Progress in Mathematics
Volume 213

Progress in Mathematics
Volume 213

The Orbit Method in Geometry and Physics

In Honor of A.A. Kirillov

Christian Duval
Laurent Guieu
Valentin Ovsienko
Editors

Springer Science+Business Media, LLC

Christian Duval
Centre de Physique Théorique
CNRS
Campus de Luminy
Marseille Cedex 9, F-13288
France

Valentin Ovsienko
Institut Girard Desargues
Université Claude Bernard Lyon 1
Villeurbanne Cedex, F-69622
France

Laurent Guieu
Université Montpellier II
Département des Sciences Mathématiques
Montpelier Cedex 5, F-34095
France

Library of Congress Cataloging-in-Publication Data

The orbit method in geometry and physics : in honor of A.A. Kirillov / C. Duval, L. Guieu, V. Ovsienko, editors.
p. cm. – (Progress in mathematics ; v. 213)
Includes bibliographical references.
ISBN 978-1-4612-6580-1 ISBN 978-1-4612-0029-1 (eBook)
DOI 10.1007/978-1-4612-0029-1
1. Orbit method. I. Kirillov, A.A. (Aleksandr Aleksandrovich), 1936- II. Duval, C. (Christian) III. Guieu, L. (Laurent) IV. Ovsienko, V. (Valentin) V. Progress in Mathematics (Boston, Mass.); v. 213

QA387.O72 2003
512'.55–dc21 2003041933

AMS Subject Classifications: 22Exx, 17Bxx, 53D50, 53B25, 53D55, 58H05, 81T30, 22E65

Printed on acid-free paper

Birkhäuser

Originally published by Birkhäuser Boston in 2003
Softcover reprint of the hardcover 1st edition 2003

ISBN 978-1-4612-6580-1 SPIN 10832700

Typeset from files by TEXniques, Inc., Cambridge, MA.

9 8 7 6 5 4 3 2 1

A.A. Kirillov

Preface

The volume is dedicated to A.A. Kirillov and emerged from an international conference which was held in Luminy, Marseille, in December 2000, on the occasion of Alexandre Alexandrovitch's 2^6th birthday. The conference was devoted to the orbit method in representation theory, an important subject that influenced the development of mathematics in the second half of the XXth century. Among the famous names related to this branch of mathematics, the name of A.A. Kirillov certainly holds a distinguished place, as the inventor and founder of the orbit method.

The research articles in this volume are an outgrowth of the Kirillov Fest and they illustrate the most recent achievements in the orbit method and other areas closely related to the scientific interests of A.A. Kirillov.

The orbit method has come to mean a method for obtaining the representations of Lie groups. It was successfully applied by Kirillov to obtain the unitary representation theory of nilpotent Lie groups, and at the end of this famous 1962 paper, it was suggested that the method may be applicable to other Lie groups as well. Over the years, the orbit method has helped to link harmonic analysis (the theory of unitary representations of Lie groups) with differential geometry (the symplectic geometry of homogeneous spaces). This theory reinvigorated many classical domains of mathematics, such as representation theory, integrable systems, complex algebraic geometry. It is now a useful and powerful tool in all of these areas.

The volume contains an overview of A.A. Kirillov's scientific work written by Jacques Dixmier. This general account provides a clear description of the orbit method, its origins and achievements, as well as an outline of other contributions of A.A. Kirillov. The other seventeen research articles published here cover a number of different topics in representation theory: the orbit method in the theory of unitary representations of Lie groups, combinatorial aspects of representation theory, quantization, harmonic analysis, Lie superalgebras and their representations, homological (super)algebras.

A. Alekseev and F. Petrov introduce a "string inspired" functional on the space of representations of the Lie algebra su(2) and prove that it achieves its minimum on the set of irreducible unitary representations. A conjecture is proposed that the same is true for any compact Lie algebra.

J. Alev and D.R. Farkas study prime and Poisson ideals in the Poisson algebra of G-invariant polynomials on a complex symplectic vector space for a given finite group G. They provide a complete classification in four dimensions or for an abelian G.

V. Baranovsky, S. Evens and V. Ginzburg analyze the Bernstein–Gelfand–

Gelfand category of representations of the quantum torus algebra. They classify its irreducible objects in terms of semistable principal bundles on elliptic curves.

R. Brylinski provides a noncommutative analog of the Kraft–Procesi model and constructs a Dixmier algebra attached to any nilpotent orbit of a complex Lie group. The associated graded algebra is proved to be a Poisson algebra isomorphic to the coordinate ring on the closure of the orbit.

P. Grozman, D. Leites and I. Shchepochkina study simple Lie superalgebras of polynomial vector fields on superspaces. They list defining relations for four of the known five exceptional Lie superalgebras of this class.

E. Gutkin considers a natural class of induced representations of the Weyl group of some classical groups. He analyzes these representation by means of the classical Schur–Weyl duality in the case of $\mathrm{GL}(n, \mathbf{C})$ and $\mathrm{SL}(n, \mathbf{C})$.

A. Joseph and A. Melnikov study a class of Lagrangian subvarieties on nilpotent coadjoint orbits of $\mathrm{SL}(n, \mathbf{C})$ called a hypersurface orbital varieties. They give a description of such varieties and construct a simple module with integral highest weight which quantizes this orbit.

D. Kazhdan and A. Polishchuk consider rational representations of a semisimple algebraic Lie group over a p-adic local field. They introduce the notion of ε-stable distributions on this representation and prove that the notions of stable and ε-stable distributions get switched by the Fourier transform.

B. Kostant and P. Michor study and describe the main properties of the (generalized) Cayley map from a connected complex reductive algebraic group to its Lie algebra. They provide an explicit decomposition of the algebra of regular functions on the group into the tensor product of the subalgebra of invariant functions, and a subspace of harmonic functions.

A.A. Kirillov introduces, via some combinatorial properties, the set of coadjoint orbits of the group of strictly upper triangular $(n \times n)$-matrices with entries in a finite field $\mathbf{F}_q$. The number of coadjoint orbits is computed in terms of a series of polynomial labeled by n, obtained by a variant of the Euler–Bernoulli triangle. The character formula for irreducible representations of the group is obtained by modification of the exponential map. Several conjectures and open problems are formulated.

V. Gorbunov, F. Malikov and V. Schechtman propose a continuation of earlier work on chiral geometry designed to geometrize the theory of vertex operator algebras. They construct chiral differential operators associated with a sheaf of commutative superalgebras over a smooth scheme.

Yu. Neretin describes the algebraic projective compactification of $\mathrm{GL}(n, \mathbf{C})$. He also constructs a universal object named the "sea urchin" for all such compactifications.

A. Okounkov discusses the concavity of the logarithm of the multiplicity of an irreducible representation as a fuction of its highest weight. He proves this property in some special cases and formulates a general conjecture that would be an analog of concavity of entropy in statistical mechanics.

G. Olshanski (with an Appendix by A. Borodin) studies a 2-parametric family of probability measures on the Thoma simplex of factor representations of the infinite symmetric groups. Probability measures on the Thoma simplex are viewed as random point processes on the real line and their correlation functions are calculated.

W. Rossmann studies smooth toric varieties of a special type, called Bott towers. In the framework of the orbit method, he associates to any such toric variety the character of a unitary representation. Inspired by the Kirillov character formula, he gives two explicit expressions of this character by means of statistical mechanics and path integral calculations.

A. Sergeev gives an interpretation of the projective Schur functions in terms of bispherical functions on certain homogeneous superspaces. These functions are realized as eigenfunctions of differential operators obtained as radial parts of Laplace operators on Lie superalgebras $\mathfrak{gl}(n|n)$ and $\mathfrak{q}(n)$.

I. Shchepochkina classifies maximal non-simple irreducible subalgebras of the classical linear Lie superalgebras, thus generalizing Dynkin's classification in the Lie algebra case.

The articles published in this volume reflect a wide range of A.A. Kirillov's research areas. Nevertheless, some topics could not be presented here, namely infinite-dimensional Lie groups, their orbits and representations, the Virasoro algebra and its applications in mathematical physics and differential geometry.

Acknowledgments. We are pleased to take this opportunity to thank the Director and the Staff of the CIRM for their efficient help, as well as the Centre de Physique Théorique (CPT-CNRS), and the Université de la Méditerranée. Special thanks are due to the Secretaries of the CIRM and CPT.

C. Duval

L. Guieu

V. Ovsienko

December 2002

Contents

Preface .... **vii**

A Principle of Variations in Representation Theory
A. Alekseev and F. Petrov **1**

Finite Group Actions on Poisson Algebras
Jacques Alev and Daniel R. Farkas **9**

Representations of Quantum Tori and G-bundles on Elliptic Curves
Vladimir Baranovsky, Sam Evens, and Victor Ginzburg **29**

Dixmier Algebras for Classical Complex Nilpotent Orbits via Kraft–Procesi Models I
Ranee Brylinski **49**

Brèves remarques sur l'oeuvre de A. A. Kirillov
Jacques Dixmier **69**

Gerbes of Chiral Differential Operators. III
Vassily Gorbounov, Fyodor Malikov, and Vadim Schechtman **73**

Defining Relations for the Exceptional Lie Superalgebras of Vector Fields
Pavel Grozman, Dimitry Leites, Irina Shchepochkina **101**

Schur–Weyl Duality and Representations of Permutation Groups
Eugene Gutkin **147**

Quantization of Hypersurface Orbital Varieties in $\mathfrak{sl}_n$
Anthony Joseph and Anna Melnikov **165**

Generalization of a Theorem of Waldspurger to Nice Representations
D. Kazhdan and A. Polischuk **197**

Two More Variations on the Triangular Theme
A. A. Kirillov **243**

The Generalized Cayley Map from an Algebraic Group to its Lie Algebra
Bertram Kostant and Peter W. Michor **259**

Geometry of $\mathrm{GL}_n(\mathbb{C})$ *at Infinity: Hinges, Complete Collineations, Projective Compactifications, and Universal Boundary*
Yurii A. Neretin **297**

Why Would Multiplicities be Log-Concave?
Andrei Okounkov **329**

Point Processes Related to the Infinite Symmetric Group
Grigori Olshanski **349**

Some Toric Manifolds and a Path Integral
W. Rossmann **395**

Projective Schur Functions as Bispherical Functions on Certain Homogeneous Superspaces
Alexander Sergeev **421**

Maximal Subalgebras of the Classical Linear Lie Superalgebras
Irina Shchepochkina **445**

The Orbit Method in Geometry and Physics

A Principle of Variations in Representation Theory

A. Alekseev
F. Petrov

Dedicated to A. A. Kirillov on the occasion of his 65th birthday

ABSTRACT We consider a certain polynomial F on the space of endomorphisms from a compact Lie algebra $\mathfrak{g}$ to $\mathrm{su}(n)$ (regarded as vector spaces). The function F was first discovered in string theory and physical considerations suggest that it is minimized on the set of irreducible representations of $\mathfrak{g}$. We prove this conjecture for $\mathfrak{g} = \mathrm{su}(2)$.

1 Introduction

The principle of variations is one of the basic tools in the theory of partial differential equations, with numerous applications in physics as well, notably as the least action principle in mechanics. In this paper, we give an example of a variational principle which singles out irreducible representations of the Lie algebra su(2).

Let $\mathfrak{g}$ be a compact Lie algebra, and let $B : \mathfrak{g} \times \mathfrak{g} \to \mathbb{R}$ be an invariant inner product on $\mathfrak{g}$. It will be convenient to choose an orthonormal basis $\{e_a\} \subset \mathfrak{g}$ such that $B(e_a, e_b) = \delta_{ab}$ and $[e_a, e_b]_{\mathfrak{g}} = \sum_c f_{abc} e_c$ with structure constants f_{abc} completely anti-symmetric in a, b, c. We shall be interested in the vector space of endomorphisms from $\mathfrak{g}$ to the Lie algebra $\mathrm{su}(n)$,

$$\mathcal{A}(\mathfrak{g}) := \{A \in \mathrm{End}(\mathfrak{g}, \mathrm{Mat}_n) | \mathrm{Tr}\, A(x) = 0,\ A(x)^* = -A(x) \text{ for all } x \in \mathfrak{g}\}.$$

In other words $\mathcal{A}$ is the space of linear maps from $\mathfrak{g}$ to the space of $n \times n$ traceless anti-Hermitian matrices. An element $A \in \mathcal{A}(\mathfrak{g})$ gives rise to a unitary representation of $\mathfrak{g}$ if

$$A([x, y]_{\mathfrak{g}}) = [A(x), A(y)] \qquad (1)$$

for all $x, y \in \mathfrak{g}$. Here on the left-hand side we use the notation $[\ ,\ ]_{\mathfrak{g}}$ for the Lie bracket of $\mathfrak{g}$ and on the right-hand side $[\ ,\ ]$ stands for the matrix commutator. Equation (1) is equivalent to a system of algebraic equations,

$$\Phi_{ab} := [A_a, A_b] - \sum_c f_{abc} A_c = 0, \qquad (2)$$

where $A_a := A(e_a)$ and $\Phi_{ab} = -\Phi_{ba}$ are traceless anti-Hermitian $n \times n$ matrices.

We shall seek a function $F : \mathcal{A}(\mathfrak{g}) \to \mathbb{R}$ which achieves its minimum exactly on the set $\mathcal{I}(\mathfrak{g}) \subset \mathcal{A}(\mathfrak{g})$ of *irreducible representations* of $\mathfrak{g}$. Note that if one drops the condition of the representation being irreducible, the task is trivially fulfilled by the function

$$f(A) = -\frac{1}{4} \sum_{ab} \mathrm{Tr}\, \Phi_{ab}^2. \tag{3}$$

Indeed, f is nonnegative and vanishes if and only if $\Phi_{ab} = 0$ for all a, b. That is, f vanishes exactly on representations of $\mathfrak{g}$.

Quite surpisingly, *string theory* provides a candidate for the function F minimized by irreducible representations

$$F(A) = -\frac{1}{4} \sum_{ab} \mathrm{Tr}\, [A_a, A_b]^2 + \frac{1}{3} \sum_{abc} f_{abc} \mathrm{Tr}\, [A_a, A_b] A_c. \tag{4}$$

Note that although formula (4) makes use of an orthonormal basis $\{e_a\}$, the function F only depends on the choice of the invariant inner product of $\mathfrak{g}$ and $\mathrm{su}(n)$, the latter given by the trace Tr. The function $F(A)$ arises in the stability analysis of certain D-branes [2, 1]. Physical considerations suggest the following conjecture about minima of the function $F(A)$:

Conjecture 1.1. *Assume that the set of irreducible representations of dimension n of the Lie algebra $\mathfrak{g}$ is nonempty. Then $F(A)$ achieves its minimum on this set.*

Note that $F(A)$ is a sum of polynomials of degrees 3 and 4. The degree 4 part is nonnegative and vanishes whenever the image of A is a commutative Lie subalgebra of $\mathrm{su}(n)$. The degree 3 part vanishes on this subset as well. A priori it is not clear whether $F(A)$ is bounded from below, for instance, the polynomial $p(x, y) = x^2y^2 + xy^2$ has similar properties and has no lower bound.

It may happen that the Lie algebra $\mathfrak{g}$ has no irreducible representations in dimension n. In this case, it seems plausible that the minimum of $F(A)$ is achieved on a $\mathfrak{g}$-representation, but we do not know how to characterize it.

In this paper, we prove Conjecture 1.1 in the simplest case of $\mathfrak{g} = \mathrm{su}(2)$. We first present an elementary proof for $n = 2$ and then proceed with a more complicated proof for arbitrary n. Our technique makes use of the concrete structure of $\mathfrak{g} = \mathrm{su}(2)$ and does not generalize to higher rank compact Lie algebras.

2 Proof of the "minimum conjecture" for $\mathfrak{g} = \mathrm{su}(2)$

In the case of $\mathfrak{g} = su(2)$ we deal with a function $F(A_1, A_2, A_3)$ of three anti-Hermitian matrices. The structure constants of $su(2)$ can be chosen as the Levi-Civita tensor ε_{abc} normalized by the convention $\varepsilon_{123} = 1$. The function (4) can be rewritten in the form

$$F(A) = -\frac{1}{2} \mathrm{Tr}\, ([A_1, A_2]^2 + [A_1, A_3]^2 + [A_2, A_3]^2) + 2\mathrm{Tr}[A_1, A_2]A_3. \tag{5}$$

2.1 The case of $n = 2$

In the standard fashion, we can assign to an anti-Hermitian 2×2 matrix A a vector $\mathbf{a} \in \mathbb{R}^3$ by the formula

$$A = -\frac{i}{2}\sum_{k=1}^{3} a_k\sigma_k,$$

where σ_k are the standard Pauli matrices. Under this identification, $\mathrm{Tr}AB = -\frac{1}{2}\mathbf{a}\cdot\mathbf{b}$ and the Lie bracket on 2×2 anti-Hermitian matrices maps to the cross product of vectors in $\mathbb{R}^3$. The function F acquires the form

$$F = \frac{1}{4}\left(|\mathbf{a_1}\times\mathbf{a_2}|^2 + |\mathbf{a_2}\times\mathbf{a_3}|^2 + |\mathbf{a_3}\times\mathbf{a_1}|^2\right) - \mathrm{Vol}(\mathbf{a_1},\mathbf{a_2},\mathbf{a_3}),$$

where

$$\mathrm{Vol}(\mathbf{a_1},\mathbf{a_2},\mathbf{a_3}) = (\mathbf{a_1}\times\mathbf{a_2})\cdot\mathbf{a_3}$$

is the volume of the parallelepiped with edges $\mathbf{a_1}, \mathbf{a_2}, \mathbf{a_3}$.

At this point it is useful to introduce $\mathbf{a_i}\times\mathbf{a_j} = \sum_k \varepsilon_{ijk}\mathbf{b}_k$. It is easy to see that

$$\mathbf{a_i}\cdot\mathbf{b_j} = \delta_{ij}\mathrm{Vol}(\mathbf{a_1},\mathbf{a_2},\mathbf{a_3}) \tag{6}$$

and

$$\mathrm{Vol}(\mathbf{b_1},\mathbf{b_2},\mathbf{b_3}) = |\mathrm{Vol}(\mathbf{a_1},\mathbf{a_2},\mathbf{a_3})|^2.$$

In terms of vectors $\mathbf{b}_k$ we obtain

$$F = \frac{1}{4}\sum_{k=1}^{3}|\mathbf{b_k}|^2 \pm |\mathrm{Vol}(|\mathbf{b_1},\mathbf{b_2},\mathbf{b_3})|^{\frac{1}{2}}.$$

The following inequality shows that F is bounded from below:

$$\begin{aligned}\tfrac{1}{4}(|\mathbf{b_1}|^2 + |\mathbf{b_2}|^2 + |\mathbf{b_3}|^2 + 1) &\geq \tfrac{1}{2}(|\mathbf{b_1}||\mathbf{b_2}| + |\mathbf{b_3}|)\\ &\geq \tfrac{1}{2}(|\mathbf{b_1}\times\mathbf{b_2}| + |\mathbf{b_3}|)\\ &\geq \sqrt{|(\mathbf{b_1}\times\mathbf{b_2})\cdot\mathbf{b_3}|}.\end{aligned}$$

Moreover, the minimum of F is achieved if and only if $|\mathbf{b_1}| = |\mathbf{b_2}|, |\mathbf{b_3}| = 1$, $\mathbf{b_1}$ and $\mathbf{b_2}$ are orthogonal to each other, and $\mathbf{b_1}\times\mathbf{b_2} = \mathbf{b_3}$. This implies that the vectors $\mathbf{b_k}$ form an oriented othonormal basis in $\mathbb{R}^3$. Equation (6) shows that $\mathbf{a_k} = \pm\mathbf{b}_k$. Since $\mathrm{Vol}(\mathbf{a_1},\mathbf{a_2},\mathbf{a_3})$ is positive for $\mathbf{a_k} = \mathbf{b_k}$, this sign choice corresponds to the minimal value of F and yields $A_i \neq 0$ and

$$[A_i, A_j] = \sum_k \varepsilon_{ijk}A_k.$$

That is, the triples (A_1, A_2, A_3) minimizing F are exactly 2-dimensional representations of su(2).

2.2 General case

Here we prove our main result:

Theorem 2.1. *Let $A_{1,2,3}$ be traceless $n \times n$ anti-Hermitian matrices, and let $F(A)$ be a polynomial function given by formula (5). The function $F(A)$ is bounded from below and achieves its minimal value $F_0 = -(n^3 - n)/24$ if and only if the triple $A_{1,2,3}$ defines an irreducible representation of* su(2).

We split the proof of this statement into a number of steps.

Lemma 2.2. *Let Γ be a connected graph with n vertices, and let d_{ij} be the distance (the length of the shortest path) between the vertices i and j. Then*

$$\sum_{i<j} d_{ij}^2 \leq N_n = \frac{n^4 - n^2}{12},$$

and the bound is achieved if and only if Γ is (up to a permutation of the vertices) a graph

$$1 - 2 - \ldots - n.$$

Proof. For $n = 2$ the statement of the lemma is obvious. Assume that it is true for graphs with the number of vertices less or equal to n. Consider a graph Γ with $n + 1$ vertex. Choose a vertex $v \in \Gamma$ such that the graph $\Gamma' = \Gamma \backslash v$ is connected. Here in $\Gamma \backslash v$ we delete the vertex v and all the edges attached to it. Since $d'_{ij} \geq d_{ij}$, we obtain

$$\sum_{i<j} d_{ij}^2 \leq N_n + \sum_i d_{vi}^2.$$

The numbers d_{vi} are bounded from above (up to permutations) by $1, 2, \ldots, n$, and $\sum_i d_{vi}^2 \leq 1^2 + 2^2 + \ldots + n^2 = n(n+1)(2n+1)/6 = M_n$. The bound is achieved if and only if Γ is of the form $v - 1 - 2 \ldots - n$ (up to permutation of vertices $1, \ldots, n$). The statement of the lemma follows from $N_{n+1} = N_n + M_n$. □

Lemma 2.3. *Let Γ be a tree with vertices enumerated by the set $\{1, 2, \ldots, n\}$, and let $X_{ij} = -X_{ji}$ be a real skew-symmetric matrix with the property that $X_{ij} \neq 0$ only if there is an edge connecting i and j. Then*

$$\sum_{i=1}^n c_i^2 \geq \frac{12}{n^3 - n} \left(\sum_{k<l} |X_{kl}| \right)^2,$$

where $c_i = \sum_{j=1}^n X_{ij}$.

Proof. Assume that $X_{kl} \neq 0$. Then there is an edge e connecting k and l. Consider the graph $\Gamma' = \Gamma \backslash e$. Since Γ is a tree, Γ' consists of two connected components, $\Gamma' = \Gamma_k \cup \Gamma_l$, where $k \in \Gamma_k$ and $l \in \Gamma_l$. Note that

$$X_{kl} = \sum_{i \in \Gamma_k} c_i \, , \; X_{lk} = \sum_{i \in \Gamma_l} c_i .$$

This observation yields $\sum_{k<l}|X_{kl}| = \sum_i m_i c_i$, where m_i is the number of edges e of Γ such that the vertex i belongs to the connected component of $\Gamma\backslash e$ with $\sum c_i > 0$. Since $\sum_{i=1}^n c_i = 0$, we obtain

$$\left(\sum_{k<l}|X_{kl}|\right)^2 = \left(\sum_i m_i c_i\right)^2 \le \sum_i c_i^2 \sum_j (m_j - \bar{m})^2 = \frac{1}{n}\sum_i c_i^2 \sum_{k<l}(m_k - m_l)^2.$$

Here $\bar{m} = \sum_j m_j/n$. The vertices k and l belong to different connected components of $\Gamma\backslash e$ if and only if e belongs to the unique shortest path from k to l. Therefore $|m_k - m_l| \le d_{kl}$. The statement of Lemma 2.3 now follows by Lemma 2.2.

Note that the equality is achieved if and only if the tree (up to permutation of vertices) is of the type $1 - 2 - \ldots - n$ and $X_{i(i+1)} = x(n-i)i/2$ with x an arbitrary constant. □

Lemma 2.4. *Let $a_i, i = 1, \ldots, n$ be real numbers, $X_{ij} = -X_{ji}$ a real skew-symmetric matrix and $c_i = \sum_{j=1}^n X_{ij}$. Then*

$$\left(2\sum_{k<l}(a_k - a_l)^2|X_{kl}| + \sum_i c_i^2\right)^3 \ge \frac{324}{n^3 - n}\left(\sum_i a_i c_i\right)^4.$$

Proof. Let Γ be a graph with vertices $\{1, 2, \ldots, n\}$ and edges between all pairs k, l such that $X_{kl} \ne 0$. Assume that Γ contains a cycle $1 - 2 - \ldots - k - 1$. Then the map $X_{12} \mapsto X_{12} + \delta, \ldots, X_{k1} \mapsto X_{k1} + \delta$ does not change c_i, and for sufficiently small δ changes $\phi := \sum_{k<l}(a_k - a_l)^2|X_{kl}|$ linearly. Therefore, one can choose δ such that ϕ does not increase and at least one of the matrix elements $X_{12}, \ldots, X_{k1}$ vanishes. With finite number of such operations one can reduce Γ to a tree without increasing ϕ and changing c_i. Hence it is sufficient to consider the case when Γ is a tree.

Note that

$$\begin{aligned}\phi &= \sum_{k<l}(a_k - a_l)^2|X_{kl}| \ge \frac{(\sum_{k<l}|a_k - a_l||X_{kl}|)^2}{\sum_{k<l}|X_{kl}|} \\ &\ge \frac{(\sum_{k<l}(a_k - a_l)X_{kl})^2}{\sum_{k<l}|X_{kl}|} = \frac{(\sum_i a_i c_i)^2}{\sum_{k<l}|X_{kl}|}\end{aligned}$$

and

$$\left(2\frac{(\sum_i a_i c_i)^2}{\sum_{kl}|X_{kl}|} + \sum_i c_i^2\right)^3 \ge 27\left(\sum_i a_i c_i\right)^4 \frac{\sum_j c_j^2}{(\sum_{k<l}|X_{kl}|)^2} \ge \frac{324}{n^3 - n}\left(\sum_i a_i c_i\right)^4.$$

Here we used the estimate of Lemma 2.3.

The equality is achieved only if the graph Γ can be reduced to a tree $1-2-\ldots-n$ (up to permutation of vertices), X_{ij} are as in Lemma 2.3, and $a_i - a_{i+1} = |x|^{1/2}$

so that $\sum_i c_i^2 = (\sum_i a_i c_i)^2 / \sum_{k<l} |X_{kl}|$. Assume that Γ was reduced to a tree by changing X_{ij} and removing some edges. It is easy to see that with our choice of a_i's, ϕ strictly decreases at the last step of this procedure (breaking the last cycle of Γ). Hence, the equality is achieved only if Γ is a tree to start with. □

Lemma 2.5. *Let $A_{1,2,3}$ be traceless anti-Hermitian $n \times n$ matrices. Then the following inequality is satisfied:*

$$\left(-\mathrm{Tr}([A_1, A_2]^2 + [A_2, A_3]^2 + [A_3, A_1]^2)\right)^3 \geq \frac{324}{n^3 - n} (\mathrm{Tr} A_1 [A_2, A_3])^4 . \quad (7)$$

Proof. Without loss of generality choose A_1 diagonal with eigenvalues $(ia_1, \ldots, ia_n)$ and denote the matrix elements of A_2 and A_3 by A^2_{ij} and A^3_{ij}, respectively. Then,

$$\begin{aligned} -\mathrm{Tr}([A_1, A_2]^2 + [A_2, A_3]^2 + [A_3, A_1]^2) \\ = 2 \textstyle\sum_{k<l} (a_k - a_l)^2 (|A^2_{ij}|^2 + |A^3_{ij}|^2) \mathrm{Tr}[A_2, A_3]^2 \\ \geq 2 \textstyle\sum_{k<l} (a_k - a_l)^2 |X_{kl}| + \sum_i c_i^2, \end{aligned}$$

where $X_{kl} = i(A^2_{kl} A^3_{lk} - A^3_{kl} A^2_{lk})$ and $c_i = \sum_j X_{ij}$. Note that we estimated the trace of $(-[A_2, A_3]^2)$ by the sum of squares of diagonal elements of $i[A_2, A_3]$. Taking into account that $\mathrm{Tr} A_1 [A_2, A_3] = \sum a_i c_i$ the inequality of Lemma 2.5 follows from that of Lemma 2.4.

In order to achieve an equality in (7) it is necessary for X_{ij} to be as in Lemma 2.3, a_i as in Lemma 2.4 and $|A^2_{kl}|^2 + |A^3_{kl}|^2 = |X_{kl}|$, which yields $A^2_{k(k+1)} = (|x|k(n-k)/2)^{1/2}$, $A^3_{k(k+1)} = i A^2_{k(k+1)}$. That is, the equality is achieved only if the matrices $A_{1,2,3}$ are positive multiples of those in the irreducible reprersentation of su(2) of dimension n. □

Now we can give a proof of Theorem 2.1. We split $F(A)$ into the cubic part $F^{(3)}(A)$ and the fourth order part $F^{(4)}(A)$. Then, on every line tA with $t \in \mathbb{R}$ we have

$$F(tA) = t^4 F^{(4)}(A) + t^3 F^{(3)}(A).$$

In our case, $F^{(4)}(A) \geq 0$ and if $F^{(4)}(A) = 0$, then $F^{(3)}(A) = 0$ as well. Hence, $F(tA)$ is always bounded from below as a function of t, and the minimum is given by

$$\min_t F(tA) = -\frac{(F^{(3)}/4)^4}{(F^{(4)}/3)^3}.$$

It is achieved at $t = -3F^{(3)}/4F^{(4)}$. According to Lemma 2.5 and formula (5), we have

$$\frac{(F^{(3)}/4)^4}{(F^{(4)}/3)^3} \leq \frac{n^3 - n}{24} = -F_0$$

which shows that F_0 is the lower bound of $F(A)$, and it is achieved on the ray tA with $t > 0$ and $A_{1,2,3}$ forming an irreducible representation of su(2). Finally, if $A_{1,2,3}$ forms a representation of su(2), $F^{(3)}/F^{(4)} = -4/3$ which yields $t = 1$ and shows that a representation always provides a minimum in t of $F(tA)$. In particular, so does an irreducible representation, which by the argument above also gives a global minimum of $F(A)$.

Acknowledgements. We are grateful to A. Bytsko and S. Smirnov for useful discussions.

References

[1] A.Yu. Alekseev, A. Recknagel, V. Schomerus, Brane Dynamics in Background Fluxes and Non-commutative Geometry, JHEP 0005 (2000) 010.

[2] R.C. Myers, Dielectric-Branes, JHEP 9912 (1999) 022.

A. Alekseev
University of Geneva
Section of Mathematics
2-4 rue du Lièvre
1211 Genève 24, Switzerland
alekseev@math.unige.ch

F. Petrov
St. Petersburg State University
Faculty of Mathematics and Mechanics
Bibliotechnaya pl. 2, 198904 Peterhof
St. Petersburg, Russia.
Fedor@FP5607.spb.edu

Finite Group Actions on Poisson Algebras

Jacques Alev
Daniel R. Farkas

Dedicated to A. A. Kirillov on the occasion of his 65th birthday

Let $\mathbf{A}_n$ denote the Weyl algebra of all differential operators on the polynomial algebra $\mathbf{C}[X_1, \ldots, X_n]$. It is well known that if G is a finite group of algebra automorphisms of $\mathbf{A}_n$, then $\mathbf{A}_n^G$ is a simple algebra. (See [M] pp. 20–23 for an algebraic proof or [Wa] Lemma 1.2 for an analytic approach.) It is natural to expect that the analogous result holds for the associated graded object. To be precise, if $\mathbf{A}_n$ is filtered by total degree, then the associated graded algebra is the larger polynomial ring $R = \mathbf{C}[X_1, \ldots, X_n, Y_1, \ldots, Y_n]$ with the Poisson bracket which describes a standard symplectic affine space. To be explicit, R is also a Lie algebra subject to

$$\{X_i, Y_j\} = \delta_{ij} \text{ and } \{X_i, X_j\} = \{Y_i, Y_j\} = 0 \text{ for all } i, j.$$

By virtue of the product or the Leibniz rule, this bracket uniquely extends a nondegenerate bilinear alternating form on the span of all of the variables. The resulting Poisson algebra R is -2-graded in the sense that

$$\{R_s, R_t\} \subseteq R_{s+t-2}.$$

The usual "differentiate until constant" argument that $\mathbf{A}_n$ is simple carries over almost verbatim to show that R is Poisson simple. That is, the only Poisson ideals of the associated graded algebra are $\{0\}$ and the algebra itself. (A *Poisson ideal* is an ideal in both the associative and Lie senses.)

Suppose that G is a finite subgroup of the symplectic group on the span of the variables. Then G extends to a group of associative and Lie algebra automorphisms of R. The algebra of invariants inherits the structure of a graded Poisson algebra. Is this fixed algebra Poisson simple?

Consider $\mathbf{C}[X, Y]$ where $\{X, Y\} = 1$ and let G be the group of order two generated by $-I$. Then $\mathbf{C}[X, Y]^G$ contains no degree one homogeneous polynomials. As a consequence of the Poisson grading, its unique maximal graded ideal is a proper Poisson ideal. This suggests that the appropriate conjecture is that the algebra of invariants has only finitely many prime and Poisson ideals.

We fix notation. Let V be a finite-dimensional complex vector space endowed with a nondegenerate alternating form. (The dimension of V is necessarily even.) Now suppose that G is a finite group which acts like a subgroup of the symplectic group $\mathbf{Sp}(V)$. That is, elements of G act linearly on V and preserve the given form. When we uniquely extend the alternating form to the polynomial ring $\mathbf{C}[V]$, it becomes the bracket $\{\cdot, \cdot\}$ for the structure of a Poisson algebra. The action of the group extends so that $g \in G$ acts as an associative algebra automorphism with

$$g(\{a, b\}) = \{g(a), g(b)\}$$

for all $a, b \in \mathbf{C}[V]$. It follows easily that the ring of invariants $\mathbf{C}[V]^G$ is also a Poisson algebra. (For this paper, $\mathbf{C}[V]$ will denote the symmetric algebra on V rather than V^*.)

A principal goal of this paper is to provide evidence for the conjecture that $\mathbf{C}[V]^G$ has only finitely many prime and Poisson ideals. In the first section we introduce special prime ideals of $\mathbf{C}[V]$ which are prime and Poisson when intersected with $\mathbf{C}[V]^G$. We use the discriminant to show that minimal prime and Poisson ideals always arise in this way.

Next we characterize those actions for which the only nonzero prime and Poisson ideal of $\mathbf{C}[V]^G$ is the unique maximal graded ideal. The rings of regular functions on the classical Kleinian singularities provide examples of this situation. Other results about maximal and Poisson ideals also appear in this section.

It has been suggested ([AL]) that $\{\mathbf{C}[V]^G, \mathbf{C}[V]^G\}$ must have finite codimension in $\mathbf{C}[V]^G$. The third section demonstrates how the theory developed earlier might be combined with techniques pioneered by Herstein to capture $\{\mathbf{C}[V]^G, \mathbf{C}[V]^G\}$.

Finally, we classify prime and Poisson ideals of $\mathbf{C}[V]^G$ when V is small or G is abelian.

These results apply to a classical context. Suppose $\mathfrak{h}$ is a Cartan subalgebra for a finite-dimensional semisimple Lie algebra $\mathfrak{g}$ and let W denote its Weyl group. In [AFSL], the Hochschild (co)homology of the algebra $\mathrm{Diff}(\mathfrak{h})^W$ was computed:

$$\dim_{\mathbf{C}} HH_{2k}(\mathrm{Diff}(\mathfrak{h})^W) = a_k,$$

where a_k equals the number of conjugacy classes of W having 1 as an eigenvalue with multiplicity k in the space $\mathfrak{h}^*$ when $1 \le k \le \mathrm{rank}(\mathfrak{g})$ and a_k is zero otherwise. Let $\mathfrak{h} \oplus \mathfrak{h}^*$ be the folded alternating form of Section 4. The results of this paper suggest that the number of prime and Poisson ideals of $\mathbf{C}[\mathfrak{h} \oplus \mathfrak{h}^*]^W$ with height $2k$ is related to a_k (checked for rank 2 in this paper). There should be a geometric way of passing from prime and Poisson ideals of $\mathbf{C}[\mathfrak{h} \oplus \mathfrak{h}^*]^W$ of height $2k$ to the Hochschild cohomology group $HH^{2k}(\mathrm{Diff}(\mathfrak{h})^W)$ in the spirit of an "extended" McKay correspondence ([EG]).

1 The discriminant theorem

Assume H is a subgroup of G. Let $\omega(H)$ denote the augmentation ideal of the group algebra $\mathbf{C}[H]$.

Theorem 1. *Set* $P_H = \mathbf{C}[V](\omega(H)V)$. *Then* P_H *is a prime ideal of* $\mathbf{C}[V]$ *satisfying*

$$\{P_H, \mathbf{C}[V]^H\} \subseteq P_H.$$

Proof. P_H is prime because $\mathbf{C}[V]/P_H$ is a polynomial ring. Also, notice that since $w \equiv h(w) \pmod{P_H}$ for all $w \in V$ and for all $h \in H$, we have $s \equiv h(s)$ $(\text{mod } P_H)$ for all $s \in \mathbf{C}[V]$.

We must show that if $f \in \mathbf{C}[V]$, $h \in H$, $v \in V$, and $a \in \mathbf{C}[V]^H$, then $\{f \cdot (1-h)(v), a\} \in P_H$. Since

$$\{f \cdot (1-h)(v), a\} = \{f, a\}(1-h)(v) + f \cdot \{(1-h)(v), a\},$$

it suffices to prove that $\{(1-h)(v), a\} \in P_H$:

$$\begin{aligned}\{(1-h)(v), a\} &= \{v, a\} - \{h(v), a\} \\ &= \{v, a\} - \{h(v), h(a)\} \\ &= \{v, a\} - h(\{v, a\}).\end{aligned}$$

According to the first paragraph, the final expression lies in P_H. □

Since $\mathbf{C}[V]^G \subseteq \mathbf{C}[V]^H$, the theorem tells us that

$$\{\mathbf{C}[V](\omega(H)V), \mathbf{C}[V]^G\} \subseteq P_H.$$

We infer that $P_H \cap \mathbf{C}[V]^G$ is a prime and Poisson ideal of $\mathbf{C}[V]^G$. As a consequence, $\mathbf{C}[V]^G$ is never Poisson simple because $P_G \cap \mathbf{C}[V]^G$ is prime and Poisson.

The point of the Discriminant Theorem which we prove next is that all minimal nonzero prime and Poisson ideals have the form P_H for a choice of cyclic H.

Theorem 2. *Assume* G *acts faithfully on* V. *If* P *is any nonzero prime ideal of* $\mathbf{C}[V]$ *such that* $\{P, \mathbf{C}[V]^G\} \subseteq P$, *then there exists a nonidentity cyclic subgroup* $C \subseteq G$ *such that*

$$\mathbf{C}[V](\omega(C)V) \subseteq P.$$

Before proceeding to the proof, we recall a classical result of Seidenberg. (A very general version can be found in [D] 3.3.) If I is an ideal in any noetherian algebra over a field of characteristic zero and $D(I) \subseteq I$ for a derivation D of the algebra, then $D(Q) \subseteq Q$ for every prime ideal minimal over I. Now consider any nonzero prime and Poisson ideal S of $\mathbf{C}[V]^G$. Then $S\mathbf{C}[V]$ is stabilized by every derivation

$$\{a, -\}$$

for $a \in \mathbf{C}[V]^G$. Thus if P is a prime ideal of $\mathbf{C}[V]$ lying over S, then Seidenberg's Lemma tells us that $\{P, \mathbf{C}[V]^G\} \subseteq P$. As a consequence, the Discriminant Theorem asserts the existence of a nonidentity group $C \subseteq G$ with

$$\mathbf{C}[V](\omega(C)V) \cap \mathbf{C}[V]^G \subseteq S.$$

In particular, all prime and Poisson ideals of $\mathbf{C}[V]^G$, minimal with respect to being nonzero, are determined by a subgroup of G, so there are only finitely many.

Lemma 3. *Let $P_1, \dots, P_m$ be nonzero prime ideals of $\mathbf{C}[V]$ such that*

$$\{P_t, \mathbf{C}[V]^G\} \subseteq P_t$$

for each t and assume these primes lie within a single G-orbit of ideals. If

$$I = P_1 \cap \cdots \cap P_m,$$

then

$$\{I, I\} \not\subseteq I.$$

Proof. It is easy to check that $P_1, \dots, P_m$ are prime ideals minimal over I. We shall assume that $\{I, I\} \subseteq I$ and reach a contradiction. Since this means that I is stabilized by the derivation $\{a, -\}$ for each $a \in I$, Seidenberg's Lemma implies that $\{I, P_t\} \subseteq P_t$ for $t = 1, \dots, m$.

Let $r \in \mathbf{C}[V]$. Because $\mathbf{C}[V]$ is integral over $\mathbf{C}[V]^G$, we may find a monic polynomial $f_t(Y) \in \mathbf{C}[V]^G[Y]$ of minimal degree such that $f_t(r) \in P_t$. Write

$$f_t(Y) = Y^d + c_{d-1}Y^{d-1} + \cdots + c_1 Y + c_0$$

with each $c_k \in \mathbf{C}[V]^G$.

Since $\{I, P_t\} \subseteq P_t$,

$$\{w, f_t(r)\} \in P_t$$

for all $w \in I$. By the product rule

$$f_t'(r)\{w, r\} + \sum_{j=0}^{d-1}\{w, c_j\}r^j \in P_t.$$

But $w \in P_t$, so $\{P_t, \mathbf{C}[V]^G\} \subseteq P_t$ implies that $\sum_j \{w, c_j\}r^j \in P_t$. Hence

$$f_t'(r)\{w, r\} \in P_t.$$

By the primality of P_t and the minimality of the degree of f we conclude that

$$\{w, r\} \in P_t.$$

Intersecting over $t = 1, \ldots, m$, we see that

$$\{I, \mathbf{C}[V]\} \subseteq I.$$

In other words, I is a nonzero proper Poisson ideal of $\mathbf{C}[V]$. This contradicts the well-known fact that $\mathbf{C}[V]$ is Poisson simple. □

Proof of the theorem. Let $P = P_1, P_2, \ldots, P_m$ be the set of all prime ideals of $\mathbf{C}[V]$ in the G-orbit of P. Clearly

$$\{\mathbf{C}[V]^G, P_t\} \subseteq P_t$$

for each t. Set $I = P_1 \cap \cdots \cap P_m$ so $\{\mathbf{C}[V]^G, I\} \subseteq I$. According to the previous lemma, there exist $a, c \in I$ such that $\{a, c\} \notin I$. As a consequence, $\{a, c\} \notin P_s$ for some choice of s.

Let $b \in \mathbf{C}[V]$ be arbitrary. Then

$$\{\sum_{\sigma \in G} \sigma(ba), c\} \in I.$$

Expanding,

$$\sum_{\sigma} \sigma(b)\{\sigma(a), c\} + \sum_{\sigma} \sigma(a)\{\sigma(b), c\} \in I.$$

Now I is a G-invariant ideal, so $\sigma(a) \in I$ for all $\sigma \in G$. Thus the second summation lies in I. We conclude that

$$\sum_{\sigma \in G} \sigma(b)\{\sigma(a), c\} \in I.$$

List the elements of G,

$$\sigma_1 = 1, \sigma_2, \ldots \sigma_n$$

for $n = |G|$ and write the previous equation n times for $b = b_1, \ldots, b_n$. If we set β to be the column vector $(b_1, \ldots, b_n)$, then Cramer's rule yields

$$\Delta(\beta)\{a, c\} \in I$$

where $\Delta(\beta) = \mathbf{det}(\sigma_1(\beta), \ldots, \sigma_n(\beta))$. (Here σ_d extends to column vectors in $\mathbf{C}[V]^n$ coordinatewise.) We have $\Delta(\beta)\{a, c\} \in P_s$ and $\{a, c\} \notin P_s$. Since P_s is prime, $\Delta(\beta) \in P_s$. It is easy to see that $\Delta(\beta)$ is a G-invariant up to sign so

$$\pm\Delta(\beta) \in P_t \text{ for all } t.$$

We conclude that $\Delta(\beta) \in P$.

Consider any $v \in V$ and focus on the special case $\beta = (1, v, v^2, \ldots, v^{n-1})$. The expansion of the Vandermonde determinant tells us that

$$\prod_{\sigma<\tau} (\sigma(v) - \tau(v)) \in P.$$

(The elements of G have been ordered according to the subscripts.) Since P is prime, there must exist distinct group elements g and h with

$$g(v) - h(v) \in P.$$

In other words, if we set

$$V_{\sigma,\tau} = \{v \in V \mid \sigma(v) - \tau(v) \in P\}$$

for each pair $(\sigma, \tau) \in G \times G$ with $\sigma \neq \tau$, then

$$V = \bigcup_{\sigma\neq\tau} V_{\sigma,\tau}.$$

It is easy to check that each $V_{\sigma,\tau}$ is a subspace of V. Since a vector space over an infinite scalar field cannot be a finite union of proper subspaces, $V = V_{g,h}$ for some single choice of distinct g and h.

We have shown that the image of V under the linear transformation $g - h$ lies in P, i.e.,

$$\mathbf{C}[V]((g-h)V) \subseteq P.$$

Since $V = g^{-1}(V)$, we have the equivalent inclusion

$$\mathbf{C}[V]((1-hg^{-1})V) \subseteq P.$$

Choose C to be the cyclic group generated by hg^{-1}. □

2 Maximal ideals

Assume that G acts faithfully on V. One says that G *acts fixed point free* on V provided the only element of the group which has a nonzero fixed point is the identity. An exhaustive description of fixed point free groups and their representations can be found in Wolf's *Spaces of Constant Curvature* ([Wo]).

The algebra of invariants $\mathbf{C}[V]^G$ is graded. If R is any graded ring, we write R_+ for its unique maximal graded ideal.

Theorem 4. *Assume G acts faithfully on V. Then $\mathbf{C}[V]^G_+$ is the unique nonzero prime and Poisson ideal of $\mathbf{C}[V]^G$ if and only if G acts fixed point free on V.*

Proof. First suppose G acts fixed point free on V. Then $1 - g$ is an invertible linear endomorphism of V for all $g \in G$ other than the identity. It follows from the Discriminant Theorem that V lies inside any prime ideal of $\mathbf{C}[V]$ lying over a nonzero prime and Poisson ideal of $\mathbf{C}[V]^G$. One direction of the corollary has been established.

Conversely, if G does not act fixed point free, there is some $g \neq 1$ in G with $(1-g)V \neq V$ and $(1-g)V \neq 0$. Thus

$$\mathbf{C}[V]((1-g)V) \cap \mathbf{C}[V]^G$$

is a bona fide intermediate prime and Poisson ideal of $\mathbf{C}[V]^G$. □

Corollary 5. *Let G be a finite subgroup of $SL(2, \mathbf{C})$ acting linearly on the polynomial algebra $\mathbf{C}[x, y]$ and let $\mathbf{C}[x, y]^G$ be the ring of regular functions on the Kleinian singularity of type G. Then the augmentation ideal is the only nonzero prime and Poisson ideal of the fixed ring.*

In general, if A is a Poisson algebra over a field of characteristic zero and $a \in A$, the Hamiltonian derivation $\{a, -\}$ is denoted by $\text{ham}(a)$. It is both an associative and Lie algebra derivation of A. The set of all Hamiltonian derivations will be written $\text{Ham}(A)$. If A is a regular affine domain, then we say A is *symplectic* provided that $\text{Der}(A)$, the collection of all associative algebra derivations of A, is generated as an A-module by $\text{Ham}(A)$ (see [FL]). For example, $F[V]$ is a symplectic Poisson algebra for any field F.

Theorem 6. *Assume that F is any field of characteristic zero and let G be a finite group of linear Poisson automorphisms of $F[V]$. Then*

$$\{F[V]^G, F[V]^G\}F[V]^G$$

has finite codimension in $F[V]^G$. More precisely, if $F[V]^G$ has a maximal ideal which is Poisson, that ideal must be the unique graded maximal ideal $F[V]^G_+$.

Proof. It suffices to prove the second statement. Let us first remark that any associative algebra ideal of a Poisson algebra which contains all brackets is automatically a Poisson ideal. Put $I = \{F[V]^G, F[V]^G\}F[V]^G$ and let $\mathcal{M}$ be a maximal ideal of $F[V]^G$ containing I; $\mathcal{M}$ is then a Poisson ideal of $F[V]^G$. Denote by $\mathcal{N}$ a maximal ideal of $F[V]$ lying over $\mathcal{M}$. By our usual argument,

$$\{\mathcal{N}, F[V]^G\} \subseteq \mathcal{N}.$$

We claim that $\{F[V], F[V]^G\} \subseteq \mathcal{N}$.

By the Nullstellensatz, $F[V]/\mathcal{N}$ is a finite field extension of the scalar field F. Given $r \in F[V]$ choose $p \in F[X]$ of minimal degree such that $p(r) \in \mathcal{N}$. If $a \in F[V]^G$, then $\{p(r), a\} \in \mathcal{N}$ so

$$p'(r)\{r, a\} \in \mathcal{N}.$$

By minimality and primality, we establish the claim that $\{r, a\} \in \mathcal{N}$.

Since $F[V]$ is symplectic, we may conclude that

$$D(a) \in \mathcal{N}$$

for all $a \in F[V]^G$ and for all associative algebra derivations D of $F[V]$. We only need to exploit the Euler derivation: $a \in \mathcal{N}$ for all homogeneous $a \in F[V]^G$ of positive degree. In other words, $F[V]^G_+ \subseteq \mathcal{N}$. It follows that $F[V]^G_+ = \mathcal{M}$. This proves that the radical of I (prime or Jacobson) is equal to $F[V]^G_+$; it then follows that there is a power of $F[V]^G_+$ included in I, which proves the theorem. □

The two theorems above also give us information about the symplectic nature of $\mathbf{C}[V]^G$. Although $\mathbf{C}[V]^G$ is never regular for nonidentity symplectic groups G, we may still ask to what extent $\mathbf{C}[V]^G\mathrm{Ham}(\mathbf{C}[V]^G)$ fills the module of all derivations of $\mathbf{C}[V]^G$. There is a bit more structure to work with. According to [F3], if A is an arbitrary Poisson algebra, then both $\mathrm{Der}(A)$ and $A\,\mathrm{Ham}(A)$ are weak Poisson A-modules. Recall that an A-module M is weak Poisson provided that there is a second action of A on M with

$$\{a, bm\}_* = \{a, b\}m + b\{a, m\}_*$$

for all $a, b \in A$ and $m \in M$. $\mathrm{Der}(A)$ is a weak Poisson A-module under

$$\{a, D\}_* = [\mathrm{ham}(a), D]$$

for $a \in A$ and $D \in \mathrm{Der}\,(A)$. It is easy to see that $A\,\mathrm{Ham}(A)$ is a weak Poisson submodule.

Lemma 7. *Let A be a Poisson algebra. If M is a weak Poisson A-module, then the annihilator of M in A is a Poisson ideal.*

Proof. Let $x \in \mathrm{ann}(M)$. If $a \in A$ and $m \in M$, then

$$\{a, x\}m = \{a, xm\}_* - x\{a, m\}_* = 0.$$

□

Theorem 8. *Let G be a finite group of symplectic automorphisms of V. Then*

$$\mathrm{Der}(\mathbf{C}[V]^G)/\mathbf{C}[V]^G\mathrm{Ham}(\mathbf{C}[V]^G)$$

is a finite-dimensional vector space if and only if G acts fixed point free on V.

Proof. Temporarily write $R = \mathbf{C}[V]$. First suppose that $R^G\mathrm{Ham}(R^G)$ has finite codimension in $\mathrm{Der}(R^G)$. These are each graded R^G-modules with finite-dimensional homogeneous components, so they agree from some degree onward. In particular, there is a positive integer N such that

$$(R^G_+)^N\mathrm{Der}(R^G) \subseteq R^G\mathrm{Ham}(R^G).$$

To exploit Theorem 4, we consider any nonzero prime ideal P of R such that $\{P, R^G\} \subseteq P$. Choose $D \in \mathrm{Der}(R)$ such that $D(P) \nsubseteq P$. Since

$$(R^G_+)^N D(P) \subseteq P,$$

the primality of P tells us that $(R^G_+)^N \subseteq P$. Hence $R^G_+ \subseteq P$. A classical result about covariants of finite groups says that $R(R^G_+)$ has finite codimension in R.

Thus P has finite codimension in R. In other words, P is a maximal ideal of R. The proof of Theorem 6 implies that $P = R_+$. Now Theorem 6 forces G to act fixed point free.

We turn to the converse and assume that G acts fixed point free. Let J be the annihilator of

$$\mathrm{Der}(R^G)/R^G\,\mathrm{Ham}(R^G).$$

By Lemma 7, J is a Poisson ideal of R^G. Let us assume, for the moment, that we know $J \neq 0$. Then Theorem 4 tells us that J has finite codimension in R^G. In other words, $\mathrm{Der}(R^G)/R^G\,\mathrm{Ham}(R^G)$ is a module over the finite-dimensional algebra R^G/J. Since $\mathrm{Der}(R^G)$ is a finitely generated R^G-module by virtue of R^G being affine, we see that the factor module is, indeed, finite dimensional.

We must show that J is nonzero. We first prove that if $r \in R$, then there is a nonzero multiple of $\mathrm{ham}(r)$ inside $R\,\mathrm{Ham}(R^G)$. Since R is integral over R^G, there is a monic polynomial $h(T) \in R^G[T]$ of minimal degree such that $h(r) = 0$. Then

$$0 = \mathrm{ham}(h(r)) = h'(r)\mathrm{ham}(r) + p$$

where $p \in R\,\mathrm{Ham}(R^G)$. By minimality, $h'(r) \neq 0$, so it is an appropriate multiplier. Now let $D \in \mathrm{Der}(R^G)$ be arbitrary. Since symplectic automorphisms of V cannot be pseudoreflections, a theorem of Levasseur ([L]) states that D can be extended to a derivation $\widehat{D}$ of $\mathrm{Der}(R)$. We use the fact that R is symplectic (i.e., $\widehat{D} \in R\,\mathrm{Ham}(R)$) and the multipliers just produced to find a nonzero $s \in R$ with

$$s\widehat{D} \in R\,\mathrm{Ham}(R^G).$$

Replace s with $\prod_{g\in G} g(s)$ so that we may assume $0 \neq s \in R^G$. Finally, apply the averaging operator

$$\frac{1}{|G|}\sum_{g\in G} g$$

to $s\widehat{D}$ and conclude that

$$sD \in R^G \mathrm{Ham} R^G.$$

Since $\mathrm{Der}(R^G)$ is a finitely generated R^G-module, we can find a single s which serves as multiplier for all derivations of R^G simultaneously. □

3 Poisson homology

Let A be any Poisson algebra. The Poisson homology group in degree zero for A is simply the vector space

$$A/\{A, A\}.$$

In [AL], the authors compute this group for $A = \mathbf{C}[V]^G$ when V is a two-dimensional space. In particular, they prove that the dimension of the 0-th Poisson homology group is the Milnor number of the corresponding Kleinian surface singularity. Their argument, which depends heavily on the special geometry available in this case, breaks into two steps. First they demonstrate that the ideal

$$\mathbf{C}[V]^G\{\mathbf{C}[V]^G, \mathbf{C}[V]^G\}$$

has finite codimension in $\mathbf{C}[V]^G$ and then they show that

$$\{\mathbf{C}[V]^G, \mathbf{C}[V]^G\} = \mathbf{C}[V]^G\{\mathbf{C}[V]^G, \mathbf{C}[V]^G\}.$$

Theorem 6 in the present paper establishes the first step in full generality. In this section, we get tantalizingly close to the complete finite dimensionality result by showing that

$$\{\mathbf{C}[V]^G, \mathbf{C}[V]^G\}^2$$

always has finite codimension in $\mathbf{C}[V]^G$.

The mathematical ideas in the next two lemmas are due to Herstein ([H], pp. 4–5) in a different context.

Lemma 9. *Let A be an arbitrary Poisson algebra. Then*

(a) $A\{A, \{A, A\}\} \subseteq \{A, A\}^2 + \{A, A\}$.

(b) *If L is any Lie ideal of A, then $\{A\{L, L\}, A\} \subseteq L$.*

Proof. For (a), let $a, b, c, d \in A$. Then

$$b\{a, \{c, d\}\} = \{a, b\{c, d\}\} - \{a, b\}\{c, d\}.$$

As to (b), choose any $a, b \in A$ and $x, y \in L$. Then

$$\begin{aligned}\{a\{x, y\}, b\} &= \{\{ax, y\} - x\{a, y\}, b\}\\ &= \{\{ax, y\}, b\} - \{x\{a, y\}, b\}.\end{aligned}$$

We show that both terms lie in L. Since L is a Lie ideal of A, $\{ax, y\} \in L$ and so $\{\{ax, y\}, b\} \in L$. As to the second term,

$$\begin{aligned}\{x\{a, y\}, b\} &= x\{\{a, y\}, b\} + \{a, y\}\{x, b\}\\ &= (\{\{a, y\}, xb\} - b\{\{a, y\}, x\}) + (\{x, \{a, y\}b\} - b\{x, \{a, y\}\})\\ &= \{\{a, y\}, xb\} + \{x, \{a, y\}b\}.\end{aligned}$$

Each summand lies in L because $\{a, y\} \in L$ and $x \in L$, respectively. □

Lemma 10. *Let A be an arbitrary Poisson algebra and assume that P is a prime ideal of A such that*

$$A\{\{A, A\}^2, \{A, A\}^2\} \subseteq P.$$

Then $\{A, A\} \subseteq P$.

Proof. We first prove something stronger: if V is a subspace of A such that $\{V, \{A, A\}^2\} \subseteq P$, then $\{V, A\} \subseteq P$.

Let $v \in V$ and $a, b, x \in A$. Then

$$\{v, \{a, b\}\{v, x^2\}\} \in P.$$

Since the characteristic of A is 0,

$$\{v, x\{a, b\}\{v, x\}\} \in P.$$

But

$$\{v, x\{a, b\}\{v, x\}\} = \{v, x\}^2\{a, b\} + x\{v, \{a, b\}\{v, x\}\}.$$

The second term on the right lies in P because P is an associative ideal. Hence the first term lies in P. Since P is prime, we conclude that either $\{a, b\} \in P$ for all $a, b \in A$ or $\{v, x\} \in P$ for all $v \in V$ and $x \in A$. The initial assertion has been established.

The hypothesis of the lemma is that $\{V, \{A, A\}^2\} \subseteq P$ for $V = \{A, A\}^2$. We conclude that either $\{A, A\} \subseteq P$ or $\{\{A, A\}^2, A\} \subseteq P$. In the second case, we have $\{V, \{A, A\}^2\} \subseteq P$ for $V = A$. Thus we may conclude that $\{A, A\} \subseteq P$ for all cases. □

Theorem 11. $\{\mathbf{C}[V]^G, \mathbf{C}[V]^G\}^2$ *has finite codimension in* $\mathbf{C}[V]^G$.

Proof. Set A to be the Poisson algebra $\mathbf{C}[V]^G$. We first show that

$$A\{\{A, A\}^2, \{A, A\}^2\}$$

has finite codimension in A. It suffices to prove that the augmentation ideal A_+ is the only prime ideal which might contain $\{\{A, A\}^2, \{A, A\}^2\}$. Let P be such a prime ideal: by Lemma 10, $\{A, A\} \subseteq P$. According to the proof of Theorem 6 about maximal ideals, the Jacobson radical and hence the prime radical of $A\{A, A\}$ is equal to A_+. This proves that $P = A_+$.

An easy application of the Leibniz property for the Poisson bracket yields

$$A\{\{A, A\}^2, \{A, A\}^2\} \subseteq A\{A, \{A, A\}^2\} \subseteq A\{A, \{A, A\}\}.$$

Now Lemma 9(a) implies that $\{A, A\}^2 + \{A, A\}$ has finite codimension in A. Thus the theorem follows once we prove that

$$\left(\{A, A\} + \{A, A\}^2\right) / \{A, A\}^2$$

is finite dimensional.

Since $A\{\{A, A\}^2, \{A, A\}^2\}$ is graded and has finite codimension in A, it contains a subspace of the form

$$_N A = \sum_{k \geq N} A_k.$$

(Here A_k denotes the k-th homogeneous component of A.) We apply Lemma 9(b) with $L = \{A, A\}^2$ – after checking that $\{A, A\}^2$ is a Lie ideal of A – to conclude that

$$\{_N A, A\} \subseteq \{A, A\}^2.$$

Hence

$$\sum_{i+j<N} \{A_i, A_j\} + \{A, A\}^2 = \{A, A\}.$$

It is clear from the grading on a polynomial ring that each A_j is finite dimensional. Since

$$\sum_{i+j<N} \{A_i, A_j\} \subseteq \sum_{t \leq 2N-4} A_t$$

we see that $\sum_{i+j<N}\{A_i, A_j\}$ is finite dimensional. □

4 Four variable invariants

We analyze the case when V is four-dimensional. As a prerequisite, we offer some general observations, so for the time being, V remains an arbitrary even dimensional vector space.

Lemma 12. *Let G be a finite group of linear Poisson automorphisms of $\mathbf{C}[V]$. Then no height one prime ideal of $\mathbf{C}[V]^G$ is Poisson. In particular the Poisson center of $\mathbf{C}[V]^G$ is $\mathbf{C}$.*

Proof. It suffices to show that no height one prime P of $\mathbf{C}[V]$ satisfies

$$\{P, \mathbf{C}[V]^G\} \subseteq P.$$

Assume such an ideal exists; Theorem 2 produces a nonidentity cyclic group C with

$$\mathbf{C}[V](\omega(C)V) \subseteq P.$$

If g is the generator of C, then V decomposes as a direct sum of two-dimensional subspaces each stabilized by g and the eigenvalues for g restricted to each subspace come in complex conjugate pairs. Thus V cannot have a one-dimensional subspace consisting of fixed points for g. Equivalently, $(1-g)V$ has codimension at least two in V. It follows that $\mathbf{C}[V](\omega(C)V)$ has height at least two. Let now z be a non unit in the Poisson center of $\mathbf{C}[V]^G$ and P a minimal prime over the principal ideal $z\mathbf{C}[V]$. The preceding argument shows that the height of P is at least 2, whereas an application of the Primidealsatz shows that it is 1. This contradiction shows that z is a unit and so $z \in \mathbf{C}$. □

Lemma 13. *Let A be any affine domain of Krull dimension one over a field of characteristic zero. If A is a Poisson algebra, then the Poisson bracket is trivial.*

Proof. Let F denote the scalar field. By Noether normalization, A is integral over a polynomial algebra $B = F[X]$. The restriction of the bracket to B is obviously trivial. Let $a \in A$ and let $f \in B[T]$ be the minimal monic polynomial satisfied by a. If $b \in B$, then

$$0 = \{f(a), b\} = f'(a)\{a, b\}.$$

By minimality, $\{a, b\} = 0$. (We usually say B lies in the *Poisson center* of A.)

Now suppose that $c \in A$ is arbitrary. Since $\{B, c\} = 0$, we see that

$$0 = \{f(a), c\} = f'(a)\{a, c\}.$$

Thus $\{a, c\} = 0$. □

The previous lemma is best understood in the context of the theory of Poisson polynomial identities ([F1],[F2]). *Let $A \supseteq B$ be Poisson domains. If A is integral over B, then A and B satisfy the same multilinear homogeneous Poisson identities.*

Before analyzing the four-variable case, we first look at prime and Poisson ideals when the dimension of V is 2. All height two primes of $\mathbf{C}[V]^G$ are maximal. So combining Lemma 12 (or 13) and Theorem 6, we see that if there is a nonzero prime and Poisson ideal then it must be $\mathbf{C}[V]^G_+$.

Theorem 14. *Let G be a finite group of linear Poisson automorphisms of*

$$\mathbf{C}[X_1, X_2, X_3, X_4],$$

then the fixed ring has finitely many prime and Poisson ideals.

Proof. We put together a number of results. Let V be the linear span of the four variables. Then $\mathbf{C}[V]^G$ has Krull dimension four. By Lemma 12, a nonzero prime and Poisson ideal must have height at least two. If its height is three or more, then it contains $\{\mathbf{C}[V]^G, \mathbf{C}[V]^G\}$ by Lemma 13. Thus Theorem 6 forces it to be the unique maximal graded ideal.

The only possible candidates left have height two. They are all minimal among prime and Poisson ideals which are nonzero. There are only finitely many of these ideals according to the remark preceding Lemma 3. □

Notice that each of the ideals described in the theorem has the form

$$\mathbf{C}[V](\omega(C)V) \cap \mathbf{C}[V]^G$$

for a cyclic subgroup C of G. We can provide even finer information in the case when G is a dihedral group. We begin by reviewing some of the classical theory of characters. Following Serre ([S] section 13.2), irreducible complex representations of a finite group G fall into three categories.

[type 1:] The character is not real valued. In this case, the underlying module cannot support any nonzero invariant bilinear form.

[type 2:] The representation is realizable over the real numbers. In this case, the module supports a nondegenerate invariant form but the form must be symmetric.

[type 3:] The character is real valued but the representation does not arise by tensoring a real representation with $\mathbf{C}$. Only in this case does there exist a nondegenerate invariant alternating form.

The description of the dihedral group as a group of plane isometries essentially shows that all of its irreducible representations are of type 2.

Now suppose V is a four-dimensional complex vector space supporting an alternating nondegenerate form and a dihedral group preserves the form. Then V is not irreducible. It is not difficult to show that if the group acts faithfully, then

$$V = W \oplus W^*$$

where W is an irreducible two-dimensional module and the form is given by

$$\langle (v, \alpha), (w, \beta) \rangle = \beta(v) - \alpha(w).$$

We shall refer to a module of this form as being *folded* and the form as being *canonical.*

Theorem 15. *Assume G is a finite dihedral group which acts faithfully on a two-dimensional complex vector space W. Let $V = W \oplus W^*$ be the folded G-module with the canonical alternating form. Then 0 and $\mathbf{C}[V]^G_+$ are prime and Poisson.*

1. *If $|G|/2$ is odd, there is one additional height two prime and Poisson ideal.*
2. *If $|G|/2$ is even, then there are exactly two additional height two prime and Poisson ideals.*

Proof. Let Q be a nonzero prime and Poisson ideal of $\mathbf{C}[V]^G$. Suppose P is a prime ideal of $\mathbf{C}[V]$ lying over Q. By the Discriminant Theorem, there exists some $1 \neq g \in G$ with

$$(1 - g)V \subseteq P.$$

If g is a rotation, then g acts without nonzero fixed points, so $1-g$ is invertible as a linear endomorphism of V. Thus in this case, $V \subseteq P$. It follows that $Q = \mathbf{C}[V]^G_+$.

Now assume $Q \neq \mathbf{C}[V]^G_+$. By the previous theorem, Q has height two in $\mathbf{C}[V]^G$, whence P has height two in $\mathbf{C}[V]$. Moreover, $(1 - g)V \subseteq P$ for some reflection $g \in G$. Notice that g has a two-dimensional subspace of fixed points in V. Hence $\mathbf{C}[V]((1 - g)V)$ is a height two prime ideal of $\mathbf{C}[V]$ contained in P. We conclude that $P = \mathbf{C}[V]((1 - g)V)$.

Conjugates of P have the same intersection with $\mathbf{C}[V]^G$ and

$$s(P) = \mathbf{C}[V]((s - sg)V) = \mathbf{C}[V]((1 - sgs^{-1})V)$$

for any $s \in G$. So there is a unique height two prime when $|G|/2$ is odd (because all reflections are conjugate in G). When $|G|/2$ is even there are two conjugacy classes of reflections. If μ_1 and μ_2 are nonconjugate reflections, then the remaining prime and Poisson ideals are

$$\mathbf{C}[V]((1 - \mu_j)V) \cap \mathbf{C}[V]^G$$

for $j = 1, 2$. We argue that these two ideals are actually different. By recalling that the prime ideals of $\mathbf{C}[V]$ lying over a prime ideal of $\mathbf{C}[V]^G$ constitute a single G-orbit, we see that it suffices to prove that

$$\mathbf{C}[V]((1-\sigma)V) \neq \mathbf{C}[V]((1-\tau)V)$$

whenever σ, τ are distinct reflections. However, if these two ideals are a common prime P, then

$$(1-\sigma)V + (1-\tau)V \subseteq P.$$

But $(1-\sigma\tau)V \subseteq (1-\sigma)V + (1-\tau)V$. As observed earlier, $1-\sigma\tau$ is invertible. It follows that $V \subseteq P$, which contradicts the fact that P has height two. □

We point out that Poisson actions of finite groups on folded modules appear naturally when studying reductive Lie algebras. Let $\mathfrak{g}$ be a finite-dimensional semisimple Lie algebra and let $\mathfrak{h}$ be a Cartan subalgebra. The theory of differential operators on $\mathfrak{g}$ invariant under the adjoint Lie group is controlled to some extent by the W-invariant differential operators on $\mathfrak{h}$, where W is the Weyl group ([Wa]). After proceeding to the associated graded algebra, the appropriate object of study turns out to be the Poisson algebra

$$\mathbf{C}[\mathfrak{h} \oplus \mathfrak{h}^*]^W.$$

Our observation about dihedral groups leads to a description of the prime and Poisson ideals of this algebra in the case when $\mathfrak{g}$ is a rank 2 semisimple Lie algebra. Indeed, the possibilities here are $A_1 \times A_1$, A_2, B_2, and G_2 with respective Weyl groups the dihedral groups of order 4, 6, 8, and 12.

5 Abelian groups

From now on we assume that G is a faithful group of linear Poisson automorphisms of $\mathbf{C}[V]$. It is plausible that every prime ideal P of $\mathbf{C}[V]^G$ which satisfies $\{P, \mathbf{C}[V]^G\} \subseteq P$ has the special form

$$P_H = \mathbf{C}[V](\omega(H)V)$$

for some subgroup $H \subseteq G$. If this is the case then not only are there only finitely many prime and Poisson ideals of $\mathbf{C}[V]^G$ but the number of such ideals is bounded by a constant dependent only on $|G|$ and not on the dimension or module structure of V. We shall show that when G is abelian, the classification of prime ideals described above is correct. Moreover, the obstruction to a similar argument for groups in general can be pinpointed.

We begin by studying the graded ideal P_H. The standard splitting of the augmentation ideal in $\mathbf{C}[H]$ implies that

$$V = \omega(H)V \oplus V^H$$

as $\mathbf{C}[H]$-modules. The calculation in Theorem 1 also shows that

$$\{\omega(H)V, V^H\} = 0.$$

It follows that the restriction of the nondegenerate form on V to V^H is also nondegenerate. One immediate consequence is that the height of P_H in $\mathbf{C}[V]$ (and of $P_H \cap \mathbf{C}[V]^G$ in $\mathbf{C}[V]^G$) is even.

The direct sum decomposition gives rise to an H-equivariant decomposition

$$\mathbf{C}[V] = \mathbf{C}[V](\omega(H)V) \oplus \mathbf{C}[V^H]$$

and an H-equivariant associative algebra projection

$$\pi : \mathbf{C}[V] \to \mathbf{C}[V^H].$$

Although π cannot be a map of Poisson algebras (because $\mathbf{C}[V]$ is Poisson simple), we have a weak substitute.

Lemma 16. *If $f \in \mathbf{C}[V]$ and $b \in \mathbf{C}[V]^H$, then*

$$\pi(\{f, b\}) = \{\pi(f), \pi(b)\}.$$

Proof. We first show that if $w \in \mathbf{C}[V^H]$, then

$$\pi(\{w, b\}) = \{w, \pi(b)\}.$$

Write $\{w, b\} = \{w, b - \pi(b)\} + \{w, \pi(b)\}$. Certainly

$$\{w, \pi(b)\} \in \mathbf{C}[V^H].$$

Now $b - \pi(b) \in \mathbf{C}[V](\omega(H)V)$ and by Theorem 1,

$$\{\operatorname{Ker} \pi, \mathbf{C}[V^H]\} \subseteq \operatorname{Ker} \pi.$$

Therefore $\{w, b - \pi(b)\} \in \operatorname{Ker} \pi$. The initial assertion follows.

To verify the identity, apply the observation for $w = \pi(f)$:

$$\pi\{\pi(f), b\} = \{\pi(f), \pi(b)\}.$$

Write $f = f_0 + \pi(f)$ with $f_0 \in P_H$. Then

$$\{f, b\} = \{f_0, b\} + \{\pi(f), b\}$$

and $\{f_0, b\} \in P_H$ (Theorem 1). Thus

$$\pi(\{f, b\}) = \pi\{\pi(f), b\}. \qquad \square$$

We now begin our analysis of an arbitrary nonzero prime ideal Q of $\mathbf{C}[V]$ which has the property that

$$\{Q, \mathbf{C}[V]^G\} \subseteq Q.$$

The relevant subgroup is

$$H = \{x \in G \mid (1 - x)V \subseteq Q\}.$$

(It is easy to check that H is a subgroup of G.) Now H is designed so that $P_H \subseteq Q$ where

$$P_H = \mathbf{C}[V](\omega(H)V).$$

Moreover, H is the "inertia group" of P_H,

$$H = \{y \in G \mid (1 - y)\mathbf{C}[V] \subseteq P_H\}.$$

Recall that the "decomposition group" of P_H is

$$D = \{g \in G \mid g \cdot P_H = P_H\}.$$

Lemma 17. *$D = N_G(H)$, the normalizer of H in G.*

Proof. It is well known that $D \subseteq N_G(H)$ because the inertia group is always normal in the decomposition group. For completeness, we include the relevant computation: if $h \in H$ and $d \in D$, then for all $v \in V$,

$$(1 - dhd^{-1})(v) = d(1 - h)(d^{-1}v) \in d \cdot P_H$$

and $d \cdot P_H \subseteq P_H \subseteq Q$.

To show that $D = N_G(H)$ it suffices to check that if $x \in N_G(H)$, then $x \cdot \omega(H)V \subseteq \omega(H)V$. But this follows because

$$x(1 - h) = (1 - xhx^{-1})x$$

with $xhx^{-1} \in H$ whenever $h \in G$. □

Set $G(Q) = N_G(H)/H$. Then $G(Q)$ is a finite group of automorphisms of $\mathbf{C}[V]/P_H$. Because

$$\mathbf{C}[V] = P_H \oplus \mathbf{C}[V^H],$$

$G(Q)$ can be regarded as a group of *linear* Poisson automorphisms of the polynomial algebra $\mathbf{C}[V^H]$.

As an application of Lemma 16, we have

$$\{\pi(Q), \pi(\mathbf{C}[V]^G)\} \subseteq \pi(Q).$$

This is almost what we need in order to carry out an inductive argument. As we shall see shortly, the stronger statement we require is

$$\{\pi(Q), (\mathbf{C}[V^H])^{G(Q)}\} \subseteq \pi(Q).$$

Since π is $G(Q)$-equivariant,

$$\pi(\mathbf{C}[V]^G) \subseteq (\mathbf{C}[V^H])^{G(Q)}.$$

The gap can be examined in light of a Galois theoretic result, Theorem 41.2 in [N]. Under the identifications of Lemma 17, it asserts that the field of fractions of $\pi(\mathbf{C}[V]^G)$ coincides with the fixed field under $G(Q)$ of the field of fractions for $(\mathbf{C}[V^H])$. Since $\mathbf{C}[V^H]$ is a polynomial algebra, it is standard that the fixed field is also the field of fractions of the integrally closed algebra $(\mathbf{C}[V^H])^{G(Q)}$. In summary,

$$(\mathbf{C}[V^H])^{G(Q)} \text{ is the integral closure of } \pi(\mathbf{C}[V]^G).$$

Proposition 18. *Assume that Q is a prime ideal of* $\mathbf{C}[V]$ *and H is the inertial subgroup of G, both as described above. If*

$$\{\pi(Q), (\mathbf{C}[V^H])^{G(Q)}\} \subseteq \pi(Q),$$

then $Q = P_H$.

Proof. First, consider the possibility that $G(Q) = 1$. Then $\pi(Q)$ is a Poisson ideal of $\mathbf{C}[V^H]$. Since $\mathbf{C}[V^H]$ is Poisson simple, $\pi(Q) = 0$. We reach the desired conclusion that $Q = P_H$.

Next, suppose $G(Q) \neq 1$. By the Discriminant Theorem applied to the action of $G(Q)$ on V^H, there is some $g \in N_G(H)$ with $g \notin H$ and

$$(1 - g)V^H \subseteq \pi(Q).$$

Since g is linear and belongs to the decomposition group of P_H,

$$(1 - g)\omega(H)V \subseteq \omega(H)V.$$

Hence the facts $V = \omega(H)V \oplus V^H$ and $P_H \subseteq Q$ force

$$(1 - g)V \subseteq Q.$$

We have reached the contradiction that $g \in H$. □

Corollary 19. *If G is an abelian group then every nonzero prime and Poisson ideal of* $\mathbf{C}[V]^G$ *has the form*

$$P_H \cap \mathbf{C}[V]^G$$

for some subgroup H of G.

Proof. We prove a slightly more general result: Suppose that Q is a nonzero prime ideal of $\mathbf{C}[V]$ such that $\{Q, \mathbf{C}[V]^G\} \subseteq Q$ and its inertial subgroup H is *normal* in G. Then

$$Q = P_H.$$

Under these hypotheses, $N_G(H) = G$. Thus

$$(\mathbf{C}[V^H])^{G(Q)} \subseteq \mathbf{C}[V^H] \cap \mathbf{C}[V]^G.$$

In other words, $(\mathbf{C}[V^H])^{G(Q)} \subseteq \pi(\mathbf{C}[V]^G)$. In this case, the gap we have discussed does not exist. □

Remark 20. After submission of this paper, the first author learned about the work of K. Brown and I. Gordon where similar questions are investigated with a different approach. Following a discussion about the finiteness of the set of prime and Poisson ideals in the invariant ring, it is established in [BG] that these are parametrized by conjugacy classes of stabilizer subgroups of G generalizing the abelian case proved in this article. In particular, this gives for the Weyl group case a parametrization by the conjugacy classes of parabolic subgroups. The last piece of information that would complete the picture is to get the number of conjugacy classes of the group which act fixed-point freely from the Poisson geometry of the quotient variety. The Kleinian singularity case ([AL]) suggests that this number could be provided in general by the dimension of the 0- th Poisson homology group of the invariant ring.

References

[AFLS] J. Alev, M. A. Farinati, T. Lambre and A. L. Solotar, Homologie des invariants d'une algèbre de Weyl sous l'action d'un groupe fini, *Journal of Algebra*, **232** (2000), 564–577.

[AL] J. Alev and T. Lambre, Comparaison de l'homologie de Hochschild et de l'homologie de Poisson pour une déformation des surfaces de Klein, 25–38 *in* Y. Khakimdjanov et al., *Algebra and Operator Theory: Proceedings of the Colloquium Tashkent, Uzbekistan*, Kluwer Academic Publ., 1998.

[BG] K. B. Brown and I. Gordon, Poisson orders, Symplectic reflection algebras and representation theory, arXiv:math.RT/0201042.

[D] J. Dixmier, *Enveloping Algebras*, North-Holland Mathematics Library **14**, Amsterdam, 1977.

[EG] P. Etingof and V. Ginzburg, Symplectic reflection algebras, Calogero-Moser space, and deformed Harish-Chandra homomorphism, *Invent. Math.*, **147** (2002), 243–348.

[F1] D. R. Farkas, Poisson polynomial identities, *Comm. in Algebra*, **26(2)** (1998), 401–416.

[F2] D. R. Farkas, Poisson polynomial identities II, *Arch. Math.*, **72** (1999), 252–260.

[F3] D. R. Farkas, Modules for Poisson algebras, *Comm. in Algebra*, **28(7)** (2000), 3293–3306.

[FL] D. R. Farkas and G. Letzter, Ring theory from symplectic geometry, *Journal of Pure and Appl. Algebra*, **125** (1998), 155–190.

[H] I. N. Herstein, *Topics in Ring Theory*, Univ. of Chicago Press, Chicago, 1969.

[L] T. Levasseur, Anneaux d'opérateurs différentiels, *Séminaire d'algèbre Malliavin*, Lecture Notes in Math. **867**, Springer-Verlag, Berlin, 1981.

[M] S. Montgomery, *Fixed Rings of Finite Automorphism Groups of Associative Rings*, Lecture Notes in Math. **818**, Springer-Verlag, Berlin, 1980.

[N] M. Nagata, *Local Rings*, Interscience: John Wiley and Sons, New York, 1962.

[S] J.-P. Serre, *Linear Representations of Finite Groups*, Grad. Texts in Math. **42**, Springer-Verlag, New York, 1977.

[Wa] N. R. Wallach, Invariant differential operators on a reductive Lie algebra and Weyl group representations, *Journal of AMS*, **6** (1993), 779–816.

[Wo] J. Wolf, *Spaces of Constant Curvature*, McGraw-Hill, New York, 1967.

Jacques Alev
Département de Mathématiques
Université de Reims CNRS UMR 6056
B. P. 1039
51687 Reims Cedex 2, France
jacques.alev@univ-reims.fr

Daniel R. Farkas
Department of Mathematics
Virginia Polytechnic Institute and State University
Blacksburg, VA 24061, USA
farkas@math.vt.edu

Representations of Quantum Tori and G-bundles on Elliptic Curves

Vladimir Baranovsky
Sam Evens
Victor Ginzburg

Dedicated to A. A. Kirillov on the occasion of his 65th birthday

ABSTRACT We study a BGG-type category of infinite-dimensional representations of $\mathcal{H}[W]$, a semidirect product of the quantum torus with parameter $\mathbf{q}$, built on the root lattice of a semisimple group G, and the Weyl group of G. Irreducible objects of our category turn out to be parametrized by semistable G-bundles on the elliptic curve $\mathbb{C}^*/\mathbf{q}^{\mathbb{Z}}$.

1 Introduction

We introduce a noncommutative deformation of the algebra of regular functions on a torus. This deformation $\mathcal{H}$, called *quantum torus algebra*, depends on a complex parameter $\mathbf{q} \in \mathbb{C}^*$. We further introduce a certain category $\mathcal{M}(\mathcal{H}, \mathcal{A})$ of representations of $\mathcal{H}$ which are locally-finite with respect to a commutative subalgebra $\mathcal{A} \subset \mathcal{H}$ whose "size" is one-half of that of $\mathcal{H}$ (our definition is modeled on the definition of the category $\mathcal{O}$ of Bernstein–Gelfand–Gelfand). We classify all simple objects of $\mathcal{M}(\mathcal{H}, \mathcal{A})$ and show that any object of $\mathcal{M}(\mathcal{H}, \mathcal{A})$ has finite length.

In §3 we consider quantum tori arising from a pair of lattices coming from a finite reduced root system. Let W be the Weyl group of this root system. We classify all simple modules over the twisted group ring $\mathcal{H}[W]$ which belong to $\mathcal{M}(\mathcal{H}, \mathcal{A})$ as $\mathcal{H}$-modules. In §4 we show that the twisted group ring $\mathcal{H}[W]$ is Morita equivalent to $\mathcal{H}^W$, the ring of W-invariants.

In §5 we establish a bijection between the set of simple modules over the algebra $\mathcal{H}[W]$ associated with a semisimple simply-connected group G, and the set of pairs (P, α), where P is a semistable principal G-bundle on the elliptic curve $\mathcal{E} = \mathbb{C}^*/\mathbf{q}^{\mathbb{Z}}$, and α is a certain "admissible representation" (cf. Definition 5.4) of the finite group $\mathrm{Aut}(P)/(\mathrm{Aut}\, P)^\circ$. Our bijection is constructed by combining the results of §3 with a bijection between $\mathbf{q}$-conjugacy classes in a loop group and G-bundles the elliptic curve $\mathcal{E}$, established earlier by some of us in [BG].

2 Holonomic modules over quantum tori

Choose to a finite rank abelian group $\mathbf{V}$, referred to as a *lattice*, and a skew symmetric $\mathbb{Z}$-valued bilinear form $\omega : \mathbf{V} \times \mathbf{V} \to \mathbb{Z}$. Associated to these data is the Heisenberg central extension

$$0 \to \mathbb{Z} \to \tilde{\mathbf{V}} \to \mathbf{V} \to 0.$$

Here $\tilde{\mathbf{V}} = \mathbf{V} \oplus \mathbb{Z}$ as a set, and the group law on $\tilde{\mathbf{V}}$ is given by

$$(v_1, z_1) \circ (v_2, z_2) = (v_1 + v_2, z_1 + z_2 + \omega(v_1, v_2)) \quad , \quad v_i \in \mathbf{V}, z_i \in \mathbb{Z}.$$

Let $\mathbb{C}\tilde{\mathbf{V}}$ denote the group algebra of $\tilde{\mathbf{V}}$ formed by all $\mathbb{C}$-linear combinations $\sum_{g \in \tilde{\mathbf{V}}} c_g[g]$. Given a complex number $\mathbf{q} \in \mathbb{C}^*$, we define a *quantum torus*, $\mathcal{H}_\mathbf{q}(\mathbf{V}, \omega)$, as the quotient of $\mathbb{C}\tilde{\mathbf{V}}$ modulo the two-sided ideal generated by the (central) element $[(0, 1)] - \mathbf{q} \cdot [(0, 0)]$. We write e^v for the image of $[(v, 0)] \in \mathbb{C}\tilde{\mathbf{V}}$ in $\mathcal{H}_\mathbf{q}(\mathbf{V}, \omega)$. The elements $\{e^v, v \in \mathbf{V}\}$ form a $\mathbb{C}$-basis of $\mathcal{H}_\mathbf{q}(\mathbf{V}, \omega)$, and we have

$$e^{v_1} \cdot e^{v_2} = \mathbf{q}^{\omega(v_1, v_2)} \cdot e^{v_1 + v_2} \quad , \quad \forall v_1, v_2 \in \mathbf{V}.$$

In particular, if $\{u_1, \dots, u_n\}$ is a $\mathbb{Z}$-basis of V, then the algebra $\mathcal{H}_\mathbf{q}(\mathbf{V}, \omega)$ is generated over $\mathbb{C}$ by the elements $X_i = e^{u_i}$ subject to the relations $X_i X_j = q_{ij}^2 X_j X_i$, where $q_{ij} = q^{\omega(u_i, u_j)}$. Therefore, $\mathcal{H}_\mathbf{q}(\mathbf{V}, \omega)$ is an iterated skew polynomial extension (cf. [MR]) and hence Noetherian ([MR, 1.2.9(iv)]).

Lemma 2.1. *If the form ω is nondegenerate, and $\mathbf{q}$ is not a root of unity, then the algebra $\mathcal{H}_q(\mathbf{V}, \omega)$ is simple.*

Proof. Suppose $h = \sum_{i=1}^{s} c_i e^{v_i}$ is an element of a two-sided ideal $J \subset \mathcal{H}_\mathbf{q}(\mathbf{V}, \omega)$, where all $v_i \in \mathbf{V}$ are distinct, and all the $c_i \in \mathbb{C}$ are nonzero. We claim that $e^{v_i} \in J$ for every i, whence $J = \mathcal{H}_\mathbf{q}(\mathbf{V}, \omega)$ since the elements e^{v_i} are invertible.

To prove the claim, we use the nondegeneracy of ω and the assumption that all the vectors v_i are distinct to find an element $v \in \mathbf{V}$ such that $\omega(v, v_i) \neq \omega(v, v_j)$, for any $i \neq j$. Hence, since q is not a root of unity, we conclude that

$$\mathbf{q}^{k \cdot \omega(v, v_i)} \neq \mathbf{q}^{k \cdot \omega(v, v_j)} \quad , \quad \forall k = 1, 2, \dots, \text{ whenever } i \neq j. \tag{2.1}$$

Now, for any $k = 0, 1, \dots,$ set $u_k := e^{k \cdot v} h e^{-k \cdot v} \in J$. We have

$$u_k = e^{k \cdot v} h e^{-k \cdot v} = \sum c_i \cdot e^{k \cdot v} e^{v_i} e^{-k \cdot v} = \sum_{i=1}^{s} c_i \cdot \mathbf{q}^{k \cdot \omega(v, v_i)} \cdot e^{v_i}.$$

Observe that the determinant of the matrix $a_{ik} := \mathbf{q}^{k \cdot \omega(v, v_i)}$ is the Vandermonde determinant $\prod_{i > j} (\mathbf{q}^{k \cdot \omega(v, v_i)} - \mathbf{q}^{k \cdot \omega(v, v_j)})$. By (2.1) this determinant is nonzero, so that the matrix is invertible. Hence, each of the elements $e^{v_1}, \dots, e^{v_s}$ can be expressed as a linear combination of the $u_0, \dots, u_{s-1} \in J$; the claim follows. $\square$

Remark. If $\mathbf{q}^m = 1$, then the elements $(1 - e^{mv})$, $v \in \mathbf{V}$ are in the center of $\mathcal{H}_\mathbf{q}(\mathbf{V}, \omega)$, hence any such element generates a nontrivial two-sided ideal.

Fix a lattice $\mathbf{X}$, let $\mathbf{Y} = \mathrm{Hom}_\mathbb{Z}(\mathbf{X}, \mathbb{Z})$, be the dual lattice and write $\langle\, , \rangle : \mathbf{X} \times \mathbf{Y} \to \mathbb{Z}$ for the canonical pairing. From now on, we take $\mathbf{V} = \mathbf{X} \oplus \mathbf{Y}$, where the form ω on $\mathbf{X} \oplus \mathbf{Y}$ is given by

$$\omega(x \oplus y, x' \oplus y') := \langle x, y' \rangle - \langle x', y \rangle \quad , \quad x, x' \in \mathbf{X},\ y, y' \in \mathbf{Y}.$$

Let $\mathcal{H} = \mathcal{H}_\mathbf{q}(\mathbf{X} \oplus \mathbf{Y}, \omega)$ denote the corresponding algebra. The elements $\{e^x, x \in \mathbf{X}\}$, resp. $\{e^y, y \in \mathbf{Y}\}$, span the commutative subalgebra $\mathbb{C}\mathbf{X} \subset \mathcal{H}$, resp. $\mathbb{C}\mathbf{Y} \subset \mathcal{H}$, and there is a natural vector space (but not *algebra*) isomorphism $\mathcal{H} \simeq \mathbb{C}\mathbf{X} \otimes_\mathbb{C} \mathbb{C}\mathbf{Y}$. The algebra structure is determined by the commutation relations

$$e^y e^x = \mathbf{q}^{<x,y>} e^x e^y \quad , \quad \forall x \in \mathbf{X},\ y \in \mathbf{Y}. \tag{2.2}$$

We introduce the complex torus $T := \mathrm{Hom}(\mathbf{X}, \mathbb{C}^*)$ so $\mathbf{X} \simeq \mathrm{Hom}_{\text{alg group}}(T, \mathbb{C}^*)$. Any element $x \in \mathbf{X}$ may be viewed as a $\mathbb{C}^*$-valued regular function $t \mapsto x(t)$ on T. Any $y \in \mathbf{Y}$ gives a well-defined element $\phi_y \in \mathrm{Hom}_{\text{alg group}}(\mathbb{C}^*, T) = Hom(\mathbf{X}, \mathbb{Z})$. We set $\mathbf{q}^y := \phi_y(\mathbf{q}) \in T$. The assignment $y \mapsto \mathbf{q}^y$ identifies the lattice $\mathbf{Y}$ with a finitely generated discrete subgroup $\mathbf{q}^\mathbf{Y} \subset T$.

Let A be a commutative $\mathbb{C}$-algebra and $\alpha : A \to \mathbb{C}$ an algebra homomorphism, referred to as a *weight*. For an A-module M, let $M(\alpha) := \{m \in M \mid am = \alpha(a) \cdot m\ ,\ \forall a \in A\}$ denote the corresponding weight subspace.

Definition. Given a $\mathbb{C}$-algebra H with a *commutative* subalgebra $A \subset H$, define

- $\mathcal{M}(H, A)$ to be the category of finitely generated H-modules M such that the H-action on M restricted to A is *locally finite*, that is, for any $m \in M$ we have $\dim_\mathbb{C} A \cdot m < \infty$.
- $\mathcal{M}^{ss}(H, A)$ to be the full subcategory of $\mathcal{M}(H, A)$ consisting of *A-diagonalizable* H-modules, i.e., H-modules M of the form

$$M = \bigoplus_{\alpha \in \text{Weights of } A} M(\alpha) \quad \text{and} \quad \dim_\mathbb{C} M(\alpha) < \infty,\ \forall \alpha.$$

Note that if $A = \mathbb{C}$, then $\mathcal{M}(H, A) = \mathcal{M}^{ss}(H, A)$ is just the category of finitely generated H-modules.

In this section we will be concerned with the special case $H = \mathcal{H}$, $A = \mathcal{A} := \mathbb{C}\mathbf{X} \subset \mathcal{H}$, (we also fix $\mathbf{q} \in \mathbb{C}^*$, not a root of unity). Since $\mathcal{H}$ is Noetherian (as explained before Lemma 2.1) any submodule N of any object $M \in \mathcal{M}(\mathcal{H}, \mathcal{A})$ is finitely generated over $\mathcal{H}$, hence $\mathcal{M}(\mathcal{H}, \mathcal{A})$ is an *abelian* category. Note the canonical algebra isomorphism $\mathbb{C}\mathbf{X} \simeq \mathbb{C}[T]$, where $\mathbb{C}[T]$ stands for the algebra of regular polynomial functions on T. Thus, the set of weights of the algebra $\mathcal{A} = \mathbb{C}\mathbf{X}$ is canonically identified with T.

For $\lambda \in T$, define an $\mathcal{H}$-module M_λ as a $\mathbb{C}$-vector space with basis $\{v_\mu\ ,\ \mu \in \lambda \cdot \mathbf{q}^{\mathbf{Y}} \subset T\}$ and with $\mathcal{H}$-action given by

$$e^y(v_\mu) = v_{\mu \cdot \mathbf{q}^y} \quad , \quad e^x(v_\mu) = x(\mu) \cdot v_\mu. \tag{2.3}$$

The module M_λ has the following interpretation. Write I_μ for the maximal ideal in $\mathbb{C}[T]$ corresponding to a point $\mu \in T$, and let $\mathbb{C}_\mu := \mathbb{C}[T]/I_\mu$ be the skyscraper sheaf at μ. Let $\mathbb{C}[\lambda \cdot \mathbf{q}^{\mathbf{Y}}] := \bigoplus_{\mu \in \lambda \cdot \mathbf{q}^{\mathbf{Y}}} \mathbb{C}_\mu$ be the (not finitely generated) $\mathbb{C}[T]$-module formed by all $\mathbb{C}$-valued, finitely supported functions on the set $\lambda \cdot \mathbf{q}^{\mathbf{Y}}$. Define an $\mathcal{H}$-action on $\mathbb{C}[\lambda \cdot \mathbf{q}^{\mathbf{Y}}]$ by the formulas

$$e^x(f) : t \mapsto x(t) \cdot f(t) \quad , \quad e^y(f) : t \mapsto f(\mathbf{q}^y \cdot t). \tag{2.4}$$

Thus, $x \in \mathbf{X}$ and $y \in \mathbf{Y}$ act via multiplication by the function $x(t)$ and shift by $\mathbf{q}^y$, respectively. It is straightforward to verify that sending $v_\mu \in \mathsf{M}_\lambda\ ,\ \mu \in \lambda \cdot \mathbf{q}^{\mathbf{Y}}$ to the characteristic function of the one-point set $\{\mu\}$ establishes an isomorphism of $\mathcal{H}$-modules $\mathsf{M}_\lambda \xrightarrow{\sim} \mathbb{C}[\lambda \cdot \mathbf{q}^{\mathbf{Y}}]$ intertwining the actions (2.3) and (2.4), respectively.

Clearly, $\mathsf{M}_\lambda \in \mathcal{M}^{ss}(\mathcal{H}, \mathcal{A})$. Moreover, it is obvious from the isomorphism $\mathsf{M}_\lambda \simeq \mathbb{C}[\lambda \cdot \mathbf{q}^{\mathbf{Y}}]$ that $\mathsf{M}_\lambda \simeq \mathsf{M}_\mu$ if $\mu \in \lambda \cdot \mathbf{q}^{\mathbf{Y}}$. Thus, the modules M_λ are effectively parametrized (up to isomorphism) by the points of the variety: $\Lambda := T/\mathbf{q}^{\mathbf{Y}}$. When $|\mathbf{q}| \neq 1$, Λ is an abelian variety. Observe that the modules corresponding to two different points of Λ have disjoint weights, hence are nonisomorphic.

Proposition 2.5.

(i) *M_λ is a simple $\mathcal{H}$-module, for any $\lambda \in \Lambda$. Moreover, the set $\{\mathsf{M}_\lambda, \lambda \in \Lambda\}$ is a complete collection of (the isomorphism classes of) simple objects of the category $\mathcal{M}(\mathcal{H}, \mathcal{A})$.*

(ii) *Any object of the category $\mathcal{M}^{ss}(\mathcal{H}, \mathcal{A})$ is isomorphic to a finite direct sum $\bigoplus_{\lambda \in \Lambda} \mathsf{M}_\lambda$, in particular, the category $\mathcal{M}^{ss}(\mathcal{H}, \mathcal{A})$ is semisimple.*

(iii) *Any object of the category $\mathcal{M}(\mathcal{H}, \mathcal{A})$ has finite length.*

Proof. Let $M \in \mathcal{M}(\mathcal{H}, \mathcal{A})$. An easy straightforward calculation shows that, for any nonzero element $m \in M(\lambda)$, the $\mathcal{H}$-submodule in M generated by m is isomorphic to M_λ. This, combined with the observation preceding the proposition, proves part (i).

Since M is finitely generated, one can find finitely many weights $\lambda_1, \ldots, \lambda_s \in T$ such that all weights of M are contained in $(\lambda_1 \cdot \mathbf{q}^{\mathbf{Y}}) \cup \ldots \cup (\lambda_s \cdot \mathbf{q}^{\mathbf{Y}})$ and, moreover, $\lambda_i \neq \lambda_j \bmod \mathbf{q}^{\mathbf{Y}}$ whenever $i \neq j$. It follows, since all weights of M are in $(\lambda_1 \cdot \mathbf{q}^{\mathbf{Y}}) \cup \ldots \cup (\lambda_s \cdot \mathbf{q}^{\mathbf{Y}})$, that M is generated by the subspace $\bigoplus_{i=1}^s M(\lambda_i)$. Furthermore, the same calculation as in the first part implies that the $\mathcal{H}$-submodule in M generated by this subspace is isomorphic to $\bigoplus_{i=1}^s \mathsf{M}_{\lambda_i} \otimes M(\lambda_i)$. This proves part (ii).

To prove (iii), suppose $M \in \mathcal{M}(\mathcal{H}, \mathcal{A})$. We use induction on the minimal dimension d of an $\mathcal{A}$-invariant subspace $V \subset M$ which generates M over $\mathcal{H}$. It follows from the definitions that if $d = 1$ then $M \simeq \mathsf{M}_\lambda$ for some λ. If $d > 1$, choose a nonzero vector $v \in V$ of some $\mathcal{A}$-weight λ and note that such a choice induces a nonzero homomorphism of $\mathcal{H}$-modules $\mathsf{M}_\lambda \to M$. Since M_λ is simple, this homomorphism is necessarily injective. The quotient M/M_λ is generated by an $\mathcal{A}$-invariant subspace $V/\langle v \rangle$, hence we can apply the assumption of induction to this $\mathcal{H}$-module, and (iii) follows. □

3 $\mathcal{H}[W]$-modules

Let $\Delta \subset \mathfrak{h}$ be a finite reduced root system. Let W be the Weyl group of Δ, write $\mathfrak{h}^\vee$ for the dual of $\mathfrak{h}$, and let $\mathbf{X} \subset \mathfrak{h}^\vee$, and $\mathbf{Y} \subset \mathfrak{h}$ be the coroot and weight lattices, respectively. The group W acts naturally on $\mathbf{X}$ and on $\mathbf{Y}$. The diagonal W-action on $\mathbf{X} \oplus \mathbf{Y}$ makes $\mathcal{H} = \mathcal{H}(\mathbf{X} \oplus \mathbf{Y})$ a left W-module with W-action $w : h \mapsto {}^w h$, $h \in \mathcal{H}$. Write $\mathcal{H}^W$ for the subalgebra of W-invariants. Further, introduce a *twisted group algebra*, $\mathcal{H}[W]$, as the complex vector space $\mathcal{H} \otimes_{\mathbb{C}} \mathbb{C}[W]$ with multiplication:

$$(f \otimes w) \cdot (g \otimes y) = (f \cdot {}^w g) \otimes (w \cdot y) \quad f, g \in \mathcal{H}, \; w, y \in W.$$

We use similar notation $\mathcal{H}[W']$ for any subgroup $W' \subset W$, and view $\mathbb{C}\mathbf{X}$, resp. $\mathbb{C}\mathbf{Y}$, as a commutative subalgebra of $\mathcal{H}[W']$ via the composition of imbeddings $\mathbb{C}\mathbf{X} \hookrightarrow \mathcal{H} \hookrightarrow \mathcal{H}[W']$.

The group W acts naturally on T and on $\Lambda = T/\mathbf{q}^\mathbf{Y}$. Given $\lambda \in T$, consider its image in Λ, and let $W^\lambda \subset W$ denote the isotropy group of the image of λ. The W^λ-action on T keeps the subset $\lambda \cdot \mathbf{q}^\mathbf{Y}$ stable, hence we may define W^λ-action on M_λ by the assignment $w : v_\mu \mapsto v_{w(\mu)}$. This way we make the twisted group algebra, $\mathcal{H}[W^\lambda]$, act on M_λ. Recall that earlier we denoted by $\mathcal{A}$ the subalgebra $\mathbb{C}\mathbf{X}$.

Theorem 3.1 (cf. [LS, 2.1]). *If $M \in \mathcal{M}^{ss}(\mathcal{H}, \mathcal{A})$, then the restriction of M to $\mathcal{H}^W$-module is semisimple, i.e., $M \in \mathcal{M}^{ss}(\mathcal{H}^W, \mathcal{A}^W)$. Furthermore,* $\operatorname{Ind}_{\mathcal{H}}^{\mathcal{H}[W]} M \in \mathcal{M}^{ss}(\mathcal{H}[W], \mathcal{A})$.

Proof. This follows from Proposition 2.5 and the twisted version of Maschke Theorem, see [M, Theorems 0.1 and 7.6(iv)]. □

Let $\mathcal{M}_\lambda(\mathcal{H}[W^\lambda], \mathcal{A})$ be the full subcategory of $\mathcal{M}^{ss}(\mathcal{H}[W^\lambda], \mathcal{A})$ formed by the modules M such that all the weights of the $\mathcal{A}$-action belong to the coset $\lambda \cdot \mathbf{q}^\mathbf{Y}$.

Let $M \in \mathcal{M}_\lambda(\mathcal{H}[W^\lambda], \mathcal{A})$. Note that the subgroup W^λ does not necessarily map the weight space $M(\lambda)$ into itself: if $w \in W^\lambda$ then, by definition of W^λ, we have $w(\lambda) \in \lambda \cdot \mathbf{q}^\mathbf{Y}$. Thus, it is possible that $w(\lambda) \neq \lambda$ so that, for $m \in M(\lambda)$, the element $w(m)$ is pushed out of the $M(\lambda)$. We define a "corrected" *dot-action* $w : m \mapsto w \cdot m$ of the group W^λ on the vector space $M(\lambda)$ as follows. As we have seen by definition, for any $w \in W^\lambda$, there exists a uniquely determined $y \in \mathbf{Y}$ such that $w(\lambda) = \lambda \cdot \mathbf{q}^y$. Then, for $m \in M(\lambda)$, put $w \cdot m = e^{-y} w(m)$. Here

$w(m) \in M$ stands for the result of w-action on m, and we claim that the element $e^{-y}w(m)$ belongs to $M(\lambda)$ (while $w(m)$ does not, in general).

Write $\mathcal{M}(W^\lambda)$ for the category of finite-dimensional $\mathbb{C}W^\lambda$-modules. With the dot-action of W^λ introduced above, we may now define a functor (cf. [LS, 2.2]) $\Phi : \mathcal{M}_\lambda(\mathcal{H}[W^\lambda], \mathcal{A}) \rightsquigarrow \mathcal{M}(W^\lambda)$ by the assignment $M \mapsto M(\lambda)$. On the other hand, given a representation N of W^λ one has an obvious $\mathcal{H}[W^\lambda]$-action on $\mathsf{M}_\lambda \otimes_{\mathbb{C}} N$ and this gives a functor $\Psi : \mathcal{M}(W^\lambda) \rightsquigarrow \mathcal{M}_\lambda(\mathcal{H}[W^\lambda], \mathcal{A})$.

Theorem 3.2. *The functors Ψ and Φ are mutually inverse equivalences.*

Proof. One has $\Phi\Psi(N) \simeq N$. If M is in $\mathcal{M}_\lambda(\mathcal{H}[W^\lambda], \mathcal{A})$ and $M(\lambda) = \Phi(M)$, then by Theorem 3.1 and Proposition 2.5, $M \simeq (\mathsf{M}_\lambda)^{\oplus m}$ as an $\mathcal{H}$-module and hence $M = \mathcal{H} \cdot M(\lambda)$. Thus, there is a morphism of $\mathcal{H}$-modules $\psi : \Psi(M(\lambda)) = \mathsf{M}_\lambda \otimes_{\mathbb{C}} M(\lambda) \to M$ given by $hv_\lambda \otimes m \mapsto h(m)$. The map ψ is injective since M_λ is simple over $\mathcal{H}$. One can easily check that ψ is actually an isomorphism of $\mathcal{H}[W^\lambda]$-modules. □

Since $\mathcal{H}$ is a subalgebra of $\mathcal{H}[W^\lambda]$ one may regard $\mathcal{H}[W^\lambda]$ as a *right* $\mathcal{H}$-module. Let $\widehat{W}^\lambda$ denote the set of isomorphism classes of simple W^λ-modules.

Proposition 3.3 (cf. [LS, 2.4]). *There is an $\mathcal{H}[W^\lambda]$-module decomposition*

$$\mathcal{H}[W^\lambda] \otimes_{\mathcal{H}} \mathsf{M}_\lambda \cong \bigoplus\nolimits_{\chi \in \widehat{W}^\lambda} (\mathsf{M}_\lambda \otimes_{\mathbb{C}} \chi)^{\oplus d_\chi} \quad , \qquad d_\chi := \dim \chi .$$

Furthermore, the $\mathcal{H}[W^\lambda]$-modules $\{\mathsf{M}_\lambda \otimes_{\mathbb{C}} \chi \,,\ \chi \in \widehat{W}^\lambda\}$ are simple and pairwise nonisomorphic.

Proof. $\Phi(\mathcal{H}[W^\lambda] \otimes_{\mathcal{H}} \mathsf{M}_\lambda)$ is the left regular representation of W^λ. □

For any $\chi \in \widehat{W}^\lambda$, put $V_\chi := \Psi(\chi) = \mathsf{M}_\lambda \otimes_{\mathbb{C}} \chi \ \in \mathcal{M}_\lambda(\mathcal{H}[W^\lambda], \mathcal{A})$. Set

$$Z_\chi \ := \ \mathrm{Ind}_{\mathcal{H}[W^\lambda]}^{\mathcal{H}[W]} V_\chi \ = \ \mathcal{H}[W] \bigotimes\nolimits_{\mathcal{H}[W^\lambda]} V_\chi \ \in \mathcal{M}^{ss}(\mathcal{H}[W^\lambda], \mathcal{A}).$$

Theorem 3.4 (cf. [LS, 2.5]). *There is an $\mathcal{H}[W]$-module isomorphism*

$$\mathcal{H}[W] \otimes_{\mathcal{H}} \mathsf{M}_\lambda \cong \bigoplus\nolimits_{\chi \in \widehat{W}^\lambda} Z_\chi{}^{\oplus d_\chi} .$$

Furthermore, Z_χ are simple pairwise nonisomorphic $\mathcal{H}[W]$-modules.

Proof. We have an obvious isomorphism:

$$\mathcal{H}[W] \otimes_{\mathcal{H}} \mathsf{M}_\lambda \ \cong \ \mathcal{H}[W] \otimes_{\mathcal{H}[W^\lambda]} \mathcal{H}[W^\lambda] \otimes_{\mathcal{H}} \mathsf{M}_\lambda .$$

The decomposition of the theorem now follows from Proposition 3.3. To prove that Z_χ are simple $\mathcal{H}[W]$-modules we write an $\mathcal{H}[W]$-module direct sum decomposition:

$$Z_\chi \cong \bigoplus\nolimits_{j=1}^{s} w_j V_\chi \quad \text{and} \quad w_j V_\chi \cong (\mathsf{M}_{w_j(\lambda)})^{\oplus d_\chi} ,$$

where $w_1 = e, \ldots, w_s$, are representatives in W of the right cosets W/W^λ. Any simple $\mathcal{H}$-submodule of Z_χ is contained in some $w_j V_\chi$.

By Theorem 3.1, the $\mathcal{H}[W]$-module $\mathcal{H}[W] \otimes_{\mathcal{H}} \mathsf{M}_\lambda$ is semisimple. Therefore, Z_χ, being a direct summand of a semisimple module, is a semisimple $\mathcal{H}[W]$-module. Hence Z_χ contains a simple submodule M with a nonzero projection from M to $w_j V_\chi$. Viewing M as an $\mathcal{H}$-module we see that $M = \bigoplus_j (M \cap w_j V_\chi)$. Since V_χ is a simple $\mathcal{H}[W^\lambda]$-module, we have $V_\chi \subset M$ and therefore $\bigoplus_j w_j V_\chi \subset M$. Hence $Z_\chi = M$.

Finally, any isomorphism $\theta : Z_\chi \xrightarrow{\sim} Z_\psi$, for some $\chi \neq \psi$, would map V_χ to V_ψ (just view it as a morphism of $\mathcal{H}$-modules). This would contradict Proposition 3.3. □

Proposition 3.5. *Any simple $\mathcal{H}[W]$-module M such that the $\mathbb{C}\mathbf{X}^W$-action on M is locally finite is isomorphic to Z_χ, for a certain $\chi \in \widehat{W^\lambda}$, $\lambda \in \Lambda/W$.*

Proof. We have $Z_\chi = \operatorname{Ind}_{\mathcal{H}[W^\lambda]}^{\mathcal{H}[W]}(V_\chi)$. By the Schur lemma and Frobenius reciprocity $\operatorname{Hom}(A, \operatorname{Res} B) = \operatorname{Hom}(\operatorname{Ind} A, B)$, it then suffices to show that $\operatorname{Res}_{\mathcal{H}[W^\lambda]}^{\mathcal{H}[W]}(M)$ has a submodule isomorphic to V_χ. But the latter follows from the proof of Theorem 3.2. □

Thus, we have reduced the classification of simple $\mathcal{H}[W]$-modules to the classification of irreducible representations of the finite group W^λ. The latter group is *not* a Weyl group, however. Therefore its representation theory is not classically known in geometric terms. In Section 5 we will develop an analogue of "Springer theory" for W^λ, relating irreducible representations of W^λ to semistable G-bundles on the elliptic curve $\mathbb{C}^*/\mathbf{q}^{\mathbb{Z}}$.

Remark 3.6. Note that one has the following alternative definition of Z_χ:

$$Z_\chi := \operatorname{Ind}_{\mathcal{A}[W^\lambda]}^{\mathcal{H}[W]}(\lambda \otimes \chi) = \mathcal{H}[W] \bigotimes_{\mathcal{A}[W^\lambda]} (\lambda \otimes \chi),$$

where λ denotes the one-dimensional $\mathcal{A}[W^\lambda]$-module, in which the group $W^\lambda \subset \mathcal{A}[W^\lambda]$ acts via the dot-action.

4 Morita equivalence

The algebra $\mathcal{H}$ may be viewed either as an $(\mathcal{H}[W], \mathcal{H}^W)$ - bimodule, $\mathcal{H}^l$, or as an $(\mathcal{H}^W, \mathcal{H}[W])$-bimodule, $\mathcal{H}^r$.

Proposition 4.1 (cf. [LS, 3.1]). (i) *$\mathcal{H}[W]$ and $\mathcal{H}^W$ are simple rings. These rings are Morita equivalent via the following functors*:

$$\mathbf{F} : \mathcal{H}[W]\text{–mod} \rightsquigarrow \mathcal{H}^W\text{–mod} \quad , \quad M \mapsto \mathcal{H}^r \otimes_{\mathcal{H}[W]} M,$$

$$\mathbf{I} : \mathcal{H}^W\text{–mod} \rightsquigarrow \mathcal{H}[W]\text{–mod} \quad , \quad N \mapsto \mathcal{H}^l \otimes_{\mathcal{H}^W} N.$$

(ii) *There are functorial isomorphisms*: $\mathbf{F}(M) \cong \mathrm{Hom}_{\mathcal{H}[W]}(\mathcal{H}^l, M) \cong M^W$.

Proof. (i) See [M, Theorems 2.3 and 2.5(a)]. (ii) Exercise. □

Similar results hold for $\mathcal{H}^{W^\lambda}$- and $\mathcal{H}[W^\lambda]$-modules, respectively. We write $\mathbf{F}_\lambda$ and $\mathbf{I}_\lambda$ for the corresponding functors.

Since $\mathcal{H}^W$ commutes with W^λ, we may regard M_λ as a left $\mathcal{H}^W \times W^\lambda$-module. Let $L_\chi = \mathrm{Hom}_{W^\lambda}(\chi^*, \mathsf{M}_\lambda)$ be the χ^*-isotypic component of the $\mathcal{H}[W^\lambda]$-module M_λ. Notice that by Proposition 4.1(ii) we have

$$L_\chi = (\mathsf{M}_\lambda \otimes_{\mathbb{C}} \chi)^{W^\lambda} = \mathbf{F}_\lambda(V_\chi) = \mathcal{H} \otimes_{\mathcal{H}[W^\lambda]} V_\chi = \mathbf{F}(Z_\chi).$$

Since $\mathsf{M}_\lambda \cong \bigoplus_{\chi \in \widehat{W}_\lambda} L_\chi \otimes V_\chi^*$ as $\mathcal{H}^W \times W^\lambda$-modules, we deduce an $\mathcal{H}[W]$-module decomposition:

$$\mathsf{M}_\lambda \cong \mathcal{H} \otimes_{\mathcal{H}[W]} \mathcal{H}[W] \otimes_{\mathcal{H}} \mathsf{M}_\lambda \cong \bigoplus\nolimits_\chi (\mathcal{H} \otimes_{\mathcal{H}[W]} Z_\chi)^{\oplus d_\chi} = \bigoplus\nolimits_\chi L_\chi^{\oplus d_\chi}. \quad (4.2)$$

Theorem 4.3 (cf. [LS, 3.4]). (i) *The $\mathcal{H}^W$-modules $\{L_\chi\ ,\ \chi \in \widehat{W}^\lambda\}$ are simple and pairwise nonisomorphic.*

(ii) *Every simple object of $\mathcal{M}(\mathcal{H}^W, \mathcal{A}^W)$ is isomorphic to L_χ, for some $\chi \in \widehat{W}^\lambda$.*

Proof. (i) Follows from Theorem 3.4 and Morita equivalence. (ii) Follows from Proposition 3.5 and Morita equvalence. □

Proposition 4.4 (cf. [LS, 3.6]). *If M_λ and M_μ have a simple $\mathcal{H}^W$-submodule in common then $\mu \in W \cdot \lambda$, in which case $\mathsf{M}_\lambda \cong \mathsf{M}_\mu$.*

Proof. By Morita equivalence and the identity $\mathsf{M}_\lambda \cong \mathcal{H} \otimes_{\mathcal{H}[W]} \mathcal{H}[W] \otimes_{\mathcal{H}} \mathsf{M}_\lambda$, it is enough to consider the $\mathcal{H}[W]$-modules $\mathcal{H}[W] \otimes_{\mathcal{H}} \mathsf{M}_\lambda$ and $\mathcal{H}[W] \otimes_{\mathcal{H}} \mathsf{M}_\mu$. Now consider these modules as $\mathcal{H}$-modules and apply the decomposition $Z_\chi \cong \bigoplus_{j=1}^s w_j V_\chi$ from the proof of Theorem 3.4. □

5 Representations and G-bundles on elliptic curves

In this section we fix G, a connected and simply-connected complex semisimple group. We write $\mathbb{T}$ for the *abstract* Cartan subgroup of G, that is, $\mathbb{T} := B/[B, B]$, for an arbitrary Borel subgroup $B \subset G$, see [CG, ch.3]. Let $\mathbb{W}$ denote the *abstract* Weyl group, the group acting on $\mathbb{T}$ and generated by the given set of simple reflections. We also fix $\mathbf{q} \in \mathbb{C}^*$ such that $|\mathbf{q}| < 1$, and set $\mathcal{E} = \mathbb{C}^*/\mathbf{q}^{\mathbb{Z}}$.

For any complex reductive group H we let $\mathfrak{M}(\mathcal{E}, H)$ denote the moduli space of topologically trivial semistable H-bundles on $\mathcal{E}$.

Definition 5.1. A G-bundle $P \in \mathfrak{M}(\mathcal{E}, G)$ is called "semisimple" if any of the following 3 equivalent conditions hold:

(i) The structure group of P can be reduced from G to a maximal torus $T \subset G$;

(ii) The automorphism group Aut$P$ is reductive;

(iii) The *substack* corresponding to the isomorphism class of P is closed in the *stack* of all G-bunles on $\mathcal{E}$.

We write $\mathfrak{M}(\mathcal{E}, G)^{ss}$ for the subspace in $\mathfrak{M}(\mathcal{E}, G)$ formed by semisimple G-bundles. To each G-bundle $P \in \mathfrak{M}(\mathcal{E}, G)$ one can assign its *semisimplification*, $P^s \in \mathfrak{M}(\mathcal{E}, G)^{ss}$. By definition, P^s corresponds to the unique closed isomorphism class in the stack of G-bundles on $\mathcal{E}$ which is contained in the closure of the isomorphism class of P. This gives the semisimplification morphism $ss : \mathfrak{M}(\mathcal{E}, G) \to \mathfrak{M}(\mathcal{E}, G)^{ss}$. It is known further (cf. for example [La]) that there are natural isomorphisms of algebraic varieties:

$$\mathfrak{M}^{\circ}(\mathcal{E}, \mathbb{T}) \simeq X_*(\mathbb{T}) \otimes_{\mathbb{Z}} \mathcal{E} \quad and \quad \mathfrak{M}(\mathcal{E}, G)^{ss} \simeq \big(X_*(\mathbb{T}) \otimes_{\mathbb{Z}} \mathcal{E}\big)/W\,, \tag{5.2}$$

where $\mathfrak{M}^{\circ}(\mathcal{E}, \mathbb{T})$ stands for the connected component of the trivial bundle in $\mathfrak{M}(\mathcal{E}, \mathbb{T})$ and $X_*(\mathbb{T}) = \mathrm{Hom}_{\text{alg group}}(\mathbb{C}^*, \mathbb{T})$. Moreover, the connected components of $\mathfrak{M}(\mathcal{E}, \mathbb{T})$ are labelled by the lattice $X_*(\mathbb{T})$, and are all isomorphic to each other.

By a B-structure on a G-bundle P we mean a reduction of its structure group from G to a Borel subgroup of G. Let $\mathcal{B}(\mathcal{E}, G)$ denote the moduli space of pairs: $\{$*G-bundle* $P \in \mathfrak{M}(\mathcal{E}, G)$, *B-structure on* $P\}$. Forgetting the B-structure gives a canonical morphism $\pi : \mathcal{B}(\mathcal{E}, G) \longrightarrow \mathfrak{M}(\mathcal{E}, G)$. On the other hand, given a B-structure on P one gets a B-bundle P_B, and push-out via the homomorphism: $B \twoheadrightarrow B/[B, B] = \mathbb{T}$ gives a $\mathbb{T}$-bundle on $\mathcal{E}$. Thus, there is a well-defined morphism of algebraic varieties $\nu : \mathcal{B}(\mathcal{E}, G) \longrightarrow \mathfrak{M}(\mathcal{E}, \mathbb{T})$. Further, set

$$\widetilde{G} = \{(x, B) \mid B \text{ is Borel subgroup in } G\,,\ x \in B\},$$

and let $\pi : \widetilde{G} \to G$ be the first projection.

We have the following two commutative diagrams, where the one on the left is the Grothendieck–Springer "universal resolution" diagram, cf. e.g., [CG, ch. 3], and the one on the right is its "analogue" for bundles on the elliptic curve $\mathcal{E}$:

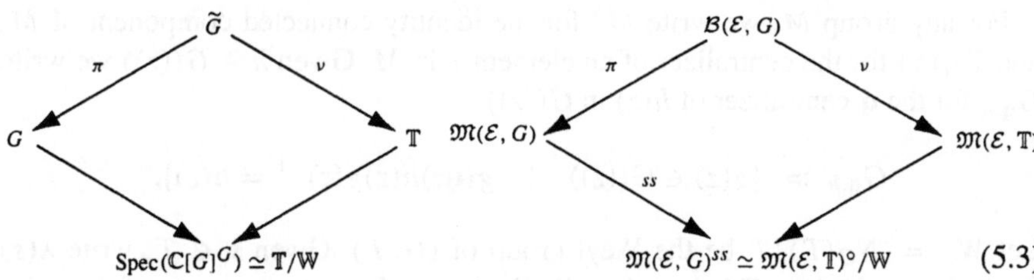

(5.3)

Observe that for any $P \in \mathfrak{M}(\mathcal{E}, G)$, the group Aut$P$ acts naturally on the set $\mathcal{B}(\mathcal{E}, G)_P := \pi^{-1}(P)$ of all B-structures on P. This induces an action of Aut $P/\mathrm{Aut}^{\circ}P$, the (finite) group of connected components, on the complex top homology group: $H_{top}(\mathcal{B}(\mathcal{E}, G)_P, \mathbb{C})$.

Definition 5.4. An irreducible representation of the group $\mathrm{Aut}\, P/\mathrm{Aut}^\circ P$ is called "admissible" if it occurs in $H_{top}(\mathcal{B}(\mathcal{E}, G)_P, \mathbb{C})$ with nonzero multiplicity.

One of the main results of this paper is the following.

Theorem 5.5. *There exists a bijection between the set of (isomorphism classes of) simple objects of $\mathcal{M}(\mathcal{H}[W], \mathcal{A})$ and the set of (isomorphism classes of) pairs (P, α), where $P \in \mathfrak{M}(\mathcal{E}, G)$, and α is an* admissible *representation of the group* $\mathrm{Aut}\, P/(\mathrm{Aut}\, P)^\circ$.

The rest of this section is devoted to the proof of the theorem. As a first approximation, recall Proposition 2.5, which states that simple objects of the category $\mathcal{M}(\mathcal{H}, \mathcal{A})$ are in one-to-one correspondence with the points of the abelian variety $\Lambda = T/\mathbf{q}^{\mathbb{Z}}$ which is, by (5.2), nothing but $\mathfrak{M}^\circ(\mathcal{E}, \mathbb{T})$. In the same spirit, it turns out that replacing algebra $\mathcal{H}$ by $\mathcal{H}[W]$ leads to the replacement of $\mathfrak{M}(\mathcal{E}, \mathbb{T})$ to $\mathfrak{M}(\mathcal{E}, G)$, as a parameter space for simple modules. Specifically, the transition from Proposition 3.5 to G-bundles will be carried out in two steps. In the first step, we reinterpret the data involved in Proposition 3.5 in terms of loop groups, and in the second step we pass from loop groups to G-bundles.

We need some notation regarding formal loop groups. Let $\mathbb{C}((z))$, $\mathbb{C}[[z]]$, $\mathbb{C}[z]$ be the field of formal Laurent series, the ring of formal Taylor series and the ring of polynomials, respectively. Let $G((z))$ be the group of all $\mathbb{C}((z))$-rational points of G, and similarly for $G[[z]]$, $G[z]$. We consider $\mathbf{q}$-conjugacy classes in $G((z))$, i.e., $G((z))$-orbits on itself under $\mathbf{q}$-conjugation: $g(z) : h(z) \mapsto g(\mathbf{q}z)h(z)g(z)^{-1}$. A $\mathbf{q}$-conjugacy class, is said to be *integral* if it contains at least one element of $G[[z]]$.

Fix a Borel subgroup $B = T \cdot U \subset G$, where T is a maximal torus of G and U is the unipotent radical of B. By [BG, Lemma 2.2] we have the following:

Jordan q-normal form for G[[z]]. *Any element $h \in G[[z]]$ is $\mathbf{q}$-conjugate to a product $s \cdot b(z)$, where $s \in T$ is a constant loop, and $b \in U[z]$ are such that:*

(J1) $b(\mathbf{q}z) \cdot s = s \cdot b(z)$,

(J2) $\mathrm{Ad}\, s(v) = \mathbf{q}^m v$, *with* $v \in \mathrm{Lie}\, G$, $m > 0$, *implies* $v \in \mathrm{Lie}\, U$.

For any group M, we write M° for the identity connected component of M, and $Z_M(x)$ for the centralizer of an element x in M. Given $h \in G((z))$ we write $G_{\mathbf{q},h}$ for the $\mathbf{q}$-centralizer of $h(z)$ in $G((z))$:

$$G_{\mathbf{q},h} := \{g(z) \in G((z)) \mid g(\mathbf{q}z)h(z)g(z)^{-1} = h(z)\}.$$

Let $W_G = N_G(T)/T$ be the Weyl group of (G, T). Given $s \in T$, write $\lambda(s)$ for its image in $\Lambda = T/\mathbf{q}^{\mathbb{Z}}$, and let $W^{\lambda(s)}$ denote the isotropy group of the point $\lambda(s) \in T/\mathbf{q}^{\mathbb{Z}}$ under the natural W-action.

Theorem 5.6. *Let $h = s \cdot b \in G[[z]]$ be written in its $\mathbf{q}$-normal form. Then we have $G_{\mathbf{q},h} = Z_{G_{\mathbf{q},s}}(b)$. Furthermore,*

(i) *$G_{\mathbf{q},s}$ is a finite-dimensional reductive group isomorphic to a (not necessarily connected) subgroup $H \subset G$ containing the maximal torus T.*

(ii) *There exists a unipotent element $u \in H$, uniquely determined up to conjugacy in H, such that under the isomorphism in* (i) *we have* $G_{\mathbf{q},h} = Z_{G_{\mathbf{q},s}}(b) \xrightarrow{\sim} Z_H(u)$.

(iii) *The group $W^{\lambda(s)}$ is isomorphic to $W_H := N_H(T)/T$, the "Weyl group" of the disconnected group H.*

The proof of the theorem will follow from Lemma 5.11 and Proposition 5.13 given later in this section.

From loop group to G-bundles. In [BG] we have constructed a bijection:

$$\mathfrak{M}(\mathcal{E}, G) \stackrel{\Theta}{\longleftrightarrow} \textit{integral } \mathbf{q}\textit{-conjugacy classes in } G((z)) . \qquad (5.7)$$

Let $P = \Theta(h)$ be the G-bundle corresponding to a $\mathbf{q}$-conjugacy class of $h \in G((z))$, and $P^s = ss(P)$ its semisimplification. Without loss of generality we may assume that h is written in its $\mathbf{q}$-normal form $h = s \cdot b$. Using Theorem 5.6 it is easy to verify that under the bijection (5.7) we have:

- $P^s = \Theta(s)$ and $\operatorname{Aut} P^s \simeq G_{\mathbf{q},s} \simeq H \subset G$. (5.8.1)
- $\operatorname{Aut} P \simeq G_{\mathbf{q},h} \simeq Z_H(u)$. (5.8.2)

Further, recall the variety $\mathcal{B}(\mathcal{E}, G)_P$ of all B-structures on P, see (5.3). Let $\mathcal{B}(\mathcal{E}, G)^\circ_P$ denote a connected component of $\mathcal{B}(\mathcal{E}, G)_P$. Write $\mathcal{B}(H)$ for the flag variety of the group H, and $\mathcal{B}(H)_u$ for the Springer fiber over u, the u-fixed point set in $\mathcal{B}(H)$. Then we have:

- $\mathcal{B}(\mathcal{E}, G)^\circ_P \simeq \mathcal{B}(H)_u$. (5.8.3)

Furthermore, the natural $Z_{H^\circ}(u)$-action on $\mathcal{B}(H)_u$ goes under the isomorphism above and the imbedding: $Z_{H^\circ}(u) \hookrightarrow Z_H(u) = \operatorname{Aut} P$ to the natural Aut P-action on $\mathcal{B}(P)$.

By isomorphism (5.8.3), one identifies the action of the finite group $Z_{H^\circ}(u)/Z^\circ_H(u)$ on $H_{top}(\mathcal{B}(H)_u, \mathbb{C})$, the top homology, with the action of the corresponding subgroup of $\operatorname{Aut} P/\operatorname{Aut}^\circ P$ on $H_{top}(\mathcal{B}(\mathcal{E}, G)^\circ_P, \mathbb{C})$. It follows that an irreducible representation of $\operatorname{Aut} P/\operatorname{Aut}^\circ P$ is admissible in the sense of Definition 5.4 if and only if the restriction of the corresponding representation of $Z_H(u)/Z^\circ_H(u)$ to the subgroup $Z_{H^\circ}(u)/Z^\circ_H(u) \subset Z_H(u)/Z^\circ_H(u)$ is isomorphic to a direct sum of irreducible representations which have nonzero multiplicity in the $Z_{H^\circ}(u)/Z^\circ_H(u)$-module $H_{top}(\mathcal{B}(H)_u, \mathbb{C})$.

Finally, we observe that the isotropy group $W^{\lambda(s)}$ occurring in part (iii) of Theorem 5.6 is exactly the group whose irreducible representations label the simple objects of the category $\mathcal{M}(\mathcal{H}[W], \mathcal{A})$, see Proposition 3.5. Thus, according to the isomorphism $W^{\lambda(s)} \simeq W_H$ of Theorem 5.6 (iii), we are interested in a parametrisation of irreducible representations of the group W_H. Such a parametrisation is provided by a version of the Springer correspondence for disconnected reductive groups, developed in the last section (Appendix) of this paper. This concludes an outline of the proof of Theorem 5.6.

We now begin a detailed exposition, and recall the Bruhat decomposition for the group $G[z, z^{-1}]$. Let $G_1[z] \subset G[z]$ denote the subgroup of loops equal to $e \in G$ at $z = 0$ and denote by $\mathcal{U}^+$ the subgroup $U \cdot G_1[z]$. Similarly, $\mathcal{U}^-$ will denote $U^- \cdot G_1[z^{-1}]$ where $U^- \subset G$ is the unipotent subgroup opposite to U and $G_1[z^{-1}]$ is the kernel of evaluation map $G[z^{-1}] \to G$ at $z = \infty$.

Proposition 5.9 (cf. [PS, Chapter 8]). *Any element of $g(z) \in G[z, z^{-1}]$ admits a* unique *representation of the form*

$$g(z) = u_1(z) \cdot \lambda(z) \cdot n_w \cdot t \cdot u_2(z)$$

where $u_1(z), u_2(z) \in \mathcal{U}^+$, $\lambda(z) \in \mathbf{Y} = Hom_{alg}(\mathbb{C}^*, T)$, $t \in T$, $w \in W$ and $u_2(z)$, in addition, satisfies $[\lambda(z)n_w] \cdot u_2(z) \cdot [\lambda(z)n_w]^{-1} \in \mathcal{U}^-$.

Corollary 5.10. *The* **q**-*conjugacy classes that intersect $T \subset G((z))$ are parametrized by Λ/W.*

Proof. Suppose that $s \in T$ is **q**-conjugate to $s' \in T$ by an element $g(z) \in G((z))$. Rewriting this in the form $g(\mathbf{q}z)s = s'g(z)$, then using the above decomposition and its uniqueness, we obtain $s' = w(s) \cdot \lambda(\mathbf{q})$. Conversely, for any $w \in W$ and $\lambda \in \mathbf{Y}$, the element s is conjugate to $w(s) \cdot \lambda(\mathbf{q})$ by the element $g(z) = \lambda(z) \cdot n_w$. □

Uniqueness of the **q**-normal Jordan form follows from

Lemma 5.11. *Suppose that two loops $s \cdot b(z)$ and $s' \cdot b'(z)$ satisfy the Jordan form conditions (J1)-(J2), and that $f(\mathbf{q}z)(s \cdot b(z))f(z)^{-1} = s' \cdot b'(z)$ for some $f(z) \in G((z))$. Then $f(\mathbf{q}z) \cdot s \cdot f(z)^{-1} = s'$ and $f(z)b(z)f(z)^{-1} = b'(z)$.*

Proof. Choose a faithful representation: $G \to GL(V)$, and a basis in V such that U maps to upper-triangular matrices and T maps to diagonal matrices. We may assume without loss of generality that the loops $s \cdot b(z)$ and $s' \cdot b'(z)$ are both maped into upper-triangular matrices, $A(z)$ and $A'(z)$, resp.

First we consider the case when all diagonal entries of $A(z)$ (resp. $A'(z)$) differ only by powers of **q**, i.e., when they are of the form $a\mathbf{q}^{m_1}, \dots, a\mathbf{q}^{m_k}$, where k is the dimension of V and $m_1 \geq m_2 \geq \dots \geq m_k$, due to Jordan form condition (J2). Further, by the Jordan form condition (J1) all entries A_{ij} above the diagonal are of the form $\alpha_{ij}z^{m_i - m_j}$, $i < j$, $\alpha_{ij} \in \mathbb{C}$. Also let $a'\mathbf{q}^{n_1}, \dots, a'\mathbf{q}^{n_k}$ be the diagonal entries of $A'(z)$ and let $F(z)$ be the matrix corresponding to $f(z)$.

We prove by descending induction on $i - j$ that $F_{i,j} = cz^l$, for an appropriate constant c and an integer l, depending on i, j. Our proof is based on the simple observation that, for any constant B and any integer l, the equation $x(\mathbf{q}z) = \mathbf{q}^l x(z) + Bz^l$ admits a solution in $\mathbb{C}((z))$ if and only if $B = 0$, in which case the solution has to be $x(z) = cz^l$, $c \in \mathbb{C}$.

The largest value of $i - j$, attained for $i = k$, $j = 1$, corresponds to the lower left corner element $F_{k,1}(z)$. From the equation $F(\mathbf{q}z)A(z) = A'(z)F(z)$ one has $F_{k,1}(\mathbf{q}z)a\mathbf{q}^{m_1} = F_{k,1}(z)a'\mathbf{q}^{n_k}$. If the ratio a/a' is not a power of **q**, this equation, as well as other equations considered below, has only zero solution (which gives a noninvertible matrix $F(z)$). Hence we can assume that $a' = a\mathbf{q}^r$ for some integer r. Then the above equation for $F_{k,1}(z)$ implies that $F_{k,1}(z) = \phi_{k,1} z^{n_k - m_1 + r}$ for some constant $\phi_{k,1}$.

We use this expression for $F_{k,1}$ to write the equations for $F_{k-1,1}$ and $F_{k,2}$, then write the equations for $F_{k-2,1}$, $F_{k-1,2}$, $F_{k,3}$, etc. In general, by descending induction on $i - j$ (ranging from $i - j = k - 1$ to $i - j = -k + 1$) one obtains equations of the type

$$F_{i,j}(\mathbf{q}z) = \mathbf{q}^{n_i - m_j + r} g_{i,j}(z) + C z^{n_i - m_j + r}$$

for some constant C depending on i, j and the previously computed values of $g_{s,t}$. As before, this leads to

$$C = 0 \quad \text{and} \quad F_{i,j}(z) = \phi_{i,j} z^{n_i - m_j + r}, \quad \phi_{ij} \in \mathbb{C}.$$

This equation implies that $f(\mathbf{q}z) \cdot s \cdot f(z)^{-1} = s'$, and $f(z)b(z)f(z)^{-1} = b'(z)$ is an immediate consequence.

In the general case, by Jordan form condition (J1) one can choose a basis of V so that $A(z)$, $A'(z)$ will have square blocks as in the first part of the proof ("**q**-Jordan blocks") along the main diagonal, and zeros everywhere else. We can assume that any two diagonal entries which differ by a power of $\mathbf{q}$, belong to the same block. A direct computation shows that, up to permutation of blocks in $A(z)$ and $A'(z)$, the conjugating matrix $F(z)$ also has square blocks along the main diagonal and zeros everywhere else. Now we apply the above argument to each individual block to obtain the result in the general case. □

Corollary 5.12. (i) *The assignment:* $s \cdot b(z) \mapsto s$ *descends to a well-defined map* $\Phi: \{\text{integral } \mathbf{q} - \text{conjugacy classes in } G((z))\} \longrightarrow \Lambda/W$.

(ii) *Let* $s \cdot b(z)$ *be a Jordan* **q***-normal form, and* $\lambda \in \Lambda/W$ *the image of* $s \in T$ *in* Λ/W. *Then the set* $\Phi^{-1}(\lambda)$ *can be identified with those (ordinary) conjugacy classes in* $G_{\mathbf{q},s}$, *the* **q***-centralizer of* s *in* $G((z))$, *which have nontrivial intersection with* $U[z]$.

Remark. We will see below that $G_{\mathbf{q},s}$ is a finite-dimensional reductive group and that $\Phi^{-1}(\lambda)$ is nothing but the set of unipotent conjugacy classes in this reductive group.

Now we begin to study the automorphism group of the G-bundle P^s associated to $s \in T$. To describe $G_{\mathbf{q},s} \simeq \operatorname{Aut} P^s$ first recall that by [BG, Lemma 2.5], $G_{\mathbf{q},s}$ consists of polynomial loops, i.e., $G_{\mathbf{q},s} \subset G[z, z^{-1}]$. Thus, there is a well-defined evaluation map $ev_{z=1} : G_{\mathbf{q},s} \to G$ sending a polynomial loop to its value at $z = 1$. Let $H \subset G$ be the image of $G_{\mathbf{q},s}$. Write $N_H(T)$ for the normalizer of T in H, and $W_H := N_H(T)/T$ for the "Weyl group" of the (generally *diconnected*) group H.

Proposition 5.13. (i) *The evaluation map* $ev_{z=1} : G_{\mathbf{q},s} \to G$ *is injective;*

(ii) *The idenity component H° of H equals the connected reductive subgroup of G corresponding to the root subsystem $\Delta_{\mathbf{q},s} \subset \Delta$ of all roots $\alpha \in \Delta \subset Hom(T, \mathbb{C}^*)$ for which $\alpha(s)$ is an integral power of $\mathbf{q}$;*

(iii) *The group W_H is isomorphic to the subgroup $W^\lambda \subset W$ of all $w \in W$ which fix $\lambda \in \Lambda = \mathbb{C}^*/\mathbf{q}^{\mathbb{Z}}$, the image of $s \in T$.*

Proof. The equation $g(\mathbf{q}z) \cdot s \cdot g(z)^{-1} = s$ can be rewritten as $g(\mathbf{q}z) = s \cdot g(z) \cdot s^{-1}$. We decompose $g(z)$ as in Proposition 5.9 and, using the uniqueness of this decomposition, obtain

$$u_1(\mathbf{q}z) = su_1(z)s^{-1}, \quad u_2(\mathbf{q}z) = su_2(z)s^{-1}, \quad s = w(s) \cdot \lambda(\mathbf{q}).$$

Rewrite $u_1(z) \in \mathcal{U}^+$ as $exp\left(\sum_{k=0}^{\infty} g_k z^k\right)$; then $\mathrm{Ad}\, s(g_k) = \mathbf{q}^k$ due to the first equation. In particular, only finitely many of g_k are nonzero and by the Jordan form condition (J2), $g_k \in \mathrm{Lie}\,(U)$. Hence $u_1(z) \in U[z]$ and, since different eigenspaces of $\mathrm{Ad}\, s$ on $\mathrm{Lie}\,(U)$ have zero intersection, $u_1(z)$ is uniquely determined by $u_1(1) = exp(\sum_{k=0}^{N} g_k)$. The same argument applies to $u_2(z)$. Moreover, since $u_2' = [\lambda(z)n_w]u_2(z)[\lambda(z)n_w]^{-1} \in U^- \cdot G_1[z^{-1}]$ and $\mathrm{Ad}\, su_2'(z) = u_2'(\mathbf{q}z)$, we can repeat the argument once more and conclude that $u_2'(1) = n_w u_2(1) n_w^{-1} \in U^-$.

Now we can show that $g(z)$ is determined by $g(1) = u(1)n_w t u_2(1)$. In fact, since $u_1(1), u_2(1) \in U$ and $n_w u_2(1) n_w^{-1} \in U^-$, the usual Bruhat decomposition for $g(1) \in G$ implies that n_w, $u_1(1)$ and $u_2(1)$ are uniquely determined by $g(1)$, hence $u_1(z)$ and $u_2(z)$ are uniquely determined by $g(1)$. The element $\lambda(z)$ can be reconstructed from a and w since $s = w(s) \cdot \lambda(\mathbf{q})$ and $\mathbf{q}$ is not a root of unity. The proposition follows. □

Example. The following example, showing that the component group H/H° can in fact be nontrivial, was kindly communicated to us by D. Vogan.

Recall that for the root system of type D_4, the coroot lattice $\mathbf{Y}$ can be identified with the subgroup of the standard Eucledian lattice $L_4 = \langle e_1, e_2, e_3, e_4 \rangle$, $(e_i, e_j) = \delta_{ij}$ formed by all vectors in L_4 with even sum of coordinates. Then the set of coroots is identified with $\pm e_i \pm e_j$, $i \neq j$, and the Weyl group W acts by permuting the e_i, and changing the sign of any even number of the basis vectors e_i. The choice of the simple coroots $\alpha_1^\vee = e_1 - e_2$, $\alpha_2^\vee = e_2 - e_3$, $\alpha_3^\vee = e_3 - e_4$, $\alpha_4^\vee = e_3 + e_4$ identifies $T = \mathbf{Y} \otimes_{\mathbb{Z}} \mathbb{C}^*$ with $(\mathbb{C}^*)^4$. Now consider the element $s = (-1, \sqrt{\mathbf{q}}, -1, -\sqrt{\mathbf{q}}) \in (\mathbb{C}^*)^4 \simeq T$. A straightforward calculation shows that, in the notation of the above proposition, $W_H = \{\pm 1\}$ while $\Delta_{\mathbf{q},s}$ is empty.

End of proof of Theorem 5.5. By Proposition 3.5 we have to establish the correspondence between the set of pairs (λ, χ) where $\lambda \in \Lambda$, $\chi \in \widehat{W^\lambda}$, and the set of pairs (P, α) as in the statement of Theorem 5.5. Take any lift $s \in T$ of the element $\lambda \in \Lambda$ and consider the G-bundle P^s corresponding to s, together with its automorphism group H. By Proposition 5.13 (iii) and the Springer Correspondence (see Appendix), the representation χ defines a unipotent H-orbit together with the admissible representation α of the centralizer of any point u in this orbit. The element $u \in H$ corresponds via Proposition 5.13 (i) to a certain loop $b(z)$, such that $s \cdot b(z)$ is a $\mathbf{q}$-normal form. The bundle P corresponds to the $\mathbf{q}$-conjugacy class of $s \cdot b(z)$. □

It will be convenient for us in the next section to reinterpret the parameters (P, α) entering Theorem 5.5 in a different way as follows. First, giving $P \in \mathfrak{M}(\mathcal{E}, G)$ is equivalent, according to (5.7), to giving the $\mathbf{q}$-conjugacy class of an element $h(z) \in G[[z]]$. Using the Jordan $\mathbf{q}$-normal form, write $h(z) = s \cdot b(z)$, where $s \in T$, is a semisimple element in G, the subgroup of constant loops. Furthermore, by Theorem 5.6 we have $\operatorname{Aut} P/\operatorname{Aut}^\circ P = Z_{G_{\mathbf{q},s}}(b)/Z^\circ_{G_{\mathbf{q},s}}(b)$.

Let $Q = ss(P)$ be the semisimplification of P. By (5.8.1), this is the G-bundle on $\mathcal{E}$ corresponding, under the bijection (5.7), to the constant loop s. Let G_Q denote the associated vector bundle on $\mathcal{E}$ corresponding to the principal G-bundle Q and the adjoint representation of the group G. By construction, $b(z)$ is a polynomial loop with unipotent values that $\mathbf{q}$-commutes with s. Hence $b(z)$ gives rise to a unipotent automorphism $\hat{b} \in \operatorname{Aut} Q$. This way one obtains a bijection:

$$\mathfrak{M}(\mathcal{E}, G) \longleftrightarrow \left\{ \begin{array}{c} \text{semisimple } G\text{-bundle } Q \in \mathfrak{M}(\mathcal{E}, G)^{ss} \\ \text{and a unipotent element } u \in \operatorname{Aut} Q \end{array} \right\}. \tag{5.14}$$

It is not difficult to show that the set $\mathcal{B}(\mathcal{E}, G)_P$, see (5.8.3) is identified, under the bijection above, with the set of u-stable B-structures on the G-bundle $ss(P)$.

Fix $\mathbf{q} \in \mathbb{C}^*$, which is not a root of unity. An element of the group $G((z))$ will be called *$\mathbf{q}$-semisimple*, resp. *$\mathbf{q}$-unipotent*, if it is $\mathbf{q}$-conjugate to a constant semisimple loop, resp. conjugate (in the ordinary sense) to an element of $U[z]$. Write $G((z))^{\mathbf{q}-\text{ss}}$ and $G((z))^{\mathbf{q}-\text{uni}}$ for the sets of $\mathbf{q}$-semisimple and $\mathbf{q}$-unipotent elements, respectively. Given $h(z) \in G((z))$, recall the notation $G_{\mathbf{q},h}$ for the $\mathbf{q}$-centralizer of h in $G((z))$, and for any $u(z) \in G((z))$, put

$$Z_{\mathbf{q},h}(u) = \{g(z) \in G((z)) \;\big|\; g(\mathbf{q}z)h(z) = h(z)g(z) \;\;\&\;\; g(z)u(z) = u(z)g(z)\}$$

a simultaneous "centralizer" of $h(z)$ and $u(z)$. If h is $\mathbf{q}$-semisimple and u $\mathbf{q}$-commutes with h, then the group $Z_{\mathbf{q},h}(u)$ acts on $\mathcal{B}(G_{\mathbf{q},h})_u$, the u-fixed point set in the flag variety of the finite-dimensional reductive group $G_{\mathbf{q},h}$, see Theorem 5.6 (i). This gives a $Z_{\mathbf{q},h}(u)/Z^\circ_{\mathbf{q},h}(u)$-action on $H_*(\mathcal{B}(G_{\mathbf{q},h})_u)$, the *total* homology. An irreducible representation of the component group $Z_{\mathbf{q},h}(u)/Z^\circ_{\mathbf{q},h}(u)$ is said to be *admissible* if it occurs in $H_*(\mathcal{B}(Z_{\mathbf{q},h})_u)$ with nonzero multiplicity. We let $Z_{\mathbf{q},h}\widehat{(u)/Z^\circ_{\mathbf{q},h}}(u)$ denote the set of admissible $Z_{\mathbf{q},h}(u)/Z^\circ_{\mathbf{q},h}(u)$-modules (cf. Definition 5.2 and the paragraph below formula (5.8.3)).

We now consider the following set:

$$\mathbf{M} = \left\{ (s, u, \chi) \;\middle|\; \begin{array}{c} s \in G((z))^{\mathbf{q}-\text{ss}}, \; u \in G((z))^{\mathbf{q}-\text{uni}} \\ s(z)u(z)s(z)^{-1} = u(\mathbf{q}z), \; \chi \in Z_{\mathbf{q},s}\widehat{(u)/Z^\circ_{\mathbf{q},s}}(u) \end{array} \right\}. \tag{5.15}$$

Thus, we can reformulate Theorem 5.5 as follows:

Theorem 5.16. *There exists a natural bijection between the set of isomorphism classes of simple objects of $\mathcal{M}(\mathcal{H}[W], \mathcal{A})$ and the set of* $\mathbf{q}$*-conjugacy classes in* $\mathbf{M}$.

Springer correspondence for disconnected groups

In this appendix we show how to extend the classical Springer Correspondence to the case of not necessarily connected reductive groups. First we recall briefly (see [CG, Chapter 3] for details) the situation for a general *connected* reductive group, such as the group H° of §5.

Let $\mathcal{N} \subset H^\circ$ be the subset of unipotent elements and $\mathcal{B}$ the variety of Borel subgroups in H°. The subvariety $\widetilde{\mathcal{N}} \subset \mathcal{B} \times \mathcal{N}$ of all pairs $\{(B_H, u) \mid u \in B_H\}$, provides an H°-equivariant smooth resolution $\pi : \widetilde{\mathcal{N}} \to \mathcal{N}$, called the *Springer resolution*.

Denote by $\mathcal{Z}$ the fiber product $\widetilde{\mathcal{N}} \times_{\mathcal{N}} \widetilde{\mathcal{N}}$, which can also be indentified with (cf. [CG]) the subvariety in $T^*(\mathcal{B} \times \mathcal{B})$ given by the union of conormal bundles to the H°-orbits on $\mathcal{B} \times \mathcal{B}$ (with respect to the diagonal action). The top Borel–Moore homology group $H(\mathcal{Z})$ is endowed with a structure of an associative algebra via the convolution product (see [CG]). Moreover, the set $\mathbb{W} \subset H(\mathcal{Z})$ of fundamental classes of irreducible components of $\mathcal{Z}$ forms a group with respect to the convolution product, called the *abstract Weyl group*, and $H(\mathcal{Z})$ can be identified with the group algebra of $\mathbb{W}$. A particular choice of a Borel subgroup $B_H \supset T$ identifies the usual Weyl group $W^\circ = N_{H^\circ}(T)/T$ with $\mathbb{W}$ by sending the class of $n_w \in N_{H^\circ}$ to the fundamental class of the conormal bundle to the H°-orbit of $(B_H, n_w B_H n_w^{-1}) \in \mathcal{B} \times \mathcal{B}$.

Consider a unipotent orbit $\mathcal{O} \subset \mathcal{N}$. The top Borel–Moore homology groups of the fibers of $\pi : \widetilde{\mathcal{N}} \to \mathcal{N}$ over $\mathcal{O}$ form an irreducible local system $L_{\mathcal{O}}$ on $\mathcal{O}$ which is equivariant with respect to $\mathbb{W} \times H^\circ$ (the action of $\mathbb{W}$ in the fibers of $L_{\mathcal{O}}$ comes from the convolution construction, cf. [CG], and the action of H° from the H°-equivariance of π). Decompose $L_{\mathcal{O}}$ into a direct sum of irreducible $\mathbb{W} \times H^\circ$-equivariant local systems $L_1, \ldots, L_k$. For any representation ϕ of $\mathbb{W}$ we can consider the local system I_i formed by the $\mathbb{W}$-invariants of the tensor product $\phi^\vee \otimes L_i$. It turns our that, for any irreducible representation ϕ, there exists a unique orbit $\mathcal{O}_\phi$ and a unique $L_\phi \in \{L_1, \ldots, L_k\}$ for which the local system I_ϕ, constructed from ϕ as above, is nonzero. Moreover, such I_ϕ is an irreducible H°-equivariant local system associated to an "admissible representation" of the component group of the centralizer Z_u of a point $u \in \mathcal{O}$. Recall (see Definition 5.4 and the discussion before Proposition 5.9) that the representation of $Z_H(u)/Z_H^\circ(u)$ is called *admissible* if its restriction to the subgroup $Z_{H^\circ}(u)/Z_H^\circ(u) \subset Z_H(u)/Z_H^\circ(u)$ is isomorphic to a direct sum of irreducible representations which have nonzero multiplicity in the $Z_{H^\circ}(u)/Z_H^\circ(u)$-module $H_{top}(\mathcal{B}(H)_u, \mathbb{C})$, where $\mathcal{B}(H)_u$ is the fiber $\pi^{-1}(u)$ over any point of the orbit $\mathcal{O}$.

Below we will use the language of equivariant local systems (which is equivalent to the language of admissible representations).

The *Springer correspondence* $\phi \mapsto (\mathcal{O}_\phi, I_\phi)$ gives a bijection between the set of irreducible representations of its Weyl group $\mathbb{W}$ and the set of pairs $(\mathcal{O}, I)$ where $\mathcal{O}$ is a unipotent orbit of H° and I is a certain H°-equivariant irreducible local system on $\mathcal{O}$ coming from an admissible representation of Z_u/Z_u°.

We proceed to representation theory of the "Weyl group" $W_H = N_H(T)/T$ of a *disconnected* reductive group H (see Proposition 5.13).

Lemma A.1. *A choice of a Borel subgroup $B_H \supset T$ in H° identifies the Weyl group W_H with the semidirect product $W^\circ \rtimes (W_H/W^\circ)$. Moreover, one has a canonical isomorphism $H/H^\circ \simeq W_H/W_H^\circ$.*

Proof. Consider the subgroup $N'(T) := N_H(B_H) \cap N_H(T)$. Then the embedding $N'(T) \subset H$ induces the isomorphisms

$$H/H^\circ \simeq N'(T)/T \simeq W_H/W^\circ.$$

Since $N'(T)$ is a subgroup of $N_H(T)$, we obtain an embedding $W_H/W^\circ \subset W_H$. Now the assertion of the lemma follows. □

Remark. A different choice of a Borel subgroup B'_H containing T gives a conjugate embedding $w(W_H/W^\circ)w^{-1} \subset W_H$, where $w \in W^\circ$ is the unique element which conjugates B_H into B'_H.

Note that H acts on $\mathcal{N}$ and on $\widetilde{\mathcal{N}}$. In particular H permutes the irreducible components of $\mathcal{Z}$. This induces an H/H°-action on $\mathbb{W}$ by group automorphisms.

Proposition A.2. *The isomorphism $\mathbb{W} = W^\circ$ (depending on the choice of B_H) and the canonical isomorphism $H/H^\circ \simeq W_H/W^\circ$ identify the above action of H/H° on $\mathbb{W}$ with the conjugation action of W_H/W° on W° arising from Lemma A.1.*

Proof. It suffices to replace the pair (H, H°) by $(N'(T), T)$. Let n_w be a lift to N_{H° of a certain element $w \in W$, and let $\mathcal{Z}_w$ be the cotangent bundle to the orbit of $(B_H, n_w B_H n_w^{-1}) \in \mathcal{B} \times \mathcal{B}$. Similarly, let $n_\sigma \in N'(T)$ be a lift of an element $\sigma \in W_H/W^\circ$. Denote $\sigma w \sigma^{-1} \in W^\circ \subset W_H$ by w^σ, then $n_{w^\sigma} = n_\sigma n_w n_\sigma^{-1}$ is a lift of w^σ to $N_{H^\circ}(T)$.

By definition of $N'(T)$ the element n_σ normalizes B_H. Hence n_w sends $(B_H, n_w B n_w^{-1})$ to $(B_H, n_{w^\sigma} B_H n_{w^\sigma}^{-1})$. Thus, $n_\sigma \cdot \mathcal{Z}_w = \mathcal{Z}_{w^\sigma}$. □

Now we recall the basic facts of Clifford theory (cf. [Hu]) which apply to any finite group W_H and its normal subroup W°, not necessarily arising as Weyl groups.

The group W_H acts by conjugation on the set $\widehat{W}^\circ$ of irreducible representations of W°. Let $\mathcal{V}_1 \dots \mathcal{V}_k$ be the orbits of its action. For any irreducible representation $\psi \in \widehat{W}_H$ we can find an orbit $\mathcal{V}_{i(\psi)}$ and a positive integer e, such that the restriction of ψ to W° is isomorphic to a multiple of the orbit sum:

$$\psi|_{W_0} \simeq e \cdot \left(\sum\nolimits_{\phi \in \mathcal{V}_{i(\psi)}} \phi\right).$$

Fix $\phi \in \mathcal{V}_{i(\psi)}$ and consider the subset $(\widehat{W}_H)_\phi \subset \widehat{W}_H$ of all representations whose restriction to W° contains an isotypical component isomorphic to ϕ (and hence automatically all representations in the orbit of ϕ). Obviously, $\widehat{W}_H$ is a disjoint union of $(\widehat{W}_H)_{\phi_i}$, where $\phi_i \in \mathcal{V}_i$ is any representative of the orbit $\mathcal{V}_i$.

To study $(\widehat{W}_H)_\phi$ we consider the stabilizer $W^\phi \subset W_H$ of $\phi \in \widehat{W}^\circ$. Then by Clifford theory (cf. [Hu]), the induction from W^ϕ to W_H establishes a bijection between $(\widehat{W}^\phi)_\phi$ and $(\widehat{W}_H)_\phi$. Moreover, any linear representation $\chi \in (\widehat{W}^\phi)_\phi$ is isomorphic to the tensor product $p_1 \otimes p_2$ of two *projective* representations p_1 and p_2 (cf. [Hu]) such that

(i) $p_1(x) = \phi(x)$, $p_2(x) = 1$ if $x \in W^\circ$

(ii) $p_1(gx) = p_1(g)\phi(x)$ and $p_1(xg) = \phi(x)p_1(g)$ if $x \in W^\circ, g \in W_H$.

Thus, p_2 is a projective representation of W_H/W° which plays the role of the multiplicity space of dimension e in terms of the above formula for $\psi|_{W^\circ}$. The second condition means that the projective cocycle of p_1 is in fact lifted from W^ϕ/W°. From now on we will fix the decomposition $\chi = p_1 \otimes p_2$.

Let $\widehat{W}_H$ denote the set of isomorphism classes of irreducible representations of W_H. Further, for any unipotent conjugacy class $\mathcal{O} \subset H$, let $\mathsf{Admiss}(\mathcal{O})$ stand for the set of (isomorphism classes of) irreducible admissible (in the sense specified before Lemma A.1) H-equivariant local systems on $\mathcal{O}$.

Theorem A.3. *There exists a bijection between the following sets*:

$$\widehat{W}_H \quad \longleftrightarrow \quad \left\{ \begin{matrix} \text{unipotent conjugacy class} \\ \mathcal{O} \subset H \text{ and } \alpha \in \mathsf{Admiss}(\mathcal{O}) \end{matrix} \right\}.$$

Proof. Take an irreducible representation ρ of W_H and let $\phi \in \widehat{W}^\circ$ be an irreducible subrepresentation of $\rho|_{W^\circ}$. By Clifford theory ρ is induced from a certain representation $\chi \in (\widehat{W}^\phi)_\phi$ as above.

Recall that by Springer Correspondence for W° the irreducible representation ϕ gives rise to a unipotent H° orbit $\mathcal{O}_\phi$ together with an H°-equivariant local system I_ϕ. We will show how to construct from $\chi = p_1 \otimes p_2$ the corresponding local system for H.

As a first step, we will construct a cerain local system $\tilde{I}_\chi$ on $\mathcal{O}_\phi$. This local system is equivariant with respect to the subgroup $H^\phi \subset H$ which corresponds to $W^\phi \subset W_H$ via the isomorphism $H/H^\circ = W_H/W^\circ$ of Lemma A.1 (it is easy to prove using Proposition A.2 that H^ϕ is the subgroup of all elements in H which preserve the orbit $\mathcal{O}$ and the local system I_ϕ). Then, imitating the induction map $(\widehat{W}^\phi)_\phi \to (\widehat{W}_H)_\phi$ we will obtain an H-equivariant local system on the unique unipotent H-orbit which contains $\mathcal{O}_\phi$ as its connected component.

We fix the choice of Borel subgroup B_H and, in particular, the isomorphisms: $\mathbb{W} \simeq W^\circ$ and $W_H \simeq W^\circ \rtimes (W_H/W^\circ)$.

Step 1. Recall that the Springer resolution $\pi : \widetilde{\mathcal{N}} \to \mathcal{N}$ is H-equivariant. It follows from the definitions that there exists an action of H^ϕ on the total space of L_ϕ, which extends the natural action of H°. However, the extended action does not commute with the W°-action any more. Instead, it satisfies the identity $h(w \cdot s) = h(w) \cdot h(s)$ where $h \in H^\phi$, $w \in W^\circ$ and s is a local section of L_ϕ. We would like to use the formula $I_\phi = (\phi^\vee \otimes L_\phi)^{W^\circ}$ to define the H^ϕ-equivariant structure on I_ϕ. To that end, we have to construct an action of H^ϕ on $\phi^\vee$ which agrees with the W°-action in the same way as before. This is done by using the composition $H^\phi \to H^\phi/H^\circ = W^\phi/W^\circ \hookrightarrow W_H$ and the projective action of W_H on $p_1^\vee$ coming from Clifford theory (recall that $p_1^\vee$ extends the W°-action on the vector space of $\phi^\vee$). By Proposition A.2 the two actions of H^ϕ on W° coincide, hence the projective action of H^ϕ on $p_1^\vee \otimes L_\phi$ indeed satisfies $h(w \cdot s) = h(w) \cdot h(s)$. Consequently, the local system $I_\phi = (p_1^\vee \otimes L_\phi)^{W^\circ}$ carries a projective action of H^ϕ.

Now we tensor the local system I_ϕ with the vector space of the projective representation $p_2^\vee$. Since H^ϕ acts on the vector space of $p_2^\vee$ via the same composition $H^\phi \to H^\phi/H^\circ \simeq W^\phi/W^\circ \hookrightarrow W_H$, the tensor product $p_2^\vee \otimes I_\phi$ carries an *a priori* projective action of H^ϕ. However, since the projective cocyles of p_1 and p_2, well-defined as functions on W^ϕ/W°, are mutually inverse, the same can be said about the projective cocycles of the H^ϕ-actions on I_ϕ and $p_2^\vee$. Therefore, these cocycles cancel out giving a *linear* action on the tensor product. This means that $p_2^\vee \otimes I_\phi$ is given the structure of an H^ϕ-equivariant local system, to be denoted by $\tilde{I}_\chi$.

Step 2. Next we consider a larger subgroup $\widehat{H}^\phi$ which preserves the unipotent orbit $\mathcal{O}_\phi$, but not necessarily the local system I_ϕ. It is easy to check that the composition $\widehat{H}^\phi \times_{H^\phi} \tilde{I}_\phi \to \widehat{H}^\phi \times_{H^\phi} \mathcal{O}_\phi \to \mathcal{O}_\phi$ defines an $\widehat{H}^\phi$-equivariant local system $\hat{I}_\chi$ over $\mathcal{O}_\phi$. Finally, $I_\chi = H \times_{\widehat{H}^\phi} \hat{I}_\chi$ is an H-equivariant local system over $H \times_{\widehat{H}^\phi} \mathcal{O}_\phi$. Note that the latter space is nothing but the union of those unipotent H°-orbits which are conjugate to each other with respect to the larger group H, i.e., a single unipotent H-orbit. It is easy to check that the assignment: $\chi \mapsto I_\chi, \chi \in (\widehat{W}^\phi)_\phi$ together with the decomposition: $\widehat{W}_H = \bigcup_{\phi_i \in \mathcal{V}_i} (\widehat{W}_H)_\phi$ and the induction isomorphisms: $(\widehat{W}^\phi)_\phi \xrightarrow{\sim} (\widehat{W}_H)_\phi$ yields the correspondence of the theorem. □

References

[BG] V. Baranovsky, V. Ginzburg. Conjugacy classes in loop groups and G-bundles on elliptic curves, *Intern. Math. Research Notices* **15** (1996), 733–751.

[CG] N. Chriss, V. Ginzburg. *Representation theory and Complex Geometry*, Birkhäuser, Boston 1997.

[Hu] B. Huppert: Character theory of finite groups. *de Gruyter Expositions in Mathematics* **25**, Walter de Gruyter & Co., Berlin, 1998.

[La] Y. Laszlo. About G-bundles over elliptic curves, *Ann. Inst. Fourier (Grenoble)* **48** (1998), no. 2, 413–424.

[LS] T. Levasseur, J.T. Stafford, Semi-simplicity of invariant holonomic systems on a reductive Lie algebra, *Amer. J. Math.* **119** (1997), 1095–1117.

[M] S. Montgomery, *Fixed Rings of Finite Automorphism Groups of Associative Rings*, LN in Math., **818** Springer-Verlag, Berlin, New York, 1980.

[MR] J.C. McConnell, J.C. Robson: *Noncommutative Noetherian Rings*, Wiley, Chichester, 1987.

[PS] A. Pressley, G. Segal: Loop groups. *Oxford Mathematical Monographs*, Clarendon Press, Oxford, 1986.

Vladimir Baranovsky
Department of Mathematics 253-37
Caltech
Pasadena CA 91125, USA
baranovs@caltech.edu

Sam Evens
Department of Mathematics
University of Notre Dame
Notre Dame IN, 46656 USA
evens.1@nd.edu

Victor Ginzburg
Department of Mathematics
University of Chicago
Chicago IL 60637, USA
ginzburg@math.uchicago.edu

Dixmier Algebras for Classical Complex Nilpotent Orbits via Kraft–Procesi Models I

Ranee Brylinski

Dedicated to Professor Alexander Kirillov on his 2^6 birthday

ABSTRACT We attach a Dixmier algebra $\mathcal{B}$ to the closure $\overline{\mathcal{O}}$ of any nilpotent orbit of G where G is $GL(n, \mathbb{C})$, $O(n, \mathbb{C})$ or $Sp(2n, \mathbb{C})$. This algebra $\mathcal{B}$ is a noncommutative analog of the coordinate ring $\mathcal{R}$ of $\overline{\mathcal{O}}$, in the sense that $\mathcal{B}$ has a G-invariant algebra filtration and $\operatorname{gr}\mathcal{B} = \mathcal{R}$.

We obtain $\mathcal{B}$ by making a noncommutative analog of the Kraft–Procesi construction which modeled $\overline{\mathcal{O}}$ as the algebraic symplectic reduction of a finite-dimensional symplectic vector space L. Indeed $\mathcal{B}$ is a subquotient of the Weyl algebra for L.

$\mathcal{B}$ identifies with the quotient of $\mathcal{U}(\mathfrak{g})$ by a two-sided ideal J, where $\mathfrak{g} = \operatorname{Lie}(G)$. Then $\operatorname{gr} J$ is the ideal $\mathfrak{I}(\overline{\mathcal{O}})$ in $S(\mathfrak{g})$ of functions vanishing on $\overline{\mathcal{O}}$. In every case where $\mathcal{O}$ is connected, J is a completely prime primitive ideal.

1 Introduction

By means of symplectic reduction in the setting of complex algebraic varieties, Kraft and Procesi ([22],[23]) constructed a model of the closure of any nilpotent coadjoint orbit $\mathcal{O}$ of G when G is one of the classical groups $GL(n, \mathbb{C})$, $O(n, \mathbb{C})$ and $Sp(2n, \mathbb{C})$. The symplectic aspect is not actually mentioned, but the construction is clearly symplectic.

In this paper we give a noncommutative analog, or quantization, of the Kraft–Procesi construction. The result is that we attach a Dixmier algebra $\mathcal{B}$ to each orbit closure $\overline{\mathcal{O}}$. Our algebra $\mathcal{B}$ has a G-invariant algebra filtration and we show that $\operatorname{gr}\mathcal{B}$ is isomorphic, as a graded Poisson algebra, to the coordinate ring $\mathcal{R}$ of $\overline{\mathcal{O}}$.

In fact, $\mathcal{B}$ identifies, as a filtered algebra, with the quotient $\mathcal{U}(\mathfrak{g})/J$ of the universal enveloping algebra $\mathcal{U}(\mathfrak{g})$ of $\mathfrak{g} = \operatorname{Lie}(G)$ by some two-sided ideal J. Then $\operatorname{gr} J$ is the ideal $\mathfrak{I}(\overline{\mathcal{O}})$ in $S(\mathfrak{g})$ defining $\overline{\mathcal{O}}$. We find that J is stable under the prin-

Keywords and phrases: nilpotent coadjoint orbit, Dixmier algebra, symplectic reduction
2000 AMS Mathematics Subject Classification: 17B35, 53D50, 22E46, 14L30

cipal anti-automorphism of $\mathcal{U}(\mathfrak{g})$, and also under the anti-linear automorphism of $\mathcal{U}(\mathfrak{g})$ defined by a Cartan involution of $\mathfrak{g}$.

The Kraft–Procesi construction attaches to $\mathcal{O}$ a complex symplectic vector space L together with a Hamiltonian action of $G \times S$ on L, where S is an auxiliary complex reductive Lie group. The actions of G and S lie inside the symplectic group $Sp(L, \mathbb{C})$. Kraft and Procesi show that $\overline{\mathcal{O}}$ is scheme-theoretically the algebraic symplectic reduction of L by S. In this way, they obtain $\mathcal{R}$ as a subquotient of the algebra $\mathcal{P}$ of polynomial functions on L. More precisely, $\mathcal{R}$ is realized as $\mathcal{P}^{inv}/\mathcal{I}^{inv}$ where $\mathcal{I}$ is an ideal in $\mathcal{P}$ and the superscript inv denotes taking S-invariants.

It is nice from the viewpoint of representation theory to regard $\mathcal{R}$ as a subquotient of the algebra $\mathcal{P}^{even}$ of even polynomials. (We can do this since $\mathcal{P}^{inv}$ lies in $\mathcal{P}^{even}$.) For $\mathcal{P}^{even}$ is the coordinate ring of the closure of the minimal nilpotent orbit $\mathcal{Y}$ of $Sp(L, \mathbb{C})$.

To make a noncommutative analog of the Kraft–Procesi construction, we start from the fact that there is a unique Dixmier algebra attached to $\overline{\mathcal{Y}}$, namely the quotient of $\mathcal{U}(\mathfrak{sp}(L, \mathbb{C}))$ by its Joseph ideal $\mathcal{J}$. We can model $\mathcal{U}(\mathfrak{sp}(L, \mathbb{C}))/\mathcal{J}$ as the even part $\mathcal{W}^{even}$ of the Weyl algebra $\mathcal{W}$ for L and then $\operatorname{gr}\mathcal{W}^{even} = \mathcal{P}^{even}$. There is an obvious quantization of the Hamiltonian S-action on L, namely the natural $(\mathfrak{s} \oplus \mathfrak{s}, S_c)$-module structure on $\mathcal{W}$. Here the subscript c denotes taking a compact real form.

We define $\mathcal{B}$ to be the coinvariants for the $(\mathfrak{s} \oplus \mathfrak{s}, S_c)$-action on $\mathcal{W}^{even}$. A priori, $\mathcal{B}$ is a $(\mathfrak{g} \oplus \mathfrak{g}, G_c)$-module with a G_c-invariant filtration, but $\mathcal{B}$ is not an algebra. However, we easily identify $\mathcal{B}$ as the quotient by a two-sided ideal of $\mathcal{W}^{inv}$ (where the superscript again indicates taking S-invariants, or equivalently, S_c-invariants). In this way, $\mathcal{B}$ becomes a filtered algebra and a subquotient of $\mathcal{W}^{even}$.

Our main result (Theorem 6.3) is to compute the associated graded algebra $\operatorname{gr}\mathcal{B}$. It is easy to see that $\operatorname{gr}\mathcal{B}$ is some quotient of $\mathcal{P}^{inv}/\mathcal{I}^{inv}$, but in fact we prove $\operatorname{gr}\mathcal{B} = \mathcal{P}^{inv}/\mathcal{I}^{inv}$. To do this, we recognize $\mathcal{B}$ as the degree zero part of the relative Lie algebra homology $H(\mathfrak{s} \oplus \mathfrak{s}, S_c; \mathcal{W}^{even})$. We consider the standard complex which computes this homology, introduce a filtration and then apply the spectral sequence for a filtered complex. We compute the E_1 term of the spectral sequence by using the fact proven by Kraft and Procesi that $\mathcal{I}$ is a complete intersection ideal. Then we easily show $E_1 = E_\infty$.

We establish some properties of $\mathcal{B}$ and the corresponding ideal J. If $\mathcal{O}$ is connected, then J is a completely prime primitive ideal (Corollary 6.5). In every case, $\mathcal{B}$ admits a unique $\mathfrak{g}^\sharp$-invariant Hermitian inner product $(\cdot|\cdot)$ such that $(1|1) = 1$, where $\mathfrak{g}^\sharp$ is a real form of $\mathfrak{g} \oplus \mathfrak{g}$ with $\mathfrak{g}^\sharp \simeq \mathfrak{g}$ (see Proposition 3.1). This prompts the question as to whether J is "good" in the sense that J is maximal and $\mathcal{U}(\mathfrak{g})/J$ is unitarizable. The latter property means (since $\mathcal{B} \simeq \mathcal{U}(\mathfrak{g})/J$) that $(\cdot|\cdot)$ is positive-definite.

Attaching "good" ideals to $\mathcal{O}$ is an important problem in representation theory and the orbit method. Quite a bit of work has been done on this (see e.g., some of the references and authors cited below) but the problem for nilpotent orbits of a complex semisimple Lie group remains unsolved.

If $G = GL(n, \mathbb{C})$, then our J is good (see Remark 6.6(i) and [8]). But $G = GL(n, \mathbb{C})$ is really a very special case for us as the geometry of $\mathcal{O}$ is incredibly nice, including but not limited to the fact that $\overline{\mathcal{O}}$ is always normal. For $G = O(n, \mathbb{C})$ or $G = Sp(2n, \mathbb{C})$, it is not the case that J is always good. Certainly if $\overline{\mathcal{O}}$ is not normal, we should not expect J to be good.

Our point of view (to be justified in [9]) is that $\mathcal{B}$ is the "canonical" quantization of the Kraft–Procesi construction, and so the failure of J to be good is really a statement about $\overline{\mathcal{O}}$. The next step is then to investigate whether we can make a modification to our quantization process in order to obtain some good ideals in $\mathcal{U}(\mathfrak{g})$ attached to $\mathcal{O}$ (and even its covers).

This paper is the first in a series. In the subsequent papers we make explicit the important role of Howe duality in our project. Indeed, $\mathcal{W}^{even}$ is the Harish-Chandra module of the (even) oscillator representation of $Sp(L, \mathbb{C})$, and the pair (G, S) constitutes a sequence of Howe dual pairs (see Remark 4.1). In taking coinvariants, we are implementing a sequence of Howe duality "operations." Each "operation" is like implementing a Howe duality correspondence, except that we do *not* pass to the irreducible quotient. In working on this project (which started in earnest in the summer of 2001 – and is part of a program we began in 1994), we have been reading the Howe duality literature. We have been influenced by especially the papers [18], [29], [1] and [25].

Our first construction of the Dixmier algebra $\mathcal{B}$ actually came out of the ideas of Howe duality and quantization by constraints. This is given in [9] and lies more in the realm of harmonic analysis than algebra. Our starting point there is the fact ([21]) that L is hyperkähler and the Kraft–Procesi construction is the algebraic analog of the hyperkähler reduction of L by S_c.

The notion of Dixmier algebra for nilpotent orbits (including their closures and their coverings) was first developed in the work of McGovern, Joseph and Vogan. See e.g., [27], [31], and [32]. The motivation for these authors and for most Dixmier algebra theorists is the search for completely prime primitive ideals. This motivation is very important for us too; we also find additional motivations coming from star products and from geometric quantization.

The results in this paper should be compared with the work in [6], [1], [3]–[7], [10, 14, 13, 16], [24]–[29] and [33]. Some of this comparison work will be done in [8] and [9].

Part of this work was carried out while I was visiting the IML and the CPT of the Université de la Méditerranée in the summer of 2001, and I thank my colleagues there for their hospitality. I especially thank Christian Duval and Valentin Ovsienko for some very valuable discussions.

It is a real pleasure to dedicate this article to Sasha Kirillov whose discoveries have opened up so many new vistas, starting of course with the Orbit Method. I warmly thank him for his friendship and his interest in my own work.

2 Dixmier algebra for the closure of a complex nilpotent orbit

Let G be a reductive complex algebraic group. Let $\mathfrak{g}$ be the Lie algebra of G and let $\mathfrak{g}^*$ be the dual space. Then G acts on $\mathfrak{g}$ and $\mathfrak{g}^*$ by, respectively, the adjoint action and the coadjoint action. The symmetric algebra $S(\mathfrak{g}) = \oplus_{p=0}^{\infty} S^p(\mathfrak{g})$ is the algebra of polynomial functions on $\mathfrak{g}^*$. The G-invariants form the graded subalgebra $S(\mathfrak{g})^G = \oplus_{p=0}^{\infty} S^p(\mathfrak{g})^G$. We can fix some nondegenerate G-invariant bilinear form $(\cdot, \cdot)$ on $\mathfrak{g}$.

The *nullcone* in $\mathfrak{g}^*$ is the set of $\lambda \in \mathfrak{g}^*$ which satisfy the following equivalent properties:

(i) the closure of the coadjoint orbit of λ contains zero,

(ii) the coadjoint orbit of λ is stable under dilations of the vector space $\mathfrak{g}^*$,

(iii) every nonconstant homogeneous G-invariant in $S(\mathfrak{g})$ vanishes on λ,

(iv) $\lambda = (x, \cdot)$ where x is a nilpotent in $\mathfrak{g}$.

The nullcone is G-stable and breaks into finitely many orbits of G, which are then called the *nilpotent coadjoint orbits*, or simply the *nilpotent orbits*, of G.

Let $\mathcal{O}$ be a nilpotent orbit of G. The closure $\overline{\mathcal{O}}$ is a complex algebraic subvariety of $\mathfrak{g}^*$; but $\overline{\mathcal{O}}$ may be reducible if G is disconnected. The coordinate ring $\mathbb{C}[\overline{\mathcal{O}}]$ of $\overline{\mathcal{O}}$ is the quotient algebra

$$\mathcal{R} = S(\mathfrak{g})/\mathfrak{I}(\overline{\mathcal{O}}) \tag{2.1}$$

where $\mathfrak{I}(\overline{\mathcal{O}})$ is the ideal of functions which vanish on $\overline{\mathcal{O}}$. Then $\mathfrak{I}(\overline{\mathcal{O}})$ is a graded ideal and $\mathcal{R} = \oplus_{p=0}^{\infty} \mathcal{R}^p$ is a graded algebra where $\mathcal{R}^p = S^p(\mathfrak{g})/\mathfrak{I}(\overline{\mathcal{O}})^p$. Each space $\mathcal{R}^p$ is a finite dimensional completely reducible representation of G. Kostant's description of $S(\mathfrak{g})$ as a module over $S(\mathfrak{g})^G$ implies that all G-multiplicities in $\mathcal{R}$ are finite.

$\mathcal{R}$ inherits from $S(\mathfrak{g})$ the structure of a graded Poisson algebra where $\{\mathcal{R}^p, \mathcal{R}^q\} \subseteq \mathcal{R}^{p+q-1}$. This Poisson bracket $\{\cdot, \cdot\}$ on $\mathcal{R}$ is G-invariant and corresponds to the holomorphic Kirillov–Kostant–Souriau symplectic form on $\mathcal{O}$.

In this situation, we define Dixmier algebras in the following way. We fix a Cartan involution ς of $\mathfrak{g}$. Then ς corresponds to a compact real form G_c of G with Lie algebra $\mathfrak{g}_c$. Let $\mathbb{N} = \{0, 1, 2, \ldots\}$.

Definition 2.1. A *Dixmier algebra* for $\overline{\mathcal{O}}$ is a quadruple $(\mathcal{D}, \xi, \tau, \vartheta)$ where

- $\mathcal{D}$ is a filtered algebra with an increasing algebra filtration $\mathcal{D} = \cup_{p\in\mathbb{N}} \mathcal{D}_p$ such that $\operatorname{gr}\mathcal{D}$ is commutative.

- $\xi : \mathfrak{g} \to \mathcal{D}_1$, $x \mapsto \xi^x$, is a homomorphism of Lie algebras and ξ induces an isomorphism of graded Poisson algebras from $S(\mathfrak{g})/\mathfrak{I}(\overline{\mathcal{O}})$ onto $\operatorname{gr}\mathcal{D}$.

- τ is a filtered algebra anti-involution of $\mathcal{D}$ such that $\tau(\xi^x) = -\xi^x$.

- ϑ is an anti-linear filtered algebra involution such that $\vartheta(\xi^x) = \xi^{\varsigma(x)}$.

Here are some explanations about the definition. First, commutativity of $\operatorname{gr}\mathcal{D}$ implies that $\operatorname{gr}\mathcal{D}$ has a natural structure of graded Poisson algebra; here the commutator in $\mathcal{D}$ induces the Poisson bracket on $\operatorname{gr}\mathcal{D}$. Second, ξ extends naturally to a filtered algebra homomorphism

$$\widetilde{\xi} : \mathcal{U}(\mathfrak{g}) \to \mathcal{D}. \tag{2.2}$$

Let J be the kernel of $\widetilde{\xi}$. Then $\operatorname{gr} J$ is a Poisson ideal of $S(\mathfrak{g})$, and $\operatorname{gr}\widetilde{\xi}$ induces a 1-to-1 homomorphism $\zeta : S(\mathfrak{g})/\operatorname{gr} J \to \operatorname{gr} D$ of graded Poisson algebras. We require that ζ be surjective and

$$\operatorname{gr} J = \mathfrak{I}(\overline{\mathcal{O}}). \tag{2.3}$$

Notice that ζ is surjective if and only if $\widetilde{\xi}$ is surjective in each filtration degree; then $\widetilde{\xi}$ induces a filtered algebra isomorphism

$$\mathcal{U}(\mathfrak{g})/J \xrightarrow{\sim} \mathcal{D}. \tag{2.4}$$

Third, τ satisfies $\tau(cA) = c\tau(A)$, $\tau(A+B) = \tau(A) + \tau(B)$, and $\tau(AB) = \tau(B)\tau(A)$ where $A, B \in \mathcal{D}$ and $c \in \mathbb{C}$. Fourth, ϑ satisfies $\vartheta(cA) = \overline{c}\vartheta(A)$, $\vartheta(A+B) = \vartheta(A) + \vartheta(B)$, and $\vartheta(AB) = \vartheta(A)\vartheta(B)$. Clearly $\tau\vartheta = \vartheta\tau$.

Notice that $\mathcal{D}$ and ξ (and ς) uniquely determine τ and ϑ, if the latter exist. Indeed, the endomorphisms $x \mapsto -x$ and $x \mapsto \varsigma(x)$ of $\mathfrak{g}$ extend uniquely to $\tau_{\mathfrak{g}}$ and $\vartheta_{\mathfrak{g}}$, where $\tau_{\mathfrak{g}}$ is an algebra anti-involution of $\mathcal{U}(\mathfrak{g})$ and $\vartheta_{\mathfrak{g}}$ is an antilinear algebra involution of $\mathcal{U}(\mathfrak{g})$. Then, via (2.4), $\tau_{\mathfrak{g}}$ and $\vartheta_{\mathfrak{g}}$ induce τ and ϑ.

We have an obvious notion of isomorphism of Dixmier algebras for $\overline{\mathcal{O}}$: $(\mathcal{D}, \xi, \tau, \vartheta)$ is isomorphic to $(\mathcal{D}', \xi', \tau', \vartheta')$ if there is a filtered algebra isomorphism $\eta : \mathcal{D} \to \mathcal{D}'$ such that $\xi' = \eta \circ \xi$, $\eta \circ \tau = \tau' \circ \eta$, and $\eta \circ \vartheta = \vartheta' \circ \eta$. We can easily classify Dixmier algebras.

Observation 2.2. *Suppose $(\mathcal{D}, \xi, \tau, \vartheta)$ is a Dixmier algebra for $\overline{\mathcal{O}}$. In addition to* (2.3), *J satisfies*

$$\tau_{\mathfrak{g}}(J) = J \quad \text{and} \quad \vartheta_{\mathfrak{g}}(J) = J. \tag{2.5}$$

In this way, we get a bijection between (isomorphism classes of) Dixmier algebras for $\overline{\mathcal{O}}$ and two-sided ideals J of $\mathcal{U}(\mathfrak{g})$ satisfying (2.3) *and* (2.5).

Proof. Clearly (2.3) and (2.5) imply that $(\mathcal{U}(\mathfrak{g})/J, \iota, \tau'_{\mathfrak{g}}, \vartheta'_{\mathfrak{g}})$ is a Dixmier algebra for $\overline{\mathcal{O}}$, where ι is the obvious map and $\tau'_{\mathfrak{g}}$ and $\vartheta'_{\mathfrak{g}}$ are induced by $\tau_{\mathfrak{g}}$ and $\vartheta_{\mathfrak{g}}$. Conversely, if $(\mathcal{D}, \xi, \tau, \vartheta)$ is given, then $(\mathcal{U}(\mathfrak{g})/J, \iota, \tau'_{\mathfrak{g}}, \vartheta'_{\mathfrak{g}})$ is isomorphic to it via (2.4). □

3 Properties of Dixmier algebras

The hopes in constructing a Dixmier algebra are (i) J will be a completely prime primitive ideal of $\mathcal{U}(\mathfrak{g})$, or even better, a completely prime maximal ideal, and (ii) $\mathcal{D}$ will be unitarizable. See [12, 3.1] for the definitions of the terms in (i).

To understand (ii), we observe that Definition 2.1 makes $\mathcal{D}$ into a $(\mathfrak{g} \oplus \mathfrak{g}, G_c)$-module. Indeed, the natural $(\mathfrak{g} \oplus \mathfrak{g}, G_c)$-module structure on $\mathcal{U}(\mathfrak{g})/J$ transfers over to $\mathcal{D}$ via (2.4). Then $\mathfrak{g} \oplus \mathfrak{g}$ acts on $\mathcal{D}$ though the representation

$$\Pi : \mathfrak{g} \oplus \mathfrak{g} \to \operatorname{End} \mathcal{D}, \qquad (x, y) \mapsto \Pi^{x,y} \tag{3.1}$$

where $\Pi^{x,y}(A) = \xi^x A - A\xi^y$. The action of G_c corresponds to the subalgebra $\{(x, x) \colon x \in \mathfrak{g}_c\}$.

Next consider the subalgebra $\mathfrak{g}^\sharp = \{(x, \varsigma(x)) : x \in \mathfrak{g}\}$ of $\mathfrak{g} \oplus \mathfrak{g}$. We say $\mathcal{D}$ is *unitarizable* if $\mathcal{D}$ admits a $\mathfrak{g}^\sharp$-invariant positive definite Hermitian inner product. In this event, by a theorem of Harish-Chandra, the operators $\Pi^{x,\varsigma(x)}$ correspond to a unitary representation of G on the Hilbert space completion of $\mathcal{D}$. This unitary representation is then a quantization of $\mathcal{O}$ in the sense of geometric quantization, if we view $\mathcal{O}$ as a *real* symplectic manifold. (If $\mathbb{C}[\overline{\mathcal{O}}] \neq \mathbb{C}[\mathcal{O}]$, then this might be just a piece of a quantization of $\mathcal{O}$.)

Notice that the following three properties are equivalent: (i) J is maximal, (ii) $\mathcal{D}$ is a simple ring, and (iii) the representation Π is irreducible.

Our formalism gives some partial results pertaining to hopes (i) and (ii). Notice that $\mathcal{D}_0 = \mathbb{C}$ by (2.4).

Proposition 3.1. *Suppose $(\mathcal{D}, \xi, \tau, \vartheta)$ is a Dixmier algebra for $\overline{\mathcal{O}}$ and $J = \ker \widetilde{\xi}$. Then*

(i) *J has an infinitesimal character.*

(ii) *If $\overline{\mathcal{O}}$ is irreducible, then J is a completely prime primitive ideal in $\mathcal{U}(\mathfrak{g})$.*

(iii) *There is a a unique G_c-invariant projection $\mathcal{T} : \mathcal{D} \to \mathbb{C}$. This map $\mathcal{T}$ is a trace, i.e., $\mathcal{T}(AB) = \mathcal{T}(BA)$.*

(iv) *$\mathcal{D}$ admits a unique $\mathfrak{g}^\sharp$-invariant Hermitian inner product $(\cdot|\cdot)$ such that $(1|1) = 1$, and it is given by*

$$(A|B) = \mathcal{T}(AB^\vartheta) \tag{3.2}$$

where $B^\vartheta = \vartheta(B)$.

Proof. (i) This means (for any proper two-sided ideal J) that J contains a maximal ideal of the center of $\mathcal{U}(\mathfrak{g})$. This happens if and only if gr J contains $S^+(\mathfrak{g})^G = \oplus_{p>0} S^p(\mathfrak{g})^G$. But gr $J = \mathfrak{I}(\overline{\mathcal{O}})$ and $\mathfrak{I}(\overline{\mathcal{O}}) \supset S^+(\mathfrak{g})^G$ since $\overline{\mathcal{O}}$ lies in the null-cone. (ii) If $\overline{\mathcal{O}}$ is irreducible, then $\mathfrak{I}(\overline{\mathcal{O}})$ is a prime ideal in $S(\mathfrak{g})$ and so J is a completely prime ideal in $\mathcal{U}(\mathfrak{g})$. This together with (i) implies, by a result of Dixmier, that J is primitive. (iii) Since $\mathfrak{I}(\overline{\mathcal{O}}) \supset S^+(\mathfrak{g})^G = S^+(\mathfrak{g})^{G_c}$, we have $\mathcal{R}^{G_c} = \mathbb{C}$ and so $\mathcal{D}^{G_c} = \mathbb{C}$. Thus we get a unique G_c-invariant projection map $\mathcal{T}$. Now G_c-invariance implies $\mathcal{T}([\xi^x, A]) = 0$ where $x \in \mathfrak{g}$. We can write this as $\mathcal{T}(\xi^x A) = \mathcal{T}(A\xi^x)$. Iteration gives $\mathcal{T}(\xi^{x_1} \cdots \xi^{x_k} A) = \mathcal{T}(A\xi^{x_1} \cdots \xi^{x_k})$. This proves $\mathcal{T}(BA) = \mathcal{T}(AB)$ since the ξ^x generate $\mathcal{D}$. (This is the same proof as in

[7, Proposition 8.4].) (iv) Suppose $(\cdot|\cdot)$ is an inner product with the desired properties. Then $(A|1) = \mathcal{T}(A)$. Now $\mathfrak{g}^\sharp$-invariance means that the operators $\Pi^{x,\varsigma(x)}$ are skew-hermitian, or equivalently, $(\xi^x A|B) = (A|B\xi^{\varsigma(x)})$. So for $B = \xi^{x_1}\cdots\xi^{x_k}$, we have $(A|B) = (\xi^{\varsigma(x_1)}\cdots\xi^{\varsigma(x_k)}A|1) = (B^\vartheta A|1) = \mathcal{T}(B^\vartheta A)$. The result is now clear. □

Corollary 3.2. *$\mathcal{D}$ is unitarizable if and only if the pairing defined by* (3.2) *is positive definite. If $\mathcal{D}$ is unitarizable, then J is maximal.*

Proof. Both statements follow from the uniqueness in Proposition 3.1(iv). □

Example 3.3. Suppose $\mathcal{O}$ is the minimal nilpotent orbit in $\mathfrak{g}$ where $\mathfrak{g}$ is simple and $\mathfrak{g} \neq \mathfrak{sl}(2,\mathbb{C})$. This is a case where $\mathbb{C}[\overline{\mathcal{O}}] = \mathbb{C}[\mathcal{O}]$. Then there is a unique Dixmier algebra for $\overline{\mathcal{O}}$. This follows by Observation 2.2 since there is exactly one choice for J satisfying (2.3) and (2.5). Moreover, (i) J is a completely prime maximal ideal of $\mathcal{U}(\mathfrak{g})$, and (ii) $\mathcal{U}(\mathfrak{g})/J$ is unitarizable if $\mathfrak{g}$ is classical.

Indeed, there is a unique two-sided ideal J satisfying (2.3) and $\tau_\mathfrak{g}(J) = J$; see [2, proof of Proposition 3.1]. Since the ideal $\mathfrak{J}(\overline{\mathcal{O}})$ is preserved by the antilinear algebra involution of $S(\mathfrak{g})$ defined by ς, it follows by the uniqueness of J that $\vartheta_\mathfrak{g}(J) = J$. Since $\overline{\mathcal{O}}$ is irreducible, Proposition 3.1(ii) implies that J is completely prime. If $\mathfrak{g} \neq \mathfrak{sl}(n,\mathbb{C})$, then J is the Joseph ideal and this is maximal by [19, Theorem 7.4]. If $\mathfrak{g} = \mathfrak{sl}(n,\mathbb{C})$ $(n \geq 3)$, then J is maximal by [30]. Finally, unitarizabilty is known; see [2, Theorem 9.1] for a uniform construction of these unitary representations on spaces of holomorphic functions on $\mathcal{O}$.

4 The Kraft–Procesi construction

In this section we recall how Kraft and Procesi in [22] and [23] constructed the closures of complex classical nilpotent orbits. We add to their construction the framework of algebraic symplectic reduction.

Let V be a complex vector space with $\mathbf{b}$ a bilinear form on V. Let G be the symmetry group of $\mathbf{b}$. We consider only the following three cases.

(i) $\mathbf{b}$ is identically zero. Then $G = GL(V,\mathbb{C})$.

(ii) $\mathbf{b}$ is nondegenerate and symmetric. Then $G = O(V,\mathbb{C})$.

(iii) $\mathbf{b}$ is nondegenerate and symplectic. Then $G = Sp(V,\mathbb{C})$.

Choose a nilpotent orbit $\mathcal{O}$ of G and an element λ in $\mathcal{O}$. Then λ corresponds (via the trace functional on End V) to $x \in \mathfrak{g}$; so x is in particular a nilpotent endomorphism of V. We note that $\mathcal{O}$ is connected, and so $\overline{\mathcal{O}}$ is irreducible, except in the following situation. If $G = O(V,\mathbb{C})$ where dim V is even and also the Jordan block size partition of $\mathcal{O}$ is very even (i.e., all parts are even and occur with even multiplicities), then $\mathcal{O}$ has two connected components and $\overline{\mathcal{O}}$ has two irreducible components. See [23], [11, Chapter 5].

Let V_d be the image of x^d. Then

$$V = V_0 \supset V_1 \supset \cdots \supset V_r \supset V_{r+1} = 0 \tag{4.1}$$

where r is the largest number such that $x^r \neq 0$. We define a complex vector space L by

$$L = L(V_0, V_1) \oplus L(V_1, V_2) \oplus \cdots \oplus L(V_{r-1}, V_r) \tag{4.2}$$

where $L(V_{d-1}, V_d)$ is obtained in the following way. If $G = GL(V, \mathbb{C})$, then

$$L(V_{d-1}, V_d) = \mathrm{Hom}(V_d, V_{d-1}) \oplus \mathrm{Hom}(V_{d-1}, V_d). \tag{4.3}$$

If $G = O(V, \mathbb{C})$ or $G = Sp(V, \mathbb{C})$, then

$$L(V_{d-1}, V_d) = \mathrm{Hom}(V_d, V_{d-1}). \tag{4.4}$$

Next we construct a complex Lie group S of the form

$$S = S_1 \times S_2 \times \cdots \times S_r \tag{4.5}$$

where S_d is obtained in the following way. To begin with, we put $\mathbf{b}_0 = \mathbf{b}$ and $S_0 = G$. If $G = GL(V, \mathbb{C})$, then for each d we put $\mathbf{b}_d = 0$ and $S_d = GL(V_d, \mathbb{C})$. If $G = O(V, \mathbb{C})$ or $G = Sp(V, \mathbb{C})$, then V_d admits an intrinsic nondegenerate complex bilinear form $\mathbf{b}_d$ and we define S_d to be the symmetry group of $\mathbf{b}_d$. In more detail, $\mathbf{b}_d$ is the bilinear form on V_d defined by $\mathbf{b}_d(x^d(u), x^d(v)) = \mathbf{b}(u, x^d(v))$. It turns out that $\mathbf{b}_d$ is nondegenerate. If $\mathbf{b}_{d-1}$ is orthogonal, then $\mathbf{b}_d$ is symplectic and we put $S_d = Sp(V_d, \mathbb{C})$. If $\mathbf{b}_{d-1}$ is symplectic, then $\mathbf{b}_d$ is orthogonal and we put $S_d = O(V_d, \mathbb{C})$.

Next we construct commuting actions of G and S on L. If $G = GL(V, \mathbb{C})$, we make G and S act by

$$\begin{aligned}(g, s_1, \ldots, s_r) \bullet (A_1, B_1, A_2, B_2, \ldots, A_r, B_r) \\ = (gA_1s_1^{-1}, s_1B_1g^{-1}, s_1A_2s_2^{-1}, s_2B_2s_1^{-1}, \ldots, s_{r-1}A_rs_r^{-1}, s_rB_rs_{r-1}^{-1})\end{aligned} \tag{4.6}$$

where $A_d \in \mathrm{Hom}(V_d, V_{d-1})$ and $B_d \in \mathrm{Hom}(V_{d-1}, V_d)$. If $G = O(V, \mathbb{C})$ or $G = Sp(V, \mathbb{C})$, we make G and S act by

$$(g, s_1, \ldots, s_r) \bullet (C_1, C_2, \ldots, C_r) = (gC_1s_1^{-1}, s_1C_2s_2^{-1}, \ldots, s_{r-1}C_rs_r^{-1}) \tag{4.7}$$

where $C_d \in \mathrm{Hom}(V_d, V_{d-1})$.

L has a (complex) symplectic form Ω given by $\Omega = \Omega_1 + \Omega_2 + \cdots + \Omega_r$ where Ω_d is the symplectic form on $L(V_{d-1}, V_d)$ defined in the following way. If $G = GL(V, \mathbb{C})$, then $\Omega_d(A+B, A'+B') = \mathrm{tr}(AB') - \mathrm{tr}(BA')$. If $G = O(V, \mathbb{C})$ or $G = Sp(V, \mathbb{C})$, then $\Omega_d(C, C') = \mathrm{tr}(C^*C)$ where $C^* \in \mathrm{Hom}(V_{d-1}, V_d)$ is the adjoint of C defined by $\mathbf{b}_{d-1}(u, C^*(v)) = \mathbf{b}_d(C(u), v)$.

Now G and S act faithfully and symplectically on L. Thereby G and S identify with commuting subgroups of the symplectic group $Sp(L, \mathbb{C})$. The action of $Sp(L, \mathbb{C})$ is Hamiltonian with canonical moment map

$$L \to \mathfrak{sp}(L, \mathbb{C})^*. \tag{4.8}$$

Hence our actions of G and S are Hamiltonian with induced moment maps (obtained by projection)

$$\gamma : L \to \mathfrak{g}^* \qquad \text{and} \qquad \sigma : L \to \mathfrak{s}^*. \tag{4.9}$$

Then γ is G-equivariant and S-invariant, and σ is S-equivariant and G-invariant.

Here are the explicit formulas for γ and σ. We may write these as $\mathfrak{g}$-valued and $\mathfrak{s}$-valued maps, with the convention that the trace functional on End L induces isomorphisms $\mathfrak{g} \xrightarrow{\sim} \mathfrak{g}^*$ and $\mathfrak{s} \xrightarrow{\sim} \mathfrak{s}^*$. If $G = GL(V, \mathbb{C})$, then γ and σ are given by

$$(-A_1 B_1) \qquad \text{and} \qquad (B_1 A_1 - A_2 B_2, \ldots, B_{r-1} A_{r-1} - A_r B_r, B_r A_r) \tag{4.10}$$

If $G = O(V, \mathbb{C})$ or $G = Sp(V, \mathbb{C})$, then γ and σ are given by

$$(-C_1 C_1^*) \qquad \text{and} \qquad (C_1^* C_1 - C_2 C_2^*, \ldots, C_{r-1}^* C_{r-1} - C_r C_r^*, C_r^* C_r). \tag{4.11}$$

Remark 4.1. For each $d = 1, \ldots, r$, S_{d-1} and S_d act on $L(V_{d-1}, V_d)$ as a Howe dual pair.

Let $\mathcal{P}$ be the algebra of polynomial functions on L. Then $\mathcal{P} = S(L^*)$ is a graded Poisson algebra with respect to the Poisson bracket $\{\cdot, \cdot\}$ defined by Ω. Our grading is $\mathcal{P} = \oplus_{j \in \frac{1}{2}\mathbb{N}} \mathcal{P}^j$ where $\mathcal{P}^j = S^{2j}(L^*)$ and then $\{\mathcal{P}^j, \mathcal{P}^k\} \subseteq \mathcal{P}^{j+k-1}$. (This choice of halving the natural degrees is convenient as the aim is to obtain $\mathcal{R}$ as a subquotient of $\mathcal{P}$.) The momentum functions γ_y ($y \in \mathfrak{g}$) and σ_x ($x \in \mathfrak{s}$) are the component functions of γ and σ; i.e., $\gamma_y(m) = \langle \gamma(m), y \rangle$ and $\sigma_x(m) = \langle \sigma(m), x \rangle$ where $m \in L$. The γ_y and σ_x lie in $\mathcal{P}^1$ and satisfy the bracket relations $\{\gamma_y, \gamma_{y'}\} = \gamma_{[y,y']}$, $\{\sigma_x, \sigma_{x'}\} = \sigma_{[x,x']}$ and $\{\gamma_y, \sigma_x\} = 0$.

Let $\mathcal{I}$ be the ideal in $\mathcal{P}$ generated by the momentum functions σ_x where $x \in \mathfrak{s}$. Then $\mathcal{I} = \oplus_{j \in \frac{1}{2}\mathbb{N}}^{\infty} \mathcal{I}^j$ is a graded ideal stable under both G and S. Hence the quotient algebra $\mathcal{P}/\mathcal{I}$ is a graded algebra on which G and S act by graded algebra automorphisms. The grading is

$$\mathcal{P}/\mathcal{I} = \oplus_{j \in \frac{1}{2}\mathbb{N}} (\mathcal{P}/\mathcal{I})^j$$

where $(\mathcal{P}/\mathcal{I})^j = \mathcal{P}^j / \mathcal{I}^j$. Kraft and Procesi proved that $\mathcal{I}$ is the full ideal of functions vanishing on $\sigma^{-1}(0)$. Thus $\mathcal{P}/\mathcal{I}$ is the coordinate ring $\mathbb{C}[\sigma^{-1}(0)]$ of the zero locus of σ.

The *algebraic symplectic reduction* L^{red} of L by S is the Mumford quotient of $\sigma^{-1}(0)$ by S. Thus L^{red} is the affine complex algebraic variety with coordinate ring

$$\mathbb{C}[L^{red}] = (\mathcal{P}/\mathcal{I})^{inv} = \mathcal{P}^{inv}/\mathcal{I}^{inv} \tag{4.12}$$

where the superscript inv denotes taking S-invariants. Moreover $\mathcal{I}^{inv}$ is a Poisson ideal in $\mathcal{P}^{inv}$. So $\mathbb{C}[L^{red}]$ inherits the structure of a graded Poisson algebra.

Notice that S contains the center $\mathbb{Z}_2 = \{1, -1\}$ of $Sp(L, \mathbb{C})$ and the action of $\mathbb{Z}_2$ induces the decomposition $\mathcal{P} = \mathcal{P}^{even} \oplus \mathcal{P}^{odd}$ where the even part is the graded Poisson algebra

$$\mathcal{P}^{even} = \oplus_{d \in \mathbb{N}} \mathcal{P}^d. \tag{4.13}$$

So $\mathcal{P}^{inv}$ lies in $\mathcal{P}^{even}$, and thus $\mathcal{P}^{inv}$ and $\mathbb{C}[L^{red}]$ are $\mathbb{N}$-graded.

The symplectic version of the Kraft–Procesi result is

Theorem 4.2. [22, Theorem 3.3], [23, Theorem 5.3] *The algebra homomorphism $\gamma^* : S(\mathfrak{g}) \to \mathcal{P}$ defined by $y \mapsto \gamma_y$ ($y \in \mathfrak{g}$) induces a G-equivariant isomorphism of $\mathbb{N}$-graded Poisson algebras from $\mathcal{R}$ onto $\mathcal{P}^{inv}/\mathcal{I}^{inv}$.*

The cited results of Kraft and Procesi are given in geometric language, and the reader who wants to read all the proofs in [22] and [23] will need some knowledge of algebraic geometry. The statements in [22, Theorem 3.3] and [23, Theorem 5.3] are easy to translate into algebra though since we are dealing with *affine* varieties. Kraft and Procesi show that γ maps $\sigma^{-1}(0)$ onto $\overline{\mathcal{O}}$, and moreover this surjection $\gamma' : \sigma^{-1}(0) \to \overline{\mathcal{O}}$ is a quotient map for the action of S. In this setting of a reductive group acting on an affine variety, "quotient map" has a very strong meaning coming from Mumford's geometric invariant theory, as explained in [22, §1.4] and [23, §0.11]. Precisely, γ' being a quotient map means that the corresponding map $\mathcal{R} \to \mathcal{P}/\mathcal{I}$ on coordinate rings is injective and has image equal to $(\mathcal{P}/\mathcal{I})^{inv}$.

In symplectic language, Kraft and Procesi proved that the moment map γ induces a G-isomorphism of affine complex algebraic varieties from L^{red} onto $\overline{\mathcal{O}}$. This isomorphism is also equivariant with respect to the natural $\mathbb{C}^*$-actions on L^{red} and $\overline{\mathcal{O}}$. Finally, since γ is a moment map it follows that γ^* preserves the Poisson brackets. Thus we get Theorem 4.2.

5 Weyl algebra $\mathcal{W}$ for L

The Kraft–Procesi construction realized $\mathcal{R}$ as a subquotient, namely $\mathcal{P}^{inv}/\mathcal{I}^{inv}$, of $\mathcal{P}^{even}$. Our aim is to make a noncommutative analog of their construction.

The image of the moment map (4.8) is the closure $\overline{\mathcal{Y}}$ of the minimal nilpotent orbit $\mathcal{Y}$ of $Sp(L, \mathbb{C})$, and $\mathcal{P}^{even} = \mathbb{C}[\overline{\mathcal{Y}}]$ as graded Poisson algebras. We know by Example 3.3 that $\overline{\mathcal{Y}}$ has a unique Dixmier algebra $(\mathcal{D}, \xi_{\mathcal{D}}, \tau_{\mathcal{D}}, \vartheta_{\mathcal{D}})$, and then $\mathcal{D}$ is the quotient of $\mathcal{U}(\mathfrak{sp}(L, \mathbb{C}))$ by its Joseph ideal. In this section we will give a more concrete model for this Dixmier algebra. Then in §6 we will perform the noncommutative analog of reduction.

Let $\mathcal{W}$ be the Weyl algebra for L^*. This means that $\mathcal{W}$ is the quotient of the tensor algebra of L^* by the two-sided ideal generated by the elements $a \otimes b - b \otimes a - \{a, b\}$ where a and b lie in L^*. Let $a \mapsto \hat{a}$ be the natural map $L^* \to \mathcal{W}$. We can identify $\mathfrak{sp}(L, \mathbb{C})$ with $S^2 L^*$ and then we have the Lie algebra embedding

$$\xi : \mathfrak{sp}(L, \mathbb{C}) \longrightarrow \mathcal{W}, \quad \xi^{ab} = \hat{a}\hat{b} + \hat{b}\hat{a}. \tag{5.1}$$

There is an increasing algebra filtration $\mathcal{W} = \cup_{j \in \frac{1}{2}\mathbb{N}} \mathcal{W}_j$ where $\mathcal{W}_j$ is the image of the space of tensors of degree at most $2j$. We have $[\mathcal{W}_j, \mathcal{W}_k] \subset \mathcal{W}_{j+k-1}$. Thus the associated graded algebra $\operatorname{gr} \mathcal{W} = \oplus_{j \in \frac{1}{2}\mathbb{N}} \mathcal{W}_j / \mathcal{W}_{j-\frac{1}{2}}$ is commutative and the commutator in $\mathcal{W}$ induces a Poisson bracket (of degree -1) on $\operatorname{gr} \mathcal{W}$. In this way gr $\mathcal{W}$ becomes a graded Poisson algebra. Then gr $\mathcal{W}$ identifies naturally with $\mathcal{P}$.

The symplectic group $Sp(L, \mathbb{C})$ acts naturally on $\mathcal{W}$ by algebra automorphisms. This action respects ξ, the filtration on $\mathcal{W}$, the Poisson bracket on gr $\mathcal{W}$, etc. The corresponding action of $\mathfrak{sp}(L, \mathbb{C})$ on $\mathcal{W}$ is given by the operators $[\xi^{ab}, \cdot]$.

We next choose a Cartan involution ς of $\mathfrak{sp}(L, \mathbb{C})$. To do this, we go back into the Kraft–Procesi construction. Recall that each space V_d in (4.1) carried a bilinear form $\mathbf{b}_d$ $(d = 0, \ldots, r)$. We can choose a positive definite hermitian form $\mathbf{h}_d$ on V_d which is compatible with $\mathbf{b}_d$ in the sense that the intersection of S_d with the unitary group of $\mathbf{h}_d$ is a maximal compact subgroup K_d of S_d. (This is an equivalent version of the setup in [21].) These $\mathbf{h}_d$ determine naturally a positive definite hermitian form $\mathbf{h}$ on L. Now we define $\varsigma(T) = -T^\dagger$ for $T \in \mathfrak{sp}(L, \mathbb{C})$, where $T^\dagger$ is the adjoint of T with respect to $\mathbf{h}$.

Corresponding to ς is a compact real form $Sp(L)$ of $Sp(L, \mathbb{C})$ with Lie algebra $\mathfrak{sp}(L)$. For later use (see §6), we notice that $G \cap Sp(L) = K_0$ and $S \cap Sp(L) = K_1 \times \cdots \times K_r$ are compact real forms of G and S, which we will denote by G_c and S_c.

Now $\mathcal{W}$ is a $(\mathfrak{sp}(L, \mathbb{C}) \oplus \mathfrak{sp}(L, \mathbb{C}), Sp(L))$-module, where the representation

$$\mathfrak{sp}(L, \mathbb{C}) \oplus \mathfrak{sp}(L, \mathbb{C}) \to \operatorname{End} \mathcal{W} \tag{5.2}$$

is given by $(x, y) \cdot A = \xi^x A - A\xi^y$ and the action of $Sp(L)$ corresponds to the subalgebra $\{(x, x) : x \in \mathfrak{sp}(L)\}$. The action of the center $\mathbb{Z}_2$ of $Sp(L)$ produces the decomposition

$$\mathcal{W} = \mathcal{W}^{even} \oplus \mathcal{W}^{odd} \tag{5.3}$$

where $\mathcal{W}^{even}$ is space of invariants for $\mathbb{Z}_2$. The induced filtration on $\mathcal{W}^{even}$ satisfies $\mathcal{W}^{even}_{p+\frac{1}{2}} = \mathcal{W}^{even}_p$ if $p \in \mathbb{N}$. So we might as well just consider the algebra filtration

$$\mathcal{W}^{even} = \cup_{d \in \mathbb{N}} \mathcal{W}^{even}_d. \tag{5.4}$$

Now we can make $\mathcal{W}^{even}$ into a Dixmier algebra.

Proposition 5.1. *The Dixmier algebra for the closure $\overline{\mathcal{Y}}$ of the minimal nilpotent orbit of $Sp(L, \mathbb{C})$ is the quadruple $(\mathcal{W}^{even}, \xi, \tau, \vartheta)$, for some unique choices of τ and ϑ.*

Proof. The map ξ induces a filtered algebra isomorphism $\pi : \mathcal{U}(\mathfrak{sp}(L, \mathbb{C}))/\mathcal{J} \to \mathcal{W}^{even}$ where $\mathcal{J}$ is the Joseph ideal (see [2, §5]). Clearly gr $\mathcal{W}^{even} = \mathcal{P}^{even}$ and so π induces a graded isomorphism $S(\mathfrak{sp}(L, \mathbb{C}))/\mathfrak{I}(\overline{\mathcal{Y}}) \xrightarrow{\sim} \mathcal{P}^{even}$. Now everything follows by Example 3.3.

6 Dixmier algebra for $\overline{\mathcal{O}}$

$\mathcal{W}^{even}$ is, by means of (5.2), both a $(\mathfrak{g} \oplus \mathfrak{g}, G_c)$-module and an $(\mathfrak{s} \oplus \mathfrak{s}, S_c)$-module, and these two actions commute.

Definition 6.1. $\mathcal{B}$ is the $(\mathfrak{g} \oplus \mathfrak{g}, G_c)$-module obtained by taking the coinvariants of $\mathcal{W}^{even}$ in the category of $(\mathfrak{s} \oplus \mathfrak{s}, S_c)$-modules.

This means that $\mathcal{B}$ is the quotient $\mathcal{W}^{even}/\mathcal{M}$ where $\mathcal{M}$ is the subspace spanned by all $\xi^x A - A\xi^y$ and $A - s \cdot A$ where $x, y \in \mathfrak{s}$, $s \in S_c$ and $A \in \mathcal{W}^{even}$ (see [20, Chapter II]). Then $\mathcal{B}$ inherits from $\mathcal{W}^{even}$ an increasing G_c-stable vector space filtration $\mathcal{B} = \cup_{d \in \mathbb{N}} \mathcal{B}_d$.

Let $\mathcal{W}^{inv}$ be the algebra of invariants for S_c. Then $\mathcal{W}^{inv}$ lies in $\mathcal{W}^{even}$ (since S_c contains $\mathbb{Z}_2$) and so $\mathcal{W}^{inv}$ inherits from $\mathcal{W}^{even}$ an algebra filtration $\mathcal{W}^{inv} = \cup_{d \in \mathbb{N}} \mathcal{W}_d^{inv}$.

Lemma 6.2. *The natural map*

$$\phi : \mathcal{W}^{inv} \to \mathcal{B} \tag{6.1}$$

is surjective in each filtration degree and its kernel is a two-sided ideal. In this way, $\mathcal{B}$ becomes a filtered algebra. The corresponding map $\operatorname{gr}\phi : \mathcal{P}^{inv} \to \operatorname{gr}\mathcal{B}$ *is a surjective homomorphism of graded Poisson algebras.*

Proof. We prove this in §7. □

Our main result is

Theorem 6.3. *We have* $\operatorname{gr}\mathcal{B} = \mathcal{P}^{inv}/\mathcal{I}^{inv}$. *So* $\operatorname{gr}\mathcal{B} \simeq \mathcal{R}$ *as graded Poisson algebras.*

Proof. The proof occupies §8. □

Corollary 6.4. *The quadruple $(\mathcal{B}, \xi_{\mathcal{B}}, \tau_{\mathcal{B}}, \vartheta_{\mathcal{B}})$ is a Dixmier algebra for $\overline{\mathcal{O}}$, where $\xi_{\mathcal{B}}$, $\tau_{\mathcal{B}}$ and $\vartheta_{\mathcal{B}}$ are the maps induced by ξ, τ, and ϑ.*

Proof. We prove this in §9. □

Let J be the kernel of the algebra homomorphism $\widetilde{\xi_{\mathcal{B}}} : \mathcal{U}(\mathfrak{g}) \to \mathcal{B}$ defined by $\xi_{\mathcal{B}}$. Proposition 3.1 gives

Corollary 6.5. *Suppose we exclude the cases where $\mathcal{O}$ is disconnected (so where $G = O(2n, \mathbb{C})$ and the Jordan block size partition of $\mathcal{O}$ is very even). Then J is a completely prime primitive ideal of $\mathcal{U}(\mathfrak{g})$ with $\operatorname{gr} J = \mathfrak{J}(\overline{\mathcal{O}})$ $\tau_{\mathfrak{g}}(J) = J$, and $\vartheta_{\mathfrak{g}}(J) = J$.*

The methods we have used thus far give no information about the excluded cases.

Remark 6.6. (i) Suppose $G = GL(n, \mathbb{C})$. Then the space L has a $G \times S$-invariant polarization, and using this we can describe $\mathcal{B}$ and ξ in the following way. Let X be the flag manifold of G of flags of the type in (4.1). Let $\mathcal{D}^{\frac{1}{2}}(X)$ be the algebra of twisted differential operators for the (locally defined) square root of the canonical bundle on X as in [7]. We can show ([8]) that $\mathcal{B}$ identifies with $\mathcal{D}^{\frac{1}{2}}(X)$ in such a way that ξ corresponds to the canonical mapping of $\mathfrak{g}$ into $\mathcal{D}^{\frac{1}{2}}(X)$. Then by [7, Corollary 8.5], J is a maximal ideal in $\mathcal{U}(\mathfrak{g})$.

(ii) Suppose $G = O(n, \mathbb{C})$ or $G = Sp(2n, \mathbb{C})$. If $\mathcal{O}$ is the minimal nilpotent orbit, then $\mathcal{B}$ is the quotient of $\mathcal{U}(\mathfrak{g})$ by its Joseph ideal. This follows by the result in Example 3.3.

7 Proof of Lemma 6.2

The action of S_c on $\mathcal{W}$ is completely reducible and locally finite, and S and S_c have the same invariants and the same irreducible subspaces. So we can form the decomposition

$$\mathcal{W}^{even} = \mathcal{W}^{inv} \oplus \mathcal{X} \tag{7.1}$$

where $\mathcal{X}$ is the sum of all non-trivial S_c-isotypic components. Then $\mathcal{X}$ is the span of the elements $A - s \cdot A$ where $A \in \mathcal{W}^{even}$ and $s \in S_c$. Then $\mathcal{M} = \mathcal{M}^{inv} \oplus \mathcal{X}$. Hence the natural map ϕ is surjective and its kernel is $\mathcal{M}^{inv}$. I.e., we have vector space isomorphisms

$$\mathcal{W}^{inv}/\mathcal{M}^{inv} \xrightarrow{\sim} \mathcal{W}^{even}/\mathcal{M} \xrightarrow{\sim} \mathcal{B}. \tag{7.2}$$

The decomposition (7.1) is compatible with the filtration on $\mathcal{W}^{even}$, since the filtration is S_c-invariant. Consequently ϕ is surjective in each filtration degree. So $\mathcal{W}^{even}$ and $\mathcal{W}^{inv}$ induce the same filtration on $\mathcal{B}$.

Next we show that $\mathcal{M}^{inv}$ is a two-sided ideal in $\mathcal{W}^{inv}$. To begin with, $\mathcal{M}^{inv}$ lies inside the subspace $\mathcal{M}'$ of $\mathcal{M}$ spanned by all $\xi^x A$ and $A\xi^x$ where $x \in \mathfrak{s}$ and $A \in \mathcal{W}^{even}$. This follows using (7.1). So it suffices to show that $D\mathcal{M}'$ and $\mathcal{M}'D$ lie in $\mathcal{M}'$ if $D \in \mathcal{W}^{inv}$. Obviously $DA\xi^x$ lies in $\mathcal{M}'$. Invariance of D gives $\xi^x D - D\xi^x = 0$ and so $D\xi^x A = \xi^x DA$ lies in $\mathcal{M}'$. Thus $D\mathcal{M}' \subseteq \mathcal{M}'$; similarly $\mathcal{M}'D \subseteq \mathcal{M}'$.

The associated graded algebra $\operatorname{gr}\mathcal{B}$ is the quotient $\operatorname{gr}\mathcal{W}^{inv}/\operatorname{gr}\mathcal{M}^{inv}$. We find $\operatorname{gr}\mathcal{W}^{inv} = \mathcal{P}^{inv}$. Now the final assertion is clear.

Remark 7.1. If we replace $\mathcal{W}^{even}$ by $\mathcal{W}$ in Definition 6.1, then we get the same thing. I.e., if $\widetilde{\mathcal{B}}$ is the module of coinvariants of $\mathcal{W}$, then $\widetilde{\mathcal{B}}$ identifies naturally with $\mathcal{B}$. Indeed $\widetilde{\mathcal{B}} = \mathcal{W}/\widetilde{\mathcal{M}}$ where $\widetilde{\mathcal{M}}$ is the subspace spanned by all $\xi^x A - A\xi^y$ and $A - s \cdot A$ where now $A \in \mathcal{W}$. But then $\widetilde{\mathcal{M}} = \mathcal{W}^{odd} \oplus \mathcal{M}$ and so the natural map $\widetilde{\phi} : \mathcal{W}^{inv} \to \widetilde{\mathcal{B}}$ is surjective with the same kernel $\mathcal{M}^{inv}$. Also the filtration $\widetilde{\mathcal{B}} = \cup_{j \in \frac{1}{2}\mathbb{N}} \widetilde{\mathcal{B}}_j$ induced by $\mathcal{W}$ reduces to the one induced by $\mathcal{W}^{inv}$ in the sense that $\widetilde{\phi}(\mathcal{W}^{inv}_d) = \widetilde{\mathcal{B}}_d = \widetilde{\mathcal{B}}_{d+\frac{1}{2}}$.

8 Proof of Theorem 6.3

We will compute $\operatorname{gr}\mathcal{B}$ by using a homology spectral sequence. We will consider the relative Lie algebra homology $H(\mathfrak{s} \oplus \mathfrak{s}, S_c; \mathcal{W}^{even})$. By definition (see [20,

Chapter II, §6-7]), $H_j(\mathfrak{s}\oplus\mathfrak{s}, S_c; \mathcal{W}^{even})$ is the jth derived functor, in the category of $(\mathfrak{s}\oplus\mathfrak{s}, S_c)$-modules, of the coinvariants. So

$$\mathcal{B} = H_0(\mathfrak{s}\oplus\mathfrak{s}, S_c; \mathcal{W}^{even}). \tag{8.1}$$

The idea is that we will introduce a filtration of the complex that computes the homology in such a way that the induced filtration on $H_0(\mathfrak{s}\oplus\mathfrak{s}, S_c; \mathcal{W}^{even})$ is the one we have already defined on $\mathcal{B}$. Then we will use the usual spectral sequence of a filtered complex to compute gr $H_0(\mathfrak{s}\oplus\mathfrak{s}, S_c; \mathcal{W}^{even})$. The computation will rely on the geometric result of Kraft and Procesi that (in the notation of §4) $\sigma^{-1}(0)$ is a complete intersection.

To begin with, we have $\mathfrak{s}\oplus\mathfrak{s} = \mathfrak{k}\oplus\mathfrak{p}$ where $\mathfrak{k} = \{(x,x) : x \in \mathfrak{s}\}$ and $\mathfrak{p} = \{(x,-x) : x\in\mathfrak{s}\}$. Then $\mathfrak{k}$ is the complexified Lie algebra of S_c. The standard complex ([20, page 163]) for computing $H(\mathfrak{s}\oplus\mathfrak{s}, S_c; \mathcal{W}^{even})$ is

$$\begin{aligned} 0 \longleftarrow \wedge^0\mathfrak{p}\otimes_{S_c}\mathcal{W}^{even} &\overset{\partial}{\longleftarrow} \wedge^1\mathfrak{p}\otimes_{S_c}\mathcal{W}^{even} \\ \overset{\partial}{\longleftarrow}\cdots &\overset{\partial}{\longleftarrow} \wedge^m\mathfrak{p}\otimes_{S_c}\mathcal{W}^{even} \longleftarrow 0. \end{aligned} \tag{8.2}$$

Here $m = \dim\mathfrak{s}$ and $\otimes_{S_c}$ denotes the S_c-coinvariants of the tensor product. We call this complex A where ${}^tA = \wedge^t\mathfrak{p}\otimes_{S_c}\mathcal{W}^{even}$.

The differential ∂ in (8.2) is given by

$$\partial(Y_1\wedge\cdots\wedge Y_t\otimes D) = \sum_{l=1}^{t}(-1)^l Y_1\wedge\cdots\wedge\widehat{Y_l}\wedge\cdots\wedge Y_t\otimes\Pi^{Y_l}(D) \tag{8.3}$$

where the Y_i lie in $\mathfrak{p}$, $D\in\mathcal{W}^{even}$, and Π is the representation (5.2). (Notice the terms involving $[Y_i, Y_j]$ are not present because $[\mathfrak{p},\mathfrak{p}]\subseteq\mathfrak{k}$.) To make the complex more transparent, we identify $\mathfrak{p}$ with $\mathfrak{s}$ so that $(x,-x)$ corresponds to x. Then (8.2) becomes

$$\partial(x_1\wedge\cdots\wedge x_t\otimes D) = \sum_{l=1}^{t}(-1)^l x_1\wedge\cdots\wedge\widehat{x_l}\wedge\cdots\wedge x_t\otimes(\xi^{x_l}D + D\xi^{x_l}). \tag{8.4}$$

Next we define an increasing filtration of A by the spaces

$${}^tA^d = \wedge^t\mathfrak{p}\otimes_{S_c}\mathcal{W}^{even}_{d-t} \tag{8.5}$$

where we set $\mathcal{W}^{even}_j = 0$ if $j<0$. Then $\partial\, {}^tA^d \subseteq {}^{t-1}A^d$. This follows since the ξ^x lie in $\mathcal{W}^{even}_1$ and so $\mathfrak{p}\cdot\mathcal{W}^{even}_d \subseteq \mathcal{W}^{even}_{d+1}$. So we have in hand a filtration of the complex (8.1). We put $A^{d,q} = {}^{d+q}A^d$; then d is the filtration degree and q is the complementary degree. The induced filtration on the homology is $H(\mathfrak{s}\oplus\mathfrak{s}, S_c; \mathcal{W}^{even}) = \cup_{d\in\mathbb{N}} F^d$ where $F^d = \oplus_{q\in\mathbb{Z}} F^{d,q}$ and $F^{d,q}$ is the dth filtration piece of $H_{d+q}(\mathfrak{s}\oplus\mathfrak{s}, S_c; \mathcal{W}^{even})$. The associated graded space gr $H(\mathfrak{s}\oplus\mathfrak{s}, S_c; \mathcal{W}^{even})$ is the direct sum of the spaces

$$\mathrm{gr}^d\, H_{d+q}(\mathfrak{s}\oplus\mathfrak{s}, S_c; \mathcal{W}^{even}) = F^{d,q}/F^{d-1,q+1} \tag{8.6}$$

Notice that ${}^0A = \mathcal{W}^{even}$ and the filtration on ${}^0A$ defined by (8.5) is the same one as in (5.4). So gr $H_0(\mathfrak{s} \oplus \mathfrak{s}, S_c; \mathcal{W}^{even}) = \text{gr } \mathcal{B}$. Our goal is to prove

$$\text{gr}^d\ H_0(\mathfrak{s} \oplus \mathfrak{s}, S_c; \mathcal{W}^{even}) = (\mathcal{P}^{inv}/\mathcal{I}^{inv})^d. \tag{8.7}$$

Now we consider the spectral sequence $E_0, E_1, \ldots$ associated to our filtered complex. (See e.g., [20, Appendix D] or [15, Chapter I,§4] for the construction of this spectral sequence in the general setting.) The E_0 term is given by $E_0^{d,q} = A^{d,q}/A^{d-1,q+1}$ and so

$$E_0^{d,q} = \wedge^{d+q}\mathfrak{p} \otimes_{S_c} \mathcal{W}^{even}_{-q}/\mathcal{W}^{even}_{-q-1}. \tag{8.8}$$

The identification gr $\mathcal{W}^{even} = \mathcal{P}^{even}$ gives

$$E_0^{d,q} = \wedge^{d+q}\mathfrak{p} \otimes_{S_c} \mathcal{P}^{-q}. \tag{8.9}$$

(Thus the E_0 term occupies the octant of the d, q plane where $q \leq 0$ and $d+q \geq 0$.) So E_0^d is the complex

$$\begin{aligned}0 \longleftarrow \wedge^0\mathfrak{p} \otimes_{S_c} \mathcal{P}^d \xleftarrow{\partial_0} \wedge^1\mathfrak{p} \otimes_{S_c} \mathcal{P}^{d-1}\\ \xleftarrow{\partial_0} \cdots \xleftarrow{\partial_0} \wedge^m\mathfrak{p} \otimes_{S_c} \mathcal{P}^{d-m} \longleftarrow 0.\end{aligned} \tag{8.10}$$

The boundary ∂_0 is induced by ∂. We can easily compute ∂_0 since the natural projection maps $\psi_d : \mathcal{W}_d \to \mathcal{P}^d$ are given by $\psi_d(\hat{a}_1 \cdots \hat{a}_{2d}) = a_1 \cdots a_{2d}$ where $a_i \in L^*$ (cf. §5). So for $D \in \mathcal{W}_d$ we have $\psi_{d+1}(\xi^x D) = \psi_{d+1}(D\xi^x) = \sigma_x\psi_d(D)$. Thus (8.4) gives

$$\partial_0(x_1 \wedge \cdots \wedge x_t \otimes f) = \sum_l (-1)^l x_1 \wedge \cdots \wedge \widehat{x_l} \wedge \cdots \wedge x_t \otimes (2\sigma_{x_l} f). \tag{8.11}$$

The total complex E_0 is

$$\begin{aligned}0 \longleftarrow \wedge^0\mathfrak{p} \otimes_{S_c} \mathcal{P}^{even} \xleftarrow{\partial_0} \wedge^1\mathfrak{p} \otimes_{S_c} \mathcal{P}^{even}\\ \xleftarrow{\partial_0} \cdots \xleftarrow{\partial_0} \wedge^m\mathfrak{p} \otimes_{S_c} \mathcal{P}^{even} \longleftarrow 0.\end{aligned} \tag{8.12}$$

The homology $H(E_0)$, together with a differential ∂_1, is the E_1 term of the spectral sequence. More precisely, $E_1^{d,q} = H_{d+q}(E_0^{d,*})$.

To compute E_1, we observe that $H(E_0)$ is the S_c-coinvariants of the homology of the complex

$$\begin{aligned}0 \longleftarrow \wedge^0\mathfrak{p} \otimes \mathcal{P}^{even} \xleftarrow{\partial_0} \wedge^1\mathfrak{p} \otimes \mathcal{P}^{even}\\ \xleftarrow{\partial_0} \cdots \xleftarrow{\partial_0} \wedge^m\mathfrak{p} \otimes \mathcal{P}^{even} \longleftarrow 0.\end{aligned} \tag{8.13}$$

Indeed, E_0 is the S_c-coinvariants of (8.13), and taking coinvariants commutes with taking homology. The latter follows because each space $\wedge^t\mathfrak{p} \otimes \mathcal{P}^{even}$ is a locally finite S_c-representation, and for any such representation $\mathcal{V}$, the natural map $\mathcal{V}^{S_c} \to \mathcal{V}_{S_c}$ from invariants to coinvariants is an isomorphism.

To compute the homology of (8.13), we recognize (8.13) as the Koszul complex K of the sequence $\sigma_{y_1}, \ldots, \sigma_{y_m}$ in $\mathcal{P}^{even}$ where $y_1, \ldots, y_m$ is any basis of $\mathfrak{s}$. Recall from §4 that $\mathcal{I}$ is the ideal in $\mathcal{P}$ generated by the σ_{y_i}. Kraft and Procesi proved in [22, Theorem 3.3] and [23, Theorem 5.3] that the subscheme $\sigma^{-1}(0)$ of L is a reduced complete intersection, i.e., $\sigma_{y_1}, \ldots, \sigma_{y_m}$ is a regular sequence in $\mathcal{P}$. Let us consider the Koszul complex $\widetilde{K}$ of this sequence in $\mathcal{P}$. By a well-known result of commutative algebra (see [17, III, Proposition 7.10A]) the homology of $\widetilde{K}$ is concentrated in degree zero and $H_0(\widetilde{K}) = \mathcal{P}/\mathcal{I}$ as graded algebras. But K is simply obtained from $\widetilde{K}$ by taking $\mathbb{Z}_2$-invariants. Hence the homology of K is concentrated in degree zero and $H_0(K) = \mathcal{P}^{even}/\mathcal{I}^{even}$ as graded algebras. Then the module $H_0(K)_{S_c}$ of coinvariants identifies with $\mathcal{P}^{inv}/\mathcal{I}^{inv}$.

Thus E_1 is the complex

$$0 \longleftarrow \mathcal{P}^{inv}/\mathcal{I}^{inv} \xleftarrow{\partial_1} 0 \xleftarrow{\partial_1} \cdots \xleftarrow{\partial_1} 0 \longleftarrow 0 \tag{8.14}$$

where the differentials $\partial_1^{d,q} : E_1^{d,q} \to E_1^{d-1,q}$ are obviously zero and

$$E_1^{d,-d} = (\mathcal{P}^{inv}/\mathcal{I}^{inv})^d \qquad \text{while} \qquad E_1^{d,q} = 0 \text{ if } q \neq -d. \tag{8.15}$$

Now we can compute the rest of the spectral sequence. We know E_{r+1} is the homology of E_r with respect to a differential ∂_r; i.e., $E_{r+1}^{d,q} = \ker \partial_r^{d,q} / \operatorname{im} \partial_r^{d+r,q-r+1}$ where $\partial_r^{d,q}$ maps $E_r^{d,q}$ to $E_r^{d-r,q+r-1}$. For $r \geq 1$, we find that $E_1^{d,q} = E_r^{d,q}$ and the differentials $\partial_r^{d,q}$ are all zero.

The E_∞ term of the spectral sequence satisfies

$$E_\infty^{d,q} = \operatorname{gr}^d \; H_{d+q}(\mathfrak{s} \oplus \mathfrak{s}, S_c; \mathcal{W}^{even}). \tag{8.16}$$

Our final step is to show our spectral sequence converges in that

$$E_1^{d,q} = E_\infty^{d,q}. \tag{8.17}$$

This will finish off the proof of Theorem 6.3 because then (8.15) gives the desired result (8.7).

The convergence (8.17) follows formally from the two properties: (i) $A^d = 0$ if $d < 0$ where $A^d = \oplus_{q \in \mathbb{Z}} A^{d,q}$ and (ii) $A^{d,q}$ has finite dimension. Indeed, following the notation in [15, I,§4.2], we have

$$\begin{aligned} E_r^{d,q} &= Z_r^{d,q}/(B_{r-1}^{d,q} + Z_{r-1}^{d-1,q+1}) \\ E_\infty^{d,q} &= Z_\infty^{d,q}/(B_\infty^{d,q} + Z_\infty^{d-1,q+1}) \end{aligned} \tag{8.18}$$

where $Z_r^{d,q} = \{z \in A^{d,q} : \partial z \in A^{d-r}\}$, $Z_\infty^{d,q} = A^{d,q} \cap \ker \partial$, $B_r^{d,q} = A^{d,q} \cap \partial A^{d+r}$ and $B_\infty^{d,q} = A^{d,q} \cap \partial A$. Suppose we fix d and q. Then (i) gives $Z_r^{d,q} = Z_\infty^{d,q}$ and $Z_{r-1}^{d-1,q+1} = Z_\infty^{d-1,q+1}$ if $r > d$, and (ii) gives $B_{r-1}^{d,q} = B_\infty^{d,q}$ if r is large enough. Therefore $E_r^{d,q} = E_\infty^{d,q}$ for r large enough. But we found $E_1^{d,q} = E_r^{d,q}$ $(r \geq 1)$ and so (8.17) follows.

We remark that it also follows that $\operatorname{gr} H_j(\mathfrak{s} \oplus \mathfrak{s}, S_c; \mathcal{W}^{even}) = 0$ for $j > 0$. Thus we have proved

Proposition 8.1. $H_j(\mathfrak{s} \oplus \mathfrak{s}, S_c; \mathcal{W}^{even}) = 0$ *if* $j > 0$.

9 Proof of Corollary 6.4

S_c and G_c are commuting subgroups of $Sp(L)$. It follows that ξ maps $\mathfrak{g}$ into $\mathcal{W}^{inv}$ and so ξ induces $\xi_{\mathcal{B}}$ where $\xi_{\mathcal{B}}$ is the composition $\mathfrak{g} \xrightarrow{\xi} \mathcal{W}^{inv} \to \mathcal{B}$. Next τ is $Sp(L)$-invariant and so in particular is S_c-invariant. Then τ preserves $\mathcal{W}^{inv}$. We see that τ preserves $\mathcal{M}$, and so τ preserves $\mathcal{M}^{inv}$. Hence τ induces a filtered algebra anti-involution $\tau_{\mathcal{B}}$ of $\mathcal{B} = \mathcal{W}^{inv}/\mathcal{M}^{inv}$. Finally, ϑ is $Sp(L)$-invariant and so in particular is S_c-invariant. Then ϑ preserves $\mathcal{W}^{inv}$. We see that ϑ preserves $\mathcal{M}$, and so ϑ induces an antilinear filtered algebra involution $\vartheta_{\mathcal{B}}$ of $\mathcal{B}$.

Thus we have in place our Dixmier algebra data for $\overline{\mathcal{O}}$. It is clear because of Proposition 5.1 that the axioms are satisfied.

References

[1] J. Adams and D. Barbasch, Reductive dual pair correspondence for complex groups, *J. Funct. Anal.* 132 (1995), 1–42.

[2] A. Astashkevich and R. Brylinski, Non-Local Equivariant Star Product on the Minimal Nilpotent Orbit, *Adv. Math* **171** (2002), 86–102.

[3] D. Barbasch, The unitary dual for complex classical Lie groups, *Inv. Math.* **96** (1989), 103–176.

[4] D. Barbasch, Orbital Integrals of Nilpotent Orbits, in *The Mathematical Legacy of Harish-Chandra*, R. S. Doran and V. S. Varadarajan, eds., Proc. Symp. Pure Math., Vol. 68, AMS, 2000, 97–110.

[5] D. Barbasch and D.A. Vogan, Primitive ideals and orbital integrals in complex classical groups, *Math. Ann.* **259** (1982), 152–199.

[6] D. Barbasch and D.A. Vogan, Unipotent representations of complex semisimple groups, *Ann. of Math.* **121** (1985), 41–110.

[7] R. Brylinski, Equivariant Deformation Quantization for the Cotangent Bundle of a Flag Manifold, *Ann. Inst. Fourier* **52**:3 (2002), 881–897.

[8] R. Brylinski, Dixmier Algebras for Classical Complex Nilpotent Orbits via Kraft–Procesi Models II, in preparation.

[9] R. Brylinski, *Quantization of Classical Complex Nilpotent Orbits*, in preparation.

[10] M. Cahen and S. Gutt, An algebraic construction of $*$ product on the regular orbits of semi simple Lie groups, in *Gravitation and Geometry,* W. Rindler and A. Trautman eds., 73–82, Bibliopolis, 1987.

[11] D.H. Collingwood and W.M. McGovern, *Nilpotent Orbits in Semisimple Lie Algebras*, Van Nostrand Reinhold, 1993.

[12] J. Dixmier, *Enveloping Algebras*, North-Holland, 1977.

[13] M. Duflo, Représentations unitaires irréductibles des groupes simples complexes de rang deux, *Bull. Soc. Math. France* **107** (1979), 55–96.

[14] C. Duval, P. Lecomte and V. Ovsienko, Conformally equivariant quantization: existence and uniqueness, *Ann. Inst. Fourier* **49:6** (1999), 1999–2029.

[15] R. Godement, *Théorie des Faisceaux*, Hermann, Paris, Third edition, 1973.

[16] K.I. Gross, The dual of a parabolic subgroup and a degenerate principal series of $Sp(n, \mathbb{C})$, *Amer. J. Math* **93** (1971), 399-428.

[17] R. Hartshorne, *Algebraic Geometry*, Graduate Texts in Math 52, Springer-Verlag New York, 1977.

[18] R. Howe, Transcending classical invariant theory, *J. Amer. Math. Soc.* **2** (1989), 535–552.

[19] A. Joseph, The minimal orbit in a simple Lie algebra and associated maximal ideal, *Ann. Scient. Ec. Norm. Sup.* **9** (1976), 1–30.

[20] A. Knapp and D.A. Vogan, *Cohomological Induction and Unitary Representations*, Princeton University Press, 1995.

[21] P.Z. Kobak and A. Swann, Classical nilpotent orbits as hyperkähler quotients, *Int. J. Math.* **7** (1996), 193–210.

[22] H. Kraft and C. Procesi, Closures of conjugacy classes of matrices are normal, *Inv. Math.* **53** (1979), 227–247.

[23] H. Kraft and C. Procesi, On the geometry of conjugacy classes in clasical groups, *Comment. Math. Helv.* **57** (1982), 539–602.

[24] T. Levasseur and J.T. Stafford, *Rings of differential operators on classical rings of invariants*, Mem. Amer. Math. Soc., no. 412, 1989.

[25] J-S Li, Singular unitary representations of classical groups, *Inv. Math.* **97** (1989), 237–255.

[26] W. McGovern, Quantization of nilpotent orbits and their covers in complex classical groups, preprint, Yale University (1989).

[27] W. McGovern, Completely prime maximal ideals and quantization, Mem. Amer. Math. Soc., no. 519, 1994.

[28] C. Moeglin, Idéaux complètement premiers de l'algèbre enveloppante de $gl_n(\mathbb{C})$, *J. Alg.* **87** (1987), 287–366.

[29] C. Moeglin, Correspondance de Howe pour les paires réductives duales, quelques calculs dans le cas Archimedien, *J. Funct. Anal.* **85** (1989), 1–85.

[30] M. Van den Bergh, Differential operators on semi-invariants for tori and weighted projective spaces, *Topics in Invariant Theory* (Paris 1989/90), 255–272, Lecture Notes in Math. 1478, Springer, Berlin 1991.

[31] D.A. Vogan, Noncommutative algebras and unitary representations, *The Mathematical Heritage of Hermann Weyl*, 35–60, Proc. Symp. Pure Math., vol. 48, Amer. Math. Soc., Providence, 1988.

[32] D.A. Vogan, Dixmier algebras, sheets and representation theory, *Operator Algebras, Unitary Representations, Enveloping Algebras and Invariant Theory*, 333–396, Progress in Math, vol. 92, Birkhäuser, 1990.

[33] D.A. Vogan, Associated varieties and unipotent representations, in *Harmonic Analysis on Reductive Lie Groups*, 315–388, Progress in Math, vol. 101, Birkhäuser, 1991.

Ranee Brylinski
Department of Mathematics
Penn State University
University Park, PA 16802, USA
`rkb@math.psu.edu` and `rkb248@yahoo.com`

Current mailing address:
P.O. Box 1089
Truro, MA 02666-1089

Brèves remarques sur l'œuvre de A. A. Kirillov

Jacques Dixmier

La thèse de Kirillov, parue en 1962, a suscité immédiatement beaucoup d'intérêt.

Soit G un groupe de Lie nilpotent simplement connexe. Soient $\mathfrak{g} = \mathrm{Lie}(G)$, $\mathfrak{g}^*$ l'espace vectoriel dual de $\mathfrak{g}$, dans lequel G opère par la représentation coadjointe. Soit $\mathfrak{g}^*/G$ l'ensemble des orbites coadjointes. Soit $\widehat{G}$ l'ensemble des représentations unitaires irréductibles de G (la plupart sont de dimension infinie). Alors Kirillov construit une bijection canonique $\mathfrak{g}^*/G \to \widehat{G}$; ainsi, les représentations unitaires irréductibles de G sont paramétrées par les orbites coadjointes. En outre, quantité de notions naturelles concernant les représentations s'interprètent géométriquement en termes d'orbites coadjointes: restriction à un sous-groupe, induction unitaire, produit tensoriel, mesure de Plancherel, topologie de $\widehat{G}$. Détaillons seulement <u>un</u> résultat, d'une élégance extraordinaire: soient $\pi \in \widehat{G}$ et Ω l'orbite associée ; soit μ la mesure G-invariante canonique sur Ω (c'est une distribution tempérée sur $\mathfrak{g}^*$) ; soit μ' sa transformée de Fourier, distribution tempérée sur $\mathfrak{g}$; transportant μ' sur G par l'application exponentielle, on obtient le caractère-distribution de π !

Kirillov s'est vite convaincu, et il a convaincu la communauté mathématique, que cette "méthode des orbites" devait être applicable à des groupes bien plus généraux que les groupes nilpotents.

Il n'a pas hésité à aborder le cas des groupes de Lie connexes quelconques. Evidemment, des difficultés considérables ont surgi immédiatement. Néanmoins, Kirillov a indiqué une voie d'accès, qui ensuite a été largement utilisée.

Depuis plus de trente ans, Kirillov applique la méthode des orbites aux groupes de Lie–Cartan de dimension infinie. Par exemple, soit M une variété C^∞ compacte. Soit G le groupe des difféomorphismes de M, ou l'un des sous-groupes obtenus en prenant les éléments de G qui conservent une forme volume, ou une structure symplectique, ou une structure de contact. On considère l'algèbre de Lie $\mathfrak{g}$ et son dual $\mathfrak{g}^*$, tous deux définis en tenant compte de la topologie. Aux orbites coadjointes sont associées des représentations unitaires irréductibles de G, du moins dans certains cas, par exemple si l'orbite est de dimension finie ou de codimension finie. Le groupe fondamental de ces orbites peut être non commutatif, par exemple être un groupe symétrique. Kirillov envisage parfois des groupes encore plus généraux (par exemple, le groupe $\mathrm{Diff}_H(E)$ où E est un fibré principal de groupe structural H), et parfois au contraire étudie de manière très approfondie des cas particuliers importants; par exemple, soient S le cercle, $\mathrm{Diff}_+(S)$ le groupe des difféomorphismes de S conservant l'orientation; alors les orbites coadjointes pour ce groupe sont liées aux fonctions holomorphes univalentes dans le disque unité, et à certaines équations différentielles linéaires; Kirillov définit sur ces orbites des structures complexes kähleriennes invariantes, d'où des représentations de l'algèbre de Virasoro et du groupe de Virasoro–Bott.

Les démonstrations précédentes ont amené Kirillov à s'intéresser, dans plusieurs articles, à des questions de pure géométrie différentielle. Soient M une variété C^∞, E un fibré en droites sur M, $\Gamma(E)$ l'ensemble des sections C^∞ de E. Un crochet de Lie $[\ ,\ ]$ sur $\Gamma(E)$ est dit local si $[s_1, s_2]$ est continue en (s_1, s_2) et si $\mathrm{supp}[s_1, s_2] \subset \mathrm{supp}\, s_1 \bigcap \mathrm{supp}\, s_2$. Kirillov classifie ces crochets. Il détermine aussi, dans des cas généraux, quels sont les opérateurs multidifférentiels invariants par difféomorphismes (tels que la dérivation extérieure, ou le crochet ordinaire de deux champs de vecteurs).

Plus récemment, Kirillov a utilisé avec succès la méthode des orbites pour des groupes très différents, les $G_n(k)$ ($n = 1, 2, \ldots$; k corps commutatif). (On note $G_n(k)$ le groupe des matrices $n \times n$ unipotentes triangulaires supérieures à éléments dans k.) La théorie est particulièrement poussée lorsque k est un corps fini. Elle amène à introduire une suite remarquable des polynômes en une variable.

Voici un autre thème longuement étudié par Kirillov en collaboration avec Gelfand. Soient G un groupe algébrique complexe, $\mathfrak{g}$ son algèbre de Lie, U l'algèbre enveloppante de $\mathfrak{g}$, D le corps enveloppant de U. Soit $D_{n,k}$ le corps engendré par des indéterminées $x_1, \ldots, x_n, y_1, \ldots, y_n, z_1, \ldots, z_k$, où tous les crochets sont nuls sauf $[x_i, y_i] = 1$. Alors Gelfand et Kirillov conjecturent en 1966 que D est isomorphe à un corps $D_{n,k}$. Dans les années suivantes, ils démontrent des cas particuliers de plus en plus nombreux de cette conjecture. Bien après, la conjecture sera reconnue comme fausse, mais elle aura suscité un grand nombre de travaux intéressants. Surtout, dès l'article de 1966, Gelfand et Kirillov introduisent diverses notions de dimensions non commutatives; l'une d'entre elles, promise à un grand avenir, sera appelée dimension de Gelfand–Kirillov.

Il n'est pas surprenant que Kirillov ait étudié, pour elles-mêmes, les algèbres de Lie de dimension infinie. Dans un série d'articles, il a mis en évidence des cas où la "croissance" de ces algèbres est strictement intermédiaire entre la croissance polynomiale et la croissance exponentielle et il a construit des identités remarquables vérifiées dans l'algèbre de Lie des champs de vecteurs sur $\mathbb{R}$, ou des champs hamiltoniens sur $\mathbb{R}^2$.

La place me manque pour analyser les nombreux articles isolés abordant des thèmes sur lesquels Kirillov n'est pas revenu par la suite. Mentionnons tout de même 3 de ces articles: 1) il a complété sur un point très important l'étude (par Gelfand–Graev) des représentations unitaires irréductibles de $SL_2(k)$ (k corps localement compact non connexe). 2) Soient $\mathcal{H}$ un espace hilbertien, G le groupe des opérateurs unitaires u dans $\mathcal{H}$ tels que $1 - u$ soit compact. Alors Kirillov a déterminé les représentations unitaires irréductibles de G. 3) Kirillov a résolu un problème essentiel de géométrie intégrale posé par Gelfand. Ces études n'utilisent pas la méthode des orbites.

Kirillov a écrit plusieurs livres. Deux d'entre eux ont été traduits en français et en anglais: 1) *Eléments de la théorie des représentations*, 2) *Théorèmes et problèmes d'analyse fonctionnelle*. Il a d'autre part rédigé beaucoup d'articles d'exposition: chacun d'eux est en réalité un court livre. La clarté des exposés fait que ces écrits ont été largement lus, et notamment, avec profit, par des chercheurs débutants.

C'est l'occasion de signaler que Kirillov a eu de nombreux élèves de thèse. Les organisateurs de ce colloque ont eu la bonne idée de faire imprimer sur un T-shirt (offert à Kirillov) la liste de ces élèves: 57 noms, un record !

Kirillov a participé, en 1967, à la fondation d'un journal célèbre: *Functional Analysis and its Applications*. Il a, pendant 4 ans, été vice-président de la Société Mathématique de Moscou. Il est membre de comités éditoriaux de plusieurs journaux. Il a suscité la traduction en russe de nombreux livres et articles édités à l'ouest.

Dans sa conversation, Kirillov mentionne le soutien constant qu'il a reçu de son épouse (qui travaille en informatique). Il est fier de l'œuvre mathématique très connue de son fils.

Ce colloque honore donc un grand mathématicien: grand mathématicien par les voies qu'il a ouvertes, grand mathématicien par l'influence qu'il exerce depuis des années.

Jacques Dixmier
11 bis rue du Val de Grâce
F-75005 Paris, France

Gerbes of Chiral Differential Operators. III

Vassily Gorbounov
Fyodor Malikov
Vadim Schechtman

Introduction

This paper is a sequel to [GII]. Its aim is to "switch on an exterior vector bundle" in the framework of [GII].

Let X be a smooth scheme over a fixed ground ring k containing $1/2$ and let E be a vector bundle (i.e., a locally free $\mathcal{O}_X$-module) of finite rank over X. Consider the exterior algebra $\Lambda E = \oplus_{i=0}^{rk(E)} \Lambda^i E$ (over $\mathcal{O}_X$); this is a sheaf of commutative superalgebras over X, where by definition $\mathcal{O}_X$ is purely even and the parity of the component $\Lambda^i E$ is equal to the parity of i.

In this paper we study the chiral counterparts of the sheaf $\mathcal{D}_{\Lambda E}$ of superalgebras of differential operators acting on ΛE. Similar to [GII], these *chiral sheaves of differential operators on* ΛE exist locally and are by no means unique; the corresponding categories form a *champ en groupoïdes* $\mathfrak{D}_{\Lambda E}$ over X, called the *gerbe of chiral differential operators on* ΛE.

Our first main result (see Theorem 5.9) says that the characteristic class $c(\mathfrak{D}_{\Lambda E})$ lies in the second hypercohomology group $H^2(X; \Omega_X^{[2,3\rangle})$ (i.e., in the same group where $c(\mathfrak{D}_X)$ lies) and is equal to

$$c(\mathfrak{D}_{\Lambda E}) = c(\Theta_{X/k}) - c(E) = c(\Omega^1_{X/k}) - c(E) \tag{0.1}$$

where $c(E)$ is the "Atiyah–Chern–Simons class" defined in [GII], 7.6. Here $\Theta_{X/k}$ is the tangent bundle. Recall that $\Omega_X^{[2,3\rangle}$ denotes the length 1 complex of sheaves $\Omega^2_{X/k} \longrightarrow \Omega^{3,closed}_{X/k}$, with $\Omega^2_{X/k}$ in degree 0.

We obtain in fact a stronger statement, namely the equality (0.1) "on the level of cocycles". As a corollary of this, we conclude that for $E = \Theta_{X/k}$ or $E = \Omega^1_{X/k}$ our gerbes admit a canonical global section. In other words, there exist canonically defined *the* sheaves of chiral differential operators $\mathcal{D}^{ch}_{\Lambda\Theta_X}$ and $\mathcal{D}^{ch}_{\Omega^{\cdot}_X}$.

Section 6 is devoted to the study of the last sheaf, which is nothing but (the underlying sheaf of) the *chiral de Rham complex* from [MSV]. We obtain the transformation laws of 4 local generators of $N = 2$ supersymmetry Q, J, G and L,

---

F. Malikov was partially supported by an NSF grant.

see Theorem 6.25. In particular, the component Q_0 of the field $Q(z)$ is a globally defined square zero derivation of $\mathcal{D}^{ch}_{\Omega^{\cdot}_X}$, which is the *chiral de Rham differential* from [GII].

This completes the alternative construction of the chiral de Rham complex sketched in Section 6 of [MSV]. Its difference from the original construction is that it does not use the Wick theorem and the arguments of "formal geometry".

In the last section we show that as a simple consequence of the Poincaré–Birkhoff–Witt theorem for $\mathcal{D}^{ch}_{\Omega^{\cdot}_X}$ and the Lefschetz fixed point theorem one gets a "moonshine style" formula, cf. Theorem 7.9.

This work was done while V.S. visited IHES. He is grateful to the Institute for the excellent working atmosphere.

1 Preliminaries

1.1. We keep the assumptions of [GII]. We assume that the ground ring k contains $1/2$. For a k-supermodule M, we denote by M^{ev} (resp. M^{odd}) the submodule of even (resp. odd) elements, so that $M = M^{ev} \oplus M^{odd}$. For a homogeneous element $a \in M$, we denote by $p(a) \in \mathbb{Z}/2\mathbb{Z}$ its parity. When we speak about graded k-supermodules $M = \oplus_{i \in I} M_i$ we imply that the I-grading is compatible with the parity, i.e., $M^x = \oplus_i M_i^x$ where $M_i^x = M^x \cap M_i$, $x = ev$ or odd.

Let A be a commutative k-superalgebra. A *Lie superalgebroid* over A is a Lie superalgebra over k equipped with a structure of an A-module, such that the identities [GII] (0.2.1) and (0.2.2) hold true.

1.2. A $\mathbb{Z}_{\geq 0}$-*graded vertex superalgebra* (over k) is a $\mathbb{Z}_{\geq 0}$-graded k-supermodule $V = \oplus_{i \geq 0} V_i$ equipped with a distinguished even vector $\mathbf{1} \in V_0$ (*vacuum vector*) and a family of bilinear operations

$$_{(n)} : V \times V \longrightarrow V, \; n \in \mathbb{Z},$$

such that

$$p(a_{(n)}b) = p(a) + p(b); \quad V_{i(n)} V_j \subset V_{i+j-n-1}. \tag{1.2.1}$$

The following properties must hold:

$$\mathbf{1}_{(n)}a = \delta_{n,-1}a; \; a_{(n)}\mathbf{1} = 0 \text{ for } n \geq 0, \; a_{(-1)}\mathbf{1} = a \tag{1.2.2}$$

and

$$\sum_{j=0}^{\infty} \binom{m}{j} (a_{(n+j)}b)_{(m+l-j)}c$$

$$= \sum_{j=0}^{\infty} (-1)^j \binom{n}{j} \{a_{(m+n-j)}b_{(l+j)}c - (-1)^{n+p(a)p(b)} b_{(n+l-j)}a_{(m+j)}c\} \tag{1.2.3}$$

for all $m, n, l \in \mathbb{Z}$. A particular case of (1.2.3) corresponding to $m = 0$ is:

$$(a_{(n)}b)_{(l)}c = \sum_{j=0}^{\infty} (-1)^j \binom{n}{j} \{a_{(n-j)}b_{(l+j)}c - (-1)^{n+p(a)p(b)} b_{(n+l-j)}a_{(j)}c\}. \tag{1.2.4}$$

Setting $n = l = -1$ we get

$$(a_{(-1)}b)_{(-1)}c = \sum_{j=0}^{\infty} \{a_{(-1-j)}b_{(-1+j)}c + (-1)^{p(a)p(b)} b_{(-2-j)}a_{(j)}c\}. \tag{1.2.5}$$

In the sequel we shall work only with $\mathbb{Z}_{\geq 0}$-graded vertex superalgebras, and call them simply vertex superalgebras. This $\mathbb{Z}_{\geq 0}$-grading will be called the grading by *conformal weight*.

1.3. Let V be a vertex superalgebra. The even operators $\partial^{(j)} : V \longrightarrow V$ of degree j ($j \in \mathbb{Z}_{\geq 0}$) are defined in the same manner as in [GII], (0.5.5), and they satisfy [GII], (0.5.7), (0.5.8), (0.5.10) and (0.5.11). The "supercommutativity" formula reads as

$$a_{(n)}b = (-1)^{n+p(a)p(b)+1} \sum_{j\geq 0} (-1)^j \partial^{(j)}(b_{(n+j)}a) \tag{1.3.1}$$

and we have the usual OPE formula

$$[a_{(m)}, b_{(n)}] = \sum_{j\geq 0} \binom{m}{j} (a_{(j)}b)_{(m+n-j)} \tag{1.3.2}$$

where on the left-hand side we have the supercommutator

$$[a_{(m)}, b_{(n)}] := a_{(m)}b_{(n)} - (-1)^{p(a)p(b)} b_{(n)}a_{(m)}. \tag{1.3.3}$$

2 Vertex superalgebroids

2.1. An *extended Lie superalgebroid* (over A) is a quintuple $\mathcal{T} = (A, T, \Omega, \partial, \langle , \rangle)$ where T is a Lie superalgebroid over A, Ω is an A-module equipped with a structure of a module over the Lie superalgebra T, $\partial : A \longrightarrow \Omega$ an even A-derivation and a morphism of T-modules, $\langle , \rangle : T \times \Omega \longrightarrow A$ is an even A-bilinear pairing.

The following identities must be true ($a \in A$, $\tau, \nu \in T$, $\omega \in \Omega$):

$$\langle \tau, \partial a \rangle = \tau(a) \tag{2.1.1}$$

$$\tau(a\omega) = \tau(a)\omega + (-1)^{p(\tau)p(a)} a\tau(\omega) \tag{2.1.2}$$

$$(a\tau)(\omega) = a\tau(\omega) + (-1)^{p(a)(p(\tau)+p(\omega))} \langle \tau, \omega \rangle \partial a \tag{2.1.3}$$

$$\tau(\langle \nu, \omega \rangle) = \langle [\tau, \nu], \omega \rangle + (-1)^{p(\tau)p(\nu)} \langle \nu, \tau(\omega) \rangle. \tag{2.1.4}$$

For example, to a Lie superalgebroid T one associates canonically an extended superalgebroid with $\Omega = \mathrm{Hom}_A(T, A)$, as in [GII], 1.2.

2.2. De Rham–Chevalley complex. Let $\mathcal{T} = (A, T, \Omega, \ldots)$ be an extended Lie A-superalgebroid. Let us define A-modules $\Omega^i = \Omega^i(\mathcal{T})$, $i \in \mathbb{Z}_{\geq 0}$, as follows. Set $\Omega^0 = A$, $\Omega^1 = \Omega$. For $i \geq 2$, Ω^i is the submodule of the module of A-polylinear homomorphisms h from T^{i-1} to Ω such that the function $\langle \tau_1, h(\tau_2, \ldots, \tau_i)\rangle$ is skew symmetric (in the graded sense) with respect to all permutations of $(\tau_1, \ldots, \tau_i)$.

For example, if $\mathcal{T}$ is associated with a Lie superalgebroid T as above, then $\Omega^i = \mathrm{Hom}_A(\Lambda^i_A T, A)$.

Let us define the even maps $d_{DR} = d^i_{DR} : \Omega^i \longrightarrow \Omega^{i+1}$ as follows. For $i = 0$ we set $d_{DR}a = -\partial a$. For $i \geq 1$ we set

$$d_{DR}h(\tau_1, \ldots, \tau_i) = d_{Lie}h(\tau_1, \ldots, \tau_i) - (-1)^{p(h)p(\tau_1)}\partial\langle \tau_1, h(\tau_2, \ldots, \tau_i)\rangle \tag{2.2.1}$$

where

$$d_{Lie}h(\tau_1, \ldots, \tau_i) = \sum_{j=1}^{i} (-1)^{j+1+p(\tau_j)(p(\tau_1)+\ldots+p(\tau_{j-1}))}\tau_j(h(\tau_1, \ldots, \widehat{\tau_j}, \ldots, \tau_i))$$

$$+ \sum_{1\leq j<l\leq i} (-1)^{j+l+p(\tau_j)(p(\tau_1)+\ldots+p(\tau_{j-1}))+p(\tau_l)(p(\tau_1)+\ldots+\widehat{p(\tau_j)}+\ldots+p(\tau_{l-1}))}$$

$$\times h([\tau_j, \tau_l], \tau_1, \ldots, \widehat{\tau_j}, \ldots, \widehat{\tau_l}, \ldots, \tau_i). \tag{2.2.2}$$

For example,

$$d_{DR}\omega(\tau) = (-1)^{p(\omega)p(\tau)}\{\tau(\omega) - \partial\langle \tau, \omega\rangle\}, \tag{2.2.3}$$

for $\omega \in \Omega^1 = \Omega$; and

$$d_{DR}h(\tau_1, \tau_2) = -h([\tau_1, \tau_2]) + (-1)^{p(h)p(\tau_1)}\tau_1(h(\tau_2))$$

$$-(-1)^{p(\tau_2)(p(h)+p(\tau_1))}\tau_2(h(\tau_1)) - (-1)^{p(h)p(\tau_1)}\partial\langle \tau_1, h(\tau_2)\rangle, \tag{2.2.4}$$

for $h \in \Omega^2$.

Let us introduce the action of the Lie algebra T on the modules Ω^i by

$$\tau(h)(\tau_1, \ldots, \tau_{i-1}) = \tau(h(\tau_1, \ldots, \tau_{i-1}))$$

$$-\sum_{j=1}^{i-1} (-1)^{p(\tau)(p(\tau_1)+\ldots+p(\tau_{j-1}))}h(\tau_1, \ldots, [\tau, \tau_j], \ldots, \tau_i). \tag{2.2.5}$$

Let us define the convolution operators $\langle \tau, \cdot\rangle : \Omega^i \longrightarrow \Omega^{i-1}$ by

$$\langle \tau, h\rangle(\tau_1, \ldots, \tau_{i-2}\rangle = (-1)^{p(\tau)p(h)}h(\tau, \tau_1, \ldots, \tau_{i-2}). \tag{2.2.6}$$

The maps $\{d^i_{DR}\}$ may be characterized as a unique collection of maps such that $d^0_{DR} = -\partial$ and the *Cartan formula*

$$\tau(h) = \langle \tau, d_{DR} h \rangle + d_{DR} \langle \tau, h \rangle \tag{2.2.7}$$

holds true.

The maps d_{DR} commute with the action of T. One checks that $d^2_{DR} = 0$, so we get a complex $(\Omega^{\bullet}(\mathcal{T}), d_{DR})$ called the *de Rham–Chevalley complex* of $\mathcal{T}$.

2.3. A *vertex superalgebroid* is a septuple $\mathcal{A} = (A, T, \Omega, \partial, \gamma, \langle\, ,\, \rangle, c)$ where A is a supercommutative k-algebra, T is a Lie superalgebroid over A, Ω is an A-module equipped with an action of the Lie superalgebra T, $\partial : A \longrightarrow \Omega$ is an even derivation commuting with the T-action,

$$\langle\, ,\, \rangle : (T \oplus \Omega) \times (T \oplus \Omega) \longrightarrow A$$

is a supersymmetric even k-bilinear pairing equal to zero on $\Omega \times \Omega$ and such that $\mathcal{T}_{\mathcal{A}} = (A, T, \Omega, \partial, \langle\, ,\, \rangle|_{T\times\Omega})$ is an extended Lie superalgebroid over A; $c : T \times T \longrightarrow \Omega$ is a skew supersymmetric even k-bilinear pairing and $\gamma : A \times T \longrightarrow \Omega$ is an even k-bilinear map.

The following axioms must hold ($a, b \in A;\ \tau, \tau_i \in T$):

$$\gamma(a, b\tau) = \gamma(ab, \tau) - a\gamma(b, \tau)$$
$$-(-1)^{p(\tau)(p(a)+p(b))}\tau(a)\partial b - (-1)^{p(a)p(b)+p(\tau)p(a)+p(\tau)p(b)}\tau(b)\partial a \tag{A1}$$

$$\langle a\tau_1, \tau_2 \rangle = a\langle \tau_1, \tau_2 \rangle + \langle \gamma(a, \tau_1), \tau_2 \rangle - (-1)^{p(a)(p(\tau_1)+p(\tau_2))}\tau_1\tau_2(a) \tag{A2}$$

$$c(a\tau_1, \tau_2) = ac(\tau_1, \tau_2) + \gamma(a, [\tau_1, \tau_2])$$
$$-(-1)^{p(\tau_2)(p(\tau_1)+p(a))}\gamma(\tau_2(a), \tau_1) + (-1)^{p(\tau_2)(p(\tau_1)+p(a))}\tau_2(\gamma(a, \tau_1))$$
$$-(-1)^{p(a)(p(\tau_1)+p(\tau_2))}\frac{1}{2}\langle \tau_1, \tau_2 \rangle \partial a + (-1)^{p(a)(p(\tau_1)+p(\tau_2))}\frac{1}{2}\partial\tau_1\tau_2(a)$$
$$-(-1)^{p(\tau_2)(p(a)+p(\tau_1))}\frac{1}{2}\partial\langle \tau_2, \gamma(a, \tau_1)\rangle \tag{A3}$$

$$\langle [\tau_1, \tau_2], \tau_3 \rangle + (-1)^{p(\tau_1)p(\tau_2)}\langle \tau_2, [\tau_1, \tau_3] \rangle = \tau_1(\langle \tau_2, \tau_3 \rangle)$$
$$-(-1)^{p(\tau_1)p(\tau_2)}\frac{1}{2}\tau_2(\langle \tau_1, \tau_3 \rangle) - (-1)^{p(\tau_3)(p(\tau_1)+p(\tau_2))}\frac{1}{2}\tau_3(\langle \tau_1, \tau_2 \rangle)$$
$$+(-1)^{p(\tau_1)p(\tau_2)}\langle \tau_2, c(\tau_1, \tau_3)\rangle + (-1)^{p(\tau_3)(p(\tau_1)+p(\tau_2))}\langle \tau_3, c(\tau_1, \tau_2)\rangle \tag{A4}$$

$$d_{Lie}c(\tau_1, \tau_2, \tau_3) = -\frac{1}{2}\partial\{\langle [\tau_1, \tau_2], \tau_3 \rangle + (-1)^{p(\tau_2)p(\tau_3)}\langle [\tau_1, \tau_3], \tau_2 \rangle$$
$$-(-1)^{p(\tau_1)(p(\tau_2)+p(\tau_3))}\langle [\tau_2, \tau_3], \tau_1 \rangle - \tau_1(\langle \tau_2, \tau_3 \rangle) + (-1)^{p(\tau_1)p(\tau_2)}\tau_2(\langle \tau_1, \tau_3 \rangle)$$
$$-(-1)^{p(\tau_3)(p(\tau_1)+p(\tau_2))}2\langle \tau_3, c(\tau_1, \tau_2)\rangle\} \tag{A5}$$

where d_{Lie} is defined by (2.2.2).

2.4. All the constructions of [GII] generalize to the $\mathbb{Z}/(2)$-graded case in an obvious manner.

3. Some formulas

3.1. Let A be a smooth k-algebra of relative dimension n, such that the A-module $T = \mathrm{Der}_k(A)$ is free and admits a base $\{\bar{\tau}_i\}$ consisting of commuting vector fields. Let E be a free A-module of rank m, with a base $\{\phi_\alpha\}$. We shall call the set $\mathfrak{g} = \{\bar{\tau}_i; \phi_\alpha\} \subset T \oplus E$ a *frame* of (T, E).

Consider a commutative A-superalgebra $\Lambda E = \oplus_{i=0}^m \Lambda_A^i(E)$ where the parity of $\Lambda_A^i(E)$ is equal to the parity of i. Each frame $\mathfrak{g}$ as above gives rise to a ΛE-base $\{\tau_i; \psi_\alpha\}$ of the Lie superalgebroid $T_{\Lambda E} = \mathrm{Der}_k(\Lambda E)$, defined as follows. We extend the fields $\bar{\tau}_i$ to derivations τ_i of the whole superalgebra ΛE by the rule

$$\tau_i(a) = \bar{\tau}_i(a); \ \tau_i(\sum a_\alpha \phi_\alpha) = \sum \bar{\tau}_i(a_\alpha)\phi_\alpha. \tag{3.1.1}$$

(Note that this extension depends on the choice of a base $\{\phi_\alpha\}$ of the module E.) The fields $\{\tau_i\}$ form a ΛE-base of the even part $T_{\Lambda E}^{ev}$.

We define the odd vector fields $\psi_\alpha \in T_{\Lambda E}^{odd}$ by

$$\psi_\alpha(\sum a_\nu \phi_\nu) = a_\alpha; \ \psi_\alpha(a) = 0. \tag{3.1.2}$$

These fields form a ΛE-base of $T_{\Lambda E}^{odd}$.

Let $\{\omega_i; \ \rho_\alpha\}$ be the dual base of the module of 1-superforms $\Omega_{\Lambda E} = \mathrm{Hom}_{\Lambda E}^{ev}(T_{\Lambda E}, \Lambda E)$, defined by

$$\langle \tau_i, \omega_j \rangle = \delta_{ij}; \ \langle \psi_\alpha, \rho_\beta \rangle = \delta_{\alpha\beta}; \ \langle \tau_i, \rho_\alpha \rangle = \langle \psi_\alpha, \omega_i \rangle = 0. \tag{3.1.3}$$

3.2. Let us describe the effect of a change of frame. Let $\mathfrak{g}' = \{\bar{\tau}_i'; \phi_\alpha'\}$ be another frame, with $\bar{\tau}_i' = g^{ij}\bar{\tau}_j; \ \phi_\alpha' = A^{\alpha\beta}\phi_\beta, \ g = (g^{ij}) \in GL_n(A), \ A = (A^{\alpha\beta}) \in GL_m(A)$.

The corresponding new bases τ_i', look as follows:

$$\tau_i' = g^{ip}\tau_p + g^{i\alpha\gamma}\phi_\gamma\psi_\alpha \tag{3.2.1}$$

where

$$g^{i\alpha\gamma} = g^{iq}\tau_q(A^{-1\alpha\mu})A^{\mu\gamma}. \tag{3.2.2}$$

Next,

$$\psi_\alpha' = A^{-1\mu\alpha}\psi_\mu \tag{3.2.3}$$

$$\omega_i' = g^{-1pi}\omega_p \tag{3.2.4}$$

$$\rho_\alpha' = \tau_i(A^{\alpha\gamma})\phi_\gamma\omega_i + A^{\alpha\mu}\rho_\mu. \tag{3.2.5}$$

Formulas for the inverse transformation:

$$\tau_q = g^{-1qi}\tau_i' + \tau_q(A^{\alpha\gamma})\phi_\gamma\psi_\alpha' \tag{3.2.6}$$

$$\psi_\beta = A^{\alpha\beta}\psi_\alpha' \tag{3.2.7}$$

$$\omega_j = g^{pj}\omega'_p \tag{3.2.8}$$

$$\rho_\beta = A^{-1\beta\alpha}\rho'_\alpha + g^{p\beta\gamma}\phi_\gamma\omega'_p. \tag{3.2.9}$$

These formulas show that T is canonically an A-module quotient of $T_{\Lambda E}$ and $\Omega = \mathrm{Hom}_A(T, A)$ is canonically an A-submodule of $\Omega_{\Lambda E}$. In fact the whole de Rham complex $\Omega^{\cdot}_A$ is canonically the subcomplex of $\Omega^{\cdot}_{\Lambda E}$.

3.3. Recall that

$$g^{ip}\tau_p(g^{jq}) = g^{jp}\tau_p(g^{iq}) \tag{3.3.1}$$

$$g^{ip}\tau_q\tau_p(g^{jq}) = g^{jq}\tau_p\tau_q(g^{ip}) \tag{3.3.2}$$

and

$$\tau_p(g^{-1qr}) = \tau_q(g^{-1pr}) \tag{3.3.3}$$

see [GII], 5.4.

It is easy to see that

$$\mathrm{tr}\{\tau_p(A)\tau_q(A^{-1})\} = \mathrm{tr}\{\tau_q(A)\tau_p(A^{-1})\}. \tag{3.3.4}$$

Using (3.3.1) and (3.3.4) one sees easily that

$$g^{ip}\tau_p(g^{j\nu\nu}) = g^{jq}\tau_q(g^{i\nu\nu}). \tag{3.3.5}$$

3.4. Let $\mathcal{A} = \mathcal{A}_{\Lambda E;\mathfrak{g}}$ be the vertex superalgebroid corresponding to the frame $\mathfrak{g}$. We have the following identities in $\mathcal{A}$:

γ-formulas

$$\gamma(a, b\tau_i) = -\tau_i(a)\partial b - \tau_i(b)\partial a \tag{3.4.1}$$

$$\gamma(a, b\phi_\nu\psi_\mu) = \delta_{\nu\mu} b\partial a \tag{3.4.2}$$

$$\gamma(a\phi_\beta, b\psi_\mu) = \delta_{\beta\mu} a\partial b. \tag{3.4.3}$$

$\langle,\rangle$-formulas

$$\langle a\tau_i, b\tau_j\rangle = -b\tau_i\tau_j(a) - a\tau_j\tau_i(b) - \tau_i(b)\tau_j(a) \tag{3.4.4}$$

$$\langle a\phi_\alpha\psi_\beta, b\tau_i\rangle = \delta_{\alpha\beta} b\tau_i(a) \tag{3.4.5}$$

$$\langle a\phi_\alpha\psi_\beta, \phi_{\alpha'}\psi_{\beta'}\rangle = ab\delta_{\beta\alpha'}\delta_{\beta'\alpha}. \tag{3.4.6}$$

c-formulas

$$c(a\tau_i, b\tau_j) = \frac{1}{2}\{\tau_i(b)\partial\tau_j(a) - \tau_j(a)\partial\tau_i(b)\} + \frac{1}{2}\partial\{b\tau_i\tau_j(a) - a\tau_j\tau_i(b)\} \tag{3.4.7}$$

$$c(a\phi_\alpha\psi_\mu, b\phi_\beta\psi_\nu) = \frac{\delta_{\mu\beta}\delta_{\nu\alpha}}{2}\{a\partial b - b\partial a\} \tag{3.4.8}$$

$$c(a\phi_\alpha\psi_\mu, b\tau_i) = -\frac{\delta_{\mu\alpha}}{2}\partial\{b\tau_i(a)\} \tag{3.4.9}$$

$$c(a\tau_i, b\psi_\alpha) = c(a\phi_\alpha\psi_\mu, b\psi_\nu) = 0. \tag{3.4.10}$$

3.5. Let $\mathfrak{g}'$ be another frame as in 3.2. We have

$$\begin{aligned}\gamma(a, \tau'_p) &= \gamma(a, g^{pq}\tau_q + g^{p\mu\nu}\phi_\nu\psi_\mu) \\ &= -\tau_q(a)\partial g^{pq} - \tau_q(g^{pq})\partial a + g^{p\mu\mu}\partial a \end{aligned} \tag{3.5.1}$$

$$\gamma(a\phi'_\mu, \psi'_\alpha) = \gamma(aA^{\mu\beta}\phi_\beta, A^{-1\nu\alpha}\psi_\nu) = aA^{\mu\beta}\partial A^{-1\beta\alpha}. \tag{3.5.2}$$

Next,

$$\begin{aligned}\langle\tau'_i, \tau'_j\rangle &= \langle g^{ip}\tau_p + g^{i\mu\alpha}\phi_\alpha\psi_\mu, g^{jq}\tau_q + g^{j\nu\beta}\phi_\beta\psi_\nu\rangle \\ &= -2g^{ip}\tau_q\tau_p(g^{jq}) - \tau_p(g^{jq})\tau_q(g^{ip}) \\ &+2g^{ip}\tau_p(g^{j\nu\nu}) + g^{ip}g^{jq}\tau_p(A^{-1\mu\beta})A^{\beta\gamma}\tau_q(A^{-1\gamma\sigma})A^{\sigma\mu}\end{aligned} \tag{3.5.3}$$

and

$$\langle\tau'_i, \psi'_\alpha\rangle = \langle\psi'_\alpha, \psi'_\beta\rangle = 0. \tag{3.5.4}$$

Finally,

$$c(\tau'_i, \tau'_j) = \frac{1}{2}\{\tau_p(g^{jq})\partial\tau_q(g^{ip}) - \tau_q(g^{ip})\partial\tau_p(g^{jq})\} + \frac{1}{2}\{g^{i\mu\nu}\partial g^{j\nu\mu} - g^{j\nu\mu}\partial g^{i\mu\nu}\} \tag{3.5.5}$$

and

$$c(\tau'_i, \psi'_\alpha) = c(\psi'_\alpha, \psi'_\beta) = 0. \tag{3.5.6}$$

4 Chern–Simons term

This section is parallel to [GII], Section 5.

4.1. We keep the setup of the previous section. Let $\mathcal{A} = \mathcal{A}_{\Lambda E;\mathfrak{g}}$, $\mathcal{A}' = \mathcal{A}_{\Lambda E;\mathfrak{g}'}$ (resp. $\mathcal{B} = \mathcal{B}_{\Lambda E;\mathfrak{g}}$, $\mathcal{B}' = \mathcal{B}_{\Lambda E;\mathfrak{g}'}$) be the vertex superalgebroids (resp. prealgebroids) corresponding to our frames.

As in [GII], 5.5 we have a canonical isomorphism

$$g = g_{\mathfrak{g},\mathfrak{g}'} = (Id_{\Lambda E}, Id_{T_{\Lambda E}}, Id_{\Omega_{\Lambda E}}, h) : \mathcal{B}' \xrightarrow{\sim} \mathcal{B} \tag{4.1.1}$$

where

$$h = h_{\mathfrak{g},\mathfrak{g}'} : T_{\Lambda E} \longrightarrow \Omega_{\Lambda E} \tag{4.1.2}$$

is defined by the condition

$$\langle x', h(y')\rangle = -\frac{1}{2}\langle x', y'\rangle, \quad x', y' \in \{\tau'_i\} \cup \{\psi'_\alpha\}. \tag{4.1.3}$$

Using (3.5.3) and (3.5.4) we find the following explicit formulas for h:

$$h(\tau'_i) = h^{ij}\omega_j; \; h(\psi'_\alpha) = 0 \tag{4.1.4}$$

where $h^{ij} = h^{ij}_{\Omega} - h^{ij}_{E}$,

$$h^{ij}_{\Omega} = \tau_p\tau_j(g^{ip}) + \frac{1}{2}\tau_q(g^{ip})\tau_p(g^{rq})g^{-1jr} \tag{4.1.5}$$

cf. [GII], (5.7.2), and

$$h^{ij}_{E} = \tau_j(g^{i\nu\nu}) + \frac{1}{2}g^{iq}\tau_j(A^{-1\mu\beta})A^{\beta\gamma}\tau_q(A^{-1\gamma\nu})A^{\nu\mu}. \tag{4.1.6}$$

The meaning of the notation h_{Ω} will become clear below, see §6.

4.2. We have

$$\mathcal{A} = g_*\mathcal{A}' + \mathfrak{b} \tag{4.2.1}$$

where the closed 3-form $\mathfrak{b} \in \Omega^{3,cl}_{\Lambda E}$ is defined by

$$\mathfrak{b}(x', y') = c(x', y') - x'(h(y')) + (-1)^{p(x')p(y')}y'(h(x')), \tag{4.2.2}$$

$x', y' \in \{\omega'_i\} \cup \{\psi'_{\alpha}\}$, cf. [GII], (5.7.3).

It is easy to see that $\psi'_{\alpha}(h(\tau'_i)) = 0$; on the other hand we know already that $h(\psi'_{\alpha}) = 0$ and $c(\psi'_{\alpha}, y') = 0$. It follows that $\mathfrak{b} \in \Omega^{3,cl} \subset \Omega^{3,cl}_{\Lambda E}$.

Next, we have

$$\tau'_i(h(\tau'_j)) = (g^{ip}\tau_p + g^{i\alpha\gamma}\phi_{\gamma}\psi_{\alpha})(h^{jq}\omega_q) =$$

(note that the second summand is zero)

$$= (g^{ip}\tau_p)(h^{jq}_{\Omega}\omega_q) + (g^{ip}\tau_p)(h^{jq}_{E}\omega_q) = g^{ip}\tau_p(h^{jq}_{\Omega})\omega_q + h^{jp}_{\Omega}\partial g^{ip}$$
$$-g^{ip}\tau_p(h^{jq}_{E})\omega_q - h^{jp}_{E}\partial g^{ip}. \tag{4.2.3}$$

It follows that $\mathfrak{b} = \mathfrak{b}_{\Omega} - \mathfrak{b}_{E}$ where $\mathfrak{b}_{\Omega}, \mathfrak{b}_{E} \in \Omega^3$ are given by

$$\mathfrak{b}_{\Omega}(\tau'_i, \tau'_j) = \frac{1}{2}\{\tau_p(g^{jq})\partial\tau_q(g^{ip}) - \tau_q(g^{ip})\partial\tau_p(g^{jq})\}$$
$$-g^{ip}\tau_p(h^{jq}_{\Omega})\omega_q - h^{jp}_{\Omega}\partial g^{ip} + g^{jp}\tau_p(h^{iq}_{\Omega})\omega_q + h^{ip}_{\Omega}\partial g^{jp} \tag{4.2.4}$$

and

$$\mathfrak{b}_{E}(\tau'_i, \tau'_j) = -\frac{1}{2}\{g^{i\mu\nu}\partial g^{j\nu\mu} - g^{j\nu\mu}\partial g^{i\mu\nu}\}$$
$$-g^{ip}\tau_p(h^{jq}_{E})\omega_q + g^{jp}\tau_p(h^{iq}_{E})\omega_q - h^{jp}_{E}\partial g^{ip} + h^{ip}_{E}\partial g^{jp}. \tag{4.2.5}$$

The form $\mathfrak{b}_{\Omega}$ has already been computed in [GII], Magic Lemma 5.6 and Theorem 6.4 (b), and is equal to

$$\mathfrak{b}_{\Omega}(\tau'_i, \tau'_j) = -\frac{1}{2}\mathrm{tr}\{g^{-1}\tau'_i(g)g^{-1}\tau'_j(g)g^{-1}\tau'_r(g) - g^{-1}\tau'_j(g)g^{-1}\tau'_i(g)g^{-1}\tau'_r(g)\}\omega'_r \tag{4.2.6}$$

cf. *loc.cit.* (5.5.3) and (6.4.2). Note that $\mathfrak{b}_\Omega$ is closed, hence $\mathfrak{b}_E$ is closed.

Magic Lemma 4.3. *We have*

$$\mathfrak{b}_E(\tau_i', \tau_j') = -\frac{1}{2}tr\{A^{-1}\tau_i'(A)A^{-1}\tau_j'(A)A^{-1}\tau_r'(A) - A^{-1}\tau_j'(A)A^{-1}\tau_i'(A)A^{-1}\tau_r'(A)\}\omega_r' \tag{4.3.1}$$

Proof. Let us denote the six terms in (4.2.5) by A, A', B, B', C and C'. We have

$$A = -\frac{1}{2}g^{iq}\tau_q(A^{-1\mu\alpha})A^{\alpha\nu}\tau_r\{g^{jp}\tau_p(A^{-1\nu\beta})A^{\beta\mu}\}\omega_r$$

$$= -\frac{1}{2}g^{iq}\tau_q(A^{-1\mu\alpha})A^{\alpha\nu}\{\tau_r(g^{jp})\tau_p(A^{-1\nu\beta})A^{\beta\mu}\omega_r + g^{jp}\tau_r\tau_p(A^{-1\nu\beta})A^{\beta\mu}\omega_r + g^{jp}\tau_p(A^{-1\nu\beta})\tau_r(A^{\beta\mu})\omega_r\}.$$

Next,

$$B = -\frac{1}{2}g^{ip}\tau_p\{g^{jq}\tau_r(A^{-1\mu\beta})A^{\beta\gamma}\tau_q(A^{-1\gamma\nu})A^{\nu\mu}\}\omega_r - g^{ip}\tau_p\tau_r(g^{j\nu\nu})\omega_r$$

$$= -\frac{1}{2}g^{ip}\tau_p(g^{jq})\tau_r(A^{-1\mu\beta})A^{\beta\gamma}\tau_q(A^{-1\gamma\nu})A^{\nu\mu}\omega_r$$
$$-\frac{1}{2}g^{ip}g^{jq}\tau_p\tau_r(A^{-1\mu\beta})A^{\beta\gamma}\tau_q(A^{-1\gamma\nu})A^{\nu\mu}\omega_r$$
$$-\frac{1}{2}g^{ip}g^{jq}\tau_r(A^{-1\mu\beta})\tau_p(A^{\beta\gamma})\tau_q(A^{-1\gamma\nu})A^{\nu\mu}\omega_r$$
$$-\frac{1}{2}g^{ip}g^{jq}\tau_r(A^{-1\mu\beta})A^{\beta\gamma}\tau_p\tau_q(A^{-1\gamma\nu})A^{\nu\mu}\omega_r$$
$$-\frac{1}{2}g^{ip}g^{jq}\tau_r(A^{-1\mu\beta})A^{\beta\gamma}\tau_q(A^{-1\gamma\nu})\tau_p(A^{\nu\mu})\omega_r - g^{ip}\tau_p\tau_r(g^{j\nu\nu})\omega_r.$$

Finally,

$$C = -\frac{1}{2}g^{jq}\tau_p(A^{-1\mu\beta})A^{\beta\gamma}\tau_q(A^{-1\gamma\nu})A^{\nu\mu}\tau_r(g^{ip})\omega_r - \tau_r(g^{ip})\tau_p(g^{j\nu\nu})\omega_r.$$

We see that $A1 = -C1'$, $A2 = -B'2$, $B1 = -B'1$ (by (3.4.6)), $B2 = -A'2$, $B4 = -B'4$ and $C1 = -A'1$. Next,

$$B6 + C2 = -\tau_r\{g^{ip}\tau_p(g^{j\nu\nu})\},$$

so $B6 + C2 = -B'6 - C'2$ by (3.3.5).

So we are left with three terms: $A3$, $B3$ and $B5$ and their primed partners. It is easy to see that

$$A3 = B3 = -B5 = -\frac{1}{2}tr\{A^{-1}\tau_i'(A)A^{-1}\tau_j'(A)A^{-1}\tau_r'(A)\}\omega_r',$$

which implies the lemma. □

5 Atiyah term

This section is parallel to [GII], Section 6.

5.1. We keep the setup of the previous section. Let us denote by $\mathcal{T}_{\Lambda E} = (\Lambda E, T_{\Lambda E}, \Omega_{\Lambda E}, \partial)$ the extended vertex superalgebroid Lie corresponding to our data.

Let $\mathfrak{g}'' = \{\bar{\tau}_i'', \phi_\alpha''\}$ be a third frame of (T, E), with $\bar{\tau}_i'' = g'^{ij}\bar{\tau}_j'$, $\phi_\alpha'' = A'^{\alpha\beta}\phi_\beta'$. We have the corresponding new bases of $T_{\Lambda E}$ and $\Omega_{\Lambda E}$ given by

$$\tau_i'' = g'^{ip}\tau_p' + g'^{i\alpha\gamma}\phi_\gamma'\psi_\alpha' \tag{5.1.1}$$

where

$$g'^{i\alpha\gamma} = g'^{iq}\tau_q'(A'^{-1\alpha\mu})A'^{\mu\gamma} \tag{5.1.2}$$

$$\psi_\alpha'' = A'^{-1\mu\alpha}\psi_\mu' \tag{5.1.3}$$

$$\omega_i'' = g'^{-1pi}\omega_p' \tag{5.1.4}$$

$$\rho_\alpha'' = \tau_i'(A'^{\alpha\gamma})\phi_\gamma'\omega_i' + A'^{\alpha\mu}\rho_\mu'. \tag{5.1.5}$$

5.2. Let $\mathcal{A}'' = \mathcal{A}_{\Lambda E,\mathfrak{g}''}$ (resp. $\mathcal{B}'' = \mathcal{B}_{\Lambda E;\mathfrak{g}''}$) be the vertex superalgebroid (resp. prealgebroid) corresponding to the third frame. We have canonical isomorphisms

$$\mathcal{B}'' \overset{g_{\mathfrak{g}',\mathfrak{g}''}}{\underset{\sim}{\longrightarrow}} \mathcal{B}' \overset{g_{\mathfrak{g},\mathfrak{g}'}}{\underset{\sim}{\longrightarrow}} \mathcal{B}$$

as well as the morphism $g_{\mathfrak{g},\mathfrak{g}''} : \mathcal{B}'' \overset{\sim}{\longrightarrow} \mathcal{B}$ over $Id_{\mathcal{T}_{\Lambda E}}$, given by functions $h_{\mathfrak{g},\mathfrak{g}'}$, $h_{\mathfrak{g}',\mathfrak{g}''}$, $h_{\mathfrak{g},\mathfrak{g}''}$, and we are aiming to compute the discrepancy

$$\mathfrak{a} = \mathfrak{a}_{\mathfrak{g},\mathfrak{g}',\mathfrak{g}''} := h_{\mathfrak{g},\mathfrak{g}'} + h_{\mathfrak{g}',\mathfrak{g}''} - h_{\mathfrak{g},\mathfrak{g}''} \in \Omega^2_{\Lambda E}. \tag{5.2.1}$$

Note that

$$\gamma(a, b\psi_\mu) = -\psi_\mu(a)\partial b - \psi_\mu(b)\partial a = 0 \tag{5.2.2}$$

(in $\mathcal{A}$), hence

$$h_{\mathfrak{g},\mathfrak{g}'}(a\psi_\mu') = ah_{\mathfrak{g},\mathfrak{g}'}(\psi_\mu') - \gamma(a, \psi_\mu') = 0. \tag{5.2.3}$$

Therefore

$$h_{\mathfrak{g},\mathfrak{g}'}(\psi_\alpha'') = h_{\mathfrak{g},\mathfrak{g}'}(A'^{-1\mu\alpha}\psi_\mu') = 0. \tag{5.2.4}$$

It follows that

$$\mathfrak{a}(\psi_\alpha'') = 0. \tag{5.2.5}$$

5.3. Let us denote for brevity $h := h_{\mathfrak{g},\mathfrak{g}'}$, $h' = h_{\mathfrak{g}',\mathfrak{g}''}$, $h'' = h_{\mathfrak{g},\mathfrak{g}''}$.

We have

$$\begin{aligned} h(a\tau_p') &= ah(\tau_p') - \gamma(a, \tau_p') \\ &= ah^{pr}\omega_r + \tau_q(a)\partial g^{pq} + \tau_q(g^{pq})\partial a - g^{p\mu\mu}\partial a \end{aligned} \tag{5.3.1}$$

and

$$h(a\phi'_\gamma\psi'_\alpha) = -aA^{\gamma\beta}\partial A^{-1\beta\alpha}. \tag{5.3.2}$$

Thus we get

$$\begin{aligned} h(\tau''_i) &= h(g'^{ip}\tau'_p + g'^{i\alpha\gamma}\phi'_\gamma\psi'_\alpha) \\ &= g'^{ip}h^{pr}_\Omega\omega_r - g'^{ip}h^{pr}_E\omega_r + \tau_q(g'^{ip})\partial g^{pq} + \tau_q(g^{pq})\partial g'^{ip} \\ &\quad - g^{p\mu\mu}\partial g'^{ip} - g'^{i\alpha\gamma}A^{\gamma\beta}\partial A^{-1\beta\alpha}. \end{aligned} \tag{5.3.3}$$

It follows that

$$\mathfrak{a} = \mathfrak{a}_\Omega - \mathfrak{a}_E \tag{5.3.4}$$

where

$$\mathfrak{a}_\Omega(\tau''_i) = g'^{ip}h^{pr}_\Omega\omega_r + \tau_q(g'^{ip})\partial g^{pq} + \tau_q(g^{pq})\partial g'^{ip} + h'^{ip}_\Omega\omega'_p - h''^{ir}_\Omega\omega_r \tag{5.3.5}$$

and

$$\mathfrak{a}_E(\tau''_i) = g'^{ip}h^{pr}_E\omega_r + g^{p\mu\mu}\partial g'^{ip} + g'^{i\alpha\gamma}A^{\gamma\beta}\partial A^{-1\beta\alpha} + h'^{ip}_E\omega'_p - h''^{ir}_E\omega_r. \tag{5.3.6}$$

The form $\mathfrak{a}_\Omega$ has already been computed in [GII], Section 6. Namely, by Theorem 6.4 (a) from *op. cit.*,

$$\mathfrak{a}_\Omega(\tau''_i) = \frac{1}{2}tr\{g'^{-1}\tau''_i(g')\tau''_s(g)g^{-1} - g'^{-1}\tau''_s(g')\tau''_i(g)g^{-1}\}\omega''_s. \tag{5.3.7}$$

5.4. Let us compute $\mathfrak{a}_E(\tau''_i)$. Let us denote the five terms on the right-hand side of (5.3.6) by $\mathfrak{A}$, $\mathfrak{B}$, $\mathfrak{C}$, $\mathfrak{D}$ and $\mathfrak{E}$. Thus,

$$\mathfrak{A} = g'^{ip}h^{pr}_E\omega_r = \frac{1}{2}g'^{ip}g^{pq}\tau_r(A^{-1\mu\beta})A^{\beta\gamma}\tau_q(A^{-1\gamma\nu})A^{\nu\mu}\omega_r + g'^{ip}\tau_r(g^{p\nu\nu})\omega_r \tag{5.4.1}$$

$$\mathfrak{B} = g^{p\mu\mu}\partial g'^{ip} = g^{pq}\tau_q(A^{-1\mu\alpha})A^{\alpha\mu}\tau_r(g'^{ip})\omega_r \tag{5.4.2}$$

$$\mathfrak{C} = g'^{i\alpha\gamma}A^{\gamma\beta}\partial A^{-1\beta\alpha} = g'^{ip}\tau'_p(A'^{-1\alpha\mu})A'^{\mu\gamma}A^{\gamma\beta}\tau_r(A^{-1\beta\alpha})\omega_r \tag{5.4.3}$$

$$\begin{aligned} \mathfrak{D} = h'^{ij}_E\omega'_j &= \frac{1}{2}g'^{is}\tau'_j(A'^{-1\mu\beta})A'^{\beta\gamma}\tau'_s(A'^{-1\gamma\nu})A'^{\nu\mu}g^{-1rj}\omega_r + \tau'_j(g'^{i\nu\nu})g^{-1rj}\omega_r \\ &= \frac{1}{2}g'^{is}\tau_r(A'^{-1\mu\beta})A'^{\beta\gamma}g^{sq}\tau_q(A'^{-1\gamma\nu})A'^{\nu\mu}\omega_r + \tau_r\{g'^{ip}g^{pq}\tau_q(A'^{-1\nu\mu})A^{\mu\nu}\}\omega_r \end{aligned} \tag{5.4.4}$$

and

$$\begin{aligned} \mathfrak{E} = -h''^{ir}\omega_r &= -\frac{1}{2}(g'g)^{iq}\tau_r((A'A)^{-1\mu\beta})(A'A)^{\beta\gamma}\tau_q((A'A)^{-1\gamma\nu})(A'A)^{\nu\mu}\omega_r \\ -\tau_r(g''^{i\nu\nu}) &= -\frac{1}{2}(g'g)^{iq}\{\tau_r(A^{-1\mu\sigma})A'^{-1\sigma\beta} + A^{-1\mu\sigma}\tau_r(A'^{-1\sigma\beta})\}A'^{\beta\rho}A^{\rho\gamma} \\ &\quad \times\{\tau_q(A^{-1\gamma\delta})A'^{-1\delta\nu} + A^{-1\gamma\delta}\tau_q(A'^{-1\delta\nu})\}A'^{\nu\epsilon}A^{\epsilon\mu}\omega_r - \tau_r(g''^{i\nu\nu}) \end{aligned}$$

$$= -\frac{1}{2}g'^{ip}g^{pq}\{\tau_r(A^{-1\mu\sigma})A^{\sigma\gamma}\tau_q(A^{-1\gamma\delta})A^{\delta\mu}\omega_r$$

$$+\tau_r(A^{-1\mu\sigma})\tau_q(A'^{-1\sigma\nu})A'^{\nu\epsilon}A^{\epsilon\mu}\omega_r$$

$$+\tau_r(A'^{-1\sigma\beta})A'^{\beta\rho}A^{\rho\gamma}\tau_q(A^{-1\gamma\sigma})\omega_r + \tau_r(A'^{-1\sigma\beta})A'^{\beta\rho}\tau_q(A'^{-1\rho\nu})A'^{\nu\sigma}\omega_r\}$$

$$-\tau_r\{(g'g)^{iq}\tau_q((A'A)^{-1\nu\mu})(A'A)^{\mu\nu}\}\omega_r. \tag{5.4.5}$$

We see that $\mathfrak{A}1 = -\mathfrak{E}1$, $\mathfrak{C} = -2\mathfrak{E}2$, $\mathfrak{D}1 = -\mathfrak{E}4$. It is easy to see that $\mathfrak{A}2 + \mathfrak{D}2 + \mathfrak{E}5 = -\mathfrak{B}$.

Finally,

$$\mathfrak{C} + \mathfrak{E}2 = \frac{1}{2}tr\{A'^{-1}\tau_i''(A')\tau_r''(A)A^{-1}\}\omega_r'' \tag{5.4.6}$$

and

$$\mathfrak{E}3 = -\frac{1}{2}tr\{A'^{-1}\tau_r''(A')\tau_i''(A)A^{-1}\}\omega_r''. \tag{5.4.7}$$

So, we have proved

Lemma 5.5. *The form* $\mathfrak{a}_E$ *is given by*

$$\mathfrak{a}_E(\tau_i'') = \frac{1}{2}tr\{A'^{-1}\tau_i''(A')\tau_r''(A)A^{-1} - A'^{-1}\tau_r''(A')\tau_i''(A)A^{-1}\}\omega_r''. \tag{5.5.1}$$

Combining 4.3 and 5.5 we get

Theorem 5.6. (a) *The cocycle* $\mathfrak{a}_{\mathfrak{g},\mathfrak{g}',\mathfrak{g}''}$ *is given by*

$$\mathfrak{a}_{\mathfrak{g},\mathfrak{g}',\mathfrak{g}''}(\tau_i'') = \frac{1}{2}tr\{g'^{-1}\tau_i''(g')\tau_r''(g)g^{-1} - g'^{-1}\tau_r''(g')\tau_i''(g)g^{-1}\}\omega_r''$$

$$-\frac{1}{2}tr\{A'^{-1}\tau_i''(A')\tau_r''(A)A^{-1} - A'^{-1}\tau_r''(A')\tau_i''(A)A^{-1}\}\omega_r''. \tag{5.6.1}$$

(b) *The* 3*-form* $\mathfrak{b}_{\mathfrak{g},\mathfrak{g}'}$ *is given by*

$$\mathfrak{b}_{\mathfrak{g},\mathfrak{g}'}(\tau_i',\tau_j') = -\frac{1}{2}tr\{g^{-1}\tau_i'(g)g^{-1}\tau_j'(g)g^{-1}\tau_r'(g)$$

$$-g^{-1}\tau_j'(g)g^{-1}\tau_i'(g)g^{-1}\tau_r'(g)\}\omega_r'$$

$$+\frac{1}{2}tr\{A^{-1}\tau_i'(A)A^{-1}\tau_j'(A)A^{-1}\tau_r'(A) - A^{-1}\tau_j'(A)A^{-1}\tau_i'(A)A^{-1}\tau_r'(A)\}\omega_r'. \tag{5.6.2}$$

Lemma 5.7. *Let* $E^* = \mathrm{Hom}_A(E, A)$ *be the dual module. We have*

$$(\mathfrak{a}_{E^*}, \mathfrak{b}_{E^*}) = (\mathfrak{a}_E, \mathfrak{b}_E). \tag{5.7.1}$$

Indeed, this follows from the easy identities

$$\mathrm{tr}\{A^t\tau_i((A^t)^{-1})A^t\tau_j((A^t)^{-1})A^t\tau_r((A^t)^{-1})\}$$
$$= -\mathrm{tr}\{A^{-1}\tau_r(A)A^{-1}\tau_j(A)A^{-1}\tau_i(A)\} \tag{5.7.2}$$

and

$$\mathrm{tr}\{A^t\tau_i((A^t)^{-1}\tau_j((B^t)^{-1})B^t\} = \mathrm{tr}\{A^{-1}\tau_i(A)\tau_j(B)B^{-1}\}. \tag{5.7.3}$$

5.8. Let us go to the global situation. Let X be a smooth variety over k and let E be a vector bundle over X. As in [GII], we define the gerbe $\mathfrak{D}_{\Lambda E}$ of chiral differential operators on ΛE over X.

Its characteristic class $c(\mathfrak{D}_{\Lambda E})$ will belong to the second hypercohomology space $H^2(X; \Omega^{[2,3\rangle}_{\Lambda E})$ (using obvious notation). Recall that we have a canonical imbedding of de Rham complexes

$$\Omega^{\cdot}_X \hookrightarrow \Omega^{\cdot}_{\Lambda E}. \tag{5.8.1}$$

We defined the "Atiyah–Chern–Simons" characteristic class $c(E) \in H^2(X; \Omega^{[2,3\rangle}_X)$ in [GII], 7.6; let us denote by $c(E)_{\Lambda E}$ its image in $H^2(X; \Omega^{[2,3\rangle}_{\Lambda E})$.

The next theorem is an immediate consequence of Theorem 5.6 and Lemma 5.7.

Theorem 5.9. *The class $c(\mathfrak{D}_{\Lambda E})$ is equal to*

$$c(\mathfrak{D}_{\Lambda E}) = c(\Theta_X) - c(E) = c(\Omega^1_X) - c(E) \tag{5.9.1}$$

where Θ_X is the tangent bundle.

6 Chiral de Rham complex

6.1. Let us return to the local situation 3.1, 4.1. Let E be equal to the module of vector fields T. Given a base $\{\bar{\tau}_i\}$ consisting of commuting vector fields, we get a frame $\mathfrak{g} = \{\bar{\tau}_i; \phi_i := \bar{\tau}_i\}$ of (A, E). Let us call such frames *natural*.

Let $\mathfrak{g}, \mathfrak{g}'$ be two natural frames, with transition matrices as in 3.2. By definition, $(A^{rs}) = (g^{rs})$. Therefore the coefficients $g^{i\alpha\gamma}$ (3.2.2) are given by

$$g^{i\alpha\gamma} = g^{iq}\tau_q(g^{-1\alpha\mu})g^{\mu\gamma} = g^{iq}\tau_\alpha(g^{-1q\mu})g^{\mu\gamma}$$
$$= -\tau_\alpha(g^{iq})g^{-1q\mu}g^{\mu\gamma} = -\tau_\alpha(g^{i\gamma}) \tag{6.1.1}$$

where we have used (3.3.3). Consequently the function h_E (4.1.6) is given by

$$h^{ij}_E = -\tau_j\tau_\nu(g^{i\nu}) + \frac{1}{2}g^{iq}\tau_j(g^{-1\mu\beta})g^{\beta\gamma}\tau_q(g^{-1\gamma\nu})g^{\nu\mu}. \tag{6.1.2}$$

The second summand is equal to

$$-\frac{1}{2}g^{iq}\tau_j(g^{-1\mu\beta})\tau_q(g^{\beta\gamma})g^{-1\gamma\nu}g^{\nu\mu} = -\frac{1}{2}g^{iq}\tau_j(g^{-1\mu\beta})\tau_q(g^{\beta\mu})$$

$$= -\frac{1}{2}g^{iq}\tau_\mu(g^{-1j\beta})\tau_q(g^{\beta\mu})$$

$$= -\frac{1}{2}\tau_\mu(g^{-1j\beta})g^{\beta q}\tau_q(g^{i\mu}) = \frac{1}{2}g^{-1j\beta}\tau_\mu(g^{\beta q})\tau_q(g^{i\mu}). \tag{6.1.3}$$

We see that the first summand in (6.1.2) is equal to minus the first summand of (4.1.5), and second summand of (6.1.2) is equal to the second summand of (4.1.5). Thus

$$h^{ij} = 2\tau_p\tau_j(g^{ip}). \tag{6.1.4}$$

If $\mathfrak{g}$, $\mathfrak{g}'$, $\mathfrak{g}''$ are natural frames of (A, T), then Theorem 5.6 says that $\mathfrak{a}_{\mathfrak{g},\mathfrak{g}',\mathfrak{g}''} = \mathfrak{b}_{\mathfrak{g},\mathfrak{g}'} = 0$. This means that

the chiral superalgebroid $\mathcal{A}_{\wedge T;\mathfrak{g}}$ does not depend, up to a canonical isomorphism, on the choice of the base $\{\bar{\tau}_i\}$. In other words, we have a canonically defined chiral algebroid $\mathcal{A}_{\wedge T}$.

Moving on to chiral envelopes, we get a canonically defined chiral (vertex) superalgebra $D^{ch}_{\wedge T}$ of *chiral differential operators on T*. It follows that:

for each smooth variety X we have a canonically defined sheaf of chiral superalgebras $\mathcal{D}^{ch}_{\wedge\Theta_X}$. These Zariski sheaves form in fact a sheaf in the étale topology.

The gluing functions for this sheaf are given explicitly by (6.1.4).

Lemma 6.2. *In the situation* 4.1, *consider the function h_E:*

$$h_E^{ij} = \tau_j(g^{i\nu\nu}) + \frac{1}{2}g^{iq}\tau_j(A^{-1\mu\beta})A^{\beta\gamma}\tau_q(A^{-1\gamma\nu})A^{\nu\mu}. \tag{6.2.1}$$

The function h_{E^*} associated with the dual module E^* is given by

$$h_{E^*}^{ij} = -\tau_j(g^{i\nu\nu}) + \frac{1}{2}g^{iq}\tau_j(A^{-1\mu\beta})A^{\beta\gamma}\tau_q(A^{-1\gamma\nu})A^{\nu\mu}. \tag{6.2.2}$$

This follows from the identities

$$\mathrm{tr}\{\tau_i(A^t)A^{-1}\} = -\mathrm{tr}\{\tau_i(A^{-1})A\} \tag{6.2.3}$$

and

$$tr\{\tau_i(A^t)A^{-1}\tau_j(A^t)A^{-1}\} = tr\{\tau_i(A^{-1})A\tau_j(A^{-1})A\} \tag{6.2.4}$$

(cf. (5.7.2), (5.7.3)).

6.3. Let E be the module of 1-forms $\Omega = \Omega^1_{A/k}$; its exterior algebra is the de Rham algebra of differential forms $\Omega^\cdot = \Omega^\cdot_{A/k}$. Frames of the form $\mathfrak{g} = \{\bar{\tau}_i, \phi_i := \omega_i\}$ will be called natural.

If $\mathfrak{g}$, $\mathfrak{g}'$ are natural frames, then formulas (6.1.2) and (6.1.3), together with the previous lemma, show that $h^{ij}_E = h^{ij}_\Omega$ where h_Ω is given by (4.1.5). (Of course one easily checks this directly. This explains the notation for h^{ij}_Ω).

In other words, we arrive at an interesting conclusion.

Theorem 6.4. *The matrices $h = (h^{ij})$ defined in* 4.1 *are equal to* 0 *if $E = \Omega$ and frames $\mathfrak{g}$, $\mathfrak{g}'$ are natural.*

6.4.1. *Warning.* The functions $h_{\mathfrak{g},\mathfrak{g}'}$ are *nonzero* since they are not linear.

6.5. On the other hand, Theorem 5.6 together with Lemma 5.7 states that $\mathfrak{a}_E$ and $\mathfrak{b}_E$ are 0 for $E = \Omega$ (for natural frames $\mathfrak{g}$, $\mathfrak{g}'$, $\mathfrak{g}''$ of (A, Ω)). This gives us a *canonically defined* chiral superalgebroid $\mathcal{A}_{\Omega^\cdot}$. Its vertex envelope will be denoted $D^{ch}_{\Omega^\cdot}$ and called *the chiral algebra of differential operators on* $\Omega^\cdot$.

This implies the following.

Theorem 6.6. *For each smooth variety X the construction* 6.2–6.5 *gives a canonically defined sheaf of chiral superalgebras $\mathcal{D}^{ch}_{\Omega^\cdot_X}$. These Zariski sheaves form a sheaf in the étale topology.*

6.7. The de Rham differential may be considered as an odd first order differential operator acting on $\Omega^\cdot_X$ (and commuting with itself).

In coordinates, if $\mathfrak{g} = \{\bar{\tau}_i; \phi_i := \omega_i\}$ is a natural frame of (A, Ω), it is given by

$$Q^{cl}_{\mathfrak{g}} = \phi_i \tau_i \tag{6.7.1}$$

(*cl* is for "classical"). Let us check the independence of (6.7.1) on the choice of a frame. Let $\mathfrak{g}' = \{\bar{\tau}'_i; \omega'_i\}$ be another natural frame, $\bar{\tau}'_i = g^{ij}\bar{\tau}_i$; $\omega'_i = g^{-1pi}\omega_p$ (cf. (3.2.4)).

Using (3.2.1) we have

$$Q^{cl}_{\mathfrak{g}'} = \phi'_i \tau'_i = g^{-1pi}\phi_p\{g^{iq}\tau_q + g^{irs}\phi_s\psi_r\} \tag{6.7.2}$$

where

$$g^{ipq} = g^{ir}\tau_r(g^{sp})g^{-1qs} = g^{sr}\tau_r(g^{ip})g^{-1qs} = \tau_q(g^{ip}) \tag{6.7.3}$$

(we used (3.3.1)). So,

$$Q^{cl}_{\mathfrak{g}'} = \phi_p\tau_p + g^{-1pi}g^{irs}\phi_s\psi_r = Q^{cl}_{\mathfrak{g}} + g^{-1pi}\tau_s(g^{ir})\phi_p\phi_s\psi_r. \tag{6.7.4}$$

Note that

$$g^{-1pi}\tau_s(g^{ir}) = -\tau_s(g^{-1pi})g^{ir},$$

which is symmetric under the permutation of p with s, due to (3.3.3); therefore the second summand in (6.7.4) is zero, i.e., $Q^{cl}_{\mathfrak{g}} = Q^{cl}_{\mathfrak{g}'}$. Thus, Q^{cl} is a correctly defined odd element of $D_{\Omega^\cdot}$. It is obvious from (6.7.1) that $[Q^{cl}, Q^{cl}] = 0$.

6.8. Let us investigate the chiral counterpart of Q^{cl}. Let us define an odd element $Q_{\mathfrak{g}}$ (of conformal weight 1) of the vertex superalgebra $D^{ch}_{\Omega^{\cdot};\mathfrak{g}} := U\mathcal{A}_{\Omega^{\cdot};\mathfrak{g}}$ by

$$Q_{\mathfrak{g}} = \phi_{i(-1)}\tau_i. \tag{6.8.1}$$

Let $\mathfrak{g}'$ be another natural frame as in 6.7. Due to Theorem 6.4, the element $Q_{\mathfrak{g}'}$ goes under the canonical isomorphism $D^{ch}_{\Omega^{\cdot};\mathfrak{g}'} = D^{ch}_{\Omega^{\cdot};\mathfrak{g}}$ to

$$Q_{\mathfrak{g}'} = \phi'_{i(-1)}\tau'_i = \phi'_i\tau'_i - \gamma(\phi'_i, \tau'_i) \tag{6.8.2}$$

cf. [GII], (3.3.1).

Lemma 6.9. *We have (in $D_{\Omega^{\cdot};\mathfrak{g}}$)*

$$\gamma(\phi'_i, \tau'_i) = -\partial\{tr(\tau_r(g)g^{-1})\phi_r\}. \tag{6.9.1}$$

6.10. Before the proof, let us write down some useful formulas:

$$\gamma(a\phi_r, b\tau_i) = -\tau_i(a)\phi_r\partial b - \tau_i(b)\partial(a\phi_r) \tag{6.10.1}$$

and

$$\gamma(a\phi_r, b\phi_s\psi_p) = -\delta_{rp}a\partial(b\phi_s) + \delta_{sp}b\partial(a\phi_r). \tag{6.10.2}$$

6.11. *Proof of 6.9.* We have

$$\gamma(\phi'_i, \tau'_i) = \gamma(g^{-1qi}\phi_q, g^{ip}\tau_p + g^{isr}\phi_r\psi_s)$$

where

$$\gamma(g^{-1qi}\phi_q, g^{ip}\tau_p) = -\tau_p(g^{-1qi})\phi_q\partial g^{ip} - \tau_p(g^{ip})\partial(g^{-1qi}\phi_q) \tag{6.11.1}$$

and

$$\gamma(g^{-1qi}\phi_q, g^{isr}\phi_r\psi_s) = -g^{-1qi}\partial(g^{iqr}\phi_r) + g^{irr}\partial(g^{-1qi})\phi_q. \tag{6.11.2}$$

Since $g^{irr} = \tau_r(g^{ir})$, the second summands in (6.11.1) and (6.11.2) cancel out. On the other hand, the first term in (6.11.1) is equal to

$$-\tau_q(g^{-1ri})\tau_s(g^{iq})\omega_s\phi_r = -\tau_r(g^{-1si})\tau_s(g^{iq})\omega_s\phi_r$$
$$= g^{-1qa}\tau_r(g^{ab})g^{-1bi}\tau_s(g^{iq})\omega_s\phi_r = -\tau_r(g^{ab})\tau_s(g^{-1ba})\omega_s\phi_r$$
$$= -\tau_r(g^{ab})\partial(g^{-1ba})\phi_r.$$

Therefore

$$\gamma(\phi'_i, \tau'_i) = -\tau_r(g^{ab})\partial(g^{-1ba})\phi_r - g^{-1ba}\partial\{\tau_r(g^{ab})\phi_r\} = -\partial\{\tau_r(g^{ab})g^{-1ba}\phi_r\},$$

□

6.12. From (6.8.1) we have $Q_{\mathfrak{g}} = \phi_i \tau_i$, and from 6.7 $\phi_i' \tau_i' = \phi_i \tau_i$. Therefore, (6.8.2) and Lemma 6.9 imply the following.

Theorem 6.13. *We have*

$$Q_{\mathfrak{g}'} = Q_{\mathfrak{g}} + \partial\{tr(\tau_r(g)g^{-1})\phi_r\}. \tag{6.13.1}$$

6.14. Consider the field $Q_{\mathfrak{g}}(z)$ acting on the vertex algebra $D^{ch}_{\Omega^{\cdot}}$. Due to (6.13.1), its zeroth component $Q_{\mathfrak{g}0}$ does not depend on the choice of the frame $\mathfrak{g}$. Therefore we get a canonical operator Q_0 acting on $D^{ch}_{\Omega^{\cdot}}$. Since it is a zeroth component of a field, it is a derivation of the vertex algebra, and it is obvious from the local definition (6.8.1) that $[Q_0, Q_0] = 0$.

Consequently, for each smooth variety X we get a canonical odd derivation Q_{0X} of the sheaf $\mathcal{D}^{ch}_{\Omega^{\cdot}_X}$, such that $[Q_{0X}, Q_{0X}] = 0$. The pair $(\mathcal{D}^{ch}_{\Omega^{\cdot}_X}, Q_{0X})$ is the *chiral de Rham complex* from [MSV].

Our Theorem 6.13 is a version of *op. cit.*, (4.1c).

6.15. In the situation 6.4, consider an even element $J_{\mathfrak{g}}$ of conformal weight 1 of the algebra $D^{ch}_{\Omega^{\cdot};\mathfrak{g}}$, given by

$$J_{\mathfrak{g}} = \phi_{i(-1)}\psi_i = \phi_i \psi_i. \tag{6.15.1}$$

After a change of frame as in *loc. cit.*, we get an element

$$J_{\mathfrak{g}'} = \phi'_{i(-1)}\psi'_i = \phi'_i\psi'_i - \gamma(\phi'_i, \psi'_i) \tag{6.15.2}$$

where we have again used Theorem 6.4. We have

$$\phi'_i\psi'_i = g^{-1pi}\phi_p g^{iq}\psi_q = \phi_p\psi_p = J_{\mathfrak{g}}$$

(see (3.2.3)). On the other hand, by (3.4.3)

$$\gamma(\phi'_i, \psi'_i) = \gamma(g^{-1pi}\phi_p, g^{iq}\psi_q) = \delta_{pq} g^{-1pi}\partial g^{iq} = g^{-1pi}\partial g^{ip} = tr(g^{-1}\partial g).$$

Thus

$$J_{\mathfrak{g}'} = J_{\mathfrak{g}} - tr(g^{-1}\partial g). \tag{6.15.3}$$

6.16. Consider an odd element $G_{\mathfrak{g}}$ of conformal weight 2 given by

$$G_{\mathfrak{g}} = \psi_{i(-1)}\omega_i. \tag{6.16.1}$$

In the frame $\mathfrak{g}'$,

$$G_{\mathfrak{g}'} = (g^{iq}\psi_q)_{(-1)}(g^{-1si}\omega_s).$$

Note that $\psi_{q(j)}a = 0$ for $j \geq 0$ (everything happens in D), hence it follows from the commutativity formula (1.3.1) that

$$a\psi_q = a_{(-1)}\psi_q = \psi_{q(-1)}a. \tag{6.16.2}$$

Therefore by "associativity" (1.2.5)

$$(a\psi_q)_{(-1)}(b\omega_s) = (\psi_{q(-1)}a)_{(-1)}(b\omega_s) = \psi_{q(-1)}a_{(-1)}b\omega_s = \psi_{q(-1)}(ab\omega_s). \tag{6.16.3}$$

Therefore

$$G_{\mathfrak{g}'} = \psi_{q(-1)}g^{iq}g^{-1si}\omega_s = \psi_{s(-1)}\omega_s = G_{\mathfrak{g}}. \tag{6.16.6}$$

6.17. Let us investigate the *Virasoro element*. Define an even element $L_{\mathfrak{g}}$ of conformal weight 2 by

$$L_{\mathfrak{g}} = L_{(b)\mathfrak{g}} + L_{(f)\mathfrak{g}} \tag{6.17.1}$$

where

$$L_{(b)\mathfrak{g}} = \omega_{i(-1)}\tau_i \tag{6.17.2}$$

((b) is for "bosonic") and

$$L_{(f)\mathfrak{g}} = \rho_{i(-1)}\psi_i \tag{6.17.3}$$

((f) is for "fermionic"), cf. [MSV], (2.3a).

6.18. We have

$$(a\omega_s)_{(-1)}(b\tau_p) = \omega_{s(-1)}\{ab\tau_p + \tau_p(a)\partial b + \tau_p(b)\partial a\} - \partial\omega_{s(-1)}b\tau_p(a). \tag{6.18.1}$$

Indeed, it follows from "associativity" (1.2.5) that

$$(a\omega_s)_{(-1)}(b\tau_p) = (\omega_{s(-1)}a)_{(-1)}(b\tau_p) = \omega_{s(-1)}a_{(-1)}(b\tau_p) + \omega_{s(-2)}a_{(0)}(b\tau_p)$$
$$+a_{(-2)}\omega_{s(0)}(b\tau_p).$$

Next,

$$a_{(-1)}(b\tau_p) = ab\tau_p - \gamma(a, b\tau_p) = ab\tau_p + \tau_p(a)\partial b + \tau_p(b)\partial a$$

(see [GII] (3.3.1));

$$a_{(0)}(b\tau_p) = -b\tau_{p(0)}a = -b\tau_p(a)$$

(see [GII] (3.3.2)). Finally

$$\omega_{s(0)}(b\tau_p) = -(b\tau_p)_{(0)}\omega_s + \partial\langle b\tau_p, \omega_s\rangle$$

where

$$(b\tau_p)_{(0)}\omega_s = (b\tau_p)(\omega_s) = \langle\tau_p, \omega_s\rangle\partial b = \delta_{ps}\partial b$$

by (1.1.3), since

$$\tau_p(\omega_s) = 0, \tag{6.18.2}$$

and $\langle b\tau_p, \omega_s\rangle = b\delta_{ps}$. This implies

$$\omega_{s(0)}(b\tau_p) = 0. \tag{6.18.3}$$

Formula (6.18.1) follows.

6.19. We have

$$(a\omega_s)_{(-1)}(b\phi_\alpha\psi_\beta) = \omega_{s(-1)}\{ab\phi_\alpha\psi_\beta - \delta_{\alpha\beta}b\partial a\}. \qquad (6.19.1)$$

Indeed, by commutativity and "associativity" (1.2.5)

$$(a\omega_s)_{(-1)}(b\phi_\alpha\psi_\beta) = (\omega_{s(-1)}a)_{(-1)}(b\phi_\alpha\psi_\beta) = \omega_{s(-1)}a_{(-1)}(b\phi_\alpha\psi_\beta).$$

On the other hand

$$a_{(-1)}(b\phi_\alpha\psi_\beta) = ab\phi_\alpha\psi_\beta - \gamma(a, b\phi_\alpha\psi_\beta) = ab\phi_\alpha\psi_\beta - \delta_{\alpha\beta}b\partial a,$$

see (3.4.2). This implies (6.19.1).

6.20. We have

$$L_{(b)\mathfrak{g}'} = L_{(b)\mathfrak{g}} + \omega_{s(-1)}\{\tau_p(g^{-1si})\partial g^{ip} + g^{-1si}\tau_\beta(g^{i\alpha})\phi_\beta\psi_\alpha\}$$
$$-\omega_{s(-2)}g^{ip}\tau_p(g^{-1si}) \qquad (6.20.1)$$

Indeed, due to Theorem 6.4

$$L_{(b)\mathfrak{g}'} = (g^{-1si}\omega_s)_{(-1)}\{g^{ip}\tau_p + g^{i\alpha\beta}\phi_\beta\psi_\alpha\}.$$

According to (6.18.1)

$$(g^{-1si}\omega_s)_{(-1)}(g^{ip}\tau_p) = \omega_{s(-1)}\{g^{-1si}g^{ip}\tau_p + \tau_p(g^{-1si})\partial g^{ip} + \tau_p(g^{ip})\partial g^{-1si}\}$$
$$-\omega_{s(-2)}g^{ip}\tau_p(g^{-1si}).$$

By (6.18.2)

$$(g^{-1si}\omega_s)_{(-1)}(g^{i\alpha\beta}\phi_\beta\psi_\alpha) = (g^{-1si}\omega_s)_{(-1)}(\tau_\beta(g^{i\alpha})\phi_\beta\psi_\alpha)$$
$$= \omega_{s(-1)}\{g^{-1si}\tau_\beta(g^{i\alpha})\phi_\beta\psi_\alpha - \delta_{\alpha\beta}\tau_\beta(g^{i\alpha})\partial g^{-1si}\}.$$

The third term in the first expression cancels out the second term in the second one, and we get (6.20.1).

6.21. We have

$$(a\rho_\mu)_{(-1)}(b\psi_\nu) = \rho_{\mu(-1)}ab\psi_\nu \qquad (6.21.1)$$

Indeed,

$$(a\rho_\mu)_{(-1)}(b\psi_\nu) = (\rho_{\mu(-1)}a)_{(-1)}(b\psi_\nu) = \rho_{\mu(-1)}ab\psi_\nu + a_{(-2)}\rho_{\mu(0)}b\psi_\nu$$

and by commutativity

$$\rho_{\mu(0)}b\psi_\nu = (b\psi_\nu)(\rho_\mu) - \partial\{(b\psi_\nu)_{(1)}\rho_\mu\} = \delta_{\nu\mu}\partial b - \delta_{\nu\mu}\partial b = 0,$$

cf. (6.18.3). This implies (6.21.1).

6.22. We have

$$(a\phi_\gamma\omega_i)_{(-1)}(b\psi_\nu) = \omega_{i(-1)}\{ab\phi_\gamma\psi_\nu - \delta_{\gamma\nu}a\partial b\} + \delta_{\nu\gamma}\omega_{i(-2)}ab. \quad (6.22.1)$$

Indeed,

$$\begin{aligned}(a\phi_\gamma\omega_i)_{(-1)}(b\psi_\nu) &= (\omega_{i(-1)}a\phi_\gamma)_{(-1)}(b\psi_\nu)\\ &= \omega_{i(-1)}(a\phi_\gamma)_{(-1)}b\psi_\nu + \omega_{i(-2)}(a\phi_\gamma)_{(0)}b\psi_\nu\end{aligned}$$

where

$$(a\phi_\gamma)_{(-1)}b\psi_\nu = a\phi_\gamma b\psi_\nu - \gamma(a\phi_\gamma, b\psi_\nu) = ab\phi_\gamma\psi_\nu - \delta_{\gamma\nu}a\partial b,$$

cf. (3.4.3), and

$$(a\phi_\gamma)_{(0)}b\psi_\nu = (b\psi_\nu)(a\phi_\gamma) = ab\delta_{\nu\gamma}.$$

6.23. We have

$$\begin{aligned}L_{(f)\mathfrak{g}'} = L_{(f)\mathfrak{g}} &+ \omega_{i(-1)}\{\tau_i(g^{-1\gamma\alpha})g^{\alpha\nu}\phi_\gamma\psi_\nu - \tau_i(g^{-1\gamma\alpha})\partial g^{\alpha\gamma}\}\\ &+ \omega_{i(-2)}\tau_i(g^{-1\gamma\alpha})g^{\alpha\gamma}.\end{aligned} \quad (6.23.1)$$

Indeed,

$$L_{(f)\mathfrak{g}'} = \rho'_{\alpha(-1)}\psi'_\alpha = \{g^{-1\mu\alpha}\rho_\mu + \tau_i(g^{-1\gamma\alpha})\phi_\gamma\omega_i\}_{(-1)}(g^{\alpha\nu}\psi_\nu).$$

By (6.21.1)

$$(g^{-1\mu\alpha}\rho_\mu)_{(-1)}(g^{\alpha\nu}\psi_\nu) = \rho_{\mu(-1)}g^{-1\mu\alpha}g^{\alpha\nu}\psi_\nu = L_{(f)\mathfrak{g}}$$

and by (6.22.1)

$$\begin{aligned}(\tau_i(g^{-1\gamma\alpha})\phi_\gamma\omega_i)_{(-1)}(g^{\alpha\nu}\psi_\nu) = \omega_{i(-1)}\{&\tau_i(g^{-1\gamma\alpha})g^{\alpha\nu}\phi_\gamma\psi_\nu - \tau_i(g^{-1\gamma\alpha})\partial g^{\alpha\gamma}\}\\ &+\omega_{i(-2)}\tau_i(g^{-1\gamma\alpha})g^{\alpha\gamma}.\end{aligned}$$

6.24. Comparing (6.20.1) and (6.23.1) we see easily that

$$L_{\mathfrak{g}'} = L_{(b)\mathfrak{g}'} + L_{(f)\mathfrak{g}'} = L_{(b)\mathfrak{g}} + L_{(f)\mathfrak{g}} = L_{\mathfrak{g}}. \quad (6.24.1)$$

Let us collect our computations of transformation rules.

Theorem 6.25. *Let $\mathfrak{g}$, $\mathfrak{g}'$ be two natural frames of (A, Ω). Consider 4 elements of the vertex superalgebra $D^{ch}_{\Omega;\mathfrak{g}}$ given by*

$$Q_{\mathfrak{g}} = \phi_{i(-1)}\tau_i \quad (6.25.1)$$

(an odd element of conformal weight 1*)*

$$J_{\mathfrak{g}} = \phi_{i(-1)}\psi_i \tag{6.25.2}$$

(an even element of conformal weight 1*)*

$$G_{\mathfrak{g}} = \psi_{i(-1)}\omega_i \tag{6.25.3}$$

(an odd element of conformal weight 2*) and*

$$L_{\mathfrak{g}} = \omega_{i(-1)}\tau_i + \rho_{i(-1)}\psi_i \tag{6.25.4}$$

(an even element of conformal weight 2*).*

After the canonical identification $D^{ch}_{\Omega^{\cdot};\mathfrak{g}'} = D^{ch}_{\Omega^{\cdot};\mathfrak{g}}$ *these elements are transformed as follows:*

$$Q_{\mathfrak{g}'} = Q_{\mathfrak{g}} + \partial\{tr(g^{-1}\tau_r(g))\phi_r\} \tag{6.25.5}$$

$$J_{\mathfrak{g}'} = J_{\mathfrak{g}} - tr(g^{-1}\partial g) \tag{6.25.6}$$

$$G_{\mathfrak{g}'} = G_{\mathfrak{g}} \tag{6.25.7}$$

and

$$L_{\mathfrak{g}'} = L_{\mathfrak{g}}. \tag{6.25.8}$$

This is a version of [MSV], Theorem 4.2.

7 Poincaré–Birkhoff–Witt

7.1 Let X be a smooth variety and let $\mathcal{D}^{ch}_{\Omega^{\cdot}_X}$ be the sheaf discussed in the previous section, cf. Theorem 6.6. It is a sheaf of $\mathbb{Z}_{\geq 0}$-graded vertex algebras, so

$$\mathcal{D}^{ch}_{\Omega^{\cdot}_X} = \oplus_{n\in\mathbb{Z}_{\geq 0}} \mathcal{D}^{ch}_{\Omega^{\cdot}_X;n} \tag{7.1.1}$$

where $\mathcal{D}^{ch}_{\Omega^{\cdot}_X;n}$ denotes the component of conformal weight n.

According to Theorem 6.25 (see (6.25.8)) we have a canonical global section L of $\mathcal{D}^{ch}_{\Omega^{\cdot}_X;2}$. Let $L(z) = \sum L_n z^{-n-2}$ be the corresponding field.

Claim 7.1.1. *A local section* $\alpha \in \mathcal{D}^{ch}_{\Omega^{\cdot}_X}$ *belongs to* $\mathcal{D}^{ch}_{\Omega^{\cdot}_X;n}$ *if and only if* $L_0(\alpha) = n\alpha$.

In a uniform notation $L(z) = \sum L_{(n)}z^{-n-1}$ we have $L_0 = L_{(1)}$. We shall check 7.1.1 simultaneously with the following:

Claim 7.1.2. *The operator* $L_{-1} = L_{(0)}$ *coincides with the canonical derivation* ∂ *of the vertex algebra* $\mathcal{D}^{ch}_{\Omega^{\cdot}_X}$.

Both statements are local, so we may assume we are in the local situation 6.17. Note that the operator $L_{(0)}$ is a derivation with respect to the operation $_{(-1)}$:

$$L_{(0)}(x_{(-1)}y) = (L_{(0)}x)_{(-1)}y + x_{(-1)}L_{(0)}y. \tag{7.1.2}$$

Therefore it suffices to check 7.1.2 on the generators $a, \tau_i, \omega_i, \psi_i, \rho_i$ of the vertex algebroid $\mathcal{A}_{\Omega^{\cdot};\mathfrak{g}}$, which is done by a simple explicit computation.

It follows from the associativity formula (1.2.4) that

$$\begin{aligned} L_{(1)}y_{(-1)}z &= (L_{(1)}y)_{(-1)}z + y_{(-1)}L_{(1)}z + L_{(0)}y_{(0)}z - y_{(0)}L_{(0)}z \\ &= (L_{(1)}y)_{(-1)}z + y_{(-1)}L_{(1)}z \end{aligned} \tag{7.1.3}$$

since

$$L_{(0)}y_{(0)}z - y_{(0)}L_{(0)}z = \partial(y_{(0)}z) - y_{(0)}\partial z = 0.$$

In other words, $L_{(1)}$ is a derivation of $_{(-1)}$. Therefore it suffices to check 7.1.1 on the generators of $\mathcal{A}_{\Omega^{\cdot};\mathfrak{g}}$ as above, which is straightforward.

7.2. In the local situation 6.1, consider the local algebra $D^{ch}_{\Omega^{\cdot};\mathfrak{g}} = U\mathcal{A}_{\Omega^{\cdot};\mathfrak{g}}$. Let us introduce a $\mathbb{Z}$-grading

$$D^{ch}_{\Omega^{\cdot};\mathfrak{g}} = \oplus_{p\in\mathbb{Z}}\, D^{ch;p}_{\Omega^{\cdot};\mathfrak{g}} \tag{7.2.1}$$

to be called *fermionic charge*. For an element $x \in D^{ch}_{\Omega^{\cdot};\mathfrak{g}}$ let us denote its fermionic charge (to be defined) by $F(x) \in \mathbb{Z}$. It is defined uniquely by the following conditions:

(a) $F(a) = F(\tau_i) = F(\omega_i) = 0;\ \ F(\phi_i) = F(\rho_i) = -F(\psi_i) = 1;$
(b) $F(x_{(-1)}y) = F(x) + F(y).$

Due to the transformation formulas (3.2.1)–(3.2.5) this grading is obviously preserved under a change of frames. Therefore in the situation of 7.1 the sheaf $\mathcal{D}^{ch}_{\Omega^{\cdot}_X}$ gets a canonical $\mathbb{Z}$-grading

$$\mathcal{D}^{ch}_{\Omega^{\cdot}_X} = \oplus_{p\in\mathbb{Z}}\, \mathcal{D}^{ch;p}_{\Omega^{\cdot}_X}. \tag{7.2.2}$$

Note that parity is equal to fermionic charge modulo 2.

Here is another way to define the grading (7.2.2). First we have the simple observation.

Lemma 7.2.1. *Let $\mathcal{A} = (A, T, \Omega, \partial, \ldots)$ be a vertex (super)algebroid. For every invertible element $a \in A$ the operator $(a^{-1}\partial a)_{(0)}$ acting on $U\mathcal{A}$ is trivial.*

Proof. Obviously this operator is trivial on $A = U\mathcal{A}_0$. Let $x \in U\mathcal{A}_1$ and $\tau \in T$ be its image under the canonical projection $U\mathcal{A}_1 \longrightarrow T$. We have

$$\begin{aligned} (a^{-1}\partial a)_{(0)}x &= -x_{(0)}a^{-1}\partial a + \partial(x_{(1)}a^{-1}\partial a) = -\tau(a^{-1}\partial a) + \partial\langle\tau, a^{-1}\partial a\rangle \\ &= a^{-2}\tau(a)\partial a - a^{-1}\partial\tau(a) + \partial(a^{-1}\tau(a)) = 0, \end{aligned}$$

so $(a^{-1}\partial a)_{(0)}$ is trivial on $U\mathcal{A}_1$. Therefore it is trivial on the whole algebra $U\mathcal{A}$ since $(?)_{(0)}$ is a derivation of the operation $_{(-1)}$. □

Applying this lemma to $a = \det(g)$ in the formula (6.25.6) we see that the component $J_{0;\mathfrak{g}}$ of the field $J_{\mathfrak{g}}(z) = \sum J_{n;\mathfrak{g}}z^{-n-1}$ is preserved under the change of frames. Consequently it gives rise to a well-defined endomorphism J_0 of the sheaf $\mathcal{D}^{ch}_{\Omega^{\cdot}_X}$.

Claim 7.2.2. *A local section $\alpha \in \mathcal{D}^{ch}_{\Omega^{\cdot}_X}$ belongs to $\mathcal{D}^{ch;p}_{\Omega^{\cdot}_X}$ if and only if $J_0(\alpha) = p\alpha$.*

Indeed, the function $F(\alpha)$ defined by $J_0(\alpha) = F(\alpha)\alpha$ obviously satisfies the condtions (a) and (b) above.

7.3. The two gradings (7.1.1) and (7.2.2) are compatible: if we denote

$$\mathcal{D}^{ch;p}_{\Omega^{\cdot}_X;n} := \mathcal{D}^{ch}_{\Omega^{\cdot}_X;n} \cap \mathcal{D}^{ch;p}_{\Omega^{\cdot}_X}, \tag{7.3.1}$$

then

$$\mathcal{D}^{ch}_{\Omega^{\cdot}_X} = \oplus_{n\in\mathbb{Z}_{\geq 0};\ p\in\mathbb{Z}}\, \mathcal{D}^{ch;p}_{\Omega^{\cdot}_X;n}. \tag{7.3.2}$$

For a fixed n, only a finite number of sheaves $\mathcal{D}^{ch;p}_{\Omega^{\cdot}_X;n}$ are nonzero.

If the ground ring k is a field of characteristic 0, then the sheaves $\mathcal{D}^{ch;p}_{\Omega^{\cdot}_X;n}$ and $\mathcal{D}^{ch;n-p}_{\Omega^{\cdot}_X;n}$ are in a certain sense dual to each other, see [MS].

7.4. From now on we assume that $k \supset \mathbb{Q}$. According to (a superversion of) the PBW theorem, [GII], Theorem 9.18, the sheaf $\mathcal{D}^{ch}_{\Omega^{\cdot}_X}$ admits a canonical filtration such that the associated graded sheaf is canonically isomorphic to

$$gr(\mathcal{D}^{ch}_{\Omega^{\cdot}_X}) = Sym_{\Omega^{\cdot}_X}\{(\oplus_{n\geq 1}\,\Theta_{\Omega^{\cdot}_X(n)}) \oplus (\oplus_{n\geq 1}\,\Omega^1_{\Omega^{\cdot}_X(n)})\}. \tag{7.4.1}$$

Here $\Theta_{\Omega^{\cdot}_X}$ (resp. $\Omega^1_{\Omega^{\cdot}_X}$) denotes the tangent (resp. the cotangent) bundle of the supervariety $(X, \Omega^{\cdot}_X)$, and $(?)_{(n)}$ means that this bundle is put into the conformal weight n.

The endomorphisms L_0 and J_0 respect the canonical filtration; hence we get a canonical finite filtration $F_{\cdot}$ on each homogeneous component $\mathcal{D}^{ch;p}_{\Omega^{\cdot}_X;n}$. The graded quotients $F_i\mathcal{D}^{ch;p}_{\Omega^{\cdot}_X;n}/F_{i+1}\mathcal{D}^{ch;p}_{\Omega^{\cdot}_X;n}$ are locally free $\mathcal{O}_X$-modules of finite rank (we shall see this in the course of computations below). This allows us to introduce the elements of the Grothendieck group $K(X)$ of vector bundles

$$[\mathcal{D}^{ch;p}_{\Omega^{\cdot}_X;n}] := \sum_i [F_i\mathcal{D}^{ch;p}_{\Omega^{\cdot}_X;n}/F_{i+1}\mathcal{D}^{ch;p}_{\Omega^{\cdot}_X;n}] \in K(X). \tag{7.4.2}$$

Here $[E]$ on the right-hand side denotes the class of a vector bundle E in $K(X)$. Consider the generating function

$$cl(\mathcal{D}^{ch}_{\Omega^{\cdot}_X})(y,q) := \sum_{p,n} [\mathcal{D}^{ch;p}_{\Omega^{\cdot}_X;n}]y^p q^n \in K(X)[y,y^{-1}][[q]]. \tag{7.4.3}$$

7.5. For a vector bundle E over X and an indeterminate x we introduce the notation

$$[S_x E] = \sum_{i=0}^{\infty} [Sym^i_{\mathcal{O}_X} E] \in K(X)[[x]] \tag{7.5.1}$$

and

$$[\Lambda_x E] = \sum_{i=0}^{\infty} [\Lambda^i_{\mathcal{O}_X} E] \in K(X)[x]. \tag{7.5.2}$$

The following fact was noticed in [BL] (cf. also [W]).

Theorem 7.6 (L. Borisov–A. Libgober). *We have*

$$cl(\mathcal{D}^{ch}_{\Omega^{\cdot}_X})(y,q) = [\Lambda_y \Omega^1_X] \cdot \prod_{n=1}^{\infty} \{[S_{q^n}\Theta_X] \cdot [S_{q^n}\Omega^1_X] \cdot [\Lambda_{y^{-1}q^n}\Theta_X] \cdot [\Lambda_{yq^n}\Omega^1_X]\}. \tag{7.6.1}$$

7.7. *Proof*. Let us look at the bundles $\Theta_{\Omega^{\cdot}_X}$ and $\Omega^1_{\Omega^{\cdot}_X}$ a little bit more attentively.

Begin with the local situation 3.1, with $E = \Omega$, so that $\Lambda E = \Omega^{\cdot}$. All our frames $\mathfrak{g}$ will be natural. Let $T_\psi \subset T_{\Omega^{\cdot}}$ be the A-submodule with the base $\{\psi_i\}$. The coordinate change formula (3.2.3) shows that it is a well-defined A-submodule of T independent of the choice of frame, canonically isomorphic to T.

We set

$$T_{\psi\Omega^{\cdot}} = \Omega^{\cdot} \otimes_A T_\psi \subset T_{\Omega^{\cdot}}.$$

We denote by $T_{\tau\Omega^{\cdot}}$ the quotient $\Omega^{\cdot}$-module $T_{\Omega^{\cdot}}/T_{\psi\Omega^{\cdot}}$. Let $T_\tau \subset T_{\tau\Omega^{\cdot}}$ be the A-submodule generated by all τ_i. The formula (3.2.1) shows that T_τ is a well-defined A-module canonically isomorphic to T, and we have

$$T_{\tau\Omega^{\cdot}} = \Omega^{\cdot} \otimes_A T_\tau.$$

Returning to our variety X, we see that we get two vector bundles Θ_ψ and Θ_τ both isomorphic to Θ_X and a canonical short exact sequence

$$0 \longrightarrow \Theta_{\psi\Omega^{\cdot}_X} \longrightarrow \Theta_{\Omega^{\cdot}_X} \longrightarrow \Theta_{\tau\Omega^{\cdot}_X} \longrightarrow 0 \tag{7.7.1}$$

with

$$\Theta_{\psi\Omega^{\cdot}_X} = \Omega^{\cdot}_X \otimes_{\mathcal{O}_X} \Theta_\psi; \quad \Theta_{\tau\Omega^{\cdot}_X} = \Omega^{\cdot}_X \otimes_{\mathcal{O}_X} \Theta_\tau. \tag{7.7.2}$$

Note that Θ_ψ has fermionic charge -1 and Θ_τ has fermionic charge 0.

Dually, we have two vector bundles Ω^1_ρ and Ω^1_ω both isomorphic to Ω^1_X and a canonical short exact sequence

$$0 \longrightarrow \Omega^1_{\omega\Omega^{\cdot}_X} \longrightarrow \Omega^1_{\Omega^{\cdot}_X} \longrightarrow \Omega^1_{\rho\Omega^{\cdot}_X} \longrightarrow 0 \tag{7.7.3}$$

with

$$\Omega^1_{\rho\Omega^{\cdot}_X} = \Omega^{\cdot}_X \otimes_{\mathcal{O}_X} \Omega^1_\rho; \quad \Omega^1_{\omega\Omega^{\cdot}_X} = \Omega^{\cdot}_X \otimes_{\mathcal{O}_X} \Omega^1_\omega. \tag{7.7.4}$$

The bundles Ω^1_ρ, Ω^1_ω have fermionic charges $1, 0$ respectively.

Note that if E is a vector bundle, then

$$Sym_{\Omega^{\cdot}_X}(\Omega^{\cdot}_X \otimes_{\mathcal{O}_X} E) = \Omega^{\cdot}_X \otimes_{\mathcal{O}_X} Sym_{\mathcal{O}_X} E. \tag{7.7.5}$$

Returning to PBW formula (7.4.1) we see that these remarks imply (7.6.1). △

7.8. From now on let us assume that $k = \mathbb{C}$. Consider the formal power series

$$\theta(y,q) = i^{-1}(y^{1/2} - y^{-1/2})q^{1/8} \prod_{n=1}^{\infty} \{(1-q^n)(1-yq^n)(1-y^{-1}q^n)\}. \tag{7.8.1}$$

This is nothing other than the theta function $\theta_1(h,z)$ as defined in [HC], II, 2, §10, formula (3), p. 204, with $q = h^2$ and $y = z^2$.

If $f \in GL(V)$ is an automorphism of a d-dimensional vector space V with eigenvalues $\lambda_1, \dots, \lambda_d$, we shall denote by $\theta_f(y,q)$ the power series

$$\theta_f(y,q) = \frac{\prod_{i=1}^d \theta(\lambda_i y, q)}{\prod_{i=1}^d \theta(\lambda_i, q)}. \tag{7.8.2}$$

Let X be a proper smooth d-dimensional algebraic variety; let $g : X \longrightarrow X$ be a *simple* automorphism, which means by definition that the graph $\Gamma_g \subset X \times X$ is transversal to the diagonal. This implies that the set X^g of its fixed points is finite.

For each $x \in X^g$ denote by g_x the induced endomorphism of the cotangent space $\Omega^1_{X;x}$. All eigenvalues of g_x are different from 1.

Theorem 7.9. *Consider the power series*

$$T_{X;g}(y,q) := y^{-d/2} \sum_{a,b,n} (-1)^{a+b} \mathrm{Tr}(g; H^a(X; \mathcal{D}^{ch;b}_{\Omega^{\cdot}_X;n}))y^b q^n. \tag{7.9.1}$$

We have

$$T_{X;g}(y,q) = \sum_{x \in X^g} \theta_{g_x}(y,q). \tag{7.9.2}$$

7.10. *Proof*. Recall that according to the Atiyah–Bott holomorphic Lefschetz fixed point formula, if E is a g-equivariant vector bundle over X, then

$$\sum_i (-1)^i \mathrm{Tr}(g; H^i(X;E)) = \sum_{x \in X^g} \frac{\mathrm{Tr}(g; E_x)}{\det(1-g_x)}; \tag{7.10.1}$$

see [AB], Theorem 4.12. Note that if (V, f) are as in 7.8, then

$$\mathrm{Tr}(f; Sym_x(V)) = \prod_{i=1}^d (1-\lambda_i x)^{-1} = \mathrm{Tr}(f; \Lambda_{-x}(V))^{-1}. \tag{7.10.2}$$

The proof of 7.6 shows that each sheaf $\mathcal{D}^{ch;p}_{\Omega^{\cdot}_X;n}$ carries a canonical filtration whose quotients are vector bundles and the associated graded sheaf is given by the formula (7.6.1). Therefore we may apply the Lefschetz formula (7.10.1).

Note that since in the expression (7.9.1) the fermionic charge $a+b$ is taken into account, we should apply the Lefschetz formula to the element $cl(\mathcal{D}^{ch}_{\Omega^{\cdot}_X})(-y,q)$. Due to (7.10.2) each fixed point x contributes

$$y^{-d/2}\prod_{i=1}^{d}\left\{\frac{1-\lambda_i y}{1-\lambda_i}\prod_{n=1}^{\infty}\frac{(1-\lambda_i y q^n)(1-\lambda_i^{-1}y^{-1}q^n)}{(1-\lambda_i q^n)(1-\lambda_i^{-1}q^n)}\right\}=\theta_{g_x}(y,q)$$

where λ_i are the eigenvalues of g_x. This implies the theorem. □

The reader may wish to compare (7.9.2) with the explicit formulas for the trace of certain automorphisms of the Frenkel–Lepowsky–Meurman Monster vertex algebra, cf. [FLM].

References

[AB] M.F. Atiyah, R. Bott, A Lefschetz fixed point formula for elliptic complexes: II. Applications, *Ann. Math.*, **88**, No. 3 (1968), 451–491.

[BL] L. Borisov, A. Libgober, Elliptic genera and applications to mirror symmetry, *Inv. Math.*, **140** (2000), 453–485; math.AG/9904126.

[FLM] I. Frenkel, J. Lepowsky, A. Meurman, Vertex operator algebras and the Monster, *Pure and Applied Mathematics*, **134**, Academic Press, Boston, 1988.

[GI] V. Gorbounov, F. Malikov, V. Schechtman, Gerbes of chiral differential operators, math.AG/9906117; *Math. Research Letters*, **7** (2000), 55–66.

[GII] V. Gorbounov, F. Malikov, V. Schechtman, Gerbes of chiral differential operators. II, math.AG/0003170.

[HC] A. Hurwitz, R. Courant, *Funktionentheorie, Vierte Auflage*, Springer-Verlag, Berlin, Göttingen, Heidelberg, New York, 1964.

[K] V. Kac, *Vertex Algebras for Beginners*, Second Edition, University Lecture Series, **10**, American Mathematical Society, Providence, Rhode Island, 1998.

[MSV] F. Malikov, V. Schechtman, A. Vaintrob, Chiral de Rham complex, *Comm. Math. Phys.*, **204** (1999), 439–473.

[MS] F. Malikov, V. Schechtman, Chiral Poincaré duality, *Math. Research Letters*, **6** (1999), 533–546.

[W] E. Witten, Elliptic genera and quantum field theory, *Comm. Math. Phys.* **109** (1987), 525–536.

Vassily Gorbounov
Department of Mathematics
University of Kentucky
Lexington, KY 40506, USA
`vgorb@ms.uky.edu`

Fyodor Malikov
Department of Mathematics
University of Southern California
Los Angeles, CA 90089, USA
`fmalikov@mathj.usc.edu`

Vadim Schechtman
Laboratoire Emile Picard, UFR MIG
Université Paul Sabatier
118, route de Narbonne
31062 Toulouse, France
`schechtman@picard.ups-tlse.fr`

Defining Relations for the Exceptional Lie Superalgebras of Vector Fields

Pavel Grozman
Dimitry Leites
Irina Shchepochkina

ABSTRACT We list defining relations for four of the five exceptional simple Lie superalgebras of vector fields with polynomial coefficients. This is done for one of many inequivalent systems of simple roots (generators). For the fifth superalgebra the result is not final: there might be infinitely many relations. On the contrary, for the same Lie superalgebra with Laurent polynomials as coefficients there are only finitely many relations. To show perspective, the complete list of simple complex vectorial Lie superalgebras is given, and their relations have been calculated (also for one system of generators only) in [GLP].

Introduction

Explicit forms of defining relations of simple Lie (super)algebras have become of acute interest. This mainly concerns algebras with Cartan matrix, but the subject is not limited to this alone; cf. e.g., [BHW] (we thank P. Etingof who suggested comparing conjecturally incomplete relations from [BHW] with complete ones [GL4]). Here we consider Lie superalgebras of vector fields with polynomial coefficients. The case of series is dealt with (for one, distinguished, set of generators) in [LP] and [GLP]. Here we consider the exceptional superalgebras introduced in [Sh5], [Sh14] and [CK2]. According to conjectures by V. Kac and D. Broadhurst [Ka2] some of the exceptional infinite-dimensional simple vectorial Lie superalgebras might be related to the Standard Model, and the Grand Unified theory, respectively. Kac laboriously elucidated and popularized his conjecture [Ka3]. Physicists are not convinced by Kac's arguments yet, but we are, and we hope our relations will be of use soon.

To make the paper of interest to a wider audience, we begin with the list of simple vectorial superalgebras and append some general background on linear algebra in superspaces, since some of the facts are not very well known and might

AMS Subject classification: 17A70, 17B66 (Primary) 17B35, 17B01 (Secondary).
Keywords and phrases: Lie superalgebra, Cartan prolongation, nonholonomic structures, defining relations.

be no less useful to some readers than the main result.

Our choice of contribution to Kirillov's Fest goes back to a 1977 prompt by A. A. Kirillov who suggested to one of us a joint project: calculating a certain Lie superalgebra cohomology. For various reasons the project was put aside. We still remember the passing wishes of our teacher, especially since we could not answer immediately; another reason was to thank one of the editors by reminding us of the problems he was busy with during his happy childhood, see [Ki].

Recall that even the conventional simple finite-dimensional Lie algebras admit several distinct presentations. For example, there are (1) the Serre relations between *Chevalley generators* (convenient and simple looking but numerous: for the non-exceptional algebras ($\mathfrak{g} = \mathfrak{sl}(n)$, $\mathfrak{sp}(2n)$, $\mathfrak{o}(n)$) the number of generators and relations grows with $\operatorname{rk}\mathfrak{g} \sim n$); (2) relations between just a pair of *Jacobson generators*, see [GL3] (the number of the relations between the Jacobson generators does not depend on n but, although indispensable in some respects, they are less convenient than the Serre relations). There are other natural choices of generators even for $\mathfrak{sl}(n)$.

Among various presentations, Chevalley generators (satisfying Serre relations) are the first choice because by using them we split the algebra into the sum of its "positive" and "negative" parts $\mathfrak{g}_\pm \simeq \mathfrak{n}$ which are nilpotent. What is so nice about this? For any nilpotent Lie algebra the very notions of "generators" and "relations" are almost transparently defined. Moreover, they admit a homological interpretation. Suppose, as will be the case in our examples, there is a set of outer derivations (the Cartan subalgebra $\mathfrak{h}$ will be such a set in our examples) acting on $\mathfrak{n}$ so that $\mathfrak{n}$ splits into the direct sum of 1-dimensional eigenspaces. Then the choice of a basis is unique up to scalar factors; moreover, usually, $\mathfrak{n} \simeq \mathfrak{n}/[\mathfrak{n}, \mathfrak{n}] \oplus [\mathfrak{n}, \mathfrak{n}]$ so any basis of the space $\mathfrak{n}/[\mathfrak{n}, \mathfrak{n}] = H_1(\mathfrak{n})$ is a set of *generators* of $\mathfrak{n}$.

To describe relations between the generators, consider the standard homology complex for $\mathfrak{n}$ with trivial coefficients ([Fu]):

$$0 \longleftarrow \mathfrak{n} \xleftarrow{d_1} \mathfrak{n}\wedge\mathfrak{n} \xleftarrow{d_2} \mathfrak{n}\wedge\mathfrak{n}\wedge\mathfrak{n} \xleftarrow{d_3} \dots .$$

By definition

$$d_1(x \wedge y) = [x, y], \quad d_2(x \wedge y \wedge z) = [x, y] \wedge z + [y, z] \wedge x + [z, x] \wedge y.$$

The condition $d_1 d_2 = 0$ expresses the Jacobi identity. Obviously, the elements of $\operatorname{Ker} d_1$ are relations. The consequences of the Jacobi identity are considered as trivial relations, and they constitute $\operatorname{Im} d_2$. Thus, for a nilpotent Lie algebra $\mathfrak{n}$, we identify a convenient basis of $H_1(\mathfrak{n})$ with the generators and that of $H_2(\mathfrak{n})$ with *defining relations*. Observe that the same arguments apply as well to *locally nilpotent* Lie (super)algebras (considered below).

In addition to $\mathfrak{g}_\pm$, there is one more piece, $\mathfrak{g}_0$, consisting of the Cartan subalgebra $\mathfrak{h}$, but the relations expressing the elements of $\mathfrak{h}$ via positive and negative generators, as well as the action of $\mathfrak{h}$ on these generators, although they are vital and introduce the Cartan matrix (or its analog), are obvious or, at least, so easy to compute that we skip them. The most difficult to describe are relations between generators "of the same sign" (positive or negative). Such presentations for Lie algebras became very fashionable when using them in connection with q-quantizations. For Lie superalgebras, even with Cartan matrix, there are some subtleties: it turned out that there are non-Serre relations even between analogs of Chevalley generators, see [FLV], [Y], [GL2].

For simple finite-dimensional Lie superalgebras, even with Cartan matrix, there are *several* non-equivalent sets of Chevalley generators and, respectively, analogs of Serre relations, see [GL2]. For the Lie superalgebras without Cartan matrix, the analogs of such sets are not described very explicitly at the moment; besides, for infinite-dimensional Lie (super)algebras there are infinitely many (non-equivalent) such sets. So we only confine ourselves to "the simplest" (or most convenient) one; for the series this corresponds to the so-called standard grading; for the exceptional vectorial Lie superalgebras the grading consistent with parity (if it exists) is the most convenient one.

In §1, we describe simple vectorial Lie superalgebras, see [LSh0, LSh1], [Sh5], [Sh14]; [CK2]. Fifteen of them are exceptional; they can be considered as regradings of only five Lie superalgebras. There are, of course, lots of gradings $\mathfrak{g} = \bigoplus\limits_{i\geq -d} \mathfrak{g}_i$, but only finitely many of the gradings are determined by a maximal subalgebra of finite codimension so that $\mathfrak{g}_{-1}$ is an irreducible $\mathfrak{g}_0$-module. Such particular (Weisfeiler) gradings are described in [Sh14], [CK2] and below. Among the 15 Weisfeiler gradings some are distinguished by the fact that the corresponding $\mathbb{Z}$-grading is consistent (with parity); and these realizations are easier to deal with in our problem of describing defining relations, so we consider them when possible.

Actually, our true goal is the description of defining relations for any realization; but, unlike in the finite-dimensional case this might be impossible: there does not seem to be a way to pass from relations in terms of one set of simple roots to relations in terms of another set.

In §2 we state our main result. To formulate it, we represent the Lie superalgebras we consider as $\mathfrak{g} = \bigoplus\limits_{i\geq -d} \mathfrak{g}_i$; we further set $\mathfrak{g}_- = \bigoplus\limits_{i<0} \mathfrak{g}_i$ and $\mathfrak{g}_+ = \bigoplus\limits_{i>0} \mathfrak{g}_i$. Observe that $\mathfrak{g}_0$ has its own $\mathbb{Z}$-grading, so it can be represented as $\mathfrak{g}_0 = \bigoplus\limits_{i} \mathfrak{g}_{0,i}$, where $\mathfrak{g}_{0,0}$ is the Cartan subalgebra in $\mathfrak{g}_0$ and in $\mathfrak{g}$. Set $\mathfrak{g}_{0-} = \bigoplus\limits_{i<0} \mathfrak{g}_{0,i}$ and $\mathfrak{g}_{0+} = \bigoplus\limits_{i>0} \mathfrak{g}_{0,i}$. Finally, set $\mathfrak{G}_+ = \mathfrak{g}_{0+} \oplus \mathfrak{g}_+$ and $\mathfrak{G}_- = \mathfrak{g}_{0-} \oplus \mathfrak{g}_-$.

Our main result is a description of defining relations for the nilpotent Lie superalgebra $\mathfrak{G}_-$ and the locally nilpotent one $\mathfrak{G}_+$.

Observe a remarkable difference in presentations for $\mathfrak{kas}$ generated by polynomials (this paper) and $\mathfrak{kas}^L$ generated by Laurent polynomials, see [GLS1]: our results (description of defining relations for $\mathfrak{G}_+$ and $\mathfrak{G}_-$) are final in all cases, except for $\mathfrak{kas}_+$; we even suspect that $\mathfrak{kas}_+$ might not be finitely presented. This never happens with other vectorial Lie superalgebras with polynomial coefficients, but it does happen with certain innocent-looking loop superalgebras, cf. [GL2].

The results are obtained with the aid of Grozman's *Mathematica*-based package SuperLie, see [GL1]. With finite-dimensional algebras $\mathfrak{G}_-$ all is clear. How reliable are our results for infinite-dimensional $\mathfrak{G}_+$? In [LP] we directly applied the general arguments of [FF1], where the homology H_i of $\mathfrak{g}_+$ for simple Lie algebras of vector fields on n indeterminates were calculated "in stable dimensions", i.e., for $n >> i$. Clearly, $\mathfrak{g}_+$ is generated by $\mathfrak{g}_1$ (except for $\mathfrak{g} = \mathfrak{vect}(1)$) and the simplest relations are of degree 2; Fuchs claimed that except for small dimensions of the manifold on which $\mathfrak{g}$ is realized, *all* the relations are of degree 2, see formula (4) below. The idea of [LP] was to proceed in the same way as for Kac–Moody algebras: add (positive) generators of $\mathfrak{g}_0$ (and, therefore, drastically diminish the number of generators of $\mathfrak{g}_+$ from $\dim \mathfrak{g}_1$ to one or two) and consider the Hochschild–Serre spectral sequence with $\mathfrak{g}_{0+}$ as a subalgebra of $\mathfrak{G}_+$. Since all homology of $\mathfrak{g}_{0+}$ (with coefficients in finite-dimensional $\mathfrak{g}_0$-modules) is known, we obtain an estimate of the number of relations from above: the space of relations is contained in the direct sum of the following spaces:

$$H_2(\mathfrak{g}_{0+}); \quad \text{(the known Serre relations)} \tag{1}$$

$$H_1(\mathfrak{g}_{0+}; H_1(\mathfrak{g}_+)) \quad \text{(calculable via the Borel–Weil–Bott theorem)} \tag{2}$$

$$H_0(\mathfrak{g}_{0+}; H_2(\mathfrak{g}_+)) = Span(\text{the } \mathfrak{g}_0\text{-lowest vectors of } H_2(\mathfrak{g}_+)); \tag{3}$$

According to [Fu] and [FF2] homologies (3) are

$$\text{the lowest weight vectors in } H_2(\mathfrak{g}_+) = [H_2(\mathfrak{g}_+)]_2 = \wedge^2(\mathfrak{g}_1)/\mathfrak{g}_2. \tag{4}$$

When homologies (1)–(3) are known, it remains to compute the differential in the spectral sequence. Usually it vanishes; if it does not vanish but we did not calculate it, we get several extra relations deducible from the remaining ones. Our relations, according to computers, are indeed the defining ones. We do not know how to distill such sets without computers, cf. [Y] with [GL2].

Notice that the relations of types (1) and (2) are of the form

$$(\mathrm{ad}X_i)^{k_{ij}}(X_j) = 0 \text{ and } (\mathrm{ad}X_i)^{m_{ij}}(Y_j) = 0, \text{ respectively,}$$

where the X_i are the generators of $\mathfrak{g}_{0+}$ and the Y_j are the lowest weight vectors of the $\mathfrak{g}_0$-module $\mathfrak{g}_1$.

Kochetkov observed that (4) (and Fuchs' proof) is true only for $\mathfrak{vect}$, $\mathfrak{svect}$ and $\mathfrak{k}$: for the Hamiltonian series there are always relations of degree 3 (but no more relations in stable dimensions). The same applies to Lie *super*algebras of Hamiltonian fields, cf. [GLP].

Related results. For description of defining relations in other cases see [GLS1] (simple stingy Lie superalgebras); [LP] (simple and close to simple vectorial Lie algebras of vector fields with polynomial coefficients), [GLP] (nonexceptional Lie superalgebras of vectorial type and some exceptional loop superalgebras); for series $\mathfrak{q}$ see [LSe].

1 Description of simple vectorial Lie superalgebras

1.1 On the setting of the problem

Selection of Lie algebras with reasonably nice properties is a matter of taste and is under the influence of the main problem considered. One of the usual choices is the class of *simple* algebras (they are relatively easy to study, have rich inner structure, and illuminate important symmetries), unless these algebras have nontirvial central extensions in which case, to satisfy our insatiable desire to quantize, we should stick to the *extensions* — and to derivations of the extensions.

For simple Lie algebras, the finite-dimensional algebras are the first to study. They can be neatly encoded by very simple graphs — Dynkin diagrams — or Cartan matrices. Next on the agenda are *$\mathbb{Z}$-graded Lie algebras of polynomial growth.* If they have a Cartan matrix, they very much resemble finite-dimensional simple Lie algebras and their theory is very similar, see [Ka1]; they have proved to be useful in various branches of mathematics and theoretical physics. Some of the algebras of polynomial growth have no Cartan matrix, but are no less useful, as for example, in the case of vectorial Lie superalgebras.

1.1.1. Types of classical Lie algebras

Let us qualitatively describe the simple Lie algebras of polynomial growth to better visualize them. They break up into the disjoint union of the following types:

1) *finite dimensional* (growth 0).

2) *loop algebras*, perhaps, twisted (growth 1); more important in applications are their "relatives" called affine *Kac–Moody* algebras, cf. [Ka1].

3) *vectorial algebras*, i.e., Lie algebras of vector fields* with polynomial coefficients (growth is equal to the number of indeterminates) or their completions with formal power series as coefficients.

4) *the stringy algebra*† $\mathfrak{vect}^L(1) = \mathfrak{der}\ \mathbb{C}[t^{-1}, t]$ (the superscript stands for Laurent). This algebra is often called *Witt algebra* $\mathfrak{witt}$ and by physicists *the centerless Virasoro algebra* because its nontrivial central extension, $\mathfrak{vir}$, is called the *Virasoro algebra*.

Strictly speaking, stringy algebras are vectorial, but we retain the generic term *vectorial* for algebras with *polynomial* or formal coefficients.

Algebras of types (1)–(4) are $\mathbb{Z}$-graded. Several algebras of types (2) and(3)

---

*These algebras are sometimes known under the recently introduced clumsy name "algebras of Cartan type": just imagine a "Cartan subalgebra in a Lie superalgebra of Cartan type with Cartan matrix" (such exist).

†The term induced by the lingo of imaginative physicists who now play with (either cherish, as we do, [GSW] or try to enforce a calmer attitude [WL]) the idea that an elementary particle is not a point but rather a slinky springy string; the term "stringy algebra" means "pertaining to string theory" but also mirrors their structure as a collection of several strings — the modules over the Witt algebra. In our papers "stringy" means "vectorial (super)algebra on a supermanifold whose base is a circle or a "relative" of such an algebra (its central extension, or an algebra of derivations, etc.), a rigorous definition follows.

(in the super case all four types) have deformations and some of the deformed algebras are not $\mathbb{Z}$-graded of polynomial growth. These deformations (studied insufficiently so far) naturally indicate one more type.

5) *simple filtered Lie algebras of polynomial growth whose associated graded ones are not simple*, e.g., the Lie algebra of matrices of complex size and its generalizations, for examples see [GL3].

1.2 Superization

After Wess and Zumino showed the importance of supersymmetries, cf. [D, WZ], it was natural to list simple "classical" Lie superalgebras.

— *Finite-dimensional Lie superalgebras.* All except for vectorial ones were classified by Scheunert, Nahm, Rittenberg [SNR] and I. Kaplansky. An unpublished preprint (1975) by Kaplansky (it appeared in a modified form as [FK], [K]) helped Kac to repair gaps in his independent proof, see [K1C]. For details of the proof (further elucidated in [Sch]) in all cases, see [Ka4], where Kac also considered representation theory, the infinite-dimensional case and real forms.

— *(Twisted) loop superalgebras.* For an intrinsic characterization of loop superalgebras without appeal to Cartan matrix (deduced from O. Mathieu's papers) together with the same for stringy superalgebras, see [GLS1]: both types of algebras $\mathfrak{g} = \bigoplus_{i=-d}^{\infty} \mathfrak{g}_i$ are of infinite depth d but in the adjoint representation they act differently:

> *for the loop algebras every root vector corresponding to every non-imaginary root acts locally nilpotently; for the stringy algebras this is not so.*

One of us, DL, conjectured that, as for Lie algebras, the simple Lie superalgebras of growth 1 are the loop superalgebras and their twisted versions correspond to outer automorphisms of $\mathfrak{g}$. Serganova listed these automorphisms and amended the conjecture, see [S1]: one should consider automorphisms modulo the connected component of the unit (it is sometimes a bit larger than the group of inner automorphisms). J. van de Leur classified twisted loop superalgebras with symmetrizable Cartan matrix [vdL] and Serganova also classified Lie superalgebras with nonsymmetrizable Cartan matrices; their lists support the conjecture [S1].

— *Stringy superalgebras.* For their conjectural list and partial proof of the completeness of this list, see [GLS1, KvdL, Ka8]. Observe that some of the stringy superalgebras possess a Cartan matrix, although a nonsymmetrizable one [GLS1].

In this paper we consider the remaining type of simple graded Lie superalgebras of polynomial growth:

— *Vectorial Lie superalgebras.* The main examples to keep in mind are the filtered Lie superalgebra $\mathcal{L} = \mathfrak{der}\, \mathbb{C}[[x]]$ of formal vector fields and the graded one associated with it, that is, $L = \mathfrak{der}\, \mathbb{C}[x]$ of polynomial vector fields with grading and filtration given by setting $\deg x_i = 1$ for all i.

1.3 Vectorial Lie superalgebras. The standard realization

The elements of the Lie algebra $\mathcal{L} = \mathfrak{der}\ \mathbb{C}[[u]]$ are considered as vector fields. The Lie algebra $\mathcal{L}$ has only one maximal subalgebra $\mathcal{L}_0$ of finite codimension (consisting of the fields that vanish at the origin). The subalgebra $\mathcal{L}_0$ determines a filtration of $\mathcal{L}$: set

$$\mathcal{L}_{-1} = \mathcal{L} \text{ and } \mathcal{L}_i = \{D \in \mathcal{L}_{i-1} \mid [D, \mathcal{L}] \subset \mathcal{L}_{i-1}\} \text{ for } i \geq 1. \tag{5}$$

The associated graded Lie algebra $L = \bigoplus\limits_{i \geq -1} L_i$, where $L_i = \mathcal{L}_i/\mathcal{L}_{i+1}$, consists of the vector fields with *polynomial* coefficients.

Superization. We consider infinite-dimensional Lie superalgebras $\mathcal{L}$ with a maximal subalgebra $\mathcal{L}_0$ of finite codimension that does not contain ideals of $\mathcal{L}$. A Lie superalgebra $\tilde{\mathcal{L}}$ with such a subalgebra $\mathcal{L}_0$ is called a *primitive Lie superalgebra*. It turns out that there are virtually as many primitive Lie algebras as there are simple vectorial algebras. On the contrary, the classification problem of primitive Lie superalgebras is wild, see [ALSh].

Weisfeiler endowed every such filtered Lie algebra $\mathcal{L}$ with the following filtration $\tilde{\mathcal{L}} = \mathcal{L}$:

$$\mathcal{L} = \mathcal{L}_{-d} \supset \mathcal{L}_{-d+1} \supset \cdots \supset \mathcal{L}_0 \supset \mathcal{L}_1 \supset \ldots$$

where $\mathcal{L}_{-1}$ is a minimal $\mathcal{L}_0$-invariant subspace strictly containing $\mathcal{L}_0$, and the other terms are defined for $i \geq 1$ by the formula

$$\mathcal{L}_{-i-1} = [\mathcal{L}_{-1}, \mathcal{L}_{-i}] + \mathcal{L}_{-i} \text{ and } \mathcal{L}_i = \{D \in \mathcal{L}_{i-1} \mid [D, \mathcal{L}_{-1}] \subset \mathcal{L}_{i-1}\}. \tag{6}$$

This d is the *depth* of $\mathcal{L}$ and of the associated graded Lie superalgebra L, where $L_k = \mathcal{L}_k/\mathcal{L}_{k+1}$.

We additionally require that for any nonzero $x \in L_k$, where $k \geq 0$, there exists $y \in L_{-1}$ such that $[x, y] \neq 0$.

An advantage of the *Weisfeiler filtrations* (6) is that for the corresponding graded Lie superalgebra L the L_0-action on L_{-1} is irreducible. These filtrations are called *Weisfeiler filtrations* even for Lie superalgebras, where Weisfeiler's construction is not literally applied; we will shortly write *W-filtrations* and call the gradings associated with W-filtrations *W-gradings*.

When the L_0-module L_{-1} is faithful, as is always the case for simple Lie superalgebras $\mathcal{L}$, such filtered Lie superalgebras $\mathcal{L}$ (and the graded ones L associated with them) can be realized by *vector fields* on $(\mathcal{L}/\mathcal{L}_0)^*$ with formal or polynomial coefficients. So, being primarily interested in simple Lie superalgebras, we will refer to W-filtered and W-graded Lie superalgebras as *vectorial* for short.

Considering the subspaces $\mathcal{L}_i$ as the basis of a topology, we can complete the graded or filtered Lie superalgebras L or $\mathcal{L}$; the elements of the completion are the vector fields with formal power series as coefficients. Although the structure of the graded algebras is easier to describe, in applications the completed Lie superalgebras are usually needed.

Remark. (1) not all filtered or graded Lie superalgebras of finite depth are *vectorial*, i.e., realizable with vector fields on a supermanifold of the same dimension as that of $\mathcal{L}/\mathcal{L}_0$; only those with faithful L_0-action on L_{-1} are; (2) unlike Lie algebras, simple vectorial *super*algebras possess *several* (finitely many, as we can show [LSh1], non-equivalent ones) nonisomorphic maximal subalgebras of finite codimension.

1) General algebras. Let $x = (u_1, \dots, u_n, \theta_1, \dots, \theta_m)$, where the u_i are even indeterminates and the θ_j are odd ones. Set $\mathfrak{vect}(n|m) = \mathfrak{der}\, \mathbb{C}[x]$; it is called *the general vectorial Lie superalgebra.*

1.3.1. Digression: the modules of tensor fields

Further, we now recall the definition of the modules of tensor fields over $\mathfrak{vect}(m|n)$ and its subalgebras, see [BL1], [L2]. For any other $\mathbb{Z}$-graded vectorial Lie superalgebra the construction is identical.

Let $\mathfrak{g} = \mathfrak{vect}(m|n)$, $\mathfrak{g}_{\geq} = \bigoplus\limits_{i\geq 0} \mathfrak{g}_i$, and $\mathfrak{g}_+ = \bigoplus\limits_{i>0} \mathfrak{g}_i$. Clearly, $\mathfrak{g}_0 \cong \mathfrak{gl}(m|n)$. Let V be the $\mathfrak{g}_0$-module with the *lowest* weight $\lambda = \mathrm{lwt}(V)$. Make V into a $\mathfrak{g}_{\geq}$-module setting $\mathfrak{g}_+ \cdot V = 0$. Let us realize $\mathfrak{g}$ by vector fields on the $m|n$-dimensional linear supermanifold $C^{m|n}$ with coordinates $x = (u, \xi)$. The superspace $T(V) = \mathrm{Hom}_{U(\mathfrak{g}_{\geq})}(U(\mathfrak{g}), V)$ is isomorphic, due to the Poincaré–Birkhoff–Witt theorem, to $\mathbb{C}[[x]] \otimes V$. Its elements have a natural interpretation as formal *tensor fields of type* V. When $\lambda = (a, \dots, a)$ we will simply write $T(\vec{a})$ instead of $T(\lambda)$. We will usually consider $\mathfrak{g}$-modules induced from irreducible $\mathfrak{g}_0$-modules.

Examples $\mathfrak{vect}(m|n)$ as $\mathfrak{vect}(m|n)$-module is $T(\mathrm{id})$. More examples: $T(\vec{0})$ is the superspace of functions; $Vol(m|n) = T(1, \dots, 1; -1, \dots, -1)$ (the semicolon separates the first m ("even") coordinates of the weight with respect to the matrix units E_{ii} of $\mathfrak{gl}(m|n)$) is the superspace of *densities* or *volume forms*. We denote the generator of $Vol(m|n)$ corresponding to the ordered set of coordinates x by $vol(x)$. The space of λ-densities is $Vol^{\lambda}(m|n) = T(\lambda, \dots, \lambda; -\lambda, \dots, -\lambda)$. In particular, $Vol^{\lambda}(m|0) = T(\vec{\lambda})$ but $Vol^{\lambda}(0|n) = T(\overrightarrow{-\lambda})$. We set: $Vol_0(0|m) = \{v \in Vol \mid \int v = 0\}$ and $T_0(\vec{0}) = \Lambda(m)/\mathbb{C} \cdot 1$.

If the generator vol of Vol is fixed, then $Vol \cong T(\vec{0})$, as $\mathfrak{svect}(m|n)$-modules. We use $T_0^0(\vec{0})$ to denote the $\mathfrak{svect}(m|n)$-module $Vol_0(0|m)/\mathbb{C}\, vol(\xi)$; let $\Lambda(0) = \mathbb{C}$.

Remark. To view the volume element as "$d^m u d^n \xi$" is totally wrong: the superdeterminant can never appear as a factor under the changes of variables. We can try to use the usual notation of differentials provided *all* the differentials anticommute. Then the linear transformations that do not intermix the even u's with the odd ξ's multiply the volume element $vol(x)$, viewed as the fraction $\frac{du_1 \cdot \ldots \cdot du_m}{d\xi_1 \cdot \ldots \cdot d\xi_n}$, by the Berezinian of the transformation. But how could we justify this? Let $X = (x, \xi)$. If we consider the usual, exterior, differential forms, then the dX_i's super anti-commute, hence, the $d\xi_i$ commute; whereas if we consider the *symmetric* product of the differentials, as in the metrics, then the dX_i's supercommute, hence, the dx_i commute. However, the $\frac{\partial}{\partial \xi_i}$ anticommute and, from the point of view of transformations, $\frac{\partial}{\partial \xi_i} = \frac{1}{d\xi_i}$. The notation, $du_1 \cdot \ldots \cdot du_m \cdot \frac{\partial}{\partial \xi_1} \cdot \ldots \cdot \frac{\partial}{\partial \xi_n}$ is nevertheless still wrong: almost any transformation $A : (u, \xi) \mapsto (v, \eta)$ sends $du_1 \cdot \ldots \cdot du_m \cdot \frac{\partial}{\partial \xi_1} \cdot \ldots \cdot \frac{\partial}{\partial \xi_n}$ to the correct element, $\mathrm{ber}(A)(du^m \cdot \frac{\partial}{\partial \xi_1} \cdot \ldots \cdot \frac{\partial}{\partial \xi_n})$, plus extra terms. Indeed, the fraction $du_1 \cdot \ldots \cdot du_m \cdot \frac{\partial}{\partial \xi_1} \cdot \ldots \cdot \frac{\partial}{\partial \xi_n}$ is the highest weight vector of an *indecomposable* $\mathfrak{gl}(m|n)$-module and $vol(x)$ is the notation for the image of this vector in the 1-dimensional quotient module modulo the invariant submodule that consists precisely of all the extra terms.

2) Special algebras. The *divergence* of the field $D = \sum\limits_i f_i \frac{\partial}{\partial u_i} + \sum\limits_j g_j \frac{\partial}{\partial \theta_j}$ is the function (in our case: a polynomial, or a series)

$$\mathrm{div} D = \sum_i \frac{\partial f_i}{\partial u_i} + \sum_j (-1)^{p(g_j)} \frac{\partial g_j}{\partial \theta_j}. \tag{7}$$

• The Lie superalgebra $\mathfrak{svect}(n|m) = \{D \in \mathfrak{vect}(n|m) \mid \mathrm{div} D = 0\}$ is called the *special* (or *divergence-free*) *vectorial superalgebra*.

It is clear that it is also possible to describe $\mathfrak{svect}(n|m)$ as $\{D \in \mathfrak{vect}(n|m) \mid L_D vol_x = 0\}$, where vol_x is the volume form with constant coefficients in coordinates x and L_D the Lie derivative with respect to D.

• The Lie superalgebra $\mathfrak{svect}_\lambda(0|m) = \{D \in \mathfrak{vect}(0|m) \mid \mathrm{div}(1 + \lambda\theta_1 \cdot \ldots \cdot \theta_m)D = 0\}$, where $p(\lambda) \equiv m \pmod 2$, — the deformation of $\mathfrak{svect}(0|m)$ — is called the *deformed special* (or *divergence-free*) *vectorial superalgebra*. Clearly, $\mathfrak{svect}_\lambda(0|m) \cong \mathfrak{svect}_\mu(0|m)$ for $\lambda\mu \neq 0$. So we briefly denote these deformations by $\widetilde{\mathfrak{svect}}(0|m)$.

Observe that for m odd the parameter of deformation, λ, is odd.

Remark. Our notation comes from differential geometry; some algebraists sometimes abbreviate $\mathfrak{vect}(n)$ to W_n (in honor of Witt) and and $\mathfrak{svect}(n)$ to S_n.

3) The algebras that preserve Pfaff equations and differential 2-forms.

• Set $u = (t, p_1, \ldots, p_n, q_1, \ldots, q_n)$; let

$$\tilde{\alpha}_1 = dt + \sum_{1 \leq i \leq n} (p_i dq_i - q_i dp_i) + \sum_{1 \leq j \leq m} \theta_j d\theta_j \quad \text{and} \quad \tilde{\omega}_0 = d\tilde{\alpha}_1 .$$

The form $\tilde{\alpha}_1$ is called *contact*, the form $\tilde{\omega}_0$ is called *symplectic*. Sometimes it is more convenient to redenote the θ's and set $\theta = \theta_{2r+1}$ if $m = 2r + 1$ and

$$\xi_j = \frac{1}{\sqrt{2}}(\theta_j - i\theta_{r+j}); \; \eta_j = \frac{1}{\sqrt{2}}(\theta_j + i\theta_{r+j}) \; \text{ for } j \leq r = [m/2]$$

(here $i^2 = -1$). In place of $\tilde{\omega}_0$ or $\tilde{\alpha}_1$ take α_1 and $\omega_0 = d\alpha_1$, respectively, where

$$\alpha_1 = dt + \sum_{1 \leq i \leq n} (p_i dq_i - q_i dp_i) + \sum_{1 \leq j \leq r} (\xi_j d\eta_j + \eta_j d\xi_j) \begin{cases} & \text{if } m = 2r \\ +\theta d\theta & \text{if } m = 2r + 1. \end{cases}$$

The Lie superalgebra that preserves the *Pfaff equation* $\alpha_1 = 0$, i.e., the superalgebra

$$\mathfrak{k}(2n+1|m) = \{D \in \mathfrak{vect}(2n+1|m) \mid L_D\alpha_1 = f_D\alpha_1 \text{ for some } f_D \in \mathbb{C}[t, p, q, \theta]\},$$

is called the *contact superalgebra.* The Lie superalgebra

$$\mathfrak{po}(2n|m) = \{D \in \mathfrak{k}(2n+1|m) \mid L_D\alpha_1 = 0\}$$

is called the *Poisson* superalgebra. (A geometric interpretation of the Poisson superalgebra: it is the Lie superalgebra that preserves the connection with form α_1 on the line bundle over a symplectic supermanifold with the symplectic form $d\alpha_1$.)

• Similarly, set $u = q = (q_1, \dots, q_n)$, let $\theta = (\xi_1, \dots, \xi_n; \tau)$ be odd. Set

$$\alpha_0 = d\tau + \sum_i(\xi_i dq_i + q_i d\xi_i), \qquad \omega_1 = d\alpha_0$$

and call these forms the *odd-contact* and *periplectic*, respectively.

The Lie superalgebra that preserves the Pfaff equation $\alpha_0 = 0$, i.e., the superalgebra

$$\mathfrak{m}(n) = \{D \in \mathfrak{vect}(n|n+1) \mid L_D\alpha_0 = f_D \cdot \alpha_0 \text{ for some } f_D \in \mathbb{C}[q, \xi, \tau]\}$$

is called the *odd-contact superalgebra.*

The Lie superalgebra

$$\mathfrak{b}(n) = \{D \in \mathfrak{m}(n) \mid L_D\alpha_0 = 0\}$$

is called the *Buttin* superalgebra ([L4]). (A geometric interpretation of the Buttin superalgebra: it is the Lie superalgebra that preserves the connection with form α_0 on the line bundle of rank $\varepsilon = (0|1)$ over a *periplectic* supermanifold, i.e., over a supermanifold with the periplectic form $d\alpha_0$.)

The Lie superalgebras

$$\mathfrak{sm}(n) = \{D \in \mathfrak{m}(n) \mid \operatorname{div} D = 0\}, \quad \mathfrak{sb}(n) = \{D \in \mathfrak{b}(n) \mid \operatorname{div} D = 0\}$$

are called the *divergence-free* (or *special*) *odd-contact* and *special Buttin* superalgebras, respectively.

Remark. A relation with finite-dimensional geometry is as follows. Clearly, $\ker\alpha_1 = \ker\tilde{\alpha}_1$. The restriction of $\tilde{\omega}_0$ to $\ker\tilde{\alpha}_1$ is the orthosymplectic form $B_{ev}(m|2n)$; the restriction of ω_0 to $\ker\alpha_1$ is $B'_{ev}(m|2n)$. Similarly, the restriction of ω_1 to $\ker\alpha_0$ is $B_{odd}(n|n)$.

1.4 Generating functions

A laconic way to describe $\mathfrak{k}$, $\mathfrak{m}$ and their subalgebras is via generating functions.

• Odd form α_1. For $f \in \mathbb{C}[t, p, q, \theta]$ set :

$$K_f = (2 - E)(f)\frac{\partial}{\partial t} - H_f + \frac{\partial f}{\partial t}E, \tag{8}$$

where $E = \sum_i y_i \frac{\partial}{\partial y_i}$ (here the y_i are all the coordinates except t) is the *Euler operator* (which counts the degree with respect to the y_i), and H_f is the hamiltonian field with Hamiltonian f that preserves $d\tilde{\alpha}_1$:

$$H_f = \sum_{i\leq n}\left(\frac{\partial f}{\partial p_i}\frac{\partial}{\partial q_i} - \frac{\partial f}{\partial q_i}\frac{\partial}{\partial p_i}\right) - (-1)^{p(f)}\sum_{j\leq m}\frac{\partial f}{\partial \theta_j}\frac{\partial}{\partial \theta_j}. \tag{9}$$

The choice of the form α_1 instead of $\tilde{\alpha}_1$ only affects the shape of H_f that we give for $m = 2k + 1$:

$$H_f = \sum_{i\leq n}\left(\frac{\partial f}{\partial p_i}\frac{\partial}{\partial q_i} - \frac{\partial f}{\partial q_i}\frac{\partial}{\partial p_i}\right) - (-1)^{p(f)}\left(\sum_{j\leq k}\left(\frac{\partial f}{\partial \xi_j}\frac{\partial}{\partial \eta_j} + \frac{\partial f}{\partial \eta_j}\frac{\partial}{\partial \xi_j}\right) + \frac{\partial f}{\partial \theta}\frac{\partial}{\partial \theta}\right).$$

• Even form α_0. For $f \in \mathbb{C}[q, \xi, \tau]$ set:

$$M_f = (2 - E)(f)\frac{\partial}{\partial \tau} - Le_f - (-1)^{p(f)}\frac{\partial f}{\partial \tau}E, \tag{10}$$

where $E = \sum_i y_i \frac{\partial}{\partial y_i}$ (here the y_i are all the coordinates except τ), and

$$Le_f = \sum_{i\leq n}\left(\frac{\partial f}{\partial q_i}\frac{\partial}{\partial \xi_i} + (-1)^{p(f)}\frac{\partial f}{\partial \xi_i}\frac{\partial}{\partial q_i}\right). \tag{11}$$

Since

$$\begin{aligned} L_{K_f}(\alpha_1) &= 2\tfrac{\partial f}{\partial t}\alpha_1 = K_1(f)\alpha_1, \\ L_{M_f}(\alpha_0) &= -(-1)^{p(f)}2\tfrac{\partial f}{\partial \tau}\alpha_0 = -(-1)^{p(f)}M_1(f)\alpha_0, \end{aligned} \tag{12}$$

it follows that $K_f \in \mathfrak{k}(2n + 1|m)$ and $M_f \in \mathfrak{m}(n)$. Observe that

$$p(Le_f) = p(M_f) = p(f) + \bar{1}.$$

• To the (super)commutators $[K_f, K_g]$ or $[M_f, M_g]$ there correspond *contact brackets* of the generating functions:

$$[K_f, K_g] = K_{\{f,\, g\}_{k.b.}}; \qquad [M_f, M_g] = M_{\{f,\, g\}_{m.b.}}$$

The explicit formulas for the contact brackets are as follows. Let us first define the brackets on functions that do not depend on t (resp. τ).

The *Poisson bracket* $\{\cdot,\cdot\}_{P.b.}$ (in the realization with the form ω_0) is given by the formula

$$\{f,g\}_{P.b.}=\sum_{i\leq n}\left(\frac{\partial f}{\partial p_i}\frac{\partial g}{\partial q_i}-\frac{\partial f}{\partial q_i}\frac{\partial g}{\partial p_i}\right)-(-1)^{p(f)}\sum_{j\leq m}\frac{\partial f}{\partial\theta_j}\frac{\partial g}{\partial\theta_j}, \tag{13}$$

and in the realization with the form ω_0 for $m=2k+1$ it is given by the formula

$$\begin{aligned}\{f,g\}_{P.b.}&=\sum_{i\leq n}\left(\frac{\partial f}{\partial p_i}\frac{\partial g}{\partial q_i}-\frac{\partial f}{\partial q_i}\frac{\partial g}{\partial p_i}\right)\\&-(-1)^{p(f)}\left(\sum_{j\leq m}\left(\frac{\partial f}{\partial\xi_j}\frac{\partial g}{\partial\eta_j}+\frac{\partial f}{\partial\eta_j}\frac{\partial g}{\partial\xi_j}\right)+\frac{\partial f}{\partial\theta}\frac{\partial g}{\partial\theta}\right).\end{aligned}$$

The *Buttin bracket* $\{\cdot,\cdot\}_{B.b.}$ is given by the formula

$$\{f,g\}_{B.b.}=\sum_{i\leq n}\left(\frac{\partial f}{\partial q_i}\frac{\partial g}{\partial\xi_i}+(-1)^{p(f)}\frac{\partial f}{\partial\xi_i}\frac{\partial g}{\partial q_i}\right). \tag{14}$$

Remark. What we call here the "Buttin bracket" was discovered in the pre-super era by Schouten; Buttin was the first to prove that this bracket establishes a Lie superalgebra structure. The interpretations of the Buttin superalgebra, similar to that of the Poisson algebra and of the elements of $\mathfrak{le}$ as analogs of Hamiltonian vector fields, was given in [L1]. The Buttin bracket and "odd mechanics" introduced in [L1] was rediscovered by Batalin with Vilkovisky; it gained a great deal of currency under the name *antibracket*, cf. [GPS]. The *Schouten bracket* was originally defined on the superspace of polyvector fields on a manifold, i.e., on the superspace of sections of the exterior algebra (over the algebra $\mathcal{F}$ of functions) of the tangent bundle $\Gamma(\Lambda^{\boldsymbol{\cdot}}(T(M)))\cong\Lambda^{\boldsymbol{\cdot}}_{\mathcal{F}}(Vect(M))$. The explicit formula of the Schouten bracket (in which the hatted slot should be ignored, as usual) is

$$\begin{aligned}&[X_1\wedge\dots\wedge\dots\wedge X_k,Y_1\wedge\dots\wedge Y_l]\\&=\textstyle\sum_{i,j}(-1)^{i+j}[X_i,Y_j]\wedge X_1\wedge\dots\wedge\hat{X}_i\wedge\dots\wedge X_k\wedge Y_1\wedge\dots\wedge\hat{Y}_j\wedge\dots\wedge Y_l.\end{aligned} \tag{15}$$

With the help of the Sign Rule we easily superize formula (15) for the case when M is replaced with a supermanifold $\mathcal{M}$. Let x and ξ be the even and odd coordinates on $\mathcal{M}$. Setting $\theta_i=\Pi(\frac{\partial}{\partial x_i})=\check{x}_i$, $q_j=\Pi(\frac{\partial}{\partial\xi_j})=\check{\xi}_j$ we get an identification of the Schouten bracket of polyvector fields on $\mathcal{M}$ with the Buttin bracket of functions on the supermanifold $\check{\mathcal{M}}$ whose coordinates are x,ξ and $\check{x}$, $\check{\xi}$ and the transformations of the checked variables are induced by those of x,ξ.

In terms of the Poisson and Buttin brackets, respectively, the contact brackets are

$$\{f,g\}_{k.b.}=(2-E)(f)\frac{\partial g}{\partial t}-\frac{\partial f}{\partial t}(2-E)(g)-\{f,g\}_{P.b.} \tag{16}$$

and

$$\{f, g\}_{m.b.} = (2-E)(f)\frac{\partial g}{\partial \tau} + (-1)^{p(f)}\frac{\partial f}{\partial \tau}(2-E)(g) - \{f, g\}_{B.b.}. \qquad (17)$$

The Lie superalgebras of *Hamiltonian fields* (or *Hamiltonian superalgebra*) and its traceless‡ subalgebra (defined only if $n = 0$) are

$$\mathfrak{h}(2n|m) = \{D \in \mathfrak{vect}(2n|m) \mid L_D\omega_0 = 0\};$$
$$\mathfrak{h}^o(m) = \{H_f \in \mathfrak{h}(0|m) \mid \int f vol_\theta = 0\}.$$

The "odd" analogues of the Lie superalgebra of Hamiltonian fields are the Lie superalgebra of vector fields Le_f introduced in [L1] and its special subalgebra:

$$\mathfrak{le}(n) = \{D \in \mathfrak{vect}(n|n) \mid L_D\omega_1 = 0\} \text{ and } \mathfrak{sle}(n) = \{D \in \mathfrak{le}(n) \mid \mathrm{div} D = 0\}.$$

It is not difficult to prove the following isomorphisms (as superspaces):

$$\begin{aligned} \mathfrak{k}(2n+1|m) &\cong Span(K_f \mid f \in \mathbb{C}[t, p, q, \xi]); \\ \mathfrak{le}(n) &\cong Span(Le_f \mid f \in \mathbb{C}[q, \xi]); \\ \mathfrak{m}(n) &\cong Span(M_f \mid f \in \mathbb{C}[\tau, q, \xi]); \\ \mathfrak{h}(2n|m) &\cong Span(H_f \mid f \in \mathbb{C}[p, q, \xi]). \end{aligned}$$

Set $\mathfrak{po}^o(m) = \{K_f \in \mathfrak{po}(0|m) \mid \int f vol_\theta = 0\}$, then $\mathfrak{h}^o(m) = \mathfrak{po}^o(m)/\mathbb{C}\cdot K_1$.

1.5 Divergence-free subalgebras

Since, as is easy to calculate,

$$\mathrm{div} K_f = (2n+2-m)K_1(f),$$

it follows that the divergence-free subalgebra of the contact Lie superalgebra either coincides with it (for $m = 2n+2$) or is the Poisson superalgebra. For the "odd" contact series the situation is more interesting: the divergence free subalgebra is simple and new (as compared with the above list).

Since

$$\mathrm{div} M_f = (-1)^{p(f)} 2\left((1-E)\frac{\partial f}{\partial \tau} - \sum_{i\leq n}\frac{\partial^2 f}{\partial q_i \partial \xi_i}\right), \qquad (18)$$

it follows that

$$\mathfrak{sm}(n) = Span\left(M_f \in \mathfrak{m}(n) \mid (1-E)\frac{\partial f}{\partial \tau} = \sum_{i\leq n}\frac{\partial^2 f}{\partial q_i \partial \xi_i}\right).$$

‡For vectorial algebras there are two analogs of trace: the divergence and a (super)trace in the original meaning of the word: the functional that vanishes on brackets. We will denote traceless subalgebras by ^{o}.

In particular,

$$\operatorname{div} Le_f = (-1)^{p(f)} 2 \sum_{i \leq n} \frac{\partial^2 f}{\partial q_i \partial \xi_i}. \tag{19}$$

Remark. The odd analog of the Laplacian, namely, the operator

$$\Delta = \sum_{i \leq n} \frac{\partial^2}{\partial q_i \partial \xi_i} \tag{20}$$

on a periplectic supermanifold appeared in physics under the name of *BRST operator*, cf. [GPS].

The divergence-free vector fields from $\mathfrak{sle}(n)$ are generated by *harmonic* functions, i.e., such that $\Delta(f) = 0$.

The lie superalgebras $\mathfrak{sle}(n)$, $\mathfrak{sb}(n)$ and $\mathfrak{svect}(1|n)$ have ideals $\mathfrak{sle}^\circ(n)$, $\mathfrak{sb}^\circ(n)$ and $\mathfrak{svect}^\circ(n)$ of codimension 1 defined from the exact sequences

$$0 \longrightarrow \mathfrak{sle}^\circ(n) \longrightarrow \mathfrak{sle}(n) \longrightarrow \mathbb{C} \cdot Le_{\xi_1 \dots \xi_n} \longrightarrow 0,$$
$$0 \longrightarrow \mathfrak{sb}^\circ(n) \longrightarrow \mathfrak{sb}(n) \longrightarrow \mathbb{C} \cdot M_{\xi_1 \dots \xi_n} \longrightarrow 0,$$
$$0 \longrightarrow \mathfrak{svect}^\circ(n) \longrightarrow \mathfrak{svect}(1|n) \longrightarrow \mathbb{C} \cdot \xi_1 \dots \xi_n \frac{\partial}{\partial t} \longrightarrow 0.$$

1.6 The Cartan prolongs

We will repeatedly use the Cartan prolong. So let us recall the definition and generalize it somewhat. Let $\mathfrak{g}$ be a Lie algebra, V a $\mathfrak{g}$-module, S^k the operator of the k-th symmetric power. Set $\mathfrak{g}_{-1} = V$, $\mathfrak{g}_0 = \mathfrak{g}$ and for any $k > 0$ define the k-th *Cartan prolong* (the result of Cartan's *prolongation*) of the pair $(\mathfrak{g}_{-1}, \mathfrak{g}_0)$ as the set $\mathfrak{g}_k$ such that for any $v_i \in \mathfrak{g}_{-1}$ we have

$$\mathfrak{g}_k = \{X \in \operatorname{Hom}(\mathfrak{g}_{-1}, \mathfrak{g}_{k-1}) \mid X(v_0)(v_1, v_2, \dots, v_k) = X(v_1)(v_0, v_2, \dots, v_k)\}.$$

Equivalently, let

$$i : S^{k+1}(\mathfrak{g}_{-1})^* \otimes \mathfrak{g}_{-1} \longrightarrow S^k(\mathfrak{g}_{-1})^* \otimes \mathfrak{g}_{-1}^* \otimes \mathfrak{g}_{-1} \tag{21}$$

be the natural embedding and

$$j : S^k(\mathfrak{g}_{-1})^* \otimes \mathfrak{g}_0 \longrightarrow S^k(\mathfrak{g}_{-1})^* \otimes \mathfrak{g}_{-1}^* \otimes \mathfrak{g}_{-1} \tag{22}$$

the natural map. Then $\mathfrak{g}_k = i(S^{k+1}(\mathfrak{g}_{-1})^* \otimes \mathfrak{g}_{-1}) \cap j(S^k(\mathfrak{g}_{-1})^* \otimes \mathfrak{g}_0)$.

The *Cartan prolong* of $(V, \mathfrak{g})$ is $(\mathfrak{g}_{-1}, \mathfrak{g}_0)_* = \bigoplus_{k \geq -1} \mathfrak{g}_k$. (In what follows $\cdot$ in superscript denotes, as is now customary, the collection of all degrees, while $*$ is reserved for dualization; in the subscripts we retain the old-fashioned $*$ instead of $\cdot$ to avoid confusion with punctuation marks.)

Suppose that the $\mathfrak{g}_0$-module $\mathfrak{g}_{-1}$ is *faithful*. Then, clearly,

$$(\mathfrak{g}_{-1}, \mathfrak{g}_0)_* \subset \mathfrak{vect}(n) = \mathfrak{der}\, \mathbb{C}[x_1, \ldots, x_n], \quad \text{where } n = \dim \mathfrak{g}_{-1} \text{ and}$$
$$\mathfrak{g}_i = \{D \in \mathfrak{vect}(n) \mid \deg D = i, \ [D, X] \in \mathfrak{g}_{i-1} \text{ for any } X \in \mathfrak{g}_{-1}\}.$$

It is easy to verify that the Lie algebra structure on $\mathfrak{vect}(n)$ induces same on $(\mathfrak{g}_{-1}, \mathfrak{g}_0)_*$.

Of the four simple vectorial Lie algebras, three are Cartan prolongs: $\mathfrak{vect}(n) = (\mathrm{id}, \mathfrak{gl}(n))_*$, $\mathfrak{svect}(n) = (\mathrm{id}, \mathfrak{sl}(n))_*$ and $\mathfrak{h}(2n) = (\mathrm{id}, \mathfrak{sp}(n))_*$. The fourth one — $\mathfrak{k}(2n+1)$ — is the result of a trifle more general construction described as follows.

1.6.1. A generalization of the Cartan prolong

Let $\mathfrak{g}_- = \bigoplus\limits_{-d \leq i \leq -1} \mathfrak{g}_i$ be a nilpotent $\mathbb{Z}$-graded Lie algebra and $\mathfrak{g}_0 \subset \mathfrak{der}_0 \mathfrak{g}_-$ a Lie subalgebra of the $\mathbb{Z}$-grading-preserving derivations of $\mathfrak{g}_-$. Let

$$i : S^{k+1}(\mathfrak{g}_-)^* \otimes \mathfrak{g}_- \longrightarrow S^k(\mathfrak{g}_-)^* \otimes \mathfrak{g}_-^* \otimes \mathfrak{g}_- \qquad (21')$$

and

$$j : S^k(\mathfrak{g}_-)^* \otimes \mathfrak{g}_0 \longrightarrow S^k(\mathfrak{g}_-)^* \otimes \mathfrak{g}_-^* \otimes \mathfrak{g}_- \qquad (22')$$

be the natural embeddings. For $k > 0$ define the k-th prolong of $(\mathfrak{g}_-, \mathfrak{g}_0)$ to be

$$\mathfrak{g}_k = \{\text{the deg} = k \text{ component of } (j(S^{\bullet}(\mathfrak{g}_-)^* \otimes \mathfrak{g}_0) \cap i(S^{\bullet}(\mathfrak{g}_-)^* \otimes \mathfrak{g}_-))\}.$$

Set $(\mathfrak{g}_-, \mathfrak{g}_0)_* = \bigoplus\limits_{i \geq -d} \mathfrak{g}_i$; then, as is easy to verify, $(\mathfrak{g}_-, \mathfrak{g}_0)_*$ is a Lie algebra.

What is the Lie algebra of contact vector fields in these terms? Denote by $\mathfrak{hei}(2n)$ the Heisenberg Lie algebra: its space is $W \oplus \mathbb{C} \cdot z$, where W is a $2n$-dimensional space endowed with a nondegenerate skew-symmetric bilinear form B and the bracket in $\mathfrak{hei}(2n)$ is given by the following relations:

$$z \text{ is in the center and } [v, w] = B(v, w) \cdot z \text{ for any } v, w \in W.$$

Clearly, $\mathfrak{k}(2n+1) \cong (\mathfrak{hei}(2n), \mathfrak{csp}(2n))_*$.

1.6.2. Lie superalgebras of vector fields as Cartan's prolongs

The superization of the constructions from Sections 1.6 and 1.6.1 are straightforward: via the Sign Rule. We thus get infinite-dimensional Lie superalgebras

$$\mathfrak{vect}(m|n) = (\mathrm{id}, \mathfrak{gl}(m|n))_*; \ \mathfrak{svect}(m|n) = (\mathrm{id}, \mathfrak{sl}(m|n))_*;$$
$$\mathfrak{h}(2m|n) = (\mathrm{id}, \mathfrak{osp}^{sk}(m|2n))_*;$$
$$\mathfrak{le}(n) = (\mathrm{id}, \mathfrak{pe}^{sk}(n))_*; \ \mathfrak{sle}(n) = (\mathrm{id}, \mathfrak{spe}^{sk}(n))_*.$$

Observe that $(\mathrm{id}, \mathfrak{osp}^{sy}(m|2n))_*$ and $(\mathrm{id}, \mathfrak{pe}^{sy}(n))_*$ are finite dimensional.

The generalization of Cartan's prolongations described in Section 1.6.1 was first defined in Tanaka's papers [T1]–[T5] and independently for Lie superalgebras in [ALSh]. Observe that the interest of Tanaka and his school (see a review

[Ym]) are quite distinct from ours: after superization there are **two** analogs associated with the contact series $\mathfrak{k}$ and $\mathfrak{m}$, respectively:

• Define the Lie superalgebra $\mathfrak{hei}(2n|m)$ on the direct sum of a $(2n, m)$-dimensional superspace W endowed with a nondegenerate skew-symmetric bilinear form and a $(1, 0)$-dimensional space spanned by z.

Clearly, we have $\mathfrak{k}(2n+1|m) = (\mathfrak{hei}(2n|m), \mathfrak{cosp}^{sk}(m|2n))_*$ and, more generally, given $\mathfrak{hei}(2n|m)$ and a subalgebra $\mathfrak{g}$ of $\mathfrak{cosp}^{sk}(m|2n)$, we call $(\mathfrak{hei}(2n|m), \mathfrak{g})_*$ the *k-prolong* of $(W, \mathfrak{g})$, where W is the identity $\mathfrak{osp}^{sk}(m|2n)$-module.

• The "odd" analog of $\mathfrak{k}$ is associated with the following "odd" analog of $\mathfrak{hei}(2n|m)$. Denote by $\mathfrak{ab}(n)$ the *antibracket* Lie superalgebra: its space is $W\oplus\mathbb{C}\cdot z$, where W is an $n|n$-dimensional superspace endowed with a nondegenerate skew-symmetric odd bilinear form B; the bracket in $\mathfrak{ab}(n)$ is given by the following relations:

$$z \text{ is odd and lies in the center; } [v, w] = B(v, w)\cdot z \text{ for } v, w \in W.$$

Clearly, $\mathfrak{m}(n) = (\mathfrak{ab}(n), \mathfrak{cpe}^{sk}(n))_*$ and, more generally, given $\mathfrak{ab}(n)$ and a subalgebra $\mathfrak{g}$ of $\mathfrak{cpe}^{sk}(n)$, we call $(\mathfrak{ab}(n), \mathfrak{g})_*$ the *m-prolong* of $(W, \mathfrak{g})$, where W is the identity $\mathfrak{pe}^{sk}(n)$-module.

Generally, given a nondegenerate form B on a superspace W and a superalgebra $\mathfrak{g}$ that preserves B, we refer to the above generalized prolongations as the *mk-prolongation* of the pair $(W, \mathfrak{g})$.

1.6.3. A partial Cartan prolong of $(\bigoplus\limits_{i\leq 0}\mathfrak{g}_i, \mathfrak{h}_1)$

Consider the generalized Cartan prolong $(\mathfrak{g}_-, \mathfrak{g}_0)_*$. Take a $\mathfrak{g}_0$-submodule $\mathfrak{h}_1$ in $\mathfrak{g}_1$ such that $[\mathfrak{g}_{-1}, \mathfrak{h}_1] = \mathfrak{g}_0$. If such $\mathfrak{h}_1$ exists (usually, the inclusion $[\mathfrak{g}_{-1}, \mathfrak{h}_1] \subset \mathfrak{g}_0$ is strict), define the ith prolongation of $(\bigoplus\limits_{i\leq 0}\mathfrak{g}_i, \mathfrak{h}_1)$ for $i \geq 2$ to be $\mathfrak{h}_i = \{D \in \mathfrak{g}_i \mid [D, \mathfrak{g}_{-1}] \in \mathfrak{h}_{i-1}\}$. Set $\mathfrak{h}_i = \mathfrak{g}_i$ for $i < 0$ and $\mathfrak{h}_* = \oplus\mathfrak{h}_i$.

Examples. $\mathfrak{vect}(1|n)$ is a subalgebra of $\mathfrak{k}(1|2n)$. The former is obtained as Cartan's prolong of the same nonpositive part as $\mathfrak{k}(1|2n)$ and a submodule of $\mathfrak{k}(1|2n)_1$, cf. Table 1.8.1. The simple exceptional superalgebra $\mathfrak{kas}$, introduced in 1.8.3, is another example.

1.7 Deformations of the Buttin superalgebra (After [ALSh])

As is clear from the definition of $\mathfrak{b}(n)$, there is a regrading (namely, $\mathfrak{b}(n; n)$ given by $\deg\xi_i = 0$, $\deg q_i = 1$ for all i) under which $\mathfrak{b}(n)$, initially of depth 2, takes the form $\mathfrak{g} = \bigoplus\limits_{i\geq -1}\mathfrak{g}_i$ with $\mathfrak{g}_0 = \mathfrak{vect}(0|n)$ and $\mathfrak{g}_{-1} \cong \Pi(\mathbb{C}[\xi])$. Replace now the $\mathfrak{vect}(0|n)$-module $\mathfrak{g}_{-1}$ of functions (with inverted parity) with the module of λ-densities, i.e., set $\mathfrak{g}_{-1} \cong \Pi(Vol(0|n)^\lambda)$, where

$$L_D(vol_\xi^\lambda) = \lambda\mathrm{div}D\cdot vol_\xi^\lambda \text{ and } p(vol_\xi^\lambda) = \bar{1}.$$

$\mathfrak{b}_\lambda(n;n)$ **as the Cartan prolong** $(\Pi(Vol(0|n)^\lambda), \mathfrak{vect}(0|n))_*$. Clearly, $\mathfrak{b}_\lambda(n)$ is a deformation of $\mathfrak{b}(n;n)$. The collection of these $\mathfrak{b}_\lambda(n;n)$ for all λ's is called the *main deformation*. (Though main, this deformation is not the quantization of the Buttin bracket, cf. [Ko1], [L4], [LSh2].)

The deformation $\mathfrak{b}_\lambda(n)$ of $\mathfrak{b}(n)$ is a regrading of $\mathfrak{b}_\lambda(n;n)$ described as follows. Set

$$\mathfrak{b}_{a,b}(n) = \{M_f \in \mathfrak{m}(n) \mid a\ \mathrm{div} M_f = (-1)^{p(f)} 2(a-bn)\frac{\partial f}{\partial \tau}\}. \tag{23}$$

Taking into account the explicit form (18) of the divergence of M_f we get

$$\begin{aligned}\mathfrak{b}_{a,b}(n) &= \{M_f \in \mathfrak{m}(n) \mid (bn - aE)\tfrac{\partial f}{\partial \tau} = a\Delta f\}\\ &= \{D \in \mathfrak{vect}(n|n+1) \mid L_D(vol^a_{q,\xi,\tau}\alpha_0^{a-bn}) = 0\}.\end{aligned}$$

It is subject to a direct check that $\mathfrak{b}_{a,b}(n)$ is isomorphic to $\mathfrak{b}_\lambda(n)$ for $\lambda = \frac{2a}{n(a-b)}$. This isomorphism shows that λ actually runs over $\mathbb{C}P^1$, not $\mathbb{C}$.

Observe that $\mathfrak{b}_{nb,b}(n) \cong \mathfrak{sm}(n)$. Observe also that $\mathfrak{b}_{a,-a}(2;2) \cong \mathfrak{h}_{1/2}(2|2)$, see (28).

As follows from the description of $\mathfrak{vect}(m|n)$-modules ([BL1]) and the criteria for simplicity of $\mathbb{Z}$-graded Lie superalgebras ([Ka4]), the Lie superalgebras $\mathfrak{b}_\lambda(n)$ are simple for $n > 1$ and $\lambda \neq 0, 1, \infty$. It is also clear that the $\mathfrak{b}_\lambda(n)$ are nonisomorphic for distinct λ's for $n > 2$.

The Lie superalgebra $\mathfrak{b}(n) = \mathfrak{b}_0(n)$ is not simple: it has an ε-dimensional, i.e., $(0|1)$-dimensional center. At $\lambda = 1$ and ∞ the Lie superalgebras $\mathfrak{b}_\lambda(n)$ are not simple either: they have an ideal of codimension ε^n and ε^{n+1}, respectively. The corresponding exact sequences are

$$0 \longrightarrow \mathbb{C}\cdot M_1 \longrightarrow \mathfrak{b}(n) \longrightarrow \mathfrak{le}(n) \longrightarrow 0,$$

$$0 \longrightarrow \mathfrak{b}_1{}^\circ(n) \longrightarrow \mathfrak{b}_1(n) \longrightarrow \mathbb{C}\cdot M_{\xi_1\ldots\xi_n} \longrightarrow 0,$$

$$0 \longrightarrow \mathfrak{b}_\infty{}^\circ(n) \longrightarrow \mathfrak{b}_\infty(n) \longrightarrow \mathbb{C}\cdot M_{\tau\xi_1\ldots\xi_n} \longrightarrow 0.$$

1.8 Detailed description of vectorial superalgebras

In [LSh0] we announced and in [LSh1] we classified simple **W-graded** vectorial Lie superalgebras. Related are the classification results of Cheng and Kac concerning simple *complete*, with respect to a x-adic topology, vectorial Lie superalgebras ([Ka7], [CK1]). Observe that Cheng and Kac only considered filtered deformations corresponding to *one* filtration; other filtrations might produce new filtered deformations.

On notation. To help grasp the general picture from the displayed formulas of the following theorem, let us immediately state that the prime example, $\mathfrak{vect}(m|n;r)$, is the Lie algebra of vector fields whose coefficients are polynomials (or formal power series, depending on the problem) in m commuting and n anticommuting indeterminates with the filtration (and grading) determined by equating the degrees of r ($0 \leq r \leq m$) odd indeterminates to 0, the degrees of the remaining indeterminates being equal to 1. The regradings of other series are determined similarly, see 1.9. We omit indicating the parameter r if $r = 0$.

In the sequel for any $\mathfrak{g}$ we write $\mathfrak{c}\mathfrak{g} = \mathfrak{g} \oplus \mathbb{C} \cdot z$ or $\mathfrak{c}(\mathfrak{g})$ to denote the trivial central extension with the 1-dimensional even center generated by z.

Let $\mathfrak{a} \subset\!\!\!\!+ \mathfrak{b}$ denote the semidirect product in which $\mathfrak{a}$ is the ideal.

If d is the operator that determines the $\mathbb{Z}$ grading of $\mathfrak{g}$ and does not belong to $\mathfrak{g}$, then the Lie algebra $\mathfrak{g} \subset\!\!\!\!+ \mathbb{C} \cdot d$ is denoted by $\mathfrak{d}(\mathfrak{g})$.

We use notation of roots in the simple finite-dimensional Lie algebras from [OV], Table 5. The standard (identity) representation of a linear Lie superalgebra $\mathfrak{g}$, i.e., a subalgebra of $\mathfrak{gl}(V)$, in V is denoted by id or, for clarity, $\mathrm{id}_{\mathfrak{g}}$.

Further elucidations. 1) In notation of the form $\mathfrak{svect}(m|n; r)$ the parentheses contain the superdimension of the superspace of indeterminates and the semicolon separates the superdimension from a shorthand description of the regrading.

When passing from one regrading to another one we take a "minimal" realization (i.e., with a minimal $\dim \mathcal{L}/\mathcal{L}_0$) as the point of reference. For the exceptional Lie superalgebras another point of reference is often more convenient, the consistent regrading K.

The regradings of the series are governed by a parameter r described in Section 1.9.1. All regradings are given in more or less suggestive notation, e.g., $\mathfrak{kas}^{\xi}(; 3\eta)$ means that having taken $\mathfrak{kas}^{\xi}$ as the point of reference, we set $\deg \eta = 0$ for each of the three η's (certainly, this imposes some conditions on the degrees of the other indeterminates). The exceptional grading E_H of $\mathfrak{h}_{\lambda}(2|2)$ is described in passing in the list of occasional isomorphisms (28).

2) Several algebras are "drop-outs" from the series, e.g., the $\mathfrak{svect}^{\circ}(1|n; r)$ are "drop-outs" from the series $\mathfrak{svect}(m|n; r)$ since the latter are not simple for $m = 1$ but contain the simple ideal $\mathfrak{svect}^{\circ}(1|n; r)$. Similarly, $\mathfrak{le}(n|n; r)$, $\mathfrak{b}_1{}^{\circ}(n|n+1; r)$ and $\mathfrak{b}_{\infty}{}^{\circ}(n|n+1; r)$ are "drop-outs" from the series $\mathfrak{b}_{\lambda}(n|n+1; r)$ corresponding to $\lambda = 0, 1$ and ∞, respectively. Though $\mathfrak{sm}$ is not a drop-out for the above reason, it is singled out by its divergence free property, hence, deserves a separate line.

The finite-dimensional Lie superalgebra $\mathfrak{h}(0|n)$ of hamiltonian vector fields is not simple either and contains a simple "traceless" ideal $\mathfrak{h}^{o}(0|n)$.

3) Observe that $\mathfrak{h}_{\lambda}(2|2; r) \simeq \mathfrak{b}_{\lambda}(2|3; r)$ ought to be considered as an exceptional series since it is a simultaneous deformation of the algebras preserving the even and the odd superanalogs of the symplectic structure.

Theorem. *The simple W-graded vectorial Lie superalgebras L constitute the following series* (24) – (26) *and five exceptional families of 15 individual algebras* (27). *They are pairwise nonisomorphic, as graded and filtered superalgebras, except for occasional isomorphisms* (28). *Parameters at which* dim *L becomes finite are marked by "FD".*

All these algebras are either Cartan prolongs or results of the generalized Cartan prolongations (described in Sections 1.6.2 and 1.6.3) and, therefore, are de-

termined by the terms L_i with $i \leq 0$ (or $i \leq 1$ in some cases). These terms are listed in Section 1.10 (and their form might drastically vary with n and r).

$\mathfrak{vect}(m\|n; r)$ for $(m\|n) \neq (0\|1)$ and $0 \leq r \leq n$; FD for $m = 0, n > 1$	
$\mathfrak{svect}(m\|n; r)$ for $m > 1, 0 \leq r \leq n$; FD for $m = 0$ and $n > 2$ $\mathfrak{svect}^{\circ}(1\|n; r)$ for $n > 1, 0 \leq r \leq n$	(24)
$\widetilde{\mathfrak{svect}}(0\|n)$ for $n > 2$ (FD)	

$\mathfrak{m}(n\|n+1; r)$ for $0 \leq r \leq n$ except $r = n-1$
$\mathfrak{sm}(n\|n+1; r)$ for $0 \leq r \leq n$ except $r = n-1$ and except $n = 3$
$\mathfrak{b}_{\lambda}(n\|n+1; r)$ for $n > 1$, where $\lambda \neq 0, 1, \infty$ and $0 \leq r \leq n$, $r \neq n-1$
$\mathfrak{b}_{\lambda}(2\|3; r)$, where $\lambda \neq 0, 1, \infty$ for $r = 0, 2$ and E (see sec. 1.9.1)
$\mathfrak{b}_1{}^{\circ}(n\|n+1; r)$ for $n > 1$ and $0 \leq r \leq n, r \neq n-1$
$\mathfrak{b}_{\infty}{}^{\circ}(n\|n+1; r)$ for $n > 1$ and $0 \leq r \leq n$ except $r = n-1$
$\mathfrak{le}(n\|n; r)$ for $n > 1$ and $0 \leq r \leq n$ except $r = n-1$
$\mathfrak{sle}^{\circ}(n\|n; r)$ for $n > 1$ and $0 \leq r \leq n, r \neq n-1$
$\widetilde{\mathfrak{sb}}_{\mu}(2^{n-1}-1\|2^{n-1})$ for $\mu \neq 0$ and $n > 2$

(25)

$\mathfrak{k}(2m+1\|n; r)$ for $0 \leq r \leq [\frac{n}{2}]$ unless $(m\|n) = (0\|2k)$ $\mathfrak{k}(1\|2k; r)$ for $0 \leq r \leq k$ except $r = k-1$	
$\mathfrak{h}(2m\|n; r)$ for $m > 0$ and $0 \leq r \leq [\frac{n}{2}]$ $\mathfrak{h}_{\lambda}(2\|2; r)$ for $\lambda \neq 0, 1, \infty$, and $r = 0, 1$ and E_H (see sec. 1.9.1) $\mathfrak{h}^{o}(0\|n)$ for $n > 3$ (FD)	(26)

Our original notation for exceptional simple vectorial superalgebras, although it reflects the way they are constructed and the geometry that is preserved, are rather long. On the other hand, to write just $\mathfrak{e}(\dim)$ is to create confusion: the superdimensions of the superspaces $(\mathcal{L}/\mathcal{L})^*$ on which the algebra $\mathcal{L}$ is realized coincide sometimes for different regradings.

Lie superalgebra	its regradings (shorthand)
$\mathfrak{vle}(4\vert 3; r)$, $r = 0, 1, K$	$\mathfrak{vle}(4\vert 3)$, $\mathfrak{vle}(5\vert 4)$, and $\mathfrak{vle}(3\vert 6)$
$\mathfrak{vas}(4\vert 4)$	$\mathfrak{vas}(4\vert 4)$
$\mathfrak{kas}(1\vert 6; r)$, $r = 0, 1, 3\xi, 3\eta$	$\mathfrak{kas}(1\vert 6)$, $\mathfrak{kas}(5\vert 5)$, $\mathfrak{kas}(4\vert 4)$, and $\mathfrak{kas}(4\vert 3)$
$\mathfrak{mb}(4\vert 5; r)$, $r = 0, 1, K$	$\mathfrak{mb}(4\vert 5)$, $\mathfrak{mb}(5\vert 6)$, and $\mathfrak{mb}(3\vert 8)$
$\mathfrak{ksle}(9\vert 6; r)$, $r = 0, 2, K, CK$	$\mathfrak{ksle}(9\vert 6)$, $\mathfrak{ksle}(11\vert 9)$, $\mathfrak{ksle}(5\vert 10)$, and $\mathfrak{ksle}(11\vert 9)$

(27)

Hereafter we inconsistently abbreviate $\mathfrak{m}(n|n+1; r)$, $\mathfrak{b}_\lambda(n|n+1; r)$, $\mathfrak{le}(n|n; r)$, etc., i.e., algebras from (S_m), to $\mathfrak{m}(n; r)$, $\mathfrak{b}_\lambda(n; r)$, etc., respectively. Sometimes instead of $\mathfrak{b}_\lambda(n; r)$, where $\lambda = \frac{2a}{n(a-b)} \in \mathbb{C} \cup \{\infty\}$, we write $\mathfrak{b}_{a,b}(n; r)$ for clarity, where parameters a, b, as natural as λ, are introduced in Section 0.7.

Occasional isomorphisms:

$$
\begin{aligned}
&\mathfrak{vect}(1|1) \cong \mathfrak{vect}(1|1; 1);\\
&\mathfrak{svect}(2|1) \cong \mathfrak{le}(2; 2);\quad \mathfrak{svect}(2|1; 1) \cong \mathfrak{le}(2)\\
&\mathfrak{sm}(n) \cong \mathfrak{b}_{2/(n-1)}(n);\ \text{ in particular, } \mathfrak{sm}(2) \cong \mathfrak{b}_2(2), \text{ and}\\
&\mathfrak{sm}(3) \cong \mathfrak{b}_1(3), \text{ hence, } \mathfrak{sm}(3) \text{ is not simple};\\
&\mathfrak{b}_{1/2}(2; 2) \cong \mathfrak{h}_{1/2}(2|2) = \mathfrak{h}(2|2);\quad \mathfrak{h}_\lambda(2|2) \cong \mathfrak{b}_\lambda(2; 2);\\
&\mathfrak{h}_\lambda(2|2; 1) \cong \mathfrak{b}_\lambda(2);\quad \mathfrak{h}_\lambda(2|2; E_H) \cong \mathfrak{b}_\lambda(2; E);\\
&\mathfrak{b}_{a,b}(2; E) \cong \mathfrak{b}_{-b,-a}(2) \cong \mathfrak{b}_{b,a}(2);\ \text{ for } a \neq b \text{ and } b \neq 0;\\
&\mathfrak{b}_{a,0}(2; E) \cong \mathfrak{le}(2) \text{ and } \mathfrak{b}_{a,a}(2; E) \cong \mathfrak{b}^\circ_\infty(2);\\
&\mathfrak{h}_\lambda(2|2) \cong \mathfrak{h}_{-1-\lambda}(2|2);\\
&\widetilde{\mathfrak{sb}}_\mu(2^{n-1} - 1|2^{n-1}) \cong \widetilde{\mathfrak{sb}}_\nu(2^{n-1} - 1|2^{n-1}) \text{ for } \mu\nu \neq 0.
\end{aligned}
\tag{28}
$$

Although $\mathfrak{b}_\lambda(2)$ and $\mathfrak{h}_\lambda(2|2)$ are isomorphic, we consider them separately since they arise as deformations of very distinct structures: of the odd and the even brackets, respectively.

Warning. Isomorphic abstract Lie superalgebras might be quite distinct as filtered: e.g., regradings provide us with isomorphisms ([ALSh])

$$\mathfrak{k}(1|2) \cong \mathfrak{vect}(1|1) \cong \mathfrak{m}(1) \text{ as abstract algebras.}$$

Observe that of the above three nonisomorphic filtered algebras only one is W-filtered.

Even if the regraded algebra can be realized by vector fields on a superspace of the same dimension, the structures preserved are totally different.

Remarks. 1) The excluded regradings $\mathfrak{k}(1|2k; k-1)$ and the ones they induce on $\mathfrak{b}_\lambda$, $\mathfrak{le}$ and $\mathfrak{sle}$ are often considered in applications: although these gradings/filtrations are not Weisfeiler ones over $\mathbb{C}$ because the $\mathfrak{g}_0$-module $\mathfrak{g}_{-1}$ is reducible, they are Weisfeiler ones for some real forms, and it is these forms that are considered.

2) $\widetilde{\mathfrak{svect}}(0|n)$, as well as $\widetilde{\mathfrak{sb}}_\mu(2^{n-1}-1|2^{n-1})$, depend on an odd parameter if n is odd.

3) The Lie superalgebras $\widetilde{\mathfrak{svect}}$ and $\mathfrak{h}^o$, as well as $\mathfrak{vect}$, $\mathfrak{svect}$ for $m=0$, are finite dimensional.

The above Lie superalgebras sometimes admit deformations that do not possess Weisfeiler filtrations. Some of these deformations are considered in detail in [L4], [Ko1], [Ko2], [LSh2].

1.9.1. Nonstandard realizations

The following are all the nonstandard gradings of the Lie superalgebras indicated. In particular, the gradings in the series $\mathfrak{vect}$ induce the gradings in the series $\mathfrak{svect}$, $\mathfrak{svect}^\circ$ and the exceptional algebras $\mathfrak{vle}(4|3)$ and $\mathfrak{vas}(4|4)$; the gradings in $\mathfrak{m}$ induce the gradings in $\mathfrak{b}_\lambda$, $\mathfrak{le}$, $\mathfrak{sle}$, $\mathfrak{sle}^\circ$, $\mathfrak{b}$, $\mathfrak{sb}$, $\mathfrak{sb}^\circ$ and the exceptional algebra $\mathfrak{mb}$; the gradings in $\mathfrak{k}$ induce the gradings in $\mathfrak{po}$, $\mathfrak{h}$ and the exceptional algebras $\mathfrak{kas}$ and $\mathfrak{ksle}$.

In what follows we consider $\mathfrak{k}(2n+1|m)$ as preserving the Pfaff equation $\tilde{\alpha}=0$, where

$$\tilde{\alpha}=dt+\sum_{i\leq n}(p_idq_i-q_idp_i)+\sum_{j\leq r}(\xi_jd\eta_j+\eta_jd\xi_j)+\sum_{k\geq m-2r}\theta_kd\theta_k.$$

The standard realizations correspond to $r=0$, and they are marked by $(*)$. Observe that the codimension of $\mathcal{L}_0$ attains its minimum in the standard realization.

Lie superalgebra	its $\mathbb{Z}$-grading
$\mathfrak{vect}(n\vert m;r)$, $0\leq r\leq m$	$\deg u_i=\deg\xi_j=1$ for any i, j $(*)$
	$\deg\xi_j=0$ for $1\leq j\leq r$; $\deg u_i=\deg\xi_{r+s}=1$ for any i, s
$\mathfrak{m}(n;r)$, $0\leq r\leq n$, $r\neq n-1$	$\deg\tau=2$, $\deg q_i=\deg\xi_i=1$ for any i $(*)$
	$\deg\tau=\deg q_i=1$, $\deg\xi_i=0$ for any i
	$\deg\tau=\deg q_i=2$, $\deg\xi_i=0$ for $1\leq i\leq r<n$; $\deg q_{r+j}=\deg\xi_{r+j}=1$ for any j
$\mathfrak{k}(2n+1\vert m;r)$, $0\leq r\leq[\frac{m}{2}]$, $r\neq k-1$ for $m=2k$ and $n=0$	$\deg t=2$, $(*)$ $\deg p_i=\deg q_i=\deg\xi_j=\deg\eta_j=\deg\theta_k=1$ for any i, j, k
	$\deg t=\deg\xi_i=2$, $\deg\eta_i=0$ for $1\leq i\leq r\leq[\frac{m}{2}]$; $\deg p_i=\deg q_i=\deg\theta_j=1$ for $j\geq 1$ and all i
$\mathfrak{k}(1\vert 2m;m)$	$\deg t=\deg\xi_i=1$, $\deg\eta_i=0$ for $1\leq i\leq m$

1.9.2. The W-regradings of exceptional algebras

Theorem ([Sh14], [CK2]). *The W-regradings of the exceptional simple vectorial Lie superalgebras are given by the following regradings of their "standard" ambients listed in Sect. 1.8; the non-Weisfeiler regradings of the ambients are underlined twice*:

1) $\mathfrak{vle}(4|3; r) = (\Pi(\Lambda(3)/\mathbb{C}\cdot 1), \mathfrak{cvect}(0|3))_* \subset \mathfrak{vect}(4|3), \quad r = 0, 1, K$;
$\underline{r=0}$: $\deg y = \deg u_i = \deg \xi_i = 1$;
$\underline{r=1}$: $\deg y = \deg \xi_1 = 0, \deg u_2 = \deg u_3 = \deg \xi_2 = \deg \xi_3 = 1, \deg u_1 = 2$;
$\underline{r=K}$: $\deg y = 0, \deg u_i = 2; \deg \xi_i = 1$;
2) $\mathfrak{vas}(4|4) = (\mathrm{spin}, \mathfrak{as})_* \subset \mathfrak{vect}(4|4)$;
3) $\mathfrak{kas} \subset \mathfrak{k}(1|6; r), \ r = 0, 1, 3\xi; \quad \mathfrak{kas}(; 3\eta) \subset \mathfrak{svect}(4|3)$;
$\underline{r=0}$: $\deg t = 2, \ \deg \eta_i = 1; \ \deg \xi_i = 1$;
$\underline{r=1}$: $\deg \xi_1 = 0, \ \deg \eta_1 = \deg t = 2, \ \deg \xi_2 = \deg \xi_3 = \deg \eta_2 = \deg \eta_3 = 1$;
$\underline{r=3\xi}$: $\deg \xi_i = 0, \ \deg \eta_i = \deg t = 1$;
$\underline{r=3\eta}$: $\deg \eta_i = 0, \deg \xi_i = \deg t = 1$;
4) $\mathfrak{mb}(4|5; r) = (\mathfrak{ab}(4), \mathfrak{cvect}(0|3))_*^m \subset \mathfrak{m}(4), \quad r = 0, 1, K$;
$\underline{r=0}$: $\deg \tau = 2, \ \deg u_i = \deg \xi_i = 1$ for $i = 0, 1, 2, 3$:
$\underline{r=1}$: $\deg \tau = \deg \xi_0 = \deg u_1 = 2, \ \deg u_2 = \deg u_3 = \deg \xi_2 = \deg \xi_3 = 1$; $\deg \xi_1 = \deg u_0 = 0$;
$\underline{r=K}$: $\deg \tau = \deg \xi_0 = 3, \ \deg u_0 = 0, \ \deg u_i = 2; \ \deg \xi_i = 1$ for $i > 0$;
5) $\mathfrak{ksle}(9|6; r) = (\mathfrak{hei}(8|6), \mathfrak{svect}_{3,4}(4))_*^k \subset \mathfrak{k}(9|6), \quad r = 0, 2, \mathrm{K}, \mathrm{CK}$:
$\underline{r=0}$: $\deg t = 2, \ \deg p_i = \deg q_i = \deg \xi_i = \deg \eta_i = 1$ for all i;
$\underline{r=2}$: $\deg t = \deg q_3 = \deg q_4 = \deg \eta_1 = 2, \ \deg q_1 = \deg q_2 = \deg p_1 = \deg p_2 = \deg \eta_2 = \deg \eta_3 = \deg \zeta_2 = \deg \zeta_3 = 1; \ \deg p_3 = \deg p_4 = \deg \zeta_1 = 0$;
$\underline{r=K}$: $\deg t = \deg q_i = 2, \ \deg p_i = 0; \ \deg \zeta_i = \deg \eta_i = 1$ for all i;
$\underline{r=CK}$: $\deg t = \deg q_1 = 3, \ \deg p_1 = 0; \ \deg q_2 = \deg q_3 = \deg q_4 = \deg \zeta_1 = \deg \zeta_2 = \deg \zeta_3 = 2; \ \deg p_2 = \deg p_3 = \deg p_4 = \deg \eta_1 = \deg \eta_2 = \deg \eta_3 = 1$.

Thus, from the point of view of classification of the W-filtered Lie superalgebras, there are five *families* of exceptional algebras consisting of 15 individual algebras.

1.9 Several first terms that determine the Cartan prolongs

To facilitate the comparison of various vectorial superalgebras, we offer the following table. The most interesting phenomena occur for extremal values of parameter r and small values of superdimension $m|n$.

The central element $z \in \mathfrak{g}_0$ is supposed to be chosen so that it acts on $\mathfrak{g}_k$ as $k \cdot \mathrm{id}$ and $\Lambda(r) = \mathbb{C}[\xi_1, \dots, \xi_r]$ is the Grassmann superalgebra generated by the ξ_i of degree 0 for all i.

Recall the range of the regrading parameter r: $0 \leq r \leq m$, where m is the number of odd indeterminates, except for the series $\mathfrak{k}$ and $\mathfrak{h}$ when $0 \leq r \leq [\frac{m}{2}]$, and $\mathfrak{m}(n)$, $\mathfrak{b}_\lambda(n)$, etc. when $0 \leq r \leq n$ with value $r = n-1$ excluded. We exclude certain values of r, namely, $r = k-1$ for $\mathfrak{k}(1|2k)$ and $r = n-1$ for $\mathfrak{m}(n)$ and their subalgebras, because for these values of r, the corresponding grading is not a W-grading: the $\mathfrak{g}_0$ module $\mathfrak{g}_{-1}$ is reducible.

N	$\mathfrak{g}$	$\mathfrak{g}_{-2}$	$\mathfrak{g}_{-1}$	$\mathfrak{g}_0$
1	$\mathfrak{vect}(n\vert m; r)$	—	$\mathrm{id}\otimes\Lambda(r)$	$\mathfrak{gl}(n\vert m-r)\otimes\Lambda(r)\ltimes\mathfrak{vect}(0\vert r)$
2	$\mathfrak{vect}(1\vert m; m)$	—	$\Lambda(m)$	$\Lambda(m)\ltimes\mathfrak{vect}(0\vert m)$
3	$\mathfrak{svect}(n\vert m; r), n\neq 1$	—	$\mathrm{id}\otimes\Lambda(r)$	$\mathfrak{sl}(n\vert m-r)\otimes\Lambda(r)\ltimes\mathfrak{vect}(0\vert r)$
4	$\mathfrak{svect}^\circ(1\vert m; r), r\neq m$	—	$\mathrm{id}\otimes Vol_0(0\vert r)$	$\mathfrak{sl}(n\vert m-r)\otimes\Lambda(r)\ltimes\mathfrak{vect}(0\vert r)$
5	$\mathfrak{svect}^\circ(1\vert m; m)$	—	$Vol_0(0\vert m)$	$\Lambda(m)\ltimes\mathfrak{svect}(0\vert m)$
5	$\mathfrak{svect}^\circ(1\vert m; m)$	—	$Vol_0(0\vert m)$	$\Lambda(m)\ltimes\mathfrak{svect}(0\vert m)$
	$\mathfrak{svect}^\circ(1\vert 2)$	—	$T_0(\bar{0})$	$\mathfrak{vect}(0\vert 2)\cong\mathfrak{sl}(1\vert 2)$
	$\mathfrak{svect}(2\vert 1)$	—	$\Pi(T_0(\bar{0}))$	$\mathfrak{vect}(0\vert 2)\cong\mathfrak{sl}(2\vert 1)$
6	$\mathfrak{h}(2n\vert m)$	—	id	$\mathfrak{osp}(m\vert 2n)$
7	$\mathfrak{h}(2n\vert m; r)$	$T_0(\bar{0})$	$\mathrm{id}\otimes\Lambda(r)$	$\mathfrak{osp}(m-2r\vert 2n)\otimes\Lambda(r)\ltimes\mathfrak{vect}(0\vert r)$
8	$\mathfrak{h}(2n\vert 2r; r)$	—	$\mathrm{id}\otimes\Lambda(r)$	$\mathfrak{sp}(2n)\otimes\Lambda(r)\ltimes\mathfrak{vect}(0\vert r)$
9	$\mathfrak{h}_\lambda(2\vert 2)$	—	$\Pi(Vol^\lambda(0\vert 2))$	$\mathfrak{osp}(2\vert 2)\cong\mathfrak{vect}(0\vert 2)$
10	$\mathfrak{h}_\lambda(2\vert 2; 1)$	—	$\mathrm{id}_{\mathfrak{sp}(2)}\otimes Vol^\lambda(0\vert 1)$	$\mathfrak{sp}(2)\otimes\Lambda(1)\ltimes\mathfrak{vect}(0\vert 1)$
11	$\mathfrak{k}(2n+1\vert m; r)$	$\Lambda(r)$	$\mathrm{id}\otimes\Lambda(r)$	$\mathfrak{cosp}(m-2r\vert 2n)\otimes\Lambda(r)\ltimes\mathfrak{vect}(0\vert r)$
12	$\mathfrak{k}(1\vert 2m; m)$	—	$\Lambda(m)$	$\Lambda(m)\ltimes\mathfrak{vect}(0\vert m)$
13	$\mathfrak{k}(1\vert 2m+1; m)$	$\Lambda(m)$	$\Pi(\Lambda(m))$	$\Lambda(m)\ltimes\mathfrak{vect}(0\vert m)$

We set $\widetilde{p(\mu)} \equiv n \pmod 2$, so μ can be odd indeterminate. The Lie superalgebras $\widetilde{\mathfrak{svect}}_\mu(0|n)$ are isomorphic for nonzero μ's; hence, so are the algebras $\widetilde{\mathfrak{sb}}_\mu(2^{n-1}-1|2^{n-1})$. So for n even we can set $\mu = 1$, while if μ is odd we should consider it as an additional indeterminate on which the coefficients depend.

N	$\mathfrak{g}$	$\mathfrak{g}_{-2}$	$\mathfrak{g}_{-1}$	$\mathfrak{g}_0$
14	$\mathfrak{m}(n; r)$	$\Pi(\Lambda(r))$	$\mathrm{id}\otimes\Lambda(r)$	$\mathfrak{cpe}(n-r)\otimes\Lambda(r)\ltimes\mathfrak{vect}(0\vert r)$
15	$\mathfrak{m}(n; n)$	—	$\Pi(\Lambda(n))$	$\Lambda(n)\ltimes\mathfrak{vect}(0\vert n)$
16	$\widetilde{\mathfrak{sb}}_\mu(2^{n-1}-1\vert 2^{n-1})$	—	$\frac{\Pi(Vol(0\vert n))}{\mathbb{C}(1+\mu\xi_1\dots\xi_n)vol(\xi)}$	$\widetilde{\mathfrak{svect}_\mu}(0\vert n)$

In what follows $\lambda = \frac{2a}{n(a-b)} \neq 0, 1, \infty$; the three exceptional cases (corresponding to the "drop-outs" $\mathfrak{le}(n)$, $\mathfrak{b}_1{}^\circ(n)$ and $\mathfrak{b}^\circ_\infty(n)$, respectively) are considered separately. The irreducibility condition of the $\mathfrak{g}_0$-module $\mathfrak{g}_{-1}$ for $\mathfrak{g} = \mathfrak{b}^\circ_\infty$ excludes $r = n-1$. The case $r = n-2$ is extra exceptional, so in the following tables, unless specified,

$$0 < r < n-2; \quad \text{additionally } a \neq b \text{ and } (a, b) \neq \alpha(n, n-2) \text{ for any } \alpha \in \mathbb{C}.$$

To further clarify the following tables, denote the superspace of the identity $k|k$-dimensional representation of $\mathfrak{spe}(k)$ by V; let $d = \mathrm{diag}(1_k, -1_k) \in \mathfrak{pe}(k)$. Let $W = V \otimes \Lambda(r)$ and $D \in \mathfrak{vect}(0|r)$. Let $\Xi = \xi_1 \cdots \xi_n \in \Lambda(\xi_1, \dots, \xi_n)$. Denote by T^r the representation of $\mathfrak{vect}(0|r)$ in $\mathfrak{spe}(n-r) \otimes \Lambda(r)$ given by the formula

$$T^r(D) = 1 \otimes D + d \otimes \frac{1}{n-r}\mathrm{div} D. \qquad (29)$$

In $\mathfrak{sle}^\circ(n; r)_0$ for $r \neq n-2$:

$$\begin{aligned}&\mathfrak{vect}(0|r) \text{ acts on the ideal } \mathfrak{spe}(n-r) \otimes \Lambda(r) \text{ via } T^r;\\ &\text{any } X \otimes f \in \mathfrak{spe}(n-r) \otimes \Lambda(r) \text{ acts in } \mathfrak{g}_{-1} \text{ as id} \otimes f \text{ and in } \mathfrak{g}_{-2} \text{ as } 0;\\ &\text{any } D \in \mathfrak{vect}(0|r) \text{ acts in } \mathfrak{g}_{-1} \text{ via } T^r \text{ and in } \mathfrak{g}_{-2} \text{ as } D.\end{aligned} \tag{30}$$

In $\mathfrak{sle}^\circ(n; n-2)_0$, observe that $\mathfrak{spe}(2) \cong \mathbb{C}(\mathrm{Le}_{q_1\xi_1 - q_2\xi_2}) \oplus \mathbb{C}\mathrm{Le}_{\xi_1\xi_2}$, whereas $\mathfrak{g}_{-2}$ and $\mathfrak{g}_{-1}$ are as above, for $r < n-2$. Set $\mathfrak{h} = \mathbb{C}(\mathrm{Le}_{q_1\xi_1 - q_2\xi_2})$. In this case

$$\mathfrak{g}_0 \cong \underline{\left(\mathfrak{h} \otimes \Lambda(n-2) \oplus \mathbb{C}\mathrm{Le}_{\xi_1\xi_2} \otimes (\Lambda(n-2) \setminus \mathbb{C}\xi_3 \cdots \xi_n)\right)} \oplus T^1(\mathfrak{vect}(0|n-2)). \tag{31}$$

The action of $\mathfrak{vect}(0|n-2)$, the quotient of $\mathfrak{g}_0$ modulo the underlined ideal, is via (30). In the subspace $\xi_1\xi_2 \otimes \Lambda(n-2) \subset \mathfrak{g}_0$ this action is the same as in the space of volume forms. So we can throw away Ξ, or, more correctly, take the irreducible submodule of forms with integral 0.

In the following table the terms "$\mathfrak{g}_{-i}$" denote the superspace isomorphic to the one in quotation marks but with the action given by formulas (30) and (31).

N	$\mathfrak{g}$	$\mathfrak{g}_{-2}$	$\mathfrak{g}_{-1}$	$\mathfrak{g}_0$
17	$\mathfrak{le}(n)$	—	id	$\mathfrak{pe}(n)$
18	$\mathfrak{le}(n; r)$	$\Pi(T_0(\vec{0}))$	$\mathrm{id} \otimes \Lambda(r)$	$\mathfrak{pe}(n-r) \otimes \Lambda(r) \oplus \mathfrak{vect}(0\|r)$
19	$\mathfrak{le}(n; n)$	—	$\Pi(T_0(\vec{0}))$	$\mathfrak{vect}(0\|n)$
20	$\mathfrak{sle}^\circ(n)$	—	id	$\mathfrak{spe}(n)$
21	$\mathfrak{sle}^\circ(n; r)$	"$\Pi(T_0(\vec{0}))$"	"$\mathrm{id} \otimes \Lambda(r)$"	$\mathfrak{spe}(n-r) \otimes$ "$\Lambda(r)$" $\oplus T^1(\mathfrak{vect}(0\|r))$
22	$\mathfrak{sle}^\circ(n; n-2)$	"$\Pi(T_0(\vec{0}))$"	"$\mathrm{id} \otimes \Lambda(r)$"	see (31)
23	$\mathfrak{sle}^\circ(n; n)$	—	$\Pi(T_0^0(\vec{0}))$	$\mathfrak{svect}(0\|n)$

Next, we consider $\mathfrak{b}_{a,b}(n; r)$ for $0 < r < n-2$ and $ar - bn \neq 0$; in particular, this excludes $\mathfrak{b}^\circ_\infty(n; n) = \mathfrak{b}^\circ_{a,a}(n; n)$ and $\mathfrak{b}^\circ_1(n; n-2) = \mathfrak{b}^\circ_{n,n-2}(n; n-2)$.

If z is the central element of $\mathfrak{cspe}(n-r)$ that acts on $\mathfrak{g}_{-1}$ as $-\mathrm{id}$, then

$$z \otimes \psi \text{ acts on } \mathfrak{g}_{-1} \text{ as } -\mathrm{id} \otimes \psi, \text{ and on } \mathfrak{g}_{-2} \text{ as } -2\mathrm{id} \otimes \psi. \tag{32}$$

Set

$$c = \frac{a}{ar - bn}.$$

Let $a \cdot \mathrm{str} \otimes \mathrm{id}$ be the representation of $\mathfrak{pe}(n-r) = \mathfrak{spe}(n-r) \oplus \mathbb{C}d$ which is $\mathrm{id}_{\mathfrak{spe}(n-r)}$ on $\mathfrak{spe}(n-r)$ and sends d to $2a\mathrm{id}$. Recall that $\mathbb{C}[k]$ is the 1-dimensional representation of $\mathfrak{g}_0$ where k is the value of the central element z from $\mathfrak{g}_0$ chosen so that $z|_{\mathfrak{g}_i} = i \cdot \mathrm{id}_{\mathfrak{g}_i}$.

N	$\mathfrak{g}$	$\mathfrak{g}_{-2}$	$\mathfrak{g}_{-1}$	$\mathfrak{g}_0$
24	$\mathfrak{b}_\lambda(n)$	$\Pi(\mathbb{C}[-2])$	id	$\mathfrak{spe}(n) \ltimes \mathbb{C}(az+bd)$
25	$\mathfrak{b}_\lambda(n;r)$	$\Pi((-c)\mathrm{str}) \otimes (Vol(0\|r))^{2c}$	$\left((-\frac{c}{2})\mathrm{str}) \otimes \mathrm{id}\right) \otimes (Vol(0\|r))^c$	$(\mathfrak{pe}(n-r) \otimes \Lambda(r)) \ltimes \mathfrak{vect}(0\|r)$
26	$\mathfrak{b}_\lambda(n;n)$	–	$\Pi(Vol^\lambda(0\|n))$	$\mathfrak{vect}(0\|n)$

In the following tables the terms "$\mathfrak{g}_i$" denote the superspace isomorphic to the one in quotation marks but with the action given by formula (32).

In the exceptional case $ar = bn$, i.e., $\lambda = \dfrac{2}{n-r}$, the $\mathfrak{vect}(0|r)$-action on the ideal $\mathfrak{cspe}(n-r) \otimes \Lambda(r) \ltimes \mathfrak{vect}(0|r)$ of $\mathfrak{g}_0$ and on $\mathfrak{g}_-$ is the same as for $\mathfrak{sle}^\circ$, see (30).

N	$\mathfrak{g}$	$\mathfrak{g}_{-2}$	$\mathfrak{g}_{-1}$	$\mathfrak{g}_0$
27	$\mathfrak{b}_{2/(n-r)}(n;r)$	$\Pi(\mathbb{C}) \otimes \Lambda(r)$	id $\otimes$ "$\Lambda(r)$"	$\mathfrak{cpe}(n-r) \otimes$ "$\Lambda(r)$" $\ltimes T^r(\mathfrak{vect}(0\|r))$
28	$\mathfrak{b}^\circ_\infty(n)$	$\Pi(\mathbb{C})$	id	$\mathfrak{spe}(n)_{a,a}$
29	$\mathfrak{b}^\circ_\infty(n;r)$	$\Pi(\mathbb{C}) \otimes \Lambda(r)$	id $\otimes \Lambda(r)$	$\left((\mathfrak{spe}(n-r)_{a,a}) \otimes \Lambda(r)\right) \ltimes \mathfrak{vect}(0\|r)$
30	$\mathfrak{b}^\circ_\infty(n;n), n>2$	–	$\Pi(\Lambda(n))$	$(\Lambda(n) \setminus \mathbb{C}\Xi) \ltimes \mathfrak{svect}(0\|n)$
31	$\mathfrak{b}^\circ_1(n)$	$\Pi(\mathbb{C})$	id	$\mathfrak{spe}(n)_{n,n-2}$
32	$\mathfrak{b}^\circ_1(n;r)$	"$\Pi(Vol_0(0\|r))$"	id $\otimes$ "$\Lambda(r)$"	$\left((\mathfrak{spe}(n-r)_{n,n-2}) \otimes \text{"}\Lambda(r)\text{"}\right) \ltimes T^r(\mathfrak{vect}(0\|r))$
33	$\mathfrak{b}^\circ_1(n;n-2)$	"$\Pi(T_0(\vec{0}))$"	id $\otimes$ "$\Lambda(r)$"	(31) for the above line with $\mathfrak{cspe}(2)$ instead of $\mathfrak{spe}(2)$
34	$\mathfrak{b}^\circ_1(n;n)$	–	$\Pi(Vol_0(0\|n))$	$\mathfrak{vect}(0\|n)$

1.10.1. The exceptional vectorial Lie subsuperalgebras

Observe that none of the simple W-graded vectorial Lie superalgebras is of depth > 3 and only two algebras are of depth 3: one is $\mathfrak{mb}(4|5;K)$, for which we have $\mathfrak{mb}(4|5;K)_{-3} \cong \Pi(\mathrm{id}_{\mathfrak{sl}(2)})$, and another, $\mathfrak{ksle}(9|6;CK) = \mathfrak{ksle}(11|9)$.

This $\mathfrak{ksle}(11|9)$ is the 15th exceptional simple vectorial Lie superalgebra; its nonpositive terms are as follows (we assume that the $\mathfrak{sl}(2)$- and $\mathfrak{sl}(3)$-modules are purely even):

$$\mathfrak{ksle}(11|9)_0 \simeq (\mathfrak{sl}(2) \oplus \mathfrak{sl}(3) \otimes \Lambda(1)) \ni \mathfrak{vect}(0|1);$$
$$\mathfrak{ksle}(11|9)_{-1} \simeq \mathrm{id}_{\mathfrak{sl}(2)} \otimes \left(\mathrm{id}_{\mathfrak{sl}(3)} \otimes \Lambda(1)\right);$$
$$\mathfrak{ksle}(11|9)_{-2} \simeq \mathrm{id}^*_{\mathfrak{sl}(3)} \otimes \Lambda(1);$$
$$\mathfrak{ksle}(11|9)_{-3} \simeq \Pi(\mathrm{id}_{\mathfrak{sl}(2)} \otimes \mathbb{C}).$$

Here are the terms $\mathfrak{g}_i$ for $i \leq 0$ of 14 of the 15 exceptional algebras, the last column gives $\dim \mathfrak{g}_-$:

$\mathfrak{g}$	$\mathfrak{g}_{-2}$	$\mathfrak{g}_{-1}$	$\mathfrak{g}_0$	$\dim\mathfrak{g}_-$
$\mathfrak{vle}(4\vert 3)$	–	$\Pi(\Lambda(3)/\mathbb{C}1)$	$\mathfrak{c}(\mathfrak{vect}(0\vert 3))$	$4\vert 3$
$\mathfrak{vle}(4\vert 3;\,1)$	$\mathbb{C}\cdot 1$	$\mathrm{id}\otimes\Lambda(2)$	$\mathfrak{c}(\mathfrak{sl}(2)\otimes\Lambda(2)\ltimes T^{1/2}(\mathfrak{vect}(0\vert 2))$	$5\vert 4$
$\mathfrak{vle}(4\vert 3;\,K)$	$\mathrm{id}_{\mathfrak{sl}(3)}$	$\mathrm{id}^*_{\mathfrak{sl}(3)}\otimes\mathrm{id}_{\mathfrak{sl}(2)}\otimes 1$	$\mathfrak{sl}(3)\oplus\mathfrak{sl}(2)\oplus\mathbb{C}z$	$3\vert 6$
$\mathfrak{vas}(4\vert 4)$	–	spin	$\mathfrak{as}$	$4\vert 4$
$\mathfrak{kas}$	$\mathbb{C}\cdot 1$	$\Pi(\mathrm{id})$	$\mathfrak{co}(6)$	$1\vert 6$
$\mathfrak{kas}(;\,1)$	$\Lambda(1)$	$\mathrm{id}_{\mathfrak{sl}(2)}\otimes(\mathrm{id}_{\mathfrak{gl}(2)}\otimes\Lambda(1))$	$(\mathfrak{sl}(2)\oplus\mathfrak{gl}(2)\otimes\Lambda(1))\ltimes\mathfrak{vect}(0\vert 1)$	$5\vert 5$
$\mathfrak{kas}(;\,3\xi)$	–	$\Lambda(3)$	$\Lambda(3)\oplus\mathfrak{sl}(1\vert 3)$	$4\vert 4$
$\mathfrak{kas}(;\,3\eta)$	–	$Vol_0(0\vert 3)$	$\mathfrak{c}(\mathfrak{vect}(0\vert 3))$	$4\vert 3$
$\mathfrak{mb}(4\vert 5)$	$\Pi(\mathbb{C}\cdot 1)$	$Vol(0\vert 3)$	$\mathfrak{c}(\mathfrak{vect}(0\vert 3))$	$4\vert 5$
$\mathfrak{mb}(4\vert 5;\,1)$	$\Lambda(2)/\mathbb{C}\cdot 1$	$\mathrm{id}_{\mathfrak{sl}(2)}\otimes\Lambda(2)$	$\mathfrak{c}(\mathfrak{sl}(2)\otimes\Lambda(2)\ltimes T^{1/2}(\mathfrak{vect}(0\vert 2))$	$5\vert 6$
$\mathfrak{mb}(4\vert 5;\,K)$	$\mathrm{id}_{\mathfrak{sl}(3)}$	$\Pi(\mathrm{id}^*_{\mathfrak{sl}(3)}\otimes\mathrm{id}_{\mathfrak{sl}(2)}\otimes\mathbb{C})$	$\mathfrak{sl}(3)\oplus\mathfrak{sl}(2)\oplus\mathbb{C}z$	$3\vert 8$
$\mathfrak{ksle}(9\vert 6)$	$\mathbb{C}\cdot 1$	$\Pi(T^0_0(\vec{0}))$	$\mathfrak{svect}(0\vert 4)_{3,4}$	$9\vert 6$
$\mathfrak{ksle}(9\vert 6;\,2)$	$\mathrm{id}_{\mathfrak{sl}(1\vert 3)}$	$\mathrm{id}_{\mathfrak{sl}(2)}\otimes\Lambda(3)$	$(\mathfrak{sl}(2)\otimes\Lambda(3))\ltimes\mathfrak{sl}(1\vert 3)$	$9\vert 11$
$\mathfrak{ksle}(9\vert 6;\,K)$	id	$\Pi(\Lambda^2(\mathrm{id}^*))$	$\mathfrak{sl}(5)$	$5\vert 10$

1.10.2. The exceptional Lie subsuperalgebra $\mathfrak{kas}$ of $\mathfrak{k}(1\vert 6)$

This Lie superalgebra is not determined by its nonpositive part and requires a closer study. The Lie superalgebra $\mathfrak{g}=\mathfrak{k}(1|2n)$ is generated by the functions from $\mathbb{C}[t,\xi_1,\dots,\xi_n,\eta_1,\dots,\eta_n]$. The standard $\mathbb{Z}$-grading of $\mathfrak{g}$ is induced by the $\mathbb{Z}$-grading of $\mathbb{C}[t,\xi,\eta]$ given by $\deg t=2$, $\deg\xi_i=\deg\eta_i=1$; namely, $\deg K_f=\deg f-2$. Clearly, in this grading $\mathfrak{g}$ is of depth 2. Let us consider the functions that generate the first several homogeneous components of $\mathfrak{g}=\bigoplus\limits_{i\geq -2}\mathfrak{g}_i$:

component	$\mathfrak{g}_{-2}$	$\mathfrak{g}_{-1}$	$\mathfrak{g}_0$	$\mathfrak{g}_1$
its generators	1	$\Lambda^1(\xi,\eta)$	$\Lambda^2(\xi,\eta)\oplus\mathbb{C}\cdot t$	$\Lambda^3(\xi,\eta)\oplus t\Lambda^1(\xi,\eta)$

As one can prove directly, the component $\mathfrak{g}_1$ generates the whole subalgebra $\mathfrak{g}_+$ of elements of positive degree. The component $\mathfrak{g}_1$ splits into two $\mathfrak{g}_0$-modules $\mathfrak{g}_{11}=\Lambda^3$ and $\mathfrak{g}_{12}=t\Lambda^1$. It is obvious that $\mathfrak{g}_{12}$ is always irreducible and the component $\mathfrak{g}_{11}$ is trivial for $n=1$.

The partial Cartan prolongs of $\mathfrak{g}_{11}$ and $\mathfrak{g}_{12}$ are well known:

$$(\mathfrak{g}_-\oplus\mathfrak{g}_0,\mathfrak{g}_{11})^{mk}_*\cong\mathfrak{po}(0|2n)\oplus\mathbb{C}\cdot K_t\cong\mathfrak{d}(\mathfrak{po}(0|2n));$$
$$(\mathfrak{g}_-\oplus\mathfrak{g}_0,\mathfrak{g}_{12})^{mk}_*=\mathfrak{g}_{-2}\oplus\mathfrak{g}_{-1}\oplus\mathfrak{g}_0\oplus\mathfrak{g}_{12}\oplus\mathbb{C}\cdot K_{t^2}\cong\mathfrak{osp}(2n|2).$$

Observe a remarkable property of $\mathfrak{k}(1|6)$: for this alone does the component $\mathfrak{g}_{11}$ split into two irreducible modules that we will denote $\mathfrak{g}^{\xi}_{11}$ and $\mathfrak{g}^{\eta}_{11}$: one is generated by $\xi_1\xi_2\xi_3$, the other by $\eta_1\eta_2\eta_3$.

Observe further that $\mathfrak{g}_0=\mathfrak{co}(6)\cong\mathfrak{gl}(4)$. As $\mathfrak{gl}(4)$-modules, $\mathfrak{g}^{\xi}_{11}$ and $\mathfrak{g}^{\eta}_{11}$ are the symmetric squares $S^2(\mathrm{id})$ and $S^2(\mathrm{id}^*)$ of the identity 4-dimensional representation and its dual, respectively.

Theorem ([Sh5], [Sh14]). *The Cartan prolong* $\mathfrak{kas}^\xi = (\mathfrak{g}_- \oplus \mathfrak{g}_0, \mathfrak{g}_{11}^\xi \oplus \mathfrak{g}_{12})_*^{mk}$ *is infinite dimensional and simple. It is isomorphic to* $\mathfrak{kas}^\eta = (\mathfrak{g}_- \oplus \mathfrak{g}_0, \mathfrak{g}_{11}^\eta \oplus \mathfrak{g}_{12})_*^{mk}$.

When it does not matter which of isomorphic algebras $\mathfrak{kas}^\xi \simeq \mathfrak{kas}^\eta$ to take we will write $\mathfrak{kas}$.

For their explicit presentation by generating functions and for related structures on the "stringy" Lie superalgebras, see [GLS1].

1.10 Grozman's theorem and a description of $\mathfrak{g}$ as $\mathfrak{g}_{\bar 0}$ and $\mathfrak{g}_{\bar 1}$

In [CK2] the exceptional algebras are described as $\mathfrak{g} = \mathfrak{g}_{\bar 0} \oplus \mathfrak{g}_{\bar 1}$. For most of the series such a description is of little value because each homogeneous component $\mathfrak{g}_{\bar 0}$ and $\mathfrak{g}_{\bar 1}$ has a complicated structure. For the exceptions (and for twisted polyvector fields) the situation is totally different!

Observe that apart from being beautiful, such a description is useful for the construction of Volichenko algebras, cf. [LS].

Recall in this relation a theorem of PG [Gr], that completely describes bilinear differential operators acting on tensor fields and invariant under all changes of coordinates. It turned out that almost all of the first order operators determine a Lie superalgebra on their domains. Some of these superalgebras are simple or close to simple. In the constructions below we use some of these invariant operators.

$\mathfrak{g} = \mathfrak{e}(5|10)$: $\mathfrak{g}_{\bar 0} = \mathfrak{svect}(5|0) \simeq d\Omega^3$, $\mathfrak{g}_{\bar 1} = \Pi(d\Omega^1)$ with the natural $\mathfrak{g}_{\bar 0}$-action on $\mathfrak{g}_{\bar 1}$ and the bracketing of odd elements being their product; we identify for any permutation $(ijklm)$ of (12345):

$$dx_i \wedge dx_j \wedge dx_k \wedge dx_l \otimes vol^{-1} = \operatorname{sign}(ijklm)\frac{\partial}{\partial x_m}.$$

$\mathfrak{g} = \mathfrak{vas}(4|4)$: $\mathfrak{g}_{\bar 0} = \mathfrak{vect}(4|0)$, and $\mathfrak{g}_{\bar 1} = \Omega^1 \otimes Vol^{-1/2}$ with the natural $\mathfrak{g}_{\bar 0}$-action on $\mathfrak{g}_{\bar 1}$ and the bracketing of odd elements being

$$[\omega_1 \otimes vol^{-1/2}, \omega_2 \otimes vol^{-1/2}] = (d\omega_1 \wedge \omega_2 + \omega_1 \wedge d\omega_2) \otimes vol^{-1}],$$

where we identify

$$dx_i \wedge dx_j \wedge dx_k \otimes vol^{-1} = \operatorname{sign}(ijkl)\frac{\partial}{\partial x_l} \text{ for any permutation } (ijkl) \text{ of } (1234).$$

$\mathfrak{g} = \mathfrak{vle}(3|6)$: $\mathfrak{g}_{\bar 0} = \mathfrak{vect}(3|0) \oplus \mathfrak{sl}(2)^{(1)}_{\geq 0}$, where $\mathfrak{g}^{(1)}_{\geq 0} = \mathfrak{g} \otimes \mathbb{C}[t]$, and $\mathfrak{g}_{\bar 1} = (\Omega^1 \otimes Vol^{-1/2}) \otimes \mathrm{id}_{\mathfrak{sl}(2)^{(1)}_{\geq 0}}$ with the natural $\mathfrak{g}_{\bar 0}$-action on $\mathfrak{g}_{\bar 1}$.

Recall that $\mathrm{id}_{\mathfrak{sl}(2)}$ is the irreducible $\mathfrak{sl}(2)$-module L^1 with highest weight 1; its tensor square splits into $L^2 \simeq \mathfrak{sl}(2)$ and the trivial module L^0; accordingly, denote by $v_1 \wedge v_2$ and $v_1 \bullet v_2$ the projections of $v_1 \otimes v_2 \in L^1 \otimes L^1$ onto the skew-symmetric and symmetric components, respectively. For $f_1, f_2 \in \Omega^0$, $\omega_1, \omega_2 \in \Omega^1$ and $v_1, v_2 \in L^1$ we set

$$\begin{aligned}&[(\omega_1 \otimes v_1)vol^{-1/2}, (\omega_2 \otimes v_2)vol^{-1/2}]\\ &= (\omega_1 \wedge \omega_2) \otimes (v_1 \wedge v_2) + d\omega_1 \wedge \omega_2 + \omega_1 \wedge d\omega_2) \otimes (v_1 \bullet v_2))\, vol^{-1},\end{aligned}$$

where we identify Ω^0 with $\Omega^3 \otimes_{\Omega^0} Vol^{-1}$ and $\Omega^2 \otimes_{\Omega^0} Vol^{-1}$ with $\mathfrak{vect}(3|0)$ by setting

$$dx_i \wedge dx_j \otimes vol^{-1} = \text{sign}(ijk)\frac{\partial}{\partial x_k} \text{ for any permutation } (ijk) \text{ of } (123).$$

$\underline{\mathfrak{g} = \mathfrak{mb}(3|8)}$: $\mathfrak{g}_{\bar 0} = \mathfrak{vect}(3|0) \oplus \mathfrak{sl}(2)^{(1)}_{\geq 0}$, and $\mathfrak{g}_{\bar 1} = \mathfrak{g}_{-1} \oplus \mathfrak{g}_1$, where

$$\mathfrak{g}_{-1} = \left(\Pi Vol^{-1/2}\right) \otimes \text{id}_{\mathfrak{sl}(2)^{(1)}_{\geq 0}} \text{ and } \mathfrak{g}_1 = \left(\Omega^1 \otimes Vol^{-1/2}\right) \otimes \text{id}_{\mathfrak{sl}(2)^{(1)}_{\geq 0}};$$

clearly, one can interchange $\mathfrak{g}_{\pm 1}$.

Multiplication is similar to that of $\mathfrak{g} = \mathfrak{vle}(3|6)$. For $f_1, f_2 \in \Omega^0, \omega_1, \omega_2 \in \Omega^1$ and $v_1, v_2 \in L^1$ we set

$$\begin{aligned}
&[(\omega_1 \otimes v_1)vol^{-1/2}, (\omega_2 \otimes v_2)vol^{-1/2}] = 0,\\
&[(f_1 \otimes v_1)vol^{-1/2}, (f_2 \otimes v_2)vol^{-1/2}] = (df_1 \wedge df_2) \otimes (v_1 \wedge v_2)vol^{-1},\\
&[(f_1 \otimes v_1)vol^{-1/2}, (\omega_1 \otimes v_2)vol^{-1/2}] =\\
&(f_1\omega_1 \otimes (v_1 \wedge v_2) + (df_1\omega_1 + f_1 d\omega_1) \otimes (v_1 \bullet v_2))\, vol^{-1}.
\end{aligned}$$

$\underline{\mathfrak{g} = \mathfrak{kas}}$: $\mathfrak{g}_{\bar 0} = \mathfrak{vect}(1|0) \oplus \mathfrak{sl}(4)^{(1)}_{\geq 0}$, and $\mathfrak{g}_{\bar 1} = \mathfrak{g}_{-1} \oplus \mathfrak{g}_1$, where $\mathfrak{g}_{-1} = \Pi\left(\Lambda^2(\text{id}_{\mathfrak{sl}(2)^{(1)}_{\geq 0}})\right)$ and $\mathfrak{g}_1 = \Pi\left(S^2(\text{id}_{\mathfrak{sl}(2)^{(1)}_{\geq 0}})\right)$; clearly, one can interchange $\mathfrak{g}_{\pm 1}$.

$\underline{\mathfrak{g} = \mathfrak{b}_\lambda(n;\bar n)}$: here $\bar n$ denotes the grading given by the formulas $\deg q_i = 0$, $\deg \xi_i = 1$ for $i = 1, \dots, n$. Then for $i = -1, 0, \dots, n-1$ we have $\mathfrak{g}_i = (\Pi^i(\Lambda^{i-1}(\mathfrak{vect}(n|0))) \otimes Vol^{-(i-1)\lambda}$.

Consider $n = 2$ more attentively. Clearly, one can interchange $\mathfrak{g}_{\pm 1}$; this possibility explains a mysterious isomorphism mentioned in (28); if $\lambda = -1 - \lambda$ we have additional automorphisms, whereas for $\lambda = \frac{1}{2}$ (and $\lambda = -\frac{3}{2}$) there is a nontrivial central extension missed in [Ko1], [Ko2] and studied in [LSh2].

2 Main result

Recall that we usually skip the wedge sign in the product of supercommuting differential forms; Π is the change of parity sign. The relations will be divided, as in the introduction: into Serre relations (S); lowest (for the positive part) or highest (for the negative part) weight relations, (LW) and (HW), respectively; other, new type relations (N_i), where i is the degree with respect to the generators z. Observe that (LW) and (HW) are of the same form as (S).

$\underline{\mathfrak{ksle}(5|10)}$ Set (positive generators):

$$X_i = x_i\frac{\partial}{\partial x_{i+1}} \quad \text{for } i = 1, 2, 3, 4, \text{ and } Z = \Pi x_5\, dx_4 dx_5.$$

Relations in $\mathfrak{ksle}_+$ are:

S	$[X_1, X_3] = 0,\quad [X_1, X_4] = 0,\quad \mathrm{ad}^2_{X_4} X_3 = 0,\quad [X_2, X_4] = 0,$
S	$\mathrm{ad}^2_{X_1} X_2 = 0,\quad \mathrm{ad}^2_{X_2} X_1 = 0,$
S	$\mathrm{ad}^2_{X_2} X_3 = 0,\quad \mathrm{ad}^2_{X_3} X_2 = 0,\quad \mathrm{ad}^2_{X_3} X_4 = 0,$
LW	$\mathrm{ad}^2_{X_3} Z = 0,\quad [X_1, Z] = 0,\quad [X_2, Z] = 0,\quad \mathrm{ad}^2_{X_4} Z = 0,$
N_2	$[Z, Z] = 0,\quad [[X_3, Z], [X_4, Z]] = 0,\quad [[X_4, [X_3, Z]], [[X_3, X_4], Z]] = 0,$
N_3	$[[[X_1, X_2], [X_3, Z]], [[Z, [X_2, X_3]], [Z, [X_3, X_4]]]] = 0.$

Set (negative generators):

$$Y_i = x_{i+1}\frac{\partial}{\partial x_i} \text{ for } i = 1, 2, 3, 4, \text{ and } Z = \Pi\, dx_1 dx_2.$$

The relations in $\mathfrak{ksle}_-$ are:

S	$[Y_1, Y_3] = 0,\quad [Y_1, Y_4] = 0,\quad [Y_2, Y_4] = 0,$
S	$\mathrm{ad}^2_{Y_1} Y_2 = 0,\quad \mathrm{ad}^2_{Y_2} Y_1 = 0,\quad \mathrm{ad}^2_{Y_2} Y_3 = 0,$
S	$\mathrm{ad}^2_{Y_3} Y_2 = 0,\quad \mathrm{ad}^2_{Y_3} Y_4 = 0,\quad \mathrm{ad}^2_{Y_4} Y_3 = 0,$
HW	$[Y_1, Z] = 0,\quad [Y_3, Z] = 0,\quad [Y_4, Z] = 0,\quad \mathrm{ad}^2_{Y_2} Z = 0,$
N_2	$[Z, Z] = 0;\quad [[[Y_2, Z], [Y_3, Y_4]], [[Z, [Y_1, Y_2]], [Z, [Y_2, Y_3]]]] = 0.$

$\mathfrak{mb}(4|5)$ Set (negative generators):

$$Y_1 = \xi_0,\quad Y_2 = q_2\xi_1,\quad Y_3 = q_3\xi_2,\quad Z = -q_0q_1 + \xi_2\xi_3.$$

The relations in $\mathfrak{mb}(4|5)_-$ are:

S	$[Y_1, Y_2] = 0,\quad [Y_1, Y_3] = 0,\quad \mathrm{ad}^2_{Y_3} Y_2 = 0,\quad \mathrm{ad}^2_{Y_2} Y_3 = 0,$
HW	$\mathrm{ad}^2_{Y_1} Z = 0,\quad \mathrm{ad}^2_{Y_2} Z = 0,\quad [Y_3, Z] = 0,$
N_2	$[Z, Z] = 0,\quad [[[Y_1, Z], [Y_2, Z]], [[Y_1, Z], [Z, [Y_2, Y_3]]]] = 0.$

Set (positive generators):

$$X_1 = {q_0}^2\xi_0 - q_0q_1\xi_1 - q_0q_2\xi_2 - q_0q_3\xi_3 + \tau q_0 + 2\xi_1\xi_2\xi_3,$$
$$X_2 = q_1\xi_2,\ X_3 = q_2\xi_3,\quad Z_1 = {q_3}^2,\quad Z_2 = \xi_0\xi_1.$$

The relations in $\mathfrak{mb}(4|5)_+$ are:

S	$[X_1, X_2]=0,\quad [X_1, X_3]=0,\quad \mathrm{ad}^2_{X_2}X_3=0,\quad \mathrm{ad}^2_{X_3}X_2=0,$
LW	$[X_2, Z_1]=0,\quad [X_3, Z_2]=0,\quad \mathrm{ad}^2_{X_1}Z_1=0,$
LW	$\mathrm{ad}^3_{X_3}Z_1=0,\quad \mathrm{ad}^2_{X_1}Z_2=0,\quad \mathrm{ad}^2_{X_2}Z_2=0,$
N_2	$[Z_1, Z_1]=0,\quad [Z_1, Z_2]=0,\quad [Z_2, Z_2]=0,$
N_2	$[Z_1, [X_3, [Z_1, X_3]]]=0,$
N_2	$[[X_1, Z_1], [X_3, Z_1]]=4\,[Z_2, [X_1, Z_1]],\quad [[X_1, Z_2], [X_2, Z_2]]=0;$
N_3	$2\,[[Z_2, [X_1, Z_1]], [[X_2, X_3], [X_3, Z_1]]]=$
	$-[[Z_1, [X_2, X_3]], [[X_1, Z_2], [X_3, Z_1]]];$
N_4	$[[[X_3, Z_1], [X_2, [X_1, Z_2]]], [[X_2, [X_1, Z_2]], [X_3, [X_3, Z_1]]]]=$
	$-4\,[[[X_1, Z_2], [X_2, X_3]], [[Z_2, [X_1, Z_1]], [Z_2, [X_2, X_3]]]]$
	$-[[[X_2, Z_2], [X_3, [X_1, Z_1]]], [[X_2, [X_1, Z_2]], [X_3, [X_3, Z_1]]]].$

<u>$\mathfrak{vle}(4|3; K)$</u> Set (negative generators):

$$\begin{gathered}
Y_1=-x_4{}^2\frac{\partial}{\partial x_4}-x_4x_5\frac{\partial}{\partial x_5}-x_4x_6\frac{\partial}{\partial x_6}-\\
x_4x_7\frac{\partial}{\partial x_7}+x_5x_6\frac{\partial}{\partial x_3}-x_5x_7\frac{\partial}{\partial x_2}+x_6x_7\frac{\partial}{\partial x_1},\\
Y_2=-x_2\frac{\partial}{\partial x_1}+x_5\frac{\partial}{\partial x_6},\quad Y_3=-x_3\frac{\partial}{\partial x_2}+x_6\frac{\partial}{\partial x_7},\quad Z=\frac{\partial}{\partial x_5}.
\end{gathered}$$

The relations in $\mathfrak{vle}(4|3; K)_-$ are:

S	$[Y_1, Y_2]=0,\quad [Y_1, Y_3]=0,\quad \mathrm{ad}^2_{Y_2}Y_3=0,\quad \mathrm{ad}^2_{Y_3}Y_2=0,$
HW	$\mathrm{ad}^2_{Y_1}Z=0,\quad \mathrm{ad}^2_{Y_2}Z=0,\quad [Y_3, Z]=0,$
N_2	$[Z, Z]=0,\quad [[Z, [Y_2, Y_3]], [[Y_1, Z], [Y_2, Z]]]=0.$

Set (positive generators):

$$\begin{gathered}
X_1=\frac{\partial}{\partial x_4},\quad X_2=-x_1\frac{\partial}{\partial x_2}+x_6\frac{\partial}{\partial x_5},\quad X_3=-x_2\frac{\partial}{\partial x_3}+x_7\frac{\partial}{\partial x_6},\\
Z_1=x_3x_4\frac{\partial}{\partial x_7}-x_3x_5\frac{\partial}{\partial x_2}+x_3x_6\frac{\partial}{\partial x_1}+x_5x_6\frac{\partial}{\partial x_7},\\
Z_2=x_2x_4\frac{\partial}{\partial x_7}-x_2x_5\frac{\partial}{\partial x_2}+x_2x_6\frac{\partial}{\partial x_1}+x_4x_5\frac{\partial}{\partial x_4}+x_5x_6\frac{\partial}{\partial x_6}-x_3x_4\frac{\partial}{\partial x_6}\\
-x_3x_5\frac{\partial}{\partial x_3}+x_3x_7\frac{\partial}{\partial x_1}+x_4x_5\frac{\partial}{\partial x_4}+x_5x_7\frac{\partial}{\partial x_7}.
\end{gathered}$$

The relations in $\mathfrak{vle}(4|3; K)_+$ are:

S	$[X_1, X_2]=0,\quad [X_1, X_3]=0,\quad \mathrm{ad}^2_{X_2}X_3=0,\quad \mathrm{ad}^2_{X_3}X_2=0,$
LW	$[X_2, Z_1]=0,\quad [X_3, Z_2]=0,\quad \mathrm{ad}^2_{X_1}Z_1=0,\quad \mathrm{ad}^2_{X_1}Z_2=0,$
LW	$\mathrm{ad}^3_{X_3}Z_1=0,\quad \mathrm{ad}^2_{X_2}Z_2=0,$

N_2	$[Z_1, Z_1]=0, \quad [Z_1, Z_2]=0, \quad [Z_2, Z_2]=0,$
N_2	$[Z_1, [X_3, [X_3, Z_1]]]=0,$
N_2	$[[X_1, Z_1], [X_3, Z_1]]=[Z_2, [X_1, Z_1]],$
N_2	$[[X_3, Z_1], [X_2, [X_1, Z_2]]]=2\,[[X_1, Z_2], [X_2, Z_2]]+$ $3\,[[X_1, Z_2], [Z_1, [X_2, X_3]]]-2\,[[X_2, X_3], [Z_2, [X_1, Z_1]]]-$ $2\,[[X_2, Z_2], [X_3, [X_1, Z_1]]];$
N_3	$[[Z_1, [X_2, X_3]], [Z_2, [X_1, Z_1]]]=-3\,[[X_2, Z_2], [Z_2, [X_1, Z_1]]];$
N_3	$[[Z_2, [X_2, X_3]], [[X_1, Z_2], [X_2, Z_2]]]=0,$
N_4	$[[Z_2, [X_1, Z_1]], [[X_1, Z_1], [X_2, Z_2]]]=0;$
N_4	$[[[X_3, Z_1], [Z_2, [X_2, X_3]]], [[X_2, [X_1, Z_2]], [X_3, [X_1, Z_1]]]]=$ $6\,[[[X_1, Z_2], [Z_1, [X_2, X_3]]], [[X_1, Z_2], [Z_2, [X_2, X_3]]]]+$ $4\,[[[X_1, Z_2], [Z_2, [X_2, X_3]]], [[X_2, X_3], [Z_2, [X_1, Z_1]]]]+$ $2\,[[[X_1, Z_2], [Z_2, [X_2, X_3]]], [[X_2, Z_2], [X_3, [X_1, Z_1]]]];$
N_7	$6\,[[[Z_2, [X_1, Z_1]], [[X_1, Z_2], [X_2, Z_2]]], [[Z_2, [X_2, X_3]], [[X_1, Z_1], [X_2, Z_2]]]]=$ $[[[X_1, Z_2], [X_2, Z_2]], [[[X_1, Z_2], [X_2, Z_2]], [[X_2, Z_2], [Z_2, [X_1, Z_1]]]]];$
N_8	$[[[Z_2, [X_1, Z_1]], [[X_1, Z_2], [X_2, Z_2]]], [[[X_1, Z_1], [X_2, Z_2]], [[X_1, Z_2], [X_2, Z_2]]]]=0.$

$\mathfrak{vas}(4|4)$ A preliminary step to the description of the presentation of $\mathfrak{vas}(4|4)$ is the description of

Generators and relations for $\mathfrak{as}$. We denote the positive generators by x and the negative by y; each even (odd) element has even (odd) index. Set

$$x_1=\eta_3, \quad x_2=\zeta_2\eta_1 \quad x_4=\zeta_3\eta_2, \qquad x_3=\zeta_1\eta_1\eta_3-\zeta_2\eta_2\eta_3$$

$$y_1=\zeta_3, \quad y_2=\zeta_2\eta_3 \quad y_4=\zeta_1\eta_2, \eta_3, \quad y_3=-\zeta_1\zeta_3\eta_1-\zeta_2\zeta_3\eta_2;$$

and let the Cartan subalgebra be spanned by

$$h_0=1, \quad h_i=\zeta_i\eta_i.$$

The relations for positive elements are:

S	$\mathrm{ad}^2_{x_2}x_4=0, \quad \mathrm{ad}^2_{x_4}x_2=0, \quad \mathrm{ad}^2_{x_4}x_1=0,$
LW	$[x_1, x_2]=0, \quad \mathrm{ad}^2_{x_4}x_3=0, \quad \mathrm{ad}^2_{x_2}x_3=0,$
N_2	$[x_1, x_1]=0, \quad [x_1, x_3]=0, \quad [x_3, x_3]=0,$
N_3	$2\,[x_3, [x_2, x_4]]=[x_4, [x_2, x_3]].$

The relations for negative elements are

$S \quad [y_1, y_1]=0, \quad [y_1, y_3]=0, \quad [y_2, y_4]=0,$

$S \quad [y_4, y_4]=0, \quad [y_4, y_3]=0, \quad [y_3, y_3]=0,$

$S \quad \mathrm{ad}^2_{y_2}y_1=0, \quad \mathrm{ad}^2_{y_2}y_3=0,$

$N \quad [[y_1, y_4], [y_4, [y_1, y_2]]]=0; \quad [[y_3, [y_1, y_2]], [[y_1, y_2], [y_1, y_4]]]=0.$

The weights with respect to the h_i are:

x_1	(0	0	0	1)	y_1	(0	0	0	−1)
x_2	(0	1	−1	0)	y_2	(0	0	0	−1)
x_4	(0	0	1	−1)	y_4	(0	−1	1	1)
x_3	(0	0	0	1)	y_3	(0	0	0	−1)

Other relations between positive and negative elements:

$$\begin{aligned}&[x_1, y_1] = -h_0, \quad [x_1, y_2] = 0, \quad [x_1, y_4] = 0,\\ &[x_1, y_3] = -h_1 - h_2, [x_2, y_1] = 0, \quad [x_2, y_2] = 0, \quad [x_2, y_4] = x_3,\\ &[x_2, y_3] = 0, \quad [x_4, y_1] = 0, \quad [x_4, y_2] = h_2 - h_3,\\ &[x_4, y_4] = 0, \quad [x_4, y_3] = 0, \quad [x_3, y_1] = -h_1 + h_2,\\ &[x_3, y_2] = 0, \quad [x_3, y_4] = 0, \quad [x_3, y_3] = 0.\end{aligned}$$

Generators and relations for $\mathfrak{vas}(4|4)$

We denote $\partial_i = \frac{\partial}{\partial x_i}$ and $\delta_i = \frac{\partial}{\partial \theta_i}$. Set (positive generators):

$$\begin{aligned}&X_1 = x_1\delta_4 - x_4\delta_1 + \theta_3\partial_2 - \theta_2\partial_3, \quad X_2 = x_2\partial_1 - \theta_1\delta_2,\\ &X_3 = x_3\delta_3, \quad X_4 = x_1\delta_4 + x_4\delta_1,\\ &Z = -x_1{}^2\partial_4 + 2x_1\theta_4\delta_1\end{aligned}$$

Relations in $\mathfrak{vas}_+$ (since only X_2 is even and the representations of $\mathfrak{g}_0$ in $\mathfrak{g}_{\pm 1}$ are not, strictly speaking either highest or lowest weight ones; all these relations are of (N) type; the same applies for $\mathfrak{vas}_-$ since only Y_2 and Y_3 are even):

$$\begin{aligned}&[X_1, X_1] = 0, \quad [X_1, X_4] = 0, \quad [X_2, X_3] = 0, \quad [X_3, X_3] = 0,\\ &[X_3, X_4] = 0, \quad [X_3, Z] = 0, \quad [X_4, X_4] = 0,\\ &[X_4, Z] = 3[X_1, Z], \quad \mathrm{ad}^2_{X_2} X_1 = 0,\\ &\mathrm{ad}^2_{X_2} X_4 = 0, \quad \mathrm{ad}^2_Z X_1 = 0,\\ &2[Z, [X_2, X_4]] = -[X_4, [X_2, Z]];\\ &\mathrm{ad}^2_Z X_2 = 0, \quad \mathrm{ad}^3_{X_2} Z = 0,\\ &[[X_1, X_3], [X_3, [X_1, X_2]]] = 0,\\ &[[X_4, [X_1, X_2]], [[X_1, X_2], [X_1, X_3]]] = 0,\\ &[[[X_1, X_2], [X_2, Z]], [[X_1, X_3], [X_2, Z]]] = 0.\end{aligned}$$

Generators in $\mathfrak{vas}(4|4)_-$. Set:

$$\begin{aligned}&Y_1 = x_2\delta_3 - x_3\delta_2 + \theta_4\partial_1 - \theta_1\partial_4, \quad Y_2 = x_2\partial_3 - \theta_3\delta_2,\\ &Y_3 = x_1\partial_2 - \theta_2\delta_1, \quad Y_4 = x_2\delta_3 + x_3\delta_2, \quad Z = \delta_4.\end{aligned}$$

Relations in $\mathfrak{vas}(4|4)_-$:

S	$\mathrm{ad}^2_{Y_2} Y_3 = 0, \quad \mathrm{ad}^2_{Y_3} Y_2 = 0,$
N	$\mathrm{ad}^2_{Y_3} Y_1 = 0, \quad \mathrm{ad}^2_{Y_2} Y_4 = 0, \quad \mathrm{ad}^2_{Y_3} Y_4 = 0,$
N	$[Y_1, Y_1] = 0, \quad [Y_1, Y_2] = 0, \quad [Y_1, Y_4] = 0, \quad [Y_2, Z] = 0,$
N	$[Y_3, Z] = 0, \quad [Y_4, Y_4] = 0, \quad [Y_4, Z] = 0, \quad [Z, Z] = 0,$
N	$2[Y_4, [Y_2, Y_3]] = [Y_3, [Y_2, Y_4]], \quad [[Y_1, Z], [Y_3, Y_4]] = [[Y_1, Y_3], [Y_1, Z]].$

$\underline{\mathfrak{kas}(4|3; K)}$ Set (negative generators):

$$Y_1 = \xi_2\eta_1, \quad Y_2 = \xi_3\eta_2, \quad Y_3 = \eta_2\eta_3, \quad Z = \xi_1.$$

The relations in $\mathfrak{kas}(4|3; K)_-$ are:

$$
\begin{array}{l|l}
S & [Y_2, Y_3] = 0, \quad \mathrm{ad}^2_{Y_2} Y_1 = 0, \quad \mathrm{ad}^2_{Y_3} Y_1 = 0, \\
S & \mathrm{ad}^2_{Y_1} Y_2 = 0, \quad \mathrm{ad}^2_{Y_1} Y_3 = 0, \\
HW & [Y_2, Z] = 0, \quad [Y_3, Z] = 0, \quad \mathrm{ad}^2_{Y_1} Z = 0, \\
N_2 & [Z, Z] = 0.
\end{array}
$$

Set (positive generators):

$$X_1 = \xi_1\eta_2, \quad X_2 = \xi_2\eta_3, \quad X_3 = \xi_2\xi_3, \quad Z_1 = t\eta_1, \quad Z_2 = \xi_3\eta_1\eta_2.$$

The relations in $\mathfrak{kas}(4|3; K)_+$ (computed up to deg ≤ 40) consist of Serre-type relations:

$$
\begin{array}{llll}
[X_1, Z_2] = 0, & [X_2, X_3] = 0, & [X_2, Z_1] = 0, & [X_3, Z_1] = 0, \\
[X_3, Z_2] = 0, & [Z_1, Z_1] = 0, & [Z_1, Z_2] = 0, & [Z_2, Z_2] = 0, \\
\mathrm{ad}^2_{X_1} X_2 = 0, & \mathrm{ad}^2_{X_1} X_3 = 0, & \mathrm{ad}^2_{X_1} Z_1 = 0, & \\
\mathrm{ad}^2_{X_2} X_1 = 0, & \mathrm{ad}^2_{X_3} X_1 = 0, & \mathrm{ad}^3_{X_2} Z_2 = 0, &
\end{array}
$$

and a new (awful) type of relations:

$[Z_2, [X_2, [X_2, Z_2]]] = 0,$

$[[[X_1, X_3], [X_2, Z_2]], [[X_1, Z_1], [X_2, Z_2]]] = -[[Z_1, [X_1, X_3]], [[X_1, Z_1], [X_2, Z_2]]],$

$4[[[Z_2, [X_1, X_2]], [Z_1, [X_3, [X_1, X_2]]]], [[Z_1, [X_1, X_3]], [[X_2, Z_2], [Z_1, [X_1, X_2]]]]] =$
$[[[Z_1, [X_1, X_2]], [Z_1, [X_1, X_3]]], [[Z_1, [X_1, X_3]], [[X_2, Z_2], [Z_1, [X_1, X_2]]]]]+$
$[[[Z_1, [X_1, X_2]], [Z_1, [X_1, X_3]]], [[Z_1, [X_3, [X_1, X_2]]], , [[X_1, Z_1], [X_2, Z_2]]]],$

$12[[[[X_1, Z_1], [X_2, Z_2]], [[Z_1, [X_1, X_2]], [Z_1, [X_1, X_3]]]], [[[X_1, Z_1], [X_3, [X_1, X_2]]], [[X_2, Z_2], [Z_1, [X_1, X_3]]]]] =$
$[[[Z_1, [X_1, X_2]], [Z_1, [X_1, X_3]]],$
$[[[Z_1, [X_1, X_2]], [Z_1, [X_1, X_3]]], [[Z_1, [X_1, X_3]], [[X_1, Z_1], [X_2, Z_2]]]]],$
$[[[[Z_1, [X_1, X_2]], [Z_1, [X_1, X_3]]], [[Z_2, [X_1, X_2]], [Z_1, [X_3, [X_1, X_2]]]]]],$

$24[[[Z_1, [X_1, X_2]], [Z_1, [X_1, X_3]]], [[Z_1, [X_1, X_3]], [[X_2, Z_2], [Z_1, [X_1, X_2]]]]]] =$
$[[[Z_1, [X_1, X_2]], [Z_1, [X_1, X_3]]], [[[Z_1, [X_1, X_2]], [Z_1, [X_1, X_3]]], [[[Z_1, [X_1, X_2]], [Z_1, [X_1, X_3]]].$

$24[[Z_1, [X_1, X_3]], [[X_2, Z_2], [Z_1, [X_1, X_2]]]]]]]+$
$[[[Z_1, [X_1, X_2]], [Z_1, [X_1, X_3]]], [[[Z_1, [X_1, X_2]], [Z_1, [X_1, X_3]]], [[[Z_1, [X_1, X_2]],$
$[Z_1, [X_1, X_3]]], [[Z_1, [X_3, [X_1, X_2]]], , [[X_1, Z_1], [X_2, Z_2]]]]]],$
$[[[[Z_1, [X_1, X_2]], [Z_1, [X_1, X_3]]], [[[X_1, Z_1], [X_2, Z_2]], [[Z_1, [X_1, X_2]], [Z_1, [X_1, X_3]]]]],$

$120[[[Z_1, [X_1, X_2]], [Z_1, [X_1, X_3]]], [[[X_1, Z_1], [X_3, [X_1, X_2]]], [[X_2, Z_2], [Z_1, [X_1, X_3]]]]]] =$
$[[[Z_1, [X_1, X_2]], [Z_1, [X_1, X_3]]], [[[Z_1, [X_1, X_2]], [Z_1, [X_1, X_3]]], [[[Z_1, [X_1, X_2]], [Z_1, [X_1, X_3]]],$
$[[[Z_1, [X_1, X_2]], [Z_1, [X_1, X_3]]], , [[Z_1, [X_1, X_3]], [[X_1, Z_1], [X_2, Z_2]]]]]]].$

$120[[[[Z_1, [X_1, X_2]], [Z_1, [X_1, X_3]]], [[[Z_1, [X_1, X_2]], [Z_1, [X_1, X_3]]], [[Z_2, [X_1, X_2]], [Z_1, [X_3, [X_1, X_2]]]]]]],$
$[[[Z_1, [X_1, X_2]], [Z_1, [X_1, X_3]]], [[[Z_1, [X_1, X_2]], [Z_1, [X_1, X_3]]], [[Z_1, [X_1, X_3]], [[X_2, Z_2], [Z_1, [X_1, X_2]]]]]]] =$
$[[[Z_1, [X_1, X_2]], [Z_1, [X_1, X_3]]], [[[Z_1, [X_1, X_2]], [Z_1, [X_1, X_3]]], [[[Z_1, [X_1, X_2]], [Z_1, [X_1, X_3]]],$

$120[[[Z_1, [X_1, X_2]], [Z_1, [X_1, X_3]]], [[[Z_1, [X_1, X_2]], [Z_1, [X_1, X_3]]], [[Z_1, [X_1, X_3]], [[X_2, Z_2], [Z_1, [X_1, X_2]]]]]]]]+$
$[[[Z_1, [X_1, X_2]], [Z_1, [X_1, X_3]]], [[[Z_1, [X_1, X_2]], [Z_1, [X_1, X_3]]], , [[[Z_1, [X_1, X_2]], [Z_1, [X_1, X_3]]];$
$[[[Z_1, [X_1, X_2]], [Z_1, [X_1, X_3]]], [[[Z_1, [X_1, X_2]], [Z_1, [X_1, X_3]]], , [[Z_1, [X_3, [X_1, X_2]]], [[X_1, Z_1], [X_2, Z_2]]]]]]]].$

Comment. The Lie superalgebra $\mathfrak{vas}$ has only one W-grading, so we consider it.

For the remaining exceptional simple Lie superalgebras $\mathfrak{g}$ we take the consistent grading. So, except for $\mathfrak{vas}$, our $\mathfrak{g}_0$ is a Lie algebra, moreover, a reductive one; $\mathfrak{g}_1$ generates the positive part of $\mathfrak{g}$, and $\mathfrak{g}_{-1}$ generates the negative part of $\mathfrak{g}$, moreover, $\mathfrak{g}_{-1}$ is an irreducible $\mathfrak{g}_0$-module. Therefore, $\mathfrak{g}$ is generated by the Chevalley generators $X_i^{\pm}$ and $H_i = [X_i^+, X_i^-]$ of $\mathfrak{g}_0$, the lowest weight vectors v_{λ_k} of $\mathfrak{g}_1$ and the highest weight vector w_Λ of $\mathfrak{g}_{-1}$. Some of the relations between these generators are already known: the Serre relations for $X_i^{\pm}$ and relations involving H_i; and also

$$(\mathrm{ad} X_i^+)^{\lambda_k(X_i^+)+1}(v_{\lambda_k}) = 0 \qquad (LW)$$

$$(\mathrm{ad} X_i^-)^{\Lambda(X_i^-)+1}(w_\Lambda) = 0 \qquad (HW)$$

and the weight relations $H_i(v_{\lambda_k}) = \lambda_k^{(i)} \cdot v_{\lambda_k}$, etc. Since finite-dimensional representations of reductive algebras are completely reducible, we have to compare $\mathfrak{g}_1 \wedge \mathfrak{g}_1$ and $\mathfrak{g}_2$ (the lowest weight vectors of $\mathfrak{g}_1 \wedge \mathfrak{g}_1 / \mathfrak{g}_2$ are relations) and similarly for the negative part. If this does not suffice, i.e., $\mathfrak{g}_1 \wedge \mathfrak{g}_1 \wedge \mathfrak{g}_1$ modulo relations is greater than $\mathfrak{g}_3$, we add new relations, etc.

For positive parts of simple vectorial Lie algebras we can use Fuchs' theorem with Kochetkov's correction (and direct calculations, as described above, for small dimensions). For the exceptional algebras we consider we have no general theory, so having obtained the whole $\mathfrak{g}_i$ and no new relations for several $i > 0$ we conjecture that we have everything. For $\mathfrak{kas}$ we keep getting new relations for every i considered.

3 Appendix

3.1 Linear algebra in superspaces. Generalities

A *superspace* is a $\mathbb{Z}/2$-graded space; for any superspace $V = V_{\bar 0} \oplus V_{\bar 1}$ denote by $\Pi(V)$ another copy of the same superspace: with the shifted parity, i.e., $(\Pi(V))_{\bar i} = V_{\bar i + \bar 1}$. The *superdimension* of V is $\dim V = p + q\varepsilon$, where $\varepsilon^2 = 1$ and $p = \dim V_{\bar 0}$, $q = \dim V_{\bar 1}$. (Usually, $\dim V$ is expressed as a pair (p, q) or $p|q$; this obscures the fact that $\dim V \otimes W = \dim V \cdot \dim W$; this fact is clear with the use of ε.)

A superspace structure in V induces a superspace structure in the space $\mathrm{End}(V)$. A *basis of a superspace* is always a basis consisting of *homogeneous* vectors; let $Par = (p_1, \dots, p_{\dim V})$ be an ordered collection of their parities. We call Par the *format* of (the basis of) V. A square *supermatrix* of format (size) Par is a $\dim V \times \dim V$ matrix whose ith row and ith column are of the same parity p_i.

One usually considers one of the simplest formats Par, e.g., Par of the form $(\bar 0, \dots, \bar 0; \bar 1, \dots, \bar 1)$ is denoted $(\dim V_{\bar 0}, \dim V_{\bar 1})$ and called *standard*. In this paper we can do without nonstandard formats. But they are vital in the classification of systems of simple roots that the reader might be interested in in connection with applications to q-quantization or integrable systems. Besides, systems of simple roots corresponding to distinct nonstandard formats are related by odd reflections

— analogs of our nonstandard regradings. (For an approach to superroots see [S2].)

The matrix unit E_{ij} is supposed to be of parity $p_i + p_j$ and the bracket of supermatrices (of the same format) is defined via the **Sign Rule**:

if something of parity p moves past something of parity q the sign $(-1)^{pq}$ *accrues; the formulas defined on homogeneous elements are extended to arbitrary ones via linearity.*

Examples. By setting $[X, Y] = XY - (-1)^{p(X)p(Y)} YX$ we get the notion of supercommutator and the ensuing notions of supercommutative superalgebra and Lie superalgebra (which in addition to superskew-commutativity satisfies the super Jacobi identity, i.e., the Jacobi identity amended with the Sign Rule). A derivation (better say, superderivation) of a superalgebra A is a linear map $D : A \longrightarrow A$ that satisfies the Leibniz rule (and the Sign rule)

$$D(ab) = D(a)b + (-1)^{p(D)p(a)} a D(b).$$

In particular, let $A = \mathbb{C}[x]$ be the free supercommutative polynomial superalgebra in $x = (x_1, \dots, x_n)$, where the superstructure is determined by the parities of the indeterminates: $p(x_i) = p_i$. Partial derivatives are defined (with the help of super Leibniz Rule) by the formulas $\frac{\partial x_i}{\partial x_j} = \delta_{i,j}$. Clearly, the collection $\mathfrak{der} A$ of all superdifferentiations of A is a Lie superalgebra whose elements are of the form $\sum f_i(x) \frac{\partial}{\partial x_i}$.

Observe that sometimes applying the Sign Rule requires some dexterity (we have to distinguish between two versions both of which turn into one, called skew- or anti-commutativity, in the nonsuper case):

$$\begin{aligned}
ba &= (-1)^{p(b)p(a)} ab && \text{(symmetry)}\\
ba &= -(-1)^{p(b)p(a)} ab && \text{(anti-symmetry)}\\
ba &= (-1)^{(p(b)+1)(p(a)+1)} ab && \text{(skew-symmetry)}\\
ba &= -(-1)^{(p(b)+1)(p(a)+1)} ab && \text{(antiskew-symmetry)}
\end{aligned}$$

In what follows the superskew-symmetric bilinear forms are named according to the above definitions.

Given the supercommutative superalgebra $\mathcal{F}$ of "functions" in the indeterminates x, define the supercommutative superalgebra Ω of differential forms as the polynomial algebra over $\mathcal{F}$ in dx, where $p(d) = \bar{1}$. Since dx is even for an odd x, we can consider not only polynomials in dx. Smooth or analytic functions in the differentials of the x are called *pseudodifferential forms* on the supermanifold with coordinates x, see [BL1]. We will need them to interpret $\mathfrak{h}_\lambda(2|2)$. The exterior differential is defined on (pseudo) differential forms by the formulas (mind Leibniz and Sign Rules):

$$d(x_i) = dx_i \text{ and } d^2 = 0.$$

The Lie derivative is defined (minding same Rules) by the formula

$$L_D(df) = (-1)^{p(D)} d(D(f)).$$

In particular, for f odd we have

$$L_D\left((df)^\lambda\right) = \lambda(-1)^{p(D)} d(D(f))(df)^{\lambda-1} \text{ for any } \lambda \in \mathbb{C}.$$

The *general linear* Lie superalgebra of all supermatrices of size Par, where $Par = (p_1, \dots)$ is the ordered collection of parities of the rows identical to that of the columns; usually, the *standard* (simplest) format $\mathfrak{gl}(\bar{0}, \dots, \bar{0}, \bar{1}, \dots, \bar{1})$ is abbreviated to $\mathfrak{gl}(\dim V_{\bar{0}} | \dim V_{\bar{1}})$. Any matrix from $\mathfrak{gl}(Par)$ can be expressed as the sum of its even and odd parts; in the standard format this is the following block expression:

$$\begin{pmatrix} A & B \\ C & D \end{pmatrix} = \begin{pmatrix} A & 0 \\ 0 & D \end{pmatrix} + \begin{pmatrix} 0 & B \\ C & 0 \end{pmatrix}, \quad p\left(\begin{pmatrix} A & 0 \\ 0 & D \end{pmatrix}\right) = \bar{0}, \; p\left(\begin{pmatrix} 0 & B \\ C & 0 \end{pmatrix}\right) = \bar{1}.$$

The *supertrace* is the map $\mathfrak{gl}(Par) \longrightarrow \mathbb{C}$, $(A_{ij}) \mapsto \sum (-1)^{p_i} A_{ii}$. Since $\operatorname{str}[x, y] = 0$, the subsuperspace of supertraceless matrices constitutes the *special linear* Lie subsuperalgebra $\mathfrak{sl}(Par)$.

There are, however, two super versions of $\mathfrak{gl}(n)$, not one. The other version is called the *queer* Lie superalgebra and is defined as the one that preserves the complex structure given by an *odd* operator J, i.e., is the centralizer $C(J)$ of J:

$$\mathfrak{q}(n) = C(J) = \{X \in \mathfrak{gl}(n|n) \mid [X, J] = 0\}, \text{ where } J^2 = -\mathrm{id}.$$

It is clear that by a change of basis we can reduce J to the form $J_{2n} = \begin{pmatrix} 0 & 1_n \\ -1 & 0 \end{pmatrix}$. In the standard format we have

$$\mathfrak{q}(n) = \left\{\begin{pmatrix} A & B \\ B & A \end{pmatrix}\right\}.$$

On $\mathfrak{q}(n)$, the *queertrace* is defined: $\operatorname{qtr} : \begin{pmatrix} A & B \\ B & A \end{pmatrix} \mapsto \operatorname{tr} B$. Denote by $\mathfrak{sq}(n)$ the Lie superalgebra of *queertraceless* matrices.

Observe that the identity representations of $\mathfrak{q}$ and $\mathfrak{sq}$ in V, though irreducible in supersetting, are not irreducible in the nongraded sense: take homogeneous (with respect to parity) and linearly independent vectors $v_1, \dots, v_n$ from V; then $Span(v_1 + J(v_1), \dots, v_n + J(v_n))$ is an invariant subspace of V which is not a subsuperspace.

A representation is *irreducible* of *general type* or just of *G-type* if there is no invariant subspace, otherwise it is called *irreducible of Q-type* (Q is after the general queer Lie superalgebra — a specifically "superish" analog of $\mathfrak{gl}$); an irreducible representation of Q-type has no invariant sub*super*space but *has* an invariant subspace.

Lie superalgebras that preserve bilinear forms: two types. To a linear map $F : V \longrightarrow W$ of superspaces there corresponds the dual map $F^* : W^* \longrightarrow V^*$ between the dual superspaces. In a basis consisting of vectors v_i of format Par, the formula $F(v_j) = \sum\limits_i v_i A_{ij}$ assigns to F the supermatrix A. In the dual bases, the *supertransposed* matrix A^{st} corresponds to F^*:

$$(A^{st})_{ij} = (-1)^{(p_i+p_j)(p_i+p(A))} A_{ji}.$$

The supermatrices $X \in \mathfrak{gl}(Par)$ such that

$$X^{st} B + (-1)^{p(X)p(B)} BX = 0 \quad \text{for an homogeneous matrix } B \in \mathfrak{gl}(Par)$$

constitute the Lie superalgebra $\mathfrak{aut}(B)$ that preserves the bilinear form B^f on V whose matrix B is given by the formula $B_{ij} = (-1)^{p(B^f)p(v_i)} B^f(v_i, v_j)$ for the basis vectors v_i.

Recall that the *supersymmetry* of the homogeneous form B^f means that its matrix B satisfies the condition $B^u = B$, where for the matrix $B = \begin{pmatrix} R & S \\ T & U \end{pmatrix}$ we set $B^u = \begin{pmatrix} R^t & (-1)^{p(B)} T^t \\ (-1)^{p(B)} S^t & -U^t \end{pmatrix}$. Similarly, *skew-supersymmetry* of B means that $B^u = -B$. Thus, we see that the *upsetting* of bilinear forms $u : \mathrm{Bil}(V, W) \longrightarrow \mathrm{Bil}(W, V)$, which for *spaces* and when $V = W$ is expressed on matrices in terms of transposition, is a new operation.

The most popular canonical forms of the nondegenerate supersymmetric form are the ones whose supermatrices in the standard format are the following canonical ones, B_{ev} or B'_{ev}:

$$B'_{ev}(m|2n) = \begin{pmatrix} 1_m & 0 \\ 0 & J_{2n} \end{pmatrix}, \quad \text{where } J_{2n} = \begin{pmatrix} 0 & 1_n \\ -1_n & 0 \end{pmatrix},$$

or

$$B_{ev}(m|2n) = \begin{pmatrix} \mathrm{antidiag}(1, \dots, 1) & 0 \\ 0 & J_{2n} \end{pmatrix}.$$

The usual notation for $\mathfrak{aut}(B_{ev}(m|2n))$ is $\mathfrak{osp}(m|2n)$ or, more precisely, $\mathfrak{osp}^{sy}(m|2n)$. Observe that the passage from V to $\Pi(V)$ sends the supersymmetric forms to superskew-symmetric ones, preserved by the "symplectico-orthogonal" Lie superalgebra, $\mathfrak{spo}(2n|m)$ or, more presisely, $\mathfrak{osp}^{sk}(m|2n)$, which is isomorphic to $\mathfrak{osp}^{sy}(m|2n)$ but has a different matrix realization.

In the standard format the matrix realizations of these algebras are

$$\mathfrak{osp}(m|2n) = \left\{ \begin{pmatrix} E & Y & X^t \\ X & A & B \\ -Y^t & C & -A^t \end{pmatrix} \right\}; \quad \mathfrak{osp}^{sk}(m|2n) = \left\{ \begin{pmatrix} A & B & X \\ C & -A^t & Y^t \\ Y & -X^t & E \end{pmatrix} \right\},$$
$$\text{where } \begin{pmatrix} A & B \\ C & -A^t \end{pmatrix} \in \mathfrak{sp}(2n), \quad E \in \mathfrak{o}(m).$$

A nondegenerate supersymmetric odd bilinear form $B_{odd}(n|n)$ can be reduced to a canonical form whose matrix in the standard format is J_{2n}. A canonical form of the superskew odd nondegenerate form in the standard format is $\Pi_{2n} = \begin{pmatrix} 0 & 1_n \\ 1_n & 0 \end{pmatrix}$. The usual notation for $\mathfrak{aut}(B_{odd}(Par))$ is $\mathfrak{pe}(Par)$. The passage from V to $\Pi(V)$ establishes an isomorphism $\mathfrak{pe}^{sy}(Par) \cong \mathfrak{pe}^{sk}(Par)$. This Lie superalgebra is called, as A. Weil suggested, *periplectic*. The matrix realizations in the standard format of these superalgebras is shorthanded to:

$$\mathfrak{pe}^{sy}(n) = \left\{ \begin{pmatrix} A & B \\ C & -A^t \end{pmatrix}, \text{ where } B = -B^t,\ C = C^t \right\};$$
$$\mathfrak{pe}^{sk}(n) = \left\{ \begin{pmatrix} A & B \\ C & -A^t \end{pmatrix}, \text{ where } B = B^t,\ C = -C^t \right\}.$$

Observe that though the Lie superalgebras $\mathfrak{osp}^{sy}(m|2n)$ and $\mathfrak{pe}^{sk}(2n|m)$, as well as $\mathfrak{pe}^{sy}(n)$ and $\mathfrak{pe}^{sk}(n)$, are isomorphic, the difference between them is sometimes crucial, see [Sh5].

The *special periplectic* superalgebra is $\mathfrak{spe}(n) = \{X \in \mathfrak{pe}(n) \mid \mathrm{str} X = 0\}$. Also of particular interest to us will be

$$\mathfrak{spe}(n)_{a,b} = \mathfrak{spe}(n) \ni \mathbb{C}(az + bd), \text{ where } z = 1_{2n}, d = \mathrm{diag}(1_n, -1_n).$$

3.2 What a Lie superalgebra is

Dealing with superalgebras it sometimes becomes useful to know their definition. Lie superalgebras were distinguished in topology in the 1930's or earlier. So when somebody offers a "better than usual" definition of a notion which seemed to have been established about 70 years ago this might look strange, to say the least. Nevertheless, the answer to the question "what is a Lie superalgebra?" is still not common knowledge. Indeed, the naive definition ("apply the Sign Rule to the definition of Lie algebra") is manifestly inadequate for considering the (singular) supervarieties of deformations and applying representation theory to mathematical physics, for example, in the study of the coadjoint representation of a Lie supergroup which can act on a supermanifold but never on a superspace (an object from another category). So, to deform Lie superalgebras and apply group-theoretical methods in a "super" setting, we must be able to recover a supermanifold from a superspace, and vice-versa.

A proper definition of Lie superalgebras is as follows, cf. [L3]–[L5]. The *Lie superalgebra* in the category of supermanifolds corresponding to the "naive" Lie superalgebra $L = L_{\bar{0}} \oplus L_{\bar{1}}$ is a linear supermanifold $\mathcal{L} = (L_{\bar{0}}, \mathcal{O})$, where the sheaf of functions $\mathcal{O}$ consists of functions on $L_{\bar{0}}$ with values in the Grassmann superalgebra on $L_{\bar{1}}^*$; this supermanifold should be such that for "any" (say, finitely generated, or from some other appropriate category) supercommutative superalgebra C, the space $\mathcal{L}(C) = \mathrm{Hom}(\mathrm{Spec} C, \mathcal{L})$, called *the space of C-points of $\mathcal{L}$*, is a Lie algebra and the correspondence $C \longrightarrow \mathcal{L}(C)$ is a functor in C. (A. Weil introduced this approach in algebraic geometry in 1954; in the super setting it is called *the language of points* or *families*, see [D], [L4].) This definition might look terribly complicated, but fortunately one can show that the correspondence $\mathcal{L} \longleftrightarrow L$ is one-to-one and the Lie algebra $\mathcal{L}(C)$, also denoted $L(C)$, admits a very simple description: $L(C) = (L \otimes C)_{\bar{0}}$.

A *Lie superalgebra homomorphism* $\rho : L_1 \longrightarrow L_2$ in these terms is a functor morphism, i.e., a collection of Lie algebra homomorphisms $\rho_C : L_1(C) \longrightarrow L_2(C)$ compatible with morphisms of supercommutative superalgebras $C \longrightarrow C'$. In particular, a *representation* of a Lie superalgebra L in a superspace V is a homomorphism $\rho : L \longrightarrow \mathfrak{gl}(V)$, i.e., a collection of Lie algebra homomorphisms $\rho_C : L(C) \longrightarrow (\mathfrak{gl}(V) \otimes C)_{\bar{0}}$.

Example. Consider a representation $\rho : \mathfrak{g} \longrightarrow \mathfrak{gl}(V)$. The tangent space of the moduli superspace of deformations of ρ is isomorphic to $H^1(\mathfrak{g}; V \otimes V^*)$. For example, if $\mathfrak{g}$ is the $0|n$-dimensional (i.e., purely odd) Lie superalgebra (with the only bracket possible: identically equal to zero), its only irreducible representations are the trivial one, $\mathbf{1}$, and $\Pi(\mathbf{1})$. Clearly, $\mathbf{1} \otimes \mathbf{1}^* \simeq \Pi(\mathbf{1}) \otimes \Pi(\mathbf{1})^* \simeq \mathbf{1}$, and because the superalgebra is commutative, the differential in the cochain complex is trivial. Therefore, $H^1(\mathfrak{g}; \mathbf{1}) = E^1(\mathfrak{g}^*) \simeq \mathfrak{g}^*$, so there are $\dim \mathfrak{g}$ odd parameters of deformations of the trivial representation. If we consider $\mathfrak{g}$ "naively" all of the odd parameters will be lost.

Which of these infinitesimal deformations can be extended to a global one is a separate much tougher question, usually solved *ad hoc*.

Examples that lucidly illustrate why one should always remember that a Lie superalgebra is not a mere linear superspace but a linear supermanifold are, e.g., deformations $\widetilde{\mathfrak{svect}}(0|2n+1)$ and $\mathfrak{sb}_\mu(2^{2n}-1|2^{2n})$ with odd parameters considered below and viewed as Lie algebras. In the category of supermanifolds these are simple Lie superalgebras.

3.3 Projectivization

If $\mathfrak{s}$ is a Lie algebra of scalar matrices, and $\mathfrak{g} \subset \mathfrak{gl}(n|n)$ is a Lie subsuperalgebra containing $\mathfrak{s}$, then the *projective* Lie superalgebra of type $\mathfrak{g}$ is $\mathfrak{pg} = \mathfrak{g}/\mathfrak{s}$. Lie superalgebras $\mathfrak{g}_1 \odot \mathfrak{g}_2$ described in sect. 3.1 are also projective.

Projectivization sometimes leads to new Lie superalgebras, for example: $\mathfrak{pgl}(n|n)$, $\mathfrak{psl}(n|n)$, $\mathfrak{pq}(n)$, $\mathfrak{psq}(n)$; whereas $\mathfrak{pgl}(p|q) \cong \mathfrak{sl}(p|q)$ if $p \neq q$.

3.4 What a semisimple Lie superalgebra is

These algebras are needed in the description of primitive Lie superalgebras of vector fields — a geometrically natural problem though wild for Lie superalgebras, see [ALSh]. Recall that a Lie superalgebra $\mathfrak{g}$ without proper ideals and of dimension > 1 is called *simple*. Examples: $\mathfrak{sl}(m|n)$ for $m > n \geq 1$, $\mathfrak{psl}(n|n)$ for $n > 1$, $\mathfrak{psq}(n)$ for $n > 2$, $\mathfrak{osp}(m|2n)$ for $mn \neq 0$ and $\mathfrak{spe}(n)$ for $n > 2$.

We will not need the remaining simple finite-dimensional Lie superalgebras of non-vectorial type. These superalgebras, first discovered by I. Kaplansky (a 1975-preprint, [K1C]) are $\mathfrak{osp}_\alpha(4|2)$, the deformations of $\mathfrak{osp}(4|2)$, and two exceptions that we denote by $\mathfrak{ag}_2$ and $\mathfrak{ab}_3$. For their description we refer to [K], [FK], [Sud], see also [GL2]. For the description of the system of simple roots see [Ka4] completed in [vdL, S1, S2].

We say that $\mathfrak{h}$ is *almost simple* if it can be sandwiched (non-strictly) between a simple Lie superalgebra $\mathfrak{s}$ and the Lie superalgebra $\mathfrak{der}\,\mathfrak{s}$ of derivations of $\mathfrak{s}$: $\mathfrak{s} \subset \mathfrak{h} \subset \mathfrak{der}\,\mathfrak{s}$.

By definition, $\mathfrak{g}$ is *semisimple* if its radical is zero. Literally following the description of semisimple Lie algebras over fields of prime characteristic, V. Kac [Ka4] gave the following description of semisimple Lie superalgebras. Let $\mathfrak{s}_1$, $\dots$, $\mathfrak{s}_k$ be simple Lie superalgebras, let $n_1, \dots, n_k$ be *pairs* of nonnegative integers $n_j = (n_j^{\bar 0}, n_j^{\bar 1})$, let $\mathcal{F}(n_j)$ be the supercommutative superalgebra of polynomials in $n_j^{\bar 0}$ even and $n_j^{\bar 1}$ odd indeterminates, and $\mathfrak{s} = \bigoplus\limits_j \left(\mathfrak{s}_j \otimes \mathcal{F}(n_j)\right)$. Then

$$\mathfrak{der}\,\mathfrak{s} = \bigoplus_j \left((\mathfrak{der}\,\mathfrak{s}_j) \otimes \mathcal{F}(n_j) \ni \mathrm{id}_{\mathfrak{s}_j} \otimes \mathfrak{vect}(n_j)\right). \tag{35}$$

Let $\mathfrak{g}$ be a subalgebra of $\mathfrak{der}\mathfrak{s}$ containing $\mathfrak{s}$. If the projection of $\mathfrak{g}$ on $\mathrm{id}_{\mathfrak{s}_j} \otimes \mathfrak{vect}(n_j)_{-1}$ *is onto for each j, then $\mathfrak{g}$ is semisimple, and all semisimple Lie superalgebras arise in the manner indicated,* i.e., as sums of *almost simple superalgebras* corresponding to the summands of (35).

3.5 A. Sergeev's central extension

In the 70's A. Sergeev proved that there is just one nontrivial central extension of $\mathfrak{spe}(n)$. It exists only for $n = 4$ and we denote it by $\mathfrak{as}$. Let us represent an arbitrary element $A \in \mathfrak{as}$ as a pair $A = x + d \cdot z$, where $x \in \mathfrak{spe}(4)$, $d \in \mathbb{C}$ and z is the central element. The bracket in $\mathfrak{as}$ is given by the formula

$$\begin{gathered}\left[\begin{pmatrix} a & b \\ c & -a^t \end{pmatrix} + d \cdot z, \begin{pmatrix} a' & b' \\ c' & -a'^t \end{pmatrix} + d' \cdot z\right] = \\ \left[\begin{pmatrix} a & b \\ c & -a^t \end{pmatrix}, \begin{pmatrix} a' & b' \\ c' & -a'^t \end{pmatrix}\right] + \operatorname{tr} c\tilde{c}' \cdot z,\end{gathered} \tag{36}$$

where $\tilde{}$ is extended via linearity from matrices $c_{ij} = E_{ij} - E_{ji}$ on which $\tilde{c}_{ij} = c_{kl}$ for any even permutation $(1234) \mapsto (ijkl)$.

The Lie superalgebra $\mathfrak{as}$ can also be described with the help of the spinor representation. For this we need several vectorial superalgebras. Consider $\mathfrak{po}(0|6)$, the Lie superalgebra whose superspace is the Grassmann superalgebra $\Lambda(\xi, \eta)$ generated by $\xi_1, \xi_2, \xi_3, \eta_1, \eta_2, \eta_3$ and whose bracket is the Poisson bracket. Recall that $\mathfrak{h}(0|6) = Span(H_f \mid f \in \Lambda(\xi, \eta))$.

Now, observe that $\mathfrak{spe}(4)$ can be embedded into $\mathfrak{h}(0|6)$. Indeed, setting $\deg \xi_i = \deg \eta_i = 1$ for all i we introduce a $\mathbb{Z}$-grading on $\Lambda(\xi, \eta)$ which, in turn, induces a $\mathbb{Z}$-grading on $\mathfrak{h}(0|6)$ of the form $\mathfrak{h}(0|6) = \underset{i\geq -1}{\oplus} \mathfrak{h}(0|6)_i$. Since $\mathfrak{sl}(4) \cong \mathfrak{o}(6)$, we can identify $\mathfrak{spe}(4)_0$ with $\mathfrak{h}(0|6)_0$.

It is not difficult to see that the elements of degree -1 in the standard gradings of $\mathfrak{spe}(4)$ and $\mathfrak{h}(0|6)$ constitute isomorphic $\mathfrak{sl}(4) \cong \mathfrak{o}(6)$-modules. It is subject to a direct verification that it is possible to embed $\mathfrak{spe}(4)_1$ into $\mathfrak{h}(0|6)_1$.

Sergeev's extension $\mathfrak{as}$ is the result of the restriction to $\mathfrak{spe}(4) \subset \mathfrak{h}(0|6)$ of the cocycle that turns $\mathfrak{h}(0|6)$ into $\mathfrak{po}(0|6)$. Quantization deforms $\mathfrak{po}(0|6)$ into $\mathfrak{gl}(\Lambda(\xi))$; the composed maps $T_\lambda : \mathfrak{as} \longrightarrow \mathfrak{po}(0|6) \longrightarrow \mathfrak{gl}(\Lambda(\xi))$ are representations of $\mathfrak{as}$ in $4|4$-dimensional modules spin_λ isomorphic to each other for all $\lambda \neq 0$. The explicit form of T_λ is as follows:

$$T_\lambda : \begin{pmatrix} a & b \\ c & -a^t \end{pmatrix} + d \cdot z \mapsto \begin{pmatrix} a & b - \lambda\tilde{c} \\ c & -a^t \end{pmatrix} + \lambda d \cdot 1_{4|4}, \tag{37}$$

where $1_{4|4}$ is the unit matrix and $\tilde{c}$ is defined in formula (36). Clearly, T_λ is an irreducible representation for any λ.

Acknowledgment. D. L. and I. S. thankfully acknowledge financial support of Université de la Méditerranée and INTAS grant 96-0538, respectively. We are also thankful to ESI for hospitality, in particular, to A. Kirillov and V. Kac, organizers of a miniprogram in 2000.

References

[ALSh] Alekseevsky D., Leites D., Shchepochkina I., New examples of simple Lie superalgebras of vector fields, *C.r. Acad. Bulg. Sci.*, **34**, no. 9 (1980), 1187–1190.

[BHW] Bernard D., Hikami K., Wadati M., The Yangian Deformation of the W-Algebras and and the Calogero–Sutherland model, hep-th/9412194 .

[BL1] Bernstein J., Leites D., Invariant differential operators and irreducible representations of Lie superalgebras of vector fields, *Serdika*, **7** (1981), 320–334 (in Russian); *Sel. Math. Sov.*, **1**, no. 2 (1981), 143–160.

[CK1] Cheng Shun-Jen; Kac V., Generalized Spencer Cohomology and filtered Deformations of $\mathbb{Z}$-graded Lie Superalgebras, math-RT/9805039; *Adv. Theor. Math. Phys.* **2**, no. 5 (1998), 1141–1182.

[CK2] Cheng S., Kac V., Structure of some $\mathbb{Z}$-graded Lie superalgebras of vector fields, *Transformation Groups*, **4** (1999), 219–272.

[D] Deligne P. et al (eds.), *Quantum Fields and Strings: a Course for Mathematicians*. Vol. 1, 2. Material from the Special Year on Quantum Field

Theory held at the Institute for Advanced Study, Princeton, NJ, 1996–1997. AMS, Providence, RI; Institute for Advanced Study (IAS), Princeton, NJ, 1999. Vol. 1: xxii+723 pp.; Vol. 2: pp. i–xxiv and 727–1501.

[FF1] Feigin B. L., Fuks D. B., Stable cohomology of the algebra W_n and relations in the algebra L_1, (Russian) *Funktsional. Anal. i Prilozhen*, **18** no. 3 (1984), 94–95.

[FF2] Feigin B. L., Fuks D. B., Cohomology of Lie groups and Lie algebras. (Russian) Current problems in mathematics. Fundamental directions, Vol. 21 (Russian), 121–209, 215, Itogi Nauki i Tekhniki, Akad. Nauk SSSR, Vsesoyuz. Inst. Nauchn. i Tekhn. Inform., Moscow, 198.

[FLV] Floreanini R., Leites D., Vinet L., On defining relations of quantum superalgebras, *Lett. Math. Phys.*, **23** (1991), 127–131.

[FK] Freund P., Kaplansky I., Simple supersymmetries, *J. Math. Phys.*, **17**, no. 2 (1976), 228–231.

[Fu] D. B. Fuks (Fuchs), *Cohomology of Infinite Dimensional Lie Algebras*. Consultants Bureau, NY, 1987.

[GPS] Gomis J., París J., Samuel S., Antibracket, antifields and gauge-theory quantization, Phys. Rept. 259 (1995), no. 1–2, 1–191.

[Gr] Grozman P., Classification of bilinear invariant operators on tensor fields (Russian), *Funktsional. Anal. i Prilozhen*, **14** no. 2, (1980), 58–59.

[GL1] Grozman P., Leites D., *Mathematica*-aided study of Lie algebras and their cohomology. From supergravity to ball bearings and magnetic hydrodynamics. In: Keränen V. (eds.), *The Second International Mathematica Symposium*, Rovaniemi, 1997, 185–192.

[GL2] Grozman P., Leites D., Defining relations for classical Lie superalgebras with Cartan matrix, hep-th 9702073; *Czech. J. Phys.*, **51**, no. 1 (2001), 1–22.

[GL3] Grozman P., Leites D., Lie superalgebras of supermatrices of complex size. Their generalizations and related integrable systems. math.RT/0202177; In: by E. Ramírez de Arellano, M. V. Shapiro, L. M. Tovar and N. L. Vasilevski (eds.) *Proc. Internatnl. Symp. Complex Analysis and Related Topics*, Mexico, 1996, Birkhäuser Verlag, 1999, 73–105.

[GL4] Grozman P., Leites D., Defining relations for simple Lie superalgebras without Cartan matrix. In: MPIM-2003, preprint, to appear.

[GLP] Grozman P., Leites D., Poletaeva E., Defining relations for simple Lie superalgebras. Lie superalgebras without Cartan matrix. In: Ivanov E. et. al. (eds.), *Supersymmetries and Quantum Symmetries* (SQS'99, 27–31 July, 1999), Dubna, JINR, 2000, 387–396; an expanded version in: *Homology, Homotopy and Applications*, vol 4 (2), 2002, 259–275.

[GLS1] Grozman P., Leites D., Shchepochkina I., Lie superalgebras of string theories, hep-th 9702120; Acta Mathematica Vietnamica, v. 26, 2001, no. 1, 27–63.

[GPS] Gomis J., Paris J., Samuel S., Antibracket, Antifields and Gauge-Theory Quantization, *Phys. Rept.* 259 (1995), 1–191.

[GSW] Green M., Schwarz J., Witten E. *Superstring theory*, vv.1, 2, 2nd ed., Cambridge Monographs on Mathematical Physics., Cambridge Univ. Press, Cambridge, 1988. x+470 pp., xii+596 pp.

[K1C] Kac V. G., Letter to editors, *Funkcional. Anal. i prilozheniya*, **10**, no. 2 (1976), 93.

[Ka1] Kac V.G., *Infinite Dimensional Lie Algebras*. 3rd ed. Cambridge Univ. Press, Cambridge, 1990, xxii+400 pp.

[Ka2] Kac V. G., Classification of infinite-dimensional simple linearly compact Lie superalgebras, ESI-preprint 605, 1997. (www.esi.ac.at)

[Ka3] Kac V. G., Classification of infinite-dimensional simple groups of supersymmetries and quantum field theory, math.QA/9912235 .

[Ka4] Kac V. G., Lie superalgebras, *Adv. Math* **26** (1977), 8–96.

[Ka7] Kac V., Classification of infinite-dimensional simple linearly compact Lie superalgebras. Adv. Math. 139 (1998), no. 1, 1–55

[KvdL] Kac, V. G., van de Leur, J. W. On classification of superconformal algebras. In: S. J. Gates, Jr., C. R. Preitschopf and W. Siegel (eds.), *Strings '88* Proceedings of the workshop held at the University of Maryland, College Park, Maryland, May 24–28, 1988. World Scientific Publishing Co., Inc., Teaneck, NJ, 1989, 77–106.

[Ka8] Kac V., Superconformal algebras and transitive group actions on quadrics, *Comm. Math. Phys.* **186**, no. 1 (1997), 233–252; Erratum: *Comm. Math. Phys.* **217**, no. 3 (2001), 697–698.

[K] Kaplansky I., Superalgebras, *Pacific J. Math.* **86**, no. 1 (1980), 93–98.

[Ki] Kirillov A. A., Ovsienko V. Yu., Udalova, O. D., Identities in the Lie algebra of vector fields on the real line. *Selecta Math. Soviet.* **10**, no. 1 (1991), 7–17.

[Ko1] Kotchetkoff Yu., Déformations de superalgébres de Buttin et quantification, *C. R. Acad. Sci. Paris,* ser. I, **299**, no. 14 (1984), 643–645.

[Ko2] Kochetkov Yu., Deformations of Lie superalgebras, VINITI Depositions, Moscow (in Russian) 1985, 384–85.

[L1] Leites D., New Lie superalgebras and mechanics, *Soviet Math. Dokl.*, **18**, no. 5 (1977), 1277–1280.

[L2] Leites D., Lie superalgebras, In: *Modern Problems of Mathematics. Recent developments*, v. 25, VINITI, Moscow, 1984, 3–49 (in Russian; English translation in: JOSMAR (J. Soviet Math.) **30**, no. 6 (1985), 2481–2512).

[L3] Leites D., *Supermanifold Theory*, Karelia Branch of the USSR Acad. Sci., Petrozavodsk, 1983, 200 pp. (in Russian; expanded in [L5]).

[L4] Leites D., Clifford algebra as a superalgebra and quantization, *Theor. Math. Phys.*, **58**, no. 2 (1984), 229–232; id., *Quantization. Supplement* 3. In: F. Berezin, M. Shubin, *Schrödinger Equation*, Kluwer, Dordrecht, 1991, 483–522.

[L5] Leites D. (ed.) *Seminar on Supermanifolds*, Reports of Stockholm University, 30/1988-13 and 31/1988-14 .

[LP] Leites D., Poletaeva E., Defining relations for classical Lie algebras of polynomial vector fields. *Math. Scand.*, **81**, no. 1 (1997), 5–19.

[LSe] Leites D., Serganova V., Defining relations for simple Lie superalgebras, I. Lie superalgebras with Cartan matrix, In: J. Mickelsson, O. Peckkonen (eds.), *Proc. of the Conf. Topological Methods in Physics, 1991*, World Scientific, 1992, 194–201.

[LS] Leites D., Serganova V., Symmetries wider than supersymmetries, In: S. Duplij and J. Wess (eds.), *Noncommutative Structures in Mathematics and Physics*, Proc. NATO Advanced Research Workshop, Kiev, 2000. Kluwer, 13–30.

[LSh0] Leites D., Shchepochkina I., Towards classification of simple vectorial Lie superalgebras. In: [L5], 31/1988-14; Leites D., Toward classification of classical Lie superalgebras. In: Nahm W., Chau L.-L. (eds.) *Differential geometric methods in theoretical physics* (Davis, CA, 1988), NATO Adv. Sci. Inst. Ser. B Phys., 245, Plenum, New York, 1990, 633–651; Leites D., Shchepochkina I., Quivers and Lie superalgebras, *Czech. J. Phys.*, **vol 47**, no. 12 (1997), 1221–1229.

[LSh1] Leites D., Shchepochkina I., Classification of simple Lie superalgebras of vector fields, MPIM-preprint, 2002, to appear.

[LSh2] Leites D., Shchepochkina I., How to quantize antibracket, preprint ESI-875 (www.esi.ac.at); *Theor. Math. Phys.*, **126**, no. 3 (2001), 339–369.

[OV] Onishchik A., Vinberg É., *Lie Groups and Algebraic Groups*, Springer-Verlag, Berlin, 1990.

[Sh3] Shchepochkina I., New exceptional simple Lie superalgebras, *C. R. Bulg. Sci.*, **36**, no. 3 (1983), 313–314.

[ShM] Shchepochkina I., Maximal subalgebras of simple Lie superalgebras. In: Leites D. (ed.), *Seminar on Supermanifolds* vols. 1–34, 1987–1990, v. 32/1988-15, Reports of Stockholm University, 1–43 (hep-th 9702120).

[Sh5] Shchepochkina I., The five exceptional simple Lie superalgebras of vector fields, hep-th 9702120; id., The five exceptional simple Lie superalgebras of vector fields, *Sov. J. Funct. Analysis*, **33**, no.3 (1999), 14 pp.

[Sh14] Shchepochkina I., Five exceptional simple Lie superalgebras of vector fields and their fourteen regradings, *Represent. Theory*, **3**, no. 3 (1999), 373–415.

[ShP] Shchepochkina I., Post G., Explicit bracket in an exceptional simple Lie superalgebra, *Internat. J. Algebra Comput.*, **8**, no. 4 (1998), 479–495; physics 9703022 .

[SNR] Scheunert M., Nahm W., Rittenberg V., Classification of all simple graded Lie algebras whose Lie algebra is reductive. I, II. Construction of the exceptional algebras. *J. Mathematical Phys.*, **17**, no. 9 (1976), 1626–1639, 1640–1644.

[Sch] Scheunert M., *The theory of Lie superalgebras. An introduction*, Lecture Notes in Mathematics, 716. Springer, Berlin, 1979.

[S1] Serganova V., Automorphisms of simple Lie superalgebras (Russian), *Izv. Akad. Nauk SSSR Ser. Mat.*, **48**, no. 3 (1984), 585–598; Feigin B. L., Leites D. A., Serganova V. V., Kac-Moody Superalgebras. In: Markov M. et al. (eds.), *Group Theoretical Methods in Physics*, Vol. 1–3 (Zvenigorod, 1982), 631–637, Harwood Academic Publ., Chur, 1985.

[S2] Serganova V., On generalizations of root systems, *Comm. Algebra*, **24**, no. 13 (1996), 4281–4299.

[Sm] Shmelev G. S., Differential operators that are invariant with respect to the Lie superalgebra $H(2,\ 2;\ \lambda)$ and its irreducible representations (Russian), *C. R. Acad. Bulgare Sci.*, **35**, no. 3 (1982), 287–290.

[St] Sternberg S., *Lectures on Differential Geometry*, Chelsey, 2nd edition, 1985.

[Sud] Sudbery A., Octonionic description of exceptional Lie superalgebras. J. Math. Phys. 24 (1983), no. 8, 1986–1988.

[T1] Tanaka N., On the pseudo-conformal geometry of hypersurfaces of the space of n complex variables, *J. Math. Soc. Japan*, **14**, no. 4 (1962), 397–429.

[T2] Tanaka N., On generalized graded Lie algebras and geometric structures, I *J. Math. Soc. Japan*, **19** no. 2 (1967), 215–254.

[T3] Tanaka N., On differential systems, graded Lie algebras and pseudo-groups, *J. Math. Kyoto Univ.*, **10** (1970), 1–82.

[T4] Tanaka N., On non-degenerate real hypersurfaces, graded Lie algebras and Cartan connections, *Japan J. Math.*, **2** (1976), 131–190.

[T5] Tanaka N., On the equivalence problems associated with simple graded Lie algebras, *Hokkaido Math. J.*, **8** (1979), 23–84.

[vdL] van de Leur J., A classification of contragredient Lie superalgebras of finite growth, *Comm. Algebra*, **17**, no. 8 (1989), 1815–1841.

[WB] Wess J., Bagger J., *Supersymmetry and Supergravity*, Princeton Series in Physics, Princeton University Press, Princeton, N.J., 1983.

[WL] Woit P., String Theory: An Evaluation, physics/0102051; Larsson T., Symmetries of Everything, math-ph/0103013 .

[WZ] Wess J., Zumino B., Supergauge transformations in four dimensions. Nuclear Phys. **B70** (1974), 39–50; Wess J., Bagger J., *Supersymmetry and Supergravity*. Second edition, Princeton University Press, Princeton, NJ, 1992.

[Y] Yamane H., On defining relations of affine Lie superalgebras and affine quantized universal enveloping superalgebras, q-alg 9603015; *Publ. Res. Inst. Math. Sci.* **35**, no. 3 (1999), 321–390.

[Ym] Yamaguchi K., Differential systems associated with simple graded Lie algebras, Progress in differential geometry, Adv. Stud. Pure Math., 22, Math. Soc. Japan, Tokyo, 1993, 413–494.

Pavel Grozman and Dimitry Leites
Department of Mathematics
University of Stockholm
Roslagsv. 101
Kräftriket hus 6
SE-106 91, Stockholm, Sweden
`mleites@matematik.su.se`

Irina Shchepochkina
Independent University of Moscow
Bolshoj Vlasievsky per.
RU-121002 Moscow, Russia
`Ira@Paramonova.mccme.ru`

Schur–Weyl Duality and Representations of Permutation Groups

Eugene Gutkin

ABSTRACT The Schur–Weyl duality provides a systematic approach to the study of representations of the classical groups. In this contribution we use the Schur-Weyl duality to analyze the natural representations of the symmetric group S_n arising from the irreducible $GL(n, \mathbf{C})$ and $SL(n, \mathbf{C})$-modules. This paper is dedicated to A. A. Kirillov on the occasion of his 2^6-th birthday.

1 Introduction

The Schur–Weyl duality [17] is a powerful tool in the study of irreducible representations of the classical groups. See [10] and [15] for contemporary developments. By a classical (complex, reductive Lie) group we mean either the full linear group $GL(n, \mathbf{C})$, or one of its classical subgroups; see [17]. By a permutation group I mean a finite group, F, acting faithfully on a finite set, X. If $|X| = N$, then F is given as a subgroup of the symmetric group S_N. The Weyl groups of the classical groups have natural realizations as permutation groups. The Weyl groups of $GL(n, \mathbf{C})$ and $SL(n, \mathbf{C})$ are isomorphic to S_n. If $G \subset GL(n, \mathbf{C})$ is either the orthogonal, the special orthogonal, or the symplectic group, then the corresponding Weyl group is a subgroup of S_n. All of them will be referred to as the classical Weyl groups.

Let us establish some notation. We will consider only complex representations, and notation such as (π, E) will be used for them. If G is the group in question, then E is the space of the representation, and $\pi : G \to GL(E)$ is the corresponding action. For brevity, we will often say that (π, E) is a G-module. It is a classical problem in representation theory to analyze the restriction of (π, E) to various subgroups of G; see, e.g., [7]. The Weyl group, W, of a complex reductive group G is not, in general, a subgroup of G. Recall that $W = N(H)/H$, where H is a Cartan subgroup, and $N(H)$ is its normalizer. However, for some groups, e.g., for $GL(n, \mathbf{C})$, the Weyl group naturally imbeds into G. Thus, for any $GL(n, \mathbf{C})$-module the restriction to W is well defined.

Let G be an arbitrary complex reductive group, and let (π, E) be a G-module. The subspace $E^H \subset E$ consists of the vectors of weight zero. The Weyl group of G naturally acts on it. Let $(\pi(W), E^H)$ be the induced representation of W.

It is the representation of the Weyl group in the space of vectors of weight zero corresponding to (π, E). In order to understand the correspondence $(\pi, E) \mapsto (\pi(W), E^H)$, it suffices to consider the case when the commutator group of G is simple, and the representation (π, E) is irreducible. Denote by $\hat{G}$ the set of (classes of) irreducible G-modules. The subset, $(\hat{G})^0 \subset \hat{G}$, of the representations (π, E) such that the space E^H is nontrivial, is well understood [12]. However, the mapping, $(\pi, E) \mapsto (\pi(W), E^H)$ for $\pi \in (\hat{G})^0$ is not understood, in general. For classical groups the problem has been studied by several authors. See [8, 6, 13, 1, 2, 14]. They were sometimes unaware of the work of their predecessors.

The results on the correspondence $(\pi, E) \mapsto (\pi(W), E^H)$ are the most complete for the groups $GL(n, \mathbf{C})$ and $SL(n, \mathbf{C})$. In this case $W = S_n$; see [8, 6, 1]. The paper [8] reduces the problem to the study of certain representations of $S_r \times S_n$, $1 \le r < \infty$. This approach raises a question about the representations of permutation groups which is of independent interest. This question is formulated below.

Let $W \subset S_n$ be a permutation group. We denote by Y the set $\{1, \ldots, n\}$. Set $X = \{1, \ldots, r\}$, and let Φ be the set of all mappings from X to Y. Then $|\Phi| = n^r$. The group $S_r \times S_n$ naturally acts on Φ. Hence the space $C(\Phi)$ of functions on Φ is a $S_r \times S_n$-module. Set $F = S_r \times W \subset S_r \times S_n$. The problem mentioned above is to analyze the induced representation of F in $C(\Phi)$. The structure of $C(\Phi)$ as a $S_r \times S_n$-module was used in [8] to identify $(\pi(S_n), E^H)$ when E is an irreducible representation of $SL(n, \mathbf{C})$ or $GL(n, \mathbf{C})$.

In this paper we extend the method of [8] to analyze the $S_r \times W$-module $C(\Phi)$ when W is the Weyl group of a classical group. The Schur–Weyl duality is then used to deduce information on the W-modules $(\pi(W), E^H)$. Here we do it in detail for $SL(n, \mathbf{C})$ and $GL(n, \mathbf{C})$. Other classical Weyl groups will be treated elsewhere [9].

A few personal remarks are in order. The decomposition of the $S_r \times S_n$-modules $C(\Phi)$ and its application to the representations (π^0, E^0) for $SL(n, \mathbf{C})$ and $GL(n, \mathbf{C})$ has been done in [8]. Unfortunately, the exposition there is extremely cryptic. This is due to the strict two-page upper bound that the journal imposed on this type of publications: Communications to the Moscow Mathematical Society. Partly because of this and partly because the paper has not been translated into English, the method of [8] remains largely unknown to researchers in the field. See [1, 2, 6, 13, 14]. Recently two mathematicians asked me independently to explain the paper [8] to them. Stimulated by their interest, I decided to extend the method of [8] to include all of the classical groups, and to give to the subject an exposition it deserves.

Finally, I find it fitting that this contribution, based on the material that I learned 2^5 years ago at the Moscow University seminar of A. A. Kirillov, will appear in the Festschrift, commemorating his 2^6th birthday. Let there be safety in numbers!

2 The setting

Recall that, classically, a partition of the set $X = \{1, \dots, r\}$ is a representation, $X = X_1 \cup \dots \cup X_m$, of it as a disjoint union of nonempty subsets. Set $k_i = |X_i|$. Then $r = k_1 + \dots + k_m$ is a partition of the number r. To distinguish between the two notions, we will speak of the set partitions and the number partitions respectively.

In the literature on partitions it is customary to assume that the parts of a partition $r = k_1 + \dots + k_m$ are positive, and go in decreasing order: $k_1 \geq \dots \geq k_m$. See, e.g., [5]. We will not follow this custom. For our purposes, a number partition is a sequence $k_1, \dots, k_n$ of nonnegative integers such that $r = k_1 + \dots + k_n$. To distinguish these partitions from the "standard" ones, we will call them the *ordered partitions*. The *size* r of the partition $r = k_1 + \dots + k_n$ will be arbitrary. On the contrary, the number of parts of an ordered number partition will be fixed, while some of them may be equal to zero. Likewise, an ordered set partition is $X = X_1 \cup \dots \cup X_n$, where some of the subsets X_i may be empty.

To every ordered partition of X there corresponds a unique ordered partition of $r = |X|$. We will denote by $P_n(X)$ (resp. $P_n(r)$) the set of ordered set (resp. number) partitions, and we will suppress the adjective "ordered" if this is not likely to cause confusion. Let me recall the notation introduced in Section 1. Thus, $Y = \{1, \dots, n\}$ and Φ is the set of mappings $\phi : X \to Y$. The symmetric groups S_r and S_n naturally act on X and Y, respectively, yielding an action of $S_r \times S_n$ on Φ. Every ϕ in Φ defines a partition $X = X_1 \cup \dots \cup X_n$ where $X_i = \phi^{-1}(i)$. This yields an $S_r \times S_n$-equivariant isomorphism of Φ and $P_n(X)$.

The relation between $P_n(X)$ and $P_n(r)$ is like the relation between Young tableaux and Young diagrams. See [5]. An (ordered) Young diagram is a sequence of n rows. The lengths k_i of the rows satisfy $k_i \geq 0$ and $\sum_{i=1}^{n} k_i = r$. Filling in the numbers from 1 to r into a Young diagram we obtain an (ordered) Young tableau. The group S_r permutes the numbers of a Young tableau. The group S_n acts by permuting the rows of a Young tableau. The two actions commute yielding an action of the direct product $S_r \times S_n$.

It is natural to identify number partitions with the points of the integer lattice $\mathbf{Z}^n$. Let $\mathbf{Z}^n_+ \subset \mathbf{Z}^n$ (resp. $\mathbf{Z}^n_{\geq} \subset \mathbf{Z}^n$) be the set of nonnegative (resp. nondecreasing) n-tuples $\mu = (k_1, \dots, k_n)$. If $A \subset \mathbf{Z}^n$, and $r \in \mathbf{Z}$, the set $A(r) \subset A$ is given by the condition $k_1 + \ldots + k_n = r$. Then $P_n(r) = \mathbf{Z}^n_+(r)$, and we will use the latter symbol in what follows. Set $\mathbf{Z}^n_{\geq +} = \mathbf{Z}^n_+ \cap \mathbf{Z}^n_{\geq}$. Then $\mathbf{Z}^n_{\geq +}(r) \subset \mathbf{Z}^n_+(r)$ is the set of standard partitions of r into at most n parts.

3 Representations of permutation groups

If G is a group, $H \subset G$ a subgroup, and (π, E) is a representation of H, then we denote by $\mathrm{Ind}(G, H, (\pi, E))$ the induced representation of G. If the representation (π, E) is inambiguously identified by either π or E, we will use the notation Ind(G, H, π) or Ind(G, H, E), respectively. For example, Ind$(G, H, 1)$ is induced by the trivial representation of H.

If G is finite, which will typically be the case, then H is arbitrary. If G is infinite, then we will mostly consider the situation when H has a finite index in G. For convenience of the reader, we recall the basic definitions. See [11] for details. The space of Ind$(G, H, (\pi, E))$ consists of the E-valued functions on G satisfying $f(gh) = \pi(h^{-1})f(g)$. The induced action of G is given by $(T(g)f)(x) = f(g^{-1}x)$. For instance, Ind$(G, H, 1)$ is the natural representation in the space of functions on the quotient G/H. Set $Q = G/H$. A choice of the cross-section $Q \to G$ identifies the space of Ind$(G, H, (\pi, E))$ with the space of E-valued functions on Q. Then the action of G becomes $(T(g)f)(q) = \pi(g, q)f(g^{-1} \cdot q)$. The multiplier $\pi(g, q)$ depends on the cross-section and on the inducing representation. The general formula for the multiplier is straightforward but not illuminating. We will need only a special case as described below.

Suppose that $G = K\,A$ is a semidirect product, where K is finite and A is abelian. Denote by $k \cdot a$ the action of K on A defining the group $K\,A$. Let $\hat{A}$ be the dual group. If $\chi \in \hat{A}$ and $a \in A$, let $< \chi, a >$ denote the pairing. We will use the same notation, $k \cdot \chi$, for the dual action of K on $\hat{A}$. Thus $\langle k \cdot \chi, k \cdot a \rangle = \langle \chi, a \rangle$. Let $K_\chi \subset K$ be the stabilizer of χ, and let $K'_\chi \subset K_\chi$ be a subgroup. Set $H = K'_\chi A \subset K_\chi A$, and let $Q = G/H$. Since H belongs to the stabilizer of χ in G, there is a G-equivariant mapping $\omega : Q \to \hat{A}$. Pulling back the characters to Q, we view elements of Q as functions on A, and set $\langle q, a \rangle = \langle \omega(q), a \rangle$. Now we write out the operators, $T(g)$, of the induced representation Ind$(K\,A, K'_\chi A, \chi)$. They act on the space of functions on Q. It suffices to know $T(k)$ and $T(a)$. Then

$$(T(k)f)(q) = f(k^{-1} \cdot q), \ (T(a)f)(q) = \langle q, a \rangle f(q). \tag{1}$$

We will use equation (1) to identify representations later on. Now we will focus the discussion further to the case at hand. Set $K = S_r \times R$, where $R \subset S_n$ is a permutation group, and let the abelian group above be $1 \times A$, where $A \subset (\mathbf{C}^*)^n$ is a group of diagonal $n \times n$ matrices, invariant under the natural action of R on $(\mathbf{C}^*)^n$.

We identify $\mathbf{Z}^n$ with the group of (complex analytic) characters of $(\mathbf{C}^*)^n$. The inclusion $A \subset (\mathbf{C}^*)^n$ yields the projection $\mathbf{Z}^n \to \hat{A} \to 1$. Hence, we associate with any Young diagram, $\mu = (k_1, \dots, k_n)$, $k_1 + \dots + k_n = r$, the character, e^μ, of A.

For convenience of the reader, we will summarize the setting. There is a permutation group $R \subset S_n$ and a R-invariant abelian subgroup $A \subset (\mathbf{C}^*)^n$. There is the set Φ of mappings $\phi : X \to Y$, where $X = \{1, \dots, r\}$, $Y = \{1, \dots, n\}$. Slightly abusing the language, we will often call the elements of Φ the Young tableaux. There is the set $\mathbf{Z}^n_+(r)$ of Young diagrams. For brevity, we will temporarily denote this set by M. Let $K = S_r \times R$. The group $S_r \times S_n$ naturally acts on Φ and on M. Let $k \cdot \phi$ and $k \cdot \mu$ denote the actions. The action of S_r on M is trivial. There is a natural projection of Φ onto M. It assigns to a Young tableau ϕ its Young diagram, $\mu = \langle \phi \rangle$. The mapping $\phi \mapsto \langle \phi \rangle$ is equivariant with respect to the actions of $S_r \times S_n$. Every $\mu \in M$ defines a character, e^{μ}, of A. Let $\phi \in \Phi$. The notation $e^{\langle\phi\rangle}$ is self-explanatory. The mappings $\phi \mapsto e^{\langle\phi\rangle}$ and $\mu \mapsto e^{\mu}$ are consistent with the actions of K on Φ, M, and $\hat{A}$. The symbol $K\,A$ means the semidirect product of the groups in question.

We will now define a representation $(\delta, C(\Phi))$ of $K\,A$. Note that the restriction of δ to $S_r \times R$ is well defined on the whole $S_r \times S_n$. It is the representation arising from the action of $S_r \times S_n$ on Φ. It suffices to define the operators $\delta(k)$ and $\delta(a)$. Then, in the notation above

$$(\delta(k)f)(\phi) = f(k^{-1}\cdot\phi), \quad (\delta(a)f)(\phi) = e^{\langle\phi\rangle}(a)f(\phi). \tag{2}$$

Let me introduce a few useful subgroups of $K\,A$ and related groups. If $\phi \in \Phi$ is a Young tableau, the group $S_\phi \subset S_r$ consists of the row-preserving permutations. Let $\mu = (k_1, \ldots, k_n) \in M$ be a Young diagram. Filling in the numbers from 1 to r in the natural order, we obtain a Young tableau, $>\mu<$. Note that $\langle(>\mu<)\rangle = \mu$, and that $S_{>\mu<} \simeq \times_{i=1}^{n} S_{k_i}$. The subgroup $R_\mu \subset R$ consists of the permutations $w \in R$ such that if $w(i) = j$, then $k_i = k_j$. Equivalently, R_μ is the stabilizer of $\mu \in \mathbf{Z}_+^n$, under the canonical action of R. For $w \in R_\mu$ let $\bar{w} \in S_r$ be the element that permutes the whole rows i and j (with $k_i = k_j$) in the Young tableau $>\mu<$. The correspondence $w \mapsto \bar{w}$ is a homomorphism $h_\mu : R_\mu \to S_r$. Let $\bar{R}_\mu \subset S_r$ be the range of h_μ. Note that $\bar{R}_\mu$ normalizes the subgroup $S_{>\mu<}$ in S_r. Also note that the kernel, $L_\mu \subset R_\mu$, of h_μ consists of those elements of R that permute only the indices i such that $k_i = 0$. Let $d = d(\mu)$ be the number of these indices. Then d is the *defect* of the partition. For example, if $R = S_n$, then $L_\mu = S_d \subset S_n$. Set $K_{>\mu<} = \{s\,\bar{w} \times w : s \in S_{>\mu<},\ w \in R_\mu\}$. The proposition below is the point of departure for our approach.

Proposition 3.1. *The following statements hold.*
1. The K-equivariant map $\phi \mapsto <\phi>$ induces a bijection between the quotients Φ/K and M/R.
2. For $\mu \in M$ set $O_\mu = K \cdot (>\mu<) \subset \Phi$. Then $O_\mu = K/K_{>\mu<}$.
3. Let δ_μ the restriction of the representation $(\delta, C(\Phi))$ of $K\,A$ to $C(O_\mu) \subset C(\Phi)$. Then

$$(\delta, C(\Phi)) = \oplus_{\mu \in M/R}(\delta_\mu, C(O_\mu)).$$

The character e^{μ} of A uniquely extends, by identity on $K_{>\mu<}$, to $K_{>\mu<}A$. Denote the extended character by e^{μ} as well. Then

$$(\delta_\mu, C(O_\mu)) = \mathrm{Ind}(K\,A, K_{>\mu<}A, e^{\mu}). \tag{3}$$

Proof. I already noted that the projection $\phi \mapsto <\phi>$ of Φ onto M is $S_r \times S_n$-equivariant. Since S_r acts trivially on $\mathbf{Z}_+^n(r)$, the first claim follows. A straightforward verification reveals that the stabilizer of $>\mu<\in \Phi$ is generated by the elements $s \in S_{>\mu<} \times 1$ and $\bar{w} \times w$, where $w \in R_\mu$. This proves the second claim. The rest of the proposition is a special case of a more general situation, described in the lemma below.

Lemma 3.2. *Let $K\ A$ be a semidirect product of a finite group with an abelian group. Suppose that K acts on a space Φ and that there is a K-equivariant mapping of Φ into $\hat{A}$. For $\phi \in \Phi$ let $e^{<\phi>}$ be the character of A, defined by this mapping. For every $\mu \in \Phi/K$ choose a representative in Φ, and denote it by $> \mu <$. Let $K_{>\mu<} \subset K$ be its stabilizer, set $O_\mu = K\cdot > \mu < \subset \Phi$. Denote by $(\delta, C(\Phi))$ the representation of $K\ A$ given by equation 2. Then :*
1. The character $e^{\langle\mu\rangle}$ of A uniquely extends to a character of $K_{>\mu<}A$, trivial on $K_{>\mu<}$. Denote it by e^μ as well ;
2. Denote by $(\delta_\mu, C(O_\mu))$ the restriction of $(\delta, C(\Phi))$ to $C(O_\mu)$. We have the direct decomposition

$$(\delta, C(\Phi)) = \oplus_{\mu\in\Phi/K}(\delta_\mu, C(O_\mu)) = \oplus_{\mu\in\Phi/K}\mathrm{Ind}(K\ A, K_{>\mu<}A, e^\mu). \quad (4)$$

The proof of the lemma is an elementary exercise, and we leave it to the reader. To finish the proof of the proposition, it suffices to show that the situation at hand is a special case of the setting of Lemma 3.2. Indeed, the semidirect product $(S_r \times R)\ A$ acts on the space Φ, and there is an equivariant mapping of Φ onto $\hat{A}$. The remaining conditions of the lemma follow. □

Remarks. When $R = S_n$ and $A = 1$, Proposition 3.1 is contained in [8]. See the proof of Theorem 1 there. In view of Proposition 3.1, the analysis of the representations of $S_r \times RA$ in $C(\Phi)$ depends on the orbit structure of the action of $R \subset S_n$ in $\mathbf{Z}^n_+(r)$. The examples below illustrate the possibilities.

Examples. 1. Let R be the full permutation group. The S_n-orbits in $\mathbf{Z}^n_+(r)$ are in 1–1 correspondence with the (standard) partitions $\mu = (k_1 \geq \cdots \geq k_n \geq 0)$.

2. Let R be the group of even permutation, $A_n \subset S_n$. Let $n > 2$. Acting by A_n, we can insure that k_1 is the largest coordinate of μ. The subgroup of A_n that fixes k_1 is A_{n-1}. If $n - 1 > 2$, we continue the inductive process. Thus the A_n-orbit of μ contains a unique partition with $k_1 \geq \cdots \geq k_{n-2} \geq \max(k_{n-1}, k_n)$. The group A_2 is trivial. Hence, we cannot insure that $k_{n-1} \geq k_n$. We say that a partition is almost standard if $k_1 \geq \cdots \geq k_{n-2} \geq k_n > k_{n-1}$. Thus the A_n-orbits in $\mathbf{Z}^n_+(r)$ are in one-to-one correspondence with the (disjoint) union of the sets of standard partitions and of almost standard partitions.

There is a classical one-to-one correspondence between the set $\hat{S}_r$ of irreducible representations of the symmetric group S_r and the set of the (standard) partitions of r. See [11, 16] or [17] for details.

If $\lambda \in \mathbf{Z}^n_{\geq+}(r)$ is such a partition, we denote by (T_λ, V_λ) the corresponding irreducible representation of S_r. Let now $\mu \in \mathbf{Z}^n_+(r)$ be another partition of r. Let $S_{>\mu<} \subset S_r$, $R_\mu \subset S_n$ and $\bar{R}_\mu \subset S_r$ be the associated subgroups. Since $\bar{R}_\mu$ normalizes $S_{>\mu<}$, the space $V^\mu_\lambda \subset V_\lambda$ of vectors fixed by $S_{>\mu<}$ is invariant under $\bar{R}_\mu$. The homomorphism $h_\mu : R_\mu \to \bar{R}_\mu$ induces a representation of R_μ in V^μ_λ. We denote this representation by $(T^\mu_\lambda, V^\mu_\lambda)$. Since the group $R \subset S_n$ will be always clear from the context, this notation should not cause any confusion. Any representation, (τ, F), of R_μ extends to a representation of $R_\mu A$, trivial on A. We denote it by the same symbol. Recall that e^{μ} is the one-dimensional representation of $R_{\mu}A$, that coincides with the character e^{μ} on A, and is trivial on R_{μ}. The representation $\tau \otimes e^{\mu}$ of $R_{\mu}A$ is τ, twisted by the character e^{μ}. It acts on the same space F, hence the notation $(\tau \otimes e^{\mu}, F)$.

If G, H are arbitrary groups, and (σ, E), (τ, F) their arbitrary representations, then $(\sigma \times \tau, E \otimes F)$ is the direct product representation of $G \times H$. When $H = G$, we will denote by $(\sigma \otimes \tau, E \otimes F)$ the tensor product representation of G. It is induced from $\sigma \times \tau$ via the diagonal imbedding of G into $G \times G$. We will also use the notation $(\sigma, E) \otimes (\tau, F)$ for the tensor product. For a representation (π, E) both $(\pi, E)^*$ and (π^*, E^*) will be used for the contragredient representation. If (π, E) is a representation of G, and $H \subset G$ is a subgroup, the restriction will often be denoted by $(\pi(H), E)$. If a subspace $F \subset E$ is invariant under $\pi(H)$, we will sometimes use $(\pi(H), F \subset E)$ for the induced representation of H.

Theorem 3.3. *Let $R \subset S_n$, S_r, $\Phi = \{\phi \ : X \to Y\}$ and $S_r \times RA$ be as in Proposition 3.1. Set $Q = \mathbf{Z}^n_+(r)/R$. The rest of the notation will be as above. Let $(\delta, C(\Phi))$ be the representation of $S_r \times RA$, given by equation 2. Then it decomposes as*

$$(\delta, C(\Phi)) = \oplus_{\lambda \in \mathbf{Z}^n_{\geq +}(r)} \oplus_{\mu \in Q} (T_\lambda, V_\lambda) \times \mathrm{Ind}(RA, R_\mu A, (T^{\mu}_{\lambda}, V^{\mu}_{\lambda})^* \otimes (e^{\mu}, \mathbf{C})).$$

Proof. For typographical reasons, I will use the symbol $\sum$ instead of $\oplus$. Slightly abusing notation, we write

$$(\delta, C(\Phi)) = \sum_{\mu \in Q} (\delta_\mu, C(O_\mu)). \tag{5}$$

Hence, in view of equations (3) and (4)

$$(\delta, C(\Phi)) = \sum_{\mu \in Q} \mathrm{Ind}(S_r \times RA, K_{>\mu<}A, e^{\mu}). \tag{6}$$

Consider the tower of inclusions $K_{>\mu<}A \subset S_r \times R_\mu A \subset S_r \times RA$. By the transitivity of the induction of representations

$$\mathrm{Ind}(S_r \times RA, K_{>\mu<}A, e^{\mu}) = \mathrm{Ind}(S_r \times RA, S_r \times R_\mu A, \mathrm{Ind}(S_r \times R_\mu A, K_{>\mu<}A, e^{\mu}). \tag{7}$$

We will use some elementary representation-theoretic facts of a general nature. Let G be any finite group. Every representation of $S_r \times G$ uniquely decomposes as $\sum_{\lambda \in \hat{S}_r} (T_\lambda, V_\lambda) \times (\tau_\lambda, F_\lambda)$, where $(\tau_\lambda, F_\lambda)$ plays the role of the multiplicity of (T_λ, V_λ) in the decomposition. In the case at hand, $G = R_\mu A$, and the remark above yields

$$\mathrm{Ind}(S_r \times R_\mu A, K_{>\mu<}A, e^{\mu}) = \sum_{\lambda \in \hat{S}_r} (T_\lambda, V_\lambda) \times (\tau_\lambda, F_\lambda). \tag{8}$$

If G is any finite group, identified with the diagonal in $G \times G$, then

$$\mathrm{Ind}(G \times G, G, 1) = \sum_{\tau \in \hat{G}} \tau \times \tau^*. \tag{9}$$

Equation (9) has natural generalizations. For example, let $H \subset G$ be a subgroup, and let us identify it with the diagonal in $H \times H \subset G \times H$. Then

$$\mathrm{Ind}(G \times H,\ H,\ 1) = \sum_{(\tau, V) \in \hat{G}} (\tau, V) \times (\tau(H), V)^*. \tag{10}$$

A slight extention of this argument, applied to the situation at hand, yields

$$\mathrm{Ind}(S_r \times R_\mu A, K_{>\mu<} A, e^\mu) = \sum_{\lambda \in \hat{S}_r} (T_\lambda, V_\lambda) \times [(T_\lambda^\mu, V_\lambda^\mu)^* \otimes (e^\mu, \mathbf{C})]. \tag{11}$$

Hence, in equation (8)

$$(\tau_\lambda, F_\lambda) = (T_\lambda^\mu, V_\lambda^\mu)^* \otimes (e^\mu, \mathbf{C}).$$

Substituting this into equations (7) and (6), we obtain the claim. □

Example. The following special case of Theorem 3.3 is contained in [8]. Let $R = S_n$ and $A = 1$. Then $Q = \mathbf{Z}^n_{\geq +}(r)$. The group $R_\mu \subset S_n$ is generated by the transpositions of the indices i and j such that $k_i = k_j$. It is the stabilizer of μ in S_n, and it will be denoted by S_n^μ. Each of the groups $S_{>\mu<}$, R_μ, $\bar{R}_\mu$ is a cartesian product of symmetric groups. Let $\lambda \in \mathbf{Z}^n_{\geq +}(r)$ be a Young diagram, and let (T_λ, V_λ) be the corresponding irreducible representation of S_r. The subspace $V_\lambda^\mu \subset V_\lambda$ consists of the vectors fixed by $S_{>\mu<}$. Since any irreducible representation τ of a symmetric group is defined over $\mathbf{Q}$ (see, e.g., [16]), we have $\tau^* = \tau$. Hence, the decomposition of Theorem 3.3 becomes

$$(\delta, C(\Phi)) = \sum_{\lambda \in \mathbf{Z}^n_{\geq +}(r)} \sum_{\mu \in \mathbf{Z}^n_{\geq +}(r)} (T_\lambda, V_\lambda) \times \mathrm{Ind}(S_n, S_n^\mu, V_\lambda^\mu). \tag{12}$$

4 Applications to the representations of the Weyl groups of $GL(n, \mathbf{C})$ and $SL(n, \mathbf{C})$

4.1 The group $GL(n, \mathbf{C})$

While establishing the necessary notation for this particular group, we introduce some general notation and terminology that will be used later on for other reductive groups. The Cartan subgroup H of $GL(n, \mathbf{C})$ is the group of diagonal $n \times n$ matrices, $H \simeq (\mathbf{C}^*)^n$. For any reductive group G let Λ (resp. Λ_+) be the set of (resp. dominant) weights of the Cartan subgroup. For $\alpha \in \Lambda$ denote by e^α the corresponding character of H. For $GL(n, \mathbf{C})$, we have $\Lambda = \mathbf{Z}^n$ and $\Lambda_+ = \mathbf{Z}^n_{\geq}$. The set $\mathbf{Z}^n_{\geq +} = \mathbf{Z}^n_{\geq} \cap \mathbf{Z}^n_+$ parametrizes the irreducible representations of $GL(n, \mathbf{C})$ that occur in the tensor algebra $\mathcal{T} = \oplus_{r=0}^{\infty} \otimes^r (\mathbf{C}^n)$.

For $\lambda \in \Lambda_+$ let (π_λ, E_λ) be the irreducible representation of G with the highest weight λ. For $\alpha \in \Lambda$ let $E_\lambda(\alpha) \subset E_\lambda$ be the subspace of weight vectors with weight α. The Weyl group, W, of a reductive group G is the quotient, $W = N(H)/H$, where $N(H) \subset G$ is the normalizer of the Cartan subgroup. For $\mu \in \Lambda_+$ set

$$E_\lambda^\mu = \oplus_{w \in W} E_\lambda(w \cdot \mu).$$

Equivalently, $E_\lambda^\mu \subset E_\lambda$ is spanned by the weight vectors in E_λ, whose weights belong to the W-orbit of μ. For $\lambda \in \Lambda_+$ denote by $\Lambda(\lambda) \subset \Lambda$ the set of weights of (π_λ, E_λ), and set $\Lambda_+(\lambda) = \Lambda(\lambda) \cap \Lambda_+$. Then

$$E_\lambda = \oplus_{\alpha \in \Lambda(\lambda)} E_\lambda(\alpha) = \oplus_{\mu \in \Lambda_+(\lambda)} E_\lambda^\mu. \tag{13}$$

The spaces $E_\lambda^\mu \subset E_\lambda$ are $N(H)$-invariant. Let $(\pi_\lambda^\mu, E_\lambda^\mu)$ denote the corresponding representation of $N(H)$. Suppose that $N(H)$ contains a subgroup RA, where $A \subset H$, and the projection $N(H) \to W$ is injective on R. Identifying R with its image under this projection, we will write $R \subset W$. Restricting $(\pi_\lambda^\mu, E_\lambda^\mu)$ to $RA \subset N(H)$, we obtain the representations $(\pi_\lambda^\mu(RA), E_\lambda^\mu)$. If the group RA is clear from the context, then it will be suppressed from the notation.

In the case at hand, i.e., $G = GL(n, \mathbf{C})$, the normalizer itself is a semidirect product : $N(H) = WH = S_n H$. The Weyl group imbeds into $GL(n, \mathbf{C})$ as the group of permutation matrices, yielding an isomorphism $W = S_n$. Let $A \subset H$ be a S_n-invariant subgroup. Any $GL(n, \mathbf{C})$-module is naturally a $S_n A$-module. In particular, let $\lambda \in \mathbf{Z}^n_{\geq +}$ and $\mu \in \Lambda_+(\lambda)$. We denote by $(\pi_\lambda^\mu, E_\lambda^\mu)$ the $S_n A$-module corresponding to the invariant subspace $E_\lambda^\mu \subset E_\lambda$. The theorem below describes these representations. Its formulation makes use of the notation of Theorem 3.3 and, more generally, of Section 3. The notation is customized to the case at hand.

For $\alpha \in \mathbf{Z}^n$ let $S_n^\alpha \subset S_n$ be the stabilizer of α. The character e^α of A extends to a character of $S_n^\alpha A$, trivial on S_n^α. Let $\lambda = (k_1 \geq \cdots \geq k_n \geq 0) \in \mathbf{Z}^n_{\geq +}(r)$. For any $\mu \in \mathbf{Z}^n_+$ the representation $(T_\lambda^\mu, V_\lambda^\mu)$ of S_n^μ is afforded by the $S_{>\mu<}$-invariant vectors of the S_r-module (T_λ, V_λ). See Section 3 for details. Twisting by the character e^μ, we obtain the $S_n^\mu A$-module $(T_\lambda^\mu \otimes e^\mu, V_\lambda^\mu)$.

Theorem 4.1. *Let $\lambda = (k_1 \geq \cdots \geq k_n \geq 0) \in \mathbf{Z}^n_{\geq +}(r)$, and let (π_λ, E_λ) be the irreducible representation of $GL(n, \mathbf{C})$ with the highest weight λ. Then for any $\mu \in \Lambda_+(\lambda)$ we have*

$$(\pi_\lambda^\mu, E_\lambda^\mu) = \mathrm{Ind}(S_n A, S_n^\mu A, (T_\lambda^\mu \otimes e^\mu, V_\lambda^\mu)). \tag{14}$$

Proof. The proof hinges on the Schur–Weyl duality. See the classical book [17]. For convenience of the reader, we formulate the Schur–Weyl theorem in a form that is more suitable for the present purpose.

For $r \geq 0$ set $\mathcal{T}_r = \otimes^r(\mathbf{C}^n)$, the space of tensors of rank r. To each $\lambda \in \mathbf{Z}^n_{\geq +}(r)$ correspond:
i) The irreducible representation (T_λ, V_λ) of S_r with the Young diagram λ;
ii) The irreducible representation (π_λ, E_λ) of $GL(n, \mathbf{C})$ with the highest weight λ.

The natural actions of the groups S_r and $GL(n, \mathbf{C})$ in $\mathcal{T}_r$ commute, yielding a representation, $(\tau_r, \mathcal{T}_r)$, of $S_r \times GL(n, \mathbf{C})$. The Schur–Weyl theorem gives the decomposition of this representation into irreducible components, namely

$$(\tau_r, \mathcal{T}_r) = \sum_{\lambda \in \mathbf{Z}^n_{\geq +}(r)} (T_\lambda, V_\lambda) \times (\pi_\lambda, E_\lambda). \tag{15}$$

Let now $G' \subset GL(n, \mathbf{C})$ be any subgroup. Let $(\pi'_\lambda, E_\lambda)$ and $(\tau'_r, \mathcal{T}_r)$ be the restrictions of (π_λ, E_λ) and $(\tau_r, \mathcal{T}_r)$ to G' and $S_r \times G'$, respectively. By (15),

$$(\tau'_r, \mathcal{T}_r) = \sum_{\lambda \in \mathbf{Z}^n_{\geq +}(r)} (T_\lambda, V_\lambda) \times (\pi'_\lambda, E_\lambda). \tag{16}$$

We apply (16) to the group $G' = S_n A \subset N(H)$. Let $e_1, \dots, e_n$ be the standard basis of $\mathbf{C}^n$. The r^n tensors $e_{i_1} \otimes \cdots \otimes e_{i_r}$, where $1 \leq i_1, \dots, i_r \leq n$, form a basis of $\mathcal{T}_r$. They are in a natural one-to-one correspondence with the mappings $\phi : X \to Y$. By equation(2), the resulting linear isomorphism $C(\Phi) \simeq \mathcal{T}_r$ is an equivalence of the representations $(\delta, C(\Phi))$ and $(\tau'_r, \mathcal{T}_r)$ of $S_r \times S_n A \subset S_r \times N(H)$. Comparing (16) with the decomposition of Theorem 3.3, and taking into account that the representation $(T^\mu_\lambda, V^\mu_\lambda)$ coincides with its contragredient, we obtain the equality

$$(\pi_\lambda(S_n A), E_\lambda) = \sum_{\mu \in \mathbf{Z}^n_{\geq +}(r)} \mathrm{Ind}(S_n A, S^\mu_n A, (T^\mu_\lambda \otimes e^\mu, V^\mu_\lambda)). \tag{17}$$

By the proof of Theorem 3.3, for any $\mu \in \mathbf{Z}^n_{\geq +}(r)$, the space of the representation $(T_\lambda, V_\lambda) \times \mathrm{Ind}(S_n A, S^\mu_n A, (T^\mu_\lambda \otimes e^\mu, V^\mu_\lambda))$ is contained in $C(O_\mu) \subset C(\Phi)$. Slightly abusing notation, we write $\mathrm{Ind}(S_n A, S^\mu_n A, (T^\mu_\lambda \otimes e^\mu, V^\mu_\lambda))$ for the subspace of E_λ corresponding to the space of $\mathrm{Ind}(S_n A, S^\mu_n A, (T^\mu_\lambda \otimes e^\mu, V^\mu_\lambda))$ under the isomorphism of (17). Therefore the weights of the weight vectors in the space $\mathrm{Ind}(S_n A, S^\mu_n A, (T^\mu_\lambda \otimes e^\mu, V^\mu_\lambda))$ belong to the orbit $S_n \cdot \mu \subset \mathbf{Z}^n_+(r)$. Thus, with our notational convention

$$\mathrm{Ind}(S_n A, S^\mu_n A, (T^\mu_\lambda \otimes e^\mu, V^\mu_\lambda)) \subset E^\mu_\lambda \subset E_\lambda.$$

This implies the inclusions

$$E_\lambda = \sum_{\mu \in \mathbf{Z}^n_{\geq +}(r)} \mathrm{Ind}(S_n A, S^\mu_n A, (T^\mu_\lambda \otimes e^\mu, V^\mu_\lambda)) \subset \sum_{\mu \in \Lambda_+(\lambda)} E^\mu_\lambda = E_\lambda.$$

Therefore the inclusion $\mathrm{Ind}(S_n A, S^\mu_n A, (T^\mu_\lambda \otimes e^\mu, V^\mu_\lambda)) \subset E^\mu_\lambda$ is, actually, the equality

$$\mathrm{Ind}(S_n A, S^\mu_n A, (T^\mu_\lambda \otimes e^\mu, V^\mu_\lambda)) = E^\mu_\lambda. \tag{18}$$

This concludes the proof. □

Setting $A = 1$ yields a natural special case of Theorem 4.1. It was obtained in [8]. For convenience of the reader, we reproduce it here.

Corollary 4.2 (Gutkin [8], Theorem 1). *Let $\lambda \in \mathbf{Z}^n_{\geq+}(r)$. Let (π_λ, E_λ) (resp. (T_λ, V_λ)) be the irreducible representation of $GL(n, \mathbf{C})$ (resp. S_r) with the highest weight λ (resp. corresponding to the Young diagram λ). Let $\mu \in \mathbf{Z}^n_{\geq+}(r)$. Recall that $(T^\mu_\lambda, V^\mu_\lambda)$ is the associated representation of the stabilizer $S^\mu_n \subset S_n$ of the weight μ. Then*

$$(\pi^\mu_\lambda, E^\mu_\lambda) = \mathrm{Ind}(S_n, S^\mu_n, (T^\mu_\lambda, V^\mu_\lambda)). \tag{19}$$

Proof. Immediate from (14). □

For applications to the group $SL(n, \mathbf{C})$ we consider an extension of another special case of Theorem 4.1. Namely, let $A = (\mathbf{Z}/2\mathbf{Z})^n$ be the group of diagonal $n \times n$ matrices with entries ± 1. We will use the notation $\chi = (\chi_1, \dots, \chi_n)$, where $\chi_i = \pm 1$. The semidirect product $S_n(\mathbf{Z}/2\mathbf{Z})^n$ consists of matrices X of 0s and ±1s such that in each row and each column of X exactly one entry is not zero. We will call $S_n(\mathbf{Z}/2\mathbf{Z})^n$ the group of signed permutation matrices. For the sake of brevity we denote it by SP_n and the group $(\mathbf{Z}/2\mathbf{Z})^n$ by X_n. There is a natural isomorphism $\hat{X}_n = X_n$. Recall that every $\mu = (m_1, \dots, m_n) \in \mathbf{Z}^n$ defines a character, e^μ, of any group A of diagonal $n \times n$ matrices. In the case at hand, this character is identified with the element $((-1)^{m_1}, \dots, (-1)^{m_n}) \in X_n$. We denote it by χ_μ.

The kernel of the homomorphism $\chi \mapsto \det(\chi)$ of SP_n into $\{\pm 1\}$ is a normal subgroup of SP_n. We will denote it by SP^+_n. For any subgroup $F \subset SP_n$ set $F^+ = F \cap SP^+_n$. Thus $F^+ = F \cap SL(n, \mathbf{C})$.

Corollary 4.3. *Let the setting and the notation be as in Corollary 4.2. Then*

$$(\pi^\mu_\lambda(SP^+_n), E^\mu_\lambda) = \mathrm{Ind}(SP^+_n, (S^\mu_n X_n)^+, (T^\mu_\lambda \otimes \chi_\mu, V^\mu_\lambda)).$$

Proof. By Theorem 4.1, the representation $(\pi^\mu_\lambda(SP^+_n), E^\mu_\lambda)$ is the restriction of $\mathrm{Ind}(SP_n, S^\mu_n X_n, (T^\mu_\lambda \otimes \chi_\mu, V^\mu_\lambda))$ to the subgroup $SP^+_n \subset SP_n$. The restriction of an induced representation to a subgroup decomposes into induced representations as well, in accordance with the double coset decomposition of the group. See, e.g., [11]. In the case at hand, there is only one double coset. □

Among partitions $\mu = (m_1, \dots, m_n) \in \mathbf{Z}^n_{\geq+}$ those with $m_1 = \cdots = m_n$ are especially relevant for me. They are characterized by the property $S^\mu_n = S_n$. We will call them the *partitions into equal parts*. Let $\mu = (m, \dots, m)$. Then the character e^μ of H satisfies $e^\mu(h) = (\det h)^m$. We will denote it by $(\det)^m$. The description of the representation of the Weyl group and related groups in the spaces V^μ_λ is especially simple when μ is a partition into equal parts.

Theorem 4.4. *Let $m \in \mathbf{Z}_+$ and let $\lambda = (l_1, \dots, l_n) \in \mathbf{Z}^n_{\geq+}$ be a partition of the size mn. Set $\mu = (m, \dots, m)$. Let (π_λ, E_λ) be the irreducible representation of $GL(n, \mathbf{C})$ with the highest weight λ and let (T_λ, V_λ) be the irreducible representation of S_{mn} with the Young diagram λ. Let $A \subset H$ be a S_n-invariant subgroup, and let the rest of the notation be as in Theorem 4.1.*

Then $E^\mu_\lambda = E_\lambda(\mu)$. The induced representation of $S_n A$ in the weight space $E_\lambda(\mu)$ is equal to $(T^\mu_\lambda \otimes (\det)^m, V^\mu_\lambda)$.

Proof. Since the weight μ is fixed under the action of the Weyl group, the subspaces E_λ^μ and $E_\lambda(\mu)$ of E_λ coincide. For the same reason, $S_n^\mu A = S_n A$. Now equation (14) implies the claim. □

The case $m = 1$ of the preceding theorem is especially simple. It is formulated below, leaving the proof to the reader. When the group A is trivial, then Theorem 4.4 and Corollary 4.5 are contained in [8].

Corollary 4.5. *Let $A \subset H$ be any S_n-invariant subgroup, and let* det *be the determinant character of A. Let $\lambda \in \mathbf{Z}^n_{\geq +}(n)$. Let (π_λ, E_λ) be the irreducible $GL(n, \mathbf{C})$-module with the highest weight λ and let (T_λ, V_λ) be the irreducible representation of S_n corresponding to the Young diagram λ.*

Then the natural representation of $S_n A$ in the subspace $E_\lambda(0)$ of E_λ of weight vectors with the weight $(1, \dots, 1)$ is the representation $T_\lambda \otimes \det$ of $S_n A$. In particular, the representation $(\pi_\lambda(S_n), E_\lambda(0))$ is irreducible. The mapping $(\pi_\lambda, E_\lambda) \mapsto (\pi_\lambda(S_n), E_\lambda(0))$ yields a one-to-one correspondence between $\mathbf{Z}^n_{\geq +}(n)$ and $\hat{S}_n$.

4.2 The group $SL(n, \mathbf{C})$

We use the general notation established in Section 4.1. The main tool of this subsection will be a reduction of the statements for $SL(n, \mathbf{C})$ to their homologues for $GL(n, \mathbf{C})$. In order to compare various constructs for these groups, we will indicate the dependence on a group in an obvious way. Thus, $\Lambda_+(GL(n))$ and $\Lambda_+(SL(n))$ are the sets of dominant weights of the respective groups, $\Lambda^{GL}(\lambda)$ and $\Lambda^{SL}(\lambda)$ are the sets of weights of the irreducible representations of the respective groups with the highest weight λ, etc.

The rank of $SL(n, \mathbf{C})$ is $n - 1$, and its lattice of (resp. set of positive) weights is naturally identified as $\Lambda = \mathbf{Z}^{n-1}$ (resp. $\Lambda_+ = \mathbf{Z}_+^{n-1}$). Let $\lambda = (l_1 \geq \cdots \geq l_n) \in \mathbf{Z}^n_{\geq} = \Lambda_+(GL(n))$, and let (π_λ, E_λ) be the corresponding irreducible representation of $GL(n, \mathbf{C})$. Its restriction to $SL(n, \mathbf{C})$ is also irreducible. Symbolically, $(\pi_\lambda(SL), E_\lambda) = (\pi_\kappa, E_\kappa)$. Let $\kappa = (k_1, \dots, k_{n-1}) \in \mathbf{Z}_+^{n-1}$ be the highest weight of the restricted representation. Then $k_i = l_i - l_{i+1}$, $1 \leq i \leq n - 1$. This yields a linear mapping, $\phi : \mathbf{Z}^n \to \mathbf{Z}^{n-1}$. The weights of the $GL(n, \mathbf{C})$-module (π_λ, E_λ) and the corresponding $SL(n, \mathbf{C})$-module (π_κ, E_κ) satisfy $\Lambda^{SL}(\kappa) = \phi(\Lambda^{GL}(\lambda))$.

Let $\lambda = (l_1, \dots, l_n) \in \mathbf{Z}^n_{\geq}$, and let $\lambda' = (l_1 + t, \dots, l_n + t) \in \mathbf{Z}^n_{\geq}$. Then $\pi_{\lambda'} = \pi_\lambda \otimes (\det)^t$. Hence, the restrictions of $\pi_{\lambda'}$ and π_λ to $SL(n, \mathbf{C})$ coincide. Every irreducible $SL(n, \mathbf{C})$-module π_κ occurs infinitely many times in the tensor algebra $\mathcal{T}$. It is convenient to realize π_κ in $\mathcal{T}_r$ with the smallest r. Define a linear mapping $\psi : \mathbf{Z}^{n-1} \to \mathbf{Z}^n$ as follows. If $\kappa = (k_1, \dots, k_{n-1}) \in \mathbf{Z}^{n-1}$ and $\lambda = \psi(\kappa) = (l_1, \dots, l_n)$, then

$$l_i = \sum_{j \geq i} k_j \ : 1 \leq i \leq n - 1, \text{ and } l_n = 0. \tag{20}$$

For $\lambda = (l_1, \dots, l_n)$ set $|\lambda| = l_1 + \ldots + l_n$. For $\kappa = (k_1, \dots, k_{n-1})$ set $\langle \kappa \rangle = |\psi(\kappa)|$. Then $\langle \kappa \rangle = k_1 + 2k_2 + \cdots + (n-1)k_{n-1}$. Let $\kappa \in \mathbf{Z}_+^{n-1}$. Then the smallest rank r such that the representation π_κ of $SL(n, C)$ occurs in $\mathcal{T}_r$ is $r = \langle \kappa \rangle$ and the corresponding highest weight is $\lambda = \psi(\kappa) \in \mathbf{Z}^n_{\geq +}$. We will need the explicit form of the one-to-one correspondence between the sets of weights $\Lambda^{GL}(\psi(\kappa)) \subset \mathbf{Z}^n$ and $\Lambda^{SL}(\kappa) \subset \mathbf{Z}^{n-1}$. In the direction from $GL(n)$ to $SL(n)$ it is given by the linear projection $\phi : \Lambda^{GL}(\psi(\kappa)) \to \Lambda^{SL}(\kappa)$. In the direction from $SL(n)$ to $GL(n)$ it is given by the proposition below.

Proposition 4.6. *Let $\kappa \in \mathbf{Z}^{n-1}_+$, and let π^{SL}_κ be the $SL(n, \mathbf{C})$-module with the highest weight κ. Let $\lambda = \psi(\kappa) \in \mathbf{Z}^n_{\geq +}$, and let π^{GL}_λ be the $GL(n, \mathbf{C})$-module with the highest weight λ. Then*
1. for any weight $\nu \in \Lambda^{SL}(\kappa)$ we have $\langle\kappa\rangle - \langle\nu\rangle \in n\mathbf{Z}_+$;
2. The corresponding weight $\mu \in \Lambda^{GL}(\lambda)$ is given by

$$\mu = \nu + \frac{\langle\kappa\rangle - \langle\nu\rangle}{n}(1, \dots, 1). \tag{21}$$

We will denote by ψ_κ the mapping given by (21). Then $\psi_\kappa : \Lambda^{SL}(\kappa) \to \Lambda^{GL}(\psi(\kappa))$ is inverse to the the projection $\phi : \Lambda^{GL}(\psi(\kappa)) \to \Lambda^{SL}(\kappa)$. We will use the isomorphisms ϕ anf ψ_κ between $\Lambda^{SL}(\kappa)$ and $\Lambda^{GL}(\psi(\kappa))$ to describe the representations of the Weyl group of $SL(n, \mathbf{C})$, and related groups in the weight spaces of the simple $SL(n, \mathbf{C})$-modules. The Cartan group H_{SL} is the group of diagonal $n \times n$ matrices with determinant one. As opposed to the case of $GL(n)$, the normalizer $N(H_{SL}) \subset SL(n, \mathbf{C})$ is not a semidirect product of the symmetric group S_n (i.e., the Weyl group of $SL(n, \mathbf{C})$) and H_{SL}. This motivates the following definition.

Definition 4.7. A subgroup $F \subset N(H_{SL})$ is an *almost semidirect product* if there exists a S_n-invariant subgroup $A \subset H_{GL}$ such that $F = (S_n A) \cap SL(n, \mathbf{C})$.

In the notation of Corollary 4.3, $F = (S_n A)^+$. The normalizer itself is an almost semidirect product: $N(H_{SL}) = (S_n H_{GL})^+$. Another natural example of an almost semidirect product is the group of signed permutation matrices with determinant one. In the notation of Corollary 4.3, this group is SP_n^+, where $SP_n = S_n X_n \subset GL(n, \mathbf{C})$ is the group of signed permutation matrices.

Theorem 4.8. *Let $\kappa = (k_1, \dots, k_{n-1}) \in \mathbf{Z}^{n-1}_+$ and let (π_κ, E_κ) be the irreducible representation of $SL(n, \mathbf{C})$ with the highest weight κ. Let $\nu \in \Lambda^{SL}_+(\kappa)$ be a weight, and let $E^\nu_\kappa \subset E_\kappa$ be the corresponding $N(H_{SL})$-invariant subspace.*

Let $F \subset N(H_{SL})$ be a subgroup. Suppose that F is an almost semidirect product : $F = (S_n A)^+$. Let $(\pi^\nu_\kappa(F), E^\nu_\kappa)$ be the natural representation of F in E^ν_κ.

Set $\lambda = \psi(\kappa) \in \mathbf{Z}^n_{\geq +}$, and let $\mu = \psi_\kappa(\nu)$. Then

$$(\pi^\nu_\kappa((S_n A)^+), E^\nu_\kappa) = \mathrm{Ind}((S_n A)^+, (S^\mu_n A)^+, (T^\mu_\lambda \otimes e^\mu, V^\mu_\lambda)). \tag{22}$$

Proof. The representation on the left-hand side of (22) is the restriction to the subgroup $(S_n A)^+$ of the natural representation of $S_n A \subset N(H_{GL})$ in the space E^μ_λ. Equation (14) of Theorem 4.1 says that the latter is equivalent to a certain induced representation. The restriction of an induced representation to a subgroup decomposes into induced representations, according to the double coset decomposition of the group. See, e.g., [11]. Arguing as in the proof of Corollary 4.3, we conclude that in the case at hand the decomposition consists of only one double coset. □

Comparing Theorem 4.8 with Theorem 4.1, we note that the description of the natural representation of the group $F \subset N(H_{SL})$ in the weight space $E_\kappa^\nu \subset E_\kappa$ is less direct in the case of $SL(n, \mathbf{C})$. In particular, it involves the auxiliary weights λ and μ in $\mathbf{Z}^n$, while the relevant parameters κ and ν belong to the lattice $\mathbf{Z}^{n-1}$. However, in specific examples the dependence on the parameters κ and ν can be made explicite. Below we consider in some detail the example of $F = SP_n^+$, where $SP_n \subset GL(n, \mathbf{C})$ is the group of signed permutation matrices.

The group SP_n is the semidirect product of the groups $S_n \subset GL(n, \mathbf{C})$ of permutation matrices, and $X_n \subset H_{GL}$ of the pure sign changes. Let $\kappa = (k_1, \ldots, k_{n-1}) \in \mathbf{Z}_+^{n-1}$ and let $\nu = (j_1, \ldots, j_{n-1}) \in \Lambda_+^{SL}(\kappa)$ be a weight in E_κ. By Proposition 4.6

$$\langle\kappa\rangle - \langle\nu\rangle = (k_1 - j_1) + 2(k_2 - j_2) + \cdots + (n-1)(k_{n-1} - j_{n-1}) = mn.$$

Then the weights $\lambda, \nu \in \mathbf{Z}^n$ of Theorem 4.8 are given by

$$\lambda = (k_1 + k_2 + \cdots + k_{n-1}, k_2 + k_3 + \cdots + k_{n-1}, \ldots, k_{n-1}, 0) \tag{23}$$

and

$$\mu = (j_1 + j_2 + \cdots + j_{n-1} + m, j_2 + j_3 + \cdots + j_{n-1} + m, \ldots, j_{n-1} + m, m). \tag{24}$$

As a consequence, the group $S_n^\mu \subset S_n$ is generated by the transpositions $s_{t,t+1}$ such that $j_t = 0$. Hence S_n^μ is directly determined by the weight ν. We will use the notation S_n^ν for it. Note that $S_n^\nu \subset S_n$ is the isotropy subgroup of ν under the natural action of S_n on $\mathbf{Z}^{n-1}$. Also note that S_n^ν is the cartesian product of symmetric groups $S_{t_1}, \ldots, S_{t_p}$, where t_i is the length of the ith block of equal coordinates in the vector ν.

Denote the elements of X_n by $\chi = (\chi_1, \ldots, \chi_n)$, where $\chi_i = \pm 1$. Let $\mu = (m_1, \ldots, m_n) \in \mathbf{Z}^n$. The character e^μ of X_n is given by $e^\mu(\chi) = \chi_1^{m_1} \cdots \chi_n^{m_n}$. Equations (23) and (24) imply

$$\begin{aligned} e^\mu(\chi) &= (\chi_1 \cdots \chi_n)^m \chi_1^{j_1 + \cdots + j_{n-1}} \chi_2^{j_2 + \cdots + j_{n-1}} \cdots \chi_{n-1}^{j_{n-1}} \\ &= (\chi_1 \cdots \chi_n)^m (\chi_1 \cdots \chi_{n-1})^{j_{n-1}} (\chi_1 \cdots \chi_{n-2})^{j_{n-2}} \cdots \chi_1^{j_1}. \end{aligned}$$

Denote by $\theta : X_n \to X_n$ the automorphism given by

$$\theta(\chi_1, \ldots, \chi_n) = (\chi_1, \chi_1\chi_2, \ldots, \chi_1 \cdots \chi_n).$$

Set $\nu_1 = (j_1, \ldots, j_{n-1}, m) \in \mathbf{Z}^n$. Then the equation above can be written as $e^\mu(\chi) = e^{\nu_1}(\theta(\chi))$. The character $\chi \mapsto e^{\nu_1}(\theta(\chi))$ of X_n depends not only on $\nu \in \mathbf{Z}^{n-1}$ but on the weight $\kappa \in \mathbf{Z}^{n-1}$ as well. Identifying $\hat{X}_n$ with X_n, we will denote it by $\chi(\kappa, \nu)$.

Corollary 4.9. *Let (π_κ, E_κ) be the representation of $SL(n, \mathbf{C})$ with the highest weight κ. Let $\nu \in \Lambda_+^{SL}(\kappa)$ and let $E_\kappa^\nu \subset E_\kappa$ be the corresponding $N(H_{SL})$-invariant subspace. Let $\lambda = \psi(\kappa)$ and let $\mu = \psi_\kappa(\nu)$ be the corresponding n-tuples. Let $r =< \kappa >$. Let (T_λ, V_λ) be the irreducible representation of S_r corresponding to the Young diagram λ. Let $(T_\lambda^\mu, V_\lambda^\mu)$ be the corresponding representation of S_n^ν — the isotropy subgroup of $\nu \in \mathbf{Z}^{n-1}$ under the natural action of S_n on $\mathbf{Z}^{n-1}$. Let $\chi(\kappa, \nu)$ be the character of X_n defined above. Denote by the same symbol its unique extension to the group $S_n^\nu X_n$.*

Let $(\pi_\kappa^\nu, E_\kappa^\nu)$ be the natural representation of $SP_n^+ = SP_n \cap SL(n, \mathbf{C})$ in E_κ^ν. Then

$$(\pi_\kappa^\nu, E_\kappa^\nu) = \mathrm{Ind}(SP_n^+, (S_n^\nu X_n)^+, (T_\lambda^\mu \otimes \chi(\kappa, \nu), V_\lambda^\mu)). \tag{25}$$

Proof. This is a direct consequence of Theorem 4.8 and the preceding discussion. □

Now we will use the preceding theorem to answer the original question. Namely, we will identify the natural representation of S_n, the Weyl group of $SL(n, \mathbf{C})$, in the space E_κ^0 of the vectors of weight zero in E_κ. The two propositions that follow are contained in the paper [8]. See Theorem 2 and its Corollary in [8].

Corollary 4.10. *Let $\kappa = (k_1, \ldots, k_{n-1}) \in \mathbf{Z}_+^{n-1}$ $(\kappa \neq 0)$, and let (π_κ, E_κ) be the irreducible representation of $SL(n, \mathbf{C})$ with the highest weight κ.*

1. Suppose that the space $E_\kappa^0 \subset E_\kappa$ is nontrivial. Then $\langle\kappa\rangle = mn$, where m is a positive integer.

2. Denote by π_κ^0 the natural representation of S_n in E_κ^0. Set $\lambda = \psi(\kappa) \in \mathbf{Z}_{\geq +}^n$ and let (T_λ, V_λ) be the representation of S_{mn} with the Young diagram λ. Let $\times^n S_m \subset S_{mn}$ be the nth cartesian power subgroup, and denote by $V_\lambda^0 \subset V_\lambda$ the subspace of vectors fixed by $T_\lambda(\times^n S_m)$. The normalizer of $\times^n S_m$ in S_{mn} is naturally identified with S_n. Denote by T_λ^0 the corresponding representation of S_n in V_λ^0. Then, for any $w \in S_n$

$$(\pi_\kappa^0(w), E_\kappa^0) = (T_\lambda^0(w)(\det w)^m, V_\lambda^0). \tag{26}$$

Proof. We will use the notation of Corollary 4.9 and Proposition 4.6. Since $\langle 0\rangle = 0$, by Proposition 4.6, $< \kappa > \in \mathbf{Z}_+$, which proves the first claim.

By Theorem 4.8, $\mu = (m, \ldots, m)$, hence the corresponding subgroup of S_{mn} is $\times^n S_m$. The subgroup S_n^ν in this case is the full group S_n. The character $\chi(\kappa, \nu)$ of Corollary 4.9 in the case at hand is given by

$$\chi = (\chi_1, \ldots, \chi_n) \mapsto (\chi_1 \cdots \chi_n)^m = (\det \chi)^m.$$

So far we identified the representation of SP_n^+ on the right-hand side of (25) with the natural representation of SP_n^+ in V_λ^0, twisted by the character $(\det \chi)^m$. The same observation holds if we replace SP_n^+ by the full group SP_n. The Weyl group S_n is the quotient of SP_n^+, and a subgroup of SP_n. Let $w \in S_n$. To obtain the operator $\pi_\kappa^0(w)$, we take the corresponding element $w \in SP_n$, multiply it by any $\chi \in X_n$, such that $w\chi \in SP_n^+$, and then take the action of $w\chi$. Equation (26) follows from the observation that $\det \chi = \det w$. □

The special case $m = 1$ of the preceding proposition yields the SL_n-version of Corollary 4.5. Recall that the set of partitions of n has a natural involution, $\lambda \mapsto \lambda'$, which interchanges the rows and the columns of the corresponding Young diagram. See, e.g., [5]. Let $\hat{S}_n$ be the set of partitions of n, viewed as the set of irreducible representations of S_n. Denote by $\tilde{S}_n \subset \hat{S}_n$ the set of all partitions except the trivial one : $n = n$. The representation corresponding to that partition is the trivial representation of S_n. Thus $\tilde{S}_n$ is the set of nontrivial irreducible representations of S_n.

Corollary 4.11. *Let $\kappa = (k_1, \dots, k_{n-1}) \in \mathbf{Z}_+^{n-1}$ be such that $k_1 + 2k_2 + \cdots (n-1)k_{n-1} = n$. Let (π_κ, E_κ) be the representation of $SL(n, \mathbf{C})$ with the highest weight κ. Let*

$$\lambda = (k_1 + k_2 + \cdots + k_{n-1}, k_2 + k_3 + \cdots + k_{n-1}, \dots, k_{n-1}, 0) \in \mathbf{Z}^n_{\geq +}$$

be the corresponding partition of n, and let $\lambda' = \lambda'(\kappa)$ be the dual partition. Then the subspace $E_\kappa^0 \subset E_\kappa$ is nontrivial, and the natural representation of S_n in E_κ^0 is the irreducible representation $T_{\lambda'}$. The mapping $(\pi_\kappa, E_\kappa) \mapsto (\pi_\kappa^0(S_n), E_\kappa^0)$ yields a one-to-one correspondence between the subset $\{\kappa \in \mathbf{Z}_+^{n-1}: <\kappa>= n\}$ and $\tilde{S}_n$.

Proof. It is immediate from Corollary 4.10 that the representation $\pi_\kappa^0(S_n)$ is equivalent to the tensor product $T_\lambda \otimes \det$. It is well known that $T_\lambda \otimes \det = T_{\lambda'}$. See, e.g., [5]. Set $K = \{\kappa \in \mathbf{Z}_+^{n-1} :<\kappa>= n\}$. The mapping $\psi : K \to \mathbf{Z}^n_{\geq +}(n)$ from K to the set of partitions of n is injective. The range consists of all partitions of n, except $\lambda_0 : n = 1 + \cdots + 1$. Since λ_0' is the trivial partition of n, the claim follows. □

Remark. The absence of the trivial representation of S_n in Corollary 4.11 is unfortunate, but accidental. It naturally corresponds to the trivial representation of $SL(n, \mathbf{C})$. This is the representation with the highest weight $\kappa = 0$. It is the only dominant weight that satisfies $<\kappa>= 0$, and the general scheme is geared toward the case $<\kappa>= mn$, $m > 0$.

Acknowledgements. This paper was written in April/June 2001. At the time I held the CNRS appointment of the Directeur de Recherche Associé at the CMAT, Ecole Polytechnique, Palaiseau. It is a pleasure to thank the faculty and the staff of the Centre de Mathématiques for the hospitable and stimulating atmosphere, and the CNRS for their financial support.

References

[1] S. Ariki, J. Matsuzawa, and I. Terada, *Representations of Weyl groups on zero weight spaces of $\mathcal{G}$-modules*, Memorial volume for Dr. T. Miyata, Kinokuniya, Tokyo, 1985.

[2] S. Ariki, *On the representations of the Weyl groups of type D corresponding to the zero weights of the representations of $SO(2n, \mathbf{C})$*, Proc. Symp. Pure Math. 47, part II, AMS, Providence, 1987, 327–342.

[3] N. Bourbaki, *Groupes et algèbres de Lie, Ch. 4–6*, Masson, Paris, 1981.

[4] C. Curtis and I. Reiner, *Representation theory of finite groups and associative algebras*, Wiley, 1962.

[5] W. Fulton, *Young Tableaux*, CUP, Cambridge, 1997.

[6] D. Gay, Characters of the Weyl group of $SU(n)$ on zero weight spaces and centralizers of permutation representations, *Rocky Mount. J. Math.* **6** (1976), 449–455.

[7] V. Guillemin, E. Lerner, S. Sternberg, *Symplectic Fibrations and Multiplicity Diagrams*, CUP, Cambridge, 1996.

[8] E. Gutkin, Representations of the Weyl group in the space of vectors of zero weight, *Uspekhi Matem. Nauk* **28** (1973), 237–238.

[9] E. Gutkin, Schur-Weyl duality and representations of the classical Weyl groups, in preparation.

[10] R. Howe, Remarks on classical invariant theory, *Trans. AMS* **313** (1989), 539–570.

[11] A. A. Kirillov, *Eléments de la théorie des représentations*, Mir, Moscow, 1974.

[12] B. Kostant, Lie group representations on polynomial rings, *Amer. J. Math.* **85** (1963), 327–404.

[13] B. Kostant, On Macdonald's η function formula, the Laplacian and generalised exponents, *Adv. Math.* **20** (1976), 179–212.

[14] J. Millson and V. Toledano, Casimir operators and monodromy representations of generalised braid groups, preprint, 2001.

[15] M. Vergne, Appendix to "Remarks on classical invariant theory" by Roger Howe (Analyse structurale d'un texte d'un mathématicien moderne), preprint, 1977.

[16] Van der Waerden, *Algebra. Vols. I, II*, Springer, New York, 1991.

[17] H. Weyl, *The classical groups. Their invariants and representations*, Princeton University Press, Princeton, 1946.

Eugene Gutkin
Department of Mathematics
Caltech
1200 California Blvd.
Pasadena, CA 91125, USA
egutkin@its.caltech.edu
and
Max-Planck-Institut for Mathematics in the Sciences
Inselstrasse 22-26
D 04103 Leipzig, Germany
gutkin@mis.mpg.de, gutkine@aol.com

Quantization of Hypersurface Orbital Varieties in $\mathfrak{sl}_n$

Anthony Joseph
Anna Melnikov

ABSTRACT Let $\mathfrak{g}$ be a semisimple Lie algebra with $\mathfrak{h}$ a Cartan subalgebra. The orbit method attempts to assign representations of $\mathfrak{g}$ to orbits in $\mathfrak{g}^*$. Orbital varieties are particular Lagrangian subvarieties of such orbits which should lead to highest weight representations of $\mathfrak{g}$. It is known that all unitary highest weight representations can be obtained by their quantization [J2]. A hypersurface orbital variety is one which is of codimension 1 in the nilradical of a parabolic. Their classification for $\mathfrak{g} = \mathfrak{sl}(n)$ is obtained from general results in [M1]. Recently Benlolo and Sanderson [BS] conjectured the form of the (nonlinear) element describing such a variety. This paper proves that conjecture and further constructs a simple module with integral highest weight which "quantizes" the variety in the precise sense that the regular functions of its closure is given a $\mathfrak{g}$ module structure compatible (up to shift by the highest weight) with its $\mathfrak{h}$ module structure.

1 Introduction

1.1. Amongst his many great contributions to mathematics, Alexander Kirillov was a founder of the "orbit method". This exploits the symplectic structure on coadjoint orbits with respect to a Lie algebra $\mathfrak{g}$. From a Lagrangian subvariety one attempts to construct a representation of $\mathfrak{g}$ associated to the given orbit. If $\mathfrak{g}$ is semisimple, the orbit is nilpotent and one wishes to construct a highest weight module, then this Lagrangian subvariety should be a so-called orbital variety. Then "quantization" of the latter leads to the required representation [J3].

In this paper we consider only the case when $\mathfrak{g} = \mathfrak{sl}_n(\mathbb{C})$. By [M4] all orbital varieties can be weakly quantized (though this fails [J3, 1.3] for arbitrary $\mathfrak{g}$). It remains to show that they can be strongly quantized. This is a rather more delicate question. We shall settle this positively for orbital varieties which are of codimension 1 in the nilradical of a parabolic. These are called hypersurface orbital varieties.

1.2. Let $\mathfrak{g} = \mathfrak{n}^- \oplus \mathfrak{h} \oplus \mathfrak{n}^+$ be a triangular decomposition. We identify $\mathfrak{n}^-$ with $(\mathfrak{n}^+)^*$ through the Killing form. Let $\mathbf{G}$ be adjoint group of $\mathfrak{g}$ and $\mathbf{B}$ the Borel subgroup with Lie algebra $\mathfrak{b} := \mathfrak{h} \oplus \mathfrak{n}^+$. For any Lie algebra $\mathfrak{a}$, let $S(\mathfrak{a})$ (resp. $U(\mathfrak{a})$) denote its symmetric (resp. enveloping) algebra. It is well known that $\mathbf{G}\mathfrak{n}^+$ is a finite

Supported in part by the Minerva Foundation, Germany, Grant No. 8466.

union (of so-called nilpotent) orbits classified by the Jordan normal form. Let $\mathcal{O}$ be such an orbit. After Spaltenstein [S] the irreducible components of $\mathcal{O} \cap \mathfrak{n}^+$ have dimension $1/2 \dim \mathcal{O}$. They are called orbital varieties and after Steinberg [St] they are classified by the standard tableaux whose shape is specified by $\mathcal{O}$. As noted in [J1, 7.3] an orbital variety $\mathcal{V}$ can be characterized as an irreducible subvariety of $\mathfrak{n}^+$ for which the ideal $I(\overline{\mathcal{V}}) \subset S(\mathfrak{n}^-)$ of definition of its closure is stable under the Poisson bracket on $S(\mathfrak{g})$ (induced by the Lie bracket).

1.3. Retain the notation of 1.2. Roughly speaking $\mathcal{V}$ can be strongly quantized if $S(\mathfrak{n}^-)/I(\overline{\mathcal{V}})$ can be given the structure of a highest weight module (not necessarily simple). Here the choice of highest weight is a crucial and delicate point. Take $\lambda \in \mathfrak{h}^*$ and let $V(\lambda)$ be a highest weight module with highest weight vector v_λ of weight λ. Let $\mathcal{F}$ be the canonical (degree) filtration on $U(\mathfrak{n}^-)$ (resp. on $S(\mathfrak{n}^-)$). We say that $V(\lambda)$ is a strong (resp. weak) quantization of $\mathcal{V}$ if $\operatorname{gr}_{\mathcal{F}} \operatorname{Ann}_{U(\mathfrak{n}^-)} v_\lambda = I(\overline{\mathcal{V}})$ (resp. if its radical equals $I(\overline{\mathcal{V}})$).

1.4. In order to exhibit a strong quantization we have to compute $I(\overline{\mathcal{V}})$ (or at least to calculate the formal character of $S(\mathfrak{n}^-)/I(\overline{\mathcal{V}})$). In [J2, Lecture 7] the general form that this should take is suggested. Let $\mathcal{N}$ be the generic matrix of $\mathfrak{n}^- + \mathrm{Id}$. Then $I(\overline{\mathcal{V}})$ should take the form $\operatorname{gr}_{\mathcal{F}}\langle a_1, a_2, \dots, a_n\rangle$, where a_i are amongst the minors of $\mathcal{N}$. This suggestion derived from an algorithm for $\operatorname{Ann}_{U(\mathfrak{n}^-)} \mathcal{V}$ based on the Enright functor [J1, 8.4] together with calculations in the thesis of E. Benlolo.

1.5. Of course the above does not tackle the difficult question as to which minors a_i to choose. Benlolo and Sanderson [BS] made an appealing conjecture for this choice concerning orbital varieties $\mathcal{V}$ having codimension 1 in the nilradical $\mathfrak{m}^+$ of a parabolic subalgebra $\mathfrak{p}$. Apart from the obvious 1×1 minors it follows from Krull's theorem that $I(\overline{\mathcal{V}})$ is generated by just one element f which can be assumed to be an $\mathfrak{h}$ weight vector. A result in [M3] describes in particular all the orbital varieties of codimension 1 in a given $\mathfrak{m}^+$. Using this, Benlolo–Sanderson [BS] show that one may reduce to the case when f cannot be expressed in terms of the generic matrices of $\mathfrak{n}^-$ obtained by deleting the first columns or last rows. Up to this reduction they show that $\mathfrak{m}^+$ is given by blocks of size $c_1, c_2, \dots, c_l$ with $c_1 = c_l = c$ and $c_i \neq c, \forall\, i : 1 < i < l$. Now let $\mathcal{M}(t)$ be the generic matrix of $\mathfrak{m}^- + t\mathrm{Id}$, where $\mathfrak{m}^-$ is the opposed algebra of $\mathfrak{m}^+$ identified with $(\mathfrak{m}^+)^*$ and t is an indeterminate. Let $M(t)$ be the $(n-c) \times (n-c)$ minor of $\mathcal{M}(t)$ lying in the bottom left-hand corner. Then $M(t)$ is a polynomial in t. They conjecture that f is the coefficient of its lowest degree term. Let β_1 be the highest root of $\mathfrak{n}^+$, β_2 the highest root of the subalgebra of $\mathfrak{n}^+$ obtained by omitting the first row and last column, and so on. They show that f has weight $-\sum_{i=1}^{c} \beta_i$.

1.6. One may note that f of 1.5 can also be described as $\operatorname{gr}_{\mathcal{F}} M(1)$. However as we shall see there is a certain computational advantage in retaining t. (Furthermore BS compute the lowest power of t in $M(t)$). One may also replace $\mathfrak{m}^- + \mathrm{Id}$ in the above by $\mathfrak{n}^- + \mathrm{Id}$ as long as one includes the obvious 1×1 minors in $I(\overline{\mathcal{V}})$. In this sense $M(1)$ only depends on the first column size c. Likewise $f = \operatorname{gr}_{\mathcal{F}} M(1)$

depends only partially on $\mathfrak{m}^-$. In particular the weight of f only depends on c. Although BS do check their conjecture in a number of cases, it is interesting to know if it holds in general since the above independence is somewhat surprising. Moreover even the relatively simple case of hypersurface varieties has some remarkable structure.

1.7. Our proof of the BS conjecture involves showing that $S(\mathfrak{m}^-)f$ is semi-prime and its zero variety is **P** stable, where $\mathbf{P} \supset \mathbf{B}$ is the parabolic subgroup with Lie algebra $\mathfrak{p}$. This holds regardless of whether $c_i \neq c$, for $i \neq 1, l$. However we show that f is irreducible if and only if this condition holds. This is a delicate point. We shall give three proofs. One is based on a careful counting of t powers in multiplying out $M(t)$ and a knowledge [M2, M3] of the orbital varieties of codimension 1 in $\mathfrak{m}^+$. The second is based on a representation-theoretic argument combined with [M4] which asserts that the associated variety of a highest weight module of *integral* highest weight is irreducible. The third is the least computational and is based on the irreducibility of the associated variety of the annihilator of a simple highest weight module.

1.8. We shall exhibit a strong quantization of $\mathcal{V}$ by determining a simple highest weight module $V(\lambda)$ whose formal character ch $V(\lambda)$ coincides with that of $S(\mathfrak{m}^-)/S(\mathfrak{m}^-)f$ and by using the linear independence of the characteristic polynomials of orbital varieties. To determine λ and compute ch $V(\lambda)$, we use the formula of Jantzen [Ja] describing the Shapovalov determinants defined with respect to $\mathfrak{p}$. Let $\mathcal{I}$ be the subset of the set of simple roots Π defining $\mathfrak{p}$. Assume $(\alpha, \alpha) = 2$ for any root. Then it suffices to take $\lambda \in \mathfrak{h}^*$ such that $(\lambda, \alpha) = 0$, $\forall\, \alpha \in \mathcal{I}$, $(\lambda, \beta_1) = c - (n-1)$ and $(\lambda, \gamma) \notin \mathbb{Z}$, for any proper sum of roots in $\Pi' := \Pi \setminus \mathcal{I}$. However we also give a choice of integral λ. This is more delicate since the condition $c_i \neq c$, $\forall\, i : 1 \leq i \leq k-1$ must be invoked. Combined with [M4] it leads to the second proof of the irreducibility of f. The non-integral case gives our third proof.

We further show that there is a choice of λ such that the annihilator of $V(\lambda)$ is maximal as well as $V(\lambda)$ being a strong quantization of $\mathcal{V}$.

1.9. The truth of the BS conjecture has the following remarkable consequence. One may specify positive integers s, t such that f is exactly the highest common divisor of the $t \times t$ minors of $\mathcal{M}(0)^s$. There seems to be no elementary proof of this purely combinatorial fact.

2 Combinatorial preliminaries and the Benlolo–Sanderson conjecture

2.1. The base field is assumed algebraically closed of characteristic zero and can be taken to be the complex field $\mathbb{C}$ without loss of generality.

Recall the notation and the hypotheses of 1.1 and 1.2. Put $\mathfrak{n} = \mathfrak{n}^+$. In particular an orbital variety $\mathcal{V}$ associated to a nilpotent orbit $\mathcal{O}$ is an irreducible component of $\mathcal{O} \cap \mathfrak{n}$. Let $\mathbf{V}$ denote the set of all orbital varieties of $\mathfrak{g}$ and W the Weyl group for the pair $(\mathfrak{g}, \mathfrak{h})$. We describe first the Steinberg map of W onto $\mathbf{V}$. Let $R \subset \mathfrak{h}^*$ denote the set of nonzero roots, R^+ the set of positive roots corresponding to $\mathfrak{n}$ in the triangular decomposition of $\mathfrak{g}$ and $\Pi \subset R^+$ the resulting set of simple roots. Let $X_\alpha = \mathbb{C}x_\alpha$ denote the root subspace corresponding to $\alpha \in R$. Then $\mathfrak{n} = \bigoplus_{\alpha \in R^+} X_\alpha$ (resp. $\mathfrak{n}^- = \bigoplus_{\alpha \in -R^+} X_\alpha$).

For each $w \in W$ set $\mathfrak{n} \cap^w \mathfrak{n} := \bigoplus_{\alpha \in R^+ \cap w(R^+)} X_\alpha$. For each closed, irreducible subgroup $\mathbf{H}$ of $\mathbf{G}$ let $\mathbf{H}(\mathfrak{n} \cap^w \mathfrak{n})$ be the union of the $\mathbf{H}$ conjugates of $\mathfrak{n} \cap^w \mathfrak{n}$. It is an irreducible locally closed subvariety. Since there are only finitely many nilpotent orbits in $\mathfrak{g}$ it follows that there exists a unique nilpotent orbit $\mathcal{O}$ such that $\overline{\mathbf{G}(\mathfrak{n} \cap^w \mathfrak{n})} = \overline{\mathcal{O}}$. A result of Steinberg [St] asserts that $\mathcal{V}_w := \overline{\mathbf{B}(\mathfrak{n} \cap^w \mathfrak{n})} \cap \mathcal{O}$ is an orbital variety and that the map $\phi : w \mapsto \mathcal{V}_w$ is a surjection of W onto $\mathbf{V}$.

2.2. It is convenient to replace $\mathfrak{sl}_n$ by $\mathfrak{g} = \mathfrak{gl}_n$. This obviously makes no difference. Note that the adjoint action is just conjugation by $\mathbf{G} = GL_n$. Let $\mathfrak{n}$ be the subalgebra of strictly upper-triangular matrices and let $\mathfrak{n}^-$ be the subalgebra of strictly lower-triangular matrices. Let $\mathbf{B}$ be the (Borel) subgroup of upper-triangular matrices in $\mathbf{G}$. All parabolic subgroups we consider further are standard, that is contain $\mathbf{B}$.

Let $e_{i,j}$ be the matrix having 1 in the ijth entry and 0 elsewhere.

Take $i < j$ and let $\alpha_{i,j}$ be the root corresponding to $e_{i,j}$. Set $\alpha_{j,i} = -\alpha_{i,j}$. We write $\alpha_{i,i+1}$ simply as α_i. Then $\Pi = \{\alpha_i\}_{i=1}^{n-1}$. Moreover $\alpha_{i,j} \in R^+$ exactly when $i < j$. For each $\alpha \in \Pi$, let $s_\alpha \in W$ be the corresponding reflection and set $s_i = s_{\alpha_i}$.

2.3. We represent every element of the symmetric group $\mathbf{S}_n$ in word form

$$w = [a_1, a_2, \ldots, a_n], \quad \text{where } a_i = w(i).$$

We identify W with $\mathbf{S}_n$ by taking s_i to be the elementary permutation interchanging $i, i+1$.

Definition. Let $w = [a_1, \ldots, a_n]$. Set $p_w(i) = j$ if $a_j = i$, that is $p_w(i)$ is the place (index) of i in the word form of w.

One has $w(p_w(i)) = w(j) = a_j = i$, that is $p_w(i) = w^{-1}(i)$. On the other hand $w(\alpha_{i,j}) = \alpha_{w(i),w(j)}$. Set $S(w) = \{\alpha \in R^+ \mid w(\alpha) \in -R^+\}$.

This gives the following result.

Lemma. *Take $i < j$. Then $\alpha_{i,j} \in S(w^{-1})$ if and only if $p_w(i) > p_w(j)$.*

Remark. In particular $\alpha_i \in S(w^{-1})$ exactly when $i+1$ comes before i in the word form of w. This is of course well known.

2.4. Let $P(n)$ denote the set of partitions of n. Then $\lambda := \{\lambda_1 \geq \lambda_2 \geq \cdots \geq \lambda_k > 0\} \in P(n)$ defines the sizes λ_i of the Jordan blocks of some element $x_\lambda \in \mathfrak{n}$. Set $\mathcal{O}_\lambda = \mathbf{G}x_\lambda$. Let $\lambda^* = \{\lambda_1^* \geq \lambda_2^* \geq \cdots \geq \lambda_l^* > 0\}$ be the dual partition. By definition $\lambda_i^* = \sharp\{j \mid \lambda_j \geq i\}$. Then $k = \lambda_1^*$, $l = \lambda_1$ and this notation will be fixed throughout. We view λ as a Young diagram D_λ with k rows of length $\lambda_1, \ldots, \lambda_k$.

The rank of a Jordan block of size r is just $r - 1$ so $\operatorname{rk} x_\lambda = \sum_{i=1}^k (\lambda_i - 1) = n - \lambda_1^*$.

For all integers $i \geq 1$, the Jordan blocks of x_λ^i are obtained by deleting the first i columns of λ. Then

(1) $\operatorname{rk} x_\lambda^i = n - \sum_{j=1}^i \lambda_j^*$.

(2) Again (cf. [H, §3.8]) $\dim \mathcal{O}_\lambda = n^2 - \sum_{i=1}^k (\lambda_i^*)^2$.

2.5. Define a partial order on $P(n)$ as follows. Let $\lambda = (\lambda_1 \geq \lambda_2 \geq \cdots \lambda_k)$ and let $\mu = (\mu_1 \geq \mu_2 \geq \cdots \mu_k)$ be partitions of n which we can assume correspond to diagrams of the same height by adding empty rows. Set $\lambda \leq \mu$ if

$$\sum_{j=1}^i \lambda_j \geq \sum_{j=1}^i \mu_j, \qquad \text{for all} \qquad i = 1, 2, \ldots, k.$$

The following result of M. Gerstenhaber (cf. [H, §3.10]) shows that this order corresponds to inclusion of nilpotent orbit closures:

Theorem. *Given two partitions λ and μ of n one has $\lambda \leq \mu$ if and only if $\overline{\mathcal{O}}_\mu \subset \overline{\mathcal{O}}_\lambda$.*

In our convention $\lambda = (n)$ is the minimal partition of n and $\lambda = (1, 1, \ldots, 1)$ is the maximal one.

2.6. For a partition λ of n one can fill the boxes of D_λ with n distinct positive integers. If the entries increase in rows from left to right and in columns from top to bottom, we call such an array a Young tableau. If the numbers in a Young tableau form a set of integers from 1 to n, we call it standard. Let $\mathbf{T}_n$ be the set of standard Young tableaux of size n.

We shall not distinguish between Young and standard tableaux. Indeed if a Young tableau has entries $t_1 < t_2 < \cdots < t_n$, then it can be identified with the standard tableau obtained by replacing t_i by i. The sequence $[t_{w(1)}, \ldots, t_{w(n)}]$: $w \in \mathbf{S}_n$ can be viewed similarly as a word form of w by replacing t_i with i. We call this process standardization. Occasionally it can cause confusion; but in such cases adequate warning will be given. We may also concatenate word forms so that then $[a_1, a_2, \ldots, a_n]$ can be written as $[a', a'']$ where $a' = [a_1, a_2, \ldots, a_i]$, $a'' = [a_{i+1}, \ldots, a_n]$.

The shape of a Young tableau T is defined to be the Young diagram from which T was built. It defines a partition of n which we denote by $\operatorname{sh} T$.

Given $T \in \mathbf{T}_n$, let $\sigma(T)$ denote the conjugate standard tableau obtained by rotation about the main diagonal. We remark that $\operatorname{sh} \sigma(T) = (\operatorname{sh} T)^*$. Again σ takes a row tableau R into a column tableau C and vice-versa.

Given any sequence S of strictly increasing integers (for example a row R or column C above) let $\widehat{S}$ denote the reversed sequence of strictly decreasing integers. For any tableau or word form S, let $\langle S\rangle$ denote the set of entries of S and $|S|$ the cardinality of $\langle S\rangle$.

The Robinson–Schensted correspondence $w \mapsto (Q(w), R(w))$ gives a bijection (see, for example [Kn] or [F]) from the symmetric group $\mathbf{S}_n$ onto the pairs of standard Young tableaux of the same shape. By R. Steinberg [St] for all $w, y \in \mathbf{S}_n$ one has $\mathcal{V}_w = \mathcal{V}_y$ iff $Q(w) = Q(y)$. This parameterizes the set of orbital varieties $\mathbf{V}$ by $\mathbf{T}_n$. Moreover sh $Q(w) = \lambda$ if and only if $\mathcal{V}_w$ is contained in $\mathcal{O}_\lambda$.

If $Q(w) = T$, we set $\mathcal{V}_T := \mathcal{V}_w$ and $T_{\mathcal{V}_w} := Q(w)$.

We define a partial order on Young tableaux and a partial order on orbital varieties analogous to the order on partitions given in 2.5. It is called the geometric order on $\mathbf{V}$ or on $\mathbf{T}_n$.

Definition.

(i) Given distinct $\mathcal{V}_1, \mathcal{V}_2 \in \mathbf{V}$, set $\mathcal{V}_2 > \mathcal{V}_1$ if $\mathcal{V}_2 \subset \overline{\mathcal{V}}_1$.

(ii) Given distinct $T_1, T_2 \in \mathbf{T}_n$, set $T_2 > T_1$ if $\mathcal{V}_{T_2} > \mathcal{V}_{T_1}$.

In general the problem of the combinatorial description of geometric order is extremely difficult. Partial results are described in [M5], which for the case studied below gives a complete answer. All that is needed here is 2.17, although a more general result is given in [M2].

2.7. Fix $T \in \mathbf{T}_n$ with sh $T = \lambda$. For all $i, j \in \{1, \dots, n\}$ let T^i_j denote the ijth entry of T when it is defined.

Definition. For any tableau T

(i) Given $s = T^i_j$, set $r_T(s) = i$ and $c_T(s) = j$.

(ii) For all $1 \le i \le \lambda^*_1 = k$, set $T^i = (T^i_1, \cdots, T^i_{\lambda_i})$. It is the ith row of T.

(iii) For all $j : 1 \le j \le \lambda_1 = l$, set $T_j = (T^1_j, \dots, T^{\lambda^*_j}_j)$. It is the jth column of T.

(iv) The hook number of the ijth entry of T is defined by

$$h(T^i_j) = 1 + (\lambda^*_j - j) + (\lambda_i - i).$$

(v) T^i_j is called a corner entry of T if $h(T^i_j) = 1$.

(vi) If $i \le j$, then $T^{i,j}$ denotes the subtableau with rows $T^i, T^{i+1}, \dots, T^j$. Otherwise it denotes the empty tableau.

A row (resp. column) of T is determined by its entries since these must increase from left to right (resp. from top to bottom).

Let $T = (T_1, T_2, \ldots, T_l)$, $S = (S_1, S_2, \ldots, S_{l'})$ be Young tableaux given by their columns. Assume that T, S have no common entries. Then we define (T, S) to be the array whose rows are the same as the rows of $(T_1, T_2, \ldots, T_l, S_1, S_2, \ldots, S_{l'})$, that is $r_{(T,S)}(T_j^i) = i$ and $r_{(T,S)}(S_j^i) = i$, and ordered in the increasing order. Of course this involves the shuffling of numbered boxes within a row.

Lemma. *(T, S) is a Young tableau.*

Proof. The proof is by induction on rows. Suppose that the integers in the first r rows have been placed in increasing order and that the resulting array of r rows obtained from (T, S) is a Young tableau. Let $i_1 < i_2 < \cdots < i_n$ be the integers of the rth row and $j_1 < j_2 < \cdots < j_m$: $m \leq n$ the ordered set of entries of $\langle T^{r+1}, S^{r+1}\rangle$. Since T, S are Young tableaux and their entries are distinct from one another, there exists an injective map $\phi : \{1, 2, \ldots, m\} \to \{1, 2, \ldots, n\}$ such that $j_t > i_{\phi(t)}$. Then $j_s \geq j_t > i_{\phi(t)}$ for all $t = 1, 2, \ldots, s$. Thus j_s exceeds s elements of the rth row. This forces $j_s > i_s$, as required. □

If the entries of S all exceed those of T, then one only needs to shift numbered boxes (to the left). In a similar fashion $\binom{T}{S}$ is defined. One has $\sigma\binom{T}{S} = (\sigma(T)\sigma(S))$.

Note that $T^{i,i+1} = \binom{T^i}{T^{i+1}}$, for example.

2.8. Take $j \in \mathbb{N}$ such that $j \notin \langle R\rangle$. Let a_i be the smallest entry of R greater than j and set

$$(R + j) := (a_1, \ldots, a_{i-1}, j, a_i, \ldots).$$

Define w_r, $w_c \in W$ by word forms

$$w_r(T) := [T^{\lambda_1^*}, \ldots, T^1], \qquad w_c(T) := [\widehat{T}_1, \ldots, \widehat{T}_{\lambda_1}].$$

Lemma [M5 or M1, 3.2.2]. *One has*

$$Q(w_r(T)) = Q(w_c(T)) = T.$$

Proof. We give a proof for completion and to clarify the nature of Robinson–Schensted correspondence.

Let T be a Young tableau and b_1 an integer > 0 not belonging to the entries of T. The Robinson–Schensted (RS) insertion of b_1 into T gives a new Young tableau $(T \downarrow b_1)$. This is an inductive procedure in which $b_1, b_2, \ldots,$ are defined and b_i is inserted into the ith row of T. In detail, write $T^1 = (a_1, a_2, \ldots, a_l)$. If $b_1 > a_l$, then b_1 is inserted into T^1 and the process stops, that is

$$(T \downarrow b_1) = \begin{pmatrix} T^1 + b_1 \\ T^{2,k} \end{pmatrix}.$$

Otherwise let i be the smallest integer such that $b_1 < a_i$. Then $(T \downarrow b_1)^1$ is obtained by replacing a_i by b_1. Set $b_2 = a_i$. Then

$$(T \downarrow b_1) = \begin{pmatrix} (T^1 \downarrow b_1)^1 \\ (T^{2,k} \downarrow b_2) \end{pmatrix}.$$

Let $w = [a_1, a_2, \dots, a_n]$ be a word form. Then $Q(w)$ is obtained inductively as follows. Take $Q_0(w)$ to be the empty tableau and set $Q_i(w) = (Q_{i-1}(w) \downarrow a_i)$, $\forall i : 1 \le i \le n$. Then $Q(w) = Q_n(w)$.

We prove the lemma by induction on rows. Let T be a Young tableau and set $T^1 = (a_1, a_2, \dots, a_l)$. Then $a_i < a_{i+1}$, $a_i < T_i^2$, $\forall i : 1 \le i \le l$ (using the convention that the undefined entries $T_i^j : j > 2$, of T are set equal to ∞). Conversely starting from $T^{2,k}$ and the sequence $R = (a_1, a_2, \dots, a_l)$ with the above properties, we obtain a Young tableau S with $S^1 = R$, $S^i = T^i$, $\forall i > 1$. One checks that $Q([w_r(T^{2,k}), R]) = S$ and this proves the first part of the lemma. The second is obtained via σ. □

2.9. Let us describe the jeu de taquin (cf. [Sch]) which removes T_j^i from T. The resulting tableau is denoted by $T - T_j^i$ and is obtained as follows.

Consider a row $R = (a_1, \cdots)$. Take $j \in \mathbb{N}$ such that $j > a_1$ and $j \notin \langle R \rangle$. Let a_i be the greatest entry of R smaller than j and set

$$(R \uparrow j) := (a_1, \dots, a_{i-1}, j, a_{i+1}, \dots).$$

One may remark that $((R \uparrow j) \downarrow a_i) = R$.

Take some element a_i of R and set

$$(R - a_i) := (a_1, \dots, a_{i-1}, a_{i+1}, \dots).$$

Similar operations are defined for a column C except that we write $(C \leftarrow j)$ for $\sigma(\sigma(C) \uparrow j)$.

With these preliminaries we write

(1) If $h(T_j^i) = 1$, then

$$(T - T_j^i) = \begin{pmatrix} T^{1,i-1} \\ (T^i - T_j^i) \\ T^{i+1,m} \end{pmatrix}$$

(2) If $h(T_j^i) > 1$, then

(i) If $T_{j+1}^i > T_j^{i+1}$ or $\lambda_i = j$, set

$$(T - T_j^i) = \begin{pmatrix} T^{1,i-1} \\ (T^i \uparrow T_j^{i+1}) \\ (T^{i+1,m} - T_j^{i+1}) \end{pmatrix}.$$

(ii) If $T^i_{j+1} < T^{i+1}_j$ or $\lambda^*_j = i$, set

$$(T - T^i_j) = (T_{1,j-1}, (T_j \leftarrow T^i_{j+1}), (T_{j+1,\lambda_1} - T^i_{j+1})).$$

Theorem [Sch]. *If T is a Young tableau, then so is $(T - T^i_j)$.*

2.10. Given $w \in W, T \in \mathbf{T}_n, \mathcal{V} \in \mathbf{V}$, a standard parabolic subgroup $\mathbf{P}$ and a standard parabolic subalgebra $\mathfrak{p}$. Let $\mathfrak{p}_{\alpha_i}$ be the minimal standard parabolic subalgebra containing $X_{-\alpha_i}$ and let $\mathbf{P}_{\alpha_i}$ be the corresponding standard parabolic subgroup. Define their τ-invariants to be

$$\begin{aligned}
\tau(w) &:= \Pi \cap S(w^{-1}),\\
\tau(T) &:= \{\alpha_i \ : r_T(i+1) > r_T(i)\},\\
\tau(\mathbf{P}) &:= \{\alpha_i \ : \mathbf{P}_{\alpha_i} \subset \mathbf{P}\},\\
\tau(\mathfrak{p}) &:= \{\alpha_i \ : \mathfrak{p}_{\alpha_i} \subset \mathfrak{p}\},\\
\tau(\mathcal{V}) &:= \{\alpha_i \ : \mathbf{P}_{\alpha_i}(\mathcal{V}) = \mathcal{V}\}.
\end{aligned}$$

Note that $\mathbf{P}$ (resp. $\mathfrak{p}$) is uniquely determined by its τ-invariant. Let $\mathbf{P}_T = \mathbf{P}_{\mathcal{V}_T}$ be the stabilizer of $\mathcal{V}_T$. It is a standard parabolic subgroup of $\mathbf{G}$ and we set $\mathfrak{p}_T = \mathfrak{p}_{\mathcal{V}_T} = \operatorname{Lie} P_T$. By [J, §9] one has

Lemma. $\tau(w) = \tau(\mathcal{V}_w) = \tau(Q(w)) = \tau(\mathbf{P}_{\mathcal{V}_w}) = \tau(\mathfrak{p}_{\mathcal{V}_w})$.

2.11. Given $\mathcal{I} \subset \Pi$, let $\mathbf{P}_\mathcal{I}$ denote the unique standard parabolic subgroup of $\mathbf{G}$ such that $\tau(\mathbf{P}_\mathcal{I}) = \mathcal{I}$. Let $\mathbf{M}_\mathcal{I}$ be the unipotent radical of $\mathbf{P}_\mathcal{I}$ and $\mathbf{L}_\mathcal{I}$ a Levi factor. Let $\mathfrak{p}_\mathcal{I}, \mathfrak{m}_\mathcal{I}, \mathfrak{l}_\mathcal{I}$ denote the corresponding Lie algebras. These notation will be conserved throughout though the subscripts may sometimes be dropped.

Given $w \in W$, let $\ell(w)$ denote its reduced length. Let $W_\mathcal{I}$ be the subgroup of W generated by the $s_\alpha : \alpha \in \mathcal{I}$ and $w_\mathcal{I}$ its unique longest element. It is also defined by condition $S(w_\mathcal{I}) = R^+ \cap \mathbb{N}\mathcal{I}$. The following is well known and easy to check.

Lemma. *$\mathcal{V}_\mathcal{I} := \mathcal{V}_{w_\mathcal{I}}$ is the unique orbital variety with closure $\mathfrak{m}_\mathcal{I}$. Moreover if $\tau(\mathcal{V}) \supset \mathcal{I}$, then $\mathcal{V} \subset \overline{\mathcal{V}_\mathcal{I}} = \mathfrak{m}_\mathcal{I}$.*

Remark. Set $\mathcal{O}_\mathcal{I} = \mathbf{G}\mathcal{V}_\mathcal{I}$. One calls $\mathcal{V}_\mathcal{I}$ the Richardson component (of $\mathcal{O}_\mathcal{I} \cap \mathfrak{n}$) defined by $\mathcal{I}$. With respect to the order relation defined in 2.6, $\mathcal{V}_\mathcal{I}$ is the unique minimal orbital variety with τ-invariant $\mathcal{I}$. In general if $\mathfrak{n} \cap^y \mathfrak{n} \subset \mathfrak{n} \cap^w \mathfrak{n}$ then trivially $\mathcal{V}_y \subset \overline{\mathbf{B}(\mathfrak{n} \cap^w \mathfrak{n})} = \overline{\mathcal{V}}_w$. By the lemma the converse holds if $w = w_\mathcal{I}$, for some $\mathcal{I} \subset \Pi$; but fails in general [M5, 4.1.1].

2.12. Given $\mathcal{I} \subset \Pi$. Write $T_{\mathcal{V}_\mathcal{I}}$ simply as $T_\mathcal{I}$. It is obtained by the following rules

(i) $r_{T_\mathcal{I}}(1) = 1$.

(ii) Given $s : 1 < s \leq n$

$$r_{T_\mathcal{I}}(s) = \begin{cases} 1, & \text{if } \alpha_{s-1} \notin \mathcal{I},\\ r_{T_\mathcal{I}}(s-1) + 1, & \text{otherwise.}\end{cases}$$

A useful way to present $T_\mathcal{I}$ is as follows. Partition $\{1, 2, \ldots, n\}$ into connected subsets $C_j := \{b_j, b_j + 1, \ldots, b_{j+1} - 1\}$ by choosing a strictly increasing sequence $1 = b_1 < b_2 < \cdots < b_{l+1} = n + 1$. Setting $\mathcal{I} = \{\alpha_i \mid i, i + 1$ belong to some $C_j\}$ defines a bijection between the set of all such partitions and the set of subsets of Π. Given $\mathcal{I} \subset \Pi$, let $\{C_i^\mathcal{I} : i = 1, 2, \ldots, l\}$ be the corresponding connected subsets which we view as columns. (Sometimes we may omit the $\mathcal{I}$ superscript.) Then in the notation of 2.7 we have $T_\mathcal{I} = (C_1^\mathcal{I}, C_2^\mathcal{I}, \ldots, C_l^\mathcal{I})$. Of course this involves some sliding of boxes to the left. However there are some advantages in this presentation. For example $T_\mathcal{I} - 1$ is obtained by simply replacing C_1 by $C_1 - 1$ and $T_\mathcal{I} - n$ is obtained by simply replacing C_l by $C_l - n$. Again one easily checks the

Proposition. *For all* $\mathcal{I} \subset \Pi$, *the word form of* $w_\mathcal{I}$ *is given by*

$$w_\mathcal{I} = [\widehat{C_1^\mathcal{I}}, \widehat{C_2^\mathcal{I}}, \ldots, \widehat{C_l^\mathcal{I}}].$$

We call $C_i^\mathcal{I}$ the i-th chain of $T_\mathcal{I}$ and $T_\mathcal{I} = (C_1^\mathcal{I}, C_2^\mathcal{I}, \ldots, C_l^\mathcal{I})$ the chain form of $T_\mathcal{I}$.

Example. Consider $\mathcal{I} = \{\alpha_1, \alpha_4, \alpha_5, \alpha_6, \alpha_9\}$ in $\mathfrak{sl}_{10}$. Then

$T_\mathcal{I} =$

1	3	4	8	9
2	5	10		
6				
7				

or in chain form $T_\mathcal{I} = (C_1, C_2, C_3, C_4, C_5)$ where $C_1 = \{1, 2\}$, $C_2 = \{3\}$, $C_3 = \{4, 5, 6, 7\}$, $C_4 = \{8\}$ and $C_5 = \{9, 10\}$.

2.13. Given an orbital variety $\mathcal{V}$, set $\mathcal{O} = \mathbf{G}\mathcal{V}$. It is a nilpotent orbit. Given a nilpotent orbit $\mathcal{O}' \subset \overline{\mathcal{O}}$ one may try to describe $\overline{\mathcal{V}} \cap \mathcal{O}'$. In general it is not even known if this intersection is equidimensional. However if $\mathcal{V} = \mathcal{V}_\mathcal{I}$ for some $\mathcal{I} \subset \Pi$ a complete answer given by [S1], [M5 or M1] and 2.11. Set $\mathcal{O}_\mathcal{I} = \mathbf{G}\mathcal{V}_\mathcal{I}$.

Theorem. *Let* $\mathcal{O}$ *be a nilpotent orbit in* $\overline{\mathcal{O}_\mathcal{I}} = \mathbf{G}\mathfrak{m}_\mathcal{I}$. *Then* $\mathfrak{m}_\mathcal{I} \cap \mathcal{O}$ *is the union of orbital varieties* $\mathcal{V}$ *in* $\mathcal{O} \cap \mathfrak{n}$ *satisfying* $\tau(\mathcal{V}) \supset \mathcal{I}$.

Proof. By [S1, p. 456, last corollary] $\mathfrak{m}_\mathcal{I} \cap \mathcal{O}$ is equidimensional.

By [M1, 4.1.8 or M5, 3.5.6] it contains at least one orbital variety and so $\dim \mathcal{V} = \frac{1}{2} \dim \mathcal{O}$, for every irreducible component $\mathcal{V}$ of $\mathcal{O} \cap \mathfrak{m}_\mathcal{I}$. Yet $\mathcal{V} \subset \mathcal{O} \cap \mathfrak{n}$, which also has dimension $\frac{1}{2} \dim \mathcal{O}$. Hence $\mathcal{V}$ is an orbital variety associated to $\mathcal{O}$. Then by 2.11 every orbital variety $\mathcal{V}$ associated to $\mathcal{O}$ lies in $\mathfrak{m}_\mathcal{I}$ if and only if $\tau(\mathcal{V}) \supset \mathcal{I}$. □

Remark. By 2.10, $\alpha_i \in \tau(\mathcal{V}_T)$ if and only if $r_T(i + 1) > r_T(i)$. In particular if $\mathcal{V} > \mathcal{V}_\mathcal{I}$ then for any i one has $r_{T_\mathcal{V}}(i) \geq r_{T_\mathcal{I}}(i)$.

2.14. Take $\mathcal{I} \subset \Pi$ and recall the notation of 2.11. Set $\mathbf{B}_{\mathcal{I}} = \mathbf{B} \cap \mathbf{L}_{\mathcal{I}}$, $\mathfrak{n}_{\mathcal{I}} = \mathfrak{n} \cap \mathfrak{l}_{\mathcal{I}}$ and $W^{\mathcal{I}} = \{w \in W \mid w\mathcal{I} \subset R^+\}$. We have decompositions $\mathbf{B} = \mathbf{M}_{\mathcal{I}} \rtimes \mathbf{B}_{\mathcal{I}}$, $\mathfrak{n} = \mathfrak{n}_{\mathcal{I}} \oplus \mathfrak{m}_{\mathcal{I}}$, $W^{\mathcal{I}} \times W_{\mathcal{I}} \tilde{\to} W$. They define projections $\mathbf{B} \to \mathbf{B}_{\mathcal{I}}$, $\mathfrak{n} \to \mathfrak{n}_{\mathcal{I}}$ and $W \to W_{\mathcal{I}}$ which we will denote by $\pi_{\mathcal{I}}$, or simply by π. Let $\mathcal{V}^{\mathcal{I}}_{\pi(w)}$ be the orbital variety of $\mathfrak{l}_{\mathcal{I}}$ with closure $\overline{\mathbf{B}_{\mathcal{I}}(\mathfrak{n}_{\mathcal{I}} \cap^{\pi(w)} \mathfrak{n}_{\mathcal{I}})}$.

Proposition [M1, 4.1.2 or M5, 4.2.2]. *For all* $w \in W$, $\mathcal{I} \subset \Pi$ *one has* $\pi_{\mathcal{I}}(\overline{\mathcal{V}}_w) = \overline{\pi(\mathcal{V}_w)} = \overline{\mathcal{V}^{\mathcal{I}}_{\pi(w)}}$.

Remark 1. It is clear that $\pi_{\mathcal{I}}$ is inclusion preserving. Thus if $\mathcal{V}_w$, $\mathcal{V}_y$ are orbital varieties of $\mathfrak{g}$ with $\overline{\mathcal{V}}_w \subset \overline{\mathcal{V}}_y$, then $\overline{\mathcal{V}}^{\mathcal{I}}_{\pi(w)} \subset \overline{\mathcal{V}}^{\mathcal{I}}_{\pi(y)}$.

Remark 2. Suppose $\mathfrak{a} \subset \mathfrak{n}$ is $\mathfrak{h}$ stable. Since root subspaces are one-dimensional we obtain $\pi_{\mathcal{I}}(\mathfrak{a}) = \mathfrak{a} \cap \mathfrak{n}_{\mathcal{I}}$. Again $\mathbf{M}_{\mathcal{I}}$ acts by 1 on $\mathfrak{n}/\mathfrak{m}_{\mathcal{I}}$. Then since $\pi_{\mathcal{I}}$ is continuous

$$\pi_{\mathcal{I}}(\mathbf{B}\mathfrak{a}) \subset \overline{\pi_{\mathcal{I}}(\mathbf{B}\mathfrak{a})} \subset \overline{\mathbf{B}\mathfrak{a}} \cap \mathfrak{n}_{\mathcal{I}} \subset \overline{\pi_{\mathcal{I}}(\mathbf{B}\mathfrak{a})}$$

and so $\overline{\mathcal{V}}^{\mathcal{I}}_{\pi(w)} = \overline{\mathcal{V}}_w \cap \mathfrak{n}_{\mathcal{I}}$.

For $1 \le i < j \le n$ set $[i, j] = \{r \in \mathbb{N} \mid i \le r \le j\}$, $\Pi_{i,j} = \{\alpha_r \mid r, r+1 \in [i, j]\}$ and $\pi_{i,j} = \pi_{\Pi_{i,j}}$. Define $\mathbf{T}_n \to \mathbf{T}_{j-i+1}$, through the jeu de taquin applied to the entries (taken in any order) of T not lying in $[i, j]$. By [M1, 2.4.16, 4.1.1 or M5 4.3.3] one has

$$\pi_{i,j}(\overline{\mathcal{V}}_T) = \overline{\mathcal{V}}_{\pi_{i,j}(T)}. \tag{$*$}$$

2.15 Definition. Let $\ge$ be an order relation on $\mathbf{V}$. Given $\mathcal{V}_{T_2} > \mathcal{V}_{T_1}$ we call $\mathcal{V}_{T_2}$ a descendant of $\mathcal{V}_{T_1}$ (respectively T_2 a descendant of T_1) if for any orbital variety $\mathcal{W}$ such that $\mathcal{V}_{T_2} \ge \mathcal{W} \ge \mathcal{V}_{T_1}$ one has $\mathcal{W} = \mathcal{V}_{T_2}$ or $\mathcal{W} = \mathcal{V}_{T_1}$. One calls $\mathcal{V}_{T_2}$ a geometric descendant of $\mathcal{V}_{T_1}$ when $\ge$ is the geometric order.

If unqualified descendant will mean geometric descendant.

The set of descendants of any $T_{\mathcal{I}} : \mathcal{I} \subset \Pi$ is described explicitly in [M2, 2.6] The simplification that results in our case can be understood as follows.

Define the Duflo (or weak Bruhat) order on W by $y \supseteq w$ if $\mathfrak{n} \cap^y \mathfrak{n} \subset \mathfrak{n} \cap^w \mathfrak{n}$. Through the map $w \mapsto Q(w)$ it induces an order relation $\supset$ on $\mathbf{T}_n$ and hence on $\mathbf{V}$ called the (induced) Duflo order. Remark 2.11 states just that $\mathcal{V}_y \subset \overline{\mathcal{V}}_w$ if $y \supseteq w$, but the converse can fail unless $w = w_{\mathcal{I}}$. The description of the induced Duflo order on $\mathbf{T}_n$ is itself a quite nontrivial problem; but the complete solution was given in [M1, 3.3.3, 3.4.5 or M5, 3.3.3, 3.4.4]. By the above remark the set of geometric descendants of a Richardson component is contained in the set of Duflo descendants in turn described by [M2, 2.6].

Present $T_{\mathcal{I}}$ as in 2.12, that is we write $T_{\mathcal{I}} = (C_1^{\mathcal{I}}, C_2^{\mathcal{I}}, \ldots, C_l^{\mathcal{I}})$. Set $c_i = |C_i|$. Then $\varsigma_i = C_i^{c_i}$ is the largest entry of C_i and moreover belongs to $T_{\mathcal{I}}^{c_i}$.

For any $1 \le i \le l$, suppose that there exists $j : 1 \le j < i$ such that $c_j \ge c_i$. Take the maximal integer $i' : 1 \le i' < i$ such that $c_{i'}$ is minimal with the property that $c_{i'} \geq c_i$. Define

$$T_{\mathcal{I}}(i) := \begin{pmatrix} T^{1,c_i-1} \\ T^{c_i} - \varsigma_i \\ T^{c_i+1,c_{i'}} \\ T^{c_{i'}+1} + \varsigma_i \\ T^{c_{i'}+2,k} \end{pmatrix}.$$

In other words we move ς_i from the c_ith to the $(c_{i'}+1)$th row according to the rules of 2.9. In Example 2.12 one has $c_2 = 1 < c_1 = 2$; $c_3 = 4$ is maximal, $c_4 = c_2 = 1$ and $c_5 = c_1 = 2$. Thus our procedure defines 3 tableaux:

$$T_{\mathcal{I}}(2) = \begin{array}{llll} 1 & 4 & 8 & 9 \\ 2 & 5 & 10 & \\ 3 & 6 & & \\ 7 & & & \end{array}, \quad T_{\mathcal{I}}(4) = \begin{array}{llll} 1 & 3 & 4 & 9 \\ 2 & 5 & 8 & 10 \\ 6 & & & \\ 7 & & & \end{array}, \quad T_{\mathcal{I}}(5) = \begin{array}{lllll} 1 & 3 & 4 & 8 & 9 \\ 2 & 5 & & & \\ 6 & 10 & & & \\ 7 & & & & \end{array}$$

Lemma. *$T_{\mathcal{I}}(i)$ is a Young tableau.*

Proof. Indeed the largest element ς_i of the ith chain is relocated on the i'th chain where it becomes the largest element since $i' < i$. Then the assertion follows from Lemma 2.7. □

2.16. As noted in 2.12 the tableau $\pi_{1,n-1}(T_{\mathcal{I}})$ is obtained from $T_{\mathcal{I}}$ by eliminating n from C_l. It corresponds to the subset $\mathcal{I}_n = \mathcal{I} \cap \Pi_{1,n-1}$ with $C_i^{\mathcal{I}_n} = C_i^{\mathcal{I}} : i < l$ and $C_l^{\mathcal{I}_n} = C_l^{\mathcal{I}} - n$. Similarly $\pi_{2,n}(T_{\mathcal{I}})$ is obtained from $T_{\mathcal{I}}$ by eliminating 1 from C_1 and corresponds to the subset $\mathcal{I}_1 = \mathcal{I} \cap \Pi_{2,n}$ with $C_i^{\mathcal{I}_1} = C_i^{\mathcal{I}} : i > 1$ and $C_1^{\mathcal{I}_1} = C_1^{\mathcal{I}} - 1$.

Lemma. *Assume that $T_{\mathcal{I}}(i)$ is defined. Then $i > 1$ and*

(i) $\pi_{1,n-1}(T_{\mathcal{I}}(i)) = T_{\mathcal{I}_n}(i)$, *if $i < l$.*

(ii) $\pi_{2,n}(T_{\mathcal{I}}(i)) = T_{\mathcal{I}_1}(i)$, *if $c_1 \neq c_i$ or there exists $j : 1 < j < i$ such that $c_j = c_i$.*

(iii) $\pi_{2,n}(T_{\mathcal{I}}(i)) = T_{\mathcal{I}_1}$, *if $c_1 = c_i$ and $c_j \neq c_i$, $\forall j : 1 < j < i$.*

Proof. (i) is an immediate consequence of $\varsigma_i < n$, for $i < l$. The hypothesis of (ii) implies that $T_{\mathcal{I}_1}(i)$ is defined. If in the choice of i' (as above) one can choose $i' : 1 < i' < i$, then both $T_{\mathcal{I}}(i)$ and $T_{\mathcal{I}_1}(i)$ are obtained by moving ς_i from the c_ith row to the $(c_{i'}+1)$th row, an operation which commutes with replacing C_1 by $C_1 - 1$. If necessarily $i' = 1$ but $c_1 > c_i$, then ς_i goes from the c_ith row to the

$(c_1 + 1)$th row in forming $T_{\mathcal{I}}(i)$, but then goes to c_1th row in forming $\pi_{2,n}(T_{\mathcal{I}}(i))$. However if $c_1 > c_i$ this is exactly what happens in forming $T_{\mathcal{I}_1}(i)$ from $T_{\mathcal{I}_1}$. Hence (ii). Under the hypothesis of (iii), ς_i is pushed back into the c_ith row in forming $\pi_{2,n}(T_{\mathcal{I}}(i))$, which is hence $T_{\mathcal{I}_1}$. □

2.17. Retain the conventions of 2.15.

Proposition. *Take $\mathcal{I} \subset \Pi$. Assume $T_{\mathcal{I}}(i)$ is defined and hence i' is defined. If $c_{i'} = c_i$, then $T_{\mathcal{I}}(i)$ is a descendant of $T_{\mathcal{I}}$ of codimension 1 and all the descendants of codimension 1 of $T_{\mathcal{I}}$ are so obtained.*

Proof. By 2.15, $T = T_{\mathcal{I}}(i)$ is a Young tableau and $\mathcal{V}_T \subset \overline{\mathcal{V}}_{\mathcal{I}}$. By 2.4 (2) we have

$$\dim \mathcal{V}_T = \dim \mathcal{V}_{\mathcal{I}} - |c_{i'} - c_i| + 1. \qquad (*)$$

Take $c_{i'} = c_i$; then $\mathcal{V}_T$ is a descendant just by dimension considerations.

Conversely assume that $T > T_{\mathcal{I}}$ with $\mathcal{V}_T$ of codimension one in $\mathfrak{m}_{\mathcal{I}}$. It remains to show that $T = T_{\mathcal{I}}(i)$, for some i.

By Remark 2.13 we have

$$r_T(s) \geq r_{T_{\mathcal{I}}}(s), \qquad \text{for all } s.$$

On the other hand by 2.4(2) and 2.5, $\operatorname{sh} T$ is obtained from $\operatorname{sh} T_{\mathcal{I}}$ by lowering exactly one box by one row. Combined with the previous result this means that there is exactly one entry, say m of $T_{\mathcal{I}}$, which is displaced on passing to T and moreover it appears exactly one row below. Suppose $\alpha_m \in \tau(T_{\mathcal{I}})$. Since $\tau(T) \supset \tau(T_{\mathcal{I}})$ by 2.11, it follows by Remark 2.13, that

$$r_T(m+1) \geq r_T(m) + 1 = r_{T_{\mathcal{I}}}(m) + 2, \quad \text{whilst} \quad r_{T_{\mathcal{I}}}(m+1) = r_{T_{\mathcal{I}}}(m) + 1.$$

This contradicts $r_T(m+1) = r_{T_{\mathcal{I}}}(m+1)$. Thus, in the previous notation $m = \varsigma_i$, the latter being the longest element of some chain C_i of $T_{\mathcal{I}}$ of length c_i. By the above we also have $T^{c_i} = T_{\mathcal{I}}^{c_i} - m$ and $T^{c_i+1} = T_{\mathcal{I}}^{c_i+1} + m$ whilst the remaining rows of T, $T_{\mathcal{I}}$ must coincide. Finally we must have some $i' < i$ with $c_{i'} = c_i$, for otherwise m would be strictly less than an entry in T above it. We conclude that $T = T_{\mathcal{I}}(i)$, as required. □

2.18. By 2.17 we conclude that the orbital varieties of codimension 1 in $\mathcal{V}_{\mathcal{I}}$ are obtained by moving ς_i from row c_i to row $(c_i + 1)$ wherever $|T_{\mathcal{I}}^{c_i}| \geq |T_{\mathcal{I}}^{c_i+1}| + 2$. Indeed if ς_i belongs to T_j (and here $j \leq i$ because some sliding of boxes from the chain form of $T_{\mathcal{I}}$ may be necessary) then the condition $c_{i'} = c_i$ implies that $|T_{j-1}| = |T_j|$.

More precisely there are exactly $t+1 := \sharp\{j \mid 1 \leq j \leq i \text{ with } c_j = c_i\}$ columns of T, namely $T_j, T_{j-1}, \dots, T_{j-t}$, having length $|T_j|$. In addition suppose that i is maximal for a given value c_i that is $c_j \neq c_i$ for any $j > i$. Then there are exactly t choices of s such that $T_{\mathcal{I}}(s)$ is defined and has the same shape as $T_{\mathcal{I}}(i)$.

Observe that the partition $\lambda_{\mathcal{I}} = \operatorname{sh} T_{\mathcal{I}}$ is defined by chain lengths, namely we have $\lambda_{\mathcal{I}}^* = \{c_i : i = 1, 2, \dots, l\}$ appropriately ordered. Fix $c \in \lambda_{\mathcal{I}}^*$ and set

$t+1 = \sharp\{i \mid c_i = c\}$. Suppose $t \geq 1$ and let $\mathcal{O}_{\mathcal{I}}(c)$ be defined through the partition $\lambda_{\mathcal{I}}(c)$ obtained from $\lambda_{\mathcal{I}}$ by replacing the appropriate pair (c, c) in $\lambda_{\mathcal{I}}^*$ by $(c+1, c-1)$. Then we may summarize the above by

Proposition. *Fix $\mathcal{I} \subset \Pi$ and $c \in \mathbb{N}^+$. Set $t = \sharp\{i \mid c_i = c\} - 1$. Assume $t \geq 1$. Then $\mathfrak{m}_{\mathcal{I}} \cap \mathcal{O}_{\mathcal{I}}(c)$ is a union of t orbital varieties of codimension 1 in $\mathfrak{m}_{\mathcal{I}}$. Moreover these orbital varieties correspond to the standard tableaux $T_{\mathcal{I}}(i) : c_i = c$, except for the minimal i that is $i : c_i = c$ and $c_j \neq c$ for all $j < i$.*

Note that orbital varieties of codimension 1 in the nilpotent radical of a parabolic appear exactly when there are repetitions in the dual partition associated to this parabolic.

2.19. Let $\mathfrak{m}^-$ be the opposed algebra of $\mathfrak{m} = \mathfrak{m}^+$ identified with $\mathfrak{m}^*$ through the Killing form. By Krull's theorem the ideal $I(\overline{\mathcal{V}})$ of definition of an orbital variety closure $\overline{\mathcal{V}}$ in the conclusion of 2.18 is principal in $S(\mathfrak{m}^-)$ and we denote by $f_{\mathcal{I}}(i)$, or simply f, the corresponding irreducible generator in $S(\mathfrak{m}^-)$.

The form of $f_{\mathcal{I}}(i)$ was conjectured by Benlolo and Sanderson in [BS]. We refer to this and their conjecture simply as BS. We prove this conjecture in Sect. 3.

As explained in 1.4, 1.5 the general form of $f_{\mathcal{I}}(i)$ is $\operatorname{gr} a$ for some minor a of a generic matrix in $\mathfrak{m}_{\mathcal{I}} + \mathrm{Id}$. They took a suitable minor lying in a bottom left-hand corner. In this they needed results similar to [M5] to reduce to the case when $\mathcal{I}$, c are chosen so that $c = c_1 = c_l$ and $c_i \neq c$ for $1 < i < l$. Then their conjecture is equivalent to saying that a is just the $(n-c) \times (n-c)$ minor lying in the bottom left-hand corner.

The BS reduction can be read off from 2.16 and 2.18 which makes precise the result of [M5] which one needs and we believe clarifies their analysis. A key point is the knowledge of the orbital varieties of codimension 1 in $\mathfrak{m}_{\mathcal{I}}$ given by 2.17 developed from the results of [M5]. We remark that the codimension 1 case is relatively easy and barely uses the full power of [M5].

For all $1 \leq i, j \leq n$, set $X_{i,j} = \mathbb{C}e_{i,j}$. Given $w \in W$, we give the affine $|R^+| - \ell(w)$ dimensional space $X(w)$ a matrix presentation through $X(w)_{i,j} = X_{i,j}$, wherever $i < j$ and $w(i) < w(j)$.

Let $x_{j,i}$ denote the coordinate function on $\mathfrak{g}$ defined by

$$x_{j,i}(e_{r,s}) = \begin{cases} -1, & \text{if } (r,s) = (i,j), \\ 0, & \text{otherwise.} \end{cases}$$

Then the Poisson bracket $\{,\}$ defined on $S(\mathfrak{g}^*)$ through the Lie bracket on $\mathfrak{g}$ satisfies

$$\{x_{i,j}, x_{r,s}\} = \delta_{j,r}x_{i,s} - \delta_{s,i}x_{r,j} \qquad (*)$$

where $\delta_{i,j}$ is the Kronecker delta. Setting $x_{i,j} = e_{i,j}$ for $i > j$ identifies $\mathfrak{m}^-$ with $\mathfrak{m}^*$.

Given $w \in W$, let $\mathcal{M}(w)$ be the matrix with entries

$$\mathcal{M}(w)_{i,j} = \begin{cases} x_{i,j}, & \text{if } X_{j,i} \subset X(w), \\ 0, & \text{otherwise.} \end{cases}$$

When $w = w_{\mathcal{I}}$, we set $\mathcal{M}(w) = \mathcal{M}_{\mathcal{I}}$.

Assume $c_1 = c_l =: c$ and let $M^c_{\mathcal{I}}(t)$ denote the $(n-c) \times (n-c)$ minor in the bottom left-hand corner of $\mathcal{M}_{\mathcal{I}} + t\mathrm{Id}$, that is

$$M^c_{\mathcal{I}}(t) := \begin{vmatrix} x_{c+1,1} & \cdots\cdots & x_{c+1,c} & t & 0 & \cdots & 0 \\ x_{c+2,1} & \cdots\cdots & \cdots & \cdots & t & \cdots & 0 \\ \vdots & \vdots & \vdots & \vdots & \vdots & \vdots & \vdots \\ x_{n-c,1} & \cdots\cdots & \cdots & \cdots & \cdots & \cdots & t \\ \vdots & \vdots & \vdots & \vdots & \vdots & \vdots & \vdots \\ x_{n,1} & \cdots\cdots & x_{n,c} & x_{n,c+1} & \cdots & \cdots & x_{n,n-c} \end{vmatrix}.$$

Note that there can be zeros in the dot places of the determinant in correspondence with the definition of the matrix $\mathcal{M}_{\mathcal{I}}$.

Example. Consider $\mathcal{I} = \{\alpha_1, \alpha_4\}$ in $\mathfrak{sl}_5$. In that case $c_1 = c_3 = 2$ and

$$\mathcal{M}_{\mathcal{I}} = \begin{bmatrix} 0 & 0 & 0 & 0 & 0 \\ 0 & 0 & 0 & 0 & 0 \\ x_{3,1} & x_{3,2} & 0 & 0 & 0 \\ x_{4,1} & x_{4,2} & x_{4,3} & 0 & 0 \\ x_{5,1} & x_{5,2} & x_{5,3} & 0 & 0 \end{bmatrix} \quad \text{and} \quad M^2_{\mathcal{I}}(t) := \begin{vmatrix} x_{3,1} & x_{3,2} & t \\ x_{4,1} & x_{4,2} & x_{4,3} \\ x_{5,1} & x_{5,2} & x_{5,3} \end{vmatrix}.$$

Developing $M^c_{\mathcal{I}}(t)$ in powers of t one obtains

$$M^c_{\mathcal{I}}(t) = m_{n-c} + m_{n-c-1}t + \cdots + m_c t^{n-2c}.$$

Set

$$d_i = \begin{cases} c_i - c & \text{if } c_i > c \\ 0 & \text{otherwise} \end{cases}, \quad d_{\mathcal{I}} = \sum_{i=0}^{k} d_i, \quad l_{\mathcal{I}} = n - c - d_{\mathcal{I}}.$$

By [BS, Lemma 2] the lowest nonzero coefficient is $m_{l_{\mathcal{I}}}$ that is the coefficient of $t^{d_{\mathcal{I}}}$. The BS conjecture states that $f_{\mathcal{I}}(l) = m_{l_{\mathcal{I}}}$.

In our example $d_{\mathcal{I}} = 0$ and $l_{\mathcal{I}} = 3$ so that

$$m_{l_{\mathcal{I}}} = x_{3,1}x_{4,2}x_{5,3} + x_{3,2}x_{4,3}x_{5,1} - x_{3,1}x_{4,3}x_{5,2} - x_{3,2}x_{4,1}x_{5,3}.$$

It is clear that $m_{l_{\mathcal{I}}} = \mathrm{gr}\, M^c_{\mathcal{I}}(1)$. We do not need to know the explicit power of t which divides $M^c_{\mathcal{I}}(t)$, though this is used in one of our three proofs of irreducibility. The BS conjecture can be expressed as the following theorem which we prove in the next section.

Theorem. *Take $\mathcal{I} \subset \Pi$ and let $T_{\mathcal{I}} = (C_1, C_2, \ldots, C_l)$ be its chain form. Set $c_i = |C_i|$, for all i.*

(i) *Assume $c_1 = c_l =: c$ and $c_j \neq c$, $\forall j : 1 < j < l$. Then $I(\overline{\mathcal{V}}_{T_{\mathcal{I}}(l)})$ is generated by $\{x_{ij} : X_{ij} \in \mathfrak{p}_{\mathcal{I}} \cap \mathfrak{n}^-\}$ and $m_{l_{\mathcal{I}}}$.*

(ii) *Suppose there exist* $i < j$ *such that* $c_i = c_j$ *and* $c_s \neq c_i$, $\forall s : i < s < j$. *Set* $u = C_i^1$, $v = C_j^{c_i}$ *and* $\mathcal{I}' = \mathcal{I} \cap \Pi_{u,v}$. *Then* $I(\overline{\mathcal{V}}_{T_{\mathcal{I}}(j)})$ *is generated by* $\{x_{ij} : X_{ij} \in \mathfrak{p}_{\mathcal{I}} \cap \mathfrak{n}^-\}$ *and* $m_{l_{\mathcal{I}'}}$ *defined with respect to* $\mathcal{V}_{T_{\mathcal{I}'}}$, *the latter viewed as an orbital variety in* $\pi_{u,v}(\mathfrak{n})$.

Note that the part (i) of the theorem is a special case of part (ii) and part (ii) will be deduced from part (i).

2.20. Let i be a positive integer $< \frac{n}{2}$ and set

$$\beta_i = \sum_{j=i}^{n-i} \alpha_j .$$

Take c as in 2.19 and set

$$\gamma_c = \sum_{i=1}^{c} \beta_i .$$

As noted in [BS, VIII] one may check that $M_{\mathcal{I}}^c(t)$, and hence $m_{l_{\mathcal{I}}}$ has weight $-\gamma_c$. Moreover with respect to the Cartan inner product (,) one has $(\gamma_c, \alpha) = 0$, $\forall \alpha \in \Pi \setminus \{\alpha_c, \alpha_{n-c}\}$. In particular

$$(\gamma_c, \alpha) = 0, \qquad \forall \alpha \in \mathcal{I}. \qquad (*).$$

3 Proof of the Benlolo–Sanderson conjecture

3.1. Let us recall some general facts about ideals of definition of orbital variety closures.

Retain the notation of 2.11 and 2.18. Let $\mathbf{P}$ be some standard parabolic subgroup of $\mathbf{G}$, $\mathfrak{p} = \text{Lie}(\mathbf{P})$ and $\mathfrak{m}$ its nilradical. Let $\mathcal{V}$ be a closed subvariety of $\mathfrak{m}$ and I (resp. J) its ideal of definition in $S(\mathfrak{m}^-)$ (resp. $S(\mathfrak{g})$). One has $J = I + S(\mathfrak{g})\mathfrak{p}$ and $J \cap S(\mathfrak{m}^-) = I$ from which one easily checks the

Lemma. *The following are equivalent*:

(i) $\{I, I\} \subset I$ *and* $\{\mathfrak{p}, I\} \subset J$.

(ii) $\{I, I\} \subset I$ *and* $\mathcal{V}$ *is* $\mathbf{P}$ *invariant.*

(iii) $\{J, J\} \subset J$.

3.2. One calls $\mathcal{V}$ involutive if any one of the above holds. If $\mathcal{V}$ is involutive, then so are its irreducible components.

Suppose $\mathcal{V} \subset \mathfrak{m}$ is irreducible. Then $\overline{\mathbf{G}\mathcal{V}}$ contains a unique dense orbit $\mathcal{O}$. If in addition $\mathcal{V}$ is involutive, then $\dim(\mathcal{V} \cap \mathcal{O}) \geq \frac{1}{2} \dim \mathcal{O}$ since $\mathcal{O}$ is a symplectic variety. Yet $\dim(\mathfrak{n} \cap \mathcal{O}) = \frac{1}{2} \dim \mathcal{O}$ and so $\mathcal{V} \cap \mathcal{O}$ must be an irreducible component of $\mathfrak{n} \cap \mathcal{O}$, hence an orbital variety associated to $\mathcal{O}$ with closure $\mathcal{V}$.

3.3. Let J_o be an ideal of $S(\mathfrak{g})$ with radical J. Suppose $\{J_o, J_o\} \subset J_o$. It is generally false that this implies $\{J, J\} \subset J$, so one cannot conclude that the zero variety of J_o is involutive. This is a delicate point in general. In the present situation, this difficulty can be avoided. Call $f \in S(\mathfrak{m}^-)$ multilinear if its degree in any one of the variables $x_{i,j}$ is at most 1. This property passes to the irreducible factors $f_1, f_2, \dots, f_t$ and we let $\mathcal{V}(f_1), \dots, \mathcal{V}(f_t)$ denote their sets of zeros in $\mathfrak{m}$. Consider the condition

$$\{\mathfrak{p}, f\} \subset S(\mathfrak{m}^-)f + S(\mathfrak{g})\mathfrak{p}. \qquad (*)$$

Proposition. *Suppose $f \in S(\mathfrak{m}^-)$ is multilinear. Then*

(i) $S(\mathfrak{m}^-)f$ *is semiprime.*

(ii) $S(\mathfrak{m}^-)f + S(\mathfrak{g})\mathfrak{p}$ *is semiprime.*

(iii) *If* $(*)$ *holds, then the* $\mathcal{V}(f_i)$ *are orbital varieties of codimension 1 in* $\mathfrak{m}$.

Proof. (i) is immediate. (ii) is easily deduced from (i). Finally (iii) follows from (ii) and 3.1, 3.2. □

3.4. Our proof of the BS conjecture involves three steps. The first is to show that $f = m_{l_\mathcal{I}}$ satisfies 3.3 $(*)$. This is rather obvious especially from the quantum viewpoint of [J2, Lecture 7]. Nevertheless a delicate point is that $M_\mathcal{I}^c(t)$ itself does not satisfy $(*)$. We may view its zero set as a deformation of $\mathcal{V}(f)$, but which is not itself orbital or **P** invariant. (Our original motivation for such deformation came from trying to define an Enright functor on orbital varieties itself inspired by the algorithm for $\operatorname{Ann}_{U(\mathfrak{n}^-)} v$ in [J1, 8.4] based on the Enright functor.) This result is given in 3.11.

The second step is to show that f is irreducible. This is a delicate point. However we can give three different proofs. The first, given in 3.15, is fairly explicit but includes some computations involving precise knowledge of $M_\mathcal{I}^c$. The second, outlined in 4.3, uses representation theory. We construct a simple highest weight module L with integral highest weight which is a strong quantization of $\mathcal{V}(f)$. This is of interest in its own right. Then we use the linear independence of the characteristic polynomials of orbital varieties and the irreducibility of the associated variety of L which holds [M4] in type A. The third method, given in 4.7, is the least computational, but the most sophisticated. Here we construct a strong quantization, but not necessary having integral highest weight. Then we use the difficult fact that $V(\operatorname{Ann} L)$ is the closure of a nilpotent orbit. In type A this has a relatively easy proof using mainly that orbit closures are normal [BK]. Finally we apply 2.18. In all these methods it is crucial to use that $c_i \neq c$, $\forall 1 < i < l$. Otherwise $\mathcal{V}(f)$ has exactly $t := \sharp\{i \mid c_i = c\} - 1$ components. A computational proof in some special cases when $m_{l_\mathcal{I}} = M_\mathcal{I}^c(0)$ was also given in [BS]. Even that is not trivial though this case can be viewed as an easy consequence of 2.4 (1) combined with 2.18.

Step three is to show that $M_{\mathcal{I}}^c$ vanishes on $X(w)$ for some $w : Q(w) = T_{\mathcal{I}}(l)$. Then $\mathbf{P}_{\mathcal{I}}$ invariance implies $\mathcal{V}(f) \supset \overline{\mathbf{B}(\mathfrak{n} \cap^w \mathfrak{n})}$. This is shown in 3.8. Since both have codimension 1 in $\mathfrak{m}_{\mathcal{I}}$, irreducibility finishes the proof. Notice this does not use involutivity, but by 3.1 the latter is equivalent to $\mathbf{P}_{\mathcal{I}}$ invariance. One may also avoid this last step by combining 3.3. and 2.18, but this is less satisfying.

3.5. Given $w \in \mathbf{S}_n$ with word form $[a_1, a_2, \dots, a_n]$, let $w - s \in \mathbf{S}_{n-1}$ be defined by deleting s and standardizing the word form. For any matrix $\mathcal{M}$ let $\mathcal{M}^{i,j}$ be the matrix obtained from $\mathcal{M}$ by deleting the ith row and jth column. Recall the definition of $X(w)$ given in 2.19.

Lemma. *For all $w \in W$, $s \in \{1, 2, \dots, n\}$ one has $X(w - s) = X(w)^{s,s}$.*

Proof. Take $1 \le i < j \le n$. Then $w(\alpha_{i,j}) \in R^+$ if and only if $w(i) < w(j)$ with a similar assertion for $y = w - s$. Yet recalling that we are standardizing $w - s$ we have up to standardization

$$y(r) = \begin{cases} a_r, & \text{if } r < w^{-1}(s), \\ a_{r+1}, & \text{if } r \ge w^{-1}(s), \end{cases}$$

From this the assertion readily follows. □

3.6. Suppose $T = Q(w)$. It is generally false that $Q(w - s) = T - s$, though it is true for the canonical elements $w_r(T)$ and $w_c(T)$ defined in 2.8. Here we shall only need to describe $Q(w_r(T) - s)$, when $s \in T^k$, that is when s lies in the last row of T. Recall that $w_r(T) = [T^k, T^{k-1}, \dots, T^1]$.

Lemma. *One has $w_r(T) - T_j^k = w_r(T - T_j^k)$ for every entry T_j^k of the last row T^k of T.*

Proof. Indeed

$$(T - T_j^k) = \begin{pmatrix} T^{1,k-1} \\ (T^k - T_j^k) \end{pmatrix}$$

which gives the required assertion. □

3.7. Fix $\mathcal{I} \subset \Pi$ and $s \in \{1, 2, \dots, n\}$. Define $\mathcal{I}' \subset \Pi \setminus \{\alpha_{s-1}\}$ by $\mathcal{I}' = \{\alpha_i \mid i < s - 1, \alpha_i \in \mathcal{I}\} \cup \{\alpha_{i-1} \mid i \ge s, \alpha_i \in \mathcal{I}\}$. Write $T_{\mathcal{I}} = (C_1^{\mathcal{I}}, C_2^{\mathcal{I}}, \dots, C_l^{\mathcal{I}})$ as in 2.12.

Lemma.

(i) $w_{\mathcal{I}} - s = w_{\mathcal{I}'}$.

(ii) $Q(w_{\mathcal{I}} - s) = T_{\mathcal{I}'}$.

(iii) *Suppose $s \in C_j^{\mathcal{I}}$. Then up to standardization $T_{\mathcal{I}'} = (C_1^{\mathcal{I}'}, C_2^{\mathcal{I}'}, \dots, C_l^{\mathcal{I}'})$ where*

$$C_i^{\mathcal{I}'} = \begin{cases} C_i^{\mathcal{I}}, & \text{if } i \ne j, \\ C_j^{\mathcal{I}} - s, & \text{if } i = j \end{cases}.$$

(iv) *If s lies in the last row of $T_\mathcal{I}$, then $T_{\mathcal{I}'} = (T_\mathcal{I} - s)$ after standardization.*

Proof. Recall from 2.15 the notation $c_i := |C_i|$ and $\varsigma_i := C_i^{c_i}$. One has

$$\mathcal{I} = \Pi \setminus \bigcup_{i=1}^{l-1} \alpha_{\varsigma_i}$$

Then (i) is clear if $s \neq \varsigma_i$ for any i. Otherwise $\alpha_s \notin \mathcal{I}$ but also $\alpha_{s-1} \notin \mathcal{I}'$. Hence (i). (ii) follows from (i) by 2.12. (iii) is obtained from 2.12 just by definition of chains. Finally (iii) implies (iv) since in that case $T_{\mathcal{I}'}$ is obtained by eliminating s and sliding the remaining entries to the left. □

3.8. Retain the notation of 2.19 and assume $c = c_1 = c_l$. Then $T_\mathcal{I}(l)$ is defined. Set $M = M_\mathcal{I}^{\underline{c}}(t)$.

Proposition. *Set $w = w_r(T_\mathcal{I}(l))$. Then $M_\mathcal{I}^{\underline{c}}(t)$ vanishes on $\mathfrak{n} \cap^w \mathfrak{n}$, that is $M(X(w)) = 0$.*

Proof. The proof is by induction on $d_\mathcal{I}$ defined in 2.19. If $d_\mathcal{I} = 0$, then $c_i \leq c$, $\forall 1 \leq i \leq l$ and so n occurs in the last row of $T_\mathcal{I}$. Moreover $T_\mathcal{I}(l)$ has $c+1$ rows with n the unique entry in row $c+1$. Consequently the word form of w starts with n. By 2.3 this forces $X_{i,n} \not\subset \mathfrak{n} \cap^w \mathfrak{n}$, $\forall i < n$. Thus the entries $x_{n,i} : i = 1, 2, \ldots, n-c$, of the last row of M vanish on $X(w)$ and consequently so does M.

Let us assume the assertion holds for $d_\mathcal{I} = d$ and take $d_\mathcal{I} = d+1$. In particular there is some $c_i > c$. Let k be the number of rows of $T := T_\mathcal{I}$. Since $c_i > c$ this is also the number of rows of $T' := T_\mathcal{I}(l)$ and so the word form of w starts with $s = T_1^k$, where $s \notin< C_1 >$. By 2.3 we conclude that $X_{i,s} \not\subset \mathfrak{n} \cap^w \mathfrak{n}$, for $i < s$ Thus the entries $x_{s,i} \ : \ i < s$ in $M_\mathcal{I}^{\underline{c}}(t)$ vanish on $X(w)$. This leaves t as the only nonzero entry of row s. Consequently

$$M(X(w)) = tM(X(w)^{s,s}) \qquad (*)$$

it being understood that n is reduced by 1 in defining the right-hand side.

By 3.5 one has $X(w)^{s,s} = X(w-s)$ whilst by 3.6 one has $X(w-s) = X(w_r(T'-s))$. Recall the definition of $\mathcal{I}'$ from 3.7. By 3.7 (iv), $T - s = T_{\mathcal{I}'}$. Then by 2.17 $T'-s$ is a descendant of $T_{\mathcal{I}'}$. Moreover $d_{\mathcal{I}'} = d_\mathcal{I} - 1 = d$ and $C_1^{\mathcal{I}'} = C_1^\mathcal{I}, C_l^{\mathcal{I}'} = C_l^\mathcal{I}$ up to standardization. In particular $|C_1^{\mathcal{I}'}| = |C_l^{\mathcal{I}'}| = c$. Set $w' = w_r(T'-s)$, $M' = M_{\mathcal{I}'}^{\underline{c}}(t)$. By the above and the induction hypothesis $M'(X(w')) = 0$. Then by $(*)$ and the above $M(X(w)) = 0$. □

3.9. Retain the notation and hypotheses of 3.8. For $1 \leq i, j \leq n$, let $M^{i,j}$ denote ijth cofactor of M. Set $d = d_\mathcal{I}$ and recall that t^d divides M. The following is clear by row expansion of the determinant.

Lemma. *Choose i, j such that $M_{i,j} = x_{c+i,j} \in \mathfrak{m}_\mathcal{I}^-$. Then $M^{i,j}$ is divisible by t^d.*

3.10. Take $f \in S(\mathfrak{m}^-)$. In computing $\{x_{i,j}, f\}$ via 2.19 $(*)$ we obtain two types of terms. The first (resp. second) is obtained from the first (resp. second) factor on the right-hand side. We call them the terms obtained from j (resp. i).

Lemma. *Suppose* $i \in \{1, 2, \dots, n-1\} \setminus \{c, n-c\}$. *Then*

$$\{x_{i,i+1}, M\} \subset S(\mathfrak{m}_{\mathcal{I}}^-)M + S(\mathfrak{g})\mathfrak{p}_{\mathcal{I}}.$$

Proof. Let $\mathcal{X}$ be a matrix with entries $\mathcal{X}_{i,j} = x_{i,j}$ and X the $(n-c) \times (n-c)$ minor in the bottom left-hand corner of $\mathcal{X} + t\mathrm{Id}$. Then $M - X \in S(\mathfrak{g})\mathfrak{p}_{\mathcal{I}}$. Since $\mathfrak{p}_{\mathcal{I}}$ is a subalgebra it is enough to prove the corresponding assertion for X.

Consider $\{x_{i,i+1}, X\}$.

(i) Suppose $i > c$. Then the sum of terms coming from $(i+1)$ forms a determinant X^{i+1} with the same rows as X, except that the entries $x_{i+1,s} + \delta_{i+1,s}t$ on the $(i+1-c)$th row of X are replaced by $x_{i,s}$. Hence this term equals $-tX^{i+1-c,i}$.

(ii) Suppose $i < n-c$. Then the sum of terms coming from i forms a determinant X_i with the same columns as X, except that the entries $x_{s,i} + \delta_{s,i}t$ on the ith column of X are replaced by $-x_{s,i+1}$. Hence this term equals $tX^{i+1-c,i}$.

If $c < i < n-c$. then terms cancel. If $i < c$ or $i > n-c$, both terms are zero. This proves the lemma. □

3.11. Set $f = m_{l_{\mathcal{I}}}$.

Lemma. *One has* $\{\mathfrak{p}_{\mathcal{I}}, f\} \subset S(\mathfrak{m}_{\mathcal{I}}^-)f + S(\mathfrak{g})\mathfrak{p}_{\mathcal{I}}$.

Proof. By 2.20 (*) f has weight zero with respect to the Cartan subalgebra $[\mathfrak{l}_{\mathcal{I}}, \mathfrak{l}_{\mathcal{I}}] \cap \mathfrak{h}$ of the semisimple Lie algebra $[\mathfrak{l}_{\mathcal{I}}, \mathfrak{l}_{\mathcal{I}}]$. Since $S(\mathfrak{g})$ and $S(\mathfrak{g})\mathfrak{l}_{\mathcal{I}}$ are locally finite $\mathfrak{l}_{\mathcal{I}}$ modules it follows from the theory of finite-dimensional $\mathfrak{l}_{\mathcal{I}}$ modules that it suffices to show $\{x_{i,i+1}, f\} \subset S(\mathfrak{m}_{\mathcal{I}}^-)f + S(\mathfrak{g})\mathfrak{p}_{\mathcal{I}}$, for all $1 \le i \le n-1$ and $\{x_{i+1,i}, f\} \subset S(\mathfrak{m}_{\mathcal{I}}^-)f + S(\mathfrak{g})\mathfrak{p}_{\mathcal{I}}$, for all $\alpha_i \in \tau(\mathfrak{p}_{\mathcal{I}})$. By 3.10 among all $\{x_{i,i+1}, f\}$ it remains to consider the cases $i = c, n-c$. Both are similar and we consider only the first. Consider $\{x_{c,c+1}, M\}$. The sum of terms coming from $c+1$ all lie in $S(\mathfrak{g})\mathfrak{p}$. The sum of terms coming from c is equal to $tM^{1,c}$ mod $S(\mathfrak{g})\mathfrak{p}$. Yet $M_{1,c} = x_{c+1,c} \in \mathfrak{m}_{\mathcal{I}}^-$ so by 3.9 this expression is divisible by t^{d+1} mod $S(\mathfrak{g})\mathfrak{p}_{\mathcal{I}}$ Thus $\{x_{c,c+1}, f\} = 0 \bmod S(\mathfrak{g})\mathfrak{p}_{\mathcal{I}}$.

As for $\{x_{i+1,i}, f\}$ we show that $\{x_{i+1,i}, M\} = 0$ exactly in the same manner as in 3.10. Let us sketch the proof. The cases $i < c$ or $i+1 > n-c$ are the same so let us show this for $i+1 > n-c$. In this case the sum of terms coming from $i+1$ is zero and the sum of terms coming from i forms a determinant M^i with the same rows as M except that the $(i-c)$th row is replaced by the $(i-c+1)$th row, so $M^i = 0$. Now if $c < i < n-c-1$, then for all $s \neq i, i+1$ one has $M_{i-c,s} = 0$ iff $M_{i+1-c,s} = 0$. Similarly for all $s \neq i-c, i+1-c$. One has $M_{s,i} = 0$ iff $M_{s,i+1} = 0$. Using this we get exactly as in 3.10 that the sum of terms coming from i equals $-tM^{i-c,i+1}$ and the sum of terms coming from $(i+1)$ equals $tM^{i-c,i+1}$, so these terms cancel. □

3.12. We may summarize the consequence of 3.3 (ii), 3.8, 3.11 as follows. Adopt the notation and hypotheses of 2.19 (i). In particular set $T' = T_{\mathcal{I}}(l)$.

Proposition. *The zero variety $\mathcal{V}(m_{l_{\mathcal{I}}})$ of $m_{l_{\mathcal{I}}}$ in $\mathfrak{m}_{\mathcal{I}}^-$ is a union of orbital variety closures of codimension 1 one of which is $\mathcal{V}_{T'}$.*

Remark. We have not yet used that $c_j \neq c$ for $1 < j < l$. This is needed for irreducibility of $\mathcal{V}(m_{l_{\mathcal{I}}})$. This would follow from 2.18 and the irreducibility of $\mathbf{G}\mathcal{V}(m_{l_{\mathcal{I}}})$; but we only have a direct proof of the latter using representation theory (cf. 4.7).

3.13. Let us show that (ii) of Theorem 2.19 results from (i). This is obtained from the following general remark. Take $\mathcal{I} \subset \Pi$ and let $T_{\mathcal{I}} = (C_1, \dots, C_l)$ be its chain form. Fix $i, j : 1 \leq i \leq j \leq l$ and set $T_{\mathcal{I}'} = (C_i, \dots, C_j)$. Let $T' \in \mathbf{T}_{|T_{\mathcal{I}'}|}$ satisfy $T' > T_{\mathcal{I}'}$ and set $T = (C_1, \dots, C_{i-1}, T', C_{j+1}, \dots, C_l)$. Then $T > T_{\mathcal{I}}$ and $\pi(T_{\mathcal{I}}) = T_{\mathcal{I}'}$, $\pi(T) = T'$, where $\pi = \pi_{\varsigma_{i-1}+1, \varsigma_j}$. Again by 2.14 (∗) $\pi(\mathcal{V}_T) = \mathcal{V}_{T'}$ and by 2.4 (2) the codimension of $\mathcal{V}_T$ in $\mathfrak{m}_{\mathcal{I}}$ equals the codimension of $\mathcal{V}_{T'}$ in $\mathfrak{m}_{\mathcal{I}'} = \pi(\mathfrak{m}_{\mathcal{I}})$. The implication (i) $\Longrightarrow$ (ii) in 2.19 follows from the

Lemma. *With the above hypotheses*

$$I(\overline{\mathcal{V}}_T) = S(\mathfrak{m}_{\mathcal{I}}^-) I(\overline{\mathcal{V}}_{T'}).$$

Proof. From the commuting diagram

$$\begin{array}{ccc} S(\pi(\mathfrak{m}_{\mathcal{I}}^-)) & \overset{\pi^*}{\hookrightarrow} & S(\mathfrak{m}_{\mathcal{I}}^-) \\ \downarrow & & \downarrow \\ R[\overline{\mathcal{V}}_{T'}] & \overset{\pi^*}{\hookrightarrow} & R[\overline{\mathcal{V}}_T] \end{array}$$

we obtain the inclusion $\supset$. On the other hand

$$S(\mathfrak{m}_{\mathcal{I}}^-)/IS(\mathfrak{m}_{\mathcal{I}}^-) \cong S(\pi(\mathfrak{m}_{\mathcal{I}}^-))/I \otimes S(\ker \pi),$$

so $IS(\mathfrak{m}_{\mathcal{I}}^-)$ is prime. Equality of codimensions finishes the proof. □

3.14. Let V be a vector space of dimension $n < \infty$. Let $\{v_i\}_{i=1}^n$ be a basis of V and set $v_0 = v_{n+1} = 0$. Define $e, f \in \operatorname{End} V$, by $ev_i = v_{i-1}$, $fv_i = v_{i+1}, \forall i$. Let r, s be integers > 0 and set $A_{r,s}^n(t) = \operatorname{Det}(te^s + f^r)$.

Lemma. *For all $r, s, n > 0$ one has*

$$A_{r,s}^n(t) = \begin{cases} ((-1)^{r+s} t^r)^h & \text{if } h(r+s) = n, \\ 0 & \text{if } r+s \text{ does not divide } n. \end{cases}$$

Proof. If $r + s > n$, then there are only zeros in the $n - s + 1$th row, so we can suppose $r + s \leq n$. Then the first r rows have exactly one entry, namely t.

Similarly the first s columns have exactly one entry, namely 1. Developing these rows and columns gives

$$A^n_{r,s}(t) = (-1)^{r+s} t^r A^{n-r-s}_{r,s}(t),$$

and hence the required result. □

3.15. The irreducibility of $m_{l_\mathcal{I}}$ is established by induction on n. In view of 3.13 it is therefore enough to consider only the situation described in 2.19 (i). Precisely we show the

Proposition. *Take $\mathcal{I} \in \Pi$ and $T_\mathcal{I} = (C_1, C_2, \dots, C_l)$ the chain form of $T_\mathcal{I}$. Set $c_i = |C_i|$. Assume $c_1 = c_l =: c$ and $c_j \neq c$, $\forall j : 1 < j < l$. Then $m_{l_\mathcal{I}}$ is irreducible.*

Proof. Otherwise by 2.17, 3.12 and 3.13, there is a subset $\mathcal{I}'$ of Π obtained from $\mathcal{I}$ exactly as in 2.19 (ii) such that the corresponding $f' := m_{l_{\mathcal{I}'}}$ is an irreducible factor of $f = m_{l_\mathcal{I}}$. We obtain a contradiction by showing that there exists a point $x \in \mathfrak{m}$ such that $f(x) = (-1)^c$, $f'(x) = 0$.

Set $a = |T^{c+1,k}_\mathcal{I}|$ and

$$\langle T^{1,c}_\mathcal{I} \rangle = \{p_1, p_2, \dots, p_{n-a}\}, \quad \langle T^{c+1,k}_\mathcal{I} \rangle = \{q_1, q_2, \dots, q_a\},$$

written in increasing order. Define $x \in \mathfrak{m}$ by

$$x_{r,s} = \begin{cases} 1, & \text{if } r = p_i, \text{ for } i : c < i \leq n-a \text{ and } s = p_{i-c} \\ 0, & \text{otherwise.} \end{cases}$$

Set $M := M^c_\mathcal{I}(t)$. Let b be an integer $0 < b \leq n - c$ and let Δ_b denote the set of subsets of $\{c+1, \dots, n\}$ of cardinality b. Given $\sigma = \{r_1, r_2, \dots, r_b\} \in \Delta_b$, let M^σ be obtained from M by deleting the $(r_i - c)$th rows and r_ith columns where $i = 1, 2, \dots, b$. Let M^σ_o be the evaluation of M^σ at $t = 0$. Developing M gives the term $(-1)^c t^b M^\sigma$ and moreover

$$m_{n-c-b} = (-1)^c \sum_{\sigma \in \Delta_b} M^\sigma_o. \tag{$*$}$$

Recall that $l_\mathcal{I} = n - c - a$. For $\sigma \in \Delta_a$ one has

$$M^\sigma_o(x) = \begin{cases} 1, & \text{if } \sigma = \{q_1, q_2, \dots, q_a\}, \\ 0, & \text{otherwise.} \end{cases}$$

Consequently

(i) $f(x) = (-1)^c$.

In the notation of Theorem 2.19 (ii) set $c' := c_i = c_j$ and $M' = M^{c'}_{\mathcal{I}'}(t)$. Then

(ii) $f'(x) = 0$, if $c' > c$.

Indeed in this case the last row of M' is indexed by the largest integer in C_j. By the hypothesis this is an entry of $T_{\mathcal{I}}^{c+1,k}$. Hence even $M'(x) = 0$.

Define u, v as in Theorem 2.19 (ii) and set $n' = v - u + 1$. Set $\sigma' = \langle T_{\mathcal{I}}^{c+1,k}\rangle \cap \{u, u+1, \dots, v\}$ and $d' = |\sigma'|$. In the notation of 3.14 we have

(iii) $M'(x) = (-1)^{c'} t^{d'} A^{n'}_{c-c',c'}(t)$, if $c' < c$.

This is obtained through a development of M' similar to that for M and noting that $M'^{\sigma}_{o}(x) = 0$ unless $\sigma \supset \sigma'$, so that $M'(x) = (-1)^{c'} t^{d'} M'^{\sigma'}(x) = (-1)^{c'} t^{d'} A^{n'}_{c-c',c'}(t)$.

Let us show finally that

(iv) $f'(x) = 0$, if $c' < c$.

By 3.14 it suffices to show that $h(c - c') + d' > d_{\mathcal{I}'}$ when $n' = hc$.

Let $C_{m_1}, C_{m_2}, \dots, C_{m_r}$ be the chains of length $> c'$ between C_i and C_j. One has $d_{\mathcal{I}'} = \sum_{s=1}^{r} (c_{m_s} - c') \le d' + r(c - c')$ and so the assertion holds if $r < h$. On the other hand $hc = n' \ge (r+2)c' + d_{\mathcal{I}'}$. Thus

$$h(c - c') + d' \ge (r+2)c' + d_{\mathcal{I}'} > d_{\mathcal{I}'} \quad \text{if} \quad r \ge h.$$

This proves (iv).

Finally the hypothesis itself excludes $c = c'$ and so the proposition follows from (i), (ii) and (iv). □

Remark. The proof of Theorem 2.19 is now complete.

3.16. Adopt the hypotheses of 2.19 (i). Let r be the number of chains C_i of length $> c$ and $s + 2c$ the sum of the lengths of the remaining chains.

Proposition. *The generator $f_{\mathcal{I}}(l)$ of $I(\overline{\mathcal{V}}_{T_{\mathcal{I}}(l)})$ in $S(\mathfrak{m}^-)$ is the highest common divisor of the (nonvanishing) $(s+c) \times (s+c)$ minors of $\mathcal{M}_{\mathcal{I}}^{r+1}$.*

Proof. Under the hypotheses of 2.19 (i), it follows from 2.18 (in the notation of 2.18) that $\mathfrak{m}_{\mathcal{I}} \cap \mathcal{O}_{\mathcal{I}}(c) = \mathcal{V}_{T_{\mathcal{I}}(l)}$. By 2.4 (i) the nilpotent orbit to which an element x belongs is determined by $\{\operatorname{rk} x^i : i = 1, 2, \dots, n\}$.

Set $\lambda = \operatorname{sh} T_{\mathcal{I}}$. Since $n - \sum_{j>r+1} \lambda_j^* = s + c$, it follows that $\operatorname{rk} x^{r+1} = s + c$, for all $x \in \mathfrak{m} \cap \mathcal{O}_{\mathcal{I}}$.

Now $\operatorname{sh} T_{\mathcal{I}}(l)$ is obtained from λ by lowering a box from the cth row to the $(c+1)$th row. Under the hypotheses of 2.19 (i) this takes it from the $(r+2)$th to the $(r+1)$th column.

From the three paragraphs above we conclude that

$$\mathfrak{m}_{\mathcal{I}} \cap \mathcal{O}_{\mathcal{I}}(c) = \{x \in \mathfrak{m} \mid \operatorname{rk} x^{r+1} = s + c - 1\}.$$

Thus this hypersurface in $\mathfrak{m}_{\mathcal{I}}$ is the set of common zeros of the $(s+c) \times (s+c)$ minors of $\mathcal{M}_{\mathcal{I}}^{r+1}$. Hence $f_{\mathcal{I}}(l)$ is exactly their largest common divisor. □

3.17. It is of course not too easy to explicitly determine $f_{\mathcal{I}}(l)$ through 3.16. Moreover these power rank conditions (i.e., 2.4 (i)) are in general insufficient to obtain ideals of definition of orbital variety closures. This was already observed by van Leeuwen [vanL, §8] and also resulted independently from [M5, §4.3] because otherwise the chain order defined in [M5, 4.3.1] would coincide with the geometric order which fails in $\mathfrak{sl}_7$, as it is shown in [M5, 4.3.6].

Combining 3.16 and 2.19 (i) (which implies that $f_{\mathcal{I}}(l) = m_{l_{\mathcal{I}}}$) gives a remarkable combinatorial fact, namely 3.16, about generic matrices. Here we note that $\deg m_{l_{\mathcal{I}}} = (r+1)c + s$, whilst the minors in the conclusion of 3.16 have degree $(r+1)(s+c)$. Thus these degrees coincide if and only if $rs = 0$. Of course when degrees coincide $f_{\mathcal{I}}(l)$ is proportional to any nonzero $(s+c) \times (s+c)$ minor of $\mathcal{M}_{\mathcal{I}}^{r+1}$. Using this, Benlolo and Sanderson were able to prove their conjecture, that is Theorem 2.19 (i), under hypothesis $r = 0$ or the hypothesis $c = 1$ (which forces $s = 0$). These cases are relatively easy up to proving irreducibility for which they developed a special trick. A purely combinatorial proof of the case $s = 0$ already appears to be rather difficult.

4 Strong quantization of hypersurface orbital varieties

4.1. We begin with a combinatorial lemma. Let $\mathfrak{R}$ be the real field.

Lemma. *Let $c, c_2, \ldots, c_l$ be positive integers with $c = c_l$. Suppose that the system of inequalities for $\{b_i\}_{i=2}^{l}$ defined by*

$$1)\ 1 + \sum_{i=2}^{s} b_i \leq \sum_{i=2}^{s-1} c_i + \max(c, c_s) : s = 2, \ldots, l-1.$$

$$2)\ 1 + \sum_{i=s+1}^{l} b_i \leq \sum_{i=s+1}^{l-1} c_i + \max(c, c_s) : s = 2, 3, \ldots, l-1.$$

$$3)\ 1 + \sum_{i=2}^{l} b_i = \sum_{i=2}^{l} c_i.$$

has a solution $b_i \in \mathfrak{R}$, $\forall i$. Then $c_s \neq c$, $\forall s : 2 \leq s < l$. Moreover, if the latter holds one can find a solution with $b_i \in \mathbb{N}$, $\forall i$.

Proof. Assume that the system has a solution $b_i \in \mathfrak{R}$, $\forall i$. Adding 1) and 2) gives

$$2 + \sum_{i=2}^{l} b_i \leq \sum_{i=2}^{l} c_i + 2\max(c, c_s) - c - c_s.$$

Then in view of 3) we obtain $1 \leq 2\max(c, c_s) - c - c_s$, that is $c \neq c_s$ for any $s : s < k$.

Conversely equality in 1) and 3) gives

$$b_i = \begin{cases} \max(c, c_2) - 1, & \text{if } i = 2 \\ c_{i-1} + \max(c, c_i) - \max(c, c_{i-1}), & \text{if } 2 < i \leq l. \end{cases} \qquad (*)$$

These imply that

$$\sum_{i=s+1}^{l} b_i = \sum_{i=s}^{l} c_i - \max(c, c_s) = \sum_{i=s+1}^{l-1} c_i + \max(c, c_s) + ((c + c_s) - 2\max(c, c_s)).$$

Then 2) holds if and only if $c + c_s \leq 2\max(c, c_s) - 1$ for all $s = 2, \ldots, l-1$. Obviously this is equivalent to $c_s \neq c$ for any $s = 2, \ldots, l-1$. Moreover, by $(*)$ these inequalities have a solution with $b_i \in \mathbb{N}$ for all i. □

4.2. Recall the notation of 2.7. Given $\mathcal{I} \subset \Pi$ let $T_\mathcal{I} = (C_1, \ldots, C_l)$ be the corresponding tableau. Assume that $c_1 = c_l = c$. One has $\Pi' := \Pi \setminus \mathcal{I} = \{\alpha_{\varsigma_i} : i = 1, 2, \ldots, l-1\}$. For $i = 1, 2, \ldots, c$ we set

$$\beta_i = \sum_{j=i}^{n-1-i} \alpha_j.$$

Set $\beta = \beta_1$. Recall that the BS element $m_{l_\mathcal{I}}$ defined in 2.18 has weight $\mathrm{wt}(m_{l_\mathcal{I}}) = -\sum_{i=1}^{c} \beta_i$. Call $\nu \in \mathfrak{h}^*$, $\mathcal{I}$-regular if $(\alpha, \nu + \rho) \neq 0$, $\forall \alpha \in R_\mathcal{I}$. Set $w.\nu := w(\nu + \rho) - \rho$. Set

$$\mathcal{S}_\nu = \{(m, \gamma) \in \mathbb{N}^+ \times R^+ \setminus R^+_\mathcal{I} \mid (\gamma^\vee, \nu + \rho) = m\}$$
$$\mathcal{S}^o_\nu = \{(m, \gamma) \in \mathcal{S}_\nu \mid s_\gamma.\nu + \rho \text{ is } \mathcal{I} \text{ regular}\}.$$

Take $c_i \neq c$ $i = 2, 3, \ldots, l-1$ in 4.1 and recall our normalization, making $(\alpha, \alpha) = 2$, $\forall \alpha \in R$. Define $\mu \in P(\Pi)$, by $(\mu + \rho, \alpha) = 1$, for $\alpha \in \mathcal{I}$ and $(\mu + \rho, \alpha_{s_i}) = -(b_{i+1} - 1)$ for $i = 1, 2, \ldots, l-1$ and where the $\{b_i\}_{i=2}^{l}$ are given by the conclusion of 4.1.

Proposition. $\mathcal{S}^o_\mu = \{(c, \beta)\}$.

Proof. One has

$$\sum_{\alpha \in \mathcal{I}} (\mu + \rho, \alpha) = \sum_{i=1}^{l} (c_i - 1) = \sum_{i=2}^{l} (c_i - 1) + c - 1$$
$$= \sum_{i=2}^{l} (b_i - 1) + c = -\sum_{\alpha \in \Pi'} (\mu + \rho, \alpha) + c$$

by 4.1 (3) and so

$$(\mu + \rho, \beta^\vee) = (\mu + \rho, \beta) = \sum_{\alpha \in \mathcal{I}} (\mu + \rho, \alpha) + \sum_{\alpha \in \Pi'} (\mu + \rho, \alpha) = c.$$

Thus $(c, \beta) \in \mathcal{S}_\mu$.

Observe that if $(c', \gamma) \in \mathcal{S}_\mu : \gamma \in R^+ \setminus R^+_\mathcal{I}$, $c' \in \mathbb{N}$, then $(c', \gamma) \in \mathcal{S}^o_\mu$ means that

$$s_\gamma(\mu + \rho) = \mu + \rho - (\gamma^\vee, \mu + \rho)\gamma = \mu + \rho - c'\gamma$$

does not vanish on any $\delta \in R^+_{\mathcal{I}}$. Note that for $\delta \in R^+_{\mathcal{I}}$ and $\gamma \in R^+ \setminus R^+_{\mathcal{I}}$ one has

$$(\delta, \gamma) = \begin{cases} 0 & \text{if } \gamma - \delta \notin R^+ \\ 1 & \text{if } \gamma - \delta \in R^+. \end{cases}$$

Thus vanishing on some $\delta \in R^+_{\mathcal{I}}$ means that $\gamma - \delta$ is a root and $(\delta, \mu + \rho) = c'$ or simply that $\gamma - \delta$ is a root such that $(\mu+\rho, \gamma-\delta) = (\mu+\rho, \gamma) - (\mu+\rho, \delta) = 0$.

Note that $\beta - \delta$ is a root for $\delta \in R^+_{\mathcal{I}}$ iff $\delta = \alpha_1 + \cdots + \alpha_i$ or $\delta = \alpha_{n-1} + \cdots + \alpha_{n-i}$ where $i < c$. Thus $(\mu + \rho, \beta - \delta) > 0$ and so $(c, \beta) \in S^o_\mu$.

Consider $\gamma = \alpha_i + \alpha_{i+1} + \cdots + \alpha_j$ for $i < c$ and $j > n - c$. One has $(\mu + \rho, \gamma) = c - (i-1) - (n - j - 1) =: c'$. Then unless $\gamma = \beta$ one can find $\delta \in R^+_{\mathcal{I}}$ starting at α_i or ending in α_j such that $(\mu + \rho, \gamma - \delta) = 0$. We conclude that $(c', \gamma) \notin S^o_\mu$.

Set $\delta_i^- = \alpha_1 + \alpha_2 + \cdots + \alpha_{\varsigma_{i-1}}$ for $2 \leq i \leq l$ and $\delta_j^+ = \alpha_{\varsigma_j} + \alpha_{\varsigma_j+1} + \cdots + \alpha_{n-1}$ for $1 \leq j \leq l-1$. One has

$$\begin{aligned}(\mu + \rho, \delta_s^-) &= \sum_{i=1}^{s-1}(c_i - 1) + \sum_{i=2}^{s} -(b_i - 1) \\ &= c + \sum_{i=2}^{s-1} c_i - \sum_{i=2}^{s} b_i \geq c + 1 - \max(c, c_s) \\ (\mu + \rho, \delta_s^+) &= \sum_{i=s+1}^{l} (c_i - 1) + \sum_{i=s+1}^{l} -(b_i - 1) \\ &= \sum_{i=s+1}^{l-1} c_i - \sum_{i=s+1}^{l} b_i + c \geq c - \max(c, c_s) + 1.\end{aligned}$$

Now consider $\gamma_{s,l} = \beta - \delta_s^-$ with $2 \leq s \leq l-1$. Then $(\mu+\rho, \gamma_{s,l}) \leq \max(c, c_s) - 1$. Thus it follows that $(c', \gamma_{s,l}) \notin S^o_\mu$. Indeed if $c_s > c$ we can subtract $\delta = \alpha_{\varsigma_{s-1}+1} + \cdots + \alpha_i$ for some $i : \varsigma_{s-1} + 1 \leq i < \varsigma_s$ from $\gamma_{s,l}$ so that $(\mu + \rho, \gamma_{s,l} - \delta) = 0$ and if $c_s < c$ we can subtract $\delta = \alpha_{n-1} + \cdots \alpha_i$ for some $i : \varsigma_{l-1} < i \leq n-1$ from $\gamma_{s,l}$ so that $(\mu + \rho, \gamma_{s,l} - \delta) = 0$. A similar assertion holds if we replace $\gamma_{s,l}$ by $\gamma = \gamma_{s,l} - (\alpha_{\varsigma_{s-1}+1} + \cdots + \alpha_i) - (\alpha_{n-1} + \cdots \alpha_j)$ for any $i : \varsigma_{s-1} + 1 \leq i < \varsigma_s$ and $j : \varsigma_{l-1} < j \leq n-1$, that is when some of these roots have already been subtracted.

A similar conclusion holds for $\gamma_{1,s} = \beta - \delta_s^+$ with $2 \leq s \leq l-1$ and for $\gamma = \gamma_{1,s} - (\alpha_1 + \cdots + \alpha_i) - (\alpha_{\varsigma_s - 1} + \cdots \alpha_j)$ for any $i : 1 \leq i < c$ and $j : \varsigma_{s-1} < j \leq \varsigma_s - 1$ (by the obvious symmetry).

Finally consider $\gamma_{i,j} = \beta - \delta_i^- - \delta_j^+$, with $2 \leq i < j \leq l-1$. We have

$$(\mu + \rho, \gamma_{i,j}) \leq \max(c, c_i) + \max(c, c_j) - c - 2.$$

Then a similar conclusion holds in this case also with slightly stronger reason. □

Remark. One may also check that the inequalities 1) and 2) of 4.1 are necessary for the conclusion of the proposition to hold with $\mu + \rho$ satisfying $(\mu + \rho, \alpha) = 1$,

$\forall \alpha \in \mathcal{I}$ and $(\mu + \rho, \beta) = c$. Consequently the proposition fails when $c = c_i$, for some $i : 1 < i < l$.

4.3. Take μ as in 4.2 and recall the definition of β_i from 2.20. Recall that $\gamma_c = \sum_{i=1}^{c} \beta_i$. Then $(\gamma_c, \alpha) = 0, \forall \alpha \in \mathcal{I}$. Moreover $\mu + \rho - \gamma_c$ is the unique $\mathcal{I}$ dominant element of $W_{\mathcal{I}}(\mu + \rho - c\beta) = W_{\mathcal{I}}(s_\beta(\mu + \rho))$.

Let $\{M^i_{\mathcal{I}}(\mu)\}_{i=0}^{\infty}$ be the Jantzen filtration of $M_{\mathcal{I}}(\mu)$. Given $(m, \gamma) \in \mathcal{S}^o_\mu$, let $\omega(s_\gamma.\mu)$ be the unique Π' dominant element in $W_{\mathcal{I}}.(s_\gamma.\mu)$ and $\operatorname{sign}(s_\gamma.\lambda) := (-1)^{l(w)}$ where $w \in W_{\mathcal{I}}$ is the unique element satisfying $w.s_\gamma.\mu = \omega(s_\gamma.\mu)$. The Jantzen sum formula combined with 4.2 gives

$$\sum_{i=1}^{\infty} \operatorname{ch} M^i_{\mathcal{I}}(\mu) = \sum_{(m,\gamma)\in\mathcal{S}^o_\mu} \operatorname{sign}(s_\gamma.\mu) \operatorname{ch} M_{\mathcal{I}}(\omega(s_\gamma.\mu)) = \operatorname{ch} M_{\mathcal{I}}(\mu - \gamma_c), \quad (*)$$

by the above noting that $\operatorname{sign}(s_\beta.\mu)$ is necessarily positive because there is only this term on the right-hand side. One may recall that the right-hand side of the Jantzen sum formula is obtained from the zeros of the Shapovalov determinants. Then the appearance of just one term on the right-hand side means that the Shapovalov determinant is nonzero on $M_{\mathcal{I}}(\mu)_\nu : \nu > \omega(s_\beta.\mu)$ and vanishes on $M_{\mathcal{I}}(\mu)_{\omega(s_\beta.\mu)}$.

(Equivalently, but more directly by Jantzen [Ja, Satz 2] the Shapovalov determinant on the ν weight subspace of $M_{\mathcal{I}(\mu)}$ takes the form

$$\prod_{m=1}^{\infty} \prod_{\gamma\in R^+\setminus R^+_{\mathcal{I}}} ((\gamma, \mu + \rho) - m)^{\chi'_\nu(\mu - m\gamma)}$$

where

$$\chi'_\nu(\lambda) = \sum_{w\in W_{\mathcal{I}}} (-1)^{l(w)} \dim M(w.\lambda)_\nu. \quad (*)$$

This has a zero if and only if some $(m, \gamma) \in \mathcal{S}_\mu$ and the sum of the corresponding exponents is strictly positive. Now by $(*)$, $\chi'_\nu(\lambda) \neq 0$ for some ν if and only if $\lambda + \rho$ is $\mathcal{I}$ regular, so we further require $(\mu, \gamma) \in \mathcal{S}^o_\mu$. Moreover in this case $\chi'_\nu(\lambda)$ equals $\operatorname{sign}(\lambda) \dim M_{\mathcal{I}}(\omega(\lambda))_\nu$. Since in addition $\mathcal{S}^o_\mu = \{(m, \beta)\}$ in the present case we must have $\chi'_\nu(s_\beta.\mu) \geq 0$ for all ν and of course no cancellations occur. Moreover from $(*)$ we obtain

$$\chi'_\nu(s_\beta.\mu) = \begin{cases} 0, & \text{if } \nu > \omega(s_\beta.\mu), \\ 1, & \text{if } \nu = \omega(s_\beta.\mu). \end{cases}$$

which gives the required assertion.)

This forces there to be a highest weight vector in $M_{\mathcal{I}}(\mu)$ of weight $\omega(s_\beta.\mu) = \mu - \gamma_c$. Since $(\alpha, \mu) = 0$ for all $\alpha \in \mathcal{I}$ it follows that $M_{\mathcal{I}}(\mu)$ is a rank 1 free $U(\mathfrak{m}^-_{\mathcal{I}})$ module. Again $(\alpha, \mu - \gamma_c) = 0$, for all $\alpha \in \mathcal{I}$. Consequently this highest weight vector must generate a submodule of $M_{\mathcal{I}}(\mu)$ isomorphic to $M_{\mathcal{I}}(\mu - \gamma_c)$

necessarily contained in the maximal submodule $M^1_{\mathcal{I}}(\mu)$ of $M_{\mathcal{I}}(\mu)$. By (*) we then conclude that $M^1_{\mathcal{I}}(\mu) = M_{\mathcal{I}}(\mu - \gamma_c)$ and so $V(\mu) = M_{\mathcal{I}}(\mu)/M_{\mathcal{I}}(\mu - \gamma_c)$ is simple.

By [M4] the associated variety of $V(\mu)$ is irreducible and hence an orbital variety closure $\overline{\mathcal{V}}$.

Set $p_{\mathcal{I}} = \prod_{\alpha \in R^+_{\mathcal{I}}} \alpha$. Since

$$\operatorname{ch} V(\mu) = (\operatorname{ch} M_{\mathcal{I}}(\mu))(1 - e^{-\gamma_c}) = \operatorname{ch} S(\mathfrak{m}^-_{\mathcal{I}})e^{\mu}(1 - e^{-\gamma_c}),$$

it follows that the characteristic polynomial of $\mathcal{V}$ is just $\gamma_c p_{\mathcal{I}}$.

Let $f_{\mathcal{I}}(l)$ be the BS element constructed in 2.18 and $\mathcal{V}_{T_{\mathcal{I}}(l)}$ the corresponding hypersurface orbital variety. By 3.12 the zero variety $\mathcal{V}(f_{\mathcal{I}}(l))$ of $f_{\mathcal{I}}(l)$ in $\mathfrak{m}^-_{\mathcal{I}}$ is a union of orbital variety closures of codimension 1, one of which is $\mathcal{V}_{T_{\mathcal{I}}(l)}$. Yet wt $f_{\mathcal{I}}(l) = -\gamma_c$ and so the characteristic polynomial of $\mathcal{V}(f_{\mathcal{I}}(l))$ is also $\gamma_c p_{\mathcal{I}}$. By the linear independence of characteristic polynomials of orbital varieties [J1] we conclude that $\mathcal{V} = \mathcal{V}(f_{\mathcal{I}}(l)) = \mathcal{V}_{T_{\mathcal{I}}(l)}$. This proves that $f_{\mathcal{I}}(l)$ is irreducible. Finally $R[\overline{\mathcal{V}}_{T_{\mathcal{I}}(l)}] = S(\mathfrak{m}^-_{\mathcal{I}})/S(\mathfrak{m}^-_{\mathcal{I}})f_{\mathcal{I}}(l)$, so $\operatorname{ch} R[\overline{\mathcal{V}}_{T_{\mathcal{I}}(l)}] = \operatorname{ch} S(\mathfrak{m}^-_{\mathcal{I}})(1-e^{-\gamma_c})$. A comparison with ch $V(\mu)$, which is the above expression up to a shift, shows that $V(\mu)$ is a strong quantization of $\mathcal{V}_{T_{\mathcal{I}}(l)}$. We have proved the

Theorem. *Every hypersurface orbital variety in $\mathfrak{sl}_n$ admits a strong quantization being a simple highest weight module with integral highest weight given by the above procedure.*

4.4. Let us consider the case where we take equality in 1) of 4.1. Using 3) we obtain

$$\sum_{i=2}^{s} b_i = \sum_{i=2}^{s-1} c_i + \max(c, c_s) - 1, \quad \sum_{i=s+1}^{l} b_i = \sum_{i=s}^{l} c_i - \max(c, c_s)$$

Then in the notation of 4.2 for $2 \leq s \leq l-1$ we obtain equalities

$$(\mu + \rho, \delta^-_s) = \sum_{i=1}^{s-1}(c_i - 1) + \sum_{i=2}^{s} -(b_i - 1) = c - \max(c, c_s) + 1$$

$$(\mu + \rho, \delta^+_s) = \sum_{i=s+1}^{l}(c_i - 1) + \sum_{i=s+1}^{l} -(b_i - 1) = \max(c, c_s) - c_s.$$

Consequently $(\mu+\rho, \gamma_{s,l}) = \max(c, c_s) - 1$ for $2 \leq s \leq l-1$ and $(\mu+\rho, \gamma_{1,s}) = c + c_s - \max(c, c_s)$ for $2 \leq s < l$. Finally for any $\gamma_{s,t}$ with $1 < s < t < l$ one has

$$\begin{aligned}(\mu+\rho, \gamma_{s,t}) &= (\mu+\rho, \gamma_{1,t}) - (\mu+\rho, \delta^-_s) = c + c_t - \max(c, c_t) - (c - \max(c, c_s) + 1) \\ &= (c_t - 1) - \max(c, c_t) + \max(c, c_s). \qquad (*)\end{aligned}$$

Now let $\gamma^-_{i,j}$ denote $\gamma_{i,j}$ with the $(c_i - 1)$ roots of the ith column and $(c_j - 1)$ of the jth column removed that is

$$\gamma^-_{i,j} = \alpha_{\varsigma_{i+1}} + \cdots + \alpha_{\varsigma_{j-1}}.$$

Then

$$(\mu+\rho,\gamma_{1,t}^-) = c + c_t - \max(c, c_t) - (c-1) - (c_t - 1) = 2 - \max(c, c_t).$$

Yet to $\gamma_{1,t}^-$ we can always add either the roots from the first column or from tth column to obtain some $\gamma := \delta + \gamma_{1,t}^- : \delta \in R_{\mathcal{I}}^+$. Such γ for which $(\mu+\rho,\gamma)$ is maximum satisfies

$$(\mu+\rho,\gamma) = 2 - \max(c, c_t) + \max(c, c_t) - 1 = 1 > 0. \qquad (**)$$

Similarly for $2 \le s < t \le l$ by $(*)$ one has

$$(\mu+\rho,\gamma_{s,t}^-) = \max(c, c_s) - \max(c, c_t) - (c_s - 1).$$

To $\gamma_{s,t}^-$ we can add either the roots from the s-th column or from t-th column. Again in the best case for such γ one has

$$(\mu+\rho,\gamma) = \max(c_s, c_t) + \max(c, c_s) - \max(c, c_t) - c_s \ge 0. \qquad (***)$$

Let us define

$$\check{S}_\mu = \{(m,\beta) \in -\mathbb{N}^+ \times R^+ \setminus R_{\mathcal{I}}^+ \mid (\beta^\vee, \mu+\rho) = m\}$$

and

$$\check{S}_\mu^o = \{(m,\beta) \in \check{S}_\mu \mid s_\beta(\mu+\rho) \text{ is } \mathcal{I} \text{ regular }\}.$$

Lemma. *With the above choice of μ one has $\check{S}_\mu^o = \emptyset$.*

Proof. As before given $\gamma \in \check{S}_\mu$ we must find $\delta \in R_{\mathcal{I}}^+$ such that $0 = (\delta, s_\gamma(\mu + \rho)) = (\delta, \mu+\rho+(\mu+\rho,\gamma)\gamma)$. It is enough to show that $\gamma+\delta$ is a root vanishing on $\mu+\rho$. For this notice for $\gamma \in \check{S}_\mu$ we must first have $r := (\gamma, \mu+\rho) \in -\mathbb{N}^+$. If for example $\gamma = \gamma_{s,t}^- : 1 \le s < t \le l$, then $(**)$, $(***)$ show that there exists $\delta \in R_{\mathcal{I}}^+$ such that $\gamma+\delta$ is a root and $(\gamma+\delta, \mu+\rho)$ takes all possible integer values from r to an integer ≥ 0. The general case is similar (and easier). □

4.5. Let $\mathbf{O}_{\mathcal{I}}$ be the subcategory of $\mathbf{O}$ in which the Levi factor defined by $\mathcal{I}$ acts finitely. By [JLT, 9.6]

Proposition. *Given $\check{S}_\mu^o = \emptyset$, then $M_{\mathcal{I}}(\mu)$ is projective in $\mathbf{O}_{\mathcal{I}}$.*

Theorem 4.6. *Define μ as in 4.4. Then* Ann $V(\mu) \in$ Max $U(\mathfrak{g})$.

Proof. Since $\mathfrak{g}$ is of type A the natural map $U(\mathfrak{g}) \to F(M_{\mathcal{I}}(\mu), M_{\mathcal{I}}(\mu))$ is surjective. Set $F^\mu = F(M_{\mathcal{I}}(\mu), M_{\mathcal{I}}(\mu)) = U(\mathfrak{g})/\text{Ann}\, M_{\mathcal{I}}(\mu)$. Let P be a maximal ideal of F^μ. By [JLT, 10.9], 4.4 and 4.5 we have

$$P = \text{Ann}_{F^\mu}(M_{\mathcal{I}}(\mu)/PM_{\mathcal{I}}(\mu)).$$

Hence $PM_{\mathcal{I}}(\mu) \subsetneq M_{\mathcal{I}}(\mu)$. Yet $M_{\mathcal{I}}(\mu-\gamma_c)$ is the maximal submodule of $M_{\mathcal{I}}(\mu)$ as a $U(\mathfrak{g})$ module. This forces $P \subset \text{Ann}\, V(\mu)$ and hence equality. □

4.7. Take $c = c_1, c_2, \ldots, c_l = c$. Proceeding as in 4.2 one may rather easily show that there exists $\mu \in \mathfrak{h}^*$ satisfying $(\mu + \rho, \alpha) = 1$, for all $\alpha \in \mathcal{I}$ and such that $\mathcal{S}^o_\mu = \{(c, \beta)\}$. Then as in 4.3, one checks that $V(\mu)$ is a strong quantization of $\mathcal{V}(f_{\mathcal{I}}(l))$. Here we do not need $c_i \neq c$, for $i : 1 < i < l$, however the resulting μ will not be integrable, nor will the associated variety of $V(\mu)$ be irreducible. However it will be contained in the associated variety of Ann $V(\mu)$ which by Borho–Kraft [BK] is just the closure of a nilpotent orbit specifically $\mathcal{O}_{T_{\mathcal{I}}(l)} := \mathbf{G}\mathcal{V}_{T_{\mathcal{I}}(l)}$ in this case. Consequently the associated variety of $V(\mu)$ is contained in $\mathfrak{m} \cap \mathcal{O}_{T_{\mathcal{I}}(l)}$ and we recall that the latter is irreducible if and only if $c_i \neq c$, for all $i : 1 < i < l$. This concludes the third proof of the irreducibility of the BS elements.

Appendix: Index of notation

Symbols appearing frequently are given below in order of appearance.

1.1. $\mathfrak{g}$

1.2. $\mathfrak{n}^-$, $\mathfrak{n}^+$, $\mathfrak{h}$, $\mathfrak{b}$, $\mathbf{G}$, $\mathbf{B}$, $S(\cdot)$, $U(\cdot)$, $\mathcal{O}$, $\mathcal{V}$, $I(\overline{\mathcal{V}})$

1.3. λ, $V(\lambda)$, $\mathcal{F}$

1.5. $\mathfrak{m}^+$, $\mathfrak{m}^-$, $\mathfrak{p}$, $\mathcal{M}(t)$, $M(t)$

1.7. $\mathbf{P}$

2.1. $\mathfrak{n}$, $\mathbf{V}$, W, R, R^+, Π, X_α, $\mathfrak{n} \cap^w \mathfrak{n}$, $\mathcal{V}_w$

2.2. $\mathfrak{g}$, $\mathbf{G}$, $\mathfrak{n}$, $\mathfrak{n}^-$, $\mathbf{B}$, $e_{i,j}$, $\alpha_{i,j}$, α_i, s_α, s_i

2.3. $\mathbf{S}_n$, $[a_1, \ldots, a_n]$, $p_w(i)$, $S(w)$

2.4. $P(n)$, λ, λ^*, k, l, D_λ

2.6. $\mathbf{T}_n$, $[a', a'']$, sh T, $\sigma(T)$, $\hat{\cdot}$, $< \cdot >$, $|\cdot|$, $Q(w)$, $\mathcal{V}_w$, $T_{\mathcal{V}}$, $\mathcal{V}_1 > \mathcal{V}_2$, $T_1 > T_2$

2.7. T^i_j, $r_T(\cdot)$, $c_T(\cdot)$, T^i, T_j, $h(T^i_j)$, $T^{i,j}$, (T, S), $\binom{T}{S}$

2.8. $(R + j)$, $w_r(T)$, $w_c(T)$, $(T \downarrow b)$

2.9. $(R \uparrow j)$, $(R - a)$, $(C \leftarrow j)$, $(T - T^i_j)$

2.10. $\tau(w)$, $\tau(T)$, $\tau(\mathbf{P})$, $\tau(\mathfrak{p})$, $\tau(\mathcal{V})$, $\mathbf{P}_T$, $\mathbf{P}_{\mathcal{V}}$, $\mathfrak{p}_T$, $\mathfrak{p}_{\mathcal{V}}$

2.11. $\mathcal{I}$, $\mathbf{P}_{\mathcal{I}}$, $\mathbf{M}_{\mathcal{I}}$, $\mathbf{L}_{\mathcal{I}}$, $\mathfrak{p}_{\mathcal{I}}$, $\mathfrak{m}_{\mathcal{I}}$, $\mathfrak{l}_{\mathcal{I}}$, $\ell(w)$, $W_{\mathcal{I}}$, $w_{\mathcal{I}}$, $\mathcal{V}_{\mathcal{I}}$, $\mathcal{O}_{\mathcal{I}}$

2.12. $T_{\mathcal{I}}$, $C^{\mathcal{I}}_i$, $(C^{\mathcal{I}}_1, C^{\mathcal{I}}_2, \ldots, C^{\mathcal{I}}_l)$

2.14. $\mathbf{B}_{\mathcal{I}},\ \mathfrak{n}_{\mathcal{I}},\ W^{\mathcal{I}},\ \pi_{\mathcal{I}},\ \pi,\ \mathcal{V}^{\mathcal{I}}_{\pi(w)},\ \pi_{i,j}$

2.15. $c_i,\ \varsigma_i,\ T_{\mathcal{I}}(i)$

2.16. $\mathcal{I}_n,\ \mathcal{I}_1$

2.18. $\lambda_{\mathcal{I}},\ \mathcal{O}_{\mathcal{I}}(c)$

2.19. $f_{\mathcal{I}}(i),\ X(w),\ x_{i,j},\ \{\ ,\ \},\ \mathcal{M}(w),\ c,\ M^c_{\mathcal{I}}(t),\ d_{\mathcal{I}},\ l_{\mathcal{I}},\ m_{l_{\mathcal{I}}}$

2.20. $\beta_i,\ \gamma_c$

3.4. f

3.5. $w-s,\ \mathcal{M}^{i,j}$

3.8. M

3.9. $M^{i,j}$

4.2. $\beta,\ \mathcal{S}_\nu,\ \mathcal{S}^o_\nu,\ \mu$

4.4. $\check{\mathcal{S}}_\mu,\ \check{\mathcal{S}}^o_\mu$

References

[BK] W. Borho and H. Kraft, Über Bahnen und deren Deformationen bei linearen Aktionen reductiver Gruppen, *Comment. Math. Helvetici* **54** (1979), 61–104.

[BS] E. Benlolo and J. Sanderson, On hypersurface orbital varieties of $\mathfrak{sl}(N,\mathbb{C})$, *J. Algebra* **245** (2001), 225–246.

[F] W. Fulton, *Young tableaux*, LMS 35, Cambridge University Press, 1997.

[H] W. Hesselink, Singularities in the nilpotent scheme of a classical group, *Trans. Am. Math. Soc.* **222** (1976), 1–32.

[Ja] J.C. Jantzen, Kontravariante Formen auf induzierten Darstellungen halbeinfacher Lie-Algebren, *Math. Ann.* **226** (1977), 53–65.

[J1] A. Joseph, On the variety of a highest weight module, *J. Algebra* **88** (1984), 238–278.

[J2] A. Joseph, Orbital varieties, Goldie rank polynomials and unitary highest weight modules, in B. Orsted and H. Schlichtkrull (eds.), *Algebraic and Analytic Methods in Representation Theory*, Perspectives in Mathematics, Vol. 17, Academic Press, London, 1997.

[J3] A. Joseph, Orbital varieties of the minimal orbit, *Ann. Ec. Norm. Sup.* **31** (1998), 17–45.

[JLT] A. Joseph, G. Letzter and D. Todoric On the Kostant–Parthasarathy–Ranga Rao–Varadarajan Determinants, III. Computation of the KPRV Determinants, *J. Algebra* **241** (2001), 67–88.

[Kn] D. E. Knuth, *The Art of computer programming*, Vol. 3, Addison-Wesley (1969), 49–72.

[M1] A. Melnikov, *On orbital variety closures in* $\mathfrak{sl}_n$. *I. Induced Duflo order*, preprint.

[M2] A. Melnikov, *On orbital variety closures in* $\mathfrak{sl}_n$. *II. Descendants of a Richardson orbital variety*, preprint.

[M3] A. Melnikov, *Robinson–Schensted procedure and combinatorial properties of geometric order* $\mathfrak{sl}(n)$, CRAS Paris **315**, série I (1992), 709–714.

[M4] A. Melnikov, *Irreducibility of the associated variety of simple highest weight modules in* $\mathfrak{sl}(n)$, CRAS, Paris **316**, série I (1993), 53–57.

[M5] A. Melnikov, *Geometric Interpretation of Robinson–Schensted procedure and related orders on Young tableaux*, thesis, Weizmann Institute 1992.

[S] N. Spaltenstein, On the fixed point set of a unipotent element on the variety of Borel subgroups, *Topology* **16** (1977), 203–204.

[S1] N.Spaltenstein, The fixed point set of a unipotent transformation on the flag manifold, *Proc. Konin. Nederl. Akad.* **79** (1976), 452–456.

[Sch] M. P. Schützenberger, *La correspondance de Robinson*, LNM Math. **597** (1976), 59–113.

[St] R. Steinberg, An occurrence of the Robinson–Schensted correspondence, *J. Algebra* **113** (1988), 523–528.

[vanL] M. A. A. van Leeuwen, *The Robinson–Schensted and Schützenberger algorithms, Part II: Geometric interpretations*, CWI report AM-R9209 (1992).

Anthony Joseph
The Donald Frey Professorial Chair
Department of Mathematics
The Weizmann Institute of Science
Rehovot 76100, Israel
joseph@wisdom.weizmann.ac.il

Anna Melnikov
Department of Mathematics
The Weizmann Institute of Science
Rehovot 76100, Israel
anna@wisdom.weizmann.ac.il

Generalization of a Theorem of Waldspurger to Nice Representations

D. Kazhdan
A. Polishchuk

This paper presents some examples of equalities of integrals over local fields of characteristic zero predicted by the stationary phase approximation. These examples should be considered in the context of algebraic integration theory, proposed by the first author in [6]. From the point of view of this theory, our main theorem gives examples of pairs of algebro-geometric data over a given local field E producing equal integrals over arbitrary finite extensions of E. Another motivation for this work is the theory of Sato's functional equations associated with prehomogeneous vector spaces over local fields. As an application of our techniques, we find an explicit form of these equations for the action of GL_n on symmetric $n \times n$ matrices in the case when n is odd (this is a generalization of a particular case of the equations obtained by W. J. Sweet Jr. in [24]).

Let $\mathbf{G}$ be a (connected) simply connected semisimple algebraic group over a local field E of characteristic zero, $\rho : \mathbf{G} \to \mathrm{Aut}(\mathbf{V})$ be a rational representation defined over E.

Definition. A representation ρ is called *nice* if the generic stabilizer subgroup $\mathbf{H}$ is connected and reductive.[1]

Note that the adjoint representation of $\mathbf{G}$ is nice. Other interesting examples can be found using Elashvili's tables of representations with positive-dimensional generic stabilizers (see [2],[3]).

By distributions on $\mathbf{V}(E)$ we always mean functionals on the Schwartz–Bruhat space $\mathcal{S}(\mathbf{V}(E))$ (if E is p-adic then $\mathcal{S}(\mathbf{V}(E))$ is the space of locally constant functions on $\mathbf{V}(E)$ with compact support). We are interested in the subspaces $\mathcal{D}^{st}$, $\mathcal{D}^{st}_{\epsilon}$ of $\mathbf{G}(E)$-invariant distributions on $\mathbf{V}(E)$ consisting of *stable* and *ϵ-stable* distributions. To define these spaces we have to introduce some notation.

Recall that for a nondegenerate quadratic form q over a local field E the Hasse–Witt invariant $\epsilon(q) = \pm 1$ is defined as follows: choose coordinates $x_1, \ldots, x_n$ in such a way that $q = a_1 x_1^2 + \ldots + a_n x_n^2$ and set $\epsilon(q) = \prod_{i<j}(a_i, a_j)$ where $(\cdot, \cdot)$ denotes the Hilbert symbol (if rank of q is equal to 1 we set $\epsilon(q) = 1$). Note that

[1] This notion is slightly more general than the one defined in Section 7 of [6].

if $E=\mathbb{R}$, then $\epsilon(q)=(-1)^{i(i-1)/2}$ where i is the number of negative squares in q (for $E=\mathbb{C}$ we have $\epsilon(q)=1$). For a pair of nondegenerate quadratic forms q and q' we define the relative Hasse–Witt invariant $\epsilon(q,q'):=\epsilon(q)\epsilon(q')$.

Let $\mathbf{V}_0\subset\mathbf{V}$ be a nonempty Zariski open affine subset such that the stabilizer of every point in $\mathbf{V}_0$ is conjugate to $\mathbf{H}$ over $\overline{E}$, and let $\mathbf{V}_0^\vee\subset\mathbf{V}^\vee$ be a similar subset in the dual representation. Let $\mathfrak{g}$ be the Lie algebra of $\mathbf{G}$, and let Q be the Killing form on $\mathfrak{g}$. For every point $x\in\mathbf{V}_0(E)$ we denote by $\mathbf{H}_x\subset\mathbf{G}$ the stabilizer of x and by $\mathfrak{h}_x\subset\mathfrak{g}$ its Lie algebra. It is easy to see that the form $Q_x:=Q|_{\mathfrak{h}_x}$ is still nondegenerate (see Lemma 1.4). Now for every $\mathbf{G}$-orbit $\mathcal{O}\subset\mathbf{V}_0$ we define the relative sign function on $\mathcal{O}(E)\times\mathcal{O}(E)$ by

$$\epsilon(x,x'):=\epsilon(Q_x,Q_{x'}).$$

Sometimes we will also use the absolute sign function

$$\epsilon(x):=\epsilon(Q_x)$$

such that $\epsilon(x,x')=\epsilon(x)\epsilon(x')$. Let us denote by $p:\mathbf{V}_0\to\mathbf{V}_0/\mathbf{G}$ the natural projection to the geometric quotient by the action of $\mathbf{G}$.

Definition. Let δ be a distribution which is given by a locally L^1 function f: $\delta=f|dv|$ where $|dv|$ is the Haar measure on $\mathbf{V}(E)$.

a) We say that δ is *stable* if the function f is constant on the fibers of the natural projection $p(E):\mathbf{V}_0(E)\to(\mathbf{V}_0/\mathbf{G})(E)$.

b) We say that δ is ϵ-stable if for (almost) any pair $x,x'\in\mathbf{V}_0(E)$ such that $p(E)(x)=p(E)(x')$ we have $f(x)=\epsilon(x,x')f(x')$.

By the definition, the space of all stable (resp. ϵ-stable) distributions on $\mathbf{V}(E)$ is the closure (with respect to the weak topology) of the space of locally L^1 stable (resp. ϵ-stable) distributions. Another definition will be given in Section 2.4.

Our main result is that in the case when E is p-adic, ρ is nice and either $\mathbf{G}$ is simple or $\mathbf{H}$ is semisimple, the notions of stable and ϵ-stable distributions get switched by the Fourier transform. Note that in the case when ρ is the adjoint representation we have $\epsilon\equiv 1$. We conjecture that similar result holds in the case $E=\mathbb{R}$. Thus, in this case our result is that stability is preserved under Fourier transform. This was proved previously by J.-L. Waldspurger (see [25], Cor. 1.6); our proof is modeled on the proof of *loc. cit.*

We also introduce the "complementary" notions of antistable and ϵ-antistable distributions and prove that they get switched by the Fourier transform in the case when ρ is nice, E is p-adic and $\mathbf{H}$ is semisimple.

The proof combines some local computations with global considerations.[2] Our main local tool is the stationary phase approximation applied to linear functionals on $\mathbf{G}$-orbits. The main global tool we use is the stabilization techniques introduced by R. Kottwitz in [11] to stabilize the elliptic semisimple part of the trace formula

[2]The sketch of the proof in the case $\epsilon\equiv 1$ was given in [6].

for $\mathbf{G}$. More precisely, we prove an analogue of the Kottwitz stabilization formula (Theorem 9.6 of [11]) for an arbitrary nice representation over a number field.

Using the same method we prove the following result concerning stable distributions and inner forms (sketched in [6] in the case $\epsilon \equiv 1$). Let $\mathbf{G}' \to \mathrm{Aut}(\mathbf{V}')$ be an inner form of a nice representation $\mathbf{G} \to \mathrm{Aut}(\mathbf{V})$ over E. Then $\mathbf{G}$-orbits on $\mathbf{V}$ are in bijection with $\mathbf{G}'$-orbits on $\mathbf{V}'$. For any $\bar{x} \in \mathbf{V}/\mathbf{G}(E)$ we denote by $\mathcal{O}_{\bar{x}} \subset \mathbf{V}, \mathcal{O}'_{\bar{x}} \subset \mathbf{V}'$ the preimages of $\bar{x}$. For any $x \in \mathcal{O}_{\bar{x}}(E), x' \in \mathcal{O}'_{\bar{x}}(E)$ we define as before $\epsilon(x, x') = \epsilon(Q_x, Q'_{x'})$ (where $Q'_{x'}$ is the restriction of the Killing form Q' on $\mathfrak{g}'$ to $\mathfrak{h}_{x'}$).

Definition. a) Let $\delta = f|dv|, \delta' = f'|dv'|$ be stable distributions on $\mathbf{V}(E), \mathbf{V}'(E)$ given by locally L^1 functions. We write $\delta \sim \delta'$ if for any $\bar{x} \in \mathbf{V}/\mathbf{G}(E), x \in \mathcal{O}_{\bar{x}}(E), x' \in \mathcal{O}'_{\bar{x}}(E)$ we have $f(x) = f'(x')$.
b) Let $\delta = f|dv|, \delta' = f'|dv'|$ be ϵ-stable distributions on $\mathbf{V}(E), \mathbf{V}'(E)$ given by locally L^1 functions. We write $\delta \sim_\epsilon \delta'$ if for any $\bar{x} \in \mathbf{V}/\mathbf{G}(E), x \in \mathcal{O}_{\bar{x}}(E), x' \in \mathcal{O}'_{\bar{x}}(E)$ we have $f(x) = f'(x')\epsilon(x, x')$.

We extend these definitions to arbitrary stable (resp. ϵ-stable) distributions and prove that in the case when E is p-adic and $\mathbf{G}$ is simple, for any stable distributions δ and δ' on $\mathbf{V}(E)$ and $\mathbf{V}'(E)$ respectively, such that $\delta \sim \delta'$, we have $\mathcal{F}(\delta) \sim_\epsilon \kappa(\mathbf{G}, \mathbf{G}')\mathcal{F}(\delta')$, where $\kappa(\mathbf{G}, \mathbf{G}')$ is the relative Hasse–Witt invariant of Killing forms on $\mathfrak{g}$ and $\mathfrak{g}'$. The latter sign coincides with the sign considered by Kottwitz in [9].

Finally, we apply our main theorem to derive the following equation for distributions on the space $\mathrm{Sym}_n(E)$ of symmetric $n \times n$ matrices over a p-adic field E in the case when n is odd:

$$\mathcal{F}(\chi(\det)) = c(\chi) \cdot \epsilon \cdot (\det, -1)^{\frac{n-1}{2}} \cdot |\det|^{-\frac{n+1}{2}} \chi^{-1}(\det), \qquad (0.0.1)$$

where χ is a generic multiplicative character of a local field E, $\chi(\det)$ is considered as a distribution on $\mathrm{Sym}_n(E)$ (defined by analytic continuation), ϵ is the function on the complement to the hypersurface ($\det = 0$) in $\mathrm{Sym}_n(E)$ that assigns to a symmetric matrix the Hasse–Witt invariant of the corresponding quadratic form, $c(\chi)$ is a nonzero constant. In the case when $\chi = |\cdot|^s\delta$, where δ is a character of $E^*/(E^*)^2$, the equation (0.0.1) follows from Proposition 4.8 of [24]. We show that in the case $E = \mathbb{R}$ the equation (0.0.1) follows from computations of Shintani in [21]. This confirms the conjecture that our main theorem holds also in the real case.

The appearence of the sign ϵ in the above results is due to the connection of the relative Hasse–Witt invariant with Weil constants for quadratic forms. First, let us introduce some more notation concerning quadratic forms. Considering a nondegenerate quadratic form q on a E-vector space L as a symmetric isomorphism $L \to L^\vee$ we can define $\det(q) \in \det(L)^{\otimes(-2)}$, where $\det(L) = \bigwedge^{\dim L} L$. By choosing a trivialization of $\det(L)$ we get an element of E^*. The corresponding element in $E^*/(E^*)^2$ does not depend on a choice of trivialization (we call it *determinant of q modulo squares*). If E is p-adic, then two nondegenerate quadratic forms q and q' are equivalent over E if and only if $\mathrm{rk} q = \mathrm{rk} q'$, $\det(q) \equiv \det(q') \bmod (E^*)^2$ and $\epsilon(q, q') = 1$.

More generally, if $E \neq \mathbb{C}$, then these invariants have the following K-theoretic meaning. Let W_E be the Witt ring of E. By the definition, W_E is the abelian group generated by pairs (L, q) where L is a finite-dimensional E-vector space and q is a nondegenerate quadratic form on L and the product in W_E comes from the operation of tensor product $(L', q') \cdot (L'', q'') := (L, q)$ where $L = L' \otimes L''$. Let $W_E^1 \subset W_E$ be the ideal of forms of even dimension, $W_E^i := (W_E^1)^i$. We define $\overline{W}_E^i := W_E^i / W_E^{i+1}$. As is well known $\overline{W}_E^0 = \mathbb{Z}/2\mathbb{Z}$,$\overline{W}_E^1 = E^*/(E^*)^2$, $\overline{W}_E^2 \simeq K_2(E)/2K_2(E) \simeq \mathbb{Z}/2\mathbb{Z}$. For a pair of nondegenerate quadratic forms (q, q') of the same rank, we have $(q) - (q') \in W_E^1$ and its image in $\overline{W}_E^1$ is identified with $\det(q)/\det(q') \bmod (E^*)^2$. If the latter element is trivial, then we have $(q) - (q') \in W_E^2$ and its image in $\overline{W}_E^2$ can be identified with the relative Hasse–Witt invariant $\epsilon(q, q')$.

On the other hand, for a fixed (nontrivial) additive character $\psi : E \to U(1)$ and a nondegenerate quadratic form q, A. Weil introduced in [26] a constant $\gamma(q, \psi)$ which is a root of unity of order 8 depending only on the equivalence class of q (and on ψ). Note that the character ψ defines a canonical Haar measure on E which is selfdual for the Fourier transform.

Let V be a vector space over E on which q is defined. For a nonzero top-degree form ν on V we have the induced Haar measure $|\nu|$ on V (corresponding to the selfdual Haar measure on E). Weil's constant is defined by the functional equation for distributions

$$\mathcal{F}(\psi(q)) = \gamma(q, \psi) \cdot |\det(q)/\nu^2|^{-\frac{1}{2}} \psi(-q^\vee), \tag{0.0.2}$$

where $q^\vee$ is the dual quadratic form on $V^\vee$, $\psi(q)$ and $\psi(-q^\vee)$ are considered as distributions on V and $V^\vee$. By the definition, the map $q \mapsto \gamma(q, \psi)$ extends to a homomorphism from W_E to roots of unity of order 8. As shown in [26] (nos. 25–28), for quadratic forms q and q' such that $(q) - (q') \in W_E^2$ one has

$$\gamma((q) - (q'), \psi) = \epsilon(q, q').$$

Here is the plan of the paper. In Section 1 we gather some algebraic facts about nice representations. In particular, we prove that the generic stabilizers of a nice representation and of its dual are conjugate to each other (Proposition 1.3). In Section 2 we draw consequences from the stationary phase approximation in the case when the ground field is p-adic. In Section 3 we prove an analogue of the stable trace formula for nice representations over number fields and combine it with local information from section 2 to derive the main result. Finally, in Section 4 we derive the equation (0.0.1).

Notation. All our fields have characteristic zero. E always denotes a local field, while F always denotes a number field. By p-adic field we mean a finite extension of $\mathbb{Q}_p$. For a vector space V over a local field E and an open subset $U \subset V$

we denote by $\mathcal{S}(U)$ the space of functions with support in U which belongs to the Schwartz–Bruhat space of V. When we work over a local field E, we fix a nontrivial additive character $\psi : E \to U(1)$. We fix the Haar measure on E which is selfdual with respect to the Fourier transform defined in terms of ψ. For a smooth variety $\mathbf{X}$ over E and a nonvanishing top-degree form ω on E, we denote by $|\omega|$ the measure on $\mathbf{X}(E)$ corresponding to ω and to the above Haar measure on E. For a variety $\mathbf{X}$ defined over a field k and an extension of fields $k \subset k'$, we denote by $\mathbf{X}_{k'}$ the variety over k' obtained from $\mathbf{X}$ by the extension of scalars. For an algebraic group $\mathbf{H}$ we denote by $Z(\mathbf{H})$ its center. The group $\mathbf{G}$ is assumed to be (connected) semisimple and simply connected, $\mathfrak{g}$ denotes the Lie algebra of $\mathbf{G}$, Q is the Killing form on $\mathfrak{g}$. When $\mathbf{G}$ acts on a vector space $\mathbf{V}$, for every point $x \in \mathbf{V}$ we denote by $\mathcal{O}_x \subset \mathbf{V}$ the $\mathbf{G}$-orbit of x and by $\mathbf{H}_x \subset \mathbf{G}$ (resp. $\mathfrak{h}_x \subset \mathfrak{g}$) the stabilizer subgroup (resp. subalgebra) of x. For a field k we denote by Γ_k the Galois group $\mathrm{Gal}(\overline{k}/k)$. For an algebraic group $\mathbf{H}$ defined over k we set $H^i(k, \mathbf{H}) = H^i(\Gamma_k, \mathbf{H}(\overline{k}))$ (where $i \leq 1$ if $\mathbf{H}$ is noncommutative).

1 Algebraic results

Throughout this section $\mathbf{G}$ is a simply connected semisimple group over a field k of characteristic zero, $\rho : \mathbf{G} \to \mathrm{Aut}(\mathbf{V})$ is a rational representation defined over k. In this section we gather results we need that can be proved after passing to an algebraic closure of k.

1.1 Representations with reductive generic stabilizer

The first restriction we impose in order for a representation ρ to be nice is the reductivity of $\mathfrak{h}$. In this subsection we discuss some consequences of this condition.

It is well known (see [13], [18]) that there exists a nonempty $\mathbf{G}$-invariant Zariski open subset $\mathbf{V}_0 \subset \mathbf{V}$ and a subgroup $\mathbf{H} \subset \mathbf{G}$ such that for every point $x \in \mathbf{V}_0(\overline{k})$ the stabilizer subgroup $\mathbf{H}_x$ of x is conjugate to $\mathbf{H}$. Furthermore, clearly we can assume that $\mathbf{V}_0$ is invariant under the natural $\mathbf{G}_m$-action on $\mathbf{V}$. The following lemma can be found in [2] (see also [17], Theorem 7.3).

Lemma 1.1. *For every $x \in \mathbf{V}_0$ one has*

$$\mathbf{V} = \mathbf{V}^{\mathfrak{h}_x} + \mathfrak{g}x.$$

Proof. This follows immediately from the surjectivity of the map $\mathbf{G}/\mathbf{H}_x \times (\mathbf{V}_0^{\mathfrak{h}_x}) \to \mathbf{V}_0$, where $\mathbf{V}_0^{\mathfrak{h}_x} = \mathbf{V}^{\mathfrak{h}_x} \cap \mathbf{V}_0$. □

Let $\mathfrak{h} \subset \mathfrak{g}$ be the Lie algebra of $\mathbf{H}$. We will call $\mathbf{H}$ (resp. $\mathfrak{h}$) the *generic stabilizer subgroup* (resp. *subalgebra*) for ρ. Henceforth, we always assume that $\mathfrak{h}$ is reductive. The theorem of V. L. Popov [16] states that in this case generic $\mathbf{G}$-orbits in $\mathbf{V}$ are closed.

Proposition 1.2. *Let $\rho^\vee : \mathbf{G} \to \mathrm{Aut}(\mathbf{V}^\vee)$ be the dual representation to ρ. Then the generic stabilizer subalgebra for $\rho^\vee$ is conjugate to $\mathfrak{h}$ over $\overline{k}$.*

Proof. Clearly, we can assume that $k = \mathbb{C}$. Let $C \subset \mathbf{G}(\mathbb{C})$ be a maximal compact subgroup. It is well known that there exists a C-invariant positive-definite Hermitian form H on $\mathbf{V}$. Let $\mathcal{O} \subset \mathbf{V}$ be a generic orbit. Let us consider the restriction of the function $x \mapsto H(x,x)$ to $\mathcal{O}(\mathbb{C})$. Since $\mathcal{O}$ is closed, there exists a vector $x \in \mathcal{O}(\mathbb{C})$ minimizing this function. In particular, we have $H(\mathfrak{g}x, x) = 0$. Let us consider the functional $x^\vee \in \mathbf{V}^\vee(\mathbb{C})$ given by $x^\vee = H(?, x)$. We claim that the stabilizer subalgebra of $x^\vee$ coincides with $\mathfrak{h}_x$. Indeed, first we note that $x^\vee \in (\mathfrak{g}x)^\perp$. According to Lemma 1.1, $\mathfrak{h}_x$ acts trivially on $V/\mathfrak{g}x$. Hence, it also acts trivially $(\mathfrak{g}x)^\perp$, so $\mathfrak{h}_x x^\vee = 0$. Let us define the $\mathbb{C}$-bilinear form on $\mathfrak{g}/\mathfrak{h}_x$ by setting

$$B(\xi_1, \xi_2) = \langle \xi_1 x^\vee, \xi_2 x\rangle = -H(\xi_1\xi_2 x, x),$$

where $\xi_1, \xi_2 \in \mathfrak{g}/\mathfrak{h}_x$. Note that since $H([\xi_1, \xi_2]x, x) = 0$, the form B is symmetric. Now let $\mathfrak{c} \subset \mathfrak{g}$ be the Lie algebra of C. We claim that the restriction of B to the real subspace $\mathfrak{c}/\mathfrak{c} \cap \mathfrak{h}_x \subset \mathfrak{g}/\mathfrak{h}_x$ is $\mathbb{R}$-valued and positive-definite. Indeed, if $\xi \in \mathfrak{c}$, then using the fact that H is $\mathfrak{c}$-invariant we get

$$B(\xi, \xi) = H(\xi x, \xi x) \in \mathbb{R}.$$

Furthermore, this implies that $B(\xi, \xi) > 0$ for $\xi \notin \mathfrak{c} \cap \mathfrak{h}_x$. Since the subspace $\mathfrak{c}/\mathfrak{c} \cap \mathfrak{h}_x$ generates $\mathfrak{g}/\mathfrak{h}_x$ over $\mathbb{C}$ this implies that the form B on $\mathfrak{g}/\mathfrak{h}_x$ is nondegenerate. In particular, the stabilizer subalgebra of $x^\vee$ is equal to $\mathfrak{h}_x$. □

Remark. The idea to look at vectors of minimal length on the orbit goes back to the work of Kempf and Ness [7]. One can give an alternative proof of Proposition 1.2 using the theorem of Mostow [14] on selfadjoint groups.

For nice representations the assertion of the above theorem holds also for generic stabilizer subgroups.

Proposition 1.3. *Assume that ρ is a nice representation. Then the dual representation $\rho^\vee$ is also nice. In this case the generic stabilizer subgroups for ρ and $\rho^\vee$ are conjugate over $\overline{k}$.*

Proof. The generic stabilizer subgroup for $\rho^\vee$ has form $\sigma(\mathbf{H})$ where σ is an automorphism of $\mathbf{G}$ which restricts to the map $t \mapsto t^{-1}$ on some maximal torus. This immediately implies that $\rho^\vee$ is nice. The second assertion follows from Proposition 1.2. □

Let Q denote the Killing form on $\mathfrak{g}$. The following lemma is well known but we include the proof since we could not find the reference (in its statement $\mathfrak{h}$ can be replaced by the Lie algebra of any reductive subgroup in $\mathbf{G}$).

Lemma 1.4. *The restriction of Q to $\mathfrak{h}$ is nondegenerate.*

Proof. We can assume that $k = \mathbb{C}$. Let $C \subset \mathbf{G}(\mathbb{C})$ be a maximal compact subgroup containing a maximal compact subgroup in $\mathbf{H}(\mathbb{C})$, and let $\mathfrak{c}$ be its Lie algebra. Then $\mathfrak{h} = \mathfrak{h} \cap \mathfrak{c} + i\mathfrak{h} \cap \mathfrak{c}$. It remains to use the fact that the restriction of Q to $\mathfrak{c}$ is $\mathbb{R}$-valued and negative definite. □

Lemma 1.5. *Let us denote* $\mathbf{W} = \mathbf{V}^{\mathbf{H}}$, $\mathbf{W}_0 = \mathbf{V}_0 \cap \mathbf{W}$.

(a) *For every point* $x_0 \in \mathbf{W}_0$ *there exists a linear subspace* $\mathbf{L} \subset \mathbf{W}_0$ *and a Zariski open neighborhood of zero* $\mathbf{S} \subset \mathbf{L}$ *such that the natural morphism*

$$a : \mathbf{G}/\mathbf{H} \times (\mathbf{S} + x_0) \to \mathbf{V}_0 : (g\mathbf{H}, x) \mapsto gx$$

is étale.

(b) *Let* $\nu_{\mathbf{V}}$ *be a nonzero translation-invariant top-degree form on* $\mathbf{V}$, *let* ω *be a nonzero* $\mathbf{G}$*-invariant top-degree form on* $\mathbf{G}/\mathbf{H}$. *Then we have* $a^*\nu_{\mathbf{V}} = \omega \wedge \nu_{\mathbf{L}}$ *for some translation-invariant top-degree form* $\nu_{\mathbf{L}}$ *on* $\mathbf{L}$.

Proof. (a) Note that we have an isomorphism

$$\mathbf{G}/\mathbf{H} \times_{N(\mathbf{H})/\mathbf{H}} \mathbf{W}_0 \to \mathbf{V}_0,$$

where $N(\mathbf{H})$ is the normalizer of $\mathbf{H}$ in $\mathbf{G}$. Let $\mathbf{L}$ to be a complement in $\mathbf{W}_0$ to the tangent space to $N(\mathbf{H})x_0$ at x_0. Then our assertion follows from Luna's results in [13] (in this situation $\mathbf{S} + x_0$ is an étale slice for the action of $N(\mathbf{H})/\mathbf{H}$ on $\mathbf{W}$).

(b) Since $\mathbf{G}$ is semisimple and connected, the form $\nu_{\mathbf{V}}$ is $\mathbf{G}$-invariant. Therefore, the pullback $a^*\nu_{\mathbf{V}}$ is also $\mathbf{G}$-invariant. On the other hand, the map a is linear in the second argument, hence $a^*\nu_{\mathbf{V}}$ is invariant with respect to translations on $\mathbf{L}$. This implies our assertion. □

Lemma 1.6. *Assume that* $\mathbf{G}$ *is simple,* ρ *is nice. Then the generic stabilizer* $\mathbf{H}$ *is either semisimple or commutative.*

Proof. This follows immediately from Tables 1 and 2 of [2]. □

1.2 Critical points

In this subsection we study critical points of the restriction of a generic linear functional $x^\vee \in \mathbf{V}_0^\vee$ to a generic orbit $\mathcal{O}$ in a representation ρ whose generic stabilizer subalgebra is reductive. Recall that a critical point x of a function ϕ is called *nondegenerate* if the quadratic form of second derivatives of ϕ at x is nondegenerate.

Lemma 1.7. *Let* x *be a critical point of* $x^\vee|_{\mathcal{O}}$. *Then:*

(a) x *has the same stabilizer subalgebra as* $x^\vee$;

(b) x *is nondegenerate if and only if* $\mathfrak{g}x \cap (\mathfrak{g}x^\vee)^\perp = 0$.

Proof. (a) Without loss of generality we can assume that $x^\vee \in \mathbf{W}_0^\vee$. The condition that x is a critical point of $x^\vee|_{\mathcal{O}}$ is equivalent to $x^\vee(\mathfrak{g}x) = 0$, i.e., $x \in (\mathfrak{g}x^\vee)^\perp$. On the other hand, by Lemma 1.1 the group $\mathbf{H}$ acts trivially on $\mathbf{V}^\vee/\mathfrak{g}x^\vee$. It follows that $(\mathfrak{g}x^\vee)^\perp$ is contained in $\mathbf{V}^{\mathbf{H}} = \mathbf{W}$, hence, x is stabilized by $\mathbf{H}$.

(b) The second derivative of $x^\vee|_{\mathcal{O}}$ at a critical point x is the following symmetric bilinear form on the tangent space $T_x\mathcal{O} = \mathfrak{g}x$: $B_{x,x^\vee}(\xi_1 x, \xi_2 x) = \langle x^\vee, \xi_1\xi_2 x\rangle$, where $\xi_1, \xi_2 \in \mathfrak{g}$. The kernel of this form is $\mathfrak{g}x \cap (\mathfrak{g}x^\vee)^\perp$. □

Let $\mathbf{I} \subset \mathbf{V}_0 \times \mathbf{V}_0^\vee$ be the subvariety consisting of $(x, x^\vee)$ such that x is a critical point of $x^\vee|_{\mathcal{O}_x}$. Let us consider the natural morphism

$$\mathbf{f} : \mathbf{I} \to \mathbf{V}_0/\mathbf{G} \times \mathbf{V}_0^\vee.$$

By the definition, $\mathbf{f}^{-1}(\mathcal{O}, x^\vee)$ consists of critical points of $x^\vee|_{\mathcal{O}}$.

Proposition 1.8. *The morphism* $\mathbf{f}$ *is dominant and there exists a nonempty* $\mathbf{G}$-*invariant open subset* $\mathbf{U} \subset \mathbf{V}_0/\mathbf{G} \times \mathbf{V}_0^\vee$ *(where* $\mathbf{G}$ *acts on the second factor) such that* $\mathbf{f}$ *is étale over* $\mathbf{U}$. *For every pair* $(\mathcal{O}, x^\vee)$ *in* $\mathbf{U}$ *the set of critical points of* $x^\vee|_{\mathcal{O}}$ *is finite and nonempty. Furthermore, all these critical points are nondegenerate.*

Proof. By the definition, the subvariety $\mathbf{I} \subset \mathbf{V}_0 \times \mathbf{V}_0^\vee$ consists of $(x, x^\vee)$ such that $\langle x^\vee, \mathfrak{g}x\rangle = 0$. Let $p_1 : \mathbf{I} \to \mathbf{V}_0$, $p_2 : \mathbf{I} \to \mathbf{V}_0^\vee$ be the natural projections. Note that the fibers of both these projections are open subsets in linear spaces of dimension $\dim \mathbf{V} - \dim \mathcal{O} = \dim \mathbf{V}/\mathbf{G}$. Hence, $\mathbf{I}$ is smooth and irreducible of dimension $\dim \mathbf{V}/\mathbf{G} + \dim \mathbf{V}$ (if nonempty). Thus, it suffices to prove that the morphism $\mathbf{f}$ is dominant. We can assume that $k = \mathbb{C}$. Let $(x, x^\vee)$ be a pair constructed in the proof of Proposition 1.2. Then $(x, x^\vee) \in \mathbf{I}$. We claim that the tangent map to $\mathbf{f}$ at $(x, x^\vee)$ is an isomorphism. Indeed, the relative tangent space of the projection $p_2 : \mathbf{I} \to \mathbf{V}_0^\vee$ at $(x, x^\vee)$ is $(\mathfrak{g}x^\vee)^\perp$. Therefore, we have to check that the linear map $(\mathfrak{g}x^\vee)^\perp \to \mathbf{V}/\mathfrak{g}x$ is an isomorphism. It remains to note that the nondegeneracy of the form B constructed in the proof of Proposition 1.2 implies that $\mathfrak{g}x \cap (\mathfrak{g}x^\vee)^\perp = 0$. This proves our claim. The last assertion of the proposition follows from Lemma 1.7 (b) and from the fact that $\mathbf{f}$ is generically étale. □

1.3 Spin-coverings

In this subsection we are going to define certain double coverings of generic stabilizers which will play an important role later. We assume that ρ is a nice representation.

Let $x \in \mathbf{V}_0$, $x^\vee \in \mathbf{V}_0^\vee$ be points such that x is a nondegenerate critical point of $x^\vee|_{\mathcal{O}_x}$. Let $B_{x,x^\vee}$ be the corresponding quadratic form of second derivatives of $x^\vee|_{\mathcal{O}_x}$ on the tangent space $T_x := T_x\mathcal{O}_x$. Then the action of the stabilizer $\mathbf{H}_x$ of x on T_x preserves $B_{x,x^\vee}$, hence, we get a homomorphism $\mathbf{H}_x \to \mathrm{Aut}(T_x, B_{x,x^\vee})$. Since $\mathbf{H}_x$ is connected, we obtain a homomorphism to the corresponding special orthogonal group

$$\iota_{x,x^\vee} : \mathbf{H}_x \to \mathrm{SO}(T_x, B_{x,x^\vee}).$$

Recall that for every vector space T equipped with a nondegenerate quadratic form B one can define the spin-covering $\mathrm{Spin}(T, B) \to \mathrm{SO}(T, B)$. It is defined in the standard way when $\dim T \geq 3$ (see e.g. [1]). If $\dim T < 3$, then we define the spin-covering as the restriction of the standard spin-covering of $SO(T \oplus U)$, where U is an orthogonal space of large dimension. The group $\mathrm{Spin}(T, B)$ is a central extension of $SO(T, B)$ by $\{\pm 1\}$. In the case $\dim(T) < 3$ this extension is trivial unless $\dim T = 2$ and the quadratic form B is anisotropic. In the latter case $SO(T, B)$ is a 1-dimensional torus of the form $R^{(1)}_{k'/k}(\mathbb{G}_m)$ for some quadratic extension $k \subset k'$, $\mathrm{Spin}(T, B)$ is isomorphic to $SO(T, B)$ and the map $\mathrm{Spin}(T, B) \to SO(T, B)$ is identified with the map $t \mapsto t^2$.

In Proposition 1.9 below we compute the pullback of the spin-covering $\mathrm{Spin}(T_x, B_{x,x^\vee}) \to SO(T_x, B_{x,x^\vee})$ by $\iota_{x,x^\vee}$. Recall that by Lemma 1.4 the restriction of Q to $\mathfrak{h}_x$ is nondegenerate. Consider the homomorphism

$$\iota_x : \mathbf{H}_x \to SO(\mathfrak{h}_x, Q|_{\mathfrak{h}_x})$$

induced by the adjoint action of $\mathbf{H}_x$. Let $\widetilde{\mathbf{H}}_x \to \mathbf{H}_x$ be the pullback of the spin-covering by ι_x. Note that if $\mathbf{H}_x$ is commutative, then ι_x is trivial, so the extension $\widetilde{\mathbf{H}}_x \to \mathbf{H}_x$ splits in this case.

Proposition 1.9. *There is a unique isomorphism of the following two central extensions of* $\mathbf{H}_x$ *by* $\{\pm 1\}$*: the pullback of the spin-covering by* $\iota_{x,x^\vee}$ *and* $\widetilde{\mathbf{H}}_x$*.*

Proof. We can reformulate the statement as follows: there exists a unique lifting of $\iota_{x,x^\vee}$ to a homomorphism $\widetilde{\mathbf{H}}_x \to \mathrm{Spin}(T_x, B_{x,x^\vee})$ which maps $\{\pm 1\} \subset \widetilde{\mathbf{H}}_x$ to $\{\pm 1\} \subset \mathrm{Spin}(T_x, B_{x,x^\vee})$ identically.

First, let us prove the uniqueness. Indeed, two such liftings differ by a homomorphism $\mathbf{H}_x \to \{\pm 1\}$. Since $\mathbf{H}_x$ is connected such a homomorphism should be trivial. Now by uniqueness it suffices to prove the existence of a lifting over $\overline{k}$. Therefore, we can assume that $k = \mathbb{C}$.

Assume first that $\dim(T_x) \geq 3$ and the twofold covering $\widetilde{\mathbf{H}}_x \to \mathbf{H}_x$ is nontrivial. Then $\widetilde{\mathbf{H}}_x$ is connected. We claim that the homomorphism

$$\pi_1(\mathbf{H}_x) \to \pi_1(SO(T_x, B_{x,x^\vee})) = \{\pm 1\} \tag{1.3.1}$$

induced by $\iota_{x,x^\vee}$ is surjective with the kernel $\pi_1(\widetilde{\mathbf{H}}_x) \subset \pi_1(\mathbf{H}_x)$. Indeed, let X be the space of $\mathbf{H}_x$-invariant nondegenerate symmetric forms on T_x. We have a continuous family of homomorphisms $\mathbf{H}_x \to SO(T_x, B)$ parametrized by $B \in X$. Since X is connected, the induced homomorphism of fundamental groups $\pi_1(\mathbf{H}_x) \to \pi_1(SO(T_x, B))$ does not depend on B. Therefore, in the above claim we can replace $B_{x,x^\vee}$ by any other form in X. Now using the natural identification of T_x with $\mathfrak{g}/\mathfrak{h}_x = \mathfrak{h}_x^\perp \subset \mathfrak{g}$ (the orthogonal complement to $\mathfrak{h}_x$ in $\mathfrak{g}$ with respect to the Killing form Q) we take $B = Q|_{\mathfrak{h}_x^\perp}$. Consider the following commutative diagram of homomorphisms

$$\begin{array}{ccc} \mathbf{H}_x & \longrightarrow & \mathrm{SO}(\mathfrak{h}_x, Q|_{\mathfrak{h}_x}) \times \mathrm{SO}(\mathfrak{h}_x^{\perp}, Q|_{\mathfrak{h}_x^{\perp}}) \\ \downarrow & & \downarrow \\ \mathbf{G} & \longrightarrow & \mathrm{SO}(\mathfrak{g}, Q) \end{array} \quad (1.3.2)$$

This diagram implies that the following composition

$$\pi_1(\mathbf{H}_x) \to \pi_1(\mathrm{SO}(\mathfrak{h}_x, Q|_{\mathfrak{h}_x})) \times \pi_1(\mathrm{SO}(\mathfrak{h}_x^{\perp}, Q|_{\mathfrak{h}_x^{\perp}})) \to \pi_1(\mathrm{SO}(\mathfrak{g}, Q))$$

is trivial (since it factors through $\pi_1(\mathbf{G}) = 1$). The assumption that the covering $\widetilde{\mathbf{H}}_x \to \mathbf{H}_x$ is nontrivial implies that $\mathbf{H}_x$ has a nontrivial semisimple component (in particular, $\dim \mathfrak{h}_x \geq 3$) and that the map $\pi_1(\mathbf{H}_x) \to \pi_1(\mathrm{SO}(\mathfrak{h}_x, Q|_{\mathfrak{h}_x}))$ is surjective with the kernel $\pi_1(\widetilde{\mathbf{H}}_x) \subset \pi_1(\mathbf{H}_x)$. On the other hand, both the maps

$$\pi_1(\mathrm{SO}(\mathfrak{h}_x, Q|_{\mathfrak{h}_x})) \to \pi_1(\mathrm{SO}(\mathfrak{g}, Q))$$

and

$$\pi_1(\mathrm{SO}(\mathfrak{h}_x^{\perp}, Q|_{\mathfrak{h}_x^{\perp}})) \to \pi_1(\mathrm{SO}(\mathfrak{g}, Q))$$

are isomorphisms. Our claim immediately follows from this, which implies that the homomorphism $\widetilde{\mathbf{H}}_x \to \mathrm{SO}(T_x, B_{x,x^\vee})$ lifts to a homomorphism $\widetilde{\mathbf{H}}_x \to \mathrm{Spin}(T_x, B_{x,x^\vee})$. It is easy to see that the latter homomorphism is nontrivial on $\{\pm 1\} \subset \widetilde{\mathbf{H}}_x$. Indeed, otherwise it would factor through $\mathbf{H}_x$, which contradicts to the nontriviality of the homomorphism (1.3.1). This finishes the proof in this case.

Now assume that the covering $\widetilde{\mathbf{H}}_x \to \mathbf{H}_x$ is trivial (and $\dim(T_x) \geq 3$). Then the map $\pi_1(\mathbf{H}_x) \to \pi_1(\mathrm{SO}(\mathfrak{h}_x, Q|_{\mathfrak{h}_x}))$ is trivial (when $\dim \mathfrak{h}_x \leq 2$ this follows from the commutativity of $\mathbf{H}_x$). As above we deduce from this that the map $\pi_1(\mathbf{H}_x) \to \pi_1(T_x, B_{x,x^\vee})$ is also trivial. Therefore, the homomorphism $\mathbf{H}_x \to \mathrm{SO}(T_x, B_{x,x^\vee})$ factors through $\mathrm{Spin}(T_x, B_{x,x^\vee})$.

Finally, if $\dim(T_x) < 3$, then we replace the orthogonal space $(T_x, B_{x,x^\vee})$ by its direct sum with a fixed orthogonal space of large dimension (on which $\mathbf{G}$ acts trivially) and apply the same argument as above. □

In the case when $k = E$ is a local field, the exact sequence $1 \to \{\pm 1\} \to \widetilde{\mathbf{H}}_x \to \mathbf{H}_x \to 1$ gives a map of Galois cohomology groups

$$\epsilon_{\mathbf{H}_x} : H^1(E, \mathbf{H}_x) \to H^2(E, \{\pm 1\}) \simeq \{\pm 1\}. \quad (1.3.3)$$

This is the sign function which will play an important role below.

2 Nice representations over local fields

In this section $\rho : \mathbf{G} \to \mathrm{Aut}(\mathbf{V})$ denotes a nice representation over a local field E. We formulate our main theorems about the behavior of $\mathbf{G}$-stable functions and distributions on $\mathbf{V}(E)$ under the Fourier transform. Also, we analyze the Fourier transform of certain stable functions using the stationary phase approximation (in the case when E is p-adic).

2.1 G-inner forms and stable G-equivalence

Let x be a point in $\mathbf{V}_0(E)$. Note that since $\mathbf{H}_x$ is reductive, the set $H^1(E, \mathbf{H}_x)$ is finite (see e.g. [15]). Consider the subset $P_x \subset \mathbf{G}(\overline{E})$ consisting of the elements g such that $g^{-1}\sigma(g) \in \mathbf{H}_x$ for all $\sigma \in \Gamma_E$. For every $g \in P_x$ the 1-cocycle

$$e_g(\sigma) = g^{-1}\sigma(g)$$

gives a class in $H^1(E, \mathbf{H}_x)$. It is easy to see that P_x is a union of right $\mathbf{H}_x(\overline{E})$-cosets and of left $\mathbf{G}(E)$-cosets, and the assignment $g \mapsto e_g$ defines a bijection

$$\mathbf{G}(E)\backslash P_x/\mathbf{H}_x(\overline{E}) \widetilde{\to} \ker(H^1(E, \mathbf{H}_x) \to H^1(E, \mathbf{G})).$$

On the other hand, we have a natural bijection

$$P_x/\mathbf{H}_x(\overline{E}) \widetilde{\to} \mathcal{O}_x(E) : g \mapsto gx.$$

In particular, we can identify the set of $\mathbf{G}(E)$-orbits on $\mathcal{O}_x(E)$ with $\ker(H^1(E, \mathbf{H}_x) \to H^1(E, \mathbf{G}))$. In the case when E is p-adic, we have $H^1(E, \mathbf{G}) = 0$ since $\mathbf{G}$ is simply connected (see [15]). Therefore, in this case we have a bijection between $\mathbf{G}(E)\backslash\mathcal{O}_x(E)$ and $H^1(E, \mathbf{H}_x)$.

Let us call two points $x, x' \in \mathbf{V}(E)$ *stably* **G**-*equivalent* if there exists an element $g \in \mathbf{G}(\overline{E})$ such that $gx = x'$, i.e., if $\mathcal{O}_x = \mathcal{O}_{x'}$. Let $\mathbf{K}, \mathbf{K}' \subset \mathbf{G}$ be subgroups defined over E. Let us say that $\mathbf{K}'$ is a **G**-*inner form* of $\mathbf{K}$ if there exists an element $g \in \mathbf{G}(\overline{E})$ such that $g\mathbf{K}(\overline{E})g^{-1} = \mathbf{K}'(\overline{E})$ and $g^{-1}\sigma(g) \in \mathbf{K}(\overline{E})$ for every $\sigma \in \Gamma_E$. It is easy to see that this defines an equivalence relation between subgroups of $\mathbf{G}$ defined over E. This definition is motivated by the following lemma.

Lemma 2.1. *Let* $x \in \mathbf{V}(E)$.

(a) *If* $x' \in \mathbf{V}(E)$ *is stably* **G**-*equivalent to* x, *then* $\mathbf{H}_{x'}$ *is a* **G**-*inner form of* $\mathbf{H}_x$.

(b) *Conversely, if* $\mathbf{H}'$ *is a* **G**-*inner form of* $\mathbf{H}_x$, *then* $\mathbf{H}'$ *is the stabilizer of some* E-*point which is stably* **G**-*equivalent to* x.

(c) *The image of the natural map*

$$a : \mathbf{G}/\mathbf{H}(E) \times \mathbf{W}_0(E) \to \mathbf{V}_0(E)$$

consists of all points in $\mathbf{V}_0(E)$ *whose stabilizer is a* **G**-*inner form of* $\mathbf{H}$.

The proof is straightforward.

Remark. It is easy to see that if $\mathbf{K}'$ is a **G**-inner form of $\mathbf{K}$, then $\mathbf{K}'$ is an inner form of $\mathbf{K}$ in the usual sense. On the other hand, if $\mathbf{K}$ is semisimple and simply connected, E is p-adic, then $\mathbf{K}'$ is a **G**-inner form of $\mathbf{K}$ if and only if $\mathbf{K}'$ is conjugate to $\mathbf{K}$ over E.

Let x and x' be a pair of stably $\mathbf{G}$-equivalent points in $\mathbf{V}_0(E)$ and let $\mathcal{O} = \mathcal{O}_x = \mathcal{O}_{x'}$ be the corresponding orbit. Then $\mathbf{G}(E)$-orbits on $\mathcal{O}(E)$ can be identified with a subset in $H^1(E, \mathbf{H}_x)$ and with a subset in $H^1(E, \mathbf{H}_{x'})$. On the other hand, $\mathbf{H}_{x'}$ is obtained from $\mathbf{H}_x$ by twisting with a cohomology class in $H^1(E, \mathbf{H}_x)$, so we have a canonical identification $H^1(E, \mathbf{H}_{x'}) \simeq H^1(E, \mathbf{H}_x)$. It is easy to see that these three identifications are compatible.

2.2 Local Kottwitz invariant and local sign function

For every connected reductive group $\mathbf{H}$ over a field k let us denote by $Z(\hat{\mathbf{H}})$ the center of the Langlands dual group (equipped with an action of the Galois group Γ_k). Note that $\hat{\mathbf{H}}$ is defined canonically up to an inner conjugation, hence, $Z(\hat{\mathbf{H}})$ is defined canonically and carries an action of Γ_k. Furthermore, an isomorphism $i : \mathbf{H} \to \mathbf{H}'$ over $\overline{k}$, such that $i^{-1}\sigma(i)$ is inner for all $\sigma \in \Gamma_k$ (an *inner twisting*), induces an isomorphism of $Z(\hat{\mathbf{H}})$ with $Z(\hat{\mathbf{H}}')$ as Γ_k-modules. Following Kottwitz we define

$$A(\mathbf{H}/k) := \pi_0(Z(\hat{\mathbf{H}})^{\Gamma_k})^D$$

where for a finite group A we denote by A^D the dual group. When $k = E$ is a local field, the local Kottwitz invariant is a functorial map

$$\mathrm{inv} = \mathrm{inv}_E : H^1(E, \mathbf{H}) \to A(\mathbf{H}/E)$$

constructed in [11]. The definition of inv is a generalization of the isomorphism derived from Tate–Nakayama duality in the case when $\mathbf{H} = \mathbf{T}$ is a torus. Indeed, this duality gives an isomorphism $H^1(E, \mathbf{T}) \simeq H^1(E, X^*(\mathbf{T}))^D$, where $X^*(\mathbf{T})$ is the module of characters of $\mathbf{T}$. Now from the exact sequence of Galois modules

$$0 \to X^*(\mathbf{T}) \to X^*(\mathbf{T}) \otimes \mathbb{C} \to X^*(\mathbf{T}) \otimes \mathbb{C}^* \to 0$$

one gets an isomorphism

$$H^1(E, X^*(\mathbf{T})) \simeq \mathrm{coker}((X^*(\mathbf{T}) \otimes \mathbb{C})^{\Gamma_E} \to (X^*(\mathbf{T}) \otimes \mathbb{C}^*)^{\Gamma_E}).$$

The latter group can be immediately identified with $\pi_0((X^*(\mathbf{T}) \otimes \mathbb{C}^*)^{\Gamma_E})$. In the general case (when $\mathbf{H}$ is not necessarily a torus), Kottwitz showed in [11] that inv_E is an isomorphism for p-adic E. In particular, for such E we obtain the structure of abelian group on $H^1(E, \mathbf{H})$.

For any 1-cocycle $e : \Gamma_E \to \mathbf{H}(\overline{E})$ we can consider the E-group $\mathbf{H}^e$ obtained from $\mathbf{H}$ by inner twisting with e. By definition $\mathbf{H}^e(\overline{E}) = \mathbf{H}(\overline{E})$ while the action of an element $\sigma \in \Gamma_E$ on $\mathbf{H}^e(\overline{E})$ differs from its action on $\mathbf{H}(\overline{E})$ by the inner automorphism associated with $e(\sigma)$. In particular, we have $Z(\hat{\mathbf{H}}^e) = Z(\hat{\mathbf{H}})$, hence $A(\mathbf{H}^e/E) = A(\mathbf{H}/E)$. According to Lemma 1.4 of [11], the following diagram is commutative

$$\begin{array}{ccc} H^1(E, \mathbf{H}^e) & \xrightarrow{\text{inv}} & A(\mathbf{H}/E) \\ \Big\downarrow i_e & & \Big\downarrow t_{\text{inv}(e)} \\ H^1(E, \mathbf{H}) & \xrightarrow{\text{inv}} & A(\mathbf{H}^e/E) = A(\mathbf{H}/E) \end{array} \qquad (2.2.1)$$

where $t_{\text{inv}(e)} : A(\mathbf{H}/E) \to A(\mathbf{H}/E)$ is the translation by $\text{inv}(e) \in A(\mathbf{H}/E)$, $i_e : H^1(E, \mathbf{H}^e) \to H^1(E, \mathbf{H})$ is the canonical identification induced by e. Thus, in the case when E is p-adic, the isomorphism i_e does not respect the group structures on the sets $H^1(E, \mathbf{H}^e)$ and $H^1(E, \mathbf{H})$, but rather respects the structures of principal homogeneous spaces over $A(\mathbf{H}/E) = A(\mathbf{H}^e/E)$.

Now assume that $\mathcal{O} \subset \mathbf{V}_0$ is an orbit. Then we have a system of compatible isomorphisms between the groups $A(\mathbf{H}_x/E)$ for $x \in \mathcal{O}(E)$.[3] Let us denote the corresponding group isomorphic to all $A(\mathbf{H}_x/E)$ by $A(\mathcal{O}/E)$. Assume for a moment that E is p-adic. Then the set of $\mathbf{G}(E)$-orbits on $\mathcal{O}(E)$ has a natural structure of a principal homogeneous space over $A(\mathcal{O}/E)$. Thus, for every pair of points $x, x' \in \mathcal{O}(E)$ we can define an element $\text{inv}(x, x') \in A(\mathcal{O}/E)$ such that $\mathbf{G}(E)x'$ is obtained from $\mathbf{G}(E)x$ by the action of $\text{inv}(x, x')$. This definition extends to the case of archimedean E as follows:

$$\text{inv}(x, gx) = \text{inv}(e_g),$$

where $g \in P_x$, e_g is the corresponding cohomology class in $H^1(E, \mathbf{H}_x)$, $\text{inv}(e_g)$ is its local Kottwitz invariant in $A(\mathbf{H}_x/E) \simeq A(\mathcal{O}/E)$. It is easy to see that the following properties are satisfied:

$$\begin{aligned} &\text{inv}(x, x') + \text{inv}(x', x) = 0, \\ &\text{inv}(x, x') + \text{inv}(x', x'') = \text{inv}(x, x''), \\ &\text{inv}(tx, tx') = \text{inv}(x, x'), \end{aligned} \qquad (2.2.2)$$

where $t \in E^*$.

Let us fix a point $x \in \mathbf{V}_0(E)$ and set $\mathbf{H} = \mathbf{H}_x$. Recall that in Section 1.3 we have defined a map

$$\epsilon_{\mathbf{H}} : H^1(E, \mathbf{H}) \to H^2(E, \{\pm 1\}) = \{\pm 1\}$$

induced by the central extension $1 \to \{\pm 1\} \to \widetilde{\mathbf{H}} \to \mathbf{H} \to 1$ (the pullback of the spin covering associated with $Q|_{\mathfrak{h}_x}$). We are going to construct a character

$$\text{sign} = \text{sign}_{\mathbf{H}} : A(\mathbf{H}/E) \to \{\pm 1\}$$

[3] For two points $x, x' \in \mathcal{O}(E)$ such that $\mathbf{H}_x = \mathbf{H}_{x'}$, the corresponding isomorphism $A(\mathbf{H}_x/E) \to A(\mathbf{H}_{x'}/E)$ is not necessarily the identity: it corresponds to the action of some element in the normalizer of $\mathbf{H}_x$.

such that $\epsilon_{\mathbf{H}} = \mathrm{sign}_{\mathbf{H}} \circ \mathrm{inv}_E$. For this we note that the above central extension is induced by the similar extension $1 \to \{\pm 1\} \to \widetilde{\mathbf{H}}_{ad} \to \mathbf{H}_{ad} \to 1$ of the adjoint group $\mathbf{H}_{ad} = \mathbf{H}/Z(\mathbf{H})$. Therefore, by functoriality of inv_E, it suffices to construct the character $\mathrm{sign}_{\mathbf{H}}$ in the case when $\mathbf{H}$ is adjoint. Thus, we can assume that $\mathbf{H}$ is semisimple. Let $u : \mathbf{H}_{sc} \to \mathbf{H}$ be the universal covering of $\mathbf{H}$ (it is defined over E), $\mathbf{C}$ be the kernel of u. Then the homomorphism u lifts uniquely to a homomorphism $\widetilde{u} : \mathbf{H}_{sc} \to \widetilde{\mathbf{H}}$. Now the restriction of $\widetilde{u}$ to $\mathbf{C}$ gives a homomorphism $\chi : \mathbf{C} \to \{\pm 1\}$ defined over E. We can consider χ as an element of order 2 in $X^*(\mathbf{C})^{\Gamma_E}$. It remains to notice that there is an isomorphism $X^*(\mathbf{C}) \simeq Z(\hat{\mathbf{H}})$ of Γ_E-modules, so we can consider χ as a character sign $=$ $\mathrm{sign}_{\mathbf{H}} : A(\mathbf{H}/E) \to \{\pm 1\}$. The following result shows that this is the character we were looking for.

Lemma 2.2. *One has the following equality of maps from $H^1(E, \mathbf{H})$ to $\{\pm 1\}$:*

$$\epsilon_{\mathbf{H}} = \mathrm{sign}_{\mathbf{H}} \circ \mathrm{inv}_E.$$

Proof. By duality for finite groups we have an isomorphism

$$H^2(E, \mathbf{C}) \simeq H^0(E, X^*(\mathbf{C}))^D \simeq A(\mathbf{H}/E).$$

According to Lemma 1.8 of [11], under this isomorphism the map $\mathrm{inv}_E : H^1(E, \mathbf{H}) \to A(\mathbf{H}/E)$ can be identified with the map $H^1(E, \mathbf{H}) \to H^2(E, \mathbf{C})$ coming from the exact sequence $1 \to \mathbf{C} \to \mathbf{H}_{sc} \to \mathbf{H} \to 1$. On the other hand, the character of $H^2(E, \mathbf{C})$ corresponding to $\mathrm{sign}_{\mathbf{H}}$ is the homomorphism on H^2 induced by the homomorphism $\chi : \mathbf{C} \to \{\pm 1\}$. Therefore, the composition $\mathrm{sign}_{\mathbf{H}} \circ \mathrm{inv}_E$ coincides with the map $H^1(E, \mathbf{H}) \to H^2(E, \{\pm 1\})$ coming from the exact sequence $1 \to \{\pm 1\} \to \widetilde{\mathbf{H}} \to \mathbf{H} \to 1$, which is the definition of $\epsilon_{\mathbf{H}}$. □

Recall that for every pair of points $x, x' \in \mathcal{O}(E)$, where $\mathcal{O} \subset \mathbf{V}_0$ is a $\mathbf{G}$-orbit, we have defined the sign $\epsilon(x, x') = \epsilon(Q|_{\mathfrak{h}_x}, Q|_{\mathfrak{h}'_x})$. It is well known that if a quadratic form B' is obtained from a nondegenerate quadratic form B by the twist with an element $e \in H^1(E, \mathrm{SO}(B))$, then the relative Hasse–Witt invariant $\epsilon(B, B')$ is equal to the image of e under the coboundary homomorphism $H^1(E, \mathrm{SO}(B)) \to H^2(E, \{\pm 1\}) = \{\pm 1\}$ coming from the spin-covering (see e.g. [22]). This implies the following relation between ϵ and the sign function (1.3.3) defined in 1.3:

$$\epsilon(x, gx) = \epsilon_{\mathbf{H}_x}(e_g), \tag{2.2.3}$$

where $x \in \mathbf{V}_0(E)$, $g \in P_x$. Comparing the definition of $\mathrm{inv}(\cdot, \cdot)$ with (2.2.3) and using Lemma 2.2 we get the following formula:

$$\epsilon(x, x') = \mathrm{sign}_{\mathbf{H}_x}(\mathrm{inv}(x, x')).$$

2.3 Critical points and stable equivalence

Assume that we have a pair of points $x \in \mathbf{V}_0(E)$, $x^\vee \in \mathbf{V}_0^\vee(E)$ such that x is a critical point of $x^\vee|_{\mathcal{O}_x}$. Proposition 1.9 implies that the following diagram is commutative:

$$\begin{array}{ccc} H^1(E, \mathbf{H}_x) & \xrightarrow{\epsilon_{\mathbf{H}_x}} & \{\pm 1\} \\ \Big\downarrow H^1(\iota_{x,x^\vee}) & & \Big\downarrow \mathrm{id} \\ H^1(E, \mathrm{SO}(T_x, B_{x,x^\vee})) & \longrightarrow & \{\pm 1\} \end{array} \tag{2.3.1}$$

where the lower horizontal arrow is the coboundary homomorphism associated with the spin-covering of $\mathrm{SO}(T_x, B_{x,x^\vee})$.

This observation leads to the following result.

Lemma 2.3. *Let ω be a $\mathbf{G}$-invariant top-degree form on $\mathcal{O}_x$. Recall that for every $g \in P_x$ one has $gx \in \mathbf{V}_0(E)$ and $gx^\vee \in \mathbf{V}_0^\vee(E)$.*

(a) *One has* $\det(B_{x,x^\vee})/\omega_x^2 = \det(B_{gx,gx^\vee})/\omega_{gx}^2$.

(b) *The quadratic forms $B_{x,x^\vee}$ and $B_{gx,gx^\vee}$ have the same determinant modulo squares. Their relative Hasse–Witt invariant is given by*

$$\epsilon(B_{gx,gx^\vee}, B_{x,x^\vee}) = \epsilon(x, gx) = \epsilon_{\mathbf{H}_x}(e_g).$$

Proof. (a) This follows from the $\overline{E}$-isomorphism of data $(T_v, B_{v,v^\vee}, \omega_v)$ and $(T_{gv}, B_{gv,gv^\vee}, \omega_{gv})$ given by the action of g.

(b) Let $e_g \in H^1(E, \mathbf{H}_v)$ be the cohomology class defined by the cocycle $\sigma \mapsto g^{-1}\sigma(g)$. Then the quadratic form $B_{gv,gv^\vee}$ is obtained from $B_{v,v^\vee}$ by twisting with the class $H^1(\iota_{v,v^\vee})(e_g) \in H^1(E, \mathrm{SO}(T_v, B_{v,v^\vee}))$. Now the diagram (2.3.1) implies that $\epsilon(B_{gx,gx^\vee}, B_{x,x^\vee}) = \epsilon_{\mathbf{H}_x}(e_g)$. □

2.4 Stable and antistable functions and distributions

Let us denote by $\mathcal{S}(\mathbf{V}(E))_{\mathbf{G}(E)}$ the space of $\mathbf{G}(E)$-coinvariants in $\mathcal{S}(\mathbf{V}(E))$. We have the natural projection $\mathcal{S}(\mathbf{V}(E)) \to \mathcal{S}(\mathbf{V}(E))_{\mathbf{G}(E)} : \phi \mapsto \overline{\phi}$. On the other hand, for every $\phi \in \mathcal{S}(\mathbf{V}(E))$ we can define a function $I(\phi)$ on $\mathbf{G}(E)\backslash\mathbf{V}_0(E)$ by the formula

$$I(\phi)(y) = \int_{x \in \mathbf{G}(E)y} \phi(x)|\omega_y|$$

where ω_y is a $\mathbf{G}$-invariant top-degree form on $\mathcal{O}_y$ (the integral is convergent since the orbit $\mathcal{O}_y$ is closed in $\mathbf{V}$). It is clear that $I(\phi)$ depends only on $\overline{\phi}$, so we will denote $I(\overline{\phi}) = I(\phi)$. Although we will not need this fact, it is worth mentioning that for a pair of functions $\phi, \phi' \in \mathcal{S}(\mathbf{V}_0(E))$ one has $\overline{\phi} = \overline{\phi'}$ if and only if $I(\phi) = I(\phi')$ (see [4]).

Definition. (i) An element $\overline{\phi} \in \mathcal{S}(\mathbf{V}(E))_{\mathbf{G}(E)}$ is called *stable* if for every $\mathbf{G}$-orbit $\mathcal{O} \subset \mathbf{V}_0$, the restriction of the function $I(\overline{\phi})$ to $\mathbf{G}(E)\backslash\mathcal{O}(E)$ is constant. We denote by $\mathcal{S}(\mathbf{V}(E))^{st} \subset \mathcal{S}(\mathbf{V}(E))_{\mathbf{G}(E)}$ the subspace of stable elements. Similarly, we define a subspace of *antistable* elements $\mathcal{S}(\mathbf{V}(E))^{as} \subset \mathcal{S}(\mathbf{V}(E))_{\mathbf{G}(E)}$. By the definition, an element $\overline{\phi} \in \mathcal{S}(\mathbf{V}(E))_{\mathbf{G}(E)}$ is antistable if for every $\mathbf{G}$-orbit $\mathcal{O} \subset \mathbf{V}_0$ the total sum of the function $I(\overline{\phi})$ over $\mathbf{G}(E)\backslash\mathcal{O}(E)$ is zero.

(ii) Let $\mathcal{D}(\mathbf{V}(E))$ denote the space of distributions on $\mathbf{V}(E)$, i.e., functionals on $\mathcal{S}(\mathbf{V}(E))$. Note that a $\mathbf{G}(E)$-invariant distribution $\alpha \in \mathcal{D}(\mathbf{V}(E))$ descends to a functional on $\mathcal{S}(\mathbf{V}(E))_{\mathbf{G}(E)}$. Now a $\mathbf{G}(E)$-invariant distribution α is called *stable* (resp. *antistable*) if $\alpha(\mathcal{S}(\mathbf{V}(E))^{as}) = 0$ (resp. $\alpha(\mathcal{S}(\mathbf{V}(E))^{st}) = 0$). We denote by $\mathcal{D}(\mathbf{V}(E))^{st} \subset \mathcal{D}(\mathbf{V}(E))$ (resp. $\mathcal{D}(\mathbf{V}(E))^{as} \subset \mathcal{D}(\mathbf{V}(E))$) the subspace of stable (resp. antistable) distributions.

For a $\mathbf{G}$-orbit $\mathcal{O} \subset \mathbf{V}_0$ and a nonzero $\mathbf{G}$-invariant top-degree form ω on $\mathcal{O}$ defined over E, we can define a stable distribution $\delta_{\mathcal{O},\omega} \in \mathcal{D}(\mathbf{V}(E))$ by the formula

$$\delta_{\mathcal{O},\omega}(\phi) = \int_{\mathcal{O}(E)} \phi|\omega|,$$

where $\phi \in \mathcal{S}(\mathbf{V}(E))$. When $\mathcal{O}(E) = \emptyset$ we set $\delta_{\mathcal{O},\omega} = 0$. When the choice of ω is clear or is not important we will abbreviate $\delta_{\mathcal{O},\omega}$ to $\delta_{\mathcal{O}}$. By the definition, an element $\overline{\phi} \in \mathcal{S}(\mathbf{V}(E))_{\mathbf{G}(E)}$ is antistable if and only if $\delta_{\mathcal{O}}(\overline{\phi}) = 0$ for all $\mathbf{G}$-orbits $\mathcal{O} \subset \mathbf{V}_0$.

More generally, for every point $x \in \mathbf{V}_0(E)$ and for a character $\kappa : A(\mathcal{O}_x/E) \to \mathbb{C}^*$ we can define a $\mathbf{G}(E)$-invariant distribution $\delta^{\kappa}_{\mathcal{O}_x} = \delta^{\kappa}_{\mathcal{O}_x,\omega}$ by the formula

$$\delta^{\kappa}_{\mathcal{O}_x,\omega}(\phi) = \int_{x' \in \mathcal{O}_x(E)} \kappa(\mathrm{inv}(x, x'))\phi(x')|\omega(x')|.$$

Note that if y is stably equivalent to x, then

$$\delta^{\kappa}_{\mathcal{O}_y,\omega} = \kappa(\mathrm{inv}(y, x)) \cdot \delta^{\kappa}_{\mathcal{O}_x,\omega}.$$

In the case when E is p-adic, an element $\overline{\phi} \in \mathcal{S}(\mathbf{V}(E))_{\mathbf{G}(E)}$ is stable if and only if $\delta^{\kappa}_{\mathcal{O}_x}(\overline{\phi}) = 0$ for all $x \in \mathbf{V}_0(E)$ and all nontrivial characters κ of $A(\mathcal{O}_x/E)$.

Also, for every G-orbit $\mathcal{O} \subset \mathbf{V}_0$ we define the distribution

$$\delta^{\epsilon}_{\mathcal{O},\omega}(\phi) = \int_{y \in \mathcal{O}(E)} \epsilon(y)\phi(y)|\omega(y)|$$

where $\epsilon : \mathbf{V}_0(E) \to \{\pm 1\}$ is the function defined in the introduction. The results of Section 2.2 show that this distribution corresponds to some character of $A(\mathcal{O}/E)$ as above. More precisely, for any point $x \in \mathbf{V}_0(E)$ we have

$$\delta^{\epsilon}_{\mathcal{O}_x,\omega} = \epsilon(x) \cdot \delta^{\mathrm{sign}_{\mathbf{H}_x}}_{\mathcal{O}_x,\omega}.$$

The Fourier transform (associated with some choice of a nontrivial additive character ψ) induces a well-defined operator

$$\mathcal{F} : \mathcal{S}(\mathbf{V}(E))_{\mathbf{G}(E)} \to \mathcal{S}(\mathbf{V}^{\vee}(E))_{\mathbf{G}(E)}.$$

We want to describe the images of the subspaces $\mathcal{S}(\mathbf{V}(E))^{st}$ and $\mathcal{S}(\mathbf{V}(E))^{as}$ under $\mathcal{F}$. This is equivalent to describing the images of the spaces of distributions $\mathcal{D}(\mathbf{V}(E))^{st}$ and $\mathcal{D}(\mathbf{V}(E))^{as}$.

Definition. (i) An element $\overline{\phi} \in \mathcal{S}(\mathbf{V}(E))_{\mathbf{G}(E)}$ is called *ϵ-stable* if for every orbit $\mathcal{O} \subset \mathbf{V}_0$ and for every pair of points $y, y' \in \mathcal{O}(E)$, one has $I(\overline{\phi})(y') = \epsilon(y, y')I(\overline{\phi})(y)$. We denote by $\mathcal{S}(\mathbf{V}(E))^{st}_{\epsilon}$ the subspace of ϵ-stable elements in $\mathcal{S}(\mathbf{V}(E))_{\mathbf{G}(E)}$. Similarly we define the subspace $\mathcal{S}(\mathbf{V}(E))^{as}_{\epsilon} \subset \mathcal{S}(\mathbf{V}(E))_{\mathbf{G}(E)}$ of *ϵ-antistable* elements. By the definition, an element $\overline{\phi} \in \mathcal{S}(\mathbf{V}(E))_{\mathbf{G}(E)}$ is ϵ-antistable if for every $y \in \mathbf{V}_0(E)$ one has

$$\sum_{y' \in \mathbf{G}(E)\backslash \mathcal{O}_y(E)} \epsilon(y, y')I(\overline{\phi})(y') = 0.$$

(ii) Dually, we define the subspace $\mathcal{D}(\mathbf{V}(E))^{st}_{\epsilon} \subset \mathcal{D}(\mathbf{V}(E))^{\mathbf{G}(E)}$ (resp. $\mathcal{D}(\mathbf{V}(E))^{as}_{\epsilon} \subset \mathcal{D}(\mathbf{V}(E))^{\mathbf{G}(E)}$) of ϵ-stable (resp. ϵ-antistable) distributions, so that $\mathcal{D}(\mathbf{V}(E))^{st}_{\epsilon}$ is the annihilator of $\mathcal{S}(\mathbf{V}(E))^{as}_{\epsilon}$ (resp. $\mathcal{D}(\mathbf{V}(E))^{as}_{\epsilon}$ is the annihilator of $\mathcal{S}(\mathbf{V}(E))^{st}_{\epsilon}$).

Note that the distributions $\delta^{\epsilon}_{\mathcal{O}}$ are ϵ-stable, and an element $\overline{\phi} \in \mathcal{S}(\mathbf{V}(E))$ is ϵ-antistable if and only if $\delta^{\epsilon}_{\mathcal{O}}(\overline{\phi}) = 0$ for all $\mathbf{G}$-orbits $\mathcal{O} \subset \mathbf{V}_0$.

We say that a function $\phi \in \mathcal{S}(\mathbf{V}(E))$ is stable (resp. antistable, ϵ-stable, ϵ-antistable) if this is true for the corresponding element $\overline{\phi} \in \mathcal{S}(\mathbf{V}(E))_{\mathbf{G}(E)}$. Clearly, a function $\phi \in \mathcal{S}(\mathbf{V}_0(E))$ is ϵ-stable (resp. ϵ-antistable) if and only if $\epsilon \cdot \phi$ is stable (resp. antistable).

Our main result is the following theorem.

Theorem 2.4. *Let $\rho : \mathbf{G} \to \mathrm{GL}(\mathbf{V})$ be a nice representation of a simply connected semisimple group $\mathbf{G}$ over a p-adic field E. Assume that either $\mathbf{G}$ is simple or the generic stabilizer is semisimple. Then $\mathcal{F}(\mathcal{D}(\mathbf{V}(E))^{st}) = \mathcal{D}(\mathbf{V}^{\vee}(E))^{st}_{\epsilon}$ (equivalently, $\mathcal{F}(\mathcal{S}(\mathbf{V}(E))^{as}) = \mathcal{S}(\mathbf{V}^{\vee}(E))^{as}_{\epsilon}$).*

Another result concerns the Fourier transform of stable functions (equivalently, antistable distributions).

Theorem 2.5. *Let $\rho : \mathbf{G} \to \mathrm{GL}(\mathbf{V})$ be a nice representation of a simply connected semisimple group $\mathbf{G}$ over a p-adic field E, such that the generic stabilizer is semisimple. Then $\mathcal{F}(\mathcal{D}(\mathbf{V}(E))^{as}) = \mathcal{D}(\mathbf{V}^{\vee}(E))^{as}_{\epsilon}$ (equivalently, $\mathcal{F}(\mathcal{S}(\mathbf{V}(E))^{st}) = \mathcal{S}(\mathbf{V}^{\vee}(E))^{st}_{\epsilon}$).*

Remark. In the case of the adjoint representation Theorem 2.4 is due to Waldspurger (for arbitrary connected reductive group $\mathbf{G}$), see [25]. We conjecture that Theorems 2.4 and 2.5 should hold under a more general assumption that E is a local field, and ρ is a nice representation of a simply connected semisimple group.

Theorem 2.4 will be deduced from the more general Theorem 2.10, which will be proved along with Theorem 2.5 in Section 3.3.

The main local ingredient of these results is the analysis of the Fourier transform of some explicit stable functions with the help of the stationary phase principle.

2.5 Stationary phase

In this subsection we assume that E is p-adic. Let $\psi : E \to \mathbb{C}^*$ be a nontrivial additive character. We will use the following easy version of the stationary phase principle over E.

Lemma 2.6. *Let* $\mathbf{X}$ *and* $\mathbf{S}$ *be smooth varieties over* E*,* $f : \mathbf{X} \times \mathbf{S} \to \mathbb{A}^1$ *be a morphism, such that for every* $s \in \mathbf{S}(\overline{E})$ *the function* $f_s = f|_{\mathbf{X}\times\{s\}}$ *on* $\mathbf{X}$ *has finitely many nondegenerate critical points. Let* $U \subset \mathbf{X}(E)$ *and* $P \subset \mathbf{S}(E)$ *be compact open subsets, and* ω *a nonvanishing top-degree form on* $\mathbf{X}$*. For every* $x_0 \in \mathrm{Cr}(f_p)$ *(where* $p \in P$*) we denote by* $q_{x_0}(x - x_0)$ *the quadratic form on the tangent space* $T_{x_0}\mathbf{X}$ *approximating* $f_p(x) - f_p(x_0)$ *near* x_0 *(the Hessian of* f_p *at* x_0*). Then there exists a positive constant* C *such that for all* $t \in (E^*)^2$ *with* $|t| > C$ *and all* $p \in P$*, we have*

$$\int_U \psi(tf_p(x))|\omega| = \sum_{x_0\in \mathrm{Cr}(f_p)} \psi(tf_p(x_0))|t|^{-n/2} \cdot c(q_{x_0}, \omega_{x_0}, \psi),$$

where $n = \dim \mathbf{X}$*,* $c(q, \nu, \psi) = \gamma(q, \psi) \cdot |\det(q)/\nu^2|^{-1/2}$*, and* $\gamma(q, \psi)$ *is the Weil constant associated with* q *and* ψ*.*

Proof. Let us assume first that $\mathbf{S}$ is a point, so that we have a function f on $\mathbf{X}$ with a finite number of nondegenerate critical points. If f has no critical points on U, then we can find a finite covering (U_i) of U by compact open subsets, such that on each U_i there exists an analytic system of coordinates $(x_1, \dots, x_n)$ with $x_1 = f$ and $|\omega| = |\lambda \cdot dx_1 \wedge \dots \wedge dx_n|$, where $\lambda \in E^*$. Furthermore, we can assume that the subsets U_i are disjoint. Therefore, the statement reduces in this case to the vanishing of the integral

$$\int_V \psi(tx_1)dx_1 \dots dx_n$$

for a compact open subset $V \in E^n$ and for sufficiently large t, which is clear. Now let $c_1, \dots, c_k$ be critical points of f contained in U. By Morse's lemma, for each point c_i there exists a neighborhood V_i of c_i and a system of coordinates $(x_1, \dots, x_n)$ on V_i, such that $f - f(c_i) = q(x_1, \dots, x_n)$ for some nondegenerate quadratic form q and $|\omega| = |\lambda \cdot dx_1 \wedge \dots \wedge dx_n|$ on V_i. Let $B_i \subset V_i$ be a small ball around c_i in this coordinate system. As we have shown above,

$$\int_{U\setminus\cup_i B_i} \psi(tf)|\omega| = 0$$

for sufficiently large t. Therefore, the statement is reduced to the case when U is an open compact subgroup in a E-vector space V, ω is translation-invariant, and

$f = q$ is a nondegenerate quadratic form. Then for any $a \in E^*$ we have

$$\mathcal{F}(\delta_{aU}) = |a|^n \cdot \mathrm{vol}(U) \cdot \delta_{a^{-1}U^\perp},$$

where $U^\perp \subset V^\vee$ is the orthogonal complement to U (with respect to ψ). Combining this with the equation (0.0.2) we obtain

$$\int_U \psi(a^2q)|\omega| = |a|^{-n} \cdot \int_{aU} \psi(q)|\omega| = \mathrm{vol}(U) \cdot c(q, \omega, \psi) \cdot \int_{a^{-1}U^\perp} \psi(-q^\vee)|\omega|^\vee$$

where $|\omega|^\vee$ is the dual measure on $V^\vee$. For sufficiently large a we have

$$\int_{a^{-1}U^\perp} \psi(-q^\vee)|\omega|^\vee = |a|^{-n} \cdot \mathrm{vol}(U^\perp).$$

By involutivity of the Fourier transform, we have $\mathrm{vol}(U)\mathrm{vol}(U^\perp) = 1$, hence we get

$$\int_U \psi(a^2q)|\omega| = c(q, \omega, \psi) \cdot |a|^{-n}.$$

One can deal with the case of a general family of functions parametrized by a compact set $P \subset \mathbf{S}(E)$ as follows. The subvariety $\mathrm{Cr}(f) \subset \mathbf{X} \times \mathbf{S}$ of critical points of f in the $\mathbf{X}$-direction is étale over $\mathbf{S}$. Thus, for every point $p \in P$ the above argument works uniformly for all p' in a sufficiently small neighborhood of p. Now our assertion follows from compactness of P. □

To analyze the result of the stationary phase approximation, it is convenient to use the following lemma.

Lemma 2.7. *Let $f_1, \ldots, f_n$ be analytic functions on a ball B in E^N centered at 0. Assume that the differentials at zero $d_0 f_1, \ldots, d_0 f_n$ are linearly independent. Then there exists a constant $a \in E^*$, such that for all $t \in E^*$ with $|t|$ sufficiently large, the functions $\psi(tf_1), \ldots, \psi(tf_n)$ on $t^{-1}aB$ are linearly independent.*

Proof. Without loss of generality we can assume that $f_i(0) = 0$. Let $C \subset E^N$ be a sufficiently large open compact containing 0, so that the restrictions of the functions $\psi(d_0 f_1), \ldots, \psi(d_0 f_n)$ to C are linearly independent. Now for sufficiently large t we have $t^{-1}C \subset B$ and $\psi(tf_i(t^{-1}x)) = \psi(d_0 f_i(x))$ for all $x \in C$. It remains to take a such that $C \subset aB$. □

2.6 Local computation

The field E is still assumed to be p-adic. We fix one of stabilizer subgroups $\mathbf{H} \subset \mathbf{G}$ and set $\mathbf{W} = \mathbf{V}^{\mathbf{H}}$, $\mathbf{W}_0 = \mathbf{W} \cap \mathbf{V}_0$, $\mathbf{W}^\vee = (\mathbf{V}^\vee)^{\mathbf{H}}$, $\mathbf{W}_0^\vee = \mathbf{W}^\vee \cap \mathbf{V}_0^\vee$. Let us consider the subset of $\mathbf{H}$-fixed points in the variety $\mathbf{I}$:

$$\mathbf{I}^{\mathbf{H}} = \mathbf{I} \cap (\mathbf{W}_0 \times \mathbf{W}_0^\vee) = \mathbf{I} \cap (\mathbf{V}_0 \times \mathbf{W}_0^\vee)$$

(the last equality follows from Lemma 1.7). Let $\mathbf{U} \subset \mathbf{V}_0/\mathbf{G} \times \mathbf{V}_0^\vee$ be a nonempty $\mathbf{G}$-invariant open subset such that the morphism $\mathbf{f} : \mathbf{I} \to \mathbf{V}_0/\mathbf{G} \times \mathbf{V}_0^\vee$ is étale over **U** (see Proposition 1.8). We have $\mathbf{I^H} = \mathbf{f}^{-1}(\mathbf{V}_0/\mathbf{G} \times \mathbf{W}_0^\vee)$. Since **U** is **G**-invariant, it has a nonempty intersection with $\mathbf{V}_0/\mathbf{G} \times \mathbf{W}_0^\vee$. Therefore, the open subset $\mathbf{f}^{-1}(\mathbf{U}) \cap \mathbf{I^H} \subset \mathbf{I^H}$ is nonempty. Note that since $\mathbf{I^H}$ is an open subset in a vector bundle over $\mathbf{W}_0^\vee$, the set of E-rational points in $\mathbf{I^H}$ is dense in Zariski topology. Therefore, there exists a point $(x_0, x_0^\vee) \in \mathbf{f}^{-1}(\mathbf{U}) \cap \mathbf{I}(E)$. Set $\mathcal{O} = \mathcal{O}_{x_0}$. Then x_0 is a critical point of $x_0^\vee|_{\mathcal{O}}$ and $(\mathcal{O}, x_0^\vee) \in \mathbf{U}$.

Recall that according to Lemma 1.5, there exists a linear subspace $\mathbf{L} \subset \mathbf{W}$ such that the morphism $p : x_0 + \mathbf{L} \to \mathbf{V}_0/\mathbf{G}$ is étale near x_0. Let $D \subset \mathbf{L}(E)$ be a small ball centered at zero, and let $U^\vee \subset \mathbf{V}_0^\vee$ be a compact open neighborhood of $x_0^\vee$. We assume that D and $U^\vee$ are small enough, so that the following two conditions are satisfied:

(i) the restriction of p to $x_0 + D$ is an isomorphism onto $p(x_0 + D) \subset \mathbf{V}_0/\mathbf{G}(E)$;
(ii) $p(x_0+D) \times U^\vee \subset \mathbf{U}(E)$ and $\mathbf{f}^{-1}(p(x_0+D) \times U^\vee)$ is analytically isomorphic to a disjoint union of open subsets mapping identically to $p(x_0 + D) \times U^\vee$.

Let $\mathcal{O}(E) = O_1 \bigsqcup \ldots \bigsqcup O_r$ be a partition of $\mathcal{O}(E)$ into $\mathbf{G}(E)$-orbits. Let us choose nonempty open compact subsets $K_i \subset O_i, i = 1, \ldots, r$ such that $\mathrm{vol}(K_i)$ does not depend on i (where the volume is computed using a **G**-invariant top-degree form on $\mathcal{O}$). In addition we assume that for every $i = 1, \ldots, r$ there exists an analytic isomorphism of K_i with a ball in E^d, where $d = \dim \mathcal{O}$. We set $K = K_1 \bigsqcup \ldots \bigsqcup K_r \subset \mathcal{O}(E)$. Using the identification $\mathbf{G}/\mathbf{H} \tilde{\to} \mathcal{O} : g\mathbf{H} \mapsto gx_0$ we can consider K as a subset of $\mathbf{G}/\mathbf{H}(E)$. By our choice of D, the restriction of the map $\mathbf{G}/\mathbf{H}(E) \times \mathbf{L}(E) \to \mathbf{V}_0(E)$ to $K \times (x_0 + D)$ is an isomorphism onto an open compact subset $K(x_0 + D) \subset \mathbf{V}_0(E)$. Now for every $t \in E^*$ with $|t| > 1$ we consider a compact open subset

$$U(t) = K(tx_0 + D) = tK(x_0 + t^{-1}D) \subset tK(x_0 + D).$$

Note that by Lemma 2.1, stabilizers of all points in $U(t)$ are **G**-inner forms of **H**. For every open compact set $C \subset \mathbf{V}(E)$ we denote by δ_C the characteristic function of C. It is clear that for every $t \in E^*$ with $|t| > 1$ the function $\delta_{U(t)} \in \mathcal{S}(\mathbf{V}_0(E))$ is stable. More generally, for every character $\kappa : A(\mathbf{H}/E) \to \mathbb{C}^*$ we define a function $\delta^\kappa_{U(t)}$ supported on $U(t)$ by

$$\delta^\kappa_{U(t)}(k(x_0 + d)) = \kappa(\mathrm{inv}(x_0, k)), \tag{2.6.1}$$

where $k \in K, d \in D$.

Lemma 2.8. *There exists a compact set $C \subset \mathbf{V}^\vee(E)$ such that for all $t \in E^*$ with $|t| > 1$ and all $\kappa \in A(\mathbf{H}/E)^D$ the support of $\mathcal{F}(\delta^\kappa_{U(t)})$ is contained in C.*

Proof. Set $U_i(t) = K_i(tx_0 + D), i = 1, \ldots, r$. Since $\delta^\kappa_{U(t)}$ is a linear combination of $\delta_{U_i(t)}$, it suffices to prove that there exists an open compact neighborhood of zero $B \subset \mathbf{V}(E)$, such that $U_i(t) + B = U_i(t)$ for all $t \in E^*$ with $|t| > 1$ and all i. Let us consider the norm on $\mathbf{V}(E)$ for which the unit ball is an integer lattice in $\mathbf{V}(E)$. Similarly, using an isomorphism of K_i with a ball in E^d and the norm on $\mathbf{L}(E)$ for which D is the unit ball, we get an (ultra)metric on $K_i \times (x_0 + D)$

such that $d((k,x),(k',x')) = \max(||k-k'||, ||x-x'||)$. Since the isomorphism $a : K_i \times (x_0 + D) \tilde{\to} U_i = K_i(x_0 + D)$ is analytic we have $d(a^{-1}(y), a^{-1}(y')) \leq c \cdot ||y - y'||$ for some constant $c > 0$, where $y, y' \in U_i$. Now for $t \in E^*$ with $|t| > 1$ the subset $t^{-1}U_i(t) = K_i(x_0 + t^{-1}D) \subset U_i$ consists of points $y \in U_i$ such that $d(a^{-1}(y), K_i \times x_0) \leq |t|^{-1}$ (since D is the unit ball for the norm on $\mathbf{L}(E)$). Now let $B \subset \mathbf{V}(E)$ be a ball of radius $< c^{-1}$ centered at zero such that $U_i + B = U_i$. We claim that $U_i(t) + B = U_i(t)$. Indeed, we have to check that for every $y \in t^{-1}U_i(t)$ one has $y + t^{-1}B \subset t^{-1}U_i(t)$. Let $y' \in y + t^{-1}B$. Then $y' \in U_i$ since $U_i + t^{-1}B = U_i$. Also, $||y - y'|| \leq c^{-1}|t|^{-1}$, hence $d(a^{-1}(y), a^{-1}(y')) \leq |t|^{-1}$. By ultrametric triangle inequality this implies that $d(a^{-1}(y'), K_i \times x_0) \leq |t|^{-1}$, i.e., $y' \in t^{-1}U_i(t)$. □

Proposition 2.9. (i) *There exists a constant $A > 0$ such that for all $t \in (E^*)^2$ with $|t| > A$, all $x^\vee \in U^\vee$ and $y^\vee \in \mathcal{O}_{x^\vee}(E)$ one has*

$$I(\mathcal{F}(\delta_{U(t)}))(y^\vee) = \epsilon(x^\vee, y^\vee) \cdot I(\mathcal{F}(\delta_{U(t)}))(x^\vee).$$

(ii) *There exists a ball $D^\vee \subset \mathbf{W}^\vee(E)$ centered at zero, such that for every sufficiently small ball D as above, the restriction of the function $y \mapsto \delta_{\mathcal{O}_y}^{\kappa^{-1}\text{sign}}(\mathcal{F}(\delta^\kappa_{U(t)}))$ to $x_0^\vee + t^{-1}D^\vee$ is not identically zero provided that $t \in (E^*)^2$ is large enough.*

Proof. (i) For $x^\vee \in \mathbf{V}_0^\vee(E)$ we have

$$\mathcal{F}(\delta_{U(t)})(x^\vee) = \int_{U(t)} \psi(\langle x^\vee, x\rangle)dx = |t|^{\dim \mathbf{V}} \cdot \int_{t^{-1}U(t)} \psi(t\langle x^\vee, x\rangle)dx,$$

where dx is the Haar measure on $\mathbf{V}(E)$ corresponding to a top-degree form defined over E. By Lemma 1.5(b) we can rewrite this integral as follows:

$$\int_{t^{-1}U(t)} \psi(t\langle x^\vee, x\rangle)dx = \int_{s\in x_0+t^{-1}D} \int_{k\in K} \psi(t\langle x^\vee, ks\rangle)|\omega(k)| \cdot |\nu(s)|,$$

where ω is a $\mathbf{G}$-invariant top-degree form on $\mathbf{G}/\mathbf{H}$, ν is a top-degree form on $\mathbf{L}$. Now the inner integral has the form

$$\int_{x\in Ks} \psi(t\langle x^\vee, x\rangle)|\omega|,$$

so we can apply the stationary phase principle to compute it. More precisely, by Lemma 2.8 we know that $\mathcal{F}(\delta_{U(t)})(x^\vee) = 0$ for $x^\vee \notin C$, where C is a compact in $\mathbf{V}^\vee(E)$. Also we know that for $x^\vee \in p^{-1}(p(U^\vee))$ and $s \in x_0 + D$, the function $x^\vee|_{\mathcal{O}_s(E)}$ has a finite number of nondegenerate critical points. Finally, we observe that the subset $p^{-1}(p(U^\vee)) \subset \mathbf{V}^\vee(E)$ is closed, hence, $C \cap p^{-1}(p(U^\vee))$ is compact. Thus, applying Lemma 2.6 we derive that there exists a constant $A > 0$ such that for $s \in x_0 + D$, $x^\vee \in C \cap p^{-1}(p(U^\vee))$, $t \in (E^*)^2$, $|t| > A$ one has

$$\int_{x\in Ks} \psi(t\langle x^\vee, x\rangle)|\omega| = |t|^{-\frac{\dim \mathcal{O}}{2}} \cdot \sum_{x\in Ks\cap \mathrm{Cr}(x^\vee|_{\mathcal{O}_s})} \psi(t\langle x^\vee, x\rangle)c(B_{x,x^\vee}, \omega_x, \psi).$$

We claim that by enlarging C if necessary we can achieve that the RHS is zero for $x^\vee \in p^{-1}(p(U^\vee)) \setminus C$. Indeed, this follows immediately from the fact that $\mathbf{I}(E) \cap (K(x_0 + D) \times p^{-1}(p(U^\vee)))$ is compact (as a preimage of the compact set $K(x_0 + D) \times p(U^\vee)$ under the map $\mathbf{I} \to \mathbf{V}_0 \times \mathbf{V}_0^\vee/\mathbf{G}$). Thus, if $t \in (E^*)^2$ is large enough, then for all $x^\vee \in p^{-1}(p(U^\vee))$ we have

$$\mathcal{F}(\delta_{U(t)})(x^\vee) = |t|^{\dim \mathbf{V} - \frac{\dim \mathcal{O}}{2}} \cdot \int_{s \in x_0 + t^{-1}D} \times \sum_{x \in Ks \cap \mathrm{Cr}(x^\vee|_{\mathcal{O}_s})} \psi(t\langle x^\vee, x\rangle) c(B_{x,x^\vee}, \omega_x, \psi)|\nu(s)|. \tag{2.6.2}$$

Now let us substitute $x^\vee$ by another point $gx^\vee \in \mathcal{O}_{x^\vee}(E)$ and make the change of variables $x \mapsto gx$ (this makes sense since x has the same stabilizer as $x^\vee$). Then we obtain

$$\mathcal{F}(\delta_{U(t)})(gx^\vee) = |t|^{\dim \mathbf{V} - \frac{\dim \mathcal{O}}{2}} \cdot \int_{s \in x_0 + t^{-1}D} \times \sum_{gx \in Ks \cap \mathrm{Cr}(gx^\vee|_{\mathcal{O}_s})} \psi(t\langle gx^\vee, gx\rangle) c(B_{gx,gx^\vee}, \omega_{gx}, \psi)|\nu(s)|.$$

Using Lemma 2.3 we can rewrite the inner sum as follows:

$$\sum_{x \in g^{-1}Ks \cap \mathrm{Cr}(x^\vee|_{\mathcal{O}_s})} \psi(t\langle x^\vee, x\rangle) \epsilon_{\mathbf{H}}(e_g) c(B_{x,x^\vee}, \omega_x, \psi),$$

where $e_g \in H^1(E, \mathbf{H})$ is the cohomology class of the cocycle $\sigma \mapsto g^{-1}\sigma(g)$. Hence, we obtain

$$\mathcal{F}(\delta_{U(t)})(gx^\vee) = |t|^{\dim \mathbf{V} - \frac{\dim \mathcal{O}}{2}} \cdot \epsilon_{\mathbf{H}}(e_g) \cdot \int_{s \in x_0 + t^{-1}D} \times \sum_{x \in g^{-1}Ks \cap \mathrm{Cr}(x^\vee|_{\mathcal{O}_s})} \psi(t\langle x^\vee, x\rangle) c(B_{x,x^\vee}, \omega_x, \psi)|\nu(s)|.$$

Now to calculate $I(\mathcal{F}(\delta_{U(t)}))$ at $gx^\vee$, we have to replace g by $g_1 g$ in the above formula where $g_1 \in \mathbf{G}(E)/(g\mathbf{H}g^{-1})(E)$ and integrate over g_1. We get

$$\begin{aligned} I(\mathcal{F}(\delta_{U(t)}))(gx^\vee) &= |t|^{\dim \mathbf{V} - \frac{\dim \mathcal{O}}{2}} \cdot \epsilon_{\mathbf{H}}(e_g) \\ &\times \int_{s \in x_0 + t^{-1}D} \int_{g_1 \in \mathbf{G}(E)/(g\mathbf{H}g^{-1})(E)} |\omega(g_1)| \\ &\cdot \sum_{x \in g^{-1}g_1^{-1}Ks \cap \mathrm{Cr}(x^\vee|_{\mathcal{O}_s})} \psi(t\langle x^\vee, x\rangle) c(B_{x,x^\vee}, \omega_x, \psi)|\nu(s)| \\ &= |t|^{\dim \mathbf{V} - \frac{\dim \mathcal{O}}{2}} \cdot \epsilon_{\mathbf{H}}(e_g) \\ &\times \int_{s \in x_0 + t^{-1}D} \sum_{x \in \mathrm{Cr}(x^\vee|_{\mathcal{O}_s})} \\ &\times \mathrm{vol}(\mathbf{G}(E)gx \cap Ks) \psi(t\langle x^\vee, x\rangle) c(B_{x,x^\vee}, \omega_x, \psi)|\nu(s)|, \end{aligned}$$

where volumes are computed using ω. By our choice of K we have $\mathrm{vol}(\mathbf{G}(E)gx \cap Ks) = \mathrm{vol}(\mathbf{G}(E)x_0 \cap K)$. Hence, we conclude that

$$I(\mathcal{F}(\delta_{U(t)}))(gx^\vee) = |t|^{\dim \mathbf{V} - \frac{\dim \mathcal{O}}{2}} \cdot \mathrm{vol}(\mathbf{G}(E)x_0 \cap K) \cdot \epsilon_{\mathbf{H}}(e_g) \qquad (2.6.3)$$
$$\times \int_{s \in x_0 + t^{-1}D} \sum_{x \in \mathrm{Cr}(x^\vee|_{\mathcal{O}_s})} \psi(t\langle x^\vee, x\rangle) c(B_{x,x^\vee}, \omega_x, \psi)|\nu(s)|.$$

Since $\epsilon_{\mathbf{H}}(e_g) = \epsilon(x^\vee, gx^\vee)$ this finishes the proof of the part (i) of the proposition.

(ii) Arguing as above, we obtain that

$$I(\mathcal{F}(\delta^\kappa_{U(t)}))(gx^\vee) = |t|^{\dim \mathbf{V} - \frac{\dim \mathcal{O}}{2}} \cdot \mathrm{vol}(\mathbf{G}(E)x_0 \cap K) \cdot \epsilon_{\mathbf{H}}(e_g)$$
$$\times \int_{s \in x_0 + t^{-1}D} \sum_{x \in \mathrm{Cr}(x^\vee|_{\mathcal{O}_s})} \kappa(\mathrm{inv}(s, gx))$$
$$\cdot \psi(t\langle x^\vee, x\rangle) c(B_{x,x^\vee}, \omega_x, \psi)|\nu(s)| \qquad (2.6.4)$$

for sufficiently large $t \in (E^*)^2$, where $x^\vee, gx^\vee \in p^{-1}(p(U^\vee))$.

By our choice of D and $U^\vee$ there exists a collection of analytic maps

$$x_i : (x_0 + D) \times U^\vee \to \mathbf{V}(E), i = 1, \ldots, n,$$

such that for every $s \in x_0 + D$ the points $x_1(s, x^\vee), \ldots, x_n(s, x^\vee)$ are disjoint and constitute the set of critical points of $x^\vee|_{\mathcal{O}_s}$. Let us set $x_i = x_i(x_0, x_0^\vee)$. Renumbering these maps if necessary we can assume that $x_1 = x_0$. We claim that the differentials at $x_0^\vee$ of the functions $x^\vee \mapsto \langle x^\vee, x_i(x_0, x^\vee)\rangle$, where $x^\vee \in U^\vee \cap \mathbf{W}^\vee(E), i = 1, \ldots, n$, are linearly independent. Note that by Lemma 1.7(a) for $x^\vee \in U^\vee \cap \mathbf{W}^\vee(E)$ we have $x_i(x_0, x^\vee) \in \mathbf{W}(E)$. Let $L_i \in \mathrm{Hom}(\mathbf{W}^\vee, \mathbf{W})$ be the differential of $x_i(x_0, x^\vee)|_{U^\vee \cap \mathbf{W}^\vee(E)}$ at $x_0^\vee$. Differentiating the condition $x_i(x_0, x^\vee) \in \mathcal{O}_{x_0}$, we get that $L_i(\mathbf{W}^\vee) \subset \mathfrak{g}x_0$. Now the differential at $x_0^\vee$ of the function $\langle x^\vee, x_i(x_0, x^\vee)\rangle$ on $U^\vee \cap \mathbf{W}^\vee$ is the functional

$$w^\vee \mapsto \langle w^\vee, x_i(x_0, x_0^\vee)\rangle + \langle x_0^\vee, L_i(w^\vee)\rangle = \langle w^\vee, x_i(x_0, x_0^\vee)\rangle$$

(here we used the equality $\langle x_0^\vee, \mathfrak{g}x_0\rangle = 0$). Since all the points $x_i(x_0, x_0^\vee) \in \mathbf{W}(E)$ are distinct, this proves our claim. By Lemma 2.7 this implies that there exists a ball $D^\vee \subset \mathbf{W}^\vee(E)$ centered at zero, such that for sufficiently large t, the functions

$$x^\vee \mapsto \psi(t\langle x^\vee, x_i(x_0, x^\vee)\rangle), \ i = 1, \ldots, n, \qquad (2.6.5)$$

are linearly independent on $x_0^\vee + t^{-1}D^\vee$. Since the functions x_i are analytic, we can choose D sufficiently small so that

$$\psi(t\langle x^\vee, x_i(s, x^\vee)\rangle) = \psi(t\langle x^\vee, x_i(x_0, x^\vee)\rangle),$$

for $s \in x_0 + t^{-1}D, x^\vee \in U^\vee, i = 1, \ldots, n$. On the other hand, if t is large enough, then setting $c_i(s, x^\vee) = c(B_{x_i(s,x^\vee),x^\vee}, \omega_{x_i(s,x^\vee)}, \psi)$ for $i = 1, \ldots, n$, we get

$$c_i(s, x^\vee) = c_i(x_0, x_0^\vee),$$
$$\kappa(\mathrm{inv}(s, gx_i(s, x^\vee))) = \kappa(\mathrm{inv}(x_0, gx_i))$$

for $s \in x_0 + t^{-1}D, x^\vee \in x_0^\vee + t^{-1}D^\vee, i = 1, \dots, n$. Finally, we have

$$\kappa(\mathrm{inv}(x_0, gx_i)) = \kappa(\mathrm{inv}(x_0, x_i)) \cdot \kappa(\tau_i(\mathrm{inv}(e_g))),$$

where $\tau_i : A(\mathbf{H}/E) \to A(\mathbf{H}/E)$ is the automorphism induced by the action of an element $g_i \in N(\mathbf{H})/\mathbf{H}$ such that $g_i x_i = x_0$. Note that since by our assumption $x_1 = x_0$, we have $\tau_1 = \mathrm{id}$. Thus, applying the formula (2.6.4) to $x^\vee \in x_0^\vee + t^{-1}D^\vee$ and large enough $t \in (E^*)^2$, and using Lemma 2.2, we get

$$\begin{aligned} I(\mathcal{F}(\delta_{U(t)}))(gx^\vee) &= |t|^{\dim \mathbf{V} - \frac{\dim \mathcal{O}}{2}} \cdot \mathrm{vol}(\mathbf{G}(E)x_0 \cap K) \\ &\times \int_{s \in x_0 + t^{-1}D} \sum_{i=1}^{n} \kappa(\mathrm{inv}(x_0, x_i))((\kappa \circ \tau_i)\mathrm{sign})(\mathrm{inv}(e_g)) \\ &\cdot \psi(t\langle x^\vee, x_i(s, x^\vee)\rangle) c_i(x_0, x_0^\vee)|\nu(s)| \\ &= |t|^{\dim \mathbf{V} - \frac{\dim \mathcal{O}}{2}} \cdot \mathrm{vol}(\mathbf{G}(E)x_0 \cap K)\mathrm{vol}(t^{-1}D) \\ &\times \sum_{i=1}^{n} \kappa(\mathrm{inv}(x_0, x_i))((\kappa \circ \tau_i)\mathrm{sign})(\mathrm{inv}(e_g)) c_i(x_0, x_0^\vee)\psi(t\langle x^\vee, x_i(x_0, x^\vee)\rangle). \end{aligned}$$

Hence, we have

$$\delta_{\mathcal{O}_{x^\vee}}^{\kappa^{-1}\mathrm{sign}}(\mathcal{F}(\delta_{U(t)})) = c(t) \cdot \sum_{gx^\vee \in \mathbf{G}(E)\backslash \mathcal{O}_{x^\vee}(E)} \sum_{i=1}^{n} \kappa(\mathrm{inv}(x_0, x_i)) \frac{\kappa \circ \tau_i}{\kappa}(\mathrm{inv}(e_g)) c_i(x_0, x_0^\vee)\psi(t\langle x^\vee, x_i(x_0, x^\vee)\rangle),$$

where $c(t) \in \mathbb{C}^*$ is a constant depending on t. Interchanging two summations and noting that $\mathrm{inv}(e_g)$ runs through the entire group $A = A(\mathcal{O}_{x^\vee}) = A(\mathcal{O}_{x_0^\vee})$, we obtain

$$\delta_{\mathcal{O}_{x^\vee}}^{\kappa^{-1}\mathrm{sign}}(\mathcal{F}(\delta_{U(t)})) = c(t)|A| \cdot \sum_{i:\kappa \circ \tau_i = \kappa} \kappa(\mathrm{inv}(x_0, x_i)) c_i(x_0, x_0^\vee)\psi(t\langle x^\vee, x_i(x_0, x^\vee)\rangle) \tag{2.6.6}$$

for $x^\vee \in x_0^\vee + t^{-1}D^\vee$ and large enough $t \in (E^*)^2$. Now our assertion follows from linear independence of the functions (2.6.5) on $x_0^\vee + t^{-1}D^\vee$ and the fact that $\tau_1 = \mathrm{id}$. □

2.7 Inner forms

Let $\mathbf{Z} \subset \mathbf{G}$ be a central subgroup defined over E, $\alpha \in H^1(E, \mathbf{G}/\mathbf{Z})$ be a cohomology class. Then α defines an inner form $\mathbf{G}'$ of $\mathbf{G}$. Note that since $\mathbf{G}$ is simply connected, the homomorphism

$$d : H^1(E, \mathbf{G}/\mathbf{Z}) \to H^2(E, \mathbf{Z})$$

is an isomorphism when E is p-adic, so in this case α is uniquely determined by $d(\alpha) \in H^2(E, \mathbf{Z})$. Assume that $\rho(\mathbf{Z}) = 1$. Then we can twist ρ by α to get a nice representation $\rho' : \mathbf{G}' \to \mathrm{Aut}(\mathbf{V}')$. Let $\mathbf{H} \subset \mathbf{G}$ (resp. $\mathbf{H}' \subset \mathbf{G}'$) be a generic stabilizer subgroup, and let $\mathbf{V}_0 \subset \mathbf{V}$ (resp. $\mathbf{V}'_0$) be the open subset consisting of points with the stabilizer conjugated to $\mathbf{H}$ (resp. $\mathbf{H}'$) over $\overline{E}$. The isomorphism $i : \mathbf{V} \to \mathbf{V}'$ defined over $\overline{E}$ induces a bijection between $\mathbf{G}$-orbits on $\mathbf{V}$ defined over E and $\mathbf{G}'$-orbits on $\mathbf{V}'$ defined over E. Let $\mathcal{O} \subset \mathbf{V}_0$ be a $\mathbf{G}$-orbit, and let $\mathcal{O}' = i(\mathcal{O}) \subset \mathbf{V}'_0$ be the corresponding $\mathbf{G}'$-orbit. If ω is a $\mathbf{G}$-invariant nonzero top degree form on $\mathcal{O}$ defined over E, then $i_*\omega$ is a $\mathbf{G}'$-invariant top degree form on $\mathcal{O}'$, also defined over E.

Definition. For a pair of functions $\phi \in \mathcal{S}(\mathbf{V}(E))$, $\phi' \in \mathcal{S}(\mathbf{V}'(E))$ we say that $\phi \sim \phi'$ (resp. $\phi \sim_\epsilon \phi'$) if for every $\mathbf{G}$-orbit $\mathcal{O} \subset \mathbf{V}_0$ defined over E one has $\delta_{\mathcal{O},\omega}(\phi) = \delta_{\mathcal{O}',i_*\omega}(\phi')$ (resp. $\delta^\epsilon_{\mathcal{O},\omega}(\phi) = \delta^\epsilon_{\mathcal{O}',i_*\omega}(\phi')$). For a pair of distributions $\delta \in \mathcal{D}(\mathbf{V}(E))$, $\delta' \in \mathcal{S}(\mathbf{V}'(E))$ we say that $\delta \sim \delta'$ (resp. $\delta \sim_\epsilon \delta'$) if for every pair of functions $\phi \in \mathcal{S}(\mathbf{V}(E))$, $\phi' \in \mathcal{S}(\mathbf{V}'(E))$ such that $\phi \sim \phi'$ (resp. $\phi \sim_\epsilon \phi'$) one has $\delta(\phi) = \delta(\phi')$.

Note that we have $\phi \sim 0$ (resp. $\phi \sim_\epsilon 0$) if and only if ϕ is antistable (resp. ϵ-antistable). Therefore, if $\delta \sim \delta'$ (resp. $\delta \sim_\epsilon \delta'$), then distributions δ and δ' are necessarily stable (resp. ϵ-stable). Also, by the definition we have $\delta_{\mathcal{O},\omega} \sim \delta_{\mathcal{O}',i_*\omega}$ (resp. $\delta^\epsilon_{\mathcal{O},\omega} \sim_\epsilon \delta^\epsilon_{\mathcal{O}',i_*\omega}$).

To formulate our result on inner forms we have to introduce a certain sign associated with $\mathbf{G}$ and $\mathbf{G}'$. Let Q (resp. Q') be the Killing form on $\mathfrak{g}$ (resp. $\mathfrak{g}'$). We set

$$\kappa(\mathbf{G}, \mathbf{G}') = \epsilon(Q, Q').$$

Note that Q and Q' have the same determinant modulo squares. Thus, in the case of p-adic E the difference between equivalence classes of these quadratic forms is measured by the sign $\kappa(\mathbf{G}, \mathbf{G}')$.

Theorem 2.10. *Let $\rho : \mathbf{G} \to \mathrm{GL}(\mathbf{V})$ be a nice representation of a simply connected semisimple group $\mathbf{G}$ over a p-adic field E, and let $\mathbf{Z} \subset \mathbf{G}$ be a central subgroup acting trivially on $\mathbf{V}$. Let $(\mathbf{G}', \mathbf{V}')$ be a twist of $(\mathbf{G}, \mathbf{V})$ by a class $\alpha \in H^1(E, \mathbf{G}/\mathbf{Z})$. Assume that either $\mathbf{G}$ is simple or the generic stabilizer is semisimple. Then for every pair of functions $\phi \in \mathcal{S}(\mathbf{V}(E))$, $\phi' \in \mathcal{S}(\mathbf{V}(E))$ one has $\phi \sim \phi'$ if and only if $\mathcal{F}(\phi) \sim_\epsilon \kappa(\mathbf{G}, \mathbf{G}')\mathcal{F}(\phi')$. The same result holds for distributions.*

The proof will be given in 3.3. The remainder of this section consists of various local ingredients of the proof.

Theorem 2.4 is an immediate consequence of Theorem 2.10. Indeed, we can take $(\mathbf{G}', \mathbf{V}') = (\mathbf{G}, \mathbf{V})$ and $\phi' = 0$. Then the condition $\phi \sim 0$ means that ϕ is antistable, while the condition $\mathcal{F}(\phi) \sim_\epsilon 0$ means that $\mathcal{F}(\phi)$ is ϵ-antistable.

Lemma 2.11. (a) *Let $\mathcal{O} \subset \mathbf{V}_0$ be an orbit, and let $x \in \mathcal{O}(E)$ be an E-point on $\mathcal{O}$. Then $\mathcal{O}'(E)$ is nonempty if and only if the class α belongs to the image of the map $H^1(E, \mathbf{H}_x/\mathbf{Z}) \to H^1(E, \mathbf{G}/\mathbf{Z})$.*

(b) *For every $\phi \in \mathcal{S}(\mathbf{V}'(E))$ the functions $x \mapsto \delta_{\mathcal{O}'_x,i_*\omega}(\phi)$ and $x \mapsto \delta^\epsilon_{\mathcal{O}'_x,i_*\omega}(\phi)$ on $\mathbf{V}_0(E)$ are locally constant.*

Proof. The proof of (a) is straightforward. To prove (b) let us fix a point $x \in \mathbf{V}_0(E)$. Set $\mathbf{H} = \mathbf{H}_x$, $\mathbf{W} = \mathbf{V}^{\mathbf{H}}$, $\mathbf{W}_0 = \mathbf{W} \cap \mathbf{V}_0$. The morphism $a : \mathbf{G}/\mathbf{H} \times \mathbf{W}_0 \to \mathbf{V}_0$ is smooth, hence $U = a(\mathbf{G}/\mathbf{H}(E) \times \mathbf{W}_0(E))$ is an open subset in $\mathbf{V}_0(E)$. Assume first that α does not belong to the image of the map $H^1(E, \mathbf{H}/\mathbf{Z}) \to H^1(E, \mathbf{G}/\mathbf{Z})$. Then for every point $y \in U$ we have $\mathcal{O}'_y(E) = \emptyset$. Thus, we can assume that α belongs to the image of this map. Then there exists an isomorphism $i : (\mathbf{G}, \mathbf{V}) \to (\mathbf{G}', \mathbf{V}')$ over $\overline{E}$, such that the subgroup $i(\mathbf{H}) \subset \mathbf{G}'$ and the morphism $i|_{\mathbf{W}}$ are defined over E. Therefore, the functions $y \mapsto \delta_{\mathcal{O}'_y, i_*\omega}(\phi)$ and $y \mapsto \delta^{\epsilon}_{\mathcal{O}'_y, i_*\omega}(\phi)$ on $\mathbf{W}_0(E)$ are locally constant. Let $U_0 \subset \mathbf{W}_0(E)$ be a neighborhood of x on which they are constant. Then $a(\mathbf{G}/\mathbf{H}(E) \times U_0)$ is an open neighborhood of x in $\mathbf{V}_0(E)$ on which a similar function is constant. □

Lemma 2.12. *Assume that E is p-adic, $\mathbf{G}$ is simple, the generic stabilizer for ρ is commutative and all irreducible components of $\rho_{\overline{E}}$ are defined over E. Then there exists a point $x \in \mathbf{V}_0(E)$ such that $\mathbf{H}_x$ is an anisotropic.*

Proof. In the case when ρ is the adjoint representation, we can use the existence of an anisotropic maximal torus $\mathbf{H}_x$ in $\mathbf{G}$ defined over E (see e.g., Theorem VI.21 of [15]). It turns out that in all other cases the action of a sufficiently big subgroup of $\mathbf{G}$ on $\mathbf{V}$ essentially reduces to the adjoint representation. Here is a more precise statement.

Claim. For every $x \in \mathbf{V}_0(E)$ there exists a semisimple subgroup $\mathbf{G}' \subset \mathbf{G}$ defined over E and a $\mathbf{G}'$-invariant decomposition $\mathbf{V} = \mathbf{V}' \oplus \mathbf{V}''$, such that the following two conditions hold:

(i) the representation of $\mathbf{G}'$ on $\mathbf{V}''$ is equivalent to the adjoint representation of $\mathbf{G}'$;

(ii) let $x = x' + x''$ where $x' \in \mathbf{V}'$, $x'' \in \mathbf{V}''$, then $\mathbf{G}'x' = x'$ and $\mathbf{H}_x$ coincides with the stabilizer of x'' in $\mathbf{G}'$ (so by (i), $\mathbf{H}_x$ is a maximal torus in $\mathbf{G}'$).

Our statement can be deduced from this claim as follows. The subset $\mathbf{V}_0 \cap (x' + \mathbf{V}'') \subset x' + \mathbf{V}''$ is nonempty and Zariski open. Therefore, we can choose $\widetilde{x}'' \in \mathbf{V}''(E)$ such that $\widetilde{x} = x' + \widetilde{x}'' \in \mathbf{V}_0(E)$ and the stabilizer of $\widetilde{x}''$ in $\mathbf{G}'$ is anisotropic. Then $\mathbf{H}_{\widetilde{x}}$ contains an anisotropic maximal torus in $\mathbf{G}'$. Since $\widetilde{x}$ is contained in $\mathbf{V}_0$, this inclusion is in fact an equality. The proof of the Claim follows from Elashvili's classification of representations of simple groups with generic stabilizers of positive dimension (see [2]). Here are the cases relevant for our situation:

(i) $\mathbf{G}_{\overline{E}} = \mathrm{SL}(W)$, $\mathbf{V}_{\overline{E}} = \mathbf{V}' \oplus \mathbf{V}''$, where $\mathbf{V}'$ and $\mathbf{V}''$ are irreducible representations of $\mathrm{SL}(W)$ in S^2W and $\bigwedge^2 W$ or in $S^2W^{\vee}$ and $\bigwedge^2 W$ (or dual to these). The stabiliser $\mathbf{G}'$ of a generic point in $\mathbf{V}'$ is the special orthogonal group $\mathrm{SO}(W)$. It is well known that the representation of $\mathrm{SO}(W)$ in $\bigwedge^2 W$ is isomorphic to the adjoint representation.

(ii) $\mathbf{G}_{\overline{E}} = \mathrm{SL}(W)$, where $\dim W = 4$, $\mathbf{V}_{\overline{E}}$ is the direct sum of 4 copies of $\bigwedge^2 W$. The stabilizer in $\mathrm{SL}(W)$ of a generic point in $(\bigwedge^2 W)^2$ is conjugate to the subgroup $\mathrm{SL}_2 \times \mathrm{SL}_2$ corresponding to a decomposition $W = W_1 \oplus W_2$, where

$\dim W_1 = \dim W_2 = 2$ (see Table 1 of [2]). The generic stabilizer in $(\bigwedge^2 W)^3$ is the subgroup $SL_2 = \{(g, g^{-1}), g \in SL_2\}$ in $SL_2 \times SL_2$ (which corresponds to choosing an isomorphism $W_1 \simeq W_2$). Decomposing the 4-th factor as a representation of $SL(W_1) \times SL(W_2)$:

$$\bigwedge^2 W = 1 \oplus 1 \oplus W_1 \otimes W_2,$$

we see that the action of the above subgroup on $\bigwedge^2 W$ is equivalent to the direct sum of the adjoint representation of SL_2 with 3 trivial representations. Thus, the above claim holds if we take $\mathbf{G}'$ to be the stabilizer of the first three components of x, $\mathbf{V}'$ to be the sum of $(\bigwedge^2 W)^3$ and of 3 trivial representations.

(iii) $\mathbf{G}_{\overline{E}} = SL(W)$, where $\dim W = 6$, $\mathbf{V}_{\overline{E}}$ is the direct sum of 2 copies of $\bigwedge^3 W$. The stabilizer in $SL(W)$ of a generic point in $\bigwedge^3 W$ is conjugate to the subgroup $SL_3 \times SL_3$ corresponding to a decomposition $W = W_1 \oplus W_2$, where $\dim W_1 = \dim W_2 = 3$ (see Table 1 of [2]). We can decompose the second copy of $\bigwedge^3 W$ as a representation of $SL(W_1) \times SL(W_2)$:

$$\bigwedge^3 W = 1 \oplus 1 \oplus W_1^\vee \otimes W_2 \oplus W_2^\vee \otimes W_1.$$

The last two factors are nonisomorphic nontrivial irreducible representations of $SL(W_1) \times SL(W_2)$, so for any E-form of the pair $(SL(W_1) \times SL(W_2), \bigwedge^3 W)$ the similar decomposition takes place. Now the stabilizer in $SL(W_1) \times SL(W_2)$ of a generic point in one of the nontrivial factors is isomorphic to SL_3, and the action of this subgroup on the second nontrivial factor is the sum of the adjoint representation and of the trivial representation. Using this we can easily construct $\mathbf{V}'$ and $\mathbf{V}''$ (with $\mathbf{G}'$ being the form of SL_3).

(iv) $\mathbf{G}_{\overline{E}} = SL(W)$, where $\dim W = 8$, $\mathbf{V}_{\overline{E}}$ is either $\bigwedge^3 W \oplus W$ or $\bigwedge^3 W \oplus W^\vee$. The stabilizer in $SL(W)$ of a generic point in $\bigwedge^3 W$ is isomorphic to SL_3. The embedding of SL_3 in $SL(W)$ corresponds to the identification of W with the adjoint representation of SL_3 (see Table 1 of [2]). Therefore, we can take $\mathbf{G}'$ to be the stabilizer of the first component of x with respect to the above decomposition of $\mathbf{V}$. □

Remark. Note that if $\mathbf{G}$ is split over E, then all irreducible components of $\rho_{\overline{E}}$ are defined over E.

Lemma 2.13. *Assume that E is p-adic and that either* $\mathbf{H}$ *is semisimple, or* $\mathbf{G}$ *is simple and all irreducible components of $\rho_{\overline{E}}$ are defined over E. Then there exists a point $x \in \mathbf{V}_0(E)$, such that the map $H^1(E, \mathbf{H}_x/\mathbf{Z}) \to H^1(E, \mathbf{G}/\mathbf{Z})$ is surjective.*

Proof. Note that the natural map $H^1(E, \mathbf{G}/\mathbf{Z}) \to H^2(E, \mathbf{Z})$ is an isomorphism (since E is p-adic). Therefore, we have to prove that the natural map

$$H^1(E, \mathbf{H}_x/\mathbf{Z}) \to H^2(E, \mathbf{Z})$$

is surjective for some $x \in \mathbf{V}_0(E)$.

Assume first that $\mathbf{H}$ is semisimple. Then we claim that this surjectivity holds for every point $x \in \mathbf{V}_0(E)$. Indeed, let $\mathbf{H}_{sc} \to \mathbf{H}$ be the universal covering of $\mathbf{H}$, $\mathbf{D} \subset \mathbf{H}_{sc}$ be the preimage of $\mathbf{Z} \subset \mathbf{H}$. Then we have an isomorphism

$$H^1(E, \mathbf{H}/\mathbf{Z}) = H^1(E, \mathbf{H}_{sc}/\mathbf{D}) \widetilde{\to} H^2(E, \mathbf{D}).$$

Since the cohomological dimension of E is equal to 2, the map $H^2(E, \mathbf{D}) \to H^2(E, \mathbf{Z})$ is surjective, which finishes the proof in this case (the same argument can be applied to any $\mathbf{H}_x$).

Now let us assume that $\mathbf{H}$ is commutative. Then by Lemma 2.12 there exists a point $x \in \mathbf{V}_0(E)$ such that $\mathbf{H}_x$ is an anisotropic torus. By Tate–Nakayama duality, for such a torus we have $H^2(E, \mathbf{H}_x) = 0$, which implies the desired surjectivity. □

In the remainder of this section we will keep the assumptions of Lemma 2.13. Furthermore, we choose a point $x_0 \in \mathbf{V}_0(E)$ as in this lemma and set $\mathbf{H} = \mathbf{H}_{x_0}$, $\mathbf{W} = \mathbf{V}^{\mathbf{H}}$, $\mathbf{W}_0 = \mathbf{W} \cap \mathbf{V}_0$, etc. We also fix a cohomology class $\alpha_0 \in H^1(E, \mathbf{H}/\mathbf{Z})$ mapping to α. This allows us to choose an isomorphism $i : (\mathbf{G}, \mathbf{V}) \to (\mathbf{G}', \mathbf{V}')$ over $\overline{E}$, such that the subgroup $\mathbf{H}' = i(\mathbf{H}) \subset \mathbf{G}'$ and the morphism $i|_{\mathbf{W}}$ are defined over E. We denote $\mathbf{W}' = (\mathbf{V}')^{\mathbf{H}'} = i(\mathbf{W})$. Also we denote by $i^\vee : \mathbf{V}^\vee \to (\mathbf{V}')^\vee$ the $\overline{E}$-isomoprhism induced by i. Note that $i^\vee|_{\mathbf{W}^\vee}$ is also defined over E.

The following lemma (generalizing Lemma 2.3) computes the sign that in the p-adic case measures the difference between the quadratic form $B_{x,x^\vee}$ introduced in 1.3 and $B_{i(x),i^\vee(x^\vee)}$. Recall that for $x \in \mathcal{O}$, $x' \in \mathcal{O}'$ we denote $\epsilon(x, x') = \epsilon(Q_x, Q'_{x'})$.

Lemma 2.14. *Let $x \in \mathbf{W}_0$ be a critical point of $x^\vee|_{\mathcal{O}_x}$, where $x^\vee \in \mathbf{W}_0^\vee$. Let ω be a nonzero $\mathbf{G}$-invariant top-degree form on $\mathcal{O}_x$ defined over E.*

(a) *One has* $\det(B_{i(x),i^\vee(x^\vee)})/(i_*\omega)^2_{i(x)} = \det(B_{x,x^\vee})/\omega_x^2$.

(b) *The quadratic forms $B_{i(x),i^\vee(x^\vee)}$ and $B_{x,x^\vee}$ have the same determinant modulo squares. Their relative Hasse–Witt invariant is given by*

$$\epsilon(B_{i(x),i^\vee(x^\vee)}, B_{x,x^\vee}) = \kappa(\mathbf{G}, \mathbf{G}')\epsilon(x, i(x)) = \kappa(\mathbf{G}, \mathbf{G}')\epsilon(x^\vee, i(x^\vee)).$$

Proof. The proof of (a) is straightforward. To prove (b) we note that there is a natural homomorphism $\iota : \mathbf{H}/\mathbf{Z} \to \mathrm{SO}(T_x, B_{x,x^\vee})$ which induces a map

$$\iota_* : H^1(E, \mathbf{H}/\mathbf{Z}) \to H^1(E, \mathrm{SO}(T_x, B_{x,x^\vee})).$$

It is easy to see that the quadratic form $B_{i(x),i^\vee(x^\vee)}$ is equivalent to the twist of $B_{x,x^\vee}$ by $\iota_*(\alpha_0)$. In particular, these forms have the same determinant modulo squares. Let

$$\delta : H^1(E, \mathrm{SO}(T_x, B_{x,x^\vee})) \to H^2(E, \{\pm 1\}) \simeq \{\pm 1\}$$

be the map induced by the spin-covering of $\mathrm{SO}(T_x, B_{x,x^\vee})$. Then the Hasse–Witt invariants of $B_{v,v^\vee}$ and of its twist by $\iota_*(\alpha_0)$ differ by $\delta(\iota_*(\alpha_0))$. It remains to prove that

$$\delta(\iota_*(\alpha_0)) = \kappa(\mathbf{G}, \mathbf{G}')\epsilon(x, i(x)). \quad (2.7.1)$$

Recall that by Proposition 1.9 the homomorphism ι lifts to a homomorphism

$$\widetilde{\mathbf{H}} \to \mathrm{Spin}(T_x, B_{x,x^\vee}),$$

where $\widetilde{\mathbf{H}} \to \mathbf{H}$ is the pullback of the spin-covering of $\mathrm{SO}(\mathfrak{h}, Q|_{\mathfrak{h}})$. Let $\widetilde{\mathbf{Z}} \subset \widetilde{\mathbf{H}}$ be the preimage of $\mathbf{Z} \subset \mathbf{H}$. Then we have an induced homomorphism

$$\chi : \widetilde{\mathbf{Z}} \to \{\pm 1\}.$$

On the other hand, we have a natural map

$$\widetilde{d} : H^1(E, \mathbf{H}/\mathbf{Z}) = H^1(E, \widetilde{\mathbf{H}}/\widetilde{\mathbf{Z}}) \to H^2(E, \widetilde{\mathbf{Z}}).$$

Now it is easy to see that

$$\delta \circ \iota_* = \chi_* \circ \widetilde{d},$$

where $\chi_* : H^2(E, \widetilde{\mathbf{Z}}) \to H^2(E, \{\pm 1\})$ is the homomorphism induced by χ. Therefore, we have

$$\delta(\iota_*(\alpha_0)) = \chi_*(\widetilde{d}(\alpha_0)). \tag{2.7.2}$$

On the other hand, we have the natural commutative diagram

$$\begin{array}{ccc} \mathbf{G} & \longrightarrow & \mathrm{Spin}(\mathfrak{g}, Q) \\ \big\downarrow & & \big\downarrow \\ \mathbf{G}/\mathbf{Z} & \longrightarrow & \mathrm{SO}(\mathfrak{g}, Q) \end{array} \tag{2.7.3}$$

so we get a homomorphism $\chi'_{\mathbf{G}} : \mathbf{Z} \to \{\pm 1\}$ such that

$$\kappa(\mathbf{G}, \mathbf{G}') = \chi'_{\mathbf{G},*}(d(\alpha)). \tag{2.7.4}$$

Composing it with the natural projection $\widetilde{\mathbf{Z}} \to \mathbf{Z}$ we get a homomorphism $\chi_{\mathbf{G}} : \widetilde{\mathbf{Z}} \to \{\pm 1\}$ such that

$$\kappa(\mathbf{G}, \mathbf{G}') = \chi_{\mathbf{G},*}(\widetilde{d}(\alpha_0)). \tag{2.7.5}$$

Also, by the definition of $\widetilde{\mathbf{H}}$ we have the following commutative diagram:

$$\begin{array}{ccc} \widetilde{\mathbf{H}} & \longrightarrow & \mathrm{Spin}(\mathfrak{h}, Q|_{\mathfrak{h}}) \\ \big\downarrow & & \big\downarrow \\ \widetilde{\mathbf{H}}/\widetilde{\mathbf{Z}} & \longrightarrow & \mathrm{SO}(\mathfrak{h}, Q|_{\mathfrak{h}}) \end{array} \tag{2.7.6}$$

which gives the homomorphism $\chi_{\mathbf{H}} : \widetilde{\mathbf{Z}} \to \{\pm 1\}$ such that

$$\epsilon(x, i(x)) = \chi_{\mathbf{H},*}(\widetilde{d}(\alpha_0)). \tag{2.7.7}$$

It remains to prove the equality

$$\chi \cdot \chi_{\mathbf{G}} \cdot \chi_{\mathbf{H}} = 1$$

of homomorphisms from $\widetilde{\mathbf{Z}}$ to $\{\pm 1\}$. Indeed, then the equation (2.7.1) would follow from (2.7.2), (2.7.5) and (2.7.7). To prove the equality of algebraic homomorphisms we can pass to the algebraic closure of E. Hence, it suffices to do this over $\mathbb{C}$. Then we can proceed similarly to the proof of Proposition 1.9. Namely, as in that proof we deduce that χ coincides with the homomorphism induced by the homomorphism $\widetilde{\mathbf{H}} \to \mathrm{Spin}(\mathfrak{h}^{\perp}, Q|_{\mathfrak{h}^{\perp}})$ (the existence of such a homomorphism also follows from the proof of Proposition 1.9). Now our statement follows from the commutativity of the diagram

$$\begin{array}{ccc} \widetilde{\mathbf{H}} & \longrightarrow & \mathrm{Spin}(\mathfrak{h}, Q|_{\mathfrak{h}}) \times \mathrm{Spin}(\mathfrak{h}^{\perp}, Q|_{\mathfrak{h}^{\perp}}) \\ \downarrow & & \downarrow \\ \mathbf{G} & \longrightarrow & \mathrm{Spin}(\mathfrak{g}, Q) \end{array} \tag{2.7.8}$$

where the left vertical arrow factors through $\mathbf{H}$. □

Remark. It is well known that the spinor representation of $\mathbf{G}$ is a multiple of the irreducible representation V_ρ corresponding to the half-sum of positive roots ρ (see [8]). It follows that the character $\chi'_{\mathbf{G}}$ concides with the restriction to $\mathbf{Z}$ of ρ considered as a character of the maximal torus of $\mathbf{G}$. Together with the equation (2.7.4) this implies that the sign $\kappa(\mathbf{G}, \mathbf{G}')$ coincides with the sign introduced by Kottwitz in [9].

The following proposition shows how to construct pairs of functions $\phi \in \mathcal{S}(\mathbf{V}_0(E))$, $\phi' \in \mathcal{S}(\mathbf{V}'_0(E))$ with $\phi \sim \phi'$ and with some control over the Fourier transforms of ϕ and ϕ'.

Proposition 2.15. *Keep the assumptions of Lemma 2.13. Then there exist $\phi \in \mathcal{S}(\mathbf{V}_0(E))$ and $\phi' \in \mathcal{S}(\mathbf{V}'_0(E))$ such that $\phi \sim \phi'$, while*

$$\delta^{\epsilon}_{\mathcal{O}_{x^\vee},\omega}(\mathcal{F}(\phi)) = \kappa(\mathbf{G}, \mathbf{G}')\delta^{\epsilon}_{\mathcal{O}'_{x^\vee}, i_*\omega}(\mathcal{F}(\phi')) \neq 0 \tag{2.7.9}$$

for some $x^\vee \in \mathbf{V}_0^\vee(E)$.

Proof. We use the notation of Section 2.6. Let us set

$$\phi = \frac{1}{\mathrm{vol}_\omega(\mathbf{G}(E)x_0 \cap K)} \cdot \delta_{U(t)}$$

for $t \in (E^*)^2$ sufficiently large. Recall that $U(t) = K(tx_0 + D) \subset \mathbf{V}_0(E)$. Let us choose some compact open subset $K' \subset \mathcal{O}_{i(x_0)}(E)$ intersecting $\mathbf{G}'(E)$-orbits at subsets of equal volumes, and let $U'(t) = K'(ti(x_0)+i(D))$ be the corresponding compact open subset of $\mathbf{V}'(E)$. Now let us set

$$\phi' = \frac{1}{\mathrm{vol}_{i_*\omega}(\mathbf{G}'(E)i(x_0) \cap K')} \cdot \delta_{U'(t)}.$$

Then clearly we have $\phi \sim \phi'$. On the other hand, using the formula (2.6.3) we obtain that for $x^\vee \in U^\vee \cap \mathbf{W}_0^\vee(E)$ one has

$$\delta^{\epsilon}_{\mathcal{O}_{x^\vee}}(\mathcal{F}(\phi)) = |t|^{\dim \mathbf{V} - \frac{\dim \mathcal{O}}{2}} \cdot |H^1(E, \mathbf{H})| \cdot \epsilon(x^\vee) \times$$
$$\int_{s \in x_0 + t^{-1}D} \sum_{x \in \mathrm{Cr}(x^\vee|_{\mathcal{O}_s})} \psi(t\langle x^\vee, x\rangle) c(B_{x,x^\vee}, \omega_x, \psi)|\nu(s)|,$$

$$\delta^{\epsilon}_{\mathcal{O}_{i(x^\vee)}}(\mathcal{F}(\phi')) = |t|^{\dim \mathbf{V} - \frac{\dim \mathcal{O}}{2}} \cdot |H^1(E, \mathbf{H}')| \cdot \epsilon(i(x^\vee)) \times$$
$$\int_{s \in x_0 + t^{-1}D} \sum_{x' \in \mathrm{Cr}(i(x^\vee)|_{\mathcal{O}_{i(s)}})} \psi(t\langle i(x^\vee), x'\rangle) c(B_{x',i(x^\vee)}, \omega_{x'}, \psi)|\nu(s)|,$$

where $\mathbf{H}' = i(\mathbf{H})$. Since $\mathbf{H}'$ is an inner form of $\mathbf{H}$, we have $|H^1(E, \mathbf{H})| = |H^1(E, \mathbf{H}')|$. On the other hand, the isomorphism $i|_{\mathbf{W}}$ sends $\mathrm{Cr}(x^\vee|_{\mathcal{O}_s})$ to $\mathrm{Cr}(i(x^\vee)|_{\mathcal{O}_{i(s)}})$. Hence, applying Lemma 2.14 we get the equality (2.7.9). It remains to note that according to Proposition 2.9(ii), we will also have $\delta^{\epsilon}_{\mathcal{O}_{x^\vee},\omega}(\mathcal{F}(\phi)) \neq 0$ for some $x^\vee \in U^\vee \cap \mathbf{W}_0^\vee$. □

3 Global methods

In this section F is a number field, $\mathbb{A}$ is the corresponding ring of adeles. For every finite set of places S we denote by $\mathbb{A}^S$ (resp. $\mathbb{A}_S$) the restricted product (resp. the usual product) of F_v over places $v \notin S$ (resp. $v \in S$) and by $a \mapsto a^S$ (resp. $a \mapsto a_S$) the corresponding projection from $\mathbb{A}$. We fix an algebraic closure $\overline{F}$ of F and denote by Γ the Galois group of $\overline{F}$ over F. For every place v of F we denote by F_v the corresponding local field and by Γ_v the local Galois group at v. For a reductive group $\mathbf{H}$ over F we denote by $\ker^1(F, \mathbf{H})$ the preimage of the trivial element under the natural map

$$H^1(F, \mathbf{H}) \to \oplus_v H^1(F_v, \mathbf{H}).$$

The Hasse principle states that for $\mathbf{H}$ semisimple and simply connected, $\ker^1(F, \mathbf{H})$ is trivial (see e.g., [15]). More generally, for arbitrary connected reductive group $\mathbf{H}$, Kottwitz constructed a bijection

$$\ker^1(F, \mathbf{H}) \simeq \ker^1(F, Z(\hat{\mathbf{H}}))^D \tag{3.0.1}$$

where $Z(\hat{\mathbf{H}})$ is the center of the Langlands dual group (see [10], (4.2.2)). We denote by $\mathcal{S}(\mathbf{V}(\mathbb{A}))$ the space of Schwartz–Bruhat functions on $\mathbf{V}(\mathbb{A})$. Let θ be the distribution on $\mathcal{S}(\mathbf{V}(\mathbb{A}))$ defined by

$$\theta(\phi) = \sum_{x \in \mathbf{V}(F)} \phi(x),$$

$\theta^{\vee}$ be the similar distribution on $\mathcal{S}(\mathbf{V}^{\vee}(\mathbb{A}))$. We set

$$\Theta(\phi) = \int_{g \in \mathbf{G}(\mathbb{A})/\mathbf{G}(F)} \theta^{g}(\phi)|dg|,$$

where $\theta \mapsto \theta^{g}$ denotes the action of g on the distribution θ (provided that the integral converges). Note that if $\mathbf{G}$ is anisotropic over F, then $\mathbf{G}(\mathbb{A})/\mathbf{G}(F)$ is compact, so $\Theta(\phi)$ is always well defined in this case. By the Poisson summation formula we have $\mathcal{F}(\theta) = \theta^{\vee}$, where $\mathcal{F}$ is the Fourier transform. Hence, for a function $\phi \in \mathcal{S}(\mathbf{V}(\mathbb{A}))$ we have

$$\Theta(\mathcal{F}(\phi)) = \Theta(\phi)$$

provided that the integral defining $\Theta(\phi)$ converges. In this section we will apply this equality to compare information about orbital integrals of a function and of its Fourier transform. In this way we will obtain global proofs of Theorems 2.5 and 2.10.

3.1 Global Kottwitz invariant

Following Kottwitz we are going to rewrite the distribution Θ evaluated on sufficiently nice functions in stably invariant terms. The main ingredient required for this is the global Kottwitz invariant defined in [10], [11].

Let $\mathbf{H}$ be a connected reductive group $\mathbf{H}$ over F. For every place v of F there is a map

$$\mathrm{inv}_v : H^1(F_v, \mathbf{H}) \to A(\mathbf{H}/F_v) = \pi_0(Z(\hat{\mathbf{H}})^{\Gamma_v})^D$$

(see 2.2). Now if we set $A(\mathbf{H}/F) = \pi_0(Z(\hat{\mathbf{H}})^{\Gamma})^D$, then for every place v we have a natural homomorphism $r_v : A(\mathbf{H}/F_v) \to A(\mathbf{H}/F)$ induced by the embedding $Z(\hat{\mathbf{H}})^{\Gamma} \to Z(\hat{\mathbf{H}})^{\Gamma_v}$. Thus, we can define a canonical map

$$\mathrm{inv} : \oplus_v H^1(F_v, \mathbf{H}) \to A(\mathbf{H}/F)$$

by setting $\mathrm{inv}((c_v)) = \prod_v r_v \mathrm{inv}_v(c_v)$.

The main result about the map inv is the exactness of the sequence

$$H^1(F, \mathbf{H}) \to \oplus_v H^1(F_v, \mathbf{H}) \xrightarrow{\mathrm{inv}} A(\mathbf{H}/F).$$

Lemma 3.1. *Assume that either* $\mathbf{H}$ *is semisimple or* $\mathbf{H}_{F_v}$ *is anisotropic. Then the homomorphism* $r_v : A(\mathbf{H}/F_v) \to A(\mathbf{H}/F)$ *is surjective.*

Proof. This follows from the fact that in both cases $Z(\hat{\mathbf{H}})^{\Gamma_v}$ is finite. □

Now let $\mathbf{V}$ be a nice representation of $\mathbf{G}$ defined over F. Then for every $x \in \mathbf{V}_0(F)$ we can identify

$$\mathbf{G}(\mathbb{A})\backslash\mathcal{O}_x(\mathbb{A})$$

with the kernel of the map

$$H^1(\Gamma, \mathbf{H}_x(\overline{\mathbb{A}})) \to H^1(\Gamma, \mathbf{G}(\overline{\mathbb{A}}))$$

where $\overline{\mathbb{A}} = \mathbb{A} \otimes_F \overline{F}$. Thus, we can restrict the global Kottwitz invariant inv : $H^1(\Gamma, \mathbf{H}_x(\overline{\mathbb{A}})) = \oplus_v H^1(F_v, \mathbf{H}) \to A(\mathbf{H}/F)$ to a map

$$\mathrm{inv}(x, \cdot) : \mathbf{G}(\mathbb{A})\backslash\mathcal{O}_x(\mathbb{A}) \to A(\mathbf{H}_x/F).$$

We claim that this function takes the value $0 \in A(\mathbf{H}_x/F)$ precisely on the set of $\mathbf{G}(\mathbb{A})$-orbits of F-rational points in $\mathcal{O}_x$. Indeed, consider the following commutative diagram:

$$\begin{array}{ccccc}
\mathbf{G}(F)\backslash\mathcal{O}_x(F) & \longrightarrow & H^1(F, \mathbf{H}_x) & \longrightarrow & H^1(F, \mathbf{G}) \\
\downarrow & & \downarrow & & \downarrow{\scriptstyle i_{\mathbf{G}}} \\
\mathbf{G}(\mathbb{A})\backslash\mathcal{O}_x(\mathbb{A}) & \longrightarrow & H^1(\Gamma, \mathbf{H}_x(\overline{\mathbb{A}})) & \longrightarrow & H^1(\Gamma, \mathbf{G}(\overline{\mathbb{A}})) \\
 & & \downarrow & & \\
 & & A(\mathbf{H}_x/F) & &
\end{array} \tag{3.1.1}$$

with exact central vertical column. The Hasse principle for $\mathbf{G}$ (which is simply connected) implies that the map $i_{\mathbf{G}}$ is injective. Now our claim follows by an easy diagram chase.

Let $\mathbf{H} = \mathbf{H}_x$ for some $x \in \mathbf{V}_0(F)$. Recall that we have defined in Section 2.2 a homomorphism $\mathrm{sign}_v : A(\mathbf{H}/F_v) \to \{\pm 1\}$ for every place v such that $\epsilon_{\mathbf{H}_{F_v}} = \mathrm{sign}_v \circ \mathrm{inv}_v$. We claim that there exists a canonical homomorphism

$$\mathrm{sign}_F : A(\mathbf{H}/F) \to \{\pm 1\}$$

such that $\mathrm{sign}_v = \mathrm{sign}_F \circ r_v$ for every place v. Indeed, since all the local homomorphisms sign_v factor through $A(\mathbf{H}_{ad}/F_v)$, it suffices to define sign_F in the case when $\mathbf{H}$ is semisimple. This is done in absolutely the same way as in the local case. Namely, we consider the exact sequence $1 \to \mathbf{C} \to \mathbf{H}_{sc} \to \mathbf{H} \to 1$ defined over F, where $\mathbf{H}_{sc}$ is simply connected. The unique morphism of this exact sequence to $1 \to \{\pm 1\} \to \widetilde{\mathbf{H}} \to \mathbf{H} \to 1$ induces a homomorphism $\mathbf{C} \to \{\pm 1\}$ defined over F, which can be considered a character $A(\mathbf{H}/F) \to \{\pm 1\}$. This is our sign_F. The compatibility with the local construction is obvious.

3.2 Stabilization

Theorem 3.2. *Assume that* $\mathbf{G}$ *is a simply connected semisimple group over* F, *and* $\rho : \mathbf{G} \to \mathrm{Aut}(\mathbf{V})$ *is a nice representation defined over* F. *Assume also that either* $\mathbf{G}$ *is anisotropic over* F, *or the generic stabilizer* $\mathbf{H}$ *is semisimple. Let* $f \in \mathcal{S}(\mathbf{V}(\mathbb{A}))$ *be a function such that* $\mathrm{supp}(f) \cap \mathbf{G}(\mathbb{A})\mathbf{V}(F) \subset \mathbf{V}_0(\mathbb{A})$. *Then*

$$\Theta(f) = \sum_{x \in \mathbf{G}(\overline{F})\backslash \mathbf{V}_0(F)} \sum_{\kappa \in A(\mathbf{H}_x/F)^D} \int_{a \in \mathcal{O}_x(\mathbb{A})} \kappa(\mathrm{inv}(x,a)) f(a) |\omega_x|, \tag{3.2.1}$$

where $\mathbf{G}(\overline{F})\backslash \mathbf{V}_0(F)$ *denotes a set of representatives of stable* $\mathbf{G}$*-equivalence classes of points in* $\mathbf{V}_0(F)$; *for every* $x \in \mathbf{V}_0(F)$ *we choose a* $\mathbf{G}$*-invariant top-degree form* ω_x *on* $\mathcal{O}_x$ *defined over* F.

Proof. For brevity we will omit $|\omega_x|$ in the integrals below. First, we claim that the infinite sum on the right-hand side of (3.2.1) is absolutely convergent. Indeed, for every point $x \in \mathbf{V}_0(F)$ there exists a Zariski neighborhood $\mathbf{U}_x \subset \mathbf{L}_x$ in a linear subspace $\mathbf{L}_x \subset \mathbf{V}^{\mathbf{H}_x}$ such that the map $\mathbf{G}/\mathbf{H}_x \times \mathbf{U}_x \to \mathbf{V}_0$ is étale. In particular, the subsets $\mathbf{G}\mathbf{U}_x$ form a Zariski open covering of $\mathbf{V}_0$, so a finite number of them cover $\mathbf{V}_0$. Therefore, it suffices to prove the absolute convergence of

$$\sum_{u \in \mathbf{U}_x(F)} \int_{a \in \mathcal{O}_u(\mathbb{A})} f(a).$$

Now from the fact that the map $\mathbf{G}/\mathbf{H}_x \times \mathbf{L}_x \to \mathbf{V}$ is linear in the second argument it is easy to deduce that the function $u \mapsto \int_{a \in \mathcal{O}_u(\mathbb{A})} f(a)$ on $\mathbf{U}_x(\mathbb{A})$ is rapidly decreasing at infinity, which implies our claim.

For every $x \in \mathbf{V}_0(F)$ we have

$$\sum_{\kappa \in A(\mathbf{H}_x/F)^D} \int_{a \in \mathcal{O}_x(\mathbb{A})} \kappa(\mathrm{inv}(x,a)) f(a) = |A(\mathbf{H}_x/F)| \cdot \int_{a \in \mathbf{G}(\mathbb{A})\mathcal{O}_x(F)} f(a)$$

since $\mathrm{inv}(x,a) = 0$ if and only if $a \in \mathbf{G}(\mathbb{A})\mathcal{O}_x(F)$. Now we have a finite covering

$$\bigsqcup_{y \in \mathbf{G}(F)\backslash \mathcal{O}_x(F)} \mathbf{G}(\mathbb{A})y \to \mathbf{G}(\mathbb{A})\mathcal{O}_x(F).$$

We claim that the degree of this covering is equal to $|\ker^1(F, Z(\hat{\mathbf{H}}))|$. Indeed, this degree is equal to the number of $y \in \mathbf{G}(F)\backslash \mathcal{O}_x(F)$ such that $\mathbf{G}(\mathbb{A})x = \mathbf{G}(\mathbb{A})y$. Such y's correspond to the classes in $\ker(H^1(F, \mathbf{H}_x) \to H^1(F, \mathbf{G}))$ that have trivial restriction at every place. Since $\mathbf{G}$ is simply connected, by the Hasse principle the elements in $H^1(F, \mathbf{G})$ with trivial restrictions at all places are trivial. Therefore, our set of cohomology classes coincides with $\ker^1(F, \mathbf{H}_x)$. Using (3.0.1) we obtain that $|\ker^1(F, \mathbf{H}_x)| = |\ker^1(F, Z(\hat{\mathbf{H}}_x))|$. It follows that

$$\sum_{\kappa \in A(\mathbf{H}_x/F)^D} \int_{a \in \mathcal{O}_x(\mathbb{A})} \kappa(\mathrm{inv}(x,a)) f(a)$$
$$= \frac{|A(\mathbf{H}_x/F)|}{|\ker^1(F, Z(\hat{\mathbf{H}}_x))|} \sum_{y \in \mathbf{G}(F)\backslash \mathcal{O}_x(F)} \int_{a \in \mathbf{G}(\mathbb{A})y} f(a).$$

Now for $y \in \mathbf{G}(F)$ we have

$$\int_{a \in \mathbf{G}(\mathbb{A})y} f(a) = \int_{g \in \mathbf{G}(\mathbb{A})/\mathbf{H}_y(\mathbb{A})} f(gy) = \tau(\mathbf{H}_y)^{-1} \int_{g \in \mathbf{G}(\mathbb{A})/\mathbf{H}_y(F)} f(gy)$$

where $\tau(\mathbf{H}_y) = \mathrm{vol}(\mathbf{H}_y(\mathbb{A})/\mathbf{H}_y(F))$ is the Tamagawa number of $\mathbf{H}_y$. As Kottwitz showed, the Tamagawa number does not change if we pass to an inner form. More precisely, we have

$$\tau(\mathbf{H}_y) = \frac{|A(\mathbf{H}_y/F)|}{|\ker^1(F, Z(\hat{\mathbf{H}}_y))|}$$

(see [10], (5.1.1) or the introduction to [12]). Since $\mathbf{H}_y$ is an inner form of $\mathbf{H}_x$, we can replace $\mathbf{H}_y$ by $\mathbf{H}_x$ in the RHS. Hence, we obtain

$$\sum_{\kappa \in A(\mathbf{H}_x/F)^D} \int_{a \in \mathcal{O}_x(\mathbb{A})} \kappa(\mathrm{inv}(x, a)) f(a) = \sum_{y \in \mathbf{G}(F)\backslash \mathcal{O}_x(F)} \int_{g \in \mathbf{G}(\mathbb{A})/\mathbf{H}_y(F)} f(gy).$$

Thus, the RHS of the formula (3.2.1) can be rewritten as follows:

$$\begin{aligned}
&\sum_{x \in \mathbf{G}(\overline{F})\backslash \mathbf{V}_0(F)} \sum_{y \in \mathbf{G}(F)\backslash \mathcal{O}_x(F)} \int_{g \in \mathbf{G}(\mathbb{A})/\mathbf{H}_y(F)} f(gy) \\
&= \sum_{y \in \mathbf{G}(F)\backslash \mathbf{V}_0(F)} \int_{g \in \mathbf{G}(\mathbb{A})/\mathbf{H}_y(F)} f(gy) \\
&= \int_{g \in \mathbf{G}(\mathbb{A})/\mathbf{G}(F)} \sum_{y \in \mathbf{G}(F)\backslash \mathbf{V}_0(F)} \sum_{x \in \mathbf{G}(F)y} f(gx) \\
&= \int_{g \in \mathbf{G}(\mathbb{A})/\mathbf{G}(F)} \sum_{x \in \mathbf{V}_0(F)} f(gx)
\end{aligned}$$

which is precisely the LHS of (3.2.1). □

3.3 Global proofs

We are going to combine the obtained local and global information about nice representations with the stabilization formula (3.2.1) to prove Theorems 2.10 and 2.5.

Proof of Theorem 2.10. We will only prove that if $\phi \sim \phi'$, then $\mathcal{F}(\phi) \sim_\epsilon \kappa(\mathbf{G}, \mathbf{G}')\mathcal{F}(\phi')$. The proof of the converse statement is absolutely analogous. For convenience we divide the proof into several steps. In step 1 we will extend our local data to the global one in an appropriate way. In the (crucial) step 2 we apply the stabilization formula to deduce the analogue of our statement for the product of local fields at two places of our global field. Finally, in step 3 we will deduce the statement for one local field. Let us rename the data $(\mathbf{G}, \mathbf{Z}, \mathbf{V}, \alpha, \mathbf{G}', \mathbf{V}')$ into $(\mathbf{G}_E, \mathbf{Z}_E, \mathbf{V}_E, \alpha_E, \mathbf{G}'_E, \mathbf{V}'_E)$ to reflect the fact that they are defined over E.

Step 1. We start by choosing a number field F, the data $(\mathbf{G}, \mathbf{Z}, \mathbf{V})$ defined over F, and a place v_0 of F such that $F_{v_0} = E$, $\mathbf{G}_{v_0} = \mathbf{G}_E$, $\mathbf{Z}_{v_0} = \mathbf{Z}_E$, $\mathbf{V}_{v_0} = \mathbf{V}_E$. In addition we can assume that there are two finite places v_1, v_2 (different from v_0) and a real place v_∞ of F such that $\mathbf{G}_{v_1}$ and $\mathbf{G}_{v_2}$ are split and $\mathbf{G}(F_{v_\infty})$ is compact. Note that the fact that $\mathbf{V}_E$ is a nice representation implies that the representations $\mathbf{V}$ and $\mathbf{V}_v$ for all places v are nice.

Let us choose a point $x_{v_0} \in \mathbf{V}_0(F_{v_0})$ such that the cohomology class α_E comes from a cohomology class $\beta_E \in H^1(\Gamma_{v_0}, \mathbf{H}_{x_{v_0}}/\mathbf{Z}_{v_0})$ (if no such point exist then the statement of the theorem is empty). In the case when the generic stabilizer is commutative we can also choose a point $x_{v_1} \in \mathbf{V}_0(F_{v_1})$ such that $\mathbf{H}_{x_{v_1}}$ is anisotropic (this is possible by Lemma 2.12). We can choose a global point $x \in \mathbf{V}_0(F)$ that approximates x_{v_0} and x_{v_1} (resp. x_{v_0} if the generic stabilizer is noncommutative) well enough, so that for $\mathbf{H} = \mathbf{H}_x$ we have a class $\beta_E \in H^1(\Gamma_{v_0}, \mathbf{H}_{v_0}/\mathbf{Z}_{v_0})$ inducing α_E, and, in the case $\mathbf{H}$ is commutative, we also have that $\mathbf{H}^0_{v_1}$ is anisotropic. Let us set $\mathbf{K} = \mathbf{H}/\mathbf{Z}$. Then $\mathbf{K}_{v_1}$ is either semisimple or an anisotropic torus, hence, by Lemma 3.1 the homomorphism $A(\mathbf{K}/F_{v_1}) \to A(\mathbf{K}/F)$ is surjective. Recall that we have an exact sequence

$$H^1(F, \mathbf{K}) \to \oplus_v H^1(F_v, \mathbf{K}) \to A(\mathbf{K}/F).$$

By surjectivity of the map $H^1(F_{v_1}, \mathbf{K}) \to A(\mathbf{K}/F)$ there exists an element $\beta \in H^1(F, \mathbf{K})$ such that $\beta_{v_0} = \beta_E$ and $\beta_v = 0$ for $v \neq v_0, v_1$. Let $\alpha \in H^1(F, \mathbf{G}/\mathbf{Z})$ be the class induced by β, and let $(\mathbf{G}', \mathbf{V}')$ be the twist of $\mathbf{G}$ by α. Then $\alpha_{v_0} = \alpha_E$, while the restrictions of α to all places other than v_0 and v_1 are trivial. Thus, we have $(\mathbf{G}'_{v_0}, \mathbf{V}'_{v_0}) = (\mathbf{G}'_E, \mathbf{V}'_E)$ and $(\mathbf{G}'_v, \mathbf{V}'_v) = (\mathbf{G}_v, \mathbf{V}_v)$ for $v \neq v_0, v_1$. In particular, $\mathbf{G}'_{v_2} = \mathbf{G}_{v_2}$ is split over F_{v_2} and $\mathbf{G}'(F_{v_\infty}) = \mathbf{G}(F_{v_\infty})$ is compact. The latter condition implies that both groups $\mathbf{G}$ and $\mathbf{G}'$ are anisotropic over F, so the distribution Θ is defined on all functions in $\mathcal{S}(\mathbf{V}(\mathbb{A}))$ (resp. $\mathcal{S}(\mathbf{V}'(\mathbb{A}))$, $\mathcal{S}(\mathbf{V}^\vee(\mathbb{A}))$ and $\mathcal{S}((\mathbf{V}')^\vee(\mathbb{A}))$).

Step 2. Set $S = \{v_0, v_1\}$. We are going to show that for every pair of functions $\phi_S \in \mathcal{S}(\mathbf{V}(\mathbb{A}_S))$, $\phi'_S \in \mathcal{S}(\mathbf{V}'(\mathbb{A}_S))$ such that $\phi_S \sim \phi'_S$, one has $\mathcal{F}(\phi_S) \sim_\epsilon \mathcal{F}(\phi'_S)$. For this we have to check that for every $x_S^\vee \in \mathbf{V}_0^\vee(\mathbb{A}_S)$ one has

$$\delta^\epsilon_{\mathcal{O}_{x_S^\vee},\omega}(\mathcal{F}(\phi_S)) = \delta^\epsilon_{\mathcal{O}'_{x_S^\vee}, i_*\omega}(\mathcal{F}(\phi'_S)).$$

Since $\mathcal{F}(\phi_S) \in \mathcal{S}(\mathbf{V}^\vee(\mathbb{A}_S))$ and $\mathcal{F}(\phi'_S) \in \mathcal{S}((\mathbf{V}')^\vee(\mathbb{A}_S))$, the functions

$$y_S \mapsto \delta^\epsilon_{\mathcal{O}_{y_S},\omega}(\mathcal{F}(\phi_S)) \text{ and } y_S \mapsto \delta^\epsilon_{\mathcal{O}'_{y_S}, i_*\omega}(\mathcal{F}(\phi'_S))$$

on $\mathbf{V}_0^\vee(\mathbb{A}_S)$ are locally constant (see Lemma 2.11). Let U_S be an open neighborhood of $x_S^\vee$ in $\mathbf{V}_0^\vee(\mathbb{A}_S)$ on which these two functions are constant. Since $\mathbf{G}_{v_2}$ is split over F_{v_2}, we can apply the construction of Section 2.6 and Proposition 2.9 to construct a stable function $\phi_{v_2} \in \mathcal{S}(\mathbf{V}_0(F_{v_2}))$ and a nonempty open subset $U_{v_2} \subset \mathbf{V}_0^\vee(F_{v_2})$ such that the restriction of $I(\mathcal{F}(\phi_{v_2}))$ to $\cup_{y \in U_{v_2}} \mathcal{O}_y(F_{v_2})$ is ϵ-stable and everywhere nonvanishing. In particular, for $y \in U_{v_2}$ we have

$\delta^{\epsilon}_{\mathcal{O}_y}(\mathcal{F}(\phi_{v_2})) \neq 0$. If $\mathbf{H}$ is commutative, then by Lemma 2.12 (and the remark after it) we can in addition assume that all points in U_{v_2} and in the support of ϕ_{v_2} have anisotropic stabilizer.

We can find a point $x^\vee \in \mathbf{V}_0^\vee(F)$ such that $x^\vee \in U_S$ and $x^\vee \in U_{v_2}$. Then we have

$$\delta^{\epsilon}_{(\mathcal{O}_{x^\vee})_{\mathbb{A}_S}}(\mathcal{F}(\phi_S)) = \delta^{\epsilon}_{\mathcal{O}_{x^\vee_S}}(\mathcal{F}(\phi_S)),$$

$$\delta^{\epsilon}_{(\mathcal{O}'_{x^\vee})_{\mathbb{A}_S}}(\mathcal{F}(\phi'_S)) = \delta^{\epsilon}_{\mathcal{O}'_{x^\vee_S}}(\mathcal{F}(\phi'_S)),$$

$$\delta^{\epsilon}_{(\mathcal{O}_{x^\vee})_{F_{v_2}}}(\mathcal{F}(\phi_{v_2})) \neq 0.$$

Let us set $S' = S \cup \{v_2\}$. We claim that for every function $\phi^{S'} \in \mathcal{S}(\mathbf{V}(\mathbb{A}^{S'}))$ one has

$$\Theta(\phi_S \otimes \phi_{v_2} \otimes \phi^{S'}) = \Theta(\phi'_S \otimes \phi_{v_2} \otimes \phi^{S'}).$$

Indeed, since ϕ_{v_2} is supported on points in $\mathbf{V}_0(F_{v_2})$ we can apply formula (3.2.1) to compute $\Theta(\phi_S \otimes \phi_{v_2} \otimes \phi^{S'})$. The RHS of this formula is the sum over $x \in \mathbf{V}_0(F)$ and over $\kappa \in A(\mathbf{H}_x/F)^D$ of terms

$$\int_{a \in \mathcal{O}_x(\mathbb{A})} \kappa(\mathrm{inv}(x, a))\phi_S(a_S)\phi_{v_2}(a_{v_2})\phi^{S'}(a^{S'})|\omega_x| = 0.$$

Since $\mathrm{inv}(x, a)$ is the product of local terms, this integral is equal to the product of the corresponding local integrals. We claim that if $\kappa \neq 0$ the local integral at v_2 is zero. Indeed, if $\mathcal{O}_x(F_{v_2})$ does not intersect the support of ϕ_{v_2} this is clear. Otherwise, by Lemma 3.1 the character $\kappa \circ r_{v_2}$ of $A(\mathbf{H}_x/F_{v_2})$ is nontrivial (here we use the fact that the stabilizers of points in the support of ϕ_{v_2} are either semisimple or anisotropic). Hence,

$$\int_{a_{v_2} \in \mathcal{O}_x(F_{v_2})} (\kappa \circ r_{v_2})(a_{v_2})\phi_{v_2}(a_{v_2})|\omega_x| = 0$$

since $\overline{\phi}_{v_2}$ is stable. Thus, the formula (3.2.1) in this case takes form

$$\Theta(\phi_S \otimes \phi_{v_2} \otimes \phi^{S'}) = \sum_{x \in \mathbf{G}(\overline{F})\backslash \mathbf{V}_0(F)} \int_{a \in \mathcal{O}_x(\mathbb{A})} \phi_S(a_S)\phi_{v_2}(a_{v_2})\phi^{S'}(a^{S'})|\omega_x|.$$

Similar formula holds for $\Theta(\phi'_S \otimes \phi_{v_2} \otimes \phi^{S'})$. Now our claim follows immediately from the condition $\phi_S \sim \phi'_S$.

Applying the Fourier transform, we obtain

$$\Theta(\mathcal{F}(\phi_S) \otimes \mathcal{F}(\phi_{v_2}) \otimes \mathcal{F}(\phi^{S'})) = \Theta(\mathcal{F}(\phi'_S) \otimes \mathcal{F}(\phi_{v_2}) \otimes \mathcal{F}(\phi^{S'})). \quad (3.3.1)$$

Let us choose one more finite place $v_3 \notin S'$ and a function ϕ_{v_3} in $\mathcal{S}(\mathbf{V}(F_{v_3}))$ such that $\mathcal{F}(\phi_{v_3})$ is supported on $\mathbf{V}_0^\vee(F_{v_3})$ and $\delta^{\epsilon}_{(\mathcal{O}_{x^\vee})_{F_{v_3}}}(\mathcal{F}(\phi_{v_3})) \neq 0$. One can define such ϕ_{v_3} by setting

$$\mathcal{F}(\phi_{v_3}) = \epsilon \cdot \delta_{KU_0},$$

where U_0 is a small compact neighborhood of $x^\vee$ in F_{v_3}-points of a linear slice for $\mathbf{G}$-action, K is a nonempty compact in $\mathcal{O}_{x^\vee}(F_{v_3})$ intersecting $\mathbf{G}(F_{v_3})$-orbits by the sets of equal volumes (see Section 2.6 for similar constructions).

Let $S'' = S' \cup \{v_3\}$. Since the function $\mathcal{F}(\phi_{v_3})$ is supported on $\mathbf{V}_0^\vee(F_{v_3})$ we can apply formula (3.2.1) to calculate $\Theta(\mathcal{F}(\phi))$ and $\Theta(\mathcal{F}(\phi'))$, where

$$\phi = \phi_S \otimes \phi_{v_2} \otimes \phi_{v_3} \otimes \phi^{S''},$$

$$\phi' = \phi'_S \otimes \phi_{v_2} \otimes \phi_{v_3} \otimes \phi^{S''},$$

for some $\phi^{S''} \in \mathcal{S}(\mathbf{V}(\mathbb{A}^{S''}))$. Now the idea is to choose $\phi^{S''}$ in such a way that all the terms in the RHS of (3.2.1) (applied to $\mathcal{F}(\phi)$ and $\mathcal{F}(\phi')$) corresponding to points of $\mathbf{V}_0^\vee(F)$ which are not stably equivalent to $x^\vee$ vanish. Indeed, let $C \subset \mathbf{V}^\vee(\mathbb{A}_{S''})$ (resp. $C' \subset \mathbf{V}^\vee(\mathbb{A}_{S''})$) be the support of $\mathcal{F}(\phi_S \otimes \phi_{v_2} \otimes \phi_{v_3})$ (resp. of $\mathcal{F}(\phi'_S \otimes \phi_{v_2} \otimes \phi_{v_3})$). Set $D = \mathbf{V}^\vee/\mathbf{G}(F) \cap p_{\mathbb{A}_{S''}}(C \cup C')$, where $p : \mathbf{V}^\vee \to \mathbf{V}^\vee/\mathbf{G}$ is the natural projection (intersection is taken in $\mathbf{V}^\vee/\mathbf{G}(\mathbb{A}_{S''})$). Since $\mathbf{V}^\vee/\mathbf{G}(F)$ is discrete in $\mathbf{V}^\vee/\mathbf{G}(\mathbb{A})$ and $C \cap C'$ is compact, it follows that D is discrete in $\mathbf{V}^\vee/\mathbf{G}(\mathbb{A}^{S''})$. Note that $p(x^\vee)$ belongs to D. Therefore, we can choose $\phi^{S''}$ in such a way that its support is disjoint from $p^{-1}_{\mathbb{A}^{S''}}(D \setminus p(x^\vee))$, while $\delta^\epsilon_{(\mathcal{O}_{x^\vee})_{\mathbb{A}^{S''}}}(\mathcal{F}(\phi^{S''})) \neq 0$. Now applying (3.2.1) we get

$$\Theta(\mathcal{F}(\phi)) = \sum_{\kappa \in A(\mathbf{H}_{x^\vee}/F)^D} \int_{a \in \mathcal{O}_{x^\vee}(\mathbb{A})} \kappa(\mathrm{inv}(x^\vee, a))\mathcal{F}(\phi_S)(a_S)$$
$$\times \mathcal{F}(\phi_{v_2})(a_{v_2})\mathcal{F}(\phi_{v_3})(a_{v_3})\mathcal{F}(\phi^{S''})(a^{S''})|\omega_{x^\vee}|.$$

Since $\mathbf{H}_{x^\vee}$ is either semisimple or anisotropic over F_{v_2}, while the function $I(\mathcal{F}(\phi_{v_2}))$ is ϵ-stable on $\mathcal{O}_{x^\vee}(F_v)$, it follows that the local integral at v_2 vanishes unless $\kappa = \epsilon_F$. The same computation works for $\Theta(\mathcal{F}(\phi'))$. Hence, the equality (3.3.1) for $\phi^{S'} = \phi_{v_2} \otimes \phi^{S''}$ reduces to

$$\delta^\epsilon_{(\mathcal{O}_{x^\vee})_{\mathbb{A}_S}}(\mathcal{F}(\phi_S)) \cdot \delta^\epsilon_{(\mathcal{O}_{x^\vee})_{F_{v_2}}}(\mathcal{F}(\phi_{v_2})) \cdot \delta^\epsilon_{(\mathcal{O}_{x^\vee})_{F_{v_3}}}(\mathcal{F}(\phi_{v_3})) \cdot \delta^\epsilon_{(\mathcal{O}_{x^\vee})_{\mathbb{A}^{S''}}}(\mathcal{F}(\phi^{S''}))$$
$$= \delta^\epsilon_{(\mathcal{O}'_{x^\vee})_{\mathbb{A}_S}}(\mathcal{F}(\phi'_S)) \cdot \delta^\epsilon_{(\mathcal{O}_{x^\vee})_{F_{v_2}}}(\mathcal{F}(\phi_{v_2})) \cdot \delta^\epsilon_{(\mathcal{O}_{x^\vee})_{F_{v_3}}}(\mathcal{F}(\phi_{v_3}))$$
$$\cdot \delta^\epsilon_{(\mathcal{O}_{x^\vee})_{\mathbb{A}^{S''}}}(\mathcal{F}(\phi^{S''})).$$

Therefore, we get

$$\delta^\epsilon_{(\mathcal{O}_{x^\vee})_{\mathbb{A}_S}}(\mathcal{F}(\phi_S)) = \delta^\epsilon_{(\mathcal{O}'_{x^\vee})_{\mathbb{A}_S}}(\mathcal{F}(\phi'_S))$$

which implies our statement.

Step 3. Applying Proposition 2.15 for the place v_1 we construct functions $\phi_{v_1} \in \mathcal{S}(\mathbf{V}_0(F_{v_1}))$ and $\phi'_{v_1} \in \mathcal{S}(\mathbf{V}'_0(F_{v_1}))$ such that $\phi_{v_1} \sim \phi'_{v_1}$, while

$$\delta^\epsilon_{\mathcal{O},\omega}(\mathcal{F}(\phi_{v_1})) = \kappa(\mathbf{G}, \mathbf{G}')\delta^\epsilon_{\mathcal{O}',i_*\omega}(\mathcal{F}(\phi'_{v_1})) \neq 0$$

for some orbit $\mathcal{O} \subset (\mathbf{V}_0)_{v_1}$. Now let $\phi_0 \in \mathcal{S}(\mathbf{V}(F_{v_0}))$, $\phi'_0 \in \mathcal{S}(\mathbf{V}'(F_{v_0}))$ be a pair of functions such that $\phi_0 \sim \phi'_0$. Then Step 2 applied to $\phi_0 \otimes \phi_{v_1}$ and $\phi'_0 \otimes \phi'_{v_1}$ implies that $\mathcal{F}(\phi_0) \sim_\epsilon \kappa(\mathbf{G}, \mathbf{G}')\mathcal{F}(\phi'_0)$. □

Remark. To generalize the above proof to the case when $\mathbf{G}$ is not necessarily simple, it would be enough to prove the following version of Lemma 2.12 for $\mathbf{G}$: if $\mathbf{G}$ is split and ρ is nice, then there exists a point $x \in \mathbf{V}_0(E)$ such that the connected component of $Z(\mathbf{H}_x)$ is anisotropic. Indeed, Lemmas 2.13 and 3.1 used in the proof can be easily generalized to the case when $Z(\mathbf{H}_x)^0$ is anisotropic. The rest of the proof does not use the assumption that $\mathbf{G}$ is simple.

Proof of Theorem 2.5. We will only prove the inclusion $\mathcal{F}(S^{st}) \subset S^{st}_\epsilon$. The proof of the inverse inclusion is absolutely analogous. Thus, we have to prove that if ϕ_0 is a stable function on $\mathbf{V}(E)$, then for any $x^\vee_E \in \mathbf{V}^\vee_0(E)$ the function $I(\epsilon \cdot \mathcal{F}(\phi_0))$ on $\mathcal{O}_{x^\vee_E}(E)$ is constant. We will split the proof into two steps which are similar to the first two steps of the previous proof: step 1 consists of constructing an appropriate global setup, while step 2 is an application of the stabilization formula. As before we rename our data $(\mathbf{G}, \mathbf{V})$ into $(\mathbf{G}_E, \mathbf{V}_E)$. We denote by $\mathbf{H}_E \subset \mathbf{G}_E$ the stabilizer of $x^\vee_E$.

Step 1. Let F be a global field, v_0 be a place of F such that $F_{v_0} = E$. We want to construct the data $(\mathbf{G}, \mathbf{V}, \mathbf{H})$ over F such that $\mathbf{G}_{v_0} = \mathbf{G}_E$, $\mathbf{V}_{v_0} = \mathbf{V}_E$, $\mathbf{H}_{v_0}$ is $\mathbf{G}(E)$-conjugate to $\mathbf{H}_E$ and such that in addition the natural homomorphism

$$A(\mathbf{H}/E) \to A(\mathbf{H}/F)$$

is an isomorphism. First, we can find the data $(\mathbf{G}, \mathbf{V})$ over F such that $\mathbf{G}_{v_0} = \mathbf{G}_E$, $\mathbf{V}_{v_0} = \mathbf{V}_E$. Now let $x^\vee \in \mathbf{V}^\vee(F)$ be a global point sufficiently close to $x^\vee_E$, and let $\mathbf{H}$ be the stabilizer of $x^\vee$. Then $\mathbf{H}_{v_0}$ is $\mathbf{G}(E)$-conjugate to $\mathbf{H}_E$. Let $\pi \subset \mathrm{Aut}(Z(\hat{\mathbf{H}}))$ be the quotient of the local Galois group Γ_{v_0}, through which it acts on $Z(\hat{\mathbf{H}})$. Let us denote by $\Gamma' \subset \Gamma$ the preimage of π under the natural homomorphism $\Gamma \to \mathrm{Aut}(Z(\hat{\mathbf{H}}))$, and let $F' \supset F$ be the finite extension corresponding to Γ'. Then Γ' contains Γ_v (which is considered as a subgroup in Γ via some fixed extension of the valuation v_0 to $\overline{F}$). Hence, there is an extension of v_0 to a place v'_0 of F' such that $F'_{v'_0} = E$. Furthermore, by construction we have $A(\mathbf{H}/E) = A(\mathbf{H}/F')$. It remains to replace (F, v_0) by (F', v'_0).

Step 2. This step is very similar to the step 2 in the previous proof. Let κ_E be a nontrivial character of $A(\mathbf{H}/E)$. We have to prove that $\delta^{\kappa_E \mathrm{sign}}_{\mathcal{O}_{x^\vee_E}}(\mathcal{F}(\phi_0)) = 0$. Let us denote $\mathbf{W}^\vee = (\mathbf{V}^\vee)^{\mathbf{H}}$. Since the function $y \mapsto \delta^{\kappa_E \mathrm{sign}}_{\mathcal{O}_y}(\mathcal{F}(\phi_0))$ of $y \in \mathbf{W}^\vee(E)$ is locally constant, we can choose a neighborhood U_{v_0} of $x^\vee_E$ on which this function is constant. By our assumption κ_E is induced by some character κ of $A(\mathbf{H}/F)$. For every place v we denote by κ_v the induced character of $A(\mathbf{H}/F_{v_1})$.

Let us choose a finite place v_1 of F (different from v_0). By Proposition 2.9 (ii) for a function of the form $\phi_{v_1} = \delta^{\kappa_{v_1}}_{U(t)} \in \mathcal{S}(\mathbf{V}_0(F_{v_1}))$, where $t \in (E^*)^2$ is large enough, one has

$$\delta^{\kappa_{v_1}\mathrm{sign}}_{\mathcal{O}_{y^\vee}}(\mathcal{F}(\phi_{v_1})) \neq 0$$

for $y^\vee \in U_{v_1}$, where $U_{v_1} \subset \mathbf{W}_0^\vee(F_{v_1})$ is a nonempty open subset. Also since the character κ_{v_1} is nontrivial by Lemma 3.1, the function ϕ_{v_1} is antistable.

Let us choose a global point $x^\vee \in \mathbf{W}_0^\vee(F)$ such that $x^\vee \in U_{v_0}$ and $x^\vee \in U_{v_1}$. Then we have

$$\delta^{\kappa_E\mathrm{sign}}_{\mathcal{O}_{x_E^\vee}}(\mathcal{F}(\phi_0)) = \delta^{\kappa_E\mathrm{sign}}_{(\mathcal{O}_{x^\vee})_{F_{v_0}}}(\mathcal{F}(\phi_0)),$$

$$\delta^{\kappa_{v_1}\mathrm{sign}}_{(\mathcal{O}_{x^\vee})_{F_{v_1}}}(\mathcal{F}(\phi_{v_1})) \neq 0.$$

Let v_2 be one more finite place of F. We can construct a function $\phi_{v_2} \in \mathcal{S}(\mathbf{V}(F_{v_2}))$ such that $\mathcal{F}(\phi_{v_2})$ has support in $\mathbf{V}_0^\vee(F_{v_2})$ and such that

$$\delta^{\kappa'}_{(\mathcal{O}_{x^\vee})_{F_{v_2}}}(\mathcal{F}(\phi_{v_2})) = 0$$

for $\kappa' \neq \kappa_{v_2}\mathrm{sign}$, while

$$\delta^{\kappa_{v_2}\mathrm{sign}}_{(\mathcal{O}_{x^\vee})_{F_{v_2}}}(\mathcal{F}(\phi_{v_2})) \neq 0.$$

Indeed, it suffices to take $\mathcal{F}(\phi_{v_2})$ to be the function of the form $\delta^{\kappa_{v_2}\mathrm{sign}}_{U(t)}$ as in Section 2.6.

Let us denote $S = \{v_0, v_1, v_2\}$. We claim that for every function $\phi^S \in \mathcal{S}(\mathbf{V}(\mathbb{A}^S))$ one has

$$\Theta(\phi_0 \otimes \phi_{v_1} \otimes \phi_{v_2} \otimes \phi^S) = 0.$$

Indeed, since ϕ_{v_1} has support in $\mathbf{V}_0(F_{v_1})$, we can apply formula (3.2.1). Now let $x \in \mathbf{V}_0(F)$ and let κ' be a character of $A(\mathbf{H}_x/F)$. If κ' is nontrivial, then by Lemma 3.1 the induced character κ'_{v_0} of $A(\mathbf{H}_x/E)$ is nontrivial. Since ϕ_0 is stable, we get

$$\int_{a\in\mathcal{O}_x(E)} \kappa'(\mathrm{inv}(x,a))\phi_0(a)|\omega_x| = 0,$$

so the corresponding term in the RHS of (3.2.1) vanishes. On the other hand, if $\kappa' = 1$, then $\kappa'_{v_1} = 1$, so the corresponding term vanishes by antistability of ϕ_{v_1}.

Applying the Fourier transform, we obtain

$$\Theta(\mathcal{F}(\phi_0) \otimes \mathcal{F}(\phi_{v_1}) \otimes \mathcal{F}(\phi_{v_2}) \otimes \mathcal{F}(\phi^S)) = 0.$$

Furthermore, since $\mathcal{F}(\phi_{v_2})$ has support in $\mathbf{V}_0^\vee(F_{v_2})$ we can apply formula (3.2.1) again. As in the proof of Theorem 2.4 we can choose ϕ^S in such a way that all the terms in the RHS of (3.2.1) corresponding to points of $\mathbf{V}_0^\vee(F)$ which are not stably conjugate to $x^\vee$ vanish, while

$$\delta_{(\mathcal{O}_{x^\vee})_{\mathbb{A}^S}}(\mathcal{F}(\phi^S)) \neq 0.$$

Now applying (3.2.1) we get

$$\begin{aligned}0 &= \Theta(\mathcal{F}(\phi_0)\otimes\mathcal{F}(\phi_{v_1})\otimes\mathcal{F}(\phi_{v_2})\otimes\mathcal{F}(\phi^S))\\ &= \sum_{\kappa'\in A(\mathbf{H}_{x^\vee}/F)^D}\int_{a\in\mathcal{O}_{x^\vee}(\mathbb{A})}\kappa'(\mathrm{inv}(x^\vee,a))\mathcal{F}(\phi_0)(a_{v_0})\\ &\quad\times\mathcal{F}(\phi_{v_1})(a_{v_1})\mathcal{F}(\phi_{v_2})(a_{v_2})\mathcal{F}(\phi^S)(a^S)|\omega_{x^\vee}|.\end{aligned}$$

By our choice of ϕ_{v_2} the local integral at v_2 vanishes unless $\kappa'_{v_2}=\kappa_{v_2}\mathrm{sign}$. By Lemma 3.1 this condition is equivalent to $\kappa'=\kappa\,\mathrm{sign}_F$. Hence, we obtain that $\delta^{\kappa_E\mathrm{sign}}_{(\mathcal{O}_{x^\vee})_{F_{v_0}}}(\mathcal{F}(\phi_0))=0$ as required. □

4 Sign function for the space of symmetric matrices

In this section we will compute the sign function $\epsilon(\cdot,\cdot)$ for the space of symmetric $n\times n$ matrices Sym_n over a local field E considered as a representation of SL_n, where $g\in\mathrm{SL}_n$ acts on Sym_n by $X\mapsto gXg^t$. As an application we will derive the formula (0.0.1) in the case of odd n.

Let us denote by $\mathrm{Sym}'_n\subset\mathrm{Sym}_n$ the complement to the hypersurface $(\det=0)$. By definition, the sign $\epsilon(A,A')$, where $A,A'\in\mathrm{Sym}'_n(E)$, is defined when A and A' belong to one $\mathrm{SL}_n(E)$-orbit, i.e., when $\det(A)=\det(A')$.

Proposition 4.1. *For a pair of symmetric $n\times n$ matrices A,A' with* $\det(A)=\det(A')\neq 0$ *one has*

$$\epsilon(A,A')=\epsilon(q_A,q_{A'})^n$$

where q_A denotes the quadratic form with matrix A.

Proof. By definition $\epsilon(A,A')$ is the product of Hasse–Witt invariants of quadratic forms obtained by restricting the Killing form Q on sl_n to stabilizer subalgebras $\mathfrak{h}_A$ and $\mathfrak{h}_{A'}$ of A and A'. Since A and A' belong to one $\mathrm{SL}_n(\overline{E})$-orbit, the forms $Q|_{\mathfrak{h}_A}$ and $Q|_{\mathfrak{h}_{A'}}$ have the same determinant modulo squares. Therefore, their relative Hasse–Witt invariant will not change if we replace Q by its scalar multiple. Thus, we can do calculation with $Q(X)=\frac{1}{2}Tr(X^2)$. We claim that

$$\epsilon(Q|_{\mathfrak{h}_A})=c(\det A)\cdot\epsilon(q_A)^n$$

where c is a sign depending only on $\det A$ modulo squares. To prove this formula we notice that both sides do not change if we replace A by gAg^t where $g\in\mathrm{GL}_n(E)$. Thus, we can assume that A is diagonal. Let $(a_1,\dots,a_n)$ be diagonal entries of A. The subalgebra $\mathfrak{h}_A\subset\mathrm{sl}_n$ consists of matrices X such that $XA+AX^t=0$. Thus, if $X=(x_{ij})$ then we should have $x_{ii}=0$ while $x_{ji}=-\frac{a_j}{a_i}x_{ij}$. Thus, the quadratic form $Q|_{\mathfrak{h}_A}$ has diagonal matrix in the natural basis on $\mathfrak{h}_A$ and we have

$$\epsilon(Q|_{\mathfrak{h}_A})=\prod_{i<j,k<l:(i,j)<(k,l)}\left(-\frac{a_j}{a_i},-\frac{a_l}{a_k}\right),$$

where $(i, j) < (k, l)$ denotes the lexicographical order. A straightforward calculation shows that the RHS is equal to

$$\prod_{i<j}(a_i, a_j)^n \cdot \prod_i (a_i, -1)^{p(n)} \cdot (-1, -1)^{q(n)}$$

where $p(n), q(n)$ are some polynomials in n. Thus, we get

$$\epsilon(Q|_{\mathfrak{h}_A}) = \epsilon(q_A)^n \cdot (\det A, -1)^{p(n)} \cdot (-1, -1)^{q(n)}$$

as required. □

Now we can derive the formula (0.0.1) in the p-adic case. It is well known (see [19],[20],[5]) that one has an equation of the form

$$\mathcal{F}(\chi(\det)) = \tau_\chi \cdot (|\cdot|^{-\frac{n+1}{2}}\chi^{-1})(\det)$$

where τ_χ is some $(\mathrm{GL}_n/\{\pm 1\})(E)$-invariant function on $\mathrm{Sym}'_n(E)$. The stabilizer subgroup $\mathbf{H} \subset (\mathrm{GL}_n/\{\pm 1\})$ of a point in $\mathrm{Sym}'_n(E)$ is the group $\mathrm{O}_n/\{\pm 1\} \simeq \mathrm{SO}_n$ (here we use the fact that n is odd). Therefore, the set of $(\mathrm{GL}_n/\{\pm 1\})(E)$-orbits on $\mathrm{Sym}'_n(E)$ can be identified with $\ker(H^1(E, \mathrm{SO}_n) \to H^1(E, \mathrm{GL}_n/\{\pm 1\}))$. Note that the homomorphism $\mathrm{SO}_n \to \mathrm{GL}_n/\{\pm 1\}$ factors through SL_n. Therefore, the map $H^1(E, \mathrm{SO}_n) \to H^1(E, \mathrm{GL}_n/\{\pm 1\})$ is trivial, so there are two $(\mathrm{GL}_n/\{\pm 1\})(E)$ orbits on $\mathrm{Sym}'_n(E)$ corresponding to two distinct elements of $H^1(E, \mathrm{SO}_n)$. These two orbits intersect the subset $\det A = 1$ by two distinct $\mathrm{SL}_n(E)$-orbits. Therefore, the function τ_χ is determined by its restriction to the subset $\det A = 1$. It remains to notice that the distribution $\chi(\det)$ is stable for the action of SL_n on Sym_n. Therefore, by Theorem 2.4 its Fourier transform is ϵ-stable. Applying Proposition 4.1 we conclude that

$$\tau_\chi(A) = c(\chi) \cdot \epsilon(q_A) \cdot f(\det A),$$

where f is a function of $\det A \bmod (E^*)^2$ (unique up to a constant) such that $A \mapsto \epsilon(q_A) \cdot f(\det A)$ is $(\mathrm{GL}_n/\{\pm 1\})(E)$-invariant. It is easy to see that $(\mathrm{GL}_n/\{\pm 1\})(E)$-orbits on $\mathrm{Sym}'_n(E)$ coincide with $\mathrm{GL}_n(E) \times E^*$-orbits on it, where E^* acts by rescaling. Now for $t \in E^*$ we have

$$\epsilon(tq_A) = (t, t)^{\frac{n(n-1)}{2}} (t, \det A)^{\frac{n-1}{2}} \epsilon(q_A) = (t, -1)^{\frac{n(n-1)}{2}} \epsilon(q_A).$$

It follows that the function $A \mapsto \epsilon(q_A)(\det A, -1)^{\frac{n-1}{2}}$ is $(\mathrm{GL}_n/\{\pm 1\})(E)$-invariant, so we should have

$$\tau_\chi(A) = c(\chi) \cdot \epsilon(q_A) \cdot (\det A, -1)^{\frac{n-1}{2}}$$

which is equivalent to the equation (0.0.1). The explicit value of the constant $c(\chi)$ can be found in [24]. Note that it can also be determined using the stationary phase approximation as in Section 2.5.

In conclusion let us show that for $E = \mathbb{R}$ the formula (0.0.1) is still true. Indeed, we have either $\chi(x) = |x|^s$ or $\chi(x) = \text{sgn}(x)|x|^s$, so our formula is equivalent to the set of two equalities

$$\int_{A \in \text{Sym}'_n(\mathbb{R})} |\det(A)|^s \hat{f}(A) dA$$
$$= c_1(s) \int_{B \in \text{Sym}'_n(\mathbb{R})} (-1)^{\frac{i_B(n-i_B)}{2}} \cdot |\det(B)|^{-s-\frac{n+1}{2}} f(B) dB, \quad (4.0.1)$$

$$\int_{A \in \text{Sym}'_n(\mathbb{R})} \text{sgn}(\det(A)) |\det(A)|^s \hat{f}(A) dA$$
$$= c_2(s) \int_{B \in \text{Sym}'_n(\mathbb{R})} (-1)^{\frac{i_B(n-i_B)}{2}} \cdot \text{sgn}(\det(B)) |\det(B)|^{-s-\frac{n+1}{2}} f(B) dB, (4.0.2)$$

where i_B is the number of negative eigenvalues of B, f is a function from the Schwartz space of $\text{Sym}_n(\mathbb{R})$, $\hat{f}$ is its Fourier transform, c_1 and c_2 are some meromorphic functions of s. These formulas can be deduced from the explicit form of Γ-matrix for $\text{Sym}_n(\mathbb{R})$ computed by T. Shintani in Lemma 15 of [21]. Indeed, let us recall the result of Shintani's computation in a form convenient for us. Let us denote by V_i the connected component of $\text{Sym}'_n(\mathbb{R})$ consisting of matrices with exactly i positive eigenvalues. Let us denote $\Phi_i(f, s) = \int_{A \in V_i} |\det(A)|^s f(A) dA$. Then one has the following equation:

$$\Phi_i(\hat{f}, s) = c(s) \cdot \sum_{j=0}^{n} v_{ij}(s) \Phi_j(f, -s - \frac{n+1}{2}),$$

where

$$v_{ij}(s) = \sum_{(\epsilon_1, \ldots, \epsilon_n)} \exp\left(\frac{\pi\sqrt{-1}}{2} [\sum_{k=1}^{j} (k+s)\epsilon_k - \sum_{k=j+1}^{n} (k-j+s)\epsilon_k]\right)$$

where the summation is taken over all n-tuples $(\epsilon_1, \ldots, \epsilon_n) = (\pm 1, \ldots, \pm 1)$ such that exactly i of the ϵ's are $+1$. Now we have

$$c_j := \sum_{i=0}^{n} v_{ij} = 2^n \cdot \prod_{k=1}^{j} \cos(\frac{\pi}{2}(k+s)) \cdot \prod_{k=1}^{n-j} \cos(\frac{\pi}{2}(k+s)).$$

Hence, for odd n we have

$$\frac{c_j}{c_{j-1}} = \frac{\cos(\frac{\pi}{2}(j+s))}{\cos(\frac{\pi}{2}(n+1-j+s))} = (-1)^{\frac{n+1}{2}+j}.$$

Therefore, the vector $(c_j)_{j=0,\ldots,n}$ is proportional to $((-1)^{j(n-j)})$ which is equivalent to the equation (4.0.1). Similarly, we have

$$c'_j := \sum_{i=0}^{n} (-1)^{n-i} v_{ij}$$

$$= \sum_{(\epsilon_1,\dots,\epsilon_n)\in\{\pm 1\}^n} \prod_{k=1}^{n} \epsilon_k$$

$$\cdot \exp(\frac{\pi\sqrt{-1}}{2}[\sum_{k=1}^{j}(k+s)\epsilon_k - \sum_{k=j+1}^{n}(k-j+s)\epsilon_k])$$

$$= (2\sqrt{-1})^n \cdot (-1)^{n-j} \cdot \prod_{k=1}^{j} \sin(\frac{\pi}{2}(k+s)) \cdot \prod_{k=1}^{n-j} \sin(\frac{\pi}{2}(k+s)).$$

Hence, for odd n the vector $(c'_j)_{j=0,\dots,n}$ is proportional to $((-1)^{(j+1)(n-j)})$ which is equivalent to (4.0.2).

Acknowledgement. Both authors were supported in part by NSF grants. We would like to thank M. Borovoi, A. Elashvili, P. Etingof, V. Kac and G. Seitz for helpful discussions.

References

[1] N. Bourbaki, *Groupes et algèbres de Lie*, Chapitres 4,5 et 6. Hermann, Paris 1968.

[2] A. G. Elashvili, Canonical form and stationary subalgebras of points of general position for simple linear Lie groups, *Funct. Anal. and its Appl.* **6** (1972), 51–62.

[3] A. G. Elashvili, Stationary subalgebras of points of the common state for irreducible linear Lie groups, *Funct. Anal. and its Appl.***6** (1972), 65–78.

[4] I. Gelfand, D. Kazhdan, Representations of the group $GL(n, K)$ where K is a local field, in *Lie groups and Their Representations (Proc. Summer School, Budapest, 1971)*, 95–118, Halsted, New York, 1975.

[5] J.-I. Igusa, Some results on p-adic complex powers, *Amer. J. Math.* **106** (1984), 1013–1032.

[6] D. Kazhdan, *An Algebraic Integration*, in *Mathematics: Frontiers and Perspectives*, AMS, Providence, 2000, 93–115.

[7] G. Kempf, L. Ness, The length of vectors in representation spaces, in *Algebraic Geometry (Proc. Summer Meeting, Copenhagen, 1978)*, 233–243, Lecture Notes in Math. 732, Springer, 1979.

[8] B. Kostant, Lie algebra cohomology and the generalized Borel–Weil theorem, *Ann. of Math.* **(2)** 74 (1961), 329–387.

[9] R. Kottwitz, Sign changes in harmonic analysis on reductive groups, *Trans. AMS* **278** (1983), 289–297.

[10] R. Kottwitz, *Stable trace formula: cuspidal tempered terms*, Duke Math. J. 51 (1984), 611–650.

[11] R. Kottwitz, Stable trace formula: elliptic singular terms, *Math. Ann.* **275** (1986), 365–399.

[12] R. Kottwitz, Tamagawa numbers, *Ann. of Math.* **(2)** 127 (1988), 629–646.

[13] D. Luna, Slices étales, in *Sur le groupes algèbriques*, 81–105. Bull. Soc. Math. France, Paris, Mémoire 33, Soc. Math. France, Paris, 1973.

[14] G. D. Mostow, Self-adjoint groups, *Ann. Math.* **62** (1955), 44–55.

[15] V. Platonov, A. Rapinchuk, *Algebraic Groups and Number Theory*, Academic Press, Boston, 1994.

[16] V. L. Popov, Criteria for the stability of the action of a semisimple group on a factorial manifold. *USSR Izvestia* **34** (1970), 523–531.

[17] V. L. Popov, E. B. Vinberg, Invariant theory, in *Algebraic Geometry IV*, Springer-Verlag, 1994.

[18] R. W. Richardson, Principal orbit types for algebraic transformation spaces in characteristic zero, *Invent. Math.* **16** (1972), 6–14.

[19] M. Sato, *Theory of prehomogeneous vector spaces (algebraic part)—the English translation of Sato's lecture from Shintani's note*. Notes by T. Shintani. Translated from the Japanese by M. Muro. *Nagoya Math. J.* **120** (1990), 1–34.

[20] M. Sato, T. Shintani, On zeta functions associated with prehomogeneous vector spaces, *Annals of Math.* **(2)** 100 (1974), 131–170.

[21] T. Shintani, On zeta-functions associated with the vector space of quadratic forms, *J. Fac. Sci. Univ. Tokyo Sect. IA Math.* **22** (1975), 25–65.

[22] T. A. Springer, On the equivalence of quadratic forms, *Indag. Math.* **21** (1959), 241–253.

[23] T. A. Springer, *Linear Algebraic Groups*, 2nd edition. Birkhäuser, Boston, 1998.

[24] W. J. Sweet, Jr., A computation of the gamma matrix of a family of p-adic zeta integrals, *J. Number Theory* **55** (1995), 222–260.

[25] J.-L. Waldspurger, Le lemme fondamental implique le transfert, *Compositio Math.* **105** (1997), 153–236.

[26] A. Weil, Sur certains groupes d'opérateurs unitaires, *Acta Math.* **111** (1964), 143–211.

D. Kazhdan
Department of Mathematics
Harvard University
Cambridge MA 02138, USA
kazhdan@math.harvard.edu

A. Polishchuk
Department of Mathematics and Statistics
Boston University
Boston MA 02215, USA
apolish@math.bu.edu

Two More Variations on the Triangular Theme

A. A. Kirillov

1 Introduction

The orbit method was created 40 years ago (see [K1]) in the attempt to describe the unitary dual $\widehat{N}$ for the group N of upper triangular matrices with units on the main diagonal.

It was immediately understood that the method has much wider applications. In fact, all main questions of representation theory of Lie groups: construction of irreducible representations, restriction-induction functors, generalized and infinitesimal characters, Plancherel measure, etc., admit a transparent description in terms of coadjoint orbits.

Ironically, the initial question: the description of the set $\mathcal{O}(N) = \mathfrak{n}^*/N \cong \widehat{N}$ remains still unanswered despite many attempts. It seems that it is really a deep problem. I refer to my paper [K2] where different aspects of this problem are discussed in more details.

In the present paper I want to propose two more approaches to this problem.

One comes essentially from the works by Adler and Kostant on integrable systems related to coadjoint orbits. The direct impulse for me was the conversation with Nicolai Reshetikhin and the recent preprint [HKKR]. In this preprint the integrable systems are considered which are related to symplectic leaves in simple complex Lie groups, their Borel subgroups and doubles. In my exposition I discuss only the geometric side of these results. Many interesting relations with integrable systems are ignored here.

The second approach had arisen during the numerous talks with my student Ning Yan. His PhD thesis is devoted to representations of the triangular group over a finite field. Ning discovered that these representations can be gathered in some "clusters" which can be explicitly classified and studied.

I feel that there is some connection between the two approaches but at this point I cannot formulate it explicitly.

This paper is addressed mainly to young researchers. T herefore, I tried to be as explicit as possible. The general theory is used only to show the place of our problem in a more general context and when it is necessary for understanding.

2 Setup

I want to discuss here two new approaches to an old problem in the Orbit Method: to classify the coadjoint orbits for the group N_n of upper triangular matrices with units on the main diagonal.

We consider N_n as an algebraic subgroup in $GL(n)$ given by the equations

$$g_{ij} = \begin{cases} 1 & \text{for } i = j, \\ 0 & \text{for } i > j. \end{cases}$$

For any field K we denote by $N_n(K)$ the group of K-points of N_n. It was understood long ago that the classification problem for $N_n(K)$-orbits in the space $\mathfrak{n}^*$ practically does not depend on the field K.

For $n \leq 6$, where the full classification is known, the set $\mathcal{O}_n$ of coadjoint orbits of N_n in $\mathfrak{n}_n^*$ is a union of simple "cells" Ω_i of different dimensions d_i which are quasiaffine manifolds of type $K^n \backslash K^m \cong \mathbb{A}^m(K) \times GL(1, K)^{n-m}$. It seems that a similar, or slightly more complicated, picture exists in higher dimensions too.

So, it is natural to conjecture that for a finite field $\mathbb{F}_q$ of q elements the number of orbits is a polynomial in q. Moreover, if we denote by $\mathcal{O}_n^m(\mathbb{F}_q)$ the set of $2m$-dimensional orbits in $\mathfrak{n}_n^*(\mathbb{F}_q)$,[1] then the sum

$$P_n(q, t) = \sum_{m \geq 0} t^m \cdot \#\mathcal{O}_n^m(\mathbb{F}_q)$$

turns out to be a polynomial in q and t. This polynomial encodes all essential information about the set of orbits.

For small n we have:

$$\begin{aligned}
P_0 &= 1, \\
P_1 &= 1, \\
P_2 &= q, \\
P_3 &= (q-1)t + q^2, \\
P_4 &= q(q-1)t^2 + (q-1)q(q+1)t + q^3, \\
P_5 &= (q-1)^2 t^4 + (q-1)q(2q-1)t^3 + (q^2-1)q(2q-1)t^2 \\
&\quad + (q-1)q^2(2q+1)t + q^4.
\end{aligned}$$

It was shown in [K2] that the beginning terms of the sequence $\{P_n\}$ written above (and even more detailed information on orbits) can be obtained by a variant of the Euler–Bernoulli triangle. But in the seventh row of the triangle the discrepancy arises. So, the general formula for P_n's is still unknown.

Note that the polynomials $Y_n(q) := P_n(q, q)$ have a still nicer behavior. To describe it we recall the definition of four remarkable sequences of polynomials

$$\{A_n\}, \quad \{B_n\}, \quad \{C_n\}, \quad \{D_n\}, \quad n \geq 0, \quad \text{in} \quad \mathbb{Z}[q]$$

---

[1] Recall that all coadjoint orbits are symplectic manifolds, hence have an even dimension.

introduced in [KM].

1. $A_n(q)$ is by definition the number of solutions of the equation $X^2 = 0$ in $n \times n$ upper-triangular matrices with elements in $\mathbb{F}_q$. More precisely, let $A_n^r(q)$ be the number of solutions which have rank r. These quantities satisfy the simple recurrence relations

$$A_{n+1}^{r+1}(q) = q^{r+1} \cdot A_n^{r+1}(q) + (q^{n-r} - q^r) \cdot A_n^r(q), \quad A_{n+1}^0(q) = 1, \quad n \geq 0.$$

We see, in particular, that $A_n^r(q)$ are polynomials in q. Thus, $A_n(q) = \sum_{r \geq 0} A_n^r(q)$ are polynomials as well.

2. Consider the triangle

$$\begin{array}{ccccccccc}
 & & & & b_{00} & & & & \\
 & & & b_{10} & \longleftarrow & b_{01} & & & \\
 & & b_{02} & \longrightarrow & b_{11} & \longrightarrow & b_{20} & & \\
 & b_{30} & \longleftarrow & b_{21} & \longleftarrow & b_{12} & \longleftarrow & b_{03} & \\
b_{04} & \longrightarrow & b_{13} & \longrightarrow & b_{22} & \longrightarrow & b_{31} & \longrightarrow & b_{40} \\
\cdots & \cdots & \cdots & \cdots & \cdots & \cdots & \cdots & \cdots & \cdots
\end{array}$$

formed by polynomials $b_{k,l}(q)$, $k \geq 0, l \geq 0$, which are labeled in a "shuttle way" and satisfy the recurrence relations

$$\begin{aligned}
b_{k,l} &= q^{-1} b_{k-1,\, l+1} + (q^{l+1} - q^l) b_{l,\, k-1} \text{ for } k > 0; \\
b_{0,l} &= q^l b_{l-1,0} \text{ for } l > 0; \quad b_{0,0} = 1.
\end{aligned}$$

Now put $B_n(q) := b_{n-1,0}(q)$, $n > 0$, $B_0(q) = 1$. This is our second sequence.

3. Define the generalized Catalan numbers $c_{k,l}$ for $k \geq 1$, $|l| \leq k$, $k \equiv l \bmod 2$, by

$$c_{k,k-2s} = \binom{k-1}{s} - \binom{k-1}{s-1}.$$

It is convenient to put $c_{k,l} = 0$ for $|l| > k$. The triangle formed by these numbers satisfies the same recurrence relation as the Pascal triangle: $c_{k,l} = c_{k-1,l} + c_{k,l-1}$, but with a different initial condition: $c_{1,-1} = -1$, $c_{1,1} = 1$ instead of $c_{0,0} = 1$.

The numbers $c_n := c_{2n+1,1}$, $n \geq 0$, form the sequence of ordinary Catalan numbers:[2] $1, 1, 2, 5, 14, 42, 132, \ldots$.

It is pertinent to remark that for a positive l the generalized Catalan number $c_{k,l}$ equals the dimension of the irreducible representation of the symmetric group $\mathbf{S}_{k-1}$ corresponding to the partition $(2^{\frac{k-l}{2}}, 1^{l-1})$. In particular, the ordinary Catalan number c_n corresponds to the rectangular diagram (2^n).

The third sequence is defined by the formula

$$C_n(q) = \sum_s c_{n+1,s} \cdot q^{\frac{n^2}{4} + \frac{1-s^2}{12}},$$

[2] Which are usually defined by the recurrence $c_{n+1} = \sum_{k=0}^{n} c_k \cdot c_{n-k}$ and the initial conditions $c_0 = c_1 = 1$.

where the sum is taken over all integers $s \in [-n-1, n+1]$ which satisfy

$$s \equiv n+1 \pmod 2, \quad s \equiv (-1)^n \pmod 3.$$

As was observed in [KM], the first 26 polynomials $C_n(q)$ coincide with $A_n(q)$ and $B_n(q)$. Later it was shown by Doron Zeilberger, using the technique from [PWZ], that $A_n = C_n$ for all $n \geq 0$.

4. The fourth sequence is defined by

$$D_n(q) := \zeta_{N_n(\mathbb{F}_q)}(-1) = \sum_{\lambda \in \widehat{N}_n(\mathbb{F}_q)} d(\lambda).$$

Here $\zeta_G(s) = \sum_{\lambda \in \widehat{G}} d(\lambda)^{-s}$ is the so-called group zeta-function (see e.g. [K2]). In other words, $D_n(q)$ is the sum of dimensions of all irreducible representations of the finite group $N_n(\mathbb{F}_q)$.

Now we can formulate the remarkable property of the sequence $\{Y_n\}$.

Proposition 1 ([KM]). *For* $0 \leq n \leq 11$ *we have*

$$A_n(q) \equiv B_n(q) \equiv C_n(q) \equiv D_n(q) \equiv Y_n(q).$$

Here are the first dozen of these polynomials:

$$\begin{aligned}
A_0 &= 1,\\
A_1 &= 1,\\
A_2 &= q,\\
A_3 &= 2q^2 - q,\\
A_4 &= 2q^4 - q^2,\\
A_5 &= 5q^6 - 4q^5,\\
A_6 &= 5q^9 - 5q^7 + q^5,\\
A_7 &= 14q^{12} - 14q^{11} + q^7,\\
A_8 &= 14q^{16} - 20q^{14} + 7q^{12},\\
A_9 &= 42q^{20} - 48q^{19} + 8q^{15} - q^{12},\\
A_{10} &= 42q^{25} - 75q^{23} + 35q^{21} - q^{15},\\
A_{11} &= 132q^{30} - 165q^{29} + 44q^{25} - 10q^{22}.
\end{aligned}$$

To finish the setting, we list here some problems which look very intriguing and are still unsolved.

1. Show that the number of coadjoint orbits for the group $N_n(\mathbb{F}_q)$ is a polynomial in q for any given n.

2. Find the asymptotics of the number of coadjoint orbits for the group $N_n(\mathbb{F}_q)$ when n goes to infinity. In particular, compute the

$$\limsup_{n\to\infty} \frac{\log\log \# \mathcal{O}_n}{\log n}.$$

3. Prove (or disprove) that the 5 sequences $A_n(q)$, $B_n(q)$, $C_n(q)$, $D_n(q)$, $Y_n(q)$ coincide for all $n \geq 0$.

4. The most intriguing problem is to interpret the formula for $\# \mathcal{O}_n(\mathbb{F}_q)$ given in [K2]:

$$\# \mathcal{O}_n(\mathbb{F}_q) = \frac{1}{\# N_n(\mathbb{F}_q)} \sum_{\substack{F\in\mathfrak{n}^*(\mathbb{F}_q)\\ X,Y\in\mathfrak{n}(\mathbb{F}_q)}} \theta(\langle F, [X, Y]\rangle)$$

as the partition function of some quantum-mechanical system, where θ is a fixed nontrivial additive character of the field K. Note that in the natural generalization of this formula to algebraic extensions of $\mathbb{F}_q$ the degree of the extension plays the role of the standard parameter β (inverse temperature) in statistical mechanics.

3 Geometry of the flag manifold

The group N_n is the unipotent radical of the semisimple algebraic group $SL(n)$. Therefore, the questions about N_n can be considered as a part of the general theory of semisimple algebraic groups. In particular, the description of coadjoint orbits for N_n (and for unipotent radicals of other semisimple algebraic groups) is related to the geometry of the so-called *flag manifolds*. We recall here some standard definitions and notation.

Let G be a complex semisimple Lie group. We denote by $\mathfrak{g}$ the Lie algebra Lie(G) and use the following notation:

$\mathfrak{h}$ – a Cartan subalgebra, $\dim \mathfrak{h} =: \operatorname{rk} \mathfrak{g} = l$;

Δ – the root system for $(\mathfrak{g}, \mathfrak{h})$, $\#\Delta = 2r$;

$\mathfrak{b} = \mathfrak{b}_+$, $\mathfrak{b}_-$ – opposite Borel subalgebras related to the splitting $\Delta = \Delta_+ \sqcup \Delta_-$;

$\Pi = (\alpha_1, \ldots, \alpha_l) \subset \Delta_+$ – the set of simple roots;

$\mathfrak{n} = \mathfrak{n}_+ = [\mathfrak{b}, \mathfrak{b}]$, $\mathfrak{n}_- = [\mathfrak{b}_-, \mathfrak{b}_-]$ – nilpotent radicals of $\mathfrak{b}_\pm$;

$(\cdot, \cdot)$ — the Cartan-Killing form on $\mathfrak{g}$ and on $\mathfrak{g}^* \cong \mathfrak{g}$;

X_α – a root vector corresponding to $\alpha \in \Delta$; the normalization condition

$$(X_\alpha, X_{-\alpha}) = 1 \qquad \text{is usually imposed;}$$

$H_\alpha \in \mathfrak{h}$ – the vector corresponding to the covector $\alpha \in \Delta \subset \mathfrak{h}^*$; we have

$$\alpha(H) = (H, H_\alpha), \qquad H_\alpha = [X_\alpha, X_{-\alpha}];$$

$C_{\alpha\beta} = \frac{2(\alpha,\beta)}{(\alpha,\alpha)}$ – the Cartan matrix;

$d_\beta = \frac{2}{(\beta,\beta)}$ – the normalization constant, making the matrix $d_\beta C_{\alpha\beta}$ symmetric;

s_α – the reflection in the hyperplane (mirror) $M_\alpha = \alpha^\perp \subset \mathfrak{h}^*$;

$W = N_G(H)/H$ – the Weyl group of G; it is usually identified with a subgroup in Aut $\mathfrak{h}^*$ generated by simple reflections $s_i = s_{\alpha_i}$, $1 \le i \le l$.

It is known that G can be viewed as the set of complex points of some algebraic group defined over $\mathbb{Z}$. So, we can introduce the group $G(K)$ of K-points of G for any field K. We do not give here the precise definitions because they are rather obvious in the case $G = SL(n)$, the only one considered below.

We denote by $B_\pm(K)$, $N_\pm(K)$, ... etc, the subgroups of $G(K)$ corresponding to Lie subalgebras $\mathfrak{b}_\pm$, $\mathfrak{n}_\pm$, ... etc. The indication to the field K is omitted when the field is fixed or immaterial.

The flag manifold $\mathcal{F}$ is defined as the set of all Borel subgroups in G. Since all such subgroups are conjugate and since B is its own normalizer in G, the flag manifold is homogeneous and can be written as the coset space $\mathcal{F} = G/B$. It is a projective algebraic manifold and the set $\mathcal{F}(K)$ of its K-points is $G(K)/B(K)$.

The famous *Bruhat Lemma* (proved first by Gelfand–Naimark for classical groups and later by Harish-Chandra in the general case) claims that double B-cosets in G are labelled by elements of W:

$$G = \bigsqcup_{w \in W} B\dot{w}B, \quad \text{where } \dot{w} \in N_G(H)$$

is a representative of $w \in W = N_G(H)/H$.

In the case $G = SL(n)$ this lemma is essentially equivalent to the Gauss method of solving a system of linear equations by successive elimination of unknowns.

Let $\mathcal{F} = G/B$ be the flag manifold for G. Under the action of the subgroup $N \subset G$ it splits into orbits

$$\mathcal{F}_w = N\dot{w}B \cong N/N_w, \quad \text{where} \quad N_w = N \bigcap \dot{w}N\dot{w}^{-1}.$$

Each orbit $\mathcal{F}_w$ is an affine algebraic submanifold in $\mathcal{F}$, isomorphic to $\mathbb{A}^{l(w)}$ where $l(w)$ is the length of w, i.e., the length of the shortest (reduced) representation of w as the product of simple reflections:

$$w = s_{i_1} \cdots s_{i_{l(w)}}. \tag{1}$$

More precisely, let $\Delta_w = \Delta_+ \bigcap w^{-1}(\Delta_-)$. Then $\# \Delta_w = l(w)$ and the general point of $\mathcal{F}_w$ can be uniquely written as

$$x(t_1, \ldots, t_{l(w)}) = \exp(t_1 X_{\alpha_1}) \cdots \exp(t_{l(w)} X_{\alpha_{l(w)}}) B \in G/B \tag{2}$$

where $\alpha_1, \ldots, \alpha_{l(w)}$ are all roots in Δ_w taken in any fixed order. In the case of the finite field $K = \mathbb{F}_q$, to make sense of (2) we can use the "fake exponential map" **exp** (see the next section) instead of ordinary exponential which is not defined.

There is another parametrization of $\mathcal{F}_w$ due to Lusztig. Namely, let us define $g_i(\tau) = \mathbf{exp}\,(\tau X_{-\alpha_i})$, $1 \le i \le l$. Then the general element $x \in \mathcal{F}_w$ can be written as

$$\widetilde{x}(\tau_1, \ldots, \tau_{l(w)}) = g_{i_1}(\tau_1)\cdots g_{i_{l(w)}}(\tau_{l(w)})B \in G/B. \qquad (2')$$

Note that (2′) is actually the parametrization of the intersection of $\mathcal{F}_w$ with the "big cell" N_-B. The transition from $(t_1, \ldots, t_{l(w)})$ to $(\tau_1, \ldots, \tau_{l(w)})$ is given by rational functions.

Let $T_B\mathcal{F}$ be the tangent space to the flag manifold at the initial point. The subgroup N is contained in $B = Stab\,B$. Therefore it acts on $T_B\mathcal{F}$ by linear transformations. The key observation is

Theorem 1. *The action of the subgroup $N \subset G$ on $T_B\mathcal{F}$ and the coadjoint action of N on $\mathfrak{n}^*$ are equivariant.*

Proof. The space $T_B\mathcal{F}$ can be identified with $\mathfrak{g}/\mathfrak{b}$ so that the action of $B = Stab(B)$ goes to the quotient of the adjoint action. Now, using the Cartan–Killing form we identify $\mathfrak{g}$ with $\mathfrak{g}^*$ and the adjoint action with the coadjoint one. Since $\mathfrak{b}$ is exactly the orthocomplement to $\mathfrak{n}$, the quotient space is identified with $\mathfrak{n}^* \cong \mathfrak{g}^*/\mathfrak{n}^\perp$. □

Our goal, consequently, is to understand the (rather complicated) N-orbit structure on $\mathfrak{n}^* \cong T_B\mathcal{F}$ by contemplating the (rather simple) N-orbit structure on $\mathcal{F}$. This problem is not yet solved and here I want to formulate some preliminary results and related conjectures.

The idea is that every N-orbit $\mathcal{F}_w$, $w \in W$, in the flag manifold has its "shadow" S_w in the tangent space which is a union of coadjoint orbits. Moreover, there is a conjectural description of orbits in a given shadow. More precisely, denote by $K^\times \cong GL(1, K)$ the set of all nonzero elements in K and put

$$r(w) = \operatorname{rk}(w - 1).$$

Then we formulate

Main Conjectures

(i) *The set of orbits in S_w is isomorphic to $(K^\times)^{r(w)}$.*

(ii) *Every orbit in S_w is isomorphic to $\mathbb{A}(K)^{l(w)-r(w)}$.*

(iii) *Any two shadows either coincide or are disjoint.*

We associate to $w \in W$ the *characteristic monomial*

$$M_w := Q^{r(w)}t^{d(w)} \in \mathbb{Z}[Q, t] \quad \text{where} \quad d(w) = \frac{l(w) - r(w)}{2}, \quad Q = q - 1. \qquad (3)$$

The first two conjectures above claim that M_w is exactly the contribution of S_w into $P_n(q, t)$. The last condition means that $P_n(q, t) = \sum_{w \in W_0} M_w$ where $W_0 \subset W$ is a certain part of W.

Thus, if the Main Conjectures are true, our problem is reduced to the question: which cells $\mathcal{F}_w$ have the same shadows in the tangent space? From all cells with the same shadow we have to include only one to W_0.

Consider in more detail the structure of the flag manifold $\mathcal{F}$ near the origin B. In the neighborhood of this point we can introduce the local chart using the action of N_-. Indeed, the N_--orbit of B is a dense open subset in $\mathcal{F}$. On the nilpotent group N_- there are several natural global coordinate systems defined over any field. Namely, if we fix somehow the order $(\alpha_1, \ldots, \alpha_r)$ of positive roots, then the general point of N_- can be written as

$$n(s_1, \ldots, s_r) = \exp(s_1 X_{-\alpha_1}) \cdots \exp(s_r X_{-\alpha_r}). \tag{4}$$

We can consider the quantities $(s_1, \ldots, s_r)$ as coordinates in $N_- B \subset \mathcal{F}$.

On the other hand, the space $\mathfrak{n}^*$ also has natural coordinates labeled by positive roots: $x_\alpha(F) = \langle F, X_\alpha \rangle$ for $F \in \mathfrak{n}^*$. Under the identification $\mathfrak{n}^* \cong \mathfrak{n}_-$ the element with coordinates $\{x_\alpha\}$ goes to $\sum_{\alpha \in \Delta_+} x_\alpha X_\alpha$, since $(X_{-\alpha}, X_\beta) = \delta_{\alpha,\beta}$.

The map $\mathbb{A}^{l(w)} \longrightarrow \mathcal{F}_w \subset \mathcal{F}$ given by (2) is rational in the coordinates chosen on N/N_w and on $N_- B$, i.e., $s_i = s_i(t_1, \ldots, t_{l(w)})$ are rational functions. Note that the initial point B has all s-coordinates zero and *is not covered* by the image of (2) for $w \neq e$.

Instead of the definition of a shadow (which is still missing) we consider several particular examples where this definition is self-suggesting.

Example 1. Let α be a simple root of $\mathfrak{g}$ and let $\sigma_\alpha \in W$ be the corresponding simple reflection. The cell $\mathcal{F}_{\sigma_\alpha} \subset \mathcal{F}$ is isomorphic to $\mathbb{A}^1(K)$ and the map (1) looks like

$$s_\alpha(t) = t^{-1}, \qquad s_\beta(q) = 0 \quad \text{for} \quad \beta \neq \alpha.$$

We define the shadow S_{σ_α} of $\mathcal{F}_{\sigma_\alpha}$ as the line $K \cdot X_{-\alpha} \subset \mathfrak{n}_- \cong \mathfrak{n}^*$ with the origin deleted. The characteristic monomial is $M_{\sigma_\alpha} = Q$. This shadow splits on $Q = q - 1 = \# GL(1, K)$ one-point orbits in $\mathfrak{n}^*$ as one can check directly.

It is in a perfect agreement with the equality $M_{\sigma_\alpha} = Q$.

Example 2. Let c be a Coxeter element in W, i.e., a product of all simple reflections in any order. It is known that there are $2^{\operatorname{rk}\mathfrak{g}}$ such elements which belong to the same conjugacy class in W. We have $l(c) = r(c) = \operatorname{rk}\mathfrak{g}$. So, the characteristic monomial is $M_c = Q^{\operatorname{rk}\mathfrak{g}}$. The computations in [HHKR] show that the images of $\mathcal{F}_c$, c a Coxeter element, in the big cell $N_- B$ have the same "tangent cones" when they approach the boundary point B. This suggests that all $\mathcal{F}_c$ have the same shadow

$$S_c = \left\{ \sum_{\alpha \in \Pi} a_\alpha X_{-\alpha} \mid a_\alpha \in K^\times \right\}.$$

Now we concentrate on the case $G = SL(n, \mathbb{C})$ where the Weyl group W is just the symmetric group S_n.

Recall that the conjugacy class of an element $w \in W$ is conveniently denoted by $1^{\alpha_1}2^{\alpha_2}\cdots n^{\alpha_n}$ where α_k is the number of cycles of length k in the permutation w. We introduce the notation:

$c(w) = \sum_k \alpha_k =$ total number of cycles;

$r(w) = \mathrm{rk}\,(w-1) = \sum_k \alpha_k(k-1) = n - c(w)$.

Lemma 1. *The number* $d(w) = \frac{l(w)-r(w)}{2}$ *is an integer.*

Proof. Consider the sign character of W given by $sgn\,(w) = \frac{\prod_{i<j}(x_i-x_j)}{\prod_{i<j}(x_{w(i)}-x_{w(j)})}$.

It is known and easy to check that $sgn\,(w) = (-1)^{l(w)}$. On the other hand, the cyclic permutation of length k has the sign $(-1)^{(k-1)}$. Therefore, $sgn\,w = (-1)^{r(w)}$ and $l(w) \equiv r(w) \bmod 2$. □

We collect here the information about cells and their shadows for $n \le 4$. Namely, we compute the map (2) and describe its image. The general scheme is easy (but rather tedious for big n): we start with a permutation matrix $\dot{w} \in S_n$, multiply it from the left by a general matrix $n \in N_n$ and rewrite the result as a product $n_- \cdot b$, $n_- \in N_-$, $b \in B$. Then we solve the system of linear equations and express the entries s_{ij} of b in terms of t_{kl} of n.

Below, the case $\dot{w} = (14) = \sigma_1\sigma_2\sigma_3\sigma_2\sigma_1$ is considered as an example. The equation looks as follows:

$$\begin{pmatrix} 1 & t_{12} & t_{13} & t_{14} \\ 0 & 1 & t_{23} & t_{24} \\ 0 & 0 & 1 & t_{34} \\ 0 & 0 & 0 & 1 \end{pmatrix} \cdot \begin{pmatrix} 0 & 0 & 0 & 1 \\ 0 & 1 & 0 & 0 \\ 0 & 0 & 1 & 0 \\ 1 & 0 & 0 & 0 \end{pmatrix}$$

$$= \begin{pmatrix} 1 & 0 & 0 & 0 \\ s_{21} & 1 & 0 & 0 \\ s_{31} & s_{32} & 1 & 0 \\ s_{41} & s_{42} & s_{43} & 1 \end{pmatrix} \cdot \begin{pmatrix} b_{11} & b_{12} & b_{13} & b_{14} \\ 0 & b_{22} & b_{23} & b_{24} \\ 0 & 0 & b_{33} & b_{34} \\ 0 & 0 & 0 & b_{44} \end{pmatrix}.$$

Solving the system of linear equations we get

$$s_{21} = t_{24}t_{14}^{-1}, \qquad s_{31} = t_{34}t_{14}^{-1}, \qquad s_{32} = t_{13}t_{24}(t_{12}t_{24}-t_{14})^{-1}, \qquad s_{41} = t_{14}^{-1},$$

$$s_{42} = t_{12}(t_{12}t_{24}-t_{14})^{-1}, \quad s_{43} = (t_{12}t_{23}-t_{13})(t_{14}-t_{12}t_{24}-t_{13}t_{34}+t_{12}t_{23}t_{34})^{-1}.$$

This means that the image of $\mathcal{F}_w$ in N_- is given by the conditions that

$$s_{41} \neq 0, \quad s_{31}s_{42} = s_{32}s_{41}.$$

Therefore it is natural to define S_w by the same conditions (identifying $\mathfrak{n}^* \cong \mathfrak{n}_-$ with N_- by the map **exp**, see the next section). Then this shadow splits on Q four-dimensional orbits given by the equations $s_{41} = const \neq 0$, $s_{31}s_{42} = s_{32}s_{41}$. Hence, the contribution of this shadow is equal to Qt^2 in accordance with the Main Conjectures.

In conclusion, we list the characteristic monomials M_w and polynomials P_n for $n \leq 4$.

$n = 1:$ $M_e = 1;$ $P_1 = 1.$

$n = 2:$ $M_e = 1,$ $M_\sigma = Q;$ $P_2 = q = Q + 1.$

$n = 3:$ $M_e = 1,$ $M_{\sigma_i} = Q,$ $M_{\sigma_i\sigma_j} = Q^2,$ $M_{\sigma_i\sigma_j\sigma_i} = Qt;$

$$P_3 = (q-1)t + q^2 = Qt + Q^2 + 2Q + 1.$$

$n = 4:$ $M_e = 1,$ $M_{\sigma_i} = Q,$ $M_{\sigma_i\sigma_j} = Q^2,$ $M_{\sigma_i\sigma_{i+1}\sigma_i} = Qt;$

$M_{\sigma_i\sigma_j\sigma_k} = Q^3,$ $M_{\sigma_i\sigma_{i+2}} = Qt,$ $M_{\sigma_i\sigma_{i+1}\sigma_i\sigma_j} = M_{\sigma_j\sigma_i\sigma_{i+1}\sigma_i} =$ Q^2t

$M_{(14)} = Qt^2,$ $M_{(1423)} = Q^3t,$ $M_{(14)(23)} = Q^2t^2.$

$$P_4 = q(q-1)t^2 + q(q^2-1)t + q^3 = Q^2t^2 + Qt^2 + Q^3t + 3Q^2t + 2Qt + Q^3 + 3Q^2 + 3Q + 1.$$

We see that indeed, for $n \leq 4$, the polynomial $P_n(q, t)$ can be composed from appropriately chosen characteristic monomials. Moreover, the cells $\mathcal{F}_w$ and $\mathcal{F}_{w'}$ seem to have the same shadow if the reduced decomposition for w' can be obtained from the one for w by replacing one Coxeter element by another.[3]

Of course, the direct computations for the cases $n = 5$ and $n = 6$ would shed more light on the situation. This work is now in progress.

4 Cluster representations

The second approach has its origin in the character formula for irreducible representations of N_n. For the real, complex or p-adic field K this formula can be written in a uniform way:

$$\chi_\Omega(\exp X) = \int_\Omega \theta(\langle F, X\rangle)dF. \tag{5}$$

Here χ_Ω is the character of an irreducible representation of N_n corresponding to an orbit $\Omega \subset \mathfrak{n}_n^*(K)$, θ is a fixed nontrivial additive character of the field K and dF is the canonical measure on Ω defined by the symplectic structure.

In the case of finite field $\mathbb{F}_q$ the following modification was suggested in [K2]:

$$\chi_\Omega(\mathbf{exp}\, X) = \frac{1}{\sqrt{\#\Omega}} \sum_{F\in\Omega} \theta(\langle F, X\rangle). \tag{6}$$

Here **exp** is a "fake exponential map" from $\mathfrak{n}_n(\mathbb{F}_q)$ to $N_n(\mathbb{F}_q)$:

$$\mathbf{exp}\, X = 1 + X. \tag{7}$$

[3]The role of Coxeter elements was brought to my attention by N. Reshetikhin.

This map is one-to-one and N_n-equivariant but fails to have other nice properties of the exponential map. In particular, $\mathbf{exp}\,(X+Y) \neq \mathbf{exp}\,(X)\mathbf{exp}\,(Y)$ even for commuting X and Y.

It soon turned out that the character formula above is overly optimistic. The computer calculations [IK] provided (indirect) evidence that in fact the character values should not always lie in the cyclotomic field suggested by (6). Nevertheless, all attempts at producing a character which directly contradicts the conjecture had failed.

In any case, the "pseudo-characters" given by the right-hand sides of (6) form a most natural orthonormal basis for the space of complex-valued class functions on the group, and it is an extremely intriguing problem to find the transition matrix which connects this basis to the actual irreducible characters.

The work by Ning Yan [Y] starts with two simple observations.

First, he proposed extending the coadjoint action of N on the dual space of its Lie algebra to an action of $N \times N$, which we shall call the *double action*:

$$(n_1, n_2) \cdot F = \left(n_2 \cdot F \cdot n_1^{-1}\right)_{low} \cdot \tag{8}$$

Here we identify $\mathfrak{n}^* \cong \mathfrak{g}^*/\mathfrak{n}^\perp \cong \mathfrak{g}/\mathfrak{b}$ with $\mathfrak{n}_-$ and use the subscript "*low*" to denote the lower part of a matrix.

Second, in the case $K = \mathbb{F}_q$ the group algebra of $N(\mathbb{F}_q)$ has a natural basis, the *Fourier basis*:

$$\phi_F(n) = \theta(\langle F, \mathbf{exp}^{-1} n\rangle) = \theta(\mathrm{tr}\,(F \cdot n)), \quad n \in N, \quad \mathbf{exp}^{-1}\, n = n - 1 \in \mathfrak{n}. \tag{9}$$

The double action and the Fourier basis are related through the regular representation of $N(\mathbb{F}_q)$ on its group algebra:

$$(n_1, n_2)\phi_F = \phi_{(n_1,n_2)\cdot F} \cdot \theta(\langle F, \mathbf{exp}^{-1}(n_1^{-1} n_2)\rangle), \qquad X \in \mathfrak{n}. \tag{10}$$

Ning shows that this relation leads to a rich theory with many of the major features one would expect from the philosophy of the orbit method. These features include a one-to-one correspondence between the *cluster characters* χ_δ and the *double orbits* O_δ, a character formula of type (6) expressing the character in terms of the orbit, and equivalence between character operations (tensor product, induction and restriction) and orbit operations.

A remarkable new feature is the canonical decomposition of the regular representation into a direct sum of *cluster modules* M_δ, the linear spans of ϕ_F, $F \in O_\delta$.

The collection of all cluster characters χ_δ forms a natural orthogonal basis for the $\mathbb{C}$-linear space of *double-invariant* functions on $N_n(\mathbb{F}_q)$. The collection of all $\mathbb{Z}$-linear combinations of cluster characters form a subring in the representation ring of $N_n(\mathbb{F}_q)$. This subring is generated by the *primary* characters, and each cluster character has a unique *primary decomposition* into a tensor product of primary characters.

The last part of Ning's work was inspired by a paper of G.I. Lehrer [L]. Lehrer was interested in the *discrete series character* of $N_n(\mathbb{F}_q)$, which is by definition the common restriction of the discrete series characters of $GL(n, \mathbb{F}_q)$. He derived a tensor product decomposition of this character, and then further decomposed each factor into a multiplicity-free sum of irreducible components. Ning proves that these irreducible components are precisely the primary characters of his theory.

To finish this introduction we draw the reader's attention to another previous work on the representation theory of $N(\mathbb{F}_q)$, by C.A.M. André [A]. Throughout his papers André makes the assumption $q \geq n$ in order to have the benefit of a well-behaved exponential map. In this case a one-to-one correspondence between the irreducible characters and the coadjoint orbits was established by D. Kazhdan [Kaz]. André defines certain "basic sums" of coadjoint orbits, and studies the characters associated to the orbit sums via the above correspondence. Note that here the popular mode of thinking is to go from $\mathbb{F}_q$ to its algebraic closure, so that the coadjoint orbits can be treated as geometric objects. It then came about that the orbit sums of André are precisely the double orbits, and the characters in his work are the cluster characters introduced by Ning.

Now we return to our main topic and discuss how the above approach is related to the classification problem for coadjoint orbits.

The idea is to gather the coadjoint orbits in some "clusters" corresponding to double orbits, i.e., $(N \times N)$-orbits in $\mathfrak{n}^*$. Note that the diagonal subgroup $N \subset N \times N$ acts on $\mathfrak{n}^*$ by the coadjoint representation. Therefore, each double orbit is a union of coadjoint orbits. The point is that the double orbits admit a rather transparent description. It turns out that they are related to the so-called *rook placements* on a triangular chess board.

Let us consider the double orbits in more details. We denote by δ the following data:

a) The collection of boxes $\{(i_1, j_1), \ldots, (i_k, j_k)\}$ satisfying the conditions:

(i) All i_p, $1 \leq p \leq k$, are different;

(ii) All j_q, $1 \leq q \leq k$, are different;

(iii) $n \geq i_p > j_p \geq 1$, $1 \leq p \leq k$.

b) The collection of nonzero elements $a_p \in K$, $1 \leq p \leq k$.

The set of all possible δ satisfying the above conditions we denote by $\Delta_n(K)$ (or simply by Δ when n and K are understood or immaterial).

Geometrically, the collection of boxes satisfying conditions (i) – (iii) is exactly the rook placement on the lower triangular chessboard with $n(n-1)/2$ boxes (i.e., such a collection which has at most one box in every row and in every column).

To a $\delta \in \Delta$ we associate three numerical characteristics. Put

$$|\delta| = k, \quad h(\delta) = \sum_{p=1}^{k} (i_p - j_p - 1), \quad i(\delta) = \#\{(p, q) \mid j_p < j_q < i_p < i_q\}. \tag{10}$$

The geometric meaning of these numbers is the following. For any box (i, j) we define its *hook* as the collection of all boxes situated to the right of (i, j) and above it. This hook consists of two parts: horizontal and vertical of the same length $h(i, j) = j - i - 1$. The number $h(\delta)$ is just the sum of lengths of all hooks corresponding to boxes in δ.

Tho hooks can intersect in at most one point. The number $i(\delta)$ counts the total amount of intersections for the hooks related to boxes in δ.

To any $\delta \in \Delta_n(K)$ we associate an element

$$F_\delta = \sum_{p=1}^{k} a_p E_{i_p j_p} \in \mathfrak{n}_-(K) \cong \mathfrak{n}^*(K) \tag{11}$$

where E_{ij} are the standard matrix units.

The very first naive attempts to classify coadjoint orbits were based on the assumption that there exists a nice set of representatives of special kind (like the hook placements). All these attempt failed for n large enough.

It turns out that for double orbits such a set indeed exists.

Theorem 2 ([Y]). a) *Every $(N_n(K)\times N_n(K))$-orbit $O \subset \mathfrak{n}^*(K)$ contains a unique element F_δ, $\delta \in \Delta_n(K)$.*

b) *The dimension of the $(N_n \times N_n)$-orbit O_δ passing through F_δ is equal to* $2h(\delta) - i(\delta)$.

c) *The self-intertwining number of the cluster representation related to O_δ is equal to $q^{i(\delta)}$. In particular, this representation is irreducible iff* $i(\delta) = 0$.

The detailed study of clusters and cluster representations has only just begun. One of the first problems here is to compute the generating function for the numbers of clusters with given characteristics $|\delta|$, $l(\delta)$ and $i(\delta)$. If we restrict ourselves by the first characteristic, the answer is known: the number of k-rook placements on the triangular chessboard with $\frac{n(n-1)}{2}$ boxes is equal to the so-called Stirling number of second kind $S(k, n)$. Therefore

$$\sum_{n\geq 0} \frac{t^n}{n!} \cdot \#\Delta_n(\mathbb{F}_q) = \sum_{n\geq 0} \left(\frac{t^n}{n!} \cdot \sum_{k=0}^{n} S(k, n) \cdot Q^k \right) = e^{Q(e^t-1)}, \quad \text{where } Q = q-1.$$

It seems plausible that the further study of clusters will help to answer the questions formulated at the end of Section 2.

I sincerely hope that the readers of this paper will contribute to this study.

Added in proof. Let σ be the permutation $(1, n+1)(2, n+2)\cdots(n, 2n) \in S_{2n}$. The structure of a double orbit of F_σ in $\mathfrak{n}_{2n}^*$ is as complicated as the whole set $\mathcal{O}(\mathcal{N}_\backslash)$.

5 Statistics of permutations

In view of the results above we are interested in the asymptotic distribution of the length function $l(s)$ on S_n for large n. Let $c_k^{(n)}$ denote the number of elements of length k in S_n. The generating function related to this sequence is given by

$$\Phi_n(q) = \sum_{k=0}^{\infty} c_k^{(n)} q^k = \prod_{m=1}^{n} \frac{1-q^m}{1-q}.$$

Consider the probability measure μ_n on $\mathbb{R}$ which is concentrated on the set $\{0, 1, \ldots, \frac{n(n-1)}{2}\}$ and the probability of the point $\{k\}$ is equal to $\frac{1}{n!}c_k^{(n)}$. The generating function shows that it is a convolution of uniform measures on the sets $\{0, 1, \ldots, k-1\}$, $1 \leq k \leq n$.

The Fourier transform of this measure is equal to

$$\Phi_n(e^{i\lambda}) = e^{i\frac{n(n-1)}{4}\lambda} \prod_{m=1}^{n} \frac{\sin\frac{m\lambda}{2}}{m \sin\frac{\lambda}{2}}.$$

Lemma. *When n goes to ∞ the quantity*

$$\left| \prod_{m=1}^{n} \frac{\sin\frac{m\lambda}{2n^{3/2}}}{m \sin\frac{\lambda}{2n^{3/2}}} - e^{-\lambda^2/144} \right|$$

goes to 0 *uniformly on every bounded interval in* $\mathbb{R}$.

Proof. For some $\theta_n \in (0, 1)$ we have

$$\sum_{m=1}^{n} \log \frac{\sin\frac{m\lambda}{2n^{3/2}}}{m \sin\frac{\lambda}{2n^{3/2}}} = \sum_{m=1}^{n} \log \frac{\frac{m\lambda}{2n^{3/2}} - \frac{(m\lambda)^3}{48n^{9/2}} + \frac{(m\theta_n\lambda)^5}{5!2^5 n^{15/2}}}{\frac{m\lambda}{2n^{3/2}} - \frac{m\lambda^3}{48n^{9/2}} + \frac{m(\theta_n\lambda)^5}{5!2^5 n^{15/2}}}$$

$$= \sum_{m=1}^{n} \log \frac{1 - \frac{m^2\lambda^2}{24n^3} + \frac{m^4\theta_n^5\lambda^4}{5!2^5 n^6}}{1 - \frac{\lambda^2}{24n^3} + \frac{\theta_n^5\lambda^4}{5!2^5 n^6}} = -\frac{\lambda^2}{144} + o\left(\frac{1}{n}\right).$$

As a corollary we get

Proposition. *Let $s \in S_n$ be a a random element. The probability that*

$$l(s) \in \left(\frac{n^2-n}{4} + \alpha\frac{n\sqrt{2n}}{12}, \frac{n^2-n}{4} + \beta\frac{n\sqrt{2n}}{12} \right)$$

is going to

$$\frac{1}{\sqrt{2\pi}} \int_{\alpha}^{\beta} e^{-\frac{x^2}{2}} dx$$

as n is going to ∞. In other words, the quantity $\frac{12l(s)-3(n^2-n)}{n\sqrt{2n}}$ *tends to the standard gaussian random variable.*

Some results can be obtained also for the probability distribution on S_n proportional to the function $q^{l(w)}$, $q > 1$. In this case we have to replace the uniform distribution on $\{0, 1, \dots, n-1\}$ by the measure $\mu_n(q)$ with probabilities $p_k = \frac{q^{k+1}-q^k}{q^n-1}$. The expectation of the coordinate x_k and its dispersion behave like

$$\langle x_k \rangle \sim k - \frac{1}{1-q^{-1}}, \qquad \langle x_k^2 \rangle - \langle x_k \rangle^2 \sim \frac{1}{q-2+q^{-1}}.$$

It means that for large n the random value $l(s)$ (which is given by $\sum_{k=1}^{n-1} x_k$) differs from its maximal value $\frac{n(n-1)}{2}$ by the quantity of order $\frac{n}{1-q^{-1}} + O(\sqrt{n})$.

The challenging problem is to compute the asymptotic behavior of $c_k^{(n)}$ when k and n both go to infinity in a prescribed way.

One question of that nature is to find the asymptotics of $c_{n-1}^{(n)}$ when $n \to \infty$. Another problem is the asymptotic behavior of the quantity $\Phi_n(q)$ for large n.

Acknowledgments. I am very thankful to my friend Nicolai Reshetikhin and to my student Ning Yan for fruitful discussions. I thank also IHES and especially Jean-Pierre Bourguignon and Maxim Kontsevich for hospitality.

References

[A] André C.A.M., The regular character of the unitriangular group, *J. Algebra* **201** (1998), 1–52.

On the coadjoint orbits of the unitriangular group, *J. Algebra* **180** (1996), 587–630.

[HHKR] Hoffmann T., Kellendonk J., Kutz N. and Reshetikhin N., Factorization dynamics and Coxeter-Toda lattices, solv-int/9906013 (1999), 1–34.

[IK] Isaacs I.M. and Karagueuzian D., Conjugacy in groups of upper triangular matrices, *J. Algebra* **202** (1998), 704–711; see also Erratum, *J. Algebra* **208**, no. 2 (1998), 722.

[Kaz] Kazhdan D., Proof of Springer's hypothesis, *Israel J. Math.* **28**, no. 4 (1977), 272–286.

[K1] Kirillov A.A., Unitary representations of nilpotent Lie groups, *Uspekhi Mat. Nauk* **35**, no. 4 (1962), 57–101.

[K2] Kirillov, A.A., Variations on the triangular theme. In *E.B. Dynkin Seminar on Lie Groups*, Amer. Math. Soc. Transl., Ser. 2, 1995, pp. 43–73.

[KM] Kirillov A.A., Melnikov A., On a remarkable sequence of polynomials. In *Algèbre Non-commutative, Groupes Quantiques et Invariants, Rencontre Franco-Belge, Reims 1995* (J. Alev, G. Cauchon, eds.), Publication SMF, No. 2, 1996, pp. 35–42.

[L] Lehrer G.I., Discrete Series and the unipotent subgroup, *Compositio Math.* **28** (1974), 9–19.

[PWZ] Petkovsek M., Wilf H. and Zeilberger D., $A = B$, A.K. Peters, 1998, p. 244.

[Y] Ning Yan, Representation theory for the finite unipotent linear groups, Ph.D. thesis, University of Pennsylvania, Philadelphia, 2001.

A. A. Kirillov
Department of Mathematics
The University of Pennsylvania
Philadelphia, PA 19104-6395, USA
kirillov@math.upenn.edu

The Generalized Cayley Map from an Algebraic Group to its Lie Algebra

Bertram Kostant
Peter W. Michor

For Alexandre Kirillov on the occasion of his 65th birthday

ABSTRACT Each infinitesimally faithful representation of a reductive complex connected algebraic group G induces a dominant morphism Φ from the group to its Lie algebra $\mathfrak{g}$ by orthogonal projection in the endomorphism ring of the representation space. The map Φ identifies the field $Q(G)$ of rational functions on G with an algebraic extension of the field $Q(\mathfrak{g})$ of rational functions on $\mathfrak{g}$. For the spin representation of Spin(V) the map Φ essentially coincides with the classical Cayley transform. In general, properties of Φ are established and these properties are applied to deal with a separation of variables (Richardson) problem for reductive algebraic groups: Find Harm(G) so that for the coordinate ring $A(G)$ of G we have $A(G) = A(G)^G \otimes \text{Harm}(G)$. As a consequence of a partial solution to this problem and a complete solution for $SL(n)$ one has in general the equality $[Q(G) : Q(\mathfrak{g})] = [Q(G)^G : Q(\mathfrak{g})^G]$ of the degrees of extension fields. Among other results, Φ yields (for the complex case) a generalization, involving generic regular orbits, of the result of Richardson showing that the Cayley map, when G is semisimple, defines an isomorphism from the variety of unipotent elements in G to the variety of nilpotent elements in $\mathfrak{g}$. In addition if G is semisimple the Cayley map establishes a diffeomorphism between the real submanifold of hyperbolic elements in G and the space of infinitesimal hyperbolic elements in $\mathfrak{g}$. Some examples are computed in detail.

1 Introduction

Let G be a connected complex reductive algebraic group and let $\mathfrak{g} = \text{Lie}(G)$. To any rational locally faithful representation $\pi : G \to \text{Aut}(V)$ one can associate a

1991 Mathematics Subject Classification: 22E50.
Keywords and phrases: Cayley mapping for representations.
BK was supported in part by the NSF and KG&G.
PWM was supported by "Fonds zur Förderung der wissenschaftlichen Forschung, Projekt P 14195 MAT."

dominant morphism

$$\Phi : G \to \mathfrak{g}$$

which (see example (3) below) we refer to as a generalized Cayley map. If $\pi' : \mathfrak{g} \to \mathrm{End}(V)$ is the differential of π, the bilinear form $(\alpha, \beta) = \mathrm{tr}(\alpha\beta)$ on $\mathrm{End}(V)$ defines a projection $pr_\pi : \mathrm{End}(V) \to \pi'(\mathfrak{g})$. The generalized Cayley map arises from the restriction of pr_π to $\pi(G)$. This paper establishes a number of striking properties of the map Φ.

The map Φ is conjugation-equivariant and consequently Φ *carries conjugacy classes in G to adjoint orbits in* $\mathfrak{g}$ and the corresponding cohomomorphism Φ^* defines a conjugation-equivariant injection

$$0 \longrightarrow A(\mathfrak{g}) \longrightarrow A(G)$$

of affine rings. On the level of quotient fields Φ^* defines $Q(G)$ as a finite algebraic extension of $Q(\mathfrak{g})$. We will write deg π for the degree of this extension.

Examples.

(1) $G = Gl(n, \mathbb{C})$, π is the standard representation, $\deg(\pi) = 1$.

(2) $G = Sl(n, \mathbb{C})$, π is the standard representation, $\deg(\pi) = n$, see (6.2).

(3) $G = \mathrm{Spin}(n, \mathbb{C})$, π is the Spin representation, $\deg(\pi) = n$ for n even and $\deg(\pi) = n - 1$ for n odd, see (7.15).

Theorem. (7.15) *In the above example (3) the map Φ is, on a Zariski open subset and up to scalar multiplication, given by the Cayley transform.*

Let $A(G)^G$ and $A(\mathfrak{g})^G$ be the subalgebras of G-invariants in $A(G)$ and $A(\mathfrak{g})$ respectively. The space $\mathrm{Harm}(g) \subset A(\mathfrak{g})$ of harmonic polynomials on $\mathfrak{g}$ was defined in [9] and the decomposition

$$A(\mathfrak{g}) = A(\mathfrak{g})^G \otimes \mathrm{Harm}(\mathfrak{g}) \tag{4}$$

was proved in [9]. As a generalization of (4), using Quillen's proof of a conjecture of Serre, Richardson in [16] proved that a G-stable subspace H of $A(G)$ exists such that

$$A(G) = A(G)^G \otimes H \tag{5}$$

holds. He also raised the question as to whether one can give an explicit construction of such a subspace H along the lines of (4). Although this problem is not solved in the present paper we do solve a weakened version of the problem. Let π be given and let $\mathrm{Sing}_\pi G$ be the subvariety of G where the differential $d\Phi$ is not invertible. Then $\mathrm{Sing}_\pi G$ is a hypersurface given as the zero set of the Jacobian $\Psi \in A(G)^G$ of the mapping Φ, see (2.1.2). Let $\mathrm{Harm}(G) = \Phi^*(\mathrm{Harm}(\mathfrak{g}))$. We localize with respect to Ψ and prove the following

Theorem. (3.2) *One has*

$$A(G)_\Psi = A(G)^G_\Psi \otimes \mathrm{Harm}(G). \tag{6}$$

The statement without localization,

$$A(G) = A(G)^G \otimes \mathrm{Harm}(G), \tag{7}$$

holds if and only if Φ maps regular orbits to regular orbits. This is the case in example (2).

Moreover (6) readily induces the following

Corollary. (3.3) *For the G-equivariant extension of the rational function fields $\Phi^* : Q(\mathfrak{g}) \to Q(G)$ the degrees satisfy*

$$[Q(G) : Q(\mathfrak{g})] = [Q(G)^G : Q(\mathfrak{g})^G].$$

Let $U \subset G$ be the unipotent variety and $N \subset \mathfrak{g}$ the nilcone. If G is semisimple and π is suitable, then Richardson and Bardsley in [1] used Φ, even in the finite characteristic case (for good primes), to establish that

$$\Phi : U \to N \tag{8}$$

is an isomorphism of algebraic varieties. We consider here the complex case and generalize this for reductive algebraic groups to

Theorem (4.5) *Let $a \in G$ be regular and assume that $d\Phi(a)$ is invertible. Then Φ restricts to an isomorphism of affine varieties*

$$\Phi : \overline{\mathrm{Conj}_G(a)} \to \overline{\mathrm{Ad}_G(\Phi(a))}. \tag{9}$$

Note that (8) is certainly not an isomorphism in Example (1).

Any $a \in G$ has a (multiplicative) Jordan decomposition $a = a_s a_u$ where a_s and a_u are respectively the semisimple and unipotent components of a. Analogously any $X \in \mathfrak{g}$ has an (additive) Jordan decomposition $X = X_s + X_n$ where X_s and X_n are the semisimple and nilpotent components. In contrast to (8) the map Φ always carries semisimple elements to semisimple elements. In fact one has

Theorem. (4.11) *For any $a \in G$ one has*

$$\Phi(a_s) = \Phi(a)_s. \tag{10}$$

An element $a \in G$ is called elliptic (resp. hyperbolic) if a is semisimple and the eigenvalues of $\pi(a)$ are of norm 1 (resp. positive) for all π. Expanding the multiplicative Jordan decomposition, every element $a \in G$ has a unique decomposition

$$a = a_e a_h a_u \tag{11}$$

where a_e and a_h are respectively elliptic and hyperbolic and all three components commute. We will say that a is of *positive type* if a_e is the identity. Let G_{pos} be the space of all elements of positive type. Analogously an element $X \in \mathfrak{g}$ is called elliptic (resp. hyperbolic) if X is semisimple and the eigenvalues of $\pi'(X)$ are pure imaginary (resp. real) for all π. Expanding the additive Jordan decomposition every element $X \in \mathfrak{g}$ has a unique decomposition

$$X = X_e + X_h + X_n \tag{12}$$

where X_e and X_h are respectively elliptic and hyperbolic and all three components commute. We will say that X is of real type if $X_e = 0$. Let $\mathfrak{g}_{real}$ be the space of all elements of real type in $\mathfrak{g}$. Given π let G_{nonsing} be the (Zariski open) complement of G_{sing} in G.

Theorem. (5.5) *If $a \in G$ is hyperbolic, then $\Phi(a)$ is hyperbolic. Furthermore one has $G_{\text{pos}} \subset G_{\text{nonsing}}$ and, extending (8), one has $\Phi(G_{\text{pos}}) \subset \mathfrak{g}_{\text{real}}$ and in fact*

$$\Phi : G_{pos} \to \mathfrak{g}_{\text{real}} \tag{13}$$

is a diffeomorphism. In particular Φ defines a bijection of the set of all hyperbolic elements in G to the set of all hyperbolic elements in $\mathfrak{g}$.

Obvious questions arise with regard to the restriction of the generalized Cayley map to subgroups of G. In this connection one readily establishes

Theorem. (2.6) (2.7) *We have*

$$\Phi_\pi | K = \Phi_{\pi|K} \tag{14}$$

in the following cases: If K is any reductive subgroup of a reductive G which is the connected centralizer of a subset $A \subset G$. Or if K is a subgroup corresponding to a simple ideal in the Lie algebra of a semisimple group G.

2 The generalized Cayley mapping and its basic properties

2.1 The setting

Let G be a (real or complex) Lie group with Lie algebra $\mathfrak{g}$, and let $\pi : G \to \text{Aut}(V)$ be a finite-dimensional representation with $\pi' : \mathfrak{g} \to \text{End}(V)$ injective. We say that G *admits a Cayley mapping* if the inner product $(A, B) \mapsto \text{tr}(AB)$ on $\text{End}(V)$ restricts to a nondegenerate inner product B_π on $\pi'(\mathfrak{g})$. Thus the orthogonal projection $\text{pr}_\pi : \text{End}(V) \to \pi'(\mathfrak{g})$ is well defined. We consider the real analytic mapping

$$\begin{aligned} &\Phi = \Phi_\pi : G \to \mathfrak{g} \quad \text{given by} \quad \pi' \circ \Phi_\pi = \text{pr}_\pi \circ \pi : G \to \text{End}(V) \to \pi'(\mathfrak{g}), \\ &B_\pi(\Phi_\pi(g), X) = \text{tr}(\pi'(\Phi_\pi(g))\pi'(X)) = \text{tr}(\pi(g)\pi'(X)), \qquad g \in G, X \in \mathfrak{g}, \end{aligned} \tag{1}$$

which we call the *generalized Cayley mapping* of the representation π. The choice of the name is justified by the fact that for the Spin representation of Spin(n) on $\mathbb{C}^n$ it coincides, up to a scalar, with the Cayley transform, see (7.15).

By $d\Phi = \mathrm{pr}_2 \circ T\Phi : TG \to T\mathfrak{g} = \mathfrak{g} \times \mathfrak{g} \to \mathfrak{g}$ we denote the differential of Φ, so that $d\Phi(g) = \mathrm{pr}_2 \circ T_g\Phi : T_gG \to \mathfrak{g}$. Let $X_1, \dots, X_n$ be a linear basis of $\mathfrak{g}$ and let $L_{X_1}, \dots, L_{X_n}$ be the corresponding left invariant vector fields on G. Let

$$\Psi_\pi(g) = \Psi(g) := \det(d\Phi(g)), \qquad g \in G \tag{2}$$

be the *Cayley determinant function* of the representation π, where the determinant is computed with respect to the bases $L_{X_i}(g)$ of T_gG and X_i of $\mathfrak{g}$, respectively. Note that Ψ does not depend on the choice of the basis (X_i) of $\mathfrak{g}$. We get the function Ψ multiplied by the modular function if we choose right invariant vector fields.

In the following we shall use the notation $\mu : G \times G \to G$ for the multiplication, $\mu(g, h) = gh = \mu_g(h) = \mu^h(g)$ for left and right translations, $T(\mu_g)$ and $T(\mu^h)$ for the corresponding tangent mappings.

2.2 Proposition

In the following cases the infinitesimally faithful representation $\pi : G \to \mathrm{Aut}(V)$ admits a Cayley mapping:

(1) *if G is a reductive complex Lie group and π is a holomorphic representation,*

(2) *if G is a real reductive Lie group and π is a real or complex representation which maps each element in the connected center to a semisimple transformation (in the complexification of V); in particular if G is a real compact Lie group.*

By a real reductive Lie group we mean one where the complexification of the Lie algebra is reductive. For abelian Lie groups there are representations acting by unipotent matrices only, and we have to exclude them.

Proof. (1) A connected reductive complex Lie group G has a compact real form, so the Lie algebra $\mathfrak{g}$ of G is the complexification of the Lie algebra $\mathfrak{k}$ of a maximal compact subgroup $K \subset G_o$. Since $\pi' : \mathfrak{g} \to \mathrm{End}(V)$ is complex linear it suffices to show that the trace form is nondegenerate on $\pi'(\mathfrak{k})$. Let us choose a K-invariant Hermitian product $(\ ,\ )$ on V by integration. Then $\pi'(X)$ is skew Hermitian with respect to this inner product for $X \in \mathfrak{k}$, so $0 \leq (\pi'(X)v, \pi'(X)v) = -(\pi'(X)^2v, v)$ and $\pi'(X)^2$ is negative definite Hermitian, so $\mathrm{tr}(\pi'(X)^2)$ is the sum of the negative eigenvalues of $\pi'(X)^2$. Thus B_π is nondegenerate on $\mathfrak{k}$.

(2) The trace form is nondegenerate on the semisimple part of $\mathfrak{g}$, and on the center since it is mapped to semisimple endomorphisms. For a compact group one can repeat the argument from the proof of (1) with a G-invariant positive definite inner product on V. $\square$

2.3 Remark

Most of the time (when not stated explicitly otherwise) G will denote a connected reductive complex algebraic group and π will be a rational representation; in particular in sections (4), (3), and (7) below.

Note that the center $\mathfrak{c}$ of the Lie algebra $\mathfrak{g}$ then belongs to any Cartan subalgebra, and its Lie group (the connected center) to the corresponding Cartan subgroup.

2.4 Proposition

Let G be a complex or real Lie group and let Φ be the generalized Cayley mapping of a representation. The Cayley mapping Φ has the following simple properties:

(1) $\Phi \circ \mathrm{Conj}_b = \mathrm{Ad}_b \circ \Phi$.

(2) *For all $g \in G$ we have (where $Z_{\mathfrak{g}}(\mathfrak{g}^g)$ denotes the centralizer of $\mathfrak{g}^g = \{X \in \mathfrak{g} : \mathrm{Ad}_g(X) = X\}$ in $\mathfrak{g}$)*

$$G^g \subseteq G^{\Phi(g)}, \qquad \mathfrak{g}^g \subseteq \mathfrak{g}^{\Phi(g)},$$
$$\Phi(G^g) \subseteq \mathfrak{g}^g, \qquad \Phi(g) \in \mathrm{Cent}(\mathfrak{g}^g) \subset Z_{\mathfrak{g}}(\mathfrak{g}^g).$$

Moreover, if G is a connected reductive complex algebraic group, then we even have $\mathrm{Cent}(\mathfrak{g}^g) = Z_{\mathfrak{g}}(\mathfrak{g}^g)$.

(3) *$d\Phi(e) : \mathfrak{g} \to \mathfrak{g}$ is the identity mapping. Thus $d\Phi(g)$ is invertible for g in the nonempty Zariski open dense subset $\{h \in G : \Psi(h) \neq 0\}$ of G, and is not invertible on the hypersurface $\{h \in G : \Psi(h) = 0\}$.*

(4) *Ψ is invariant under conjugation.*

(5) *Let $H \subset G$ be a Cartan subgroup with Cartan Lie subalgebra $\mathfrak{h} \subset \mathfrak{g}$. Then $\Phi(H) \subset \mathfrak{h}$.*

(6) *Let χ_π be the character of the representation π, given by $\chi_\pi(g) = \mathrm{tr}(\pi(g))$. Then $d\chi_\pi(g)(T_e(\mu_g)X) = \mathrm{tr}(\pi'(\Phi_\pi(g))\pi'(X)) = B_\pi(\Phi_\pi(g), X)$.*

(7) *The differential $d\Phi_\pi(g).T_e(\mu_g).X \in \mathfrak{g}$ is given by the implicit equation $\mathrm{tr}(\pi'(d\Phi_\pi(g)\,T_e(\mu_g)\,X)\pi'(Y)) = \mathrm{tr}(\pi(g)\pi'(X)\pi'(Y))$ for $Y \in \mathfrak{g}$.*

(8) *If $\Phi(e) = 0$ and $a \in G$ is such that $\pi(a) \in \pi'(\mathfrak{g})$, then $d\Phi(a^{-1})$ is not invertible.*

Proof. (1) follows by the invariance of the trace.

(2) Most of it is obvious. Let G be a connected reductive complex algebraic group. We claim that g then lies in the identity component of G^g; this is immediate if g is semisimple (see (4.1) below) since then g lies in a Cartan subgroup. Using the Jordan decomposition (see (2.6) below) it is true in general. But then if $X \in \mathfrak{g}$

commutes with $\mathfrak{g}^g$ it commutes with g, hence $X \in \mathfrak{g}^g$ and consequently $X \in \mathrm{Cent}(\mathfrak{g}^g)$, establishing $Z_{\mathfrak{g}}(\mathfrak{g}^g) = \mathrm{Cent}(\mathfrak{g}^g)$.

(3) $T_e\pi(G) = \pi'(\mathfrak{g})$.

(4) follows from (1) and the form of the determinant (2.1.2).

(5) Let $a \in H$ be regular in G so that $G^a = H$ and $\mathfrak{g}^a = \mathfrak{h}$. Then use (2).

(6) Insert the definitions.

(7) This follows from

$$\begin{aligned}
&\mathrm{tr}(\pi'(d\Phi_\pi(g).T_e(\mu_g).X)\pi'(Y)) \\
&\quad = \tfrac{d}{dt}|_0 \mathrm{tr}(\pi'(\Phi_\pi(g\exp(tX)))\pi'(Y)) \\
&\quad = \tfrac{d}{dt}|_0 \mathrm{tr}(\pi(g\exp(tX))\pi'(Y)) = \tfrac{d}{dt}|_0 \mathrm{tr}(\pi(g)\pi(\exp(tX))\pi'(Y)) \\
&\quad = \mathrm{tr}(\pi(g)\pi'(X)\pi'(Y)).
\end{aligned}$$

(8) Take $X \in \mathfrak{g}$ with $\pi(a) = \pi'(X)$ so that $\pi(a^{-1})\pi'(X) = \mathrm{Id}_V$. Then we have $\mathrm{tr}(\pi(a^{-1})\pi'(X)\pi'(Y)) = \mathrm{tr}(\pi'(Y)) = 0$ for all $Y \in \mathfrak{g}$ so that $d\Phi(a^{-1})$ has a nontrivial kernel by (7). □

2.5 Proposition

Let G be a (real or complex) Lie group and let $\pi : G \to \mathrm{Aut}(V)$ be a representation which admits a Cayley mapping.

(1) *For $a \in G$ we have*

$$\begin{aligned}
d\Phi(a)(T_a(\mathrm{Conj}_G(a))) &= T_{\Phi(a)}(\mathrm{Ad}_G(\Phi(a))) = [\mathfrak{g}, \Phi(a)], \\
d\Phi(a)(T_a(G^a)) &\subseteq \mathfrak{g}^a \subseteq \mathfrak{g}^{\Phi(a)}, \\
T_a(\mathrm{Conj}_G(a)) &= \{T_e(\mu^a)X - T_e(\mu_a)X : X \in \mathfrak{g}\} \\
&= \{T_e(\mu^a)(X - \mathrm{Ad}_a X) : X \in \mathfrak{g}\}, \\
T_a(G^a) &= T_e(\mu_a)\mathfrak{g}^a = T_e(\mu^a)\mathfrak{g}^a.
\end{aligned}$$

Moreover, if G is a reductive complex algebraic group and a is a semisimple element in G, then we have

$$T_aG = T_a(\mathrm{Conj}_G(a)) \oplus T_a(G^a), \qquad \mathfrak{g} = [\mathfrak{g}, \Phi(a)] \oplus \mathfrak{g}^{\Phi(a)}.$$

(2) *Suppose that $d\Phi(a) : T_aG \to \mathfrak{g}$ is invertible for $a \in G$. Then $\mathfrak{g}^a = \mathfrak{g}^{\Phi(a)}$. In particular, the G-orbit $\mathrm{Conj}_G(a)$ of a in G has the same dimension as its Φ-image, $\mathrm{Ad}_G(\Phi(a))$; consequently, $\Phi : \mathrm{Conj}_G(a) \to \mathrm{Ad}_G(\Phi(a))$ is a covering. If in addition G is connected reductive and a is semisimple, then the centralizer $G_{\Phi(a)}$ is connected, so Φ is a diffeomorphism between the orbits.*

Proof. (1) Most of it follows easily from (2.4). Let us now suppose that G is a reductive complex algebraic group and that $a \in G$ is semisimple (4.1) so that a is contained in a Cartan subgroup with Cartan Lie subalgebra $\mathfrak{h}$. By (2.4.5) we get $\Phi(a) \in \mathfrak{h}$. We claim that then $\mathfrak{g} = [\mathfrak{g}, \Phi(a)] \oplus \mathfrak{g}^{\Phi(a)}$. By dimension it suffices to check that $[\mathfrak{g}, \Phi(a)] \cap \mathfrak{g}^{\Phi(a)} = 0$. This follows from the root space decomposition: Put $\Phi(a) = H \in \mathfrak{h}$, let $Y = [X, H] \in \mathfrak{g}^H$, let $X = X_0 + \sum_{\alpha\in R} X_\alpha$ be the root space decomposition. Then $Y = [X, H] = 0 + \sum_{\alpha\in R} \alpha(H) X_\alpha$ and $0 = [Y, H] = \sum_{\alpha\in R} \alpha(H)^2 X_\alpha$ whence either $\alpha(H) = 0$ or $X_\alpha = 0$ so that $Y = 0$.

Similarly it suffices to see that $\mathfrak{g}^a \cap (\mathrm{Id} - \mathrm{Ad}_a)\mathfrak{g} = 0$. So let $a = \exp(H)$ for some $H \in \mathfrak{h}$ and suppose that $Y = X - \mathrm{Ad}_a X \in \mathfrak{g}^a$. Consider again the root space decomposition $X = X_0 + \sum_{\alpha\in R} X_\alpha$. Then we get $Y = X - \mathrm{Ad}_{\exp(H)} X = 0 + \sum_{\alpha\in R}(1 - e^{\alpha(H)}) X_\alpha$ and $0 = Y - \mathrm{Ad}_{\exp(H)} Y = \sum_{\alpha\in R}(1 - e^{\alpha(H)})^2 X_\alpha$ whence either $1 - e^{\alpha(H)} = 0$ or $X_\alpha = 0$ so that $Y = 0$.

(2) By (2.4.2) we have $\mathfrak{g}^a \subseteq \mathfrak{g}^{\Phi(a)}$, so it suffices to prove the converse. Let $X \in \mathfrak{g}^{\Phi(a)}$ so that $[X, \Phi(a)] = 0$. Then

$$\begin{aligned} d\Phi(a) T_e(\mu^a)(X - \mathrm{Ad}_a X) &= \tfrac{d}{dt}|_0 \Phi(\exp(tX) a \exp(-tX)) \\ &= \tfrac{d}{dt}|_0 \mathrm{Ad}_{\exp(tX)} \Phi(a) = [X, \Phi(a)] = 0 \end{aligned}$$

so that $X - \mathrm{Ad}_a X = 0$ since $d\Phi(a)$ is invertible. If G is connected reductive and a is semisimple (see (4.1)) then by (2.4.5) $\Phi(a)$ is also semisimple, thus $G^{\Phi(a)}$ is connected. □

2.6 Theorem

Let G be a (real or complex) Lie group and let $\pi : G \to \mathrm{Aut}(V)$ be a representation which admits a Cayley mapping. Let $H = (\bigcap_{a\in A} G^a)_o = (G^A)_o \subseteq G$ be a subgroup which is the connected centralizer of a subset $A \subseteq G$ and suppose that H is itself reductive.

Then the representation $\pi|H : H \to \mathrm{End}(V)$ admits a Cayley mapping and we have

$$\Phi_\pi|H = \Phi_{\pi|H} : H \to \mathfrak{h}.$$

Note that this result proves again (2.4.5) by choosing A the Cartan subgroup. If we choose $A = G$, then H is the connected center of G. The assumption of this theorem holds in the following two cases:

(1) if $H = (G^g)_o$ is the connected centralizer of a semisimple element g in a complex algebraic group G (see (4.1) below), since then the connected centralizer $(G^g)_o = Z_G(g)_o$ is again reductive, by Proposition 13.19 in [2] on p. 321,

(2) if $H = (G^A)_o$ is the connected centralizer of a reductive subgroup $A \subseteq G$ of a complex reductive Lie group G, since it is well known that H is then itself reductive.

Proof. By (2.2.1) both Cayley mappings Φ_π and $\Phi_{\pi|H}$ exist. By (2.1.2) we have $\Phi_\pi(G^a) \subseteq \mathfrak{g}^a$ for all $a \in A$, thus also

$$\Phi_\pi(H) \subseteq \bigcap_{a\in A} \Phi_\pi(G^a) \subseteq \bigcap_{a\in A} \mathfrak{g}^a = \mathfrak{h}.$$

Moreover, for $h \in H$ and $X \in \mathfrak{h}$ we have

$$\operatorname{tr}(\pi'(\Phi_\pi(h)\pi'(X)) = \operatorname{tr}(\pi(h)\pi'(X)) = \operatorname{tr}(\pi'(\Phi_{\pi|(G^g)_o}(h)\pi'(X))).$$

Thus $\Phi_\pi(h) = \Phi_{\pi|H}(h)$. □

2.7 Theorem

Let G be a semisimple real or complex Lie group, let $\pi : G \to \operatorname{Aut}(V)$ be an infinitesimally effective representation. Let $\mathfrak{g} = \mathfrak{g}_1 \oplus \ldots \oplus \mathfrak{g}_k$ be the decomposition into the simple ideals $\mathfrak{g}_i$. Let $G_1, \ldots, G_k$ be the corresponding connected subgroups of G. Then

$$\Phi_\pi | G_i = \Phi_{\pi|G_i} \quad \text{for } i = 1, \ldots, k.$$

This result cannot be extended to reductive groups as the standard representation of GL_n shows where Φ is the embedding $GL_n \to \mathfrak{gl}_n$.

Proof. Let $g_1 \in G_1$ and suppose that $\Phi_\pi(g_1) = \sum_{i=1}^k \Phi_\pi(g_1)_i \in \bigoplus_{i=1}^k \mathfrak{g}_i$ with $\Phi_\pi(g_1)_i \neq 0 \in \mathfrak{g}_i$ for some $i \neq 1$. Then for each $g_i \in G_i$ we have

$$\Phi_\pi(g_1) = \Phi_\pi(g_i\, g_1\, g_i^{-1}) = \operatorname{Ad}_{g_i} \Phi_\pi(g_1)$$

so that $\Phi_\pi(g_1)$ contains a nontrivial G_i-orbit which is absurd. This shows that $\Phi_\pi(G_j) \subseteq \mathfrak{g}_j$ for $j = 1, \ldots, k$. Moreover for $g_j \in G_j$ and $X_j \in \mathfrak{g}_j$ we have

$$\operatorname{tr}(\pi'(\Phi_\pi(g_j))\pi'(X_j)) = \operatorname{tr}(\pi(g_j)\pi'(X_j)) = \operatorname{tr}((\pi|G_j)'(\Phi_{\pi|G_j}(g_j)(\pi|G_j)'(X_j))$$

which implies the result. □

2.8 Theorem

Let G be a simple real or complex Lie group, and let $\pi_i : G \to \operatorname{End}(V_i)$ be nontrivial representations for $i = 1, 2$. The inner product B_{π_i} on $\mathfrak{g}$ from (2.1) is a multiple of the Cartan Killing form B, and we write $B_{\pi_i} = j_{\pi_i} B$. Then we have

(1) *For the direct sum representation $\pi_1 \oplus \pi_2 : G \to \operatorname{End}(V_1 \oplus V_2)$ we have*

$$\Phi_{\pi_1\oplus\pi_2}(g) = \frac{j_{\pi_1}}{j_{\pi_1\oplus\pi_2}} \Phi_{\pi_1}(g) + \frac{j_{\pi_2}}{j_{\pi_1\oplus\pi_2}} \Phi_{\pi_2}(g) \in \mathfrak{g}.$$

(2) *For the tensor product representation* $\pi_1 \otimes \pi_2 : G \to \mathrm{End}(V_1 \otimes V_2)$ *we have*

$$\Phi_{\pi_1\otimes\pi_2}(g) = \frac{j_{\pi_1}\chi_{\pi_2}(g)}{j_{\pi_1\otimes\pi_2}}\,\Phi_{\pi_1}(g) + \frac{\chi_{\pi_1}(g)j_{\pi_2}}{j_{\pi_1\otimes\pi_2}}\,\Phi_{\pi_2}(g) \in \mathfrak{g}.$$

(3) *For the n-fold tensor product representation* $\otimes^n\pi : G \to \mathrm{End}(\otimes^n V)$ *we have*

$$\Phi_{\otimes^n\pi}(g) = \left(\frac{\chi_\pi(g)}{\dim(V)}\right)^{n-1}\Phi_\pi(g).$$

(4) *For the contragredient representation* $\pi^\top : G \to \mathrm{End}(V^*)$ *given by* $\pi^\top(g) = \pi(g^{-1})^\top$ *we have*

$$\Phi_{\pi^\top}(g) = -\Phi_\pi(g^{-1}).$$

For a complex simple Lie group the number j_{π_i} is a multiple of the *Dynkin index* of the representation π_i. It is nonnegative and satisfies

$$\begin{aligned} j_{\pi_1\oplus\pi_2} &= j_{\pi_1} + j_{\pi_2}, \\ j_{\pi_1\otimes\pi_2} &= \dim(V_2)j_{\pi_1} + \dim(V_1)j_{\pi_2}, \\ j_\pi &= \frac{\dim(V)}{\dim(\mathfrak{g})}B(\lambda_\pi, \lambda_\pi + \rho), \end{aligned} \tag{5}$$

for an irreducible representation π with highest weight λ_π, where ρ is half the sum of all positive roots. This is due to Dynkin [4] and can be found in [5, p.100]. Using this, equation (2) becomes

$$\begin{aligned} \Phi_{\pi_1\otimes\pi_2}(g) = &\frac{B(\lambda_{\pi_1}, \lambda_{\pi_1} + \rho)}{B(\lambda_{\pi_1}, \lambda_{\pi_1} + \rho) + B(\lambda_{\pi_2}, \lambda_{\pi_2} + \rho)}\frac{\chi_{\pi_2}(g)}{\dim(V_2)}\Phi_{\pi_1}(g) \\ &+ \frac{B(\lambda_{\pi_2}, \lambda_{\pi_2} + \rho)}{B(\lambda_{\pi_2}, \lambda_{\pi_2} + \rho) + B(\lambda_{\pi_1}, \lambda_{\pi_1} + \rho)}\frac{\chi_{\pi_1}(g)}{\dim(V_1)}\Phi_{\pi_2}(g) \end{aligned} \tag{2'}$$

Proof. (1), (2) and (4) are easy computations.

(3) By induction and (5) we check that $j_{\otimes^n\pi} = n\dim(V)^{n-1}j_\pi$ which via (2) leads quickly to the result. □

2.9 Proposition

For the Cayley mapping of a rational representation of a connected reductive complex algebraic group G, the pullback mapping $\Phi^* : A(\mathfrak{g}) = S^*(\mathfrak{g}^*) \to A(G)$ *between the algebras of regular functions is injective, equivariant, and maps the subalgebras of invariant regular functions to each other,* $\Phi^* : A(\mathfrak{g})^G \to A(G)^G$. *Consequently,* $\Phi : G \to \mathfrak{g}$ *is a dominant algebraic morphism. By the algebraic*

Peter–Weyl theorem we have $A(G) = \bigoplus_{\lambda \in D} A_\lambda$ where D is the set of all dominant integral highest weights, and where

$$A_\lambda = \{f \in A(G) : f(g) = \operatorname{tr}(\pi_\lambda(g)B) \text{ for some } B \in \operatorname{End}(V_\lambda)\}.$$

For an irreducible representation π we thus have $\Phi^*(\mathfrak{g}^*) \subset A_\lambda$ where λ is the highest weight of π.

Proof. We prove the last statement. Let λ be the highest weight of π. A_λ is a vector space of dimension $(\dim(V_\lambda))^2$ of functions on G. Take $X \in \mathfrak{g}$ and consider the function $f \in A_\lambda$ given by $f(g) = \operatorname{tr}(\pi(g)\pi'(X))$. Then $f = \Phi^*(B_\pi(\ ,X))$ where $B_\pi(\ ,X) \in \mathfrak{g}^*$. □

3 A separation of variables theorem for reductive algebraic groups

In this section G is a connected reductive complex algebraic group and Φ is the Cayley mapping of a rational representation.

3.1 Rational vector fields on G

Let $x_i : \mathfrak{g} \to \mathbb{C}$ be the coordinate functions for the basis $X_1, \ldots, X_n$ of $\mathfrak{g}$, and consider the constant vector fields $\partial_{x_i} = \frac{\partial}{\partial x_i}$.

Theorem. *Let G be a connected reductive complex algebraic group and let Φ be the Cayley mapping of a rational representation.*

Then the pullback vector fields $Y_i := \Phi^*\partial_{x_i}$ for $i = 1, \ldots, n$ are commuting vector fields with rational coefficients on G. The fields $\Psi.Y_i$ are regular vector fields (with algebraic coefficients). The fields Y_i induce an G-equivariant injective algebra homomorphism from the symmetric algebra $S^*\mathfrak{g}$ into the algebra of differential operators on G with rational coefficients with polar divisors contained in the hypersurface $\{h \in G : \Psi(h) = 0\}$.

Proof. For a matrix A we denote by $C(A)$ the classical adjoint or the matrix of algebraic complements which satisfies $A.C(A) = C(A).A = \det(A).\mathrm{Id}$. Applying this to $d\Phi(g)$ we can write the pull backfields as

$$\begin{aligned}(\Phi^*\partial_{x_i})(g) &= d\Phi(g)^{-1}.\partial_{x_i}|_{\Phi(g)} = \frac{1}{\det(d\Phi(g))}.C(d\Phi(g)).\partial_{x_i} \\ &= \frac{1}{\Psi(g)}.\sum_{j=1}^{n} C(d\Phi(g))_{ij} L_{X_j}(g).\end{aligned}$$

These are well-defined algebraic vector fields on the Zariski open set $\{\Psi \neq 0\}$ and are Φ-related to the constant fields ∂_{x_i} on $\mathfrak{g}$, thus they commute. □

3.2 Invariants and harmonic functions

For the algebra of regular functions we have $A(\mathfrak{g}) = A(\mathfrak{g})^G \otimes \mathrm{Harm}(\mathfrak{g})$ by [9], Theorem 0.2, where the space $\mathrm{Harm}(\mathfrak{g})$ is by definition the space of all regular functions that are killed by all invariant differential operators with constant coefficients. We define $\mathrm{Harm}_\pi(G) := \Phi_\pi^*(\mathrm{Harm}(\mathfrak{g}))$. It is a G-module. Let us denote by $A(G)_\Psi$ the localization at Ψ.

Theorem. *Let G be a connected reductive complex algebraic group and let Φ be the Cayley mapping of a rational representation. Then*

$$A(G)_\Psi = A(G)_\Psi^G \otimes \mathrm{Harm}_\pi(G).$$

Moreover, we have

$$A(G) = A(G)^G \otimes \mathrm{Harm}_\pi(G)$$

if and only if $\Phi : G \to \mathfrak{g}$ maps regular orbits in G to regular orbits in $\mathfrak{g}$.

Note that for the standard representation π of $SL_n(\mathbb{C})$ the Cayley mapping Φ_π carries regular orbits to regular orbits, see (6.2) below. In general this is wrong. If there exists a closed orbit $O = \mathrm{Conj}_G(a)$ in G for a in a Cartan subgroup H, such that $\dim(\Phi(O)) < \dim(O)$, then there exists already a regular orbit with that property, $\mathrm{Ad}(G)(a\,u) \subset \mathfrak{g}$ where u is a principal unipotent element in $(G^a)_o$, see (4.8.3). The orbits through

$$\mathrm{diag}(1, 1, 1, 1, -1, -1, -1, -1) \in SO(8) \cap \mathfrak{so}(8)$$

have this property.

Proof. By [15], Theorem A, we have

$$A(G) = A(G)^G \otimes H \tag{1}$$

for some G-submodule H of $A(G)$. Then H is by restriction isomorphic to the affine ring $A(\mathrm{Conj}_G(a))$ for any regular orbit $\mathrm{Conj}_G(a)$ in G.

Let λ be any highest weight of G such that the corresponding irreducible G-module V_λ has a nonzero weight space V_λ^0 with weight 0 of dimension $d(\lambda)$. Let H_λ be the primary component of H of type V_λ so that one has the direct sum

$$H = \bigoplus_\lambda H_\lambda.$$

By [9], p 348, Proposition 8, the multiplicity of V_λ in H_λ is $d(\lambda)$, since H restricts bijectively to any regular orbit in G; see also [15], theorem A. Thus we may write as a direct sum

$$H_\lambda = \bigoplus_{j=1}^{d(\lambda)} H_\lambda^j \tag{2}$$

where H_λ^j is irreducible and hence equivalent to V_λ.

Now let $\mathrm{Harm}(G)_\lambda$ be the primary component of the G-module $\mathrm{Harm}(G) = \Phi^*(\mathrm{Harm}(\mathfrak{g})) \subset A(G)$ of type V_λ. By [9] the multiplicity of V_λ in $\mathrm{Harm}(G)$ is again $d(\lambda)$ so that we have

$$\mathrm{Harm}(G)_\lambda = \bigoplus_{j=1}^{d(\lambda)} \mathrm{Harm}(G)_\lambda^j \tag{3}$$

where each $\mathrm{Harm}(G)_\lambda^j$ is an irreducible G-submodule of $A(G) = A(G)^G \otimes H$ of highest weight λ and hence equivalent to V_λ. Thus there exists a $(d(\lambda) \times d(\lambda))$-matrix $S = (s_{ij})$ with entries in $A(G)^G$ so that

$$\mathrm{Harm}(G)_\lambda^i = \sum_{j=1}^{d(\lambda)} s_{ij} H_\lambda^j. \tag{4}$$

One would like to replace the abstract G-module H_λ by the equivalent and explicit G-module $\mathrm{Harm}(G)_\lambda$. For that one needs that the matrix S is invertible in $A(G)^G$. The determinant $\det(S)$ is nonzero in $A(G)^G$ by the independence in (2) and (3). One would need that $\det(S)$ is a constant. Let $Z = \mathrm{Zero}(\det(S))$ be the zero set of $\det(S)$. By Steinberg [18], Theorem 1.3, the set of irregular elements in G is of codimension 3. Since $\det(S) \in A(G)^G$ its zero set Z contains full orbits and is of codimension 1, so it is the union of the closures of all regular orbits $\mathcal{O}$ in Z. But $\Phi(\mathcal{O})$ cannot be a regular adjoint orbit in $\mathfrak{g}$ since $\mathrm{Harm}(\mathfrak{g})$ restrict faithfully to $A(\mathrm{Ad}(G).X)$ for every regular orbit in $\mathfrak{g}$. Consequently, if one knew that Φ carried each regular orbit to a regular orbit, then Z was empty and hence $\det(S)$ a nonzero constant which proves part of the second assertion.

In the general case, if we localize $A(G)$ and $A(G)^G$ at the function Ψ from (2.4), then we restrict Φ to the Zariski open affine subvariety $G_{\Psi \neq 0}$ where Ψ does not vanish. But there Φ is locally biholomorphic and thus carries regular orbits to regular orbits in $\mathfrak{g}$, see (2.5.2). Hence $\det(S)$ does not vanish on $G_{\Psi \neq 0}$ and is thus invertible in $A(G)_\psi^G$. Consequently we have $A(G)_\Psi = A(G)_\Psi^G.\mathrm{Harm}(G)$.

It remains to prove that $A(G)_\Psi = A(G)_\Psi^G.\mathrm{Harm}(G)$ implies that $A(G)_\Psi = A(G)_\Psi^G \otimes \mathrm{Harm}(G)$. Assume for contradiction that this is false. Then there exist linearly independent $b_i \in \mathrm{Harm}(G)$ and linearly independent $a_i \in A(G)_\Psi^G$ for $1 \le i \le k$ such that $\sum_i a_i b_i = 0$. Since $G_{\Psi \neq 0}$ is Zariski open in G there exists a regular orbit $\mathcal{O}$ in $G_{\Psi \neq 0}$ such that $a_i(g) \neq 0$ for all i and any $g \in \mathcal{O}$. On the other hand $\sum_i a_i b_i | \mathcal{O} = \sum_i a_i(g) b_i | \mathcal{O}$ vanishes on $\mathcal{O}$. But $\Phi(\mathcal{O})$ is a regular orbit in $\mathfrak{g}$. Let $b_i = \Phi^*(c_i)$ for $c_i \in \mathrm{Harm}(\mathfrak{g})$; then the c_i are linearly independent. Thus the vanishing of $\sum_i a_i(g) c_i$ on the regular orbit $\Phi(\mathcal{O})$ contradicts the fact that $\mathrm{Harm}(\mathfrak{g})$ restricts faithfully onto each regular orbit.

Finally, if we had $A(G) = \mathrm{Harm}_\pi(G) \otimes A(G)^G$ then $\mathrm{Harm}_\pi(G) = \Phi^*\mathrm{Harm}(\mathfrak{g})$ would restrict faithfully to each regular orbit in G. Since the same is true for $\mathrm{Harm}(\mathfrak{g})$ the mapping Φ had to map regular orbits to regular orbits. □

3.3 Corollary

Let G be a connected reductive complex algebraic group and let Φ be the Cayley mapping of a rational representation with $\Phi(e) = 0 \in \mathfrak{g}$. Then for the G-equivariant extension of the rational function fields $\Phi^* : Q(\mathfrak{g}) \to Q(G)$ the degrees satisfy

$$[Q(G) : Q(\mathfrak{g})] = [Q(G)^G : Q(\mathfrak{g})^G].$$

See (6.2) below for the explicit extension in the case of the standard representation of $SL_n(\mathbb{C})$.

Proof. Note that $Q(\mathfrak{g})^G$ is the quotient field of $A(\mathfrak{g})^G$, and $Q(G)^G$ is the quotient field of $A(G)^G$. We have $A(\mathfrak{g}) = A(\mathfrak{g})^G \otimes \mathrm{Harm}(\mathfrak{g})$ by [9], Theorem 0.2, and $A(G)_\Psi = A(G)_\Psi^G \otimes \mathrm{Harm}_\pi(G)$ by (3.2), where $\mathrm{Harm}_\pi(G) := \Phi^*(\mathrm{Harm}(\mathfrak{g}))$ is isomorphic to $\mathrm{Harm}(\mathfrak{g})$.

Let $k = [Q(G), Q(\mathfrak{g})]$. Then any $q \in Q(G)$ satisfies a unique monic polynomial of degree $\le k$ with coefficients in $Q(g)$. Choose $q \in Q(G)^G$. Then the coefficients must be in $Q(g)^G$ since otherwise by conjugating by an element in G we would obtain a new minimal polynomial which contradicts uniqueness. Thus $[Q(G)^G, Q(\mathfrak{g})^G] \le k$.

On the other hand, if $m = [Q(G)^G, Q(\mathfrak{g})^G]$, then there exists q in $Q(G)^G$ which satisfies an equation of degree m over $Q(\mathfrak{g})^G$ and $Q(G)^G = (Q(\mathfrak{g})^G)[q]$. But $A(G)_\Psi^G$ is contained in $Q(G)^G$. Thus $A(G)_\Psi$ is contained in $Q(\mathfrak{g})[q]$ by (3.2). Hence the quotient field $Q(G)$ of $A(G)_\Psi$ is contained in $Q(\mathfrak{g})[q]$. Thus $m = k$. □

4 The behavior of the Jordan decomposition under the Cayley map

In this section G is a connected reductive complex algebraic group and Φ is the generalized Cayley mapping of a rational representation.

4.1 The Jordan decomposition

For references about the Jordan decomposition (additive in the Lie algebra and multiplicative in the algebraic group) see [2] or Chapter IV of [6]. In our case given $a \in G$ we write a_s for the semisimple part of a and a_u for the unipotent part of a so that $a = a_s\, a_u = a_u\, a_s$.

Recall that semisimple means that a_s is G-conjugate to an element in the Cartan subgroup H and a_u unipotent means that a_u is conjugate to an element in the unipotent variety U. We shall use the decomposition $H = T\, H_{\mathbb{R}}$ where T is a maximal torus. An element $a \in G$ is called elliptic if a is semisimple and the eigenvalues of $\pi(a)$ are of norm 1 for all π; equivalently, a is conjugated to an element in T. Likewise, an element $a \in G$ is called hyperbolic if a is semisimple and the eigenvalues of $\pi(a)$ are real for all π; or equivalently, if a is conjugated to an element in $H_{\mathbb{R}}$. Expanding the multiplicative Jordan decomposition every element $a \in G$ has a unique decomposition

$$a = a_e a_h a_u$$

where a_e and a_h are respectively elliptic and hyperbolic and all three components commute. We say that a is *of positive type* if $a_e = e \in G$.

Analogously, for $X \in \mathfrak{g}$ there is the unique (additive) Jordan decomposition $X = X_s + X_n$ where $[X_s, X_n] = 0$ and where X_s is semisimple (conjugate to an element in $\mathfrak{h}$) and X_n is nilpotent (conjugate to an element in the nilcone N). Expanding the additive Jordan decomposition every element $X \in \mathfrak{g}$ has a unique decomposition

$$X = X_e + X_h + X_n$$

where X_e and X_h are respectively elliptic and hyperbolic and all three components commute. We will say that X is *of real type* if $X_e = 0$. Let $\mathfrak{g}_{real}$ be the space of all elements of real type in $\mathfrak{g}$. For information on hyperbolic elements in algebraic groups see [11], especially Section 2 on p. 418.

4.2 Ad-complete modules

Let $D \subset h^*$ denote the set of dominant integral weights for G (relative to some fixed Borel subgroup) and for each $\lambda \in D$ let $\pi_\lambda : G \to \operatorname{Aut} V_\lambda$ be a fixed irreducible representation with highest weight λ.

A completely reducible G-module M will be said to Ad-complete if one has an equivalence

$$M \cong \oplus_{\lambda \in D} \dim(V_\lambda^H)\, V_\lambda,$$

i.e., each irreducible component occurs with multiplicity equal to the dimension of its zero weight space.

Let $\operatorname{Reg}(\mathfrak{g})$ (resp. $\operatorname{Reg}(G)$) be the set of regular elements in $\mathfrak{g}$ (respectively G). We recall the following results.

4.3 Theorem

([9]). *For $X \in \operatorname{Reg}(\mathfrak{g})$ one has $A(\operatorname{Ad}_G(X)) = A(\overline{\operatorname{Ad}_G(X)})$ and as a G-module $A(\overline{\operatorname{Ad}_G(X)})$ is* Ad-*complete. Furthermore there exists a section $\operatorname{Reg}_\#(\mathfrak{g})$ of the map $\operatorname{Reg}(\mathfrak{g}) \to \operatorname{Reg}(\mathfrak{g})/G$ which in addition has the property that*

$$\operatorname{Reg}_\#(\mathfrak{g}) \to \mathbb{C}^\ell, \qquad X \mapsto (I_1(X), \ldots, I_\ell(X)) \tag{1}$$

is an algebraic isomorphism, where the I_k form a basis of $A(\mathfrak{g})^G$. Finally,

$$\mathfrak{g} = \bigcup_{X \in \operatorname{Reg}_\#(\mathfrak{g})} \overline{\operatorname{Ad}_G(X)} \tag{2}$$

is a disjoint union.

Subsequently Steinberg proved the following group-theoretic analogue:

4.4 Theorem

([18]). *Assume G is simply-connected semisimple. For $a \in \operatorname{Reg}(G)$ one has*

$$A(\operatorname{Conj}_G(a)) = A(\overline{\operatorname{Conj}_G(a)})$$

and as a G-module $A(\overline{\operatorname{Conj}_G(a)})$ is Ad*-complete. Furthermore there exists a section $\operatorname{Reg}_\#(G)$ of the map $\operatorname{Reg}(G) \to \operatorname{Reg}(G)/G$ which in addition has the property that*

$$\operatorname{Reg}_\#(G) \to \mathbb{C}^\ell, \qquad a \mapsto (\chi_1(a), \dots, \chi_\ell(a)) \tag{1}$$

is an algebraic isomorphism. Here $\{\chi_j\}$ are the characters of the fundamental representations. Finally

$$G = \bigcup_{a \in \operatorname{Reg}_\#(G)} \overline{\operatorname{Conj}_G(a)} \tag{2}$$

is a disjoint union.

4.5 Theorem

Let G be a reductive complex algebraic group and let $\pi : G \to \operatorname{Aut}(V)$ be a locally faithful rational representation of G. Let $a \in G$ be regular. Assume that a is nonsingular with respect to the Cayley map $\Phi = \Phi_\pi$ so that $\Phi(a)$ is regular in $\mathfrak{g}$ by (2.5.2). Then Φ restricts to an isomorphism

$$\Phi : \overline{\operatorname{Conj}_G(a)} \to \overline{\operatorname{Ad}_G(\Phi(a))} \tag{1}$$

of affine varieties.

Proof. Let G^s be the simply connected covering group of the commutator (semisimple) subgroup G' of G. Let $\gamma : G^s \to G'$ be the covering map. We may write $a = bc$ where $b \in \operatorname{Cent}(G)$ and $c \in G'$. Let $g \in G^s$ be such that $\gamma(g) = c$; note that g is regular. Clearly the mapping $G^s \to G$, given by $h \mapsto b\gamma(h)$, restricts to a surjective G^s-equivariant morphism $\beta : \operatorname{Conj}_{G^s}(g) \to \operatorname{Conj}_G(a)$; thus by continuity β also restricts to a dominant morphism

$$\beta : \overline{\operatorname{Conj}_{G^s}(g)} \to \overline{\operatorname{Conj}_G(a)}. \tag{2}$$

But also by continuity one has a dominant morphism

$$\Phi : \overline{\operatorname{Conj}_G(a)} \to \overline{\operatorname{Ad}_G(\Phi(a))} \tag{3}$$

and hence if α is the composite of (2) and (3) it defines a dominant morphism

$$\alpha : \overline{\operatorname{Conj}_{G^s}(g)} \to \overline{\operatorname{Ad}_G(\Phi(a))}. \tag{4}$$

But then the cohomomorphism of (4) is injective. However since the affine algebras in question are Ad-complete by Theorems 4.5 and 4.4, it follows that (4) must be an isomorphism. But then obviously (2) and (3) are isomorphisms. □

4.6

Richardson proved that for semisimple groups the generalized Cayley map defines an isomorphism of the unipotent variety in G with the nilcone in $\mathfrak{g}$. His theorem is very general and includes the case of G defined over fields of finite characteristic as long as the prime is good. An application of Theorem 4.5 yields Richardson's theorem for the complex case.

Theorem. *Let $a \in G$ be a principal unipotent element. Then a is nonsingular. Let $U \subset G$ be the unipotent variety $U = \overline{\mathrm{Conj}_G(a)}$. Then*

$$\Phi : U \to \overline{\mathrm{Ad}_G(\Phi(a))} \tag{1}$$

is an isomorphism of affine varieties. Furthermore if $\Phi(e) = 0$ (e.g., G is semisimple), then $\Phi(a)$ is principal nilpotent so that $\overline{\mathrm{Ad}_G(\Phi(a))}$ is the nilcone $N \subset \mathfrak{g}$ and hence (1) is an isomorphism

$$\Phi : U \to N. \tag{2}$$

Proof. Since $e \in U$ and $d\Phi(e)$ is invertible it is immediate that $d\Phi(a)$ is invertible. Thus (1) follows from Theorem 4.5. If $\Phi(e) = 0$ then $0 \in \overline{\mathrm{Ad}_G(\Phi(a))}$. But this implies that $\Phi(a)$ is principal nilpotent and hence one has (2). □

4.7 Corollary

If G is semisimple and X is a principal nilpotent element in $\mathfrak{g}$ and if π is irreducible, then $\Phi^{-1}(X) \subset G$ consists of $|Z(G)|$ many elements.

Proof. Let $b \in \Phi^{-1}(X)$. Since $d\Phi(b)$ is invertible we have $b \in G^b = G^X = Z(G) \times (G^X)_0$. Also by (4.6) there is a unique $a \in \Phi^{-1}(X) \cap U$. Since π is irreducible and $Z(G)$ is a finite group, for $c \in Z(G)$ we have $\pi(c) = k_c.1_V$ where $k_c \in S^1 \subset \mathbb{C}$. Thus

$$\begin{aligned} \mathrm{tr}(\pi'\Phi(c.g)\pi'(Y)) &= \mathrm{tr}(\pi(c)\pi(g)\pi'(Y)) = k_c.\mathrm{tr}(\pi(g)\pi'(Y)) \\ &= \mathrm{tr}(\pi'\Phi(g)\pi'(k_c.Y)) \end{aligned}$$

for all $Y \in \mathfrak{g}$. This implies $\Phi(c.g) = k_c.\Phi(g)$ for each $g \in G$. Choose the unique elements $a_c \in \Phi^{-1}(\frac{1}{k_c}X) \cap U$ for all $c \in Z(G)$. Then $c.a_c \in G^X$ is in the coset $c.(G^X)_0$ and $\Phi(c.a_c) = X$. So $\Phi^{-1}(X)$ consists of $|Z(G)|$ many elements. □

4.8

For any $a \in G$ (resp. $x \in \mathfrak{g}$) let G^a (resp. G^x) denote the centralizer of a (resp. x) in G. Let $\mathfrak{g}^a = \mathrm{Lie}\, G^a$ (resp. $\mathfrak{g}^x = \mathrm{Lie}\, G^x$). If a is semisimple and G^a_o is the identity component of G^a then G^a_o is a reductive subgroup of G (see Proposition 13.19 in [2] on p. 321). If x is semisimple then G^x is connected (see e.g., Theorem

22.3, p. 140 in [6]) and reductive (since $\mathfrak{g}^x$ is clearly reductive). For any semisimple element $b \in G$ let U_b be the unipotent variety in G_o^b and N_b the nilpotent cone in $\mathfrak{g}^b$. Moreover, let $l = \operatorname{rank} G$ and recall that an element $a \in G$ (resp. $x \in \mathfrak{g}$) is called regular if dim $G^a = l$ or equivalently dim $\mathfrak{g}^a = l$ (resp. dim $G^x = l$ or equivalently dim $\mathfrak{g}^x = l$).

For $a \in G$ with Jordan decomposition $a = a_s a_u$ let $\mathfrak{c} = \mathit{cent}\, \mathfrak{g}^{a_s}$ and let $\mathfrak{s} = [\mathfrak{g}^{a_s}, \mathfrak{g}^{a_s}]$ so that $\mathfrak{s}$ is semisimple and

$$\mathfrak{g}^{a_s} = \mathfrak{c} \oplus \mathfrak{s}. \tag{1}$$

Proposition. *Let G be a connected reductive algebraic group.*

(2) *Let $a \in G$ so that $a = a_s a_u$. Then a_u is in the unipotent variety U_{a_s} of the reductive subgroup $G_o^{a_s}$ of G. Conversely if $b \in U_{a_s}$, then $g = a_s b$ is the Jordan decomposition of g so that*

$$a_s U_{a_s} = \{b \in G \mid b_s = a_s\}$$

(3) *If $a \in G$, then a is regular if and only if a_u is principal unipotent in $G_o^{a_s}$.*

(4) *If $X \in \mathfrak{g}$, then X is regular if and only if X_n is principal nilpotent in $\mathfrak{g}^{X_s}$.*

(5) *If $a \in G$, let H be a Cartan subgroup of G which contains a_s and let $\mathfrak{h} = \operatorname{Lie} H$. Let C be the center of $G_o^{a_s}$. Then $G_o^{a_s} = C\,S$ and $a_s \in C$ so that a_s and hence a are in $G_o^{a_s}$. Furthermore $C \subset H$ and $\mathfrak{c} \subset \mathfrak{h}$ and in addition* $\operatorname{Lie} C = \mathfrak{c}$.

In spite of (5) it is not true in general that C is connected. In fact if a_s corresponds to a vertex of the fundamental simplex, then $\mathfrak{c} = 0$ and C is finite.

Proof. (2) It clearly suffices to show that a_u is in the identity component of G^{a_s}. But this is immediate from the bijection between U and N defined by the exponential map. The latter implies that if $x = \log(a_u)$, then $x \in \mathfrak{g}^{a_s}$. Hence the one parameter subgroup defined by x is in $G_o^{a_s}$ so that $a_u \in G_o^{a_s}$.

(3) and (4) It suffices by an identical argument to consider only the group case. Clearly by the uniqueness of the Jordan decomposition and (2) we have

$$\mathfrak{g}^a = \mathfrak{g}^{a_s} \cap \mathfrak{g}^{a_u} = (\mathfrak{g}^{a_s})^{a_u}. \tag{6}$$

Of course rank $\mathfrak{g}^{a_s} = l$ since a_s is conjugate to an element in H. Thus

$$\operatorname{rank} \mathfrak{c} + \operatorname{rank} \mathfrak{s} = l = \operatorname{rank} G. \tag{7}$$

But if S is the semisimple subgroup corresponding to $\mathfrak{s}$, then clearly $U_{a_s} \subset S$ so that $a_u \in S$ and hence by (6) one has

$$\mathfrak{g}^a = \mathfrak{c} \oplus \mathfrak{s}^{a_u} \tag{8}$$

so that $\dim \mathfrak{g}^a = \operatorname{rank} \mathfrak{c} + \dim \mathfrak{s}^{a_u}$. But $\dim \mathfrak{s}^{a_u} \geq \operatorname{rank} \mathfrak{s}$ and by definition of principal unipotent one has $\dim \mathfrak{s}^{a_u} = \operatorname{rank} \mathfrak{s}$ if and only if a_u is principal unipotent in S or equivalently in $G_o^{a_s}$. Now the result follows from (7).

(5) That $G_o^{a_s} = C\,S$ follows from (1). Obviously $H \subset G_o^{a_s}$ and hence H is a Cartan subgroup of $G_o^{a_s}$. But since $H \subset G_o^{a_s}$ it follows that $a_s \in G_o^{a_s}$. But then obviously $a_s \in C$. But since the center of a connected reductive group lies in every Cartan subgroup one has $C \subset H$. We get $\operatorname{Lie} C = \mathfrak{c}$ since the centers correspond to each other under the Lie subgroup – Lie subalgebra correspondence. □

4.9

Let $z_a = \log(a_u)$ so that $z_a \in N_{a_s}$. By the theorem of Jacobson–Morosov, see [7], p. 983, there exists $h_a \in \mathfrak{s}$ so that $[h_a, z_a] = 2z_a$. But then if $r_a(t) = \exp(t\,h_a)$ one has clearly has that

$$\lim_{t \to -\infty} r_a(t)\, a_u\, r_a(t)^{-1} = 1.$$

But of course $r_a(t)$ commutes with a_s since $a_s \in C$. Thus

$$\lim_{t \to -\infty} r_a(t)\, a\, r_a(t)^{-1} = a_s. \tag{1}$$

4.10 Corollary

Let G be a connected reductive complex algebraic group and let Φ be the Cayley mapping of a rational representation.

If for some $a \in G$ the differential $d\Phi(a_s) : T_{a_s}G \to \mathfrak{g}$ is invertible, then also $d\Phi(a) : T_a G \to \mathfrak{g}$. This is the case if a_s is hyperbolic, by (5.4) below.

Proof. This is an immediate consequence of (4.9.1). □

4.11

We will begin to establish results leading to the main theorem on the commutativity of the generalized Cayley mapping and the operation of taking the semisimple part for Jordan decompositions. We will use the notation of (4.8).

Theorem. *Let G be a connected reductive complex algebraic group and let Φ be the Cayley mapping of a rational representation. Let $a \in G$.*

Then for the semisimple parts we have $\Phi(a_s) = \Phi(a)_s$ and the Jordan decomposition is the decomposition into components with respect to (4.8.8)

$$\Phi(a) = \Phi(a_s) + \Phi(a)_n \in \mathfrak{g}^a = \mathfrak{c} \oplus \mathfrak{s}^{a_u}.$$

Proof. Let $\Phi(a) = Z + F \in \mathfrak{g}^a = \mathfrak{c} \oplus \mathfrak{s}^{a_u}$ be the decomposition into components with respect to (4.8.8). Recall from (4.9) the curve $r_a : \mathbb{R} \to S$ satisfying

$\lim_{t\to-\infty} r_a(t)\, a\, r_a(t)^{-1} = a_s$ by (4.9.1). Hence by the continuity of Φ one has

$$\Phi(a_s) = \lim_{t\to-\infty} \mathrm{Ad}_{r_a(t)}(\Phi(a)) = \lim_{t\to-\infty} \mathrm{Ad}_{r_a(t)}(Z+F) = Z + \lim_{t\to-\infty} \mathrm{Ad}_{r_a(t)} F.$$

By (2.4.2) and (4.8.1) we have $\Phi(a_s) \in \mathrm{Cent}(\mathfrak{g}^{a_s}) = \mathfrak{c}$. But also $\mathrm{Ad}_{r_a(t)} F \in \mathfrak{s}$ by (4.9) so that $\lim_{t\to-\infty} \mathrm{Ad}_{r_a(t)} F = 0$ which implies that F is nilpotent, and $Z = \Phi(a_s)$ which is semisimple since $\mathfrak{c} \subseteq \mathfrak{h}$ by (4.8.5). Finally note that $[Z, F] = 0$ since $\mathfrak{c} = \mathrm{Cent}(g^{a_s})$ so that the result follows. □

4.12

We now consider the nilpotent part of $\Phi(a)$. The situation is more complicated. Let $b \in G$ be semisimple. If $w \in U_b$, then of course

$$(bw)_s = b \quad \text{and} \quad (bw)_u = w. \tag{1}$$

For any $w \in U_b$ one has $\Phi(bw)_n \in N_b$ since $\Phi(bw)_n$ is, by the uniqueness of the Jordan decomposition, clearly invariant under $Ad\, b$. One thus obtains a map

$$\Phi_b : U_b \to N_b, \quad \Phi_b(w) = \Phi(bw)_n. \tag{2}$$

Proposition. *Let $b \in G$ be semisimple. The map $\Phi_b : U_b \to N_b$ is a regular morphism which commutes with the adjoint actions of G^b.*

Proof. The commutation of Φ_b with the adjoint action of G^b is obvious. The rationality is an immediate consequence of Theorem 4.11 since it clearly implies that

$$\Phi_b(w) = \Phi(bw) - \Phi(b). \quad \square \tag{3}$$

4.13 Theorem

Let G be a connected reductive complex algebraic group and let Φ be the Cayley mapping of a rational representation with $\Phi(e) = 0 \in \mathfrak{g}$. Let $b \in G$ be semisimple and suppose that $d\Phi(b) : T_bG \to \mathfrak{g}$ is invertible. This holds for b hyperbolic, see (5.4).

Then $\Phi_b : U_b \to N_b$ is an isomorphism of algebraic varieties.

This is generalization of Theorem 4.6. It also follows directly from the 'lemme fondamental' in [14], as in Richardsons proof of (4.6).

Proof. The proof of (4.6) that $\Phi : U \to N$ is an algebraic isomorphism is a consequence of the fact that Φ carries a principal unipotent orbit to a principal nilpotent orbit. Replacing G by the semisimple algebraic group S (using the notation of (4.8) where $b = a_s$) the same argument yields the isomorphism of Φ_b as soon as we demonstrate that Φ_b carries a principal unipotent orbit in U_b to a principal nilpotent orbit in N_b. Let $w \in U_b$ be a principal unipotent element in S. But bw is a regular element in G by (4.8.3). Using the notation of the proof of Proposition 4.8 where $a = bw$, $a_s = b$, $a_u = w$ one has dim $\mathfrak{g}^{bw} = l$ and $\mathfrak{g}^{bw} = \mathfrak{c} \oplus \mathfrak{s}^w$ where if $c = \dim \mathfrak{c}$ and $s = \mathfrak{s}^w$ then $l = c + s$ and $s = \operatorname{rank} \mathfrak{s}$. One the other hand, if $\mathfrak{v} = \mathfrak{s}^{\Phi_b(w)}$ and $v = \dim \mathfrak{v}$, then $v \geq s$ and $v = s$ if and only if $\Phi_b(w)$ is principal nilpotent in S. But, by Theorem 4.11 clearly

$$\mathfrak{c} \oplus \mathfrak{v} \subset \mathfrak{g}^{\Phi(bw)}. \tag{1}$$

But now by assumption $d\Phi(b)$ is invertible. Thus $d\Phi(bw)$ is invertible by Proposition 4.10. But then $\mathfrak{g}^{bw} = \mathfrak{g}^{\Phi(bw)}$ by Theorem 1.8. In particular dim $\mathfrak{g}^{\Phi(bw)} = l$. But the left side of (1) has dimension $c + v$. Thus one must have $v = s$ (and equality in (1). Hence $\Phi_b(w)$ is a principal nilpotent in S. □

5 The mapping degree of the Cayley mapping

5.1 The mapping degree of Φ for compact G

Now let G be a compact group and let the ground field be the reals. Then $\Phi(G)$ is compact in $\mathfrak{g}$, so $\Phi : G \to \mathfrak{g}$ is not surjective. Let us embed $\mathfrak{g}$ into the one-point-compactification $\mathfrak{g} \cup \infty = S^n$; then the topological mapping degree of $\Phi : G \to \mathfrak{g} \cup \infty$ is 0 since Φ is not surjective. Thus, over a regular value of Φ which exists by Sard's theorem, there is an even number of sheets: on one half of them Φ is orientation-preserving, on the other half, it is orientation-reversing.

5.2 The mapping degree in the complex case

Since in general there are conjugacy orbits on G which map to points in $\mathfrak{g}$, the mapping Φ is not proper in the sense of (Hausdorff) topology (which means that compact sets have compact inverse images; in the usual topology on $\mathfrak{g}$). But by (2.4.1) the mapping Φ induces a mapping between the algebraic orbit spaces $\bar{\Phi}$: $G//\mathrm{Conj}_G \to \mathfrak{g}//\mathrm{Ad}_G$, i.e., the affine varieties with coordinate rings $A(G)^G$ and $A(\mathfrak{g})^G$, respectively.

Theorem. *Let G be a connected reductive complex Lie group and let Φ be the Cayley mapping of a representation. Let $\mathfrak{h} \subset \mathfrak{g}$ be a Cartan subalgebra with Cartan subgroup H.*

If the Cayley mapping $\Phi : H \to \mathfrak{h}$ is proper:

(1) The mapping degree of $\Phi : H \to \mathfrak{h}$ is positive and consequently the mapping is surjective.

(2) $\bar{\Phi} : G//\mathrm{Conj}_G \to \mathfrak{g}//\mathrm{Ad}_G$ is a proper mapping for the Hausdorff topologies on the affine varieties. Thus $\bar{\Phi}$ has positive mapping degree and is surjective.

Proof. Let $\mathfrak{h}$ be a Cartan subalgebra with Cartan subgroup H. By (2.4.5) we have $\Phi : H \to \mathfrak{h}$.

(1) $\Phi : H \to \mathfrak{h}$ is a proper smooth mapping; thus its mapping degree is defined as the value of the induced mapping in the top De Rham cohomology with compact supports which is isomorphic to $\mathbb{R}$ via integration:

$$\deg(\Phi) = \Phi^* : H^l_c(\mathfrak{h}; \mathbb{R}) = \mathbb{R} \to H^l_c(H, \mathbb{R}) = \mathbb{R}, \quad l = \dim(\mathfrak{h}).$$

Choose a regular value Y of Φ. Then for each $g \in \Phi^{-1}(Y)$ the tangent mapping $T_g\Phi$ is invertible and orientation preserving, since it is complex holomorphic. Thus the mapping degree is the number of sheets over a regular point, which is positive. But then $\Phi : H \to \mathfrak{h}$ has to be surjective: if not, its image is closed, and an n-form with support in the complement is pulled backed to 0 on H, in contradiction to the positivity of the mapping degree.

(2) Since $G//\mathrm{Conj}_G \cong H/W$ and $\mathfrak{g}//\mathrm{Ad}_G \cong \mathfrak{h}/W$ where W is the Weyl group, the result follows directly from (1). □

5.3 Lemma

Suppose that for reals r_i we have $\sum_{i=1}^N r_i = 0$. Then

$$\sum_{i=1}^N r_i e^{r_i} \geq \frac{1}{2N}\sum_{i=1}^N r_i^2.$$

Proof. We separate negative and positive summands and consider for $s_i > 0$ and $t_j > 0$ the expression

$$-\sum_{i=1}^n s_i e^{-s_i} + \sum_{j=1}^m t_j e^{t_j}, \qquad \sum_{i=1}^n s_i = A = \sum_{j=1}^m t_j.$$

The function $f(x) = x\,e^x$ is strictly convex for $x > 0$; thus

$$\frac{1}{m}\sum_{j=1}^m f(t_j) \geq f\Big(\frac{1}{m}\sum_{j=1}^m t_j\Big) = f(A/m)$$

with equality only if all t_j are equal. Moreover, $-\sum_{i=1}^n s_i e^{-s_i} \geq -\sum_{i=1}^n s_i = -A$. For the right-hand side we have $A^2 = \Big(\sum_{j=1}^m t_j\Big)^2 \geq \sum_{j=1}^m t_j^2$ and similarly $A^2 \geq \sum_{i=1}^n s_i^2$. Finally

$$-\sum_{i=1}^n s_i e^{-s_i} + \sum_{j=1}^m t_j e^{t_j} \geq -A + m\frac{A}{m}e^{A/m} \geq -A + A\big(1 + \frac{A}{m}\big) = \frac{A^2}{m} = \frac{1}{2m}2A^2$$

$$\geq \frac{1}{2m}\Big(\sum_{i=1}^n s_i^2 + \sum_{j=1}^m t_j^2\Big).$$

□

5.4 Theorem

Let G be a connected reductive complex algebraic group and let Φ be the Cayley mapping of a rational representation π. Let $G_{\text{hyp}} \subset G$ be the subset of all semisimple hyperbolic elements in G, and let $\mathfrak{g}_{\text{hyp}}$ be the set of all semisimple hyperbolic elements in the Lie algebra $\mathfrak{g}$.

Then for each $g \in G_{\text{hyp}}$ the differential $d\Phi(g) : T_gG \to \mathfrak{g}$ is invertible.

Moreover, the mapping $\Phi_\pi : G_{\text{hyp}} \to \mathfrak{g}_{\text{hyp}}$ is bijective and a diffeomorphism between real subvarieties if $\Phi_\pi(e) = 0$, or if the representation π is also self-contragredient.

Proof. Let $\mathfrak{h} \subset \mathfrak{g}$ be a Cartan subalgebra with Cartan subgroup H. Let $\mathfrak{h} = \mathfrak{h}_{\mathbb{R}} + i\mathfrak{h}_{\mathbb{R}}$. Let $g \in G$ be a hyperbolic element which we assume to lie in H. Then $g = \exp X$ where $X \in \mathfrak{h}_{\mathbb{R}}$, so $g \in H_{\mathbb{R}}$. We have $\Phi(g) \in \mathfrak{h}_{\mathbb{R}}$ since

$$\operatorname{tr}(\pi'(\Phi(g))\pi'(Y)) = \operatorname{tr}(\pi(g)\pi'(Y)) \in \mathbb{R}$$

for all $Y \in \mathfrak{h}_{\mathbb{R}}$. We choose a maximal compact subgroup $G_u \subset G$ such that $i\mathfrak{h}_{\mathbb{R}} \subset \operatorname{Lie}(G_u) =: \mathfrak{g}_u$. Then there exists a Hermitian inner product on V (a Hilbert space structure) such that all elements of $\pi(G_u)$ are unitary. Then $\pi'(\mathfrak{g}_u)$ consists of skew Hermitian operators, and $\pi'(i\mathfrak{g}_u)$ consists of Hermitian operators. Thus $\pi'(\mathfrak{h}_{\mathbb{R}})$ and $\pi(g)$ are Hermitian operators. We use $\operatorname{tr}(AB)$ as real positive inner product on the the space of Hermitian operators on V.

We have $g = h^2$ for $h = \exp(\frac{1}{2}\log(g)) \in H_{\mathbb{R}}$. We claim that $\pi(g) : V \to V$ is a positive definite Hermitian operator, i.e., $(X, Y) \mapsto \operatorname{tr}(\pi(g)\pi'(X)\pi'(Y)^*)$ is a positive definite Hermitian form on $\mathfrak{g}$. Namely, for $X \in \mathfrak{g}$ arbitrary we have $X = X_1 + X_2$ for unique $X_1 \in \mathfrak{g}_u$ and $X_2 \in i\mathfrak{g}_u$. Let $X^* := X_1 - X_2$; then $\pi'(X^*)$ equals the adjoint operator $\pi'(X)^*$. We have

$$\begin{aligned}\operatorname{tr}(\pi(g)\pi'(X)\pi'(X)^*) &= \operatorname{tr}(\pi(h)^2\pi'(X)\pi(X)^*)\\ &= \operatorname{tr}(\pi(h)\pi'(X)(\pi(h)\pi'(X))^*) = \|\pi(h)\pi'(X)\|^2 > 0.\end{aligned}$$

By (2.4.7) this implies that $d\Phi(g) : T_gG \to \mathfrak{g}$ is invertible.

For the second assertion, we claim that in both cases we have for $C > 0$

$$\|\pi'(\Phi(g))\| \geq C\|\pi'(X)\|. \tag{1}$$

If $\Phi(e) = 0$, we have

$$\begin{aligned}0 &= \operatorname{tr}(\pi'(\Phi(e))\pi'(X)) = \operatorname{tr}(\pi(e)\pi'(X)) = \operatorname{tr}(\pi'(X))\\ &= \sum_{\lambda\in\text{weight}(\pi)} \operatorname{tr}(\pi'(X)|V_\lambda) = \sum_{\lambda\in\text{weight}(\pi)} \dim(V_\lambda)\lambda(X).\end{aligned}$$

Thus by Lemma 5.3 we have

$$\sum_{\lambda\in\text{weight}(\pi)} \dim(V_\lambda)\lambda(X)e^{\lambda(X)} \geq C \sum_{\lambda\in\text{weight}(\pi)} \dim(V_\lambda)\lambda(X)^2$$

for a positive constant C. Then we have by Cauchy–Schwarz,

$$\begin{aligned}\|\pi'\Phi(g)\|\,\|\pi'(X)\| \geq \operatorname{tr}(\pi(g)\pi'(X)) &= \sum_{\lambda\in\operatorname{weight}(\pi)} \operatorname{tr}(\pi(\exp(X))\pi'(X)|V_\lambda)\\ &= \sum_{\lambda\in\operatorname{weight}(\pi)} \dim(V_\lambda)\lambda(X)e^{\lambda(X)} \geq C\sum_\lambda \lambda(X)^2\dim(V_\lambda)\\ &= C\sum_\lambda \operatorname{tr}(\pi'(X)^2|V_\lambda) = C\operatorname{tr}(\pi'(X)^2) = C\|\pi'(X)\|^2.\end{aligned}$$

If the representation π is self-contragredient, a similar argument works using $re^r - re^{-r} \geq 2r$ for $r > 0$ instead of the inequality (5.3).

Now we may finish the proof. Note that $\exp : \mathfrak{h}_\mathbb{R} \to H_\mathbb{R}$ is a diffeomorphism with inverse $\log : H_\mathbb{R} \to \mathfrak{h}_\mathbb{R}$. Thus estimate (1) implies that $\Phi : H_\mathbb{R} \to \mathfrak{h}_\mathbb{R}$ is a proper mapping. It is also a local diffeomorphism, thus a covering mapping and a diffeomorphism since $\mathfrak{h}_\mathbb{R}$ is vector space.

Finally, each hyperbolic element $g \in G$ is contained in $H_\mathbb{R}$ for a suitable Cartan subgroup H, and the above arguments show that $\Phi : G_{\text{hyp}} \to \mathfrak{g}_{\text{hyp}}$ is locally a diffeomorphism and is surjective. It is also injective: Let $\Phi(g_1) = Y = \Phi(g_2)$. Then g_1 and some conjugate $h\,g_2\,h^{-1}$ lie in the same Cartan subgroup $H_\mathbb{R}$ on which Φ is a diffeomorphism, thus $g_1 = h.g_2.h^{-1}$. By equivariancy we have $\operatorname{Ad}(h)Y = Y$. Since Φ is a local diffeomorphism near g_1, the orbits have the same dimension; thus g_1 and Y have the same connected component of the isotropy group. But isotropy groups of semisimple elements of the Lie algebra are connected, as mentioned in (4.8). Thus $g_1 = g_2$. □

5.5 Theorem

Let G be a connected reductive complex algebraic group and let Φ be the Cayley mapping of a rational representation with $\Phi(e) = 0 \in \mathfrak{g}$.

Then $\Phi : G_{\text{pos}} \to \mathfrak{g}_{\text{real}}$ is bijective and a fiber respecting isomorphism of real algebraic varieties, where G_{pos} and $\mathfrak{g}_{\text{real}}$ are defined in (4.1).

Proof. Let $g \in G_{\text{pos}}$, then $g_s = g_h$, thus by (5.4) $d\Phi(g_s) : T_{g_s}G \to \mathfrak{g}$ is invertible and thus by theorem 4.13 $\Phi_{g_s} : U_{g_s} \to N_{g_s}$ is an automorphism of algebraic varieties, where U_{g_s} is the unipotent variety in the reductive group $G_o^{g_s}$, see (4.8). The set $G_{\text{pos}} = \bigsqcup_{h\in G_{\text{hyp}}} h.U_h \to G_{\text{hyp}}$ is a fibration with complex algebraic varieties as fibers and the real algebraic variety G_{hyp} as base. Likewise $\mathfrak{g}_{\text{real}} = \bigsqcup_{h\in G_{\text{hyp}}}(\Phi(h) + N_h) \to \mathfrak{g}_{\text{hyp}}$ is a is a fibration with the nilpotent cones N_h as fibers. $\Phi : G_{\text{hyp}} \to \mathfrak{g}_{\text{hyp}}$ is given by $\Phi(g) = \Phi(g_s) + \Phi_{g_s}(g_u) = \Phi(g)_s + \Phi(g)_n$ and is a fiber respecting isomorphism by theorem (4.13) and (5.4). □

6 Examples

6.1 The Cayley mappings for the representations of $SL_2(\mathbb{C})$

The standard representation $SL_2(\mathbb{C}) \subset \mathrm{End}(\mathbb{C}^2)$. Here $\Phi(A) = A - \frac{1}{2}\mathrm{tr}(A)1_{\mathbb{C}^2}$ and $\Psi(A) = \frac{1}{2}\mathrm{tr}(A)$ for $A \in SL_2$. Φ is surjective and proper and has mapping degree 2. On the Cartan subgroup we get

$$\Phi_{\pi_1}\begin{pmatrix} a & 0 \\ 0 & a^{-1} \end{pmatrix} = \begin{pmatrix} \frac{1-a^2}{2a} & 0 \\ 0 & \frac{a^2-1}{2a} \end{pmatrix} \qquad \Psi_{\pi_1}\begin{pmatrix} a & 0 \\ 0 & a^{-1} \end{pmatrix} = \frac{a^2+1}{2a}.$$

For the $(n+1)$-dimensional representation we computed

$$\Phi_{\pi_n}\begin{pmatrix} a & 0 \\ 0 & a^{-1} \end{pmatrix} = \frac{3}{n^3+3n^2+2n}\sum_{p=0}^{n}(n-2p)a^{n-2p}\begin{pmatrix} 1 & 0 \\ 0 & -1 \end{pmatrix}$$

$$\Psi_{\pi_n}\begin{pmatrix} a & 0 \\ 0 & a^{-1} \end{pmatrix} = \frac{-108}{(n^3+3n^2+2n)^3}\sum_{p_1,p_2,p_3=0}^{n}(n-2p_1)^2(n-p_2)na^{3n-2(p_1+p_2+p_3)}$$

6.2 Example: The standard representation of $SL_n(\mathbb{C})$

The standard representation $SL_n(\mathbb{C}) \subset \mathrm{End}(\mathbb{C}^n)$. Here $\Phi(A) = A - \frac{1}{n}\mathrm{tr}(A)1_{\mathbb{C}^n}$, and Φ has mapping degree n, which coincides with the degree of the smooth hypersurface $SL_n \subset \mathrm{End}(\mathbb{C}^n)$. Thus Φ maps regular orbits in SL_n to regular orbits in $\mathfrak{sl}_n$, and by (3.2) we have

$$A(SL_n) = A(SL_n)^{SL_n} \otimes \mathrm{Harm}(SL_n)).$$

The set G_{sing} where the differential of the Cayley transform is singular is given by $\Psi^{-1}(0) = (SL_n \cap \mathfrak{sl}_n)^{-1} = \{A \in SL_n : \mathrm{tr}(A^{-1}) = 0\}$.

Let $\chi \in A(SL_n)$ be given by $\chi(A) = \frac{1}{n}\mathrm{tr}(A)$. Then

$$\begin{aligned} A(SL_n)^{SL_n} &= A(\mathfrak{sl}_n)^{SL_n}[\chi], \\ Q(SL_n)^{SL_n} &= Q(\mathfrak{sl}_n)^{SL_n}[\chi], \\ Q(SL_n) &= Q(\mathfrak{sl}_n)[\chi], \end{aligned}$$

in terms of (3.3).

This can be shown as follows: $\mathrm{End}(\mathbb{C}^n) = \mathfrak{Sl}_n \oplus \mathbb{C}.1_n$ is an orthogonal direct sum, thus the orthogonal projection is given by $\mathrm{pr}(A) = A - \frac{1}{n}\mathrm{tr}(A)$. Of course $\Phi = \mathrm{pr}|SL_n$. For $Y \in \mathfrak{sl}_n$ we have $A \in \Phi^{-1}(Y)$ if and only if $A - Y = t.1_n$ for some $t \in \mathbb{C}$. The set of all these t is given by the equation $\det(Y + t.1_n) = 1$. Generically there are n solution, thus the degree of Φ is n, and Φ is surjective.

For $A \in SL_n$ and $X \in \mathfrak{sl}_n$ then $A.X$ is a typical tangent vector in $T_A(SL_n)$. Now $d\Phi(A) = \Phi : T_A(SL_n) \to \mathfrak{sl}_n$ is not invertible if its kernel $\mathbb{C}.1_n \cap T_A(SL_n)$

is non trivial, so there exists an $X \in \mathfrak{sl}_n$ with $A.X = 1_n$ (by scaling X appropriately). But then $X = A^{-1} \in \mathfrak{sl}_n$ and thus $\operatorname{tr}(A^{-1}) = 0$.

Let us discuss the adjunction now. The fiber over any $X \in \mathfrak{sl}_n$ consists of all $A = X + t.1_n$ where t runs through the roots of the polynomial $f_X(t) = \operatorname{tr}(t.1_n + X) - 1$, and moreover $t = \chi(A)$, by the formula for Φ. Let us consider the expansion

$$f_X(t) = \operatorname{tr}(t.1_n + X) - 1 = \sum_{j=0}^{n} p_j(X)\, t^j,$$

where $p_j \in A(\mathfrak{sl}_n)^{SL_n}$ form a system of generators with $p_0(X) = \det(X) - 1$, $p_{n-1}(X) = \operatorname{tr}(X) = 0$, and $p_n(X) = 1$. We have

$$f_X(\chi) = \operatorname{tr}(t.1_n + X) - 1 = \sum_{j=0}^{n} p_j(X)\, \chi^j = 0,$$

and this is the minimal polynomial for χ over $A(\mathfrak{sl}_n)^{SL_n}$. Finally note that χ is $1/n$ times the character of the standard representation and Φ^* pulls back the elements $p_j \in A(\mathfrak{sl}_n)^{SL_n}$ to the characters of the remaining fundamental representations of SL_n. Thus $A(SL_n)^{SL_n} = A(\mathfrak{sl}_n)^{SL_n}[\chi]$. The other assertions follow from (3.3).

6.3 The standard representation of $O_n(\mathbb{C})$

In the standard representation $O_n(\mathbb{C}) \to \operatorname{End}(\mathbb{C}^n)$ we have

$$\begin{aligned} O_n(\mathbb{C}) &= \{A \in \operatorname{End}(\mathbb{C}^n) : AA^\top = 1_n\} \\ \mathfrak{so}_n(\mathbb{C}) &= \{X \in \operatorname{End}(\mathbb{C}^n) : X + X^\top = 0\} \\ \Phi(A) &= \tfrac{1}{2}(A - A^\top). \end{aligned}$$

7 Spin representations and Cayley transforms

In this section we treat only the complex group $\operatorname{Spin}(n, \mathbb{C})$. Some of the results here were inspired by [13]. We first recall notations and results from [12].

7.1 Clifford multiplication in terms of exterior algebra operations

Let V be a complex vector space of finite dimension $n > 0$. Assume that $B_V = (\ ,\ )$ is a fixed nonsingular symmetric bilinear form on V. Its extends naturally to a symmetric nonsingular bilinear form $B_{\wedge V}$ on the exterior algebra $\wedge V$. The graded structure on $\wedge V$ induces a $\mathbb{Z}$-graded structure on the operator algebra $\operatorname{End}(\wedge V)$ and also on the $\mathbb{Z}$-graded super Lie algebra $\operatorname{Der}(\wedge V)$ of graded super derivations. For $\beta \in \operatorname{End}(\wedge V)$ we denote its transpose by $\beta^t \in \operatorname{End}(\wedge V)$ so that $(\beta u, v) = (u, \beta^t v)$ for any $u, v \in \wedge V$.

For any $u \in \wedge V$ let $\epsilon(u) \in \mathrm{End}(\wedge V)$ (left exterior multiplication) be defined so that $\epsilon(u)w = u \wedge w$ for any w. For any $u \in \wedge V$ let $\iota(u) = \epsilon(u)^t$. If $u \in \wedge^k V$ then $\epsilon(u) \in \mathrm{End}^k(\wedge V)$ and $\iota(u) \in \mathrm{End}^{-k}(\wedge V)$. Furthermore if $x \in V$ then $\iota(x) \in \mathrm{Der}^{-1} \wedge V$. In fact $\iota(x)$ is the unique element of $\mathrm{Der}^{-1}(\wedge V)$ such that $\iota(x)y = (x, y)$ where we have identified $\mathbb{C} = \wedge^0 V$. For a basis $\{z_j\}$ of V we have $\mathrm{Der}(\wedge V) = \bigoplus_j \epsilon(\wedge V)\iota(z_j)$. Let $\kappa \in \mathrm{End}^0(\wedge V)$ be defined so that $\kappa = (-1)^k$ on $\wedge^k V$. Let $\wedge^{\mathrm{even}} V$ and $\wedge^{\mathrm{odd}} V$ be the eigenspaces for κ corresponding, respectively, to the eigenvalues 1 and -1.

We recall that the Clifford algebra $C(V)$ over V with respect to $B_V|V$ is the tensor algebra $T(V)$ over V modulo the ideal generated by all elements in $T(V)$ of the form $x \otimes x - (x, x)$ where $x \in V$ and we regard $\mathbb{C} = T^0(V)$. We note that $\epsilon(x)\iota(y) + \iota(y)\epsilon(x) = (x, y) \in \mathrm{End}(\wedge V)$. For any $x \in V$ let $\gamma(x) = \epsilon(x) + \iota(x)$ then $\gamma(x)^2 = (x, x)$. Thus the correspondence $x \mapsto \gamma(x)$ extends to a homomorphism $C(V) \to \mathrm{End}(\wedge V)$, denoted by $u \mapsto \gamma(u)$, defining the structure of a $C(V)$-module on $\wedge V$. This leads to the map $u \mapsto \gamma(u)1$, denoted by $\psi : C(V) \to \wedge V$. By Theorem II.1.6, p. 41 in [3] this map ψ is bijective and we will identify $C(V)$ with $\wedge V$, using ψ. We consequently then recognize that the linear space $\wedge V = C(V)$ has 2 multiplicative structures. If $u, v \in \wedge V$ there is exterior multiplication $u \wedge v$ and Clifford multiplication $uv = \gamma(u)v$. Also we will write $C^0(V) = \wedge^{\mathrm{even}} V$ and $C^1(V) = \wedge^{\mathrm{odd}} V$ so that $C^i(V)C^j(V) \subset C^{i+j}(V)$ for $i, j \in \mathbb{Z}/(2)$. That is, the parity automorphism κ of the exterior algebra is also an automorphism of the Clifford algebra. In addition if $\alpha \in \mathrm{End} \wedge V$ is the unique involutory antiautomorphism of $\wedge V$, as an exterior algebra, such that $\alpha(x) = x$ for all $x \in V$, then α is also an antiautomorphism of the Clifford algebra structure; explicitly, $\alpha = (-1)^{\frac{k(k-1)}{2}}$ on $\wedge^k V$. Using α and κ one can show easily that for $x \in V$ and $u \in C(V)$ we have $ux = (\varepsilon(x) - \iota(x))\kappa(u)$.

Let $\mathrm{Spin}(V)$ be the set of all $g \in C^0(V)$ such that $g\alpha(g) = 1$ and $gx\alpha(g) \in V$ for all $x \in V$. This is an algebraic group under Clifford multiplication. For any $g \in \mathrm{Spin}\,V$ let $T(g) \in GL(V)$ be the mapping given by $T(g)x = gx\alpha(g)$ for $x \in V$.

The Lie algebra $\mathrm{LieSpin}\,V$ is a Lie subalgebra of $C^0(V)$, in fact $\mathrm{LieSpin}\,V = \wedge^2 V$ as explained on p. 68 in [3]. We need this in detail: Let $u \in \wedge^2 V$. For any $x \in V$ one has $\iota(x)u \in V$ and one can define $\tau(u) \in \mathfrak{so}(V) \subset \mathrm{End}(V)$ by $\tau(u)x = -2\iota(x)u$.

For any $z \in \mathrm{End}\,V$ there exists a unique operator $\xi(z) \in \mathrm{Der}^0(\wedge V)$ which extends the action of z on V. Clearly $\xi : \mathrm{End}\,V \to \mathrm{Der}^0(\wedge V)$ is a Lie algebra isomorphism.

7.2 Theorem

$\mathrm{Spin}\,V \subset C^0(V)$ is a connected Lie group. If $g \in \mathrm{Spin}\,V$ then $T(g) \in SO(V)$ and $T : \mathrm{Spin}\,V \to SO(V)$ is an epimorphism with kernel $\{\pm 1\}$. In particular T defines $\mathrm{Spin}\,V$ as a double cover of $SO(V)$.

The subspace $\wedge^2 V$ is a Lie subalgebra of $C^0(V)$. In fact $\wedge^2 V = \mathrm{LieSpin}\,V$ so *that* $\mathrm{Spin}V$ *is the group generated by all* $\exp(u)$ *for* $u \in \wedge^2 V$ *where exponentiation is with respect to Clifford multiplication.*

Furthermore the map $\tau : \wedge^2 V \to \mathrm{Lie}SO(V)$ *is a Lie algebra isomorphism and is the differential of the double cover* $T : \mathrm{Spin}V \to SO(V)$. *Finally if* $u \in \wedge^2 V$ *then* $\mathrm{ad}(u) = \xi(\tau(u))$ *so that not only is* $\mathrm{ad}(u)$ *a derivation of both exterior algebra and Clifford algebra structures on* $C(V)$ *but also* $\mathrm{ad} : \wedge^2 V \to \mathrm{Der}^0 \wedge V$ *is a faithful representation of the Lie algebra* $\wedge^2 V$ *on* $\wedge V$.

7.3

Let $\{z_i\}$ be a B_V-orthonormal basis of V. On these elements the Clifford product equals the exterior product, $z_i z_j = z_i \wedge z_j$. Let $N = \{1, \dots, n\}$ and let $\mathcal{I}$ be the set of all subsets of N. We regard any subset I as ordered: $I = \{i_1 < \dots < i_k\}$. Let

$$z_I = z_{i_1} \cdots z_{i_k} = z_{i_1} \wedge \cdots \wedge z_{i_k}.$$

The set of elements $\{z_I\}$ with $|I| = k$, is a $B_{\wedge V}$-orthonormal basis of $\wedge^k V$ so that the set $\{z_I\}$, $I \in \mathcal{I}$, $|I|$ is even (resp. odd) is a $B_{\wedge V}$-orthonormal basis of $C^0(V)$ (resp. $C^1(V)$) and the full set $\{z_I\}$, $I \in \mathcal{I}$ is a $B_{\wedge V}$-orthonormal basis of $C(V)$. Moreover for $I, J \in \mathcal{I}$ let $I \diamond J \in \mathcal{I}$ be the symmetric difference $(I \setminus J) \cup (J \setminus I)$. Then

$$z_I z_J = c_{I,J} z_{I \diamond J}, \qquad \text{where } c_{I,J} \in \{1, -1\},$$
$$(z_I)^{-1} = \alpha(z_I) = (-1)^{|I|(|I|-1)/2} z_I.$$

7.4 The spin module

The following description of the spin module S is uniform in n and the spin representation will always be a direct sum of 2 (possibly equivalent) irreducible representations.

Let $Z = \wedge^0 V + \wedge^n V$ so that Z is a 2-dimensional subspace of $C(V)$ and let $C^Z(V)$ be the centralizer of Z in $C(V)$. There exists (uniquely up to sign) an element $\mu \in \wedge^n V$ such that $\mu^2 = 1$. Hence if we put $e_+ = \frac{1}{2}(1 + \mu)$ and $e_- = \frac{1}{2}(1 - \mu)$ then $\{e_+, e_-\} \subset Z$ are orthogonal idempotents in the sense of ring theory and $1 = e_+ + e_-$. In particular, Z is a subalgebra of $C(V)$ and is isomorphic to $e_+ Z \oplus e_- Z = \mathbb{C} \oplus \mathbb{C} \subset C^Z(V)$. We have $x\mu = (-1)^{n+1}\mu v$ for any $v \in V$ so that $C^0(V) \subset C^Z(V)$ and in fact $C^Z(V) = ZC^0(V)$. Let $C^Z_+(V) = e_+ C^Z(V)$ and $C^Z_-(V) = e_- C^Z(V)$. Let $n_1 = [\frac{n+1}{2}]$. It is well known (see e.g., II.2.4 and II.2.6 in [3]) that both $C^Z_+(V)$ and $C^Z_-(V)$ are each isomorphic to a $2^{n_1-1} \times 2^{n_1-1}$ matrix algebra so that $C^Z(V)$ is a semisimple associative algebra of dimension 2^{2n_1-1} and the unique decomposition of $C^Z(V)$ into simple ideals is given by $C^Z(V) = C^Z_+(V) \oplus C^Z_-(V)$. In particular $Z = \mathrm{Cent}(C^Z(V))$ and e_+ and e_- are, respectively, the identity elements of $C^Z_+(V)$ and $C^Z_-(V)$. Thus there exists a $C^Z(V)$-module S (the spin module) S of dimension 2^{n_1}, defined by a representation $\sigma : C^Z(V) \to \mathrm{End}S$ characterized uniquely, up to equivalence, by the condition that if $S_+ = e_+ S$ and $S_- = e_- S$ then $S = S_+ \oplus S_-$ is the unique decomposition of S into irreducible $C^Z(V)$-modules. Let $\sigma_+ : C^Z(V) \to \mathrm{End} S_+$ and $\sigma_- : C^Z(V) \to \mathrm{End} S_-$ be the corresponding irreducible representations.

Recall $\mathrm{Spin}(V) \subset C^0(V) \subset C^Z(Z)$. The restriction of σ to $\mathrm{Spin}(V)$ will be denoted by Spin so that Spin is a representation of $\mathrm{Spin}(V)$ on S. Its differential, also denoted by Spin, is the restriction σ to $\wedge^2 V$. Thus one has a Lie algebra representation $\mathrm{Spin} : \wedge^2 V \to \mathrm{End} S$. Replacing σ by $\sigma_\pm$ and S by $S_\pm$ one similarly has representations $\mathrm{Spin}_\pm$ of $\mathrm{Spin} V$ and $\wedge^2 V$ on $S_\pm$. Since $\wedge^2 V$ generates $C^0(V)$, both $\mathrm{Spin}_\pm$ are irreducible representations.

Finally, let $n_0 = [\frac{n}{2}]$ so that one has $n = n_0 + n_1$. Note that if n is even then $n_1 = n_0 = \frac{1}{2} n$, whereas if n is odd then $n_1 = n_0 + 1$ so that $n_1 = \frac{1}{2}(n+1)$ and $n_0 = \frac{1}{2}(n-1)$.

7.5

Let $\mathrm{pr}_k : C(V) \to \wedge^k V$ be the projection defined by the graded structure of $\wedge V$. We identify $\wedge^0 V$ with $\mathbb{C}$ so that the image of pr_0 is scalar-valued.

The Clifford algebra $C(V)$ as a module over itself by left multiplication (γ) is equivalent to 2^{n_0} copies of the $C(V)$ spin module S with respect to σ (see (7.4)). One has that $2^n = 2^{n_0} 2^{n_1}$

Proposition. *For any $w \in C(V)$ one has*

$$\mathrm{pr}_0(w) = \frac{1}{2^n} \mathrm{tr}\gamma(w) = \frac{1}{2^{n_1}} \mathrm{tr}\sigma(w).$$

Proof. Of course for $c \in \wedge^0 V$ one obviously has that $c = \frac{1}{2^n} \mathrm{tr}\gamma(c)$. With the notation of (7.3), it suffices to prove that $\mathrm{tr}\gamma(z_I) = 0$ if $k > 0$. But this follows from (7.3). □

7.6

We now note that the bilinear form given by $\mathrm{tr}\sigma$ is essentially given by $B_{\wedge V}$.

Proposition. *Let $u, w \in C(V)$ Then*

$$\mathrm{tr}\sigma(u)\sigma(w) = 2^{n_1}(u, \alpha(w))$$

Proof. The left side is just $\mathrm{tr}\sigma(uw)$. By proposition (7.5) this is just $2^{n_1}\mathrm{pr}_0(uw)$. Using the basis z_I of (7.3) and writing $u = \sum_I u_I z_I$ and $w = \sum_I w_I z_I$ it is immediate from (7.3) that $\mathrm{pr}_0(uw) = \sum_I u_I w_I \mathrm{pr}_0(z_I^2)$. But clearly by (7.3) one has $(z_I, \alpha(z_J)) = 0$ for $I \neq J$ and $z_I^2 = \mathrm{pr}_0(z_I^2) = (z_I, \alpha(z_I))$. □

7.7 The Cayley mapping

The generalized Cayley mapping $\Phi_T : \mathrm{Spin}(V) \to \mathrm{LieSpin}(V)$ for the representation $T : \mathrm{Spin}(V) \to SO(V)$ is given by $\mathrm{tr}(\sigma(g)\sigma(y)) = \mathrm{tr}(\sigma(\Phi_T(g))\sigma(y))$. By (7.2) one has $\mathrm{LieSpin}(V) = \wedge^2 V$.

Proposition. *Let* $g \in \mathrm{Spin}(V)$. *Then* $\Phi_T(g) = \mathrm{pr}_2(g)$.

Proof. Let $y \in \wedge^2 V$. Let $x = \Phi_T(g)$ and let $z = \mathrm{pr}_2(g)$. Note that $\alpha(y) = -y$. By (7.6) one has

$$\mathrm{tr}(\sigma(g)\sigma(y)) = -2^{n_1}(g, y) = -2^{n_1}(\mathrm{pr}_2(g), y) = \mathrm{tr}(\sigma(\mathrm{pr}_2(g))\sigma(y)). \qquad \square$$

7.8

Let $\theta : SO(V) \to \mathrm{Aut}(\wedge V)$ be the representation so that if $a \in SO(V)$ then $\theta(a)$ is the unique exterior algebra automorphism which extends the action of a on V. Clearly

$$\mathrm{tr}\theta(a) = \det(1 + a)$$

Proposition. *For any* $g \in \mathrm{Spin}(V)$ *and* $w \in C(V)$ *then*

$$\theta(T(g))w = g\, wg^{-1}$$

In particular for $\theta(a)$, *for* $a \in SO(V)$ *is an automorphism of both the Clifford and exterior algebra structures on* $C(V)$.

Proof. Using the notation of (7.1) and (7.2) it is clear that $\xi|\mathrm{Lie}SO(V)$ is the differential of θ. To prove the proposition it suffices to establish its infinitesimal analogue. But this is stated in (7.2). $\square$

7.9 Proposition

The tensor product representation Spin ⊗ Spin *of* Spin(V) *on* $S \otimes S$ *descends to* $SO(V)$ *and is equivalent to* θ *if* n *is even and 2 copies of* θ *if* n *is odd.*

This is well known; we include a proof in the conventions used above.

Proof. Assume that M is a matrix algebra and β is an antiautomorphism of M. Let L be a minimal left ideal of M so that $(\dim L)^2 = \dim M$. Then clearly $R = \beta(L)$ is a minimal right ideal of M. There exists $v \in M$ such that $LvR \neq 0$ so that $LvR = M$ since LvR is a 2-sided ideal and M is simple. This implies that the map

$$\mu : L \otimes L \to M, \qquad \mu(a \otimes b) = a\, v\, \beta(b)$$

is surjective and thus a linear isomorphism by dimension. If $g \in M$, then clearly $\mu(ga \otimes gb) = g\mu(a \otimes b)\beta(g)$, which by Proposition 7.8 proves the assertion in case n is even by choosing $M = C(V)$, $\beta = \alpha$, $L = S$ and $g \in \mathrm{Spin}(V)$.

If n is odd, then $C(V) = M_1 \oplus M_2$ where M_i, $i = 1, 2$, are matrix algebras. Then $S = L_1 \oplus L_2$ where L_i, $i = 1, 2$, are respectively minimal left ideals in M_i (and left ideals in $C(V)$). Then

$$S \otimes S = L_1 \otimes L_1 \oplus L_1 \otimes L_2 \oplus L_2 \otimes L_1 \oplus L_2 \otimes L_2. \tag{1}$$

We have $\mathrm{Spin}(V) \subset C^0(V)$ which implies that all 4 summands on the right side of (1) define $\mathrm{Spin}(V)$-equivalent submodules of $\mathrm{Spin} \otimes \mathrm{Spin} : \mathrm{Spin}(V) \to \mathrm{Aut}(S \otimes S)$, since the parity automorphism $\kappa \in \mathrm{Aut}C(V)$ from (7.1) of course fixes $C^0(V)$ but interchanges M_1 and M_2. This is clear from the definition of $e_\pm$ in (7.4) since $\kappa = -1$ on $\wedge^n V$. Now if $i \in \{1, 2\}$ let $i' \in \{1, 2\}$ be defined so that $\alpha(M_i) = M_{i'}$. But then $\alpha(L_i) = R_{i'}$ is a minimal right ideal (and right ideal in $C(V)$) in $M_{i'}$. But then

$$C(V) = L_1 M_1 R_{1'} \oplus L_1 M_1 R_{2'} \oplus L_2 M_2 R_{1'} \oplus L_2 M_2 R_{2'},$$

where 2 of the 4 components are identically zero and the remaining 2 are M_1 and M_2. Note that the 2-sided ideal M_i in $C(V)$ is stable under θ by Proposition 7.8. The argument in the even case then readily implies that θ is equivalent to 2 of the 4 components in (1) and hence $\mathrm{Spin} \otimes \mathrm{Spin}$ is equivalent to 2 copies of θ. □

7.10 Theorem

Let $g \in \mathrm{Spin}(V)$. *Then* $\mathrm{pr}_0(g) = 0$ *if* $g \notin \mathrm{Spin}(V)^* := \{g \in \mathrm{Spin}(V) : \det(1 + T(g)) \neq 0\}$. *On the other hand, if* $g \in \mathrm{Spin}(V)^*$, *then*

$$\mathrm{pr}_0(g) = \frac{1}{2^{n/2}}\sqrt{\det(1 + T(g))}$$

for one of two choices of square root. Both choices are taken for the two elements $g, g' \in \mathrm{Spin}(V)$ *such that* $T(g) = T(g')$.

Proof. Apply Proposition 7.5 to the case where $w = g \in \mathrm{Spin}(V)$. Then

$$\mathrm{pr}_0(g)^2 = \frac{1}{2^{2n_1}}(\mathrm{trSpin}(g))^2.$$

But $(\mathrm{trSpin}(g))^2 = \mathrm{tr}(\mathrm{Spin} \otimes \mathrm{Spin})(g)$. If n is even, then $2n_1 = n$ and we have $\mathrm{tr}(\mathrm{Spin} \otimes \mathrm{Spin})(g) = \det(1 + T(g))$ by (7.8) and Proposition 7.9. Thus in this case

$$\mathrm{pr}_0(g)^2 = \frac{1}{2^n}\det(1 + T(g)). \tag{1}$$

Now assume that n is odd. Then $\mathrm{tr}(\mathrm{Spin} \otimes \mathrm{Spin})(g) = 2\det(1 + T(g))$ by (7.8) and Proposition 7.9. But since $2n_1 - 1 = n$ equation (1) holds for all n. The theorem now follows: If $g_1, g_2 \in \mathrm{Spin}(V)$, then $T(g_1) = T(g_2)$ if and only if $g_1 = \pm g_2$. □

7.11

For any $u \in \wedge^2 V$ let $\hat{e}^u$ denote the exponential of u using exterior (and hence commutative since $\wedge^{\text{even}} V$ is commutative) multiplication. If $w \in \wedge^2 V$ and $x \in V$, then by (7.1) and (7.2) one has, by differentiating an exponential,

$$-\iota(x)\hat{e}^{2w} = \varepsilon([w, x])e^{2w},$$

where the bracket is Clifford commutation. But then also

$$(-\varepsilon([w, x]) - \iota(x))\hat{e}^{2w} = (\iota(x) + \varepsilon([w, x]))\hat{e}^{2w}.$$

Adding $(\varepsilon(x) + \iota([w, x]))\hat{e}^{2w}$ to both sides of the last equation yields

$$(\varepsilon(x - [w, x]) - \iota(x - [w, x]))\hat{e}^{2w} = (\varepsilon(x + [w, x]) + \iota(x + [w, x]))\hat{e}^{2w}.$$

But this is just Clifford multiplication. If $y \in V$ and $u \in C(V)$ then $(\varepsilon(y) + \iota(y))u = yu$ and, if $u \in C^0(V)$, $(\varepsilon(y) - \iota(y))u = uy$, both by (7.1). We have proved:

Proposition. *Let $x \in V$ and $w \in \wedge^2 V$. Then*

$$\hat{e}^{2w}(x - [w, x]) = (x + [w, x])\hat{e}^{2w}.$$

7.12

Let $s \in \text{Lie}SO(V)$ and let V_s^1 and V_s^{-1} be, respectively, the s-eigenspaces for the eigenvalues ± 1. For $x, y \in V$ the equation

$$(sx, y) = -(x, sy)$$

readily implies the following.

Lemma. *The subspaces V_s^1 and V_s^{-1} are isotropic with respect to the symmetric bilinear form B_V and are nonsingularly paired by B_V. In particular*

$$\dim V_s^1 = \dim V_s^{-1}$$

Consequently $1 + s$ is invertible if and only if $1 - s$ is invertible.

7.13

Using Lemma 7.12 we define the Zariski open subset of $\text{Lie}SO(V)$ and $SO(V)$ by

$$\begin{aligned} \text{Lie}SO(V)^{(*)} &= \{s \in \text{Lie}SO(V) : \dim(V_s^1) = \dim(V_s^{-1}) = 0\} \\ &= \{s \in \text{Lie}SO(V) : 1 + s \text{ and } 1 - s \text{ are invertible}\}, \\ SO(V)^* &= \{a \in SO(V) \mid \det(1 + a) \neq 0\}, \end{aligned}$$

so that in the notation of Theorem 7.2 and Theorem 7.10 one has

$$\mathrm{Spin}(V)^* = T^{-1}(SO(V)^*).$$

Proposition. *The subsets* $\mathrm{Lie}SO(V)^{(*)}$ *and* $SO(V)^*$ *are algebraically isomorphic via the mapping*

$$\mathrm{Lie}SO(V)^{(*)} \to SO(V)^*, \qquad s \mapsto a = \frac{1-s}{1+s}, \qquad s = \frac{1-a}{1+a} \leftarrow a \tag{1}$$

for $s \in \mathrm{Lie}SO(V)^{(*)}$ *and* $a \in SO(V)^*$. *In addition one has the relation* $(1+a)(1+s) = 2$.

Proof. If $b \in \mathrm{End}V$ let b^t be the transpose endomorphism with respect to B_V. Since $s^t = -s$ for $s \in \mathrm{Lie}SO(V)$ it is immediate from (1) that $a^t = a^{-1}$ so that a is orthogonal. Since $\mathrm{Lie}SO(V)^{(*)}$ is Zariski open it is connected and $0 \in \mathrm{Lie}SO(V)^{(*)}$. It follows then that $a \in SO(V)$. Furthermore $(1+a)(1+s) = 2$ is immediate from (1) and hence $a \in SO(V)^*$. The map $s \mapsto a$ is injective since one recovers s from a by (1). Conversely if $a \in SO(V)^*$, then $s \in \mathrm{Lie}SO(V)$ since $a^t = a^{-1}$ implies $s^t = -s$. But this formula yields $(1+a)(1+s) = 2$ so that $s \in \mathrm{Lie}SO(V)^{(*)}$. Hence the mapping (1) is bijective. □

7.14

In general if $b \in \mathrm{End}V$ is such that $1+b$ is invertible we will put

$$\Gamma(b) = \frac{1-b}{1+b}$$

Also let $\wedge^2 V^{(*)}$ be the inverse image of $\mathrm{Lie}SO(V)^{(*)}$ under the isomorphism $\tau : \wedge^2 V \to \mathrm{Lie}SO(V)$.

Theorem. *Let* $g \in \mathrm{Spin}(V)^*$. *Let the sign of square root be chosen so that*

$$\mathrm{pr}_0(g) = \frac{1}{2^{n/2}}\sqrt{\det(1+T(g))}$$

(see Theorem 7.10). Then putting $c = \mathrm{pr}_0(g)$ *one has*

$$g = c\,\hat{e}^{-2w}$$

where $w = \tau^{-1}(\Gamma(T(g))) \in \wedge^2 V$. *In particular* $w \in \wedge^2 V^{(*)}$.

Proof. We have $\Gamma(\tau(w)) = T(g)$. But by Proposition 7.11, for any $x \in V$, one has $\hat{e}^{-2w}((1+\tau(w))x) = ((1-\tau(w))x)\hat{e}^{-2w}$. Putting $y = (1+\tau(w))x)$ one therefore has

$$\hat{e}^{-2w}\; y = T(g)(y)\; \hat{e}^{-2w}$$

On the other hand $g\,y\,g^{-1} = T(g)\,(y)$ so that

$$y\,g^{-1} = g^{-1}T(g)(y).$$

But the last two equations imply that $g^{-1} e^{-2w}$ commutes with y. Since $y \in V$ is arbitrary this implies that $g^{-1} e^{-2w} \in \mathrm{Cent}(C(V)) \cap C^0(V)$. But $\mathrm{Cent}(C(V)) \cap C^0(V) = \mathbb{C}$. Hence there exists $d \in \mathbb{C}$ such that $d\, e^{-2w} = g$. But by applying pr_0 it follows that $d = c$. □

7.15

We determine the generalized Cayley mapping $\Phi : \mathrm{Spin}(V) \to \mathrm{LieSpin}(V)$, corresponding to to the representation Spin, for elements g in the Zariski open set $\mathrm{Spin}(V)^*$.

Theorem. *Let $g \in \mathrm{Spin}(V)^*$. Let the sign of square root be chosen so that*

$$\mathrm{pr}_0(g) = \frac{1}{2^{n/2}} \sqrt{\det(1 + T(g))},$$

according to theorem 7.10. Then

$$\Phi_{\mathrm{Spin}}(g) = -2\,\mathrm{pr}_0(g)\,\tau^{-1}(\Gamma(T(g))) \in \wedge^2 V.$$

Thus the generalized Cayley mapping $\Phi_{\mathrm{Spin}} : \mathrm{Spin}(V) \to \wedge^2 V$ factors to the classical Cayley transform $\Gamma : SO(V)^* \to \mathrm{LieSpin}(V)^{(*)}$, up to multiplication by a regular function, via the natural identifications.

For the degree of the spin representation we have

$$\deg(\mathrm{Spin}) = \begin{cases} n & \text{if } n \text{ is even,} \\ n-1 & \text{if } n \text{ is odd.} \end{cases}$$

Let $\chi \in A(\mathrm{Spin}(n, \mathbb{C}))$ be given by $\chi(a) = \sqrt{\det(1 + T(a))} = 2^{n/2}\mathrm{pr}_0(a)$. Then

$$\begin{aligned} Q(\mathrm{Spin}(n))^{\mathrm{Spin}(n)} &= Q(\mathrm{LieSpin}(n))^{\mathrm{Spin}(n)}[\chi], \\ Q(\mathrm{Spin}(n)) &= Q(\mathrm{LieSpin}(n))[\chi], \end{aligned}$$

in terms of (3.3).

Proof. By (7.7) one has $\Phi_{\mathrm{Spin}}(g) = \mathrm{pr}_2(g)$ but since the exponential in Theorem 7.14 is an exponential for exterior multiplication the result follows from this theorem.

Let us now determine deg(Spin). Consider the fiber of Φ over a generic X in $\mathrm{LieSpin}(n)$. Let $G_o = \{a \in \mathrm{Spin}(n) \mid \det(1 + T(a)) = 0\}$. We may assume that X is in the complement of $\Phi(G_o)$. Then modulo some fixed scalar (which can be ignored in determining the cardinality of the fiber) if $a \in \mathrm{Spin}(n)$ is in the fiber then

$$\sqrt{\det(1 + T(a))}(1 - T(a))(1 + T(a))^{-1} = X.$$

Put $t = \sqrt{\det(1 + T(a))}$ so that $(1 - T(a))(1 + T(a))^{-1} = X/t$. Recall that an equation of matrices $(1 - c)(1 + c)^{-1} = d$ is symmetric in c and d and that $(1 + c)(1 + d) = 2$. Thus

$$(1 + T(a))(1 + X/t) = 2. \tag{1}$$

Taking the determinant of both sides of (1) one has

$$t^2 \det(1 + X/t) = 2^n \tag{2}$$

or

$$f_X(t) = \det(t + X) - 2^n t^{n-2} = 0. \tag{3}$$

Comparing (1) and (2) we get $\det(1+T(a)) = t^2$ which checks with the definition of t. Given a nonzero root t of (3) the corresponding $T(a)$ is given by $T(a) = (1 - X/t)(1 + X/t)^{-1}$. But t is a specific square root of $\det(1 + T(a))$ and hence a itself is uniquely defined.

Now (3) is a polynomial of degree n in t with leading term t^n and constant term $\det(X)$. If n is even then $\det(X) \neq 0$ for generic X and we also may assume that the polynomial $\det(t + X)$ has pairwise different roots. Replacing X by a large multiple rX and considering

$$\tfrac{1}{r^n} f_{rX}(t) = \tfrac{1}{r^n}\left(\det(t + rX) - 2^n t^{n-2}\right) = \det(\frac{t}{r} + X) - \tfrac{2^n}{r^2}(\tfrac{t}{r})^{n-2}$$

which approaches $\det(t/r + X)$ we see that also the polynomial (3) has pairwise distinct roots and they are all nonzero since $\det(X) \neq 0$.

If n is odd, then $\det(X) = 0$ but generically the rank of X is $n - 1$; by the argument above (3) has again pairwise distinct roots for generic X, but one of them is 0 so that there are $n - 1$ different nonzero roots.

Let us discuss the adjunction now. We use the above description of the fiber of Φ over a generic X. Let us consider the expansion

$$f_X(t) = \det(t + X) - 2^n t^{n-2} = \sum_{j=0}^{n} p_j(X)\, t^j, \tag{4}$$

where $p_j \in A(\mathrm{LieSpin}(n))^{\mathrm{Spin}(n)}$ form a system of generators with $p_0(X) = \det(X)$ which is 0 for odd n, and $p_n(X) = 1$. We have

$$f_X(\chi) = \det(\chi + X) - 2^n t^{n-2} = \sum_{j=0}^{n} p_j(X)\, \chi^j = 0,$$

since t in the discussion above corresponds to $\chi(a)$, and this is the minimal polynomial for χ over $Q(\mathrm{LieSpin}(n))^{\mathrm{Spin}(n)}$ for n even. For n odd the minimal polynomial is $f_X(\chi)/\chi$. The reason for this is that $f_X(t)$ has exactly deg(Spin) many pairwise nonzero roots for generic X, in a Zariski dense open set which we may determine as the complement to $\Phi(\{g : \Psi(g) = 0\})$. □

7.16 The Cayley transform for the ρ-representation of any semisimple Lie group

Let G be any simply connected semisimple Lie group. One has a homomorphism

$$G \to \mathrm{Spin}(\mathfrak{g}) \tag{1}$$

which lifts the adjoint representation. The Lie algebra of Spin($\mathfrak{g}$) is $\wedge^2\mathfrak{g}$. Thus for any $a \in G$ there exists $u = u(a) \in \wedge^2\mathfrak{g}$ where

$$u = \sqrt{\det(1 + Ada)}(1 - Ada)/(1 + Ada) \tag{2}$$

On the other hand $\wedge\mathfrak{g}$ is a differential chain complex with respect to a boundary operator c where $c : \wedge^2\mathfrak{g} \to \mathfrak{g}$ is given by $c(X \wedge Z) = [X, Z]$ for $X, Z \in \mathfrak{g}$.

Theorem. *Let G be any simply connected semisimple Lie group. For the ρ-representation one has up to a fixed scalar multiple*

$$\Phi(a) = c(u). \tag{3}$$

Proof. Actually the first reference establishing that the restriction (using (1)) of the spin representation s of Spin($\mathfrak{g}$) to G is a multiple of the ρ-representation is in reference 9 in [12] (see top paragraph on p. 358 where ρ has been written as g). But our result (2.8.1) on direct sums (here a multiple of the same representation, thus we may use (2.8.1) for G semisimple) we may use the restriction $\pi = s|G$ to compute Φ_ρ. But then $\pi' : \mathfrak{g} \to \wedge^2\mathfrak{g}$ is an injection. Let $\mathfrak{m}$ be the orthocomplement of $\pi'(\mathfrak{g})$ in $\wedge^2\mathfrak{g}$ so that one has a direct sum $\wedge^2\mathfrak{g} = \pi'(\mathfrak{g}) + \mathfrak{m}$ with corresponding projection $p : \wedge^2\mathfrak{g} \to \pi'(\mathfrak{g})$. Let $a \in G$ and put $X = \Phi(a)$. In the notation of (3) we must prove that $X = c(u)$ up to a scalar multiple. But one clearly has

$$\pi'(X) = p(u) \tag{4}$$

That is for some $v \in \mathfrak{m}$ one has

$$u = v + \pi'(X) \tag{5}$$

But now besides the boundary operator c on $\wedge\mathfrak{g}$ one has a coboundary operator d on $\wedge\mathfrak{g}$. See section 2 in [10] where ∂ in [10] is written as c here. One has $d : \mathfrak{g} \to \wedge^2\mathfrak{g}$. The homomorphism π' is δ in (69) of [12]. By (106) in [12] one has $\pi' = d/2$, so that (4) becomes

$$d(X)/2 = p(u) \tag{6}$$

But by (94) in [12] one has that $dc + cd$ is 1/2 times the Casimir operator on $\wedge\mathfrak{g}$. But the Casimir is normalized so that it takes the value 1 on $\mathfrak{g}$. However c vanishes on $\mathfrak{g} = \wedge^1\mathfrak{g}$. Thus $cd = 1/2$ on $\mathfrak{g}$. Thus upon applying c to (6) yields

$$X/4 = c(pu). \tag{7}$$

On the other hand, $\mathfrak{m}$ is the kernel of c on $\wedge^2\mathfrak{g}$ by (4.4.4) in [10]. Thus c vanishes on v in (5) and by applying c to both sides of (5) one has $c(u) = c(pu)$ and hence $X/4 = c(u)$. But this is just (3) up to a scalar. □

Acknowledgements. We thank N. Wallach, V.L. Popov, T.A. Springer, D. Luna, D.V. Alekseevsky, J. Hofbauer, and E. Vinberg for helpful comments.

References

[1] Bardsley, P., Richardson, R.W., Etale slices for algebraic transformation groups in characteristic p, *Proc. Lond. Math. Soc., I. Ser.* **51** (1985), 295–317.

[2] Borel, A., *Linear Algebraic Groups*, W. A. Benjamin, 1969.

[3] Chevalley, C., *The Algebraic Theory of Spinors*, Columbia Univ. Press, New York, 1954.

[4] Dynkin, E.B., Semisimple subalgebras of semisimple Lie algebras, *Mat. Sbornik Nov. Ser.* **30** (1952), 349–462. English transl. in *AMS Transl. II. Ser* **6** (1957), 111–243.

[5] Gorbatsevich, V.V., Onishchik, A.L., Vinberg, E.B., Structure of Lie groups and Lie algebras. In: *Lie Groups and Lie Algebras III* (eds., A.L. Onishchik, E.B. Vinberg), Encyclopedia of Mathematical Sciences **41**, Springer-Verlag, Berlin, 1994.

[6] Humphrey, J., *Linear Algebraic Groups*, GTM 21, Springer-Verlag, 1975, 1981.

[7] Kostant, B., The principal three-dimensional subgroup and the Betti numbers of a complex simple Lie group, *Amer. J. Math.* **81** (1959), 973–1032.

[8] Kostant, B., Lie group representations on polynomial rings, *Bull. Amer. Math. Soc.* **69** (1963), 518–526.

[9] Kostant, B., Lie group representations on polynomial rings, *Amer. J. Math.* **85** (1963), 327–404.

[10] Kostant, B., Eigenvalues of a Laplacian and commutative Lie subalgebras, *Topology* **3** (1965), 147–159.

[11] Kostant, B., On Convexity, the Weyl group and the Iwasawa decomposition, *Ann. Scient. Ec. Norm. Sup.* **6** (1973), 413–455.

[12] Kostant, B., Clifford Algebra Analogue of the Hopf–Koszul–Samelson Theorem, the ρ-decomposition $C(\mathfrak{g}) = \operatorname{End} V_\rho \otimes C(P)$, and the $\mathfrak{g}$-module structure of $\wedge \mathfrak{g}$, *Adv. of Math.* **125** (1997), 275–350.

[13] Lipschitz, R., Correspondence, *Ann. Math.* **69** (1959), 247–251.

[14] Luna, D., Slices étales, *Bull. Soc. Math. France*, Mémoire **33** (1973), 81–105.

[15] Richardson, R. W., The conjugating representation of a semisimple group, *Invent. Math.* **54** (1979), 229–245.

[16] Richardson, R. W., An application of the Serre conjecture to semisimple algebraic groups. In *Algebra, Carbondale 1980*, Lecture Notes in Math. 848, Springer-Verlag, 1981, pp. 141–151.

[17] Springer, T. A., The unipotent variety of a semisimple group. In *Algebr. Geom., Bombay Colloq. 1968*, 1969, pp. 373–391.

[18] Steinberg, R., Regular elements of semisimple algebraic groups, *Inst. Hautes Études Sci. Publ. Math.* **25** (1965), 49–80.

[19] Steinberg, R., *Conjugacy Classes in Algebraic Groups. Notes by Vinay V. Deodhar*, Lecture Notes in Mathematics 36, Springer-Verlag, 1974, pp. vi+159.

Bertram Kostant
MIT
Department of Mathematics
Cambridge, MA 02139-4307, USA
kostant@math.mit.edu

Peter W. Michor
Erwin Schrödinger Institute of Mathematical Physics
Boltzmanngasse 9
A-1090 Wien, Austria
and
Institut für Mathematik
Universität Wien
Strudlhofgasse 4, A-1090 Wien, Austria
Peter.Michor@esi.ac.at

Geometry of $\mathrm{GL}_n(\mathbb{C})$ at Infinity: Hinges, Complete Collineations, Projective Compactifications, and Universal Boundary

Yurii A. Neretin

Compactifications of semisimple groups and of homogeneous spaces arise in natural ways in different branches of mathematics (enumerative algebraic geometry, noncommutative harmonic analysis, automorphic forms, etc.). The most important construction of this kind is the family of objects called the *Satake boundary*, or the *De Concini–Procesi boundary* or the *wonderful compactification*; see [26], [7], [4], [12], and [22]. For the first time, such compactifications of the group $\mathrm{PGL}_n(\mathbb{C})$ (the *complete collineations*) and of the symmetric space $\mathrm{PGL}_n(\mathbb{C})/\mathrm{PO}_n(\mathbb{C})$ (the *complete quadrics*) were discovered by Semple [27], [28], [29].

In both the geometry and analysis of symmetric spaces, there arise some more complicated compactifications such as the Karpelevich compactification ([11], [15], [8]) and the Martin compactification ([6], [23], [8]).

There exists also a wide theory of algebraic equivariant compactifications of reductive groups and their homogeneous spaces, see [31], [16], [12], [24], [2] (if a reductive group is a torus, then this theory becomes the theory of toric varieties).

This paper has two purposes. The first aim is to give an explicit description in elementary geometric terms of all algebraic projective compactifications (see below) of $\mathrm{GL}_n(\mathbb{C})$. The second aim is to construct a universal object (the sea urchin) for all the compactifications of this type.

0.1 Projective compactifications

Consider the group $\mathrm{GL}_n(\mathbb{C})$ of all complex invertible $n \times n$ matrices. Consider a polynomial (generally speaking, reducible) representation ζ of the group $\mathrm{GL}_n(\mathbb{C})$ in an N-dimensional complex linear space Z.

Denote by $\mathrm{Mat}(Z)$ the space of all linear operators in Z. Consider the space $\mathbb{P}\mathrm{Mat}(Z)$ consisting of nonzero operators defined up to a nonzero scalar factor.

Supported by the grants RFBR 98-01-00303 and NWO 047-008-009.

* The wonderful compactification of $\mathrm{PGL}(3, \mathbb{C})/\mathrm{PO}(3, \mathbb{C})$ was constructed by E. Study [30] in 1886.

This space is the complex projective space $\mathbb{CP}^{N^2-1}$. Consider the maps

$$\mathrm{GL}_n(\mathbb{C}) \xrightarrow{\zeta} \mathrm{Mat}(Z) \to \mathbb{P}\mathrm{Mat}(Z).$$

Denote by $[\mathrm{GL}_n]_\zeta$ the closure of the image of $\mathrm{GL}_n(\mathbb{C})$ in $\mathbb{P}\mathrm{Mat}(Z)$. The spaces $[\mathrm{GL}_n]_\zeta$ are called *projective compactifications* of $\mathrm{GL}_n(\mathbb{C})$; see [25].

Remark. Let ζ be an irreducible representation with a signature $\nu = (\nu_1, \dots, \nu_n)$ (see below 2.12) satisfying the condition

$$\nu_1 > \nu_2 > \dots > \nu_n.$$

In this case, the space $[\mathrm{GL}_n]_\zeta$ is called the Semple complete collineation space or the De Concini–Procesi compactification of $\mathrm{PGL}_n(\mathbb{C})$.

Remark. Let us give the definition of a projective compactification of a homogeneous space G/H, where G and H are Lie groups (this will not be used below). Let ρ be a finite-dimensional projective representation of G in a linear space V, or equivalently, let G act on the projective space $\mathbb{P}V$ by projective transformations. Assume that there exists a point $w \in \mathbb{P}V$ such that H is the stabilizer of w. Then the G-orbit Gw of w is the homogeneous space G/H. The closure of the orbit Gw in $\mathbb{P}V$ is a *projective compactification* of G/H.

If we vary an action of G on a projective space and a point w, we obtain various projective compactifications of G/H. In our case, $G = \mathrm{GL}_n(\mathbb{C}) \times \mathrm{GL}_n(\mathbb{C})$, $H = \mathrm{GL}_n(\mathbb{C})$ is the diagonal subgroup in G, and $V = \mathrm{Mat}(Z)$.

We emphasize that the Martin and Karpelevich compactifications of Riemannian noncompact symmetric spaces are not projective compactifications.

0.2 Abstract characterization of projective compactifications

Let the group $\mathrm{GL}_n(\mathbb{C}) \times \mathrm{GL}_n(\mathbb{C})$ act on a projective algebraic variety M. Denote by $\Delta \simeq \mathrm{GL}_n(\mathbb{C})$ the diagonal subgroup in $\mathrm{GL}_n(\mathbb{C}) \times \mathrm{GL}_n(\mathbb{C})$. Let G have an open $\mathrm{GL}_n(\mathbb{C}) \times \mathrm{GL}_n(\mathbb{C})$-orbit in M, and the stabilizer of the orbit contain Δ (any proper subgroup of $\mathrm{GL}_n(\mathbb{C}) \times \mathrm{GL}_n(\mathbb{C})$ containing Δ is a product of Δ and a subgroup in the center $\mathbb{C}^* \times \mathbb{C}^*$ of $\mathrm{GL}_(\mathbb{C}) \times \mathrm{GL}_(\mathbb{C})$).

The Kambayashi theorem ([9], see also the exposition in [3], 5.1) implies that all normal $\mathrm{GL}(\mathbb{C}) \times \mathrm{GL}(\mathbb{C})$-varieties M satisfying these conditions are of the form $[\mathrm{GL}]_\zeta$.

We explore only the constructive definition 0.1.

0.3 Sea urchin

The sea urchin $\mathfrak{S}\mathfrak{U}_n$ is the universal object for all the projective compactifications of $\mathrm{GL}_n(\mathbb{C})$ in the following sense.

a) For each projective compactification $[\mathrm{GL}_n]_\zeta$, there exists a canonical $\mathrm{GL}_(\mathbb{C}) \times \mathrm{GL}_n(\mathbb{C})$-equivariant surjective map

$$\pi_\zeta : \mathfrak{S}\mathfrak{U}_n \to [\mathrm{GL}_n]_\zeta.$$

b) Denote by $\dot{D}_\varepsilon$ the punctured disk $0 < |z| < \varepsilon$ in $\mathbb{C}$. Consider the germ at 0 of an algebraic curve

$$\dot{D}_\varepsilon \to GL_n(\mathbb{C}).$$

Any such germ has a limit as $z \to 0$ in the sea urchin $\mathfrak{SU}_n$. Two germs γ_1, γ_2 have the same limit in the sea urchin iff for each projective compactification $[GL_n(\mathbb{C})]_\zeta$ the limits $\lim_{z\to 0} \gamma_1(z)$ and $\lim_{z\to 0} \gamma_2(z)$ in $[GL_n(\mathbb{C})]_\zeta$ coincide.

c) $\pi_\zeta(\lim_{z\to 0} \gamma(z)) = \lim_{z\to 0} \pi_\zeta(\gamma(z))$ for all germs γ and all $[GL_n]_\zeta$.

The existence of the sea urchin is obvious. Indeed, the condition b) defines an equivalence of germs and a point of the sea urchin is a class of equivalent germs.

Strangely enough, this object admits an explicit parametrization in elementary terms. Points of the sea urchin are enumerated by collections of integers $(m_1, \ldots, m_n)$ (defined up to a common factor) and some special collections of subspaces (hinges) $P_1, \ldots, P_\tau \subset \mathbb{C}^n \oplus \mathbb{C}^n$.

Remark. The sea urchin admits a natural topology (see 3.11). Nevertheless, it is not a compact topological space. By its definition, any meromorphic curve in $GL_n(\mathbb{C})$ has a limit in the sea urchin; but there are sequences $g_j \in GL_n(\mathbb{C})$ that have no limit points in the sea urchin.

Remark. There exists a universal object for all projective compactifications in the category of compact topological spaces. This object (for the case of Riemannian noncompact symmetric spaces) was investigated by Kushner [13], [14]. The sea urchin is not a compact space in the usual sense, hence the sea urchin differs from the Kushner compactification. Also, the sea urchin is simpler.

Remark. The sea urchin is not a projective algebraic variety.

0.4 Basic observations

Denote by Gr_n the Grassmannian of n-dimensional subspaces in $\mathbb{C}^n \oplus \mathbb{C}^n$. Consider the canonical embedding $GL \to Gr_n$ taking any operator to its graph.

Consider the germ at $z = 0$ of an algebraic (or meromorphic) map $\gamma : \dot{D}_\varepsilon \to GL_n(\mathbb{C})$. For any integer k, consider the following limit in the Grassmannian

$$R_k = \lim_{z\to 0} z^{-k}\gamma(z).$$

We obtain the bilateral sequence

$$\ldots, R_{-2}, R_{-1}, R_0, R_1, R_2, \ldots \tag{0.1}$$

Consider all k such that R_k is not a sum of a horizontal subspace and a vertical subspace, i.e.,

$$R_k \neq \Big[R_k \cap (\mathbb{C}^n \oplus 0)\Big] \oplus \Big[R_k \cap (0 \oplus \mathbb{C}^n)\Big].$$

---

†Let$M_1, M_2, \ldots$ be all projective compactifications. Consider the diagonal embedding $GL_n \to M_1 \times M_2 \times \ldots$ (the product is equipped with the topology of pointwise convergence = the Tihonov topology). The closure of the image of this embedding is the required universal object.

Thus we obtain some *finite* collection of integers

$$k_1 > k_2 > \ldots > k_\tau.$$

Then we select the corresponding R_{k_j} from the sequence (0.1). Thus we obtain the finite ($\tau \leqslant n$) family of subspaces in $\mathbb{C}^n \oplus \mathbb{C}^n$

$$\mathcal{R} = (R_{k_1}, \ldots, R_{k_\tau}).$$

All possible families $\mathcal{R}$ can easily be described; see the definition of hinges in 2.1.

Thus, for each meromorphic germ $\gamma(z)$, we associate the following data

$$(k_1, \ldots, k_\tau;\; R_{k_1}, \ldots, R_{k_\tau}). \tag{0.2}$$

Our paper contains two observations.

1. For any polynomial representation ζ of $\mathrm{GL}_n(\mathbb{C})$, the limit of $\zeta(\gamma(z))$ in $[\mathrm{GL}_n]_\zeta$ is completely determined by the data (0.2).

2. The operator

$$\lim_{z\to 0} \zeta(\gamma(z)) \in \mathbb{P}\mathrm{Mat}(Z)$$

admits a simple explicit description in terms of the data (0.2).

This gives the explicit description of all the spaces $[\mathrm{GL}_n]_\zeta$ and also the description of the sea urchin.

0.5 Structure of this paper

Section 1 contains preliminaries regarding the category GA of linear relations and the fundamental representation of GA. These objects appeared in [17]; a detailed exposition is contained in [18] and [21], 2.5.

Section 2 contains the preliminaries concerning the hinges and the hinge semigroup. It is mainly based on [19], except for Subsections 2.10, 2.11; a detailed exposition of the work [19] is presented in [22]. We also explain here relations between the hinge semigroup and some well-known constructions.

Section 3 contains the construction of the sea urchin.

0.6 Other classical groups and symmetric spaces

In this paper, we consider only the groups $\mathrm{GL}_n(\mathbb{C})$. The hinge language is common for all the classical groups, all the classical symmetric varieties, and for their real forms (i.e., Riemannian and pseudo-Riemannian symmetric spaces), see [19], [22]. The results of this paper extend to this general situation more or less automatically.

Notation

We denote by $\mathbb{C}^*$ the multiplicative group of nonzero complex numbers.

Let V be an n-dimensional complex linear space. We denote by $\mathrm{GL}_n(\mathbb{C})$ or GL(V) the group of all invertible linear operators in $V \simeq \mathbb{C}^n$.

By $\mathrm{PGL}_n(\mathbb{C})$ we denote the quotient group $\mathrm{GL}_n(\mathbb{C})/\mathbb{C}^*$ of $\mathrm{GL}_n(\mathbb{C})$ by the subgroup $\mathbb{C}^*$ of all scalar matrices.

For a linear space Z, we denote by $\mathbb{P}Z$ the projective space $(Z \setminus 0)/\mathbb{C}^*$.

By $\mathrm{Mat}(Z)$ we denote the space of all linear operators in a linear space Z. By $\mathbb{P}\mathrm{Mat}(Z)$ we denote the quotient space $(\mathrm{Mat}(Z) \setminus 0)/\mathbb{C}^*$.

1 Category GA and its fundamental representation

1.1 Linear relations

Let V, W be finite-dimensional linear spaces over $\mathbb{C}$. A *linear relation* $P : V \rightrightarrows W$ is a linear subspace in $V \oplus W$.

Example. Let $A : V \to W$ be a linear operator. Its *graph* $\mathrm{graph}(A)$ consists of all vectors of the form $v \oplus Av \in V \oplus W$. Thus $\mathrm{graph}(A)$ is a linear relation $V \rightrightarrows W$. *Below we do not distinguish linear operators and their graphs.*

Let V, W, Y be linear spaces, and let $P : V \rightrightarrows W$, $Q : W \rightrightarrows Y$ be linear relations. Their product $S = QP$ is the linear relation $S : V \rightrightarrows Y$ consisting of all $v \oplus y \in V \oplus Y$ such that there exists $w \in W$ satisfying the conditions

$$v \oplus w \in P, \qquad w \oplus y \in Q.$$

For any linear relation $P : V \rightrightarrows W$, we define

a) the *kernel* $\mathrm{Ker}\, P \subset V$ is $P \cap (V \oplus 0)$;

b) the *image* $\mathrm{Im}\, P \subset W$ is the image of the projection of $P \subset V \oplus W$ on W along V,

c) the *domain* $\mathrm{Dom}\, P \subset V$ is the image of the projection of $P \subset V \oplus W$ on V along W,

d) the *indefiniteness* $\mathrm{Indef}\, P \subset W$ is $P \cap (0 \oplus W)$,

e) the *dimension* $\dim P$ is the dimension of P,

f) the *rank*

$$\begin{aligned} \mathrm{rk}\, P &:= \dim P - \dim \ker P - \dim \mathrm{Indef}\, P \\ &= \dim \mathrm{Dom}\, P - \dim \mathrm{Ker}\, P = \dim \mathrm{Im}\, P - \dim \mathrm{Indef}\, P. \end{aligned}$$

Remark. For any linear operators $A : V \to W$, $B : W \to Y$,

$$\mathrm{graph}(BA) = \mathrm{graph}(B)\, \mathrm{graph}(A);$$

$$\mathrm{Ker}\, \mathrm{graph}(A) = \mathrm{Ker}\, A; \qquad \mathrm{Im}\, \mathrm{graph}(A) = \mathrm{Im}\, A$$

$$\mathrm{rk}\, \mathrm{graph}(A) = \mathrm{rk}\, A.$$

For $P : V \rightrightarrows W$ we define the *pseudoinverse* linear relation $P^{\square} : W \rightrightarrows V$. It is the same subspace $P \subset V \oplus W$ regarded as a subspace in $W \oplus V$.

For $P : V \rightrightarrows W$ and $c \in \mathbb{C}^*$ we define the linear relation $c \cdot P$ consisting of all vectors $v \oplus cw$, where $v \oplus w$ ranges in P.

Remark. Let A be a linear operator. Then $c \cdot \mathrm{graph}(A) = \mathrm{graph}(cA)$. For an invertible linear operator $A : V \to V$, we have

$$(\mathrm{graph}\, A)^{\square} = \mathrm{graph}(A^{-1}).$$

1.2 Category GA

The objects of the category GA are finite-dimensional linear spaces over $\mathbb{C}$. The set of morphisms $\mathrm{Mor}(V, W) = \mathrm{Mor}_{\mathrm{GA}}(V, W)$ consists of all linear relations $P : V \rightrightarrows W$ and the formal element $\mathrm{null}_{V,W}$.

Remark. The dimension of P is an arbitrary number $0, 1, 2, \dots, \dim V + \dim W$. The element $\mathrm{null}_{V,W}$ cannot be identified with a linear relation.

Let $P : V \rightrightarrows W$, $Q : W \rightrightarrows Y$ be linear relations. If

$$\mathrm{Im}\, P + \mathrm{Dom}\, Q = W; \tag{1.1}$$

$$\mathrm{Indef}\, P \cap \mathrm{Ker}\, Q = 0, \tag{1.2}$$

then the product QP in the category GA coincides with the product of linear relations. Otherwise,

$$QP = \mathrm{null}_{V,Y}.$$

The product of null and any morphism is null.

Proposition 1.1 (see [21], 2.7). a) *For any linear spaces V, W, Y, Z and any morphisms $P \in \mathrm{Mor}(V, W)$, $Q \in \mathrm{Mor}(W, Y)$, $R \in \mathrm{Mor}(Y, Z)$, associativity holds*

$$(RQ)P = R(QP).$$

b) *Let $P \in \mathrm{Mor}(V, W)$, $Q \in \mathrm{Mor}(W, Y)$. If $QP \neq \mathrm{null}$, then*

$$\dim(QP) = \dim Q + \dim P - \dim W.$$

Remark. The group $\mathrm{Aut}(V)$ of automorphisms of an object V is the group GL(V).

1.3 Semigroup $\Gamma(V)$

Denote by $\Gamma(V)$ the subset in $\mathrm{Mor}(V, V)$ consisting of $\mathrm{null}_{V,V}$ and all linear relations $R : V \rightrightarrows V$ such that $\dim R = \dim V$.

By Proposition 1.1, $\Gamma(V)$ is closed with respect to the multiplication. Also, for a linear relation $R \in \Gamma(V)$

$$\dim \mathrm{Dom}\, R + \dim \mathrm{Indef}\, R = \dim V; \qquad \dim \mathrm{Ker}\, R + \dim \mathrm{Im}\, R = \dim V.$$

1.4 Exterior algebras

Let V be a complex linear space. Denote by $\Lambda(V)$ the exterior algebra of the space V. Recall that $\Lambda(V)$ is the associative algebra with generators v, where v ranges in V, and relations

$$v \wedge w = -w \wedge v,$$
$$(\mu v_1 + \nu v_2) \wedge w = \mu(v_1 \wedge w) + \nu(v_2 \wedge w),$$

where $\mu, \nu \in \mathbb{C}$, $v, w \in V$ and the sign $\wedge$ denotes the multiplication in $\Lambda(V)$.

We denote by $\Lambda^k V$ the linear subspace in $\Lambda(V)$ spanned by all vectors having the form

$$v_1 \wedge v_2 \wedge \ldots \wedge v_k, \qquad v_j \in V.$$

The space $\Lambda^k V$ is called *the k-th exterior power* of V. If $e_1, \ldots, e_n$ is a basis in V, then the collection $e_{i_1} \wedge \ldots \wedge e_{i_k}$, where $i_1 < i_2 < \ldots < i_k$, is a basis in $\Lambda^k V$.

Let $A : V \to W$ be a linear operator. We define the linear *operator of change of variables*

$$\lambda_{\mathrm{cha}}(A) : \Lambda V \to \Lambda W$$

by

$$\lambda_{\mathrm{cha}}(A) v_1 \wedge \ldots \wedge v_k = A v_1 \wedge \ldots \wedge A v_k.$$

If A is an operator $V \to W$, and B is an operator $W \to Y$, then

$$\lambda_{\mathrm{cha}}(B)\lambda_{\mathrm{cha}}(A) = \lambda_{\mathrm{cha}}(BA).$$

The operators $\lambda_{\mathrm{cha}}(A)$ preserve the degree k, and hence we also obtain the operators in the exterior powers

$$\lambda^k_{\mathrm{cha}}(A) : \Lambda^k V \to \Lambda^k W.$$

1.5 Fundamental representation of the category GA

Let $S : V \rightrightarrows W$ be a linear relation. Then there exist a basis

$$f_1, \ldots, f_\alpha, g_1, \ldots, g_\beta, h_1, \ldots, h_\gamma \tag{1.3}$$

in V and a basis

$$F_1, \ldots, F_\mu, G_1, \ldots, G_\beta, H_1, \ldots, H_\nu \tag{1.4}$$

in W such that $S \subset V \oplus W$ is spanned by the vectors

$$0 \oplus F_1, \ldots, 0 \oplus F_\mu, \quad g_1 \oplus G_1, \ldots, g_\beta \oplus G_\beta, \quad h_1 \oplus 0, \ldots h_\gamma \oplus 0. \tag{1.5}$$

Remark. Thus,

– the vectors h_i form a basis in Ker S;
– the vectors h_i and g_j form a basis in Dom S;
– the vectors F_k form a basis in Indef S;

– the vectors F_k, G_j form a basis in Im S.

Remark. Let $V = W$ and $S \in \Gamma(V)$. Then $\alpha = \mu$, $\gamma = \nu$.

We define the linear operator

$$\lambda(S) : \Lambda(V) \to \Lambda(W)$$

by

$$\lambda(S) f_1 \wedge f_2 \wedge \ldots \wedge f_\alpha \wedge g_{i_1} \wedge \ldots \wedge g_{i_k} = F_1 \wedge F_2 \wedge \ldots \wedge F_\mu \wedge G_{i_1} \wedge \ldots \wedge G_{i_k},$$

and

$$\lambda(S)\xi = 0$$

for all other basic vectors ξ in ΛV.

Remark. The bases (1.3), (1.4) are not uniquely determined by the linear relation S. If we change the bases, then the operator $\lambda(S)$ is multiplied by a nonzero constant.

Let $A \in$ GL(W), $B \in$ GL(V). Then

$$\lambda(ASB) = \lambda_{\text{cha}}(A)\lambda(S)\lambda_{\text{cha}}(B).$$

We also set

$$\lambda(\text{null}) = 0.$$

Remark. Let S be the graph of a linear operator A. Then $\lambda(S) = \lambda_{\text{cha}}(A)$. Nevertheless, $\lambda_{\text{cha}}(A)$ is a well-defined operator in ΛV and the operator $\lambda(S)$ is defined up to a nonzero scalar factor. For this reason, we preserve both notations $\lambda(\cdot)$ and $\lambda_{\text{cha}}(\cdot)$, since their meanings slightly differ.

Theorem 1.3 (see [21], II.7 or [22], §1). a) *Let $P : V \rightrightarrows W$, $Q : W \rightrightarrows Y$ be linear relations. Then*

$$\lambda(Q)\lambda(P) = c(Q, P)\lambda(QP),$$

where $c(Q, P) \in \mathbb{C}$.

b) $c(Q, P) \neq 0$ *iff* $QP \neq$ null.

Remark. For the intrinsic definition of the operators $\lambda(P)$, see [18], [21].

Remark. By construction, the operator $\lambda(P)$ takes homogeneous vectors to homogeneous vectors. Thus we obtain the family of operators

$$\lambda^k(P) : \Lambda^k V \to \Lambda^{k+\dim P - \dim V} W. \tag{1.6}$$

1.6 Fundamental representations of the semigroup $\Gamma(V)$

The main tool below is the semigroup $\Gamma(V)$ defined in 1.3. By (1.6), for $P \in \Gamma(V)$ we have

$$\lambda^m(P) : \Lambda^m V \to \Lambda^m V.$$

Thus we obtain the collection of projective representations λ^m of the semigroup $\Gamma(V)$ in the spaces $\Lambda^m V$.

Lemma 1.3. *Consider* $P \in \Gamma(V)$.

a) $\lambda^m(P) \neq 0$ *iff* $\dim \operatorname{Indef} P \leqslant m \leqslant \dim \operatorname{Im} P$

b) *If* $m = \dim \operatorname{Indef} P$, *then* $\operatorname{rk} \lambda^m(P) = 1$, *and*

$$\lambda^m(P) = \lambda^m(Q)$$

for any $Q \in \Gamma(V)$ *such that* $\operatorname{Indef} Q = \operatorname{Indef} P$, $\operatorname{Dom} Q = \operatorname{Dom} P$. *In particular, we can choose* $Q = \operatorname{Dom} P \oplus \operatorname{Indef} P$.

c) *If* $m = \dim \operatorname{Im} P$, *then* $\operatorname{rk} \lambda^m(P) = 1$, *and*

$$\lambda^m(P) = \lambda^m(R)$$

for any $R \in \Gamma(V)$ *such that* $\operatorname{Im} R = \operatorname{Im} P$, $\operatorname{Ker} R = \operatorname{Ker} P$. *In particular, we can choose* $R = \operatorname{Ker} P \oplus \operatorname{Im} P$.

Proof. Consider the canonical form described in 1.5. The conditions $\dim V = \dim W = \dim P$ imply $\alpha = \mu$, $\gamma = \nu$. Now all statements become obvious, see also [22]. □

Proposition 1.4. *The map* $\lambda(P)$ *is a continuous map from the Grassmannian* Gr_n *of all n-dimensional subspaces in* $V \oplus V$ *to* $\mathbb{P}\mathrm{Mat}(\Lambda V)$. *In particular, its image is a closed subset in* $\mathbb{P}\mathrm{Mat}(\Lambda V)$.

This follows from [21], Theorems 2.2 and 6.5.

2 Hinges

This section contains the preliminaries on the hinges with short explanations and sketches of proofs. For more details, see [22].

In this Section, V is an n-dimensional complex linear space and Gr_n is the Grassmannian of all n-dimensional linear subspaces in $V \oplus V \simeq \mathbb{C}^n \oplus \mathbb{C}^n$.

2.1 Hinges

A hinge

$$\mathcal{P} = (P_1, \ldots, P_k) : V \rightrightarrows V$$

is a family of linear relations $P_j : V \rightrightarrows V$ such that $\dim P_j = \dim V = n$ (hence $P_j \in \Gamma(V)$) and

$$\operatorname{Ker} P_j = \operatorname{Dom} P_{j+1}, \qquad \text{where} \quad j = 1, 2, \dots, k-1, \tag{2.1}$$

$$\operatorname{Im} P_j = \operatorname{Indef} P_{j+1}, \qquad \text{where} \quad j = 1, 2, \dots, k-1, \tag{2.2}$$

$$\operatorname{Dom} P_1 = V, \tag{2.3}$$

$$\operatorname{Im} P_k = V, \tag{2.4}$$

$$P_j \neq \operatorname{Ker} P_j \oplus \operatorname{Indef} P_j, \qquad \text{where} \quad j = 1, 2, \dots, k. \tag{2.5}$$

2.2 Comments

2.3.a. *Remarks on the definition*

Remark. The conditions (2.3)–(2.4) are the interpretation of the conditions (2.1)–(2.2) for $j = 0$ and $j = k$. The condition (2.3) means that the first term P_1 of a hinge is an operator. The condition (2.4) means that the last term P_k is a linear relation pseudoinverse to an operator.

Remark. The condition (2.5) is equivalent to the condition $\operatorname{rk} P_j > 0$.

Example. The graph of an *invertible* operator is a hinge ($k = 1$). The graph of a noninvertible operator is not a hinge.

Example. Let $A : V \to V$, $B : V \to V$ be linear operators such that

$$\operatorname{Im} A = \operatorname{Ker} B; \qquad \operatorname{Ker} A = \operatorname{Im} B.$$

Then

$$(\operatorname{graph}(A), \operatorname{graph}(B)^{\square})$$

is a hinge. Any hinge consisting of two terms ($k = 2$) has this form.

Remark. We have

$$V \supset \operatorname{Ker} P_1 \supset \operatorname{Ker} P_2 \supset \dots \supset \operatorname{Ker} P_k = 0,$$
$$0 \subset \operatorname{Im} P_1 \subset \operatorname{Im} P_2 \subset \dots \subset \operatorname{Im} P_k = V.$$

By (2.5), $\operatorname{Ker} P_{j+1} \neq \operatorname{Ker} P_j$. This implies $k \leqslant n$.

2.2.b. *Completed hinges*

For each hinge $\mathcal{P}$ we define the *completed hinge*

$$\widehat{\mathcal{P}} := (Q_0, P_1, Q_1, P_2, Q_2, \dots, P_k, Q_k), \tag{2.6}$$

where

$$\begin{aligned} Q_0 &= V \oplus 0, \\ Q_j &= \operatorname{Ker} P_j \oplus \operatorname{Im} P_j = \operatorname{Dom} P_{j+1} \oplus \operatorname{Indef} P_{j+1}, \\ Q_k &= 0 \oplus V, \end{aligned} \tag{2.7}$$

Obviously, $\mathcal{P}$ is uniquely determined by $\widehat{\mathcal{P}}$. It is easy to give an abstract definition of a completed hinge. It is a family (2.6) of n-dimensional linear relations $\mathbb{C}^n \rightrightarrows \mathbb{C}^n$ satisfying the conditions (2.7) and the condition $\operatorname{rk} P_j > 0$.

Evidently, the space of all hinges and the space of all completed hinges coincide.

2.2.c. Rephrasing of definition

We have

$$\operatorname{Ker} P_j \oplus \operatorname{Indef} P_j \subset P_j \subset \operatorname{Dom} P_j \oplus \operatorname{Im} P_j.$$

The image of P_j under the natural projection

$$\operatorname{Dom} P_j \oplus \operatorname{Im} P_j \longrightarrow \big(\operatorname{Dom} P_j / \operatorname{Ker} P_j\big) \oplus \big(\operatorname{Im} P_j / \operatorname{Indef} P_j\big)$$

is a graph of an invertible operator

$$\operatorname{Dom} P_j \big/ \operatorname{Ker} P_j \to \operatorname{Im} P_j \big/ \operatorname{Indef} P_j.$$

Thus, hinges can be defined in the following way. Consider two flags

$$0 = Y_0 \subset Y_1 \subset \ldots \subset Y_k = V,$$
$$V = Z_0 \supset Z_1 \supset \ldots \supset Z_k = 0,$$

such that

$$\dim Y_j / Y_{j-1} = \dim Z_{j-1} / Z_j \qquad \text{for all} \quad j.$$

For each j, we fix an invertible linear operator

$$A_j : Y_j / Y_{j-1} \to Z_{j-1} / Z_j.$$

By the previous remark, these data define a hinge.

2.3 Notation

We denote by $\operatorname{Hinge}(V) = \operatorname{Hinge}_n$ the space of all hinges $\mathcal{P} : V \Rrightarrow V$.

We denote by $\operatorname{Hinge}^*(V) = \operatorname{Hinge}_n^*$ the space of all hinges $\mathcal{P} = (P_1, \ldots, P_k) : V \Rrightarrow V$ defined up to the equivalence

$$(P_1, \ldots, P_k) \sim (c_1 P_1, \ldots, c_k P_k), \qquad \text{where} \quad c_j \in \mathbb{C}^*.$$

Considering 1-term hinges ($k = 1$), we obtain

$$\operatorname{Hinge}_n \supset GL_n(\mathbb{C}); \qquad \operatorname{Hinge}_n^* \supset PGL_n(\mathbb{C}).$$

2.4 Topology on Hinge_n^*

Below we define a structure of an irreducible smooth projective algebraic variety on Hinge_n^*. In this subsection, we define the topology on Hinge_n^*.

2.4.a. Convergence of sequences $g_j \in \mathrm{PGL}_n$ to points of Hinge_n^*

A sequence $g_j \in \mathrm{PGL}_n$ converges to $\mathcal{P} = (P_1, \ldots, P_k) \in \mathrm{Hinge}_n^*$, if there exist k sequences

$$\beta_1^{(1)}, \beta_2^{(1)}, \beta_3^{(1)}, \ldots; \qquad \beta_1^{(2)}, \beta_2^{(2)}, \beta_3^{(2} \ldots; \qquad \ldots\ldots; \qquad \beta_1^{(k)}, \beta_2^{(k)}, \beta_3^{(k)}, \ldots, \tag{2.8}$$

($\beta_j^{(\sigma)} \in \mathbb{C}^*$) such that $\beta_j^{(\sigma)} g_j$ converge to P_σ in Gr_n for all $\sigma = 1, \ldots, k$.

Remark. This implies

$$\lim_{j\to\infty} \beta_j^{(\sigma)} / \beta_j^{(\tau)} = \infty \qquad \text{for } \sigma > \tau.$$

Example. Let

$$g_j = \begin{pmatrix} 4^j & 0 & 0 \\ 0 & 2^j & 0 \\ 0 & 0 & 2^{-j} \end{pmatrix}.$$

Then we can choose

$$\beta_j^{(1)} = 4^{-j}; \qquad \beta_j^{(2)} = 2^{-j}; \qquad \beta_j^{(3)} = 2^j.$$

The limits P_1, P_2, P_3 are the subspaces in $\mathbb{C}^3 \oplus \mathbb{C}^3$ consisting of all the vectors of the following type

$$\begin{aligned} P_1 &: \quad (x, y, z) \oplus (x, 0, 0) \\ P_2 &: \quad (0, y, z) \oplus (x, y, 0) \\ P_3 &: \quad (0, 0, z) \oplus (x, y, z), \end{aligned}$$

where x, y, z range in $\mathbb{C}$. For the sequences

$$\mu_j^{(0)} = 8^{-j}; \quad \mu_j^{(1)} = 2^{-3/2\, j}; \quad \mu_j^{(2)} = 1; \quad \mu_j^{(3)} = 4^j$$

the limits $Q_j = \lim_{j\to\infty} \mu_j^{(\sigma)} g_j$ in Gr_n also exist, they consist of the vectors

$$\begin{aligned} Q_0 &: \quad (x, y, z) \oplus (0, 0, 0) \\ Q_1 &: \quad (0, y, z) \oplus (x, 0, 0) \\ Q_2 &: \quad (0, 0, z) \oplus (x, y, 0) \\ Q_3 &: \quad (0, 0, 0) \oplus (x, y, z) \end{aligned}$$

All limits $Q_0, \ldots, Q_3$ have rank 0. The collection $(Q_0, P_1, Q_1, P_2, \ldots, Q_4)$ is a completed hinge, see (2.6).

Lemma 2.1. *Any sequence $g_j \in \mathrm{PGL}_n(\mathbb{C})$ contains a subsequence convergent in our sense.*

Proof. We represent g_j in the form

$$g_j = A_j \begin{pmatrix} u_j^{(1)} & 0 & \dots \\ 0 & u_j^{(2)} & \dots \\ \vdots & \vdots & \ddots \end{pmatrix} B_j,$$

where A_j, B_j are unitary matrices, and

$$u_j^{(1)} \geqslant u_j^{(2)} \geqslant \dots > 0$$

Selecting a subsequence, we can assume that

1) the sequences A_j, B_j are convergent

2)$\exists \lim_{j\to\infty} u_j^{(m)}/u_j^{(1)} = \alpha_m$ for all $m > 1$.

We have $1 \geqslant \alpha_2 \geqslant \alpha_3 \geqslant \dots \geqslant 0$. Consider τ such that $\alpha_{\tau-1} \neq 0$, $\alpha_\tau = 0$. After the next selection of a subsequence, we can assume

3) $\exists \lim_{j\to\infty} u_j^{(m)}/u_j^{(\tau)} = \beta_m$ for all $m > \tau$. Obviously, $1 \geqslant \beta_{\tau+1} \geqslant \beta_{\tau+2} \geqslant \dots \geqslant 0$.

Then we repeat the same argument again, again, and again.

Now we assume $\beta_j^{(1)} = (u_j^{(1)})^{-1}$, $\beta_j^{(2)} = (u_j^{(\tau)})^{-1}$, etc.

Thus the space $\text{Hinge}_n^* \setminus PGL_n(\mathbb{C})$ is some kind of boundary of $PGL_n(\mathbb{C})$.

2.4.b. *Convergence on the boundary*

A sequence $\mathcal{P}^j = (P_1^j, \dots P_k^j)$ converges to a hinge $\mathcal{Q} = (Q_1, \dots, Q_l)$, if for any Q_u, there exist $v = 1, \dots, k$ and a sequence $\mu_j \in \mathbb{C}^*$ such that $\mu_j P_v^j$ converges to Q_u in Gr_n.

2.4.c. *Formal description of the structure of a compact metric space on* Hinge_n^*

Consider the action of the group $\mathbb{C}^*$ on Gr_n given by $P \mapsto c \cdot P$. Fixed points of $\mathbb{C}^*$ in Gr_n are linear relations of rank 0. In other words, the fixed points have the form

$$Q = \text{Ker}\, Q \oplus \text{Indef}\, Q.$$

All other orbits of $\mathbb{C}^*$ in Gr_n have trivial stabilizers. If $P \in \text{Gr}_n$ is not a fixed point of the $\mathbb{C}^*$-action, then the closure of the orbit $\mathbb{C}^* \cdot P$ consists of the orbit itself and of the pair of the points

$$\text{Dom}\, P \oplus \text{Indef}\, P, \qquad \text{Ker}\, P \oplus \text{Im}\, P.$$

For a hinge $\mathcal{P} = (P_1, \dots, P_k)$, we define the subset $\Omega(\mathcal{P})$ in Gr_n by

$$\Omega(\mathcal{P}) := Q_0 \cup \mathbb{C}^* P_1 \cup Q_1 \cup \mathbb{C}^* P_2 \cup Q_2 \cup \dots \cup Q_{k-1} \cup \mathbb{C}^* P_k \cup Q_k,$$

where the points $Q_j = \operatorname{Ker} P_j \oplus \operatorname{Im} P_j = \operatorname{Dom} P_{j+1} \oplus \operatorname{Indef} P_{j+1}$ are the elements of the completed hinge, see (2.6).

We emphasize that the closure of $\mathbb{C}^* P_j$ contains Q_{j-1} and Q_j. Hence the subset $\Omega(\mathcal{P})$ is closed and connected.

Denote by $\operatorname{Close}(\mathrm{Gr}_n)$ the space of all closed subsets of the Grassmannian Gr_n. Consider an arbitrary metric ρ on Gr_n compatible with the topology. For $x \in \mathrm{Gr}_n$ and $A \in \operatorname{Close}(\mathrm{Gr}_n)$, we define the distance

$$\rho(x, A) = \min_{y \in A} \rho(x, y).$$

The *Hausdorff metric* in $\operatorname{Close}(\mathrm{Gr}_n)$ is defined by

$$d(A, B) = \max\Big[\max_{x \in A} d(x, B),\ \max_{y \in B} d(y, A)\Big].$$

Theorem 2.2. a) *The image of the embedding* $\mathrm{Hinge}_n^* \to \operatorname{Close}(\mathrm{Gr}_n)$ *given by* $\mathcal{P} \mapsto \Omega(\mathcal{P})$ *is a compact subset in* $\operatorname{Close}(\mathrm{Gr}_n)$. *Thus we obtain a topology of a compact metric space on* Hinge_n^*.

b) *The group* $\mathrm{PGL}_n(\mathbb{C})$ *is dense in* Hinge_n^*.

Remark. The space of orbits of $\mathbb{C}^*$ on Gr_n is a non-Hausdorff topological space. The construction described above is the result of an application of the construction of the Hausdorff quotient described in [20],[22]. In algebraic geometry, there exist also more delicate constructions of separated quotients, i.e., the Hilbert scheme quotient and the Chow scheme quotient, see [1],[10]; in our case, these constructions are equivalent to the Hausdorff quotient.

2.5 Orbits of the group $\mathrm{GL}_n(\mathbb{C}) \times \mathrm{GL}_n(\mathbb{C})$ in Hinge_n

The group $\mathrm{GL}_n(\mathbb{C}) \times \mathrm{GL}_n(\mathbb{C})$ acts on $V \oplus V \simeq \mathbb{C}^n \oplus \mathbb{C}^n$ in the obvious way. Hence it acts on the spaces Hinge_n and Hinge_n^*.

Lemma 2.3. *The group* $\mathrm{GL}_n(\mathbb{C}) \times \mathrm{GL}_n(\mathbb{C})$ *has* 2^n *orbits in the space* Hinge_n. *These orbits are enumerated by the number* $k = 1, 2, \ldots, n$ *of terms of a hinge and the positive integers*

$$\alpha_1 = \operatorname{rk} P_1, \ldots, \alpha_k = \operatorname{rk} P_k, \qquad \text{where} \quad \alpha_1 + \ldots + \alpha_k = n.$$

Proof. Obvious. □

We denote these orbits by

$$\mathcal{O}[\alpha] = \mathcal{O}[\alpha_1, \ldots, \alpha_k].$$

Fix a basis $e_1, \ldots, e_n \in V$. For a given collection $\alpha_1, \ldots, \alpha_k$, we define the *canonical hinge* $\mathcal{P}_{\alpha_1,\ldots,\alpha_k} \in \mathcal{O}[\alpha_1, \ldots, \alpha_k]$ by

$$\mathcal{P}_{\alpha_1,\ldots,\alpha_k} = (P_1, \ldots, P_k) : V \rightrightarrows V,$$

where the linear relation $P_j \subset V \oplus V$ is spanned by the vectors

$$\begin{aligned} 0 \oplus e_\sigma, &\quad \text{where} \quad \sigma \leqslant \alpha_1 + \ldots + \alpha_{j-1}, \\ e_\tau \oplus e_\tau, &\quad \text{where} \quad \alpha_1 + \ldots + \alpha_{j-1} < \tau \leqslant \alpha_1 + \ldots + \alpha_j, \\ e_\mu \oplus 0, &\quad \text{where} \quad \mu > \alpha_1 + \ldots + \alpha_j. \end{aligned}$$

Remark. Denote by $\mathcal{O}^*[\alpha]$ the image of $\mathcal{O}[\alpha]$ in Hinge_n^*. We have

$$\dim \mathcal{O}[\alpha] = n^2; \qquad \dim \mathcal{O}^*[\alpha_1, \ldots, \alpha_k] = n^2 - k.$$

2.6 Alternative

Theorem 2.4. *Let $\mathcal{P} = (P_1, \ldots, P_k) : V \rightrightarrows V$ be a hinge. Fix $m = 0, 1, \ldots, n$. Consider the family of operators*

$$\lambda^m(P_1), \lambda^m(P_2), \ldots, \lambda^m(P_k) : \Lambda^m V \to \Lambda^m V. \tag{2.9}$$

Then there are only two possibilities.

1) There exists a unique j such that $\lambda^m(P_j) \neq 0$.

2) There exists j such that $\lambda^m(P_j) \neq 0$, $\lambda^m(P_{j+1}) \neq 0$, and $\lambda^m(P_\tau) = 0$ for all $\tau \neq j, j+1$. In this case, $\lambda^m(P_j)$ and $\lambda^m(P_{j+1})$ have rank 1 and coincide up to a nonzero factor. They also coincide with $\lambda(Q_j)$, where $Q_j = \mathrm{Ker}\, P_j \oplus \mathrm{Im}\, P_j$ is the term of the completed hinge $\widehat{\mathcal{P}}$, see (2.6).

Proof. This is a consequence of Lemma 1.3. □

Now for each hinge $\mathcal{P} = (P_1, \ldots, P_k) : V \rightrightarrows V$ and for each $m = 0, 1, \ldots, n$, we define the operator

$$\lambda^m(\mathcal{P}) : \Lambda^m V \to \Lambda^m V$$

as the unique nonzero term of the sequence (2.9). By construction, this operator is defined up to a nonzero factor.

Remark. Obviously, for any $g_1, g_2 \in \mathrm{GL}_n(\mathbb{C})$,

$$\lambda^m(g_1 \mathcal{P} g_2) = \lambda^m_{\mathrm{cha}}(g_1)\lambda^m(\mathcal{P})\lambda^m_{\mathrm{cha}}(g_2).$$

2.7 Example: The operators $\lambda^m(\mathcal{P})$ for canonical hinges

Consider the canonical hinge $\mathcal{P}_\alpha = \mathcal{P}_{\alpha_1, \ldots, \alpha_k}$. We intend to describe the operator $\mathcal{L}^m := \lambda^m(\mathcal{P}_\alpha)$. Set

$$u_j = \alpha_1 + \ldots + \alpha_{j-1}.$$

Let $u_j \leqslant m \leqslant u_{j+1}$, let $s := m - u_j$ and $u_j < l_1 < \ldots < l_s \leqslant u_{j+1}$. Then for each element

$$h = e_1 \wedge e_2 \wedge e_3 \wedge \ldots \wedge e_{u_j} \wedge e_{l_1} \wedge \ldots \wedge e_{l_s}$$

of the standard basis, we have

$$\lambda^m(\mathcal{P}_\alpha)h = h;$$

and $\lambda^m(\mathcal{P}_\alpha)$ annihilates all other elements $e_{t_1} \wedge \ldots \wedge e_{t_m}$ of the standard basis in $\Lambda^m V$.

2.8 The projective embedding of Hinge_n^*

In 2.6, for any $\mathcal{P} \in \mathrm{Hinge}_n$, we constructed the family of nonzero linear operators

$$\lambda^\circ(\mathcal{P}) := \big(\lambda^1(\mathcal{P}), \lambda^2(\mathcal{P}), \ldots, \lambda^{n-1}(\mathcal{P})\big) \tag{2.10}$$

defined up to nonzero factors. Consider two hinges

$$\mathcal{P} = (P_1, \ldots, P_k); \qquad \mathcal{R} = (c_1 P_1, \ldots, c_k P_k), \quad \text{where} \quad c_j \in \mathbb{C}^*.$$

By the construction 2.6, the operators $\lambda^m(\mathcal{P})$ and $\lambda^m(\mathcal{R})$ coincide up to a nonzero factor. Thus we obtain the map

$$\lambda^\circ : \mathrm{Hinge}_n^* \to \bigtimes_{m=1}^{n-1} \mathbb{P}\mathrm{Mat}(\Lambda^m V).$$

Consider also the map

$$\lambda^\circ : \mathrm{PGL}_n(\mathbb{C}) \to \bigtimes_{m=1}^{n-1} \mathbb{P}\mathrm{Mat}(\Lambda^m V)$$

given by

$$\lambda^\circ(g) := \big(\lambda^1_{\mathrm{cha}}(g), \ldots, \lambda^{n-1}_{\mathrm{cha}}(g)\big). \tag{2.11}$$

Theorem 2.5. ([19],[22]) *The map* (2.10) *from* Hinge_n^* *to* $\bigtimes_{m=1}^{n-1} \mathbb{P}\mathrm{Mat}(\Lambda^m V)$ *is continuous.*

Corollary 2.6. a) *The image of the map* (2.10) *is compact.*

b) *The image of the map* (2.10) *is the closure of* $\lambda^\circ\big(\mathrm{PGL}_n(\mathbb{C})\big)$.

c) *The space* Hinge_n^* *is an irreducible projective variety.*

Sketch of proof of Theorem 2.5. We shall prove the implication

$$g_j \in \mathrm{GL}_n(\mathbb{C}) \text{ converges to } \mathcal{P} \implies \lambda^\circ(g_j) \text{ converges to } \lambda^\circ(\mathcal{P}). \tag{2.12}$$

Let us represent g_j in the form

$$g_j = A_j D_j B_j,$$

where A_j, B_j are unitary matrices and D_j are diagonal matrices with decreasing eigenvalues. We can assume that the sequences $A_1, A_2, \ldots$ and $B_1, B_2, \ldots$ are convergent. Thus the question is reduced to the case $A_j = 1$, $B_j = 1$. For this case, the statement can easily be checked.

Corollary 2.6 implies the following consequence.

Corollary 2.7. a) *The space* Hinge_n^* *is an irreducible smooth projective variety.*

b) *The variety* Hinge_n^* *coincides with the Semple complete collineation space.*

c) *The variety* Hinge_n^* *coincides with the De Concini–Procesi* [4] *compactification of* $\mathrm{PGL}_n(\mathbb{C})$.

By definition (see [28]), the Semple complete collineation space is the closure of the image of the map (2.11). By the Semple theorem, the complete collineation space is a smooth projective variety, and by [4] it coincides with the De Concini–Procesi compactification, see also below 2.14.

2.9 Semigroup of hinges

Let us define one more variation of the space Hinge_n. Denote by $\widetilde{\mathrm{Hinge}}_n$ the set of all elements of $\times_{m=0}^{n} \mathrm{Mat}(\Lambda^m V)$ having the form

$$\mathcal{A} = (A_0, \dots, A_n) = \big(c_0 \cdot \lambda^0(\mathcal{P}), c_1 \cdot \lambda^1(\mathcal{P}), c_2 \cdot \lambda^2(\mathcal{P}), \dots, c_n \cdot \lambda^n(\mathcal{P})\big),$$

where $\mathcal{P}$ is a hinge and $c_0, \dots, c_n \in \mathbb{C}$. We say that $\mathcal{A}$ *lies over* $\mathcal{P}$. If all c_j are nonzero, we say that $\mathcal{A}$ is *nondegenerate*.

Proposition 2.8. $\widetilde{\mathrm{Hinge}}_n$ *is a subsemigroup in* $\times_{m=0}^{n} \mathrm{Mat}(\Lambda^m V)$.

This easily follows from Theorem 2.5.

Now we intend to give a constructive description of the product in $\widetilde{\mathrm{Hinge}}_n$.

Let $\mathcal{R} = (R_1, \dots, R_s)$ be a family of linear relations $V \rightrightarrows V$, and $\dim R_j = n$. We say that $\mathcal{R}$ is a *weak hinge* if for each j

$$\mathrm{Ker}\, R_j \supset \mathrm{Dom}\, R_{j+1},$$
$$\mathrm{Im}\, R_j \subset \mathrm{Indef}\, R_{j+1}.$$

Remark. Let $\widehat{\mathcal{P}} = (Q_0, P_1, Q_1, P_2, \dots, P_k, Q_k)$ be a completed hinge (see (2.6)). Then any subcollection of $\widehat{\mathcal{P}}$ is a weak hinge, and each weak hinge can be obtained in this way.

For any weak hinge $\mathcal{R} = (R_1, \dots, R_s)$ and any $m = 0, 1, \dots, n$, we intend to construct the canonical operator

$$\lambda^m(\mathcal{R}) : \Lambda^m V \to \Lambda^m V$$

defined up to a scalar factor. For this, we consider the sequence

$$\lambda^m(R_1), \lambda^m(R_2), \dots, \lambda^m(R_s).$$

If this family contains a nonzero term $\lambda^m(R_j)$, then we set

$$\lambda^m(\mathcal{R}) := \lambda^m(R_j).$$

Otherwise,

$$\lambda^m(\mathcal{R}) := 0.$$

Theorem 2.9 ([19], [22]). *Let* $\mathcal{R} = (R_1, \dots, R_s)$, $\mathcal{T} = (T_1, \dots, T_\tau)$ *be weak hinges. Then the family of all*

$$T_i R_j \neq \text{null}$$

is a weak hinge.

Sketch of proof. Let $T_i R_j \neq$ null. It can easily be checked that the segments

$$\dim \operatorname{Indef} R_j \leqslant m \leqslant \dim \operatorname{Im} R_j, \tag{2.13}$$

$$\dim \operatorname{Indef} T_i \leqslant m \leqslant \dim \operatorname{Im} T_i \tag{2.14}$$

have nonzero intersection. After this remark, Theorem 2.9 can easily be checked. □

Theorem 2.10 ([19], [22]). *For each m,*

$$\lambda^m(\mathcal{T})\lambda^m(\mathcal{R}) = c \cdot \lambda^m(\mathcal{T}\mathcal{R}),$$

where $c \in \mathbb{C}^*$.

Proof. Assume m satisfies the equations (2.13), (2.14). Then

$$\lambda^m(\mathcal{T}) = \lambda^m(T_i); \qquad \lambda^m(\mathcal{R}) = \lambda^m(R_j).$$

Thus,

$$\lambda^m(\mathcal{T})\lambda^m(\mathcal{R}) = \lambda^m(T_i)\lambda^m(R_j) = \lambda^m(T_i R_j) = \lambda^m(\mathcal{T}\mathcal{R}).$$ □

Thus, for $\mathcal{A}$ lying over a weak hinge $\mathcal{P}$ and $\mathcal{B}$ lying over a weak hinge $\mathcal{Q}$, the product $\mathcal{A}\mathcal{B}$ lies over $\mathcal{P}\mathcal{Q}$.

2.10 Canonical embedding $\mathrm{Hinge}_n \to \widetilde{\mathrm{Hinge}}_n$

Let $\mathcal{P} \in \mathrm{Hinge}_n$. The construction 2.6 defines the operators $\lambda^m(\mathcal{P}) : \Lambda^m V \to \Lambda^m V$ up to nonzero factors. We intend to define these operators in a canonical way.

Fix $\alpha = (\alpha_1, \dots \alpha_k)$. Consider $\mathcal{P} = (P_1, \dots, P_k) \in \mathcal{O}[\alpha]$. In particular, $\dim \operatorname{Im} P_j = \alpha_1 + \dots + \alpha_j$, and $\dim \operatorname{Indef} P_j = \alpha_1 + \dots + \alpha_{j-1}$

We have the family of operators

$$\lambda(P_1), \lambda(P_2), \dots, \lambda(P_k) : \Lambda(V) \to \Lambda(V)$$

defined up to nonzero factors. We have

$$\lambda^m(P_j) \neq 0 \qquad \text{iff} \quad \dim \operatorname{Indef} P_j \leqslant m \leqslant \dim \operatorname{Im} P_j,$$

and

$$\lambda^{\alpha_1}(P_2) = c_1 \cdot \lambda^{\alpha_1}(P_1),$$
$$\lambda^{\alpha_1+\alpha_2}(P_3) = c_2 \cdot \lambda^{\alpha_1+\alpha_2}(P_2),$$
$$\lambda^{\alpha_1+\alpha_2+\alpha_3}(P_4) = c_3 \cdot \lambda^{\alpha_1+\alpha_2+\alpha_3}(P_3),$$

and so on. The linear relation P_1 is a graph of some linear operator A. Thus the operator $\lambda(P_1) := \lambda_{\mathrm{cha}}(A)$ is well defined. After this we define the operator $\lambda(P_2)$ by the condition $c_1 = 1$, and then we define the operator $\lambda(P_3)$ by the condition $c_2 = 1$, and so on.

We denote by $\mathcal{L}^m(\mathcal{P}) : \Lambda^m V \to \Lambda^m V$ the unique nonzero operator among

$$\lambda^m(P_1), \dots, \lambda^m(P_k).$$

We denote by $\mathcal{L}(\mathcal{P})$ the collection $(\mathcal{L}^0(\mathcal{P}), \mathcal{L}^1(\mathcal{P}), \dots, \mathcal{L}^n(\mathcal{P}))$. Thus we obtain the embedding

$$\mathcal{L} : \mathrm{Hinge}_n \to \widetilde{\mathrm{Hinge}}_n.$$

Remark. For the canonical hinge $\mathcal{P}_\alpha$, the family of operators $\mathcal{L}^m = \mathcal{L}^m(\mathcal{P}_\alpha)$ in the exterior powers $\Lambda^m(V)$ was described in 2.7.

Lemma 2.11. *For any* $g_1, g_2 \in GL_n(\mathbb{C})$,

$$\mathcal{L}^m(g_1 \mathcal{P} g_2) = \lambda^m_{\mathrm{cha}}(g_1)\mathcal{L}^m(\mathcal{P})\lambda^m_{\mathrm{cha}}(g_2).$$

Proof. Indeed, the multiplication by $\lambda_{\mathrm{cha}}(g)$ does not change the "gluing conditions" $c_1 = c_2 = \dots = 1$. □

2.11 Reduced hinge semigroup

Consider the image $\mathcal{L}(\mathrm{Hinge}_n)$ of the embedding $\mathcal{L}$ described in the previous subsection. Denote by $\overline{\mathrm{Hinge}}_n$ the closure of this image.

The set $\overline{\mathrm{Hinge}}_n$ is the union of 2^{n-1} affine algebraic varieties $\overline{\mathcal{L}(\mathcal{O}_\alpha)}$, the dimension of all these varieties is n^2.

The set $\overline{\mathrm{Hinge}}_n$ admits the following explicit description. Let $\mathcal{R} = (R_1, \dots, R_k)$ be a weak hinge. Consider the family of operators

$$c_j \cdot \lambda(R_j) : \Lambda(V) \to \Lambda(V),$$

where $c_j \in \mathbb{C}^*$. We say that the family is *well glued* if

1. the condition $\lambda^m(R_j) \neq 0$, $\lambda^m(R_{j+1}) \neq 0$ (or, equivalently, $\operatorname{Ker}(R_j) = \operatorname{Dom}(R_{j+1})$, and their dimension is m) implies

$$c_j \cdot \lambda^m(R_j) = c_{j+1} \cdot \lambda^m(R_{j+1});$$

2. if R_1 is an operator, then $c_1\lambda(R_1) = \lambda_{\text{cha}}(R_1)$.

The set $\overline{\text{Hinge}_n}$ coincides with the set of all well-glued families.

It can be easily checked that the multiplication of hinges preserves the gluing condition. This implies the following statement:

Proposition 2.12. *The set* $\overline{\text{Hinge}_n}$ *is a subsemigroup in the semigroup* $\widetilde{\text{Hinge}_n}$.

2.12 Representations of the semigroup of hinges

Recall the construction of irreducible polynomial finite-dimensional representations of $\text{GL}_n(\mathbb{C})$. Consider a collection of integers

$$\nu_1 \geqslant \nu_2 \geqslant \ldots \geqslant \nu_n \geqslant 0. \tag{2.15}$$

We call such collections *signatures*. Denote by π the standard representation of $\text{GL}_n(\mathbb{C})$ in $V = \mathbb{C}^n$. Fix a basis $e_1, \ldots, e_n$ in $\mathbb{C}^n$. Denote by ξ_s the vector

$$\xi_s = e_1 \wedge_2 \wedge \ldots \wedge e_s \in \Lambda^s V.$$

Consider the space

$$\mathfrak{H}_\nu := \bigotimes_{j=1}^{n} \Lambda^j(V)^{\otimes(\nu_j - \nu_{j+1})}$$

(we set $\nu_{n+1} = 0$). Consider the representation $\mathfrak{r}_\nu$ of the group $\text{GL}_n(\mathbb{C})$ in the space $\mathfrak{H}_\nu$ given by

$$\mathfrak{r}_\nu(g) = \bigotimes_{j=1}^{n} \lambda^j_{\text{cha}}(\pi(g))^{\otimes(\nu_j - \nu_{j+1})}.$$

Consider also the vector

$$\Xi_\nu := \bigotimes_{j=1}^{n} \xi_j^{\otimes(\nu_j - \nu_{j+1})} \in \mathfrak{H}_\nu.$$

Denote by H_ν the $\text{GL}_n(\mathbb{C})$-cyclic span of the vector Ξ_ν. We denote by ρ_ν the representation of the group $\text{GL}_n(\mathbb{C})$ in the space H_ν.

It is well known (see, for instance [32]) that the representations ρ_ν are irreducible and all polynomial irreducible representations of the group $\text{GL}_n(\mathbb{C})$ have this form.

Let us define the representations ρ_ν of the semigroup $\widetilde{\text{Hinge}_n}$. Let

$$\mathcal{A} := (A_0, A_1, A_2, \ldots, A_n) \in \widetilde{\text{Hinge}_n}; \qquad A_j \in \text{Mat}(\Lambda^j V).$$

We define the operator $\mathfrak{r}_\nu(\mathcal{A})$ in $\mathfrak{H}_\nu$ by

$$\mathfrak{r}_\nu(\mathcal{A}) := \bigotimes_{j=1}^{n} A_j^{\otimes(\nu_j - \nu_{j+1})}.$$

Lemma 2.13. *The subspace $H_\nu \subset \mathfrak{H}_\nu$ is invariant with respect to the operators $\mathfrak{r}_\nu(\mathcal{A})$.*

Proof. Assume $\mathfrak{r}_\nu(\mathcal{A}) \neq 0$. Then the operator $\mathfrak{r}_\nu(\mathcal{A})$ depends (up to a scalar factor) only on the hinge $\mathcal{P}$ lying under $\mathcal{A}$. But any hinge can be approximated by elements of $\mathrm{GL}_n(\mathbb{C})$. □

We define the operator

$$\rho_\nu(\mathcal{A})$$

as the restriction of the operator $\mathfrak{r}_\nu(\mathcal{A})$ to the subspace H_ν. Obviously, ρ_ν is a linear representation of the semigroup Hinge_n:

$$\rho_\nu(\mathcal{A})\rho_\nu(\mathcal{B}) = \rho_\nu(\mathcal{AB}).$$

Lemma 2.14. *Let $\mathcal{A}$ be a nondegenerate element of $\widetilde{\mathrm{Hinge}_n}$ lying over the canonical hinge $\mathcal{P}_{\alpha_1,\dots,\alpha_k}$. Then*

$$\rho_\nu(\mathcal{A})\Xi_\nu = c\cdot \Xi_\nu,$$

where $c \in \mathbb{C}^*$.

Proof. Indeed,

$$\lambda^m(\mathcal{P}_{\alpha_1,\dots,\alpha_k})e_1\wedge\dots\wedge e_m = e_1\wedge\dots\wedge e_m,$$

see 2.7. This implies the required statement. □

Corollary 2.15. *Let $\mathcal{A}$ be a nondegenerate element of $\widetilde{\mathrm{Hinge}_n}$. Then $\rho_\nu(\mathcal{A}) \neq 0$.*

Proof. Each hinge $\mathcal{Q}$ is of the form $g_1\mathcal{P}_\alpha g_2$, there $g_1, g_2 \in \mathrm{GL(V)}$. Thus,

$$\rho_\nu(\mathcal{A}) = c\cdot\rho_\nu(g_1)\rho_\nu(\mathcal{P}_\alpha)\rho_\nu(g_2).$$

The first and the third factors are invertible and the factor in the middle is nonzero. □

2.13 Extension of reducible representations

Let ζ be a finite-dimensional polynomial representation of $\mathrm{GL}_n(\mathbb{C})$ in the space Z. Then

$$\zeta = \oplus\rho_{\nu(j)};\qquad Z = \oplus H_{\nu(j)},$$

where $\nu^{(1)}, \nu^{(2)}\dots$ are collections of signatures satisfying (2.15), and H_ν are the corresponding spaces. We define the representation $\zeta(\mathcal{A})$ of the semigroup $\widetilde{\mathrm{Hinge}_n}$ by

$$\zeta(\mathcal{A}) = \oplus\rho_{\nu(j)}(\mathcal{A}).$$

2.14 Identification of Hinge_n^* with the De Concini–Procesi compactification of $\mathrm{GL}_n(\mathbb{C})$

Consider an *irreducible* representation ρ_ν of the group $\mathrm{GL}_n(\mathbb{C})$. Let $\mathcal{A} \in \widetilde{\mathrm{Hinge}_n}$ lie over a hinge $\mathcal{P}$. Then the operator $\rho_\nu(\mathcal{A})$ is determined up to a factor by the underlying hinge $\mathcal{P}$. Moreover, it is determined by $\mathcal{P}$ considered as an element of Hinge_n^*. By Corollary 2.15, ρ_ν determines the map

$$\mathrm{Hinge}_n^* \to \mathbb{P}\mathrm{Mat}(H_\nu).$$

It can easily be checked that this map is continuous.

Theorem 2.16. a) *Let ρ_ν be an irreducible polynomial representation of* $\mathrm{GL}_n(\mathbb{C})$. *Then the image of* Hinge_n^* *in* $\mathbb{P}\mathrm{Mat}(H_\nu)$ *coincides with the projective compactification* $[\mathrm{GL}_n]_{\rho_\nu}$ *defined in* 0.1.

b) *If $\nu_1 > \nu_2 > \ldots > \nu_n$, then the map $\rho_\nu : \mathrm{Hinge}_n^* \to \mathbb{P}\mathrm{Mat}(H_\nu)$ is an embedding.*

The statement b) identifies Hinge_n^* with the De Concini–Procesi construction, [4].

3 Sea urchin

In this section, we show that the hinge language is sufficient for a description of all the projective compactifications of $\mathrm{GL}_n(\mathbb{C})$.

3.1 Meromorphic matrices

Denote by D_ε the disk $|z| < \varepsilon$ on $\mathbb{C}$. Denote by $\dot{D}_\varepsilon$ the punctured disk $0 < |z| < \varepsilon$ on $\mathbb{C}$.

Denote by $\mathrm{GL}_n(\mathcal{F})$ the group of germs of holomorphic maps $D_\varepsilon \to \mathrm{GL}_n(\mathbb{C})$. The elements of this group are $N \times N$ matrices

$$\gamma(z) = \begin{pmatrix} \gamma_{11}(z) & \ldots & \gamma_{1N}(z) \\ \vdots & \ddots & \vdots \\ \gamma_{N1}(z) & \ldots & \gamma_{NN}(z) \end{pmatrix},$$

where $\gamma_{ij}(z)$ are functions holomorphic in a neighborhood of 0, and $\gamma(0)$ is invertible.

We say that a map

$$\gamma : \dot{D}_\varepsilon \to \mathrm{GL}_n(\mathbb{C})$$

is a *meromorphic family* if all matrix elements γ_{ij} are holomorphic functions in some punctured disk $\dot{D}_\varepsilon$ with poles or removable singularities at 0, and $\gamma(z)$ is invertible for z lying in some punctured disk $\dot{D}_{\varepsilon'}$.

We define the order $\mathrm{ord}(\gamma)$ of the pole of γ as the maximal order of poles at 0 of the matrix elements γ_{ij}. In this definition, we admit a negative order of a pole (a function has a pole of negative order $-k$ at a point 0, if it has a zero of order k at 0).

The value $\mathrm{ord}(\gamma)$ coincides with the minimal k such that the map

$$z \mapsto z^k \gamma(z)$$

from D_ε to $\mathrm{Mat}(\mathbb{C}^N)$ is holomorphic.

3.2 Exponents of meromorphic families

Lemma 3.1. a) *Any meromorphic family* $\gamma(z) : \dot{D}_\varepsilon \to GL_n(\mathbb{C})$ *can be represented in the form*

$$\gamma(z) = a(z) \begin{pmatrix} z^{-m_1} & 0 & 0 & \dots \\ 0 & z^{-m_2} & 0 & \dots \\ 0 & 0 & z^{-m_3} & \dots \\ \vdots & \vdots & \vdots & \ddots \end{pmatrix} b(z), \tag{3.1}$$

where $a(z), b(z) \in GL_n(\mathcal{F})$ *and*

$$m_1 \geqslant m_2 \geqslant \dots \geqslant m_n. \tag{3.2}$$

b) *Consider the* j*-th exterior power* $\lambda^j_{\mathrm{cha}}(\gamma(z))$ *of the matrix* $\gamma(z)$. *Then*

$$\mathrm{ord}\, \lambda^j_{\mathrm{cha}}(\gamma(z)) = m_1 + m_2 + \dots + m_j.$$

In particular, the numbers m_j *are uniquely determined by the function* $\gamma(z)$.

Proof. a) It is sufficient to apply the Gauss elimination algorithm.
b) Obvious. □

We say that the numbers m_i are the *exponents* of the meromorphic family $\gamma(z)$. We fix the notation

$$\mathbf{m} := (m_1, \dots, m_n)$$

for the exponents. We also define the numbers

$$k_1 > k_2 > \dots > k_\tau$$

to be all the pairwise different exponents m_i. We denote by α_j the number of copies k_j in the collection $(m_1, m_2, \dots, m_n)$. We say that α_j, k_j are the *numbers associated with* m_i.

From now on, the meaning of numbers m_i, α_j, k_j *is fixed.*

3.3 Limits of meromorphic families in Hinge_n

Let $\gamma(z)$ be a meromorphic family of $n \times n$ matrices. Let $m_1, \dots, m_n$ be its exponents, and let $k_1, \dots, k_\tau, \alpha_1, \dots, \alpha_\tau$ be the associated numbers.

Proposition 3.2. a) *For each j the following limit in* Gr_n

$$P_j := \lim_{z \to 0} z^{k_j} \operatorname{graph}(\gamma(z)) \tag{3.3}$$

exists, and $\operatorname{rk} P_j = \alpha_j$.

b) *The collection $\mathcal{P}^\gamma := (P_1, \dots, P_\tau)$ is a hinge.*

Remark. In 2.4, we defined the convergence of sequences $g_j \in \mathrm{PGL}_n(\mathbb{C})$ to elements of Hinge_n^*. It is impossible to define a convergence of sequences $g_j \in \mathrm{GL}_n(\mathbb{C})$ to points of Hinge_n. But in Proposition 3.2, we consider holomorphic families instead of sequences.

Proof. Let $\gamma(z)$ be the diagonal matrix with eigenvalues $z^{-m_1}, \dots, z^{-m_n}$. In this case, the limit of $\gamma(z)$ is the canonical hinge $\mathcal{P}_\alpha$. For a general family (3.1), the limit is

$$a(0)\mathcal{P}_\alpha b(0).$$

3.4 Limits of meromorphic families in $\overline{\mathrm{Hinge}_n}$

Let γ be a meromorphic family, let m_j be its exponents.

Proposition 3.3 a) *For each $j = 0, 1, \dots, n$ there exists a nonzero limit*

$$\mathcal{L}^j = \lim_{z \to 0} z^{m_1 + \dots + m_j} \lambda^j_{\mathrm{cha}}(\gamma(z)) \tag{3.4}$$

in $\mathrm{Mat}(\Lambda^j V)$.

b) *The collection $\mathcal{L}_j$ coincides with the the collection $\mathcal{L}^j(\mathcal{P}^\gamma)$, where $\mathcal{P}^\gamma$ is the hinge constructed in Proposition* 3.2 *and the operators $\mathcal{L}^j(\mathcal{P})$ were described in* 2.10.

Example. Consider the diagonal matrix $\delta_{\mathbf{m}}(z)$ with eigenvalues $z^{-m_1}, \dots, z^{-m_n}$. Then $\mathcal{P}^{\delta_{\mathbf{m}}}$ is the canonical hinge $\mathcal{P}_\alpha$. The operator $\lambda^j_{\mathrm{cha}}(\delta_{\mathbf{m}}(z))$ is the diagonal operator in the basis $e_{p_1} \wedge \dots \wedge e_{p_j}$. The eigenvalues are $z^{-m_{p_1} - \dots - m_{p_j}}$. The maximal absolute value of the eigenvalues is $|z|^{-m_1 - \dots - m_j}$.

The eigenspace $W_j \subset \Lambda^j V$ corresponding to the maximal eigenvalue is spanned by the vectors

$$e_1 \wedge e_2 \wedge \dots \wedge e_u \wedge e_{q_1} \wedge \dots \wedge e_{q_{j-u}}, \tag{3.5}$$

where u is the largest i such that $m_i > m_j$ and $m_{q_1} = \dots = m_{q_{j-u}} = m_j$. The operator

$$\mathcal{L}^j = \lim_{z \to 0} z^{m_1 + \dots + m_j} \lambda_{\mathrm{cha}}(\delta_{\mathbf{m}}(z))$$

is the identity operator on the subspace spanned by the vectors (3.5), and $\mathcal{L}^j$ annihilates all other basic vectors. Thus, $\mathcal{L}^j$ coincides with the operator $\mathcal{L}^j(\mathcal{P}_\alpha)$ described in 2.7.

Proof of Proposition 3.3. Let us represent $\gamma(z)$ in the form (3.1). Then, by the example given above,

$$\mathcal{P}^\gamma = a(0)\mathcal{P}_\alpha b(0); \qquad \mathcal{L}^j = \lambda^j_{\mathrm{cha}}\big(a(0)\big)\mathcal{L}^j(\mathcal{P}_\alpha)\lambda^j_{\mathrm{cha}}\big(b(0)\big).$$

Now we apply Lemma 2.11. □

3.5 Limits of meromorphic families in irreducible representations

Let ρ_ν be an irreducible holomorphic representation of $\mathrm{GL}_n(\mathbb{C})$. We extend ρ_ν to the representation of $\widetilde{\mathrm{Hinge}}_n$ by the procedure 2.12.

Lemma 3.4. *Let γ be a meromorphic family of $n \times n$ matrices, let m_i be its exponents. Let $\mathcal{L} : \mathrm{Hinge}_n \to \widetilde{\mathrm{Hinge}}_n$ be the embedding defined in* 2.10.

a) *There exists the nonzero limit*

$$[\rho_\nu(\gamma)] := \lim_{z\to 0} z^{\sum m_i \nu_i}\rho_\nu(\gamma(z)). \tag{3.6}$$

b) *This limit coincides with the operator*

$$\rho_\nu\big(\mathcal{L}(\mathcal{P}^\gamma)\big).$$

Proof. Obvious. □

3.6 Change of the variable z in meromorphic families

We intend to consider a limit of a meromorphic curve independently of its parametrization.

a) If we change the variable z by the formula

$$z \mapsto z + c_2z^2 + c_3z^3 + \dots,$$

then nothing changes. The exponents m_j, the hinge $\mathcal{P}^\gamma$, the limit (3.4) in $\bigoplus \mathrm{Mat}(\Lambda^j V)$, and the limit (3.6) in $\mathrm{Mat}(H_\nu)$ remain the same.

b) If we change the variable z by the formula

$$z \mapsto z^p,$$

then the exponents m_j are replaced by pm_j. All other data (i.e., $\mathcal{P}^\gamma$ and $\mathcal{L}^j$) remain the same as above.

c) Let us change the variable z by

$$c \mapsto cz, \qquad \text{where} \quad c \in \mathbb{C}^*.$$

Then m_j's do not change. The hinge $\mathcal{P}^\gamma$ transforms by the rule

$$(P_1, \ldots, P_\tau) \mapsto (c^{-k_1} P_1, \ldots, c^{-k_\tau} P_\tau).$$

The collection $\mathcal{L}(\mathcal{P}^\gamma) = (\mathcal{L}^0(\mathcal{P}^\gamma), \ldots, \mathcal{L}^n(\mathcal{P}^\gamma)) \in \widetilde{\mathrm{Hinge}_n}$ transforms by the rule

$$(\mathcal{L}^0, \mathcal{L}^1, \mathcal{L}^2, \ldots) \mapsto (c^0 \cdot \mathcal{L}^0, c^{-m_1}\mathcal{L}^1, c^{-m_1-m_2}\mathcal{L}^2, \ldots).$$

The operators $[\rho_\nu(\gamma)]$ transform by the rule

$$[\rho_\nu(\gamma)] \mapsto c^{-\sum m_j \nu_j}[\rho_\nu(\gamma)].$$

3.7 Sea urchin

We define the *sea urchin* $\mathfrak{SU}_n$ as the union of $\mathrm{GL}_n(\mathbb{C})$ and all the spikes $\mathfrak{sp}[\mathbf{m}]$.

The *spikes* $\mathfrak{sp}[\mathbf{m}]$ are enumerated by the collections of integers

$$\mathbf{m} : m_1 \geqslant m_2 \geqslant \ldots \geqslant m_n$$

such that $m_1, \ldots, m_n$ have no common divisor. Let k_j, α_j be the associated numbers (see 3.2). Points of the spike $\mathfrak{sp}[\mathbf{m}]$ are hinges $\mathcal{P} \in \mathcal{O}[\alpha_1, \ldots, \alpha_\tau]$ defined up to the equivalence

$$(P_1, \ldots, P_\tau) \sim (c^{k_1} P_1, \ldots, c^{k_\tau} P_\tau); \qquad c \in \mathbb{C}^*. \tag{3.7}$$

Remark. $\dim \mathrm{GL} = n^2$, and the dimension of any spike is $n^2 - 1$. □

3.8 Limits of meromorphic curves in the sea urchin

Let γ be a meromorphic curve in $\mathrm{GL}_n(\mathbb{C})$. Let $\hat{m}_1, \ldots, \hat{m}_n$ be its exponents, let u be the greatest (positive) common divisor of the numbers $\hat{m}_j$. Then we define the collection $\mathbf{m}^\gamma$ by

$$m_j = m_j^\gamma = \hat{m}_j / u.$$

The limit $\mathcal{P}^\gamma$ of the curve γ in the spike $\mathfrak{sp}[\mathbf{m}]$ is defined by Proposition 3.2. By 3.6, the collection $\mathbf{m}^\gamma$ and the limit of the curve γ in $\mathfrak{sp}[\mathbf{m}]$ do not depend on the parametrization of the curve γ.

3.9 Projective compactifications of $\mathrm{GL}_n(\mathbb{C})$

Let $\rho_{\nu(1)}, \ldots, \rho_{\nu(\sigma)}$ be irreducible holomorphic representations of $\mathrm{GL}_n(\mathbb{C})$, where

$$\nu^{(l)} : \nu_1^{(l)} \geqslant \ldots \geqslant \nu_n^{(l)}$$

are signatures (see 2.11). Let $H_{\nu^{(l)}}$ be the spaces of the representations $\rho_{\nu^{(l)}}$. Consider the direct sum ζ of the representations $\rho_{\nu^{(l)}}$ and the space Z of the representation ζ:

$$\zeta := \bigoplus_{l=1}^{\sigma} \rho_{\nu^{(l)}}; \qquad Z = \bigoplus_{l=1}^{\sigma} H_{\nu^{(l)}}.$$

We have the maps

$$\mathrm{GL}_n(\mathbb{C}) \xrightarrow{\zeta} \mathrm{Mat}(Z) \setminus 0 \to \mathbb{P}\mathrm{Mat}(Z). \tag{3.8}$$

Denote by $[\mathrm{GL}_n]_\zeta$ the closure of the image of $\mathrm{GL}_n(\mathbb{C})$ in $\mathbb{P}\mathrm{Mat}(Z)$. The spaces $\mathbb{P}\mathrm{Mat}(Z)$ are called *projective compactifications* of $\mathrm{GL}_n(\mathbb{C})$.

3.10 The canonical maps $\pi_\zeta : \mathfrak{S}\mathfrak{U}_n \to \mathbb{P}\mathrm{Mat}(Z)$

We must extend the map (3.8) to the spikes $\mathfrak{sp}[\mathbf{m}]$ of the sea urchin $\mathfrak{S}\mathfrak{U}_n$.

Let us fix a spike $\mathfrak{sp}[\mathbf{m}]$. Consider the numbers

$$v^{(l)} := \sum_j m_j \nu_j^{(l)}.$$

Let

$$\mathbf{v} := \max_l v^{(l)}.$$

For a hinge $\mathcal{P} \in \mathfrak{sp}[\mathbf{m}]$, we define the operator

$$\zeta(\mathbf{m}; \mathcal{P}) = \bigoplus B_l(\mathbf{m}, \mathcal{P}) : \bigoplus H_{\nu^{(l)}} \to \bigoplus H_{\nu^{(l)}},$$

where the operators

$$B_l = B_l(\mathbf{m}; \mathcal{P}) : H_{\nu^{(l)}} \to H_{\nu^{(l)}}$$

are given by

$$B_l = \begin{cases} \rho_{\nu^{(l)}}(\mathcal{L}(\mathcal{P}^\gamma)), & \text{if } \; v^{(l)} = \mathbf{v} \\ 0, & \text{if } \; v^{(l)} < \mathbf{v}. \end{cases}$$

Remark. The hinge $\mathcal{P} \in \mathfrak{sp}[\mathbf{m}]$ is defined up to the equivalence (3.7). Hence the operators $[\rho_{\nu^{(l)}}(\mathcal{P}^\gamma)]$ are defined up to the factors $c^{\sum_i m_j \nu_j^{(l)}} = c^{v^{(l)}}$. But for all nonzero B_l, these factors coincide, and hence $\zeta(\mathbf{m}; \mathcal{P})$ is a nonzero operator defined up to a factor $c^{\mathbf{v}}$. □

Theorem 3.5. *The image of the sea urchin under the map* $(\mathbf{m}, \mathcal{P}) \mapsto \zeta(\mathbf{m}, \mathcal{P})$ *coincides with the projective compactification* $[\mathrm{GL}_n]_\zeta$.

Lemma 3.6. *Let* $z \mapsto \gamma(z)$ *be a meromorphic family. Then the limit of* $\zeta(\gamma(z))$ *in* $[\mathrm{GL}_n]_\zeta$ *exists and coincides with* $\zeta(\mathbf{m}^\gamma, \mathcal{P}^\gamma)$.

Proof of Lemma 3.6. We have

$$\operatorname{ord} \rho_{\nu^{(l)}}(\gamma(z)) = \sum_i \nu_i^{(l)} m_i = v^{(l)}.$$

Thus our limit is

$$\lim_{z\to 0} z^{\mathbf{v}} \zeta(\gamma(z)) = \lim_{z\to 0} z^{\mathbf{v}} \bigoplus_l \rho_{\nu^{(l)}}(\gamma(z))$$

and Lemma 3.4 implies the required result. □

Proof of Theorem 3.5. The set $[\mathrm{GL}_n]_\zeta \setminus \mathrm{GL}_n(\mathbb{C})$ is a subvariety of the projective variety $[\mathrm{GL}_n]_\zeta$. Thus any point of this set can be attained along an algebraic curve $\gamma(z) \subset \mathrm{GL}_n(\mathbb{C})$. By Lemma 3.6, this point is contained in the image of the sea urchin.

Conversely, each point $(\mathbf{m}, \mathcal{P})$ of the sea urchin can be attained along an algebraic curve $\gamma(z)$. By Lemma 3.6, the image of the point $(\mathbf{m}, \mathcal{P})$ can be attained along the curve $\zeta(\gamma(z))$ and thus $\zeta(\mathbf{m}, \mathcal{P}) \in [\mathrm{GL}_n]_\zeta$. □

Remark. Let us give a description of the sea urchin in formal terms. The spikes are homogeneous spaces $\mathrm{GL}_n(\mathbb{C}) \times \mathrm{GL}_n(\mathbb{C})/G_{\mathbf{m}}$, where the family of subgroups $G_{\mathbf{m}}$ can be obtained by the following algorithm.

Consider two opposite parabolic subgroups P_+ and P_- in $\mathrm{GL}_n(\mathbb{C})$. Denote by N_+ and N_- the unipotent radicals in P_+, P_-. Denote by Q the Levi subgroup ($Q \subset P_+$, $Q \subset P_-$). Denote by Z the center of the Levi subgroup. Let S be a one-parameter subgroup in Z. Then say that S is positive if all root lying in N_+ are positive on the generator of S. Now we consider the subgroup

$$G^S \subset \mathrm{GL}_n(\mathbb{C}) \times \mathrm{GL}_n(\mathbb{C})$$

consisting of all pairs

$$(g_1, g_2) = (n_- q_1, q_2 n_+)$$

where $n_- \in N_-$, $n_+ \in N_+$, $q_1, q_2 \in Q$ and $q_1^{-1} q_2 \in S$. Then the family of subgroups G^S coincides with the family $G_{\mathbf{m}}$.

This remark allows one to extend the sea urchin construction to an arbitrary complex semisimple group. For all classical groups the De Concini–Procesi compactification can be described in the hinge language (see [19], [22]) and the sea urchins also can easily be described in these terms. □

3.11 Convergence in the sea urchin

Let $\mathcal{P} \in \mathcal{O}(\alpha_1, \ldots, \alpha_\tau)$ be a point of a spike $\mathfrak{sp}[\mathbf{m}]$. The sequence $g_j \in \mathrm{GL}_n(\mathbb{C})$ converges to $\mathcal{P} = (P_1, \ldots, P_\tau)$ if there exists a sequence of nonzero complex numbers $z_j \to 0$, such that for each $s = 1, 2, \ldots, \tau$ the sequence $z_j^{-k_s} g_j$ converges to P_s in the Grassmannian.

It is possible to define a topology in $\mathfrak{S}\mathfrak{U}_n$ compatible with this convergence; we do not discuss this.

Example. The sequence

$$g_j = \begin{pmatrix} e^j & 0 & 0 \\ 0 & j & 0 \\ 0 & 0 & 1 \end{pmatrix} \in GL_3(\mathbb{C})$$

does not converge in the sea urchin (the curve $(e^z, z, 1)$ has an essential singularity at infinity). Moreover, the sequence g_j has no convergent subsequences. It can be easily checked that the sequence g_j has a limit in Hinge_3^*; also it has a limit in any projective compactification.

Thus the sea urchin is not a compact topological space.

Acknowledgments. I am grateful to C. De Concini, E.B. Vinberg and V.L. Popov for meaningful discussion on this subject. I thank the administrators of the Erwin Schrödinger Institute for Mathematical Physics, where this work was done, for their hospitality.

References

[1] Bialynicki-Birula, A., Sommese, A.J., A conjecture about compact quotient by tori, in *Complex Analytic Singularities,* Adv. Stud. Pure Math., 8, North-Holland, 1987, pp. 59–68.

[2] Brion, M., Luna, D., Vust, Th., Espaces homogènes sphériques, *Invent. Math.* **84** (1986), 617–632.

[3] Chriss, N., Ginzburg, V., *Representations and Complex Geometry,* Birkhäuser, 1997.

[4] De Concini, C., Procesi, C., *Complete symmetric varieties.* Lect. Notes Math., vol. 996, 1983, 1–44.

[5] De Concini, C., Springer, T., Betti numbers of complete symmetric varieties, in *Geometry Today*, Birkhäuser (1985), 87–107.

[6] Dynkin, E.B., Nonnegative eigenfunctions of the Laplace–Beltrami operator and Brownian motion on certain symmetric spaces, *Izv. Akad. Nauk SSSR, ser. Mat.* **30** (1966); 455–478; English transl., Amer. Math.Soc. Translations, vol. 72, Amer. Math. Soc., Providence, 1968, 203–228.

[7] Furstenberg, H., Poisson formula for semisimple Lie groups. *Ann. of Math.* **(2)** 77 (1963), 335–386.

[8] Guivarch, Y., Ji., L., Taylor, J.C., *Compactifications of Symmetric Spaces.* Birkhäuser, 1998.

[9] Kambayashi, T., Projective representations of algebraic groups of transformations, *Amer. J. Math.* **88** (1966), 199–205.

[10] Kapranov, M.M., Chow quotients of Grassmannians, I, in *I.M. Gelfand Seminar*, S.I.Gelfand and S.G. Gindikin, eds., Adv. in Sov.Math., 16, Amer. Math. Soc. (1991), 29–110.

[11] Karpelevich, F.I., The geometry of geodesics and eigenfunctions of the Laplace–Beltrami operator on the symmetric spaces, *Trudy Mosk. Mat. Obshch.* **14** (1965), 48–185; English transl. *Trans. Moscow Math. Soc.* **14** (1965), 51–199.

[12] Knop, F., Automorphisms, root systems, and compactifications of homogeneous varieties, *J. Amer. Math. Soc.* **9** (1996), 153–174.

[13] Kushner, G.F., A certain compactification of noncompact symmetric spaces, *Dokl. Akad. Nauk.* **190** (1970), 1282–1285; English transl., *Sov. Math. Dokl.* **11** (1970), 284–287.

[14] Kushner, G.F., The compactification of noncompact symmetric Riemannian spaces, *Trudy Sem. Vektor. Tenzor. Anal.* **16** (1972), 99–152. (Russian)

[15] Kushner, G.F., F.I. Karpelevich compactification is homeomorphic to a ball, *Trudy Sem. Vektor. Tenzor. Anal.* **19** (1979), 96–111; English transl., Amer.Math. Soc. Translations, ser.2, 134 (1987), 119–133.

[16] Luna, D., Vust, Th., Plongements d'espaces homogènes, *Comment. Math. Helv.* **58** (1983), 186–245.

[17] Neretin, Yu. A., Spinor representation of the infinite-dimensional orthogonal semigroup and the Virasoro algebra, *Funkts. Anal. Prilozh.* **23** (1989), no. 3, 32–44; English transl., *Funct. Anal. Appl.* **23** (1990), 196–207.

[18] Neretin, Yu. A., Extensions of representations of classical groups to representations of categories, *Algebra i Analiz* **3** (1991), 176–202; English transl., *St-Petersburg Math. J.* **3** (1992), 147–169.

[19] Neretin, Yu. A., On universal completions of complex classical groups, *Funkt. Anal. Prilozh.* **26** (1992), no. 4, 30–44; English transl., *Funct. Anal. Appl.* **26** (1993), 254–265.

[20] Neretin, Yu. A., Hausdorff metrics, construction of a separated quotient space and boundaries of symmetric spaces, *Funkt. Anal. Prilozh.* **31** (1997), no.1, 83–86; English transl., *Funct. Anal. Appl.* **31** (1997), 65–67.

[21] Neretin, Yu. A., *Categories of Symmetries and Infinite-dimensional Groups*, Oxford University Press, 1996; Russian edition: Moscow, URSS, 1998.

[22] Neretin, Yu. A., *Hinges and Study–Semple–Satake–Furstenberg–De Concini–Procesi–Oshima boundary*, Amer. Math. Soc. Transl., 181 (1998), 165–230.

[23] Olshanetsky, M.A., Martin boundaries for real semisimple groups, *J. Funct. Anal.* **126** (1994), 169–216.

[24] Popov, V.L., Contractions of actions of reductive algebraic groups, *Mat. Sb.* **130** (1986), no. 3, 310–334; English transl., *Math. USSR.–Sbornik* **58** (1987), 2, 311–335.

[25] Rashevskii, P.K., Global projective models of complex homogeneous spaces, *Trudy Moscow Math. Soc.* **25** (1971), 3–14, English transl. *Trans. Moscow Math. Soc.* **25** (1971), 1–12.

[26] Satake, I., On representations and compactifications of symmetric Riemannian spaces, *Ann. of Math.* **71** (1960), 77–110.

[27] Semple, J.G., On complete quadrics, *J. Lond. Math. Soc.* **23** (1948), 258–267.

[28] Semple, J.G., The variety whose points represent complete collineations of S_r on S'_r, *Rend. Math. Univ. Roma* **(5)** 10 (1951), 201–208.

[29] Semple, J.G., On complete quadrics, II, *J. Lond. Math. Soc.* **27** (1952), 280–287.

[30] Study, E., Über die Geometrie der Kogelschnitte, insbesondere deren Charakteristikeproblem, *Math. Ann.* **17** (1886), 58–101.

[31] Vinberg, E.B., Popov, V.L., *A certain class of quasihomogeneous affine varieties,* Izv. Acad. Nauk SSSR, Ser. Math., 36 (1972), 749–764, English transl., Math. USSR Izv. 6 (1972), 743–758.

[32] Želobenko, D.P., *Compact Lie groups and their representations,* Nauka, Moscow, 1970 (Russian); English translation Amer. Math. Soc., 1973.

Yurii A. Neretin
Institute of Theoretical and Experimental Physics
Bolshaya Cheremushkinskaya
25 Moscow 117259, Russia
neretin@main.mccme.rssi.ru; neretin@gate.itep.ru

Why Would Multiplicities be Log-Concave?

Andrei Okounkov

To A. A. Kirillov on his 64th birthday

ABSTRACT A basic property of entropy in statistical physics is that it is concave as a function of energy. The analog of this in representation theory would be the concavity of the logarithm of the multiplicity of an irreducible representation as a function of its highest weight. We discuss various situations where such concavity can be established or reasonably conjectured and consider some implications of this concavity.

0 Introduction

The aim of this paper is to discuss some heuristic arguments, conjectures, and rigorous results related to the following phenomenon. The physical analogy, explained in Section 1, suggests that under certain circumstances the logarithms of multiplicities of irreducible representations can be expected to be concave as a function of the highest weight. In Section 2 we discuss some cases when this is known or expected to be the case and explore various implications of this concavity. In Section 3 we discuss the classical limit, in which much more general results can be established.

This text is not a survey, but is rather based on the various talks I have given and my personal point of view on the subject. I hope that the somewhat informal style of here will make the basic ideas easier to explain. For missing details, the reader is referred to the original papers [16, 8, 17, 12, 18]. For surveys on log-concavity in general, see [4, 22].

I very much benefited from discussions with my colleagues in the Kirillov seminar and from the Institute of Problems of Information Transmission, especially A. A. Kirillov, R. Dobrushin, V. Ginzburg, G. Olshanski, S. Pirogov, and also in talks with W. Graham and A. Khovanskii.

I would like to thank A. Buch for providing me with a program for computation of tensor product multiplicities. Partial financial supported by NSF (DMS-9801466) and by a Sloan foundation fellowship is gratefully acknowledged.

1 Physical motivation: entropy and its concavity

1.1

Consider a quantum mechanical system, that is, a selfadjoint operator H on a Hilbert space V. For simplicity, we assume that V is spanned by the eigenvectors of H.

The multiplicity $\Omega(E)$ of an eigenvalue $E \in \mathrm{spec}(H)$ measures how many states of our system have the energy E. In other words, fixing an energy level E does not determine the state of the system uniquely: there remain $\Omega(E)$ possibilities. The size of this indeterminacy equals $\log_2 \Omega(E)$ bits of information.

In statistical physics, there is the basic relation[1]

$$\boxed{S = k \log \Omega}$$

where Ω is the number of states with given values of macroscopic parameters such as energy, k is the Boltzmann constant, and S is the *entropy*, which measures the degree of disorder in the system or, in other words, the lack of information about the precise state of the system. We are thus led to think of

$$S(E) = \log \Omega(E)$$

as of the entropy of the energy level E.

1.2

In statistical mechanics, the entropy is always a concave function of all additive macroscopic parameters such as energy E, volume V, or the number of particles N. There is a simple physical argument for this concavity which goes as follows. Suppose we have two systems with parameters $(E_1, V_1, \dots)$ and $(E_2, V_2, \dots)$, respectively, contained in two reservoirs separated by an impervious wall:

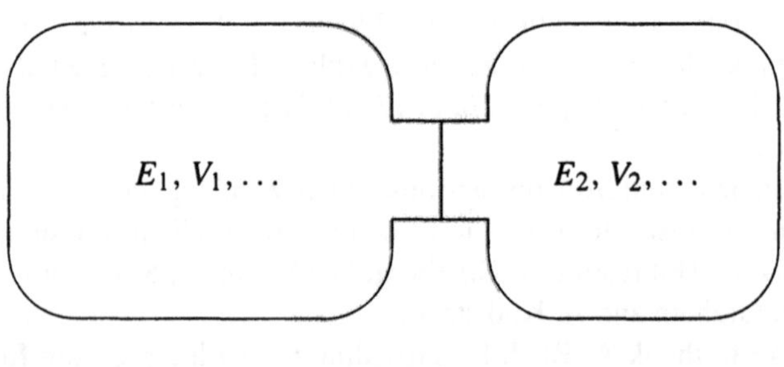

[1]This relation, in the form $S = k \log W$, is written on Boltzmann's tombstone.

Let us now bring them in contact by removing this wall. The energy and the volume of the new system will be $E_1 + E_2$ and $V_1 + V_2$, respectively, whereas the entropy will increase

$$S(E_1 + E_2, \ldots) \geq S(E_1, \ldots) + S(E_2, \ldots) \tag{1.1}$$

because of the additional disorder introduced by allowing the systems to mix. The net increase in entropy is called the *entropy of mixing* and its positivity reflects the irreversibility of mixing.

There is, however, one case when the mixing is clearly reversible and that is when the two systems were identical to begin with, that is, when

$$(E_1, V_1, \ldots) = (E_2, V_2, \ldots),$$

in which case we can simply reinsert the wall to recover the original situation. Thus, in this case the entropy of mixing vanishes. In other words,

$$S(2E, \ldots) = 2S(E, \ldots). \tag{1.2}$$

Combining (1.1) with (1.2) we get the concavity of the entropy.[2]

1.3

Of course, in order to apply statistical considerations one needs the system in question to have a very large or infinite number of degrees of freedom. Still, it is natural to ask whether in some interesting cases one can expect or, even better, prove the concavity of $S(E)$. Fortunately, interesting examples do exists.

An obvious limitation for the entropy concavity principle is that the concavity of $S(E)$ is clearly not preserved under direct sums. Hence our system has be in some sense *irreducible*. The concrete meaning of this irreducibility will be different in a different context. In Section 2, the space V will be an irreducible module of some ambient group. In Section 3, we will be dealing with group actions on irreducible algebraic varieties.

1.4

First, however, one has to modify the definition of concavity. Indeed, the support $\operatorname{spec}(H)$ of the function $\Omega(E)$ is countable and hence $S(E)$ cannot be concave in the usual sense.

We suppose that $\operatorname{spec}(H)$ is contained in a lattice, which without loss of generality we can take to be $\mathbb{Z} \subset \mathbb{R}$, and we define concavity to mean

$$S(\alpha E_1 + (1-\alpha)E_2) \geq \alpha S(E_1) + (1-\alpha)S(E_2), \quad \alpha \in [0, 1],$$

[2]In thermodynamics, one has the relation $\frac{\partial S}{\partial E} = \frac{1}{T}$ where T is the temperature. Therefore $\frac{\partial^2 S}{\partial E^2} < 0$ means that temperature rises when energy increases.

whenever the middle point $\alpha E_1 + (1-\alpha)E_2$ lies in the lattice $\mathbb{Z}$.

The abstract form of this convention is the following:

Definition 1. Let $F : \mathbb{A} \to \mathbb{O}$ be a function from an Abelian semigroup $\mathbb{A}$ to an ordered Abelian semigroup $\mathbb{O}$. We say that this function is *concave* if

$$(p+q)\, F(C) \geq p\, F(A) + q\, F(B)$$

for any $A, B, C \in \mathbb{A}$ satisfying

$$(p+q)\, C = p\, A + q\, B\,, \quad p, q \in \mathbb{Z}_{\geq 0}\,.$$

In the case when

$$\mathbb{O} = (\mathbb{R}_{\geq 0}, \times)$$

is the multiplicative semigroup of nonnegative real numbers with the usual ordering, we also call the function F logarithmically concave, or *log-concave* for short.

In our examples, $\mathbb{A}$ will be usually isomorphic to $\mathbb{Z}^n$ or $\mathbb{R}^n$, whereas the target semigroup $\mathbb{O}$ will occasionally be something more interesting.

1.5

Since the eigenvalues of H are now integers, the time evolution e^{itH} defines a representation of the standard circle $\mathbb{T}^1$ on V.

More generally, for any compact group[3] K, one can ask whether for some interesting representation V of K the multiplicities $\Omega(\lambda)$ of irreducible representations V^λ

$$V = \bigoplus_{\lambda \in K^\wedge} \Omega(\lambda)\, V^\lambda$$

form a log-concave function on the weight lattice of $K^\wedge$.

Examples of such representations will be discussed in Section 2. They are, in a sense, related to the "thermodynamics" of classical groups.

1.6

Now consider a Hamiltonian system of classical mechanics, that is, a manifold M^{2n} with symplectic form ω and with an energy function

$$h : M^{2n} \to \mathbb{R}\,.$$

The form $\frac{\omega^n}{n!}$ is a volume form on M^{2n} which defines a measure (called the *Liouville measure*). Let $\Omega(E)$ be the density of the push-forward of this measure under h

$$h_*\left(\frac{\omega^n}{n!}\right) = \Omega(E)\, dE\,, \quad E \in \mathbb{R}\,.$$

---

[3]Here and in what follows we assume all compact groups to be *connected*.

In other words, $\Omega(E)$ tells us how many states of our system have the energy E. Again, we think of

$$S(E) = \log \Omega(E)$$

as of entropy[4] of the energy level E.

As in Section 1.5, this can be generalized to the situation of a Hamiltonian action of a compact group K action on M^{2n}. Let

$$\phi : M^{2n} \to \mathrm{Lie}(K)^*$$

be the moment map for this action. For any $\xi \in \mathrm{Lie}(K)^*$ the volume

$$\Omega(\xi) = \mathrm{Vol}\phi^{-1}(\xi)$$

measures how many points of M^{2n} have the energy ξ. This function is clearly invariant under the coadjoint action of K on $\mathrm{Lie}(K)^*$, so we can and will assume that ξ lies in the positive Weyl chamber $\mathfrak{h}_+$.

We can ask whether for some actions the function $\log \Omega(\xi)$ is concave on $\mathfrak{h}_+$. Observe that such a concavity implies, in particular, that the set

$$\mathrm{supp}\Omega(\xi) = \phi(M^{2n}) \cap \mathfrak{h}_+ ,$$

is convex, which is a famous classical result [1, 10, 11, 14].

It turns out that the supply of cases where $\log \Omega(\xi)$ is concave is now much richer than in the quantum situation. As shown by W. Graham in [8], it includes all torus actions on compact Kähler manifolds. It also includes [17] all actions on projective varieties, possibly singular. It was conjectured by V. Ginzburg and A. Knutson that it is true for any symplectic M^{2n}. This was shown to be not the case by Y. Karshon in [12].

We will discuss this classical situation in an algebraic setting in Section 3.

2 Some results and conjectures on log-concavity of multiplicities

2.1

Again, we begin with a motivation, this time a historical one. Here is how the question of logarithmic concavity of multiplicities arose in the "thermodynamics" of classical groups.

Let $U(\infty)$ denote the inductive limit of $U(n)$ with respect to standard embeddings $U(n) \subset U(n+1)$ which can be visualized as follows:

---

[4]This entropy is not to be confused with the entropy of the dynamical system defined on $h^{-1}(E)$ by the Hamiltonian flow $\dot{x} = \{h, x\}$.

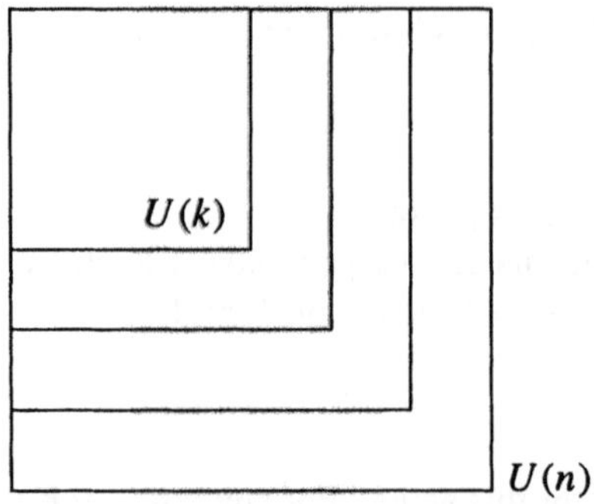

The description of the characters[5] of $U(\infty)$ is a fundamental result with nontrivial history. Voiculescu in [25] proved that functions of the form

$$g \mapsto \det \left[e^{\gamma^+(g-1)+\gamma^-(g^{-1}-1)} \prod \frac{1+\beta_i^+(g-1)}{1-\alpha_i^+(g-1)} \frac{1+\beta_i^-(g^{-1}-1)}{1-\alpha_i^-(g^{-1}-1)} \right] \tag{2.1}$$

where

$$0 \le \alpha_i^{\pm}, \quad 0 \le \beta_i^{\pm} \le 1, \quad 0 \le \gamma^{\pm}$$

are parameters satisfying the conditions

$$\sum (\alpha_i^+ + \alpha_i^- + \beta_i^+ + \beta_i^-) < \infty, \quad \beta_1^+ + \beta_1^- \le 1,$$

are characters of $U(\infty)$ and conjectured that there are no other characters. It was observed by Boyer [3] and, independently, by Vershik and Kerov [24] that this conjecture is equivalent to Schoenberg's conjecture concerning the so-called totally positive sequences (see below) which was already established by Edrei in [6] using some deep results about entire functions.

2.2

Vershik and Kerov also outlined a different and more direct proof which uses approximation of characters of $U(\infty)$ by normalized characters of $U(n)$. It follows from a general principle due to Vershik (see [23] and also [20]), that any character χ of $U(\infty)$ is a limit of a sequence of normalized characters χ_n of $U(n)$ as $n \to \infty$ in the sense that

$$\chi_n \Big|_{U(k)} \xrightarrow{\text{uniformly}} \chi \Big|_{U(k)} \tag{2.2}$$

---

[5] An abstract definition of characters is: indecomposable central continuous positive definite functions. More concretely, they are spherical functions of the Gelfand pair

$$U(\infty) \times U(\infty) \supset \operatorname{diag} U(\infty)$$

or, equivalently, traces of factor representations of type I_n or II_1.

for any fixed $k = 1, 2, 3, \ldots$. In [24], Vershik and Kerov gave necessary and sufficient conditions for the convergence of $\{\chi_n\}$ and identified the corresponding limits with functions (2.1).

This approximation principle is a materialization of certain general ergodic theory ideas and is closely akin to some standard constructions in statistical physics such as construction of Gibbsian measures in an infinite volume by a thermodynamic limit transition. In this case, one chooses a sequence of boxes which fill up the space (just as in the above visualization of $U(\infty)$); for each box one picks some boundary condition which specifies a Gibbsian measure (in our case, χ_n), and one requires convergence of the induced measures on all compact sets.

2.3

Note that the formula (2.1) is multiplicative in the eigenvalues of $g \in U(\infty)$. This multiplicativity can be established a priori; as shown by Olshanski, see for example [21], this multiplicativity and more general multiplicativity are very characteristic for representations of infinite-dimensional classical groups.

Any character χ of $U(\infty)$ is therefore uniquely determined by its restriction to $U(1)$

$$x(z) = \chi\left(\begin{bmatrix} z & & \\ & 1 & \\ & & \ddots \end{bmatrix}\right) = \sum_{k\in\mathbb{Z}} x_k\, z^k .$$

Conversely, any function $g \mapsto \det x(g)$ is a character of $U(\infty)$ provided that it is positive definite, which means that its restriction to any $U(n)$ is a nonnegative linear combination of the characters of $U(n)$, that is, of the rational Schur functions s_λ.

The identity

$$\prod_{i=1}^{n}\left(\sum_{k\in\mathbb{Z}} x_k\, z_i^k\right) = \sum_{\lambda=(\lambda_1\geq\cdots\geq\lambda_n)\in\mathbb{Z}^n} \det\left[x_{\lambda_i-i+j}\right]_{i,j=1\ldots n} s_\lambda(z_1,\ldots,z_n),$$

shows that this positivity is equivalent to the positivity of some (in fact, all) minors of the infinite Toeplitz matrix $\left[x_{j-i}\right]_{i,j\in\mathbb{Z}}$, which is precisely Schoenberg's definition of a *totally positive* sequence.

In particular, the positivity of 2×2 minors means that

$$x_n^2 \geq x_{n-1}\, x_{n+1} .$$

Thus, one knows a priori that the restriction of any character of $U(\infty)$ to $U(1)$ has log-concave multiplicities.

2.4

The question whether the same is true before the limit, that is, whether the restriction of any irreducible representation of $U(n)$ to standard $U(1)$ has log-concave multiplicities, surfaced when we were working with G. Olshanski on a generalization of the Vershik–Kerov theorem [24]. Originally, this log-concavity was needed to replace the uniform convergence in (2.2) by convergence of Taylor series, see Section 3 in [16]. Eventually, in [19] it was replaced by a more elementary argument, but nonetheless this log-concavity is a valid question with interesting answer.

As it turns out, for any representation V^λ of $U(n)$ the multiplicity of the irreducible representation V^μ of the standard $U(k) \subset U(n)$ is a log-concave function of the pair

$$(\lambda, \mu) \in U(n)^\wedge \oplus U(k)^\wedge .$$

In fact, one can say more. Without loss of generality, let us assume that λ is a partition and consider the space

$$V^{\lambda/\mu} = \mathrm{Hom}_{U(k)}(V^\mu \to V^\lambda)$$

whose dimension is the multiplicity in question. The space $V^{\lambda/\mu}$ is an $U(n-k)$ module with character given by the skew Schur function $s_{\lambda/\mu}$

$$\mathrm{tr}_{V^{\lambda/\mu}} \left(\begin{bmatrix} z_1 & & \\ & z_2 & \\ & & \ddots \end{bmatrix} \right) = s_{\lambda/\mu}(z_1, z_2, \dots) .$$

One has the following.

Theorem 1 ([16]). *Suppose* (λ_i, μ_i), $i = 1, 2, 3$, *are partitions such that*

$$(\lambda_2, \mu_2) = \tfrac{1}{2}(\lambda_1, \mu_1) + \tfrac{1}{2}(\lambda_3, \mu_3) .$$

Then the following polynomial has nonnegative coefficients:

$$s^2_{\lambda_2/\mu_2} - s_{\lambda_1/\mu_1} s_{\lambda_3/\mu_3} \in \mathbb{Z}_{\geq 0}[z_1, z_2, \dots] . \tag{2.3}$$

The coefficients of the polynomial $s_{\lambda/\mu}$ correspond to standard tableaux of shape λ/μ. In the proof of Theorem 1, one constructs a certain transformation on pairs of standard tableaux and proves that it is injective.

Similar results for orthogonal and symplectic groups are also established in [16].

2.5

It is likely that (2.3) is actually a nonnegative linear combination of Schur functions. One can propose a conjecture which would, among other things, imply this property.

Recall that the Littlewood–Richardson coefficients $c_{\lambda\mu\nu}$ are defined by

$$c_{\lambda\mu\nu} = \dim \left(V^{\lambda} \otimes V^{\mu} \otimes V^{\nu}\right)^{G}$$

where the superscript G stands for invariants of $G = U(n)$. If either of the arguments of $c_{\lambda\mu\nu}$ is not a dominant weight, we set $c_{\lambda\mu\nu} = 0$ by definition. Often, one uses the numbers

$$c^{\lambda}_{\mu\nu} = c_{\lambda^*\mu\nu}$$

where λ^* is the highest weight of the dual module $\left(V^{\lambda}\right)^*$

$$(\lambda_1, \ldots, \lambda_n)^* = (-\lambda_n, \ldots, -\lambda_1) .$$

The numbers $c^{\lambda}_{\mu\nu}$ are coefficients in the expansions

$$V^{\lambda/\mu} = \sum_{\nu} c^{\lambda}_{\mu\nu} V^{\nu} ,$$

$$V^{\mu} \otimes V^{\nu} = \sum_{\lambda} c^{\lambda}_{\mu\nu} V^{\lambda} .$$

Conjecture 1. *The function*

$$(\lambda, \mu, \nu) \to \log c_{\lambda\mu\nu}$$

is concave.

If true, this concavity would have some interesting applications. In particular, since

$$c_{\lambda\mu\nu} = c_{\lambda\nu\mu}$$

we conclude that

$$c_{\lambda, \frac{\mu+\nu}{2}, \frac{\mu+\nu}{2}} \overset{?}{\geq} c_{\lambda\nu\mu} ,$$

provided $\frac{\mu+\nu}{2}$ is an integral weight. This is equivalent to the inclusion of representations

$$V^{\nu} \otimes V^{\mu} \overset{?}{\subset} \left(V^{\frac{\mu+\nu}{2}}\right)^{\otimes 2} , \tag{2.4}$$

which can be interpreted as saying that the representation valued function

$$V : \lambda \mapsto V^{\lambda} \tag{2.5}$$

is concave with respect to the natural ordering and tensor multiplication of representations.[6]

[6]Remark that it follows from Weyl's dimension formula that

$$\lambda \mapsto \log \dim V^{\lambda}$$

is a concave function. That is, the function (2.5) considered as a function into just vector spaces without group action is concave with respect to the tensor product.

If (2.4) is true, then we certainly have the following inclusion of $U(n)$-modules:

$$\left(V^{\lambda_1} \otimes V^{\mu_1}\right) \otimes \left(V^{\lambda_3} \otimes V^{\mu_3}\right) \overset{?}{\subset} \left(V^{\lambda_2} \otimes V^{\mu_2}\right)^{\otimes 2} \tag{2.6}$$

for (λ_i, μ_i) as in Theorem 1. The last inclusion is equivalent to

$$V^{\lambda_1/\mu_1} \otimes V^{\lambda_3/\mu_3} \overset{?}{\subset} \left(V^{\lambda_2/\mu_2}\right)^{\otimes 2}. \tag{2.7}$$

Indeed, the equation (2.7) is equivalent to

$$\left(\sum_{\nu_1} c_{\lambda_1^* \mu_1 \nu_1} V^{\nu_1}\right) \otimes \left(\sum_{\nu_3} c_{\lambda_3^* \mu_3 \nu_3} V^{\nu_3}\right) \overset{?}{\subset} \left(\sum_{\nu_2} c_{\lambda_2^* \mu_2 \nu_2} V^{\nu_2}\right) \otimes \left(\sum_{\nu_4} c_{\lambda_2^* \mu_2 \nu_4} V^{\nu_4}\right), \tag{2.8}$$

whereas (2.6) says that

$$\left(\sum_{\nu_1} c_{\lambda_1 \mu_1 \nu_1^*} V^{\nu_1}\right) \otimes \left(\sum_{\nu_3} c_{\lambda_3 \mu_3 \nu_3^*} V^{\nu_3}\right) \overset{?}{\subset} \left(\sum_{\nu_2} c_{\lambda_2 \mu_2 \nu_2^*} V^{\nu_2}\right) \otimes \left(\sum_{\nu_4} c_{\lambda_2 \mu_2 \nu_4^*} V^{\nu_4}\right). \tag{2.9}$$

To get (2.8) from (2.9), take the dual space of everything, which will replace V^{ν_i} by $V^{\nu_i^*}$, and then replace λ_i by λ_i^* and ν_i^* by ν_i. The inclusion (2.7) is equivalent to Schur positivity of (2.3).

Similarly, the conjecture and symmetry

$$c_{\lambda\mu\nu} = c_{\nu\lambda\mu}$$

imply that

$$c_{\lambda',\mu',\nu'} \overset{?}{\geq} c_{\lambda\mu\nu}$$

provided the weight

$$\begin{pmatrix} \lambda' \\ \mu' \\ \nu' \end{pmatrix} = \begin{pmatrix} \alpha & 0 & 1-\alpha \\ 1-\alpha & \alpha & 0 \\ 0 & 1-\alpha & \alpha \end{pmatrix} \begin{pmatrix} \lambda \\ \mu \\ \nu \end{pmatrix},$$

is an integral weight and $0 \leq \alpha \leq 1$.

2.6

Here is another implication of Conjecture 1 which is actually known to be true. Concavity of $\log c_{\lambda\mu\nu}$ implies that the support

$$\operatorname{supp} c_{\lambda\mu\nu} = \{(\lambda, \mu, \nu),\ c_{\lambda\mu\nu} \neq 0\}$$

is convex. In particular, since it contains the origin (0, 0, 0), it is *saturated*, meaning that

$$c_{k\lambda,k\mu,k\nu} \neq 0 \Rightarrow c_{\lambda,\mu,\nu} \neq 0\,, \tag{2.10}$$

for any $k = 2, 3, \ldots$. In fact, since $c_{0,0,0} = 1$, Conjecture 1 implies that

$$c_{k\lambda,k\mu,k\nu} \overset{?}{\leq} (c_{\lambda,\mu,\nu})^k\,.$$

The saturation (2.10) turns out to be a very important property, see [7]. It has been established by A. Knutson and T. Tao in [15], see also [2].

2.7

As already pointed out in Section 1.3, log-concavity is (in contrast to so many things in representation theory) not an additive property: it is totally destroyed by direct sums.

It seems however likely that log-concavity should be a *multiplicative* property, that is, it should behave nicely with respect to tensor products. For example, recall that it is well known and easy to prove that the convolution of two log-concave sequences is again log-concave. This is equivalent to saying that the set of $U(1)$-modules with log-concave multiplicities is closed under tensor products.

This multiplicativity principle fits together nicely with the above conjecture about tensor product multiplicities.

3 Log-concavity in the classical limit

3.1

Dealing with actual multiplicities may be a subtle business. Fortunately, many of these subtleties disappear in the classical limit and much more general results can be established.

Let us assume that the phase space of our classical system is an irreducible projective algebraic variety $X \in \mathbb{P}^N$ over $\mathbb{C}$ which is stable under the action of a compact group $K \subset GL(N+1)$. We write X in place of M^{2n} to stress the fact that we are now working with projective algebraic varieties which are allowed to be singular.

Even for singular X, the moment map

$$\phi : X \to \text{Lie}(K)^*$$

is still well defined as the restriction of the moment map for the K-action on $\mathbb{P}^N$. It is well known (see e.g. Theorem 6.5 in [9]) that the function $\Omega(\xi)$ from Section 1.6 describes the asymptotics of the multiplicities of K-modules in polynomials of very large degree on X.

More concretely, let

$$\mathbb{C}[X] = \oplus_{d=0}^{\infty} \mathbb{C}[X]_d$$

be the homogeneous coordinate ring of X. The space $\mathbb{C}[X]_d$ of degree d polynomials on X decomposes as a K-module

$$\mathbb{C}[X]_d = \bigoplus_{\lambda \in K^\wedge} \Omega_d(\lambda)\, V^\lambda .$$

One can view $\Omega_k(\lambda)$ as a measure on $K^\wedge \subset \mathfrak{h}_+$. After proper normalization, the measures $\Omega_k(k\lambda)$ converge weakly to $\Omega(\xi)\, d\xi$ as $k \to \infty$ where $d\xi$ is the Lebesgue measure on $\mathfrak{h}_+$.

In other words, for any $A \subset \mathfrak{h}_+$, the integral $\int_A \Omega(\xi)\, d\xi$ describes the leading asymptotics of the sum $\sum_{\lambda \in kA} \Omega_k(\lambda)$ as $k \to \infty$. Hence, informally, $\Omega(\xi)$ is the multiplicity $\Omega_k(k\lambda)$ averaged over some infinitesimal neighborhood of ξ. Such an averaging over infinitesimally close energy levels is a very natural thing to do from the statistical physics perspective.

3.2

The function $\Omega(\xi)$ depends not only on the K-action on X as such but also on the embedding $X \subset \mathbb{P}^N$, where $\mathbb{C}^{N+1}$ is a representation space of K or, equivalently, of the complexification G of K.

In intrinsic terms, such an embedding is a very ample invertible sheaf L in the G-linearized Picard group $\operatorname{Pic}^G(X)$ of X. Write $\Omega(\xi, L)$ to stress the dependence on both ξ and L. Because L enters the definition of $\Omega(\xi, L)$ only via $L^{\otimes n}$, $n \to \infty$, the function $\Omega(\xi, L)$ is well defined for any

$$L \in \operatorname{Pic}^G(X) \otimes_{\mathbb{Z}} \mathbb{Q} .$$

Since ample sheafs form a semigroup in $\operatorname{Pic}^G(X)$ it makes sense to ask whether $\log \Omega(\xi, L)$ is concave as a function of the pair (ξ, L).

In fact, we already saw an example of such a bivariate concavity in Section 2.4 where the multiplicities for restrictions from $U(n)$ to $U(k)$ turned out to be concave in the pair of highest weights.

3.3

In this setting, the log-concavity of $\Omega(\xi)$ and $\Omega(\xi, L)$ was established in [17] and [18], respectively, by using the classical Brunn–Minkowski inequality of convex analysis.

Here we want to use the same ideas to approach the problem from a slightly different angle. Instead of looking at the weak limit of measures $\Omega_k(k\lambda)$, which involves averaging over infinitesimally close energy levels, we want to look at the asymptotics of the sequence $\Omega_k(k\lambda)$ for some fixed λ. This is a natural thing to do from the representation theory point of view.

3.4

There is a standard trick which allows one to dispose of the first variable ξ in $\Omega(\xi, L)$ by enlarging the variety X. Indeed, by definition of the multiplicities Ω_k we have

$$\begin{aligned}\Omega_k(k\lambda, L) &= \dim \left(H^0(X, L^{\otimes k}) \otimes V^{k\lambda^*}\right)^G ,\\ &= \dim H^0\left(X \times G/B, (L \boxtimes L_{\lambda^*})^{\otimes k}\right)^G ,\end{aligned}$$

where $\lambda^* \in K^\wedge$ is the highest weight of $(V^\lambda)^*$, $G/B = K/T$ is the flag variety of K, the sheaf $L_{\lambda^*} \in \mathrm{Pic}^G(G/B)$ corresponds to the map of G/B onto the orbit of the highest vector in $P(V^\lambda)$, and superscript G denotes G-invariants. So, without loss of generality, we can assume that $\xi = 0$.

3.5

Recall that, by definition,

$$\mathbb{C}[X/\!\!/_L G] = \bigoplus_k H^0(X, L^{\otimes k})^G$$

is the homogeneous coordinate ring of the Geometric Invariant Theory quotient $X/\!\!/_L G$ corresponding to $L \in \mathrm{Pic}^G(X)$. We write $X/\!\!/_L G$ in place of the standard $X/\!\!/G$ to stress the dependence on L.

The sequence

$$\Omega_k(0, L) = \dim H^0(X, L^{\otimes k})^G , \quad k = 0, 1, 2, \dots \tag{3.1}$$

may fail to have a $k \to \infty$ asymptotics for the following trivial reason. Consider the set

$$\{k \,,\, H^0(X, L^{\otimes k})^G \neq 0\} \subset \mathbb{Z}_{\geq 0} \,.$$

Since X is irreducible, it is a semigroup and it either contains all sufficiently large integers or lies in a proper subgroup of $\mathbb{Z}$. We want to avoid the latter case because in that case the sequence (3.1) does not have any asymptotics. So we will replace L by a suitable power of L in that case.

By replacing L by its power one can also achieve that $\mathbb{C}[X/\!\!/_L G]$ is generated by its degree 1 graded component and so we can assume this as well. Thus, we have an embedding

$$X/\!\!/_L G \subset \mathbb{P}\left(\left(H^0(X, L)^G\right)^*\right)$$

and we denote by $\deg X/\!\!/_L G$ the degree of this embedding. It follows that in this case

$$\Omega_k(0, L) \sim \deg X/\!\!/_L G \,\frac{k^d}{d!} , \quad d = \dim X/\!\!/_L G ,$$

as k goes to ∞. We now want to show that $\log \deg X/\!\!/_L G$ is a concave function of L.

3.6

In fact, one can establish a more general fact. Let Y be an irreducible algebraic variety of dimension d and let $\mathbb{K} = \mathbb{C}(Y)$ be the field of rational functions on Y. Let $S \subset \mathbb{K}$ be a $\mathbb{C}$-linear subspace such that $1 \in S$ and which generates $\mathbb{K}$ as a field. The embedding $S \subset \mathbb{K}$ corresponds to a subvariety $Y_S \subset S^*$ which is birationally isomorphic to Y. Let $\deg Y_S$ denote its degree.

Given two such subspaces S_1 and S_2, denote by $S_1 S_2$ the subspace generated by all products $f_1 f_2$, where $f_i \in S_i$. We will show that

$$\sqrt[d]{\deg Y_{S_1 S_2}} \geq \sqrt[d]{\deg Y_{S_1}} + \sqrt[d]{\deg Y_{S_2}} \tag{3.2}$$

for any such pair S_1 and S_2. Since, clearly,

$$\deg Y_{S^2} = 2^d \deg Y_S$$

the inequality (3.2) implies that $\sqrt[d]{\deg Y_S}$, and consequently, $\log \deg Y_S$ is a concave function of S.

3.7

In particular, (3.2) would imply the concavity of $\log \deg X /\!\!/_L G$. Indeed, although the varieties $X /\!\!/_L G$ may not be isomorphic for different L, they are always birationally isomorphic. Their common field of fractions is the field $\mathbb{K} = \mathbb{C}(X)^G$ of rational G-invariants.

Given some L_1 and L_2, pick some $\phi_i \in H^0(X, L_i)^G$. Replacing the L_i's if necessary by their multiples, we can assume that $\mathbb{C}[X /\!\!/_{L_i} G]$ is generated by $S_i = \phi_i^{-1} H^0(X, L_i)^G \subset \mathbb{K}$. Since the algebra $\mathbb{C}[X /\!\!/_{L_1 \otimes L_2} G]$ contains the algebra generated by $S_1 S_2$, we get from (3.2) the desired lower bound on the asymptotics of the dimensions of the graded components of $\mathbb{C}[X /\!\!/_{L_1 \otimes L_2} G]$.

3.8

Now, in order to establish (3.2), we will construct convex sets $\Delta_S \in \mathbb{R}^d$ of dimension $d = \dim Y$ such that

$$\deg Y_S = d! \operatorname{vol} \Delta_S$$

and

$$\Delta_{S_1 S_2} \supset \Delta_{S_1} + \Delta_{S_2}\,.$$

The inequality (3.2) will then follow immediately from the classical Brunn–Minkowski inequality, see e.g. [5]. See also e.g. the appendix by A. Khovanskii in [5] for a discussion of the relationship between classical inequalities of the convex analysis and algebraic geometry. For example, the Alexandrov–Fenchel inequality, which is stronger than the Brunn–Minkowski inequality, corresponds to the Hodge index theorem for surfaces.

3.9

The construction of $\Delta_S \in \mathbb{R}^d$ is similar to the definition of a Newton polytope.

Choose a smooth point $y \in Y$ which lies away from the singularities of the maps from Y to Y_{S_i} and their inverses. Choose a flag of subvarieties

$$Y \supset Y^1 \supset \cdots \supset Y^d = y, \quad \operatorname{codim} Y^k = k,$$

which are all smooth at y. Fix some local equation u_k of Y^k in Y^{k-1}.

These data give rise to a map

$$\mathbb{K} \setminus 0 \ni f \mapsto \mathfrak{v}(f) = (\mathfrak{v}_1(f), \ldots, \mathfrak{v}_d(f)) \in \mathbb{Z}^d,$$

where

$$\begin{aligned}
\mathfrak{v}_1(f) &= \operatorname{ord}_{Y^1} f, \\
\mathfrak{v}_2(f) &= \operatorname{ord}_{Y^2} \left(f u_1^{-\mathfrak{v}_1(f)} \right)\Big|_{Y^1}, \\
\mathfrak{v}_3(f) &= \operatorname{ord}_{Y^3} \left(\left(f u_1^{-\mathfrak{v}_1(f)} \right)\Big|_{Y^1} u_2^{-\mathfrak{v}_2(f)} \right)\Big|_{Y^2} \cdots
\end{aligned}$$

and so on. It is clear that $\mathfrak{v}$ is a valuation, that is,

$$\begin{aligned}
\mathfrak{v}(fg) &= \mathfrak{v}(f) + \mathfrak{v}(g), \\
\mathfrak{v}(f+g) &\geq \min\{\mathfrak{v}(f), \mathfrak{v}(g)\},
\end{aligned} \tag{3.3}$$

where the ordering on $\mathbb{Z}^d$ is lexicographic.

It is also clear that the residue field of $\mathfrak{v}$ is isomorphic to $\mathbb{C}$ and hence for any $\mathbb{C}$-linear subspace of $S \subset \mathbb{K}$ we have

$$\dim_{\mathbb{C}} S = |\mathfrak{v}(S \setminus 0)|. \tag{3.4}$$

3.10

By definition, set

$$\Gamma_S = \left\{ (k, \mathfrak{v}(f)),\ f \in S^k \setminus 0 \right\} \subset \mathbb{Z}^{1+d}.$$

It follows from (3.3) that Γ_S is a semigroup.

Denote by $\Lambda_S \subset \mathbb{Z}^{1+d}$ the lattice generated by Γ_S. Let $\nabla_S \subset \mathbb{R}^{1+d}$ be the closed convex cone generated by Γ_S and let Δ_S be the intersection of ∇_S with the subspace $(1, \mathbb{R}^d) \subset \mathbb{R}^{1+d}$. It is clear that

$$\Delta_S = \overline{\left\{ \frac{\mathfrak{v}(f)}{k},\ f \in S^k \setminus 0 \right\}},$$

where bar denotes closure.

Since the point y corresponds to a smooth point of Y_S, there exist

$$f_0, f_1, \ldots, f_d = 1 \in \mathbb{C}[Y_S]$$

such that

$$\mathfrak{v}(f_k) = (\underbrace{0, \ldots, 0}_{k \text{ times}}, 1, \ldots).$$

Thus, Δ_S contains a d-dimensional simplex and so

$$\dim \Delta_S = d. \tag{3.5}$$

3.11

Let us now prove that

$$\operatorname{vol} \Delta_S = d! \deg Y_S .$$

On the one hand, it follows that (3.4) and the definition of Γ_S that

$$\left|\Gamma_S \cap (k, \mathbb{Z}^d)\right| = \dim_{\mathbb{C}} S^k \sim \deg Y_S \frac{k^d}{d!}, \quad k \to \infty .$$

Since $\Gamma_S \subset \nabla_S \cap \Lambda_S$ we have

$$\frac{\deg Y_S}{d!} \leq \operatorname{vol} \Delta_S ,$$

where the normalization of Lebesgue measure is given by the intersection of the lattice Λ_S with the subspace $(1, \mathbb{R}^d) \subset \mathbb{R}^{1+d}$.

The inverse inequality will be deduced from the following results of Khovanskii [13]. Choose a sequence of finitely generated subgroups

$$\Gamma_1 \subset \Gamma_2 \subset \cdots \subset \Gamma_S$$

such that $\Gamma_S = \bigcup \Gamma_i$ and each Γ_i generates the lattice Λ_S. Let ∇_i denote the cone generated by Γ_i and let Δ_i be the corresponding hyperplane section of ∇_i. It is clear that

$$\Delta_1 \subset \Delta_2 \subset \cdots \subset \Delta_S = \overline{\bigcup \Delta_i} . \tag{3.6}$$

It is a theorem of Khovanskii, see Proposition 3 in Section 3 of [13], that there exists vectors $\gamma_i \in \Gamma_i$ such that

$$(\nabla_i + \gamma_i) \cap \Lambda_S \subset \Gamma_i .$$

It follows that

$$\frac{\left|\Gamma_i \cap (k, \mathbb{Z}^d)\right|}{k^d} \to \operatorname{vol} \Delta_i , \quad k \to \infty .$$

Hence $\operatorname{vol} \Delta_i \leq (d!)^{-1} \deg Y_S$ and (3.6) implies that

$$\operatorname{vol} \Delta_S \leq \frac{\deg Y_S}{d!} ,$$

as was to be shown.

3.12

The relation

$$\Delta_{S_1 S_2} \supset \Delta_{S_1} + \Delta_{S_2}\,. \tag{3.7}$$

follows immediately from (3.3) and the definition of Δ_S. Now we are almost in position to finish the proof of (3.2) by applying the Brunn–Minkowski inequality. One remaining detail is that our normalization of the volume vol Δ_S depends on the lattice Λ_S. It is, however, clear that

$$\Lambda_{S_1 S_2} \subset \Lambda_{S_1}, \Lambda_{S_2}\,.$$

Therefore, if we normalize the volume according to the lattice $\Lambda_{S_1 S_2}$ we have

$$\operatorname{vol} \Delta_{S_1 S_2} = \frac{\deg Y_{S_1 S_2}}{d!}\,, \quad \operatorname{vol} \Delta_{S_i} \geq \frac{\deg Y_{S_i}}{d!}\,.$$

This and the Brunn–Minkowski inequality applied to (3.7)

$$\sqrt[d]{\operatorname{vol},\, \Delta_{S_1 S_2}} \geq \sqrt[d]{\operatorname{vol} \Delta_{S_1}} + \sqrt[d]{\operatorname{vol} \Delta_{S_2}}$$

completes the proof of (3.2).

References

[1] M. F. Atiyah, Convexity and commuting Hamiltonians, *Bull. London Math. Soc.* **14** no. 1, (1982), 1–15.

[2] A. Buch, *The saturation conjecture (after A. Knutson and T. Tao)*, math.CO/9810180.

[3] R. Boyer, Infinite traces of AF-algebras and characters of $U(\infty)$, *J. Operator Theory* **9** (1983), 205–236.

[4] F. Brenti, *Unimodal, log-concave, and Polya frequency sequences in combinatorics*, Mem. AMS 81, no. 413, (1989).

[5] Yu. Burago and V. Zalgaller, *Geometric Inequalities*, Grundlehren der Mathematischen Wissenschaften 285, Springer-Verlag, Berlin-New York, 1988.

[6] A. Edrei. On the generation function of a doubly infinite, totally positive sequence, *Trans. AMS* **74** (1953), 367–383.

[7] W. Fulton, *Eigenvalues, invariant factors, highest weights, and Schubert calculus*, math.AG/9908012, to appear in *Bull. AMS*.

[8] W. Graham, *Logarithmic convexity of push-forward measures*, Invent. Math., **123** (1996), 315–322.

[9] V. Guillemin and S. Sternberg, Geometric quantization and multiplicities of group representations, *Invent. Math.* **67** (1982), no. 3, 515–538.

[10] V. Guillemin and S. Sternberg, *Convexity properties of the moment mapping*, Invent. Math. **67** (1982), no. 3, 491–513.

[11] V. Guillemin and S. Sternberg, Convexity properties of the moment mapping. II, *Invent. Math.* **77** (1984), no. 3, 533–546.

[12] Y. Karshon, Example of a non-log-concave Duistermaat–Heckman measure, *Math. Res. Lett.* **3** no. 4, (1996), 537–540.

[13] A. Khovanskii, The Newton polytope, the Hilbert polynomial and sums of finite sets, *Functional Anal. Appl.* **26**, no. 4, 276–281.

[14] F. Kirwan, Convexity properties of the moment mapping. III, *Invent. Math.* **77** no. 3, (1984), 547–552.

[15] A. Knutson and T. Tao, The honeycomb model of $\mathrm{GL}_n(C)$ tensor products. I. Proof of the saturation conjecture, *Journal of Amer. Math. Soc.* **12** no. 4, (1999), 1055–1090.

[16] A. Okounkov, Log-Concavity of multiplicities with Application to Characters of $U(\infty)$, *Adv. Math.* **127** no. 2, (1997), 258–282.

[17] A. Okounkov, Brunn-Minkowski inequality for multiplicities, *Invent. Math.* **125** (1996), 405-411.

[18] A. Okounkov, *Multiplicities and Newton polytopes*, Kirillov's seminar on representation theory, 231–244, AMS Transl. Ser. 2, **181**, AMS, Providence, RI, 1998.

[19] A. Okounkov and G. Olshanski, Asymptotics of Jack polynomials as the number of variables goes to infinity, *Internat. Math. Res. Notices* **13** (1998), 641–682.

[20] G. Olshanski, *Unitary representations of infinite-dimensional pairs* (G, K) *and the formalism of R. Howe*, Representation of Lie groups and related topics, 269–463, Adv. Stud. Contemp. Math. **7**, Gordon and Breach, New York, 1990.

[21] G. Olshanski, *On semigroups related to infinite-dimensional groups*, Topics in representation theory, 67–101, Adv. Soviet Math. **2**, Amer. Math. Soc., Providence, RI, 1991.

[22] R. Stanley, *Log-concave and unimodal sequences in algebra, combinatorics, and geometry*, Graph theory and its applications: East and West (Jinan, 1986), 500–535, Ann. New York Acad. Sci. **576**, 1989.

[23] A. Vershik, Description of invariant measures for the action of some infinite-dimensional groups, *Soviet Math. Doklady* **15** (1974), 1396–1400.

[24] A. Vershik and S. Kerov, Characters and factor representations of the infinite unitary group, *Soviet Math. Doklady* **26** (1982), 570–574.

[25] D. Voiculescu, Représentations factorielles de type II_1 de $U(\infty)$, *J. Math. Pures Appl.* **55** no. 1, (1976), 1–20.

Andrei Okounkov
Department of Mathematics
University of California at Berkeley
Evans Hall #3840
Berkeley, CA 94720-3840, USA
okounkov@math.berkeley.edu

Point Processes Related to the Infinite Symmetric Group

Grigori Olshanski

To my teacher A. A. Kirillov on the occasion of his 65th birthday

ABSTRACT We study a two-parametric family of probability measures on an infinite-dimensional simplex (the Thoma simplex). These measures originate in harmonic analysis on the infinite symmetric group (S. Kerov, G. Olshanski and A. Vershik, *Comptes Rendus Acad. Sci. Paris I* **316** (1993), 773–778). Our approach is to interpret them as probability distributions on a space of point configurations, i.e., as certain point stochastic processes, and to find the correlation functions of these processes.

In this paper we relate the correlation functions to the solutions of certain multidimensional moment problems. Then we calculate the first correlation function which leads to a conclusion about the support of the initial measures. In the appendix, we discuss a parallel but more elementary theory related to the well-known Poisson–Dirichlet distribution.

The paper is part of a joint project with Alexei Borodin and can serve as an introduction to the subject. The goal of our work is to understand new phenomena in noncommutative harmonic analysis which arise when the irreducible representations depend on countably many continuous parameters.

We begin with a study of a remarkable family of stochastic point processes. These processes live on the punctured interval $[-1, 1] \setminus \{0\}$ and depend on two real parameters. Our purpose is to calculate their correlation functions which supply substantial information about the processes. The present paper is the first in a series of papers and contains an introduction to the subject, a description of the method, a calculation of the first correlation function (also called the density function) and an application concerning the "support" of the processes. The higher correlation functions are calculated in the papers [B1, B2] by Alexei Borodin. An overview of the results is given in the note [BO1] (it is based on a series of electronic preprints, [P.I]–[P.V]). Another approach is contained in [BO2]. For further developments of the subject, see [BO3, BO4, BOO, B3, Ok2, Ok3].

The origins of the point processes in question come from harmonic analysis on the infinite symmetric group [KOV]: these processes govern the decomposition of the so-called generalized regular representations. I shall briefly discuss the link with representation theory, as this is the main motivation of the work. I believe that this new type of relationship between representations and probabilities is interesting. However, in the body of the paper, we are not dealing with representations, and I have tried to make the exposition formally independent of [KOV] and accessible to non-experts in representation theory.

Starting with the infinite symmetric group $S(\infty) = \varinjlim S(n)$ (the union of the finite symmetric groups), we form a "(G, K)-pair," where G is the product $S(\infty) \times S(\infty)$ and K is the diagonal subgroup in G. (Let us emphasize at once that irreducible representations of G, generally speaking, are *not* tensor products of two irredicible representations of $S(\infty)$ as would be the case for a "tame" group in place of $S(\infty)$; but $S(\infty)$ is *not tame*.) A *spherical representation* of (G, K) is a couple (T, v), where T is a unitary representation of G and v is a distinguished K-invariant unit vector in the space of T. Note that if T is irreducible then v is unique up to a scalar factor. Note also that the spherical representations are a particular case of more general "admissible representations" of (G, K), see [Ol1, Ok1].

There exists a parametrization $\omega \leftrightarrow (T^{(\omega)}, v^{(\omega)})$ of irreducible spherical representations by points ω of an infinite-dimensional simplex Ω; the latter is called the *Thoma simplex*, see [T, VK, Ol1, Ok1]. Any cyclic representation (T, v) can be decomposed into a direct integral,

$$T = \int_{\Omega} T^{(\omega)} P(d\omega), \quad v = \int_{\Omega} v^{(\omega)} P(d\omega). \tag{0.1}$$

Here P is a probability measure on Ω, which is uniquely defined; it is called the *spectral measure* for (T, v).

In [KOV], we constructed a family $\{T_z\}$ of admissible representations of (G, K) depending on a complex parameter z. Each T_z is realized in a L^2 space on a compactification $\mathfrak{S}$ of the discrete space $S(\infty)$. When z tends to infinity, T_z approaches the conventional biregular representation T_∞ of G in the ℓ^2 space on $S(\infty)$, so that the representations T_z form a deformation of T_∞; we call them the *generalized regular representations.*

As is well known, the representation T_∞ is irreducible. However, the representations T_z are highly reducible (as "true" regular representations should be). Their decomposition can be viewed as a model problem of noncommutative harmonic analysis in the situation when the irreducible representations depend on infinitely many parameters.

The construction of [KOV] shows that each T_z possesses a distinguished K-invariant vector v; moreover, v is cyclic provided that z is not integral. Applying the abstract decomposition (0.1), we arrive at a family $\{P_z\}$ of spectral measures on the simplex Ω. The probability measures P_z are the main object of the paper.[1] Note that they are pairwise disjoint ([KOV, Theorem 5.3]).

To study the measures P_z we propose the following approach:

1) We define an embedding of the Thoma simplex Ω to the space Ξ of configurations in the locally compact space $I = [-1, 1] \setminus \{0\}$ (a configuration in I is an

[1] Actually, we also consider a "complementary series" of spectral measures. In the text, we use the notation $P_{zz'}$, where either $z' = \bar{z}$ (the "principal series") or both z and z' are real and satisfy some extra condition (the "complementary series").

unordered collection of points which can accumulate only at 0). Then any probability measure on Ω will define a *random* configuration in I, i.e., a stochastic point process. Thus, the measures P_z can be interpreted as certain point processes $\mathcal{P}_z$.

2) As a characteristic of the point processes $\mathcal{P}_z$ we choose the correlation functions. Let $n = 1, 2, \ldots$ and $x_1, \ldots, x_n$ be an arbitrary n-tuple of points in I. Given a point process, the probability of the event that a random configuration intersects each of the infinitely small intervals $x_1 + dx_1, \ldots, x_n + dx_n$ has the form $\rho_n(x_1, \ldots, x_n)dx_1 \ldots dx_n$, and the density ρ_n is called the nth *correlation function*. All point processes coming from probability measures on Ω are uniquely determined by their correlation functions.

3) We show that the correlation functions of $\mathcal{P}_z$ can be obtained from a sequence $\sigma_1, \sigma_2, \ldots$ of probability measures called the *controlling measures*. The nth controlling measure lives on the n-dimensional cube $[-1, 1]^n$, and one can write down all its moments. This reduces the problem of calculating the functions to a certain multidimensional moment problem.

The paper is organized as follows.

§1 contains preliminaries on symmetric functions, the Thoma simplex, coherent systems of distributions on the Young graph, and their spectral measures. Using this formalism, we introduce in §2 the so-called coherent z-systems on the Young graph, by means of which we define the spectral measures P_z. In §3 we discuss the controlling measures and their moments, and in §4 we pass to the point processes.

The technical part of the work begins in §5. Here we solve a moment problem and calculate the density function of $\mathcal{P}_z$. We present an integral representation of the density function, Theorem 5.8, and an explicit expression in terms of a multivariate hypergeometric function (the Lauricella function F_B in three variables), Theorem 5.12.

In §6 we show that the first controlling measure has no atom at 0, which implies that the measures P_z are concentrated on a distinguished face of the simplex Ω (this result was announced in [KOV]).

There are two appendices.

In the first appendix (§7) we discuss the Poisson–Dirichlet distributions $PD(t)$. These distributions were extensively studied in the literature and they play an important role in the construction [KOV] of the representations T_z. Our purpose is to derive the Watterson [W] formula for the correlation functions of $PD(t)$ by employing the general formalism of §§1–4.

The second appendix (§8) contains a simple direct proof, due to A. Borodin, of Theorem 2.1 asserting the existence of the coherent z-systems.

I am deeply grateful to Alexei Borodin for numerous discussions which substantially influenced the final version of the paper. The work was supported by RFBR grant 98-01-00303.

1 Coherent systems of distributions on the Young graph

Symmetric functions [M]

Let Λ denote the algebra of symmetric functions over the base field $\mathbb{R}$. Formally, Λ may be defined as $\mathbb{R}[p_1, p_2, \dots]$, the algebra of polynomials over infinitely many indeterminates $p_1, p_2, \dots$, called the *power sums*. Another system of generators of Λ is formed by the *complete symmetric functions* $h_1, h_2, \dots$, which are expressed through $p_1, p_2, \dots$ via the the following relation of generating series with formal parameter t:

$$1 + \sum_{n \geq 1} h_n t^n = \exp\Big(\sum_{n \geq 1} (p_n/n) t^n\Big).$$

A *partition* is a weakly decreasing sequence $\lambda = (\lambda_1, \lambda_2, \dots)$ of nonnegative integers with finitely many nonzero terms. The number of nonzero terms is called the *length* of λ and denoted as $\ell(\lambda)$. Partitions are also viewed as Young diagrams. By $|\lambda|$ we denote the sum $\lambda_1 + \lambda_2 + \dots$ or, equivalently, the number of boxes in the diagram λ. The zero partition (or the empty Young diagram) is denoted as $\varnothing$.

The elements

$$p_\rho = p_{\rho_1} p_{\rho_2} \cdots p_{\rho_l}, \quad l = \ell(\rho),$$

where ρ ranges over the set of partitions, form a basis in Λ. Another distinguished basis in Λ is formed by the *Schur functions*, which are also indexed by arbitrary partitions and can be defined, in terms of $h_1, h_2, \dots$, by the *Jacobi–Trudi formula*

$$s_\lambda = \det[h_{\lambda_i - i + j}],$$

where the order of determinant is $\ell(\lambda)$ and it is assumed that $h_0 = 1$ and $h_n = 0$ when $n < 0$.

We shall need two important formulas:

$$s_\mu \cdot p_1 = \sum_{\lambda:\, \lambda \searrow \mu} s_\lambda, \tag{1.1}$$

$$p_\rho = \sum_{\lambda:\, |\lambda| = |\rho|} \chi^\lambda_\rho s_\lambda, \tag{1.2}$$

see [M, I.5.16 and I.7.8]. Here and in what follows the notation $\lambda \searrow \mu$ (or, equivalently, $\mu \nearrow \lambda$) means that the diagram λ contains the diagram μ and differs from it by a single box (in particular, we have $|\lambda| = |\mu| + 1$). By χ^λ we denote the irreducible character indexed by λ (it is a character of the symmetric group of degree $|\lambda|$), and χ^λ_ρ is the value of χ^λ on the conjugacy class indexed by ρ, see [M, §I.7].

In the customary realization of the algebra Λ, its generators p_n are identified with the expressions $x_1^n + x_2^n + \dots$ in indeterminates $x_1, x_2, \dots$, so that elements of

Λ become symmetric functions in $x_1, x_2, \ldots$. But we shall need another realization, obtained by specializing the generators $p_n \in \Lambda$ to the following expressions in the indeterminates $\alpha = (\alpha_1, \alpha_2, \ldots)$, $\beta = (\beta_1, \beta_2, \ldots)$, and γ:

$$\begin{aligned} p_1 &\mapsto \widetilde{p}_1(\alpha, \beta, \gamma) := \sum_{i\ge1} \alpha_i + \sum_{i\ge1} \beta_i + \gamma \\ p_n &\mapsto \widetilde{p}_n(\alpha, \beta, \gamma) := \sum_{i\ge1} \alpha_i^n + (-1)^{n-1} \sum_{i\ge1} \beta_i^n, \quad n \ge 2, \end{aligned} \tag{1.3}$$

which is equivalent to

$$1 + \sum_{n\ge1} h_n t^n \mapsto e^{\gamma t} \prod_{i\ge1} \frac{1+\beta_i t}{1-\alpha_i t}.$$

This is a generalization of the well-known "super" realization of Λ; indeed, setting $\gamma = 0$ converts the above expressions to "supersymmetric" functions in α and $-\beta$, see [S] and [M, §I.3, Ex. 23].

The Thoma simplex [VK, KV, KOO]

We shall abbreviate

$$\alpha = (\alpha_1, \alpha_2, \ldots), \qquad \beta = (\beta_1, \beta_2, \ldots).$$

Let Ω be the set of the triples $\omega = (\alpha, \beta, \gamma)$ such that

$$\alpha_1 \ge \alpha_2 \ge \cdots \ge 0, \quad \beta_1 \ge \beta_2 \ge \cdots \ge 0, \quad \gamma \ge 0,$$

$$\sum_{i\ge1} (\alpha_i + \beta_i) + \gamma = 1. \tag{1.4}$$

Since γ is determined by α and β, we shall sometimes omit γ and write $\omega = (\alpha, \beta)$.

The set Ω is an infinite-dimensional simplex; its faces of codimension 1 have the form

$$\Omega_i = \{\omega \in \Omega \mid \alpha_i = \alpha_{i+1}\}, \quad \Omega_{-i} = \{\omega \in \Omega \mid \beta_i = \beta_{i+1}\}, \quad i \ge 1, \tag{1.5}$$

$$\Omega_0 = \{\omega \in \Omega \mid \gamma = 0\}. \tag{1.6}$$

The simplex Ω is called the *Thoma simplex* in connection with the pioneering work of Thoma [T]. We equip Ω with the weakest topology in which the coordinates α_i and β_i (but not γ) are continuous functions. In this topology, Ω is a metrizable compact space, and the face Ω_0 is a dense subset.

We define the functions $\widetilde{p}(\omega)$ on Ω by setting $\omega = (\alpha, \beta, \gamma) \in \Omega$ in the formulas (1.3). Note that $\widetilde{p}_1(\omega) \equiv 1$. It is readily verified ([KOO, Lemma 5.2]) that these functions are continuous on Ω.[2] Consequently, any element $f \in \Lambda$ can be

---

[2]Observe that the expression $\sum(\alpha_i + \beta_i)$ is not continuous on Ω.

converted to a continuous function $\widetilde{f}$ on Ω by writing f as a polynomial in the generators p_n and then replacing each p_n by $\widetilde{p}_n$. We shall call $\widetilde{f}$ the *extended version* of f, cf. [KV, KOO]. In particular, we shall deal with the *extended power sums* $\widetilde{p}_n$ and the *extended Schur functions* $\widetilde{s}_\lambda$.

Note that the functions $\widetilde{f}$, where f ranges over Λ, form a dense subalgebra in the Banach algebra $C(\Omega)$ of (real) continuous functions on the compact space Ω, see [KOO, Lemma 5.3].

The algebra Λ possesses an involutive automorphism (denoted as ω in [M]) such that

$$p_n \mapsto (-1)^{n-1} p_n, \quad s_\lambda \mapsto s_{\lambda'},$$

where λ' means the transposed diagram. In terms of the realization (1.3) this involution exactly corresponds to the symmetry $\alpha \leftrightarrow \beta$. The latter symmetry also defines a symmetry of the Thoma simplex, which will be denoted as $\omega \mapsto \omega'$. It follows that

$$\widetilde{s}_{\lambda'}(\omega) = \widetilde{s}_\lambda(\omega').$$

Harmonic functions and coherent systems of distributions on the Young graph [V, VK, KV, Ke2, KOO]

By definition, the vertices of the *Young graph* $\mathbb{Y}$ are arbitrary Young diagrams (including $\varnothing$), and its (oriented) edges are formed by the couples μ, λ such that $\mu \nearrow \lambda$. The number of oriented paths from $\varnothing$ to λ is called the *dimension* of λ and denoted as $\dim\lambda$ (we agree that $\dim\varnothing = 1$). The function dim on the vertices of $\mathbb{Y}$ satisfies the *recurrence relation*

$$\dim\lambda = \sum_{\mu:\, \mu\nearrow\lambda} \dim\mu.$$

These definitions are inspired by the *Young branching rule* for the characters of the finite symmetric groups $S(n)$, $n = 1, 2, \ldots,$

$$\chi^\lambda\mid_{S(n-1)} = \sum_{\mu:\, \mu\nearrow\lambda} \chi^\mu, \quad n = |\lambda|.$$

(See [V, JK, OV].) The above two relations show that the dimension of a diagram λ coincides with the dimension of the character χ^λ, i.e., with the number $\chi^\lambda_{(1^n)}$, $n = |\lambda|$.

A (real) function $\varphi(\lambda)$ on the vertices of $\mathbb{Y}$ is called *harmonic* if it satisfies the "harmonicity condition"

$$\varphi(\mu) = \sum_{\lambda:\, \lambda\searrow\mu} \varphi(\lambda)$$

for each diagram μ. Let $\mathbb{Y}_n$ denote the set of Young diagrams with n boxes (equivalently, the set of partitions of the number n). The sets $\mathbb{Y}_n$ define a *grading* of the graph $\mathbb{Y}$. Clearly, the knowledge of a harmonic function on a "floor" $\mathbb{Y}_n$ determines it on the preceding floors $\mathbb{Y}_{n-1}, \mathbb{Y}_{n-2}, \ldots$.

Let $M(\lambda)$ be a function on the vertices of $\mathbb{Y}$ and M_n denote its restriction to the nth floor $\mathbb{Y}_n$, $n = 0, 1, 2, \ldots$. We call $M = (M_n)$ a *coherent system of distributions* on the Young graph (coherent system, for short) if M is nonnegative, normalized at $\lambda = \varnothing$ (i.e., $M(\varnothing) = M_0(\varnothing) = 1$), and the function $M(\lambda)/\dim\lambda$ is harmonic, i.e.,

$$M(\mu) = \sum_{\lambda:\, \lambda \searrow \mu} \frac{\dim\mu}{\dim\lambda} M(\lambda).$$

The latter formula and the recurrence relation for the dimension function imply that the M_n's are probability distributions on the floors of the Young graph (see [KOO, Lemma 8.1]).

By a measure on Ω we shall always mean a Borel measure with respect to the canonical Borel structure of the topological space Ω. According to [KOO, Theorem B], there is a bijective correspondence $M \leftrightarrow P$ between the coherent systems M on the Young graph and the probability measures P on the Thoma simplex. This correspondence is characterized by the relation

$$\frac{M(\lambda)}{\dim\lambda} = \int_{\Omega} \widetilde{s}_\lambda(\omega)\, P(d\omega), \tag{1.7}$$

where λ is an arbitrary Young diagram and $\widetilde{s}_\lambda$ is the extended Schur function as defined above. Moreover, as is shown in the proof of this result, the measure P is approached, in a certain sense, by the finite probability distributions M_n as $n \to \infty$. We shall call P the *spectral measure* of M.

If M is a coherent system on $\mathbb{Y}$, then the function $M^t(\lambda) = M(\lambda^t)$ is a coherent system as well (indeed, this follows from the fact that the involution $\lambda \mapsto \lambda^t$ is a symmetry of the Young graph). On the other hand, if P is a measure on Ω, let P^t denote the image of P under the symmetry $\omega \mapsto \omega^t$ defined above. Now, we have

$$\text{if} \quad M \leftrightarrow P \quad \text{then} \quad M^t \leftrightarrow P^t. \tag{1.8}$$

Indeed, this follows from (1.7) and the equality $\widetilde{s}_{\lambda^t}(\omega) = \widetilde{s}_\lambda(\omega^t)$ mentioned above.

Connection with representations [T, VK, KV, Ol1, Ol2]

The constructions described above are inspired by the representation theory of the infinite symmetric group $S(\infty) = \varinjlim S(n)$. Let M be a coherent system on $\mathbb{Y}$ and $\varphi(\lambda) = M(\lambda)/\dim\lambda$ the corresponding harmonic function. For each n, the linear combination of irreducible characters

$$\chi_n = \sum_{\lambda:\, |\lambda|=n} \varphi(\lambda)\chi^\lambda$$

is a central positive definite function on $S(n)$, normalized at the unit element. By the harmonicity condition, the functions $\chi_1, \chi_2, \ldots$ are compatible with the embeddings $S(n-1) \hookrightarrow S(n)$. Consequently, they define a central, positive definite, normalized function χ on the group $S(\infty)$. Let X denote the set of all such functions. This is a convex set; its extremal points are called the *characters* of the group $S(\infty)$ (in the sense of von Neumann).

By virtue of the Gelfand–Naimark–Segal construction, characters generate *finite factor representations* of the group $S(\infty)$. They also correspond to *irreducible unitary spherical representations* of the Gelfand pair (G, K), where G stands for the "bisymmetric group" $S(\infty) \times S(\infty)$ and K is the diagonal subgroup of G, see [Ol1, Ol2].

The integral representation (1.7) has the following meaning. First, via the correspondence $M \leftrightarrow \chi$, (1.7) implies that the characters are parametrized by the points $\omega \in \Omega$ (we shall write them as $\chi^{(\omega)}$).[3] This result is known as *Thoma's theorem* [T]; see also [VK]. The extreme coherent system $M^{(\omega)} \leftrightarrow \chi^{(\omega)}$ is given by the formula

$$M^{(\omega)}(\lambda) = \dim \lambda \cdot \widetilde{s}_\lambda(\omega),$$

which is equivalent to *Thoma's formula* [T]

$$\chi^{(\omega)}(\rho) = \widetilde{p}_{\rho_1}(\omega)\widetilde{p}_{\rho_2}(\omega) \dots .$$

In the latter expression, the left-hand side is the value of $\chi^{(\omega)}$ at the conjugacy class in $S(\infty)$ of an arbitrary cycle-type $\rho = (\rho_1, \rho_2, \dots, 1, 1, \dots)$, and the right-hand side is correctly defined, because almost all cycles have length 1 and $\widetilde{p}_1 \equiv 1$.

Next, (1.7) implies that any function $\chi \in X$ is uniquely decomposed into a continual convex combination of the characters $\chi^{(\omega)}$. As χ generates a cyclic spherical representation T of the pair (G, K), this also means that the spectral measure P governs the decomposition of T in a direct integral of irreducible spherical representations.

2 The coherent z-systems

Set

$$M_{zz'}(\lambda) = \frac{\prod_{(ij)\in\lambda} (z+j-i)(z'+j-i)}{(t)_{|\lambda|}} \cdot \frac{\dim^2 \lambda}{|\lambda|\,!}. \tag{2.1}$$

Here λ is an arbitrary Young diagram, z and z' are complex parameters, $(ij) \in \lambda$ is an arbitrary box of λ (i and j are the numbers of the row and the column containing the box), $t = zz'$, and an expression of type $(t)_n$ means the Pochhammer symbol:

$$(t)_n = \Gamma(t+n)/\Gamma(t) = t(t+1)\dots(t+n-1);$$

(2.1) is correctly defined if $t \neq 0, -1, -2, \dots$. We also agree that $M_{zz'}(\varnothing) = 1$.

Theorem 2.1. *The function $M_{zz'}(\lambda)/\dim\lambda$ is harmonic.*

---

[3]Thus, the Thoma simplex may be viewed as the *spherical dual* of the Gelfand pair (G, K).

Proof. This follows from [KOV, Theorem 3.1]. There exists also a number of direct proofs [Ke2, P, BO3] (the papers [Ke2, BO3] actually contain a more general result). One more direct proof is given below in §8. □

Proposition 2.2. *Assume that z, z' are not integers and are such that $t = zz'$ is not equal to* $0, -1, -2, \ldots$. *Then $M_{zz'}(\lambda) \neq 0$ for any λ. Moreover, $M_{zz'}$ is strictly positive if and only if the parameters z, z' satisfy one of the following conditions*:

(i) *$z' = \bar{z}$ and $z \notin \mathbb{Z}$;*

(ii) *z and z' are real and are both contained in an open interval of the form $(m, m+1)$ with $m \in \mathbb{Z}$.*

Proof. The first claim is obvious from (2.1), let us check the second claim. Assume that (i) or (ii) holds. Then $t > 0$ and $(z+k)(z'+k) > 0$ for any $k \in \mathbb{Z}$, which implies $M_{zz'}(\lambda) > 0$.

Conversely, assume that $M_{zz'}(\lambda) > 0$ for any λ. Let $\mu \nearrow \lambda$ be an edge of $\mathbb{Y}$, $n = |\mu|$, (ij) be the box $\lambda \setminus \mu$, and $k = j - i$. Comparing $M_{zz'}(\mu)$ and $M_{zz'}(\lambda)$, we see that

$$\frac{(z+k)(z'+k)}{(t+n)} > 0.$$

Clearly, this inequality holds for any number $k \in \mathbb{Z}$ and $n = 1, 2, \ldots$ that correspond to edges of $\mathbb{Y}$, that is, for any k, n such that either $k \neq 0$ and $n \geq |k|$, or $k = 0$ and $n \geq 3$.

Now, we fix k and let $n \to \infty$. From the above inequality we conclude that the numerator $(z+k)(z'+k)$ must be real and strictly positive. Since this holds for any $k \in \mathbb{Z}$, both zz' and $z + z'$ are real. It follows that either z, z' are complex conjugate or both real. It remains to examine the latter possibility. Using the fact that $(z+k)(z'+k)$ is not only real but strictly positive, we get condition (ii) of the proposition. □

Corollary 2.3. *Under assumptions* (i) *or* (ii) *of Proposition* 2.2, *$M_{zz'}$ is a coherent system of distributions on the Young graph.*

From now on we shall assume that the parameters z, z' satisfy one of the conditions (i), (ii) of Proposition 2.2 (note then that $t \neq 0, -1, -2, \ldots$ holds automatically). We shall call the coherent systems $M_{zz'}$ the (coherent) *z-systems*. To distinguish between (i) and (ii), we shall speak about the *principal series* and *complementary series* of z-systems, respectively. A motivation for such a terminology is that the difference $z - z'$ ranges, respectively, over the imaginary axis and the open interval $(-1, 1)$, like the parameters of the principal or the complementary series for $SL(2, \mathbb{R})$.

The principal series of z-systems first appeared in [KOV]: in that form we described the spherical functions of the generalized regular representations of the pair (G, K). The existence of the complementary series was observed in 1995 by Borodin. The z-systems can be characterized as the only coherent systems of distributions on the Young graph satisfying a "multiplicativity condition," see [R].

In the present paper we do not deal with a "degenerate series" of coherent systems which arises when one of the parameters z, z' is integral. The "degenerate" coherent systems live, in essence, on truncated versions of the Young graph. See [KOV] and [BO4] for more about them.

Remark 2.4. The expression (2.1) for $M_{zz'}$ has two evident symmetries, each of which has a representation-theoretic meaning. First, (2.1) does not change under the transposition $z \leftrightarrow z'$, which leads to certain intertwining operators for the generalized regular representations. Second, we have

$$M_{zz'}(\lambda') = M_{-z,-z'}(\lambda), \tag{2.2}$$

where $\lambda \mapsto \lambda'$ denotes transposition of Young diagrams. This symmetry reflects the well-known fact that, for irreducible characters χ^λ, transposing λ is equivalent to tensoring χ^λ with the one-dimensional sign character.

In the next two propositions, we rewrite the expression (2.1) by making use of two explicit formulas for $\dim\lambda$.

Proposition 2.5. *The expression* (2.1) *can be written as*

$$M_{zz'}(\lambda) = \frac{|\lambda|!}{(t)_{|\lambda|}} \prod_{i=1}^{l} (z-i+1)_{\lambda_i}(z'-i+1)_{\lambda_i} \times \frac{\prod\limits_{1\le i,j\le l} (\lambda_i - \lambda_j + j - i)^2}{\prod\limits_{1\le i\le l} ((\lambda_i + l - i)!)^2}, \tag{2.3}$$

where $l \ge \ell(\lambda)$ may be taken arbitrarily.

Proof. The first product is exactly the product over the boxes $(ij) \in \lambda$ in formula (2.1), and the remaining terms come from the formula

$$\frac{\dim\lambda}{|\lambda|!} = \frac{\prod\limits_{1\le i,j\le l} (\lambda_i - \lambda_j + j - i)}{\prod\limits_{1\le i\le l} (\lambda_i + l - i)!}, \qquad l \ge \ell(\lambda). \tag{2.4}$$

It is readily seen that the right-hand side is stable on $l \ge \ell(\lambda)$. Hence, it suffices to check it for a particular value of l. But for $l = |\lambda|$, this coincides with the formula of [M, §I.7, Ex. 6]. □

The explicit expression (2.3) is not quite satisfactory, because it does not reflect the symmetry (2.2). A symmetric expression can be obtained with the help of the Frobenius notation for Young diagrams [M, §I.1]:

$$\lambda = (p_1, \dots, p_d \,|\, q_1, \dots, q_d), \tag{2.5}$$

where d is the length of the diagonal in λ, and

$$p_i = \lambda_i - i, \quad q_i = (\lambda^t)_i - i, \qquad 1 \le i \le d,$$

are the *Frobenius coordinates* of λ. Note that

$$|\lambda| = |p| + |q| + d,$$

where

$$|p| = p_1 + \cdots + p_d, \quad |q| = q_1 + \cdots + q_d.$$

Proposition 2.6. *In the Frobenius notation* (2.5), *the expression* (2.1) *can be written as follows:*

$$M_{zz'}(\lambda) = \frac{(|p|+|q|+d)!\, t^d}{(t)_{|p|+|q|+d}} \prod_{i=1}^{d} \frac{(z+1)_{p_i}(z'+1)_{p_i}(-z+1)_{q_i}(-z'+1)_{q_i}}{(p_i!)^2\,(q_i!)^2}$$

$$\times \prod_{i=1}^{d}\prod_{j=1}^{d}(p_i+q_j+1)^{-2} \prod_{1\le i<j\le d}(p_i-p_j)^2(q_i-q_j)^2. \tag{2.6}$$

Proof. Given a box $(ij) \in \lambda$, its *hook* is defined as the shape formed by the boxes $(kl) \in \lambda$ such that either $k = i, l \ge j$ or $k > i, l = j$. The total number of boxes in the hook is called the *hook-length* and denoted as $h(i, j)$.

Let us represent the shape λ as the union of the diagonal hooks. Then the contribution of the kth diagonal hook ($k = 1, \dots, d$) to the product

$$\prod_{(ij)\in\lambda} (z+j-i)(z'+j-i)$$

is equal to

$$\begin{aligned}
&(z-q_k)\dots(z-1)z(z+1)\dots(z+p_k)\\
&\quad\times (z'-q_k)\dots(z'-1)z'(z'+1)\dots(z'+p_k)\\
&= zz'(z+1)_{p_k}(z'+1)_{p_k}(-z+1)_{q_k}(-z'+1)_{q_k}\\
&= t(z+1)_{p_k}(z'+1)_{p_k}(-z+1)_{q_k}(-z'+1)_{q_k},
\end{aligned}$$

which explains the term t^d and the first product in (2.6). The remaining terms in (2.6) come from the following formula expressing $\dim\lambda$ in the Frobenius notation:

$$\frac{\dim\lambda}{|\lambda|!} = \frac{\prod\limits_{1\le i,j\le d}(p_i-p_j)(q_i-q_j)}{\prod\limits_{1\le i\le d}\ \prod\limits_{1\le j\le d}(p_i+q_j+1)\cdot \prod\limits_{1\le i\le d}(p_i!\,q_i!)}. \tag{2.7}$$

To check (2.7) we start with the well-known *hook formula*

$$\frac{\dim\lambda}{|\lambda|!} = \prod_{(ij)\in\lambda} h(i,j)^{-1}, \tag{2.8}$$

which, by virtue of formula (2.4), is equivalent to the identity

$$\prod_{(ij)\in\lambda} h(i,j)^{-1} = \frac{\prod\limits_{1\le i,j\le l}(\lambda_i-\lambda_j+j-i)}{\prod\limits_{1\le i\le l}(\lambda_i+l-i)!}, \qquad l\ge \ell(\lambda), \tag{2.9}$$

see [M, §I.1, Ex. 1].

Now, let us divide the shape λ into three pieces: the square shape of size $d\times d$, the diagram λ^+ formed by the boxes (ij) with $j>d$, and the diagram λ^- formed by the boxes (ij) with $i>d$. The hook-length of a box (ij) from the square shape is equal to p_i+q_j+1. Consequently, the product of the hook-lengths over the boxes entering the square shape is equal to the double product

$$\prod_{1\le i\le d}\ \prod_{1\le j\le d}(p_i+q_j+1)$$

in the denominator of (2.7). To explain the remaining terms in (2.7), we express the products of the hook-lengths in the diagrams λ^+ and λ^- via the identity (2.9), where we substitute $l=d$ and $\lambda=\lambda^+$ or $\lambda=(\lambda^-)'$. It should also be noted that for any box in $\lambda^\pm$, its hook with respect to λ is the same as the hook with respect to $\lambda^\pm$. □

Formula (2.6) plays a fundamental role in our theory, see [B2, BO1, BO2, BO3, BO4]. In the present paper, it will be used for calculations in §5 below.

3 Controlling measures

Given a point $\omega=(\alpha,\beta,\gamma)\in\Omega$, we define the corresponding *Thoma measure* on $[-1,1]$ as

$$\nu^{(\omega)}=\sum_{i\ge1}\alpha_i\delta(\alpha_i)+\sum_{i\ge1}\beta_i\delta(-\beta_i)+\gamma\delta(0),$$

where $\delta(x)$ stands for the Dirac mass at $x\in[-1,1]$. Clearly, $\nu^{(\omega)}$ is a probability measure.

Proposition 3.1. *The moments of $\nu^{(\omega)}$ are given by the formula*

$$\int_{-1}^{1} x^l\,\nu^{(\omega)}(dx)=\widetilde{p}_{l+1}(\omega), \qquad l=0,1,2,\dots,$$

where $\widetilde{p}_{l+1}$ are the extended power sums as defined in §1.

Proof. This is a direct consequence of the definition of the Thoma measure and that of $\widetilde{p}_n$. □

Let Prob$[-1,1]$ denote the set of probability Borel measure on $[-1,1]$; this set has a natural Borel structure [DVJ]. The map

$$\Omega \to \mathrm{Prob}[-1,1], \qquad \omega \mapsto \nu^{(\omega)},$$

is Borel-measurable (this is a routine exercise). Therefore, equipping Ω with a probability measure P, we obtain a *random measure* on $[-1,1]$. Since the above map is injective, we may interpret probability measures P on Ω as random measures on $[-1,1]$.

The next construction looks rather natural from the point of view of the theory of random measures (see, e.g., [DVJ]) or the exchangeability theory [A]. We take the infinite product

$$\nu_\infty^{(\omega)} = \nu^{(\omega)} \times \nu^{(\omega)} \times \dots, \qquad \omega \in \Omega,$$

which is a probability measure on the infinite-dimensional cube

$$[-1,1]^\infty = [-1,1] \times [-1,1] \times \dots,$$

and we average $\nu_\infty^{(\omega)}$ with respect to a given probability measure P on Ω:

$$\sigma = \int_\Omega \nu_\infty^{(\omega)}\, P(d\omega).$$

In other words, viewing (Ω, P) as a probability space, we take the expectation of the random measure $\nu_\infty^{(\omega)}$. We shall call σ the *controlling measure* (of infinite order) for the measure P. Clearly, σ is a symmetric probability measure on the infinite-dimensional cube. Its projection on the n-dimensional cube $[-1,1]^n$ will be called the nth *controlling measure* of P and denoted as σ_n:

$$\sigma_n = \int_\Omega \underbrace{\nu^{(\omega)} \times \dots \times \nu^{(\omega)}}_{n \text{ times}} P(d\omega).$$

For a Young diagram λ, let $d(\lambda)$ be the number of diagonal boxes. By $\chi^\lambda_{(r_1,\dots,r_n)}$, where $r_1,\dots,r_n$ are (not necessarily decreasing) numbers ≥ 1 such that $r_1+\dots+r_n = |\lambda|$, we denote the value of the irreducible character at any permutation with n cycles of length $r_1,\dots,r_n$.

Proposition 3.2. *Let M be a coherent system on the Young graph, let P be its spectral measure on Ω, and let σ_n be the nth controlling measure of P.*

(i) *The moments of σ_n satisfy the relations*

$$\int\limits_{[-1,1]^n} x_1^{l_1} \dots x_n^{l_n}\, \sigma_n(dx_1 \dots dx_n) = \int_\Omega \widetilde{p}_{l_1+1}(\omega) \dots \widetilde{p}_{l_n+1} P(d\omega) \qquad (3.1a)$$

$$= \sum_{\substack{\lambda:\, d(\lambda)\leq n \\ |\lambda| = l_1+\dots+l_n+n}} \chi^\lambda_{(l_1+1,\dots,l_n+1)} \frac{M(\lambda)}{\dim \lambda}, \qquad (3.1b)$$

where $l_1, \dots, l_n = 0, 1, 2, \dots$.

Proof. By definition of σ_n, the left-hand side is the expectation (with respect to (Ω, P)) of the integral

$$\int_{[-1,1]^n} x_1^{l_1} \dots x_n^{l_n} \, \nu^{(\omega)}(dx_1) \dots \nu^{(\omega)}(dx_n).$$

By Proposition 3.1, this integral is equal to

$$\widetilde{p}_{l_1+1}(\omega) \dots \widetilde{p}_{l_n+1}(\omega).$$

Integrating over $\omega \in \Omega$ with respect to P, we get (3.1a).

According to the identity (1.2), the right-hand side of (3.1a) is equal to

$$\int_\Omega \sum_{\lambda:\ |\lambda|=l_1+\dots+l_n+n} \chi^\lambda_{(l_1+1,\dots,l_n+1)} \, \widetilde{s}_\lambda(\omega) P(d\omega).$$

The Murnaghan–Nakayama rule ([M, §I.7, Ex. 5]) implies that $\chi^\lambda_{(l_1+1,\dots,l_n+1)}$ vanishes when $d(\lambda) > n$, so we may introduce the supplementary requirement $d(\lambda) \le n$ into the above sum. Finally, transposing integration and summation and applying (1.7) we get (3.1b). □

Thus, when M is known, we can, in principle, find the controlling measures σ_n from the moment problem (3.1). The latter has a unique solution because the support of σ_n is bounded.

The restriction $d(\lambda) \le n$ that appears in formula (3.1) will play a crucial role in what follows. It means that the nth controlling measure is completely determined by the restriction of M to the set of diagrams λ contained in the Γ-like shape formed by the boxes (ij) with $\min(i, j) \le n$. In particular, for calculating σ_1, it suffices to know the values of M on the hook diagrams.

Let $P_{zz'}$ be the spectral measure of $M_{zz'}$; we shall refer to the family $\{P_{zz'}\}$ as to that of *spectral z-measures*. The nth controlling measure of $P_{zz'}$ will be denoted by $\sigma_n^{(zz')}$.

Proposition 3.3. *The moments of the measure* $\sigma_n^{(zz')}$ *are given by the formula*

$$\begin{aligned}
&\int_{[-1,1]^n} x_1^{l_1} \dots x_n^{l_n} \, \sigma_n^{(zz')}(dx_1 \dots dx_n) \\
&\quad = \sum_{d=1}^{n} \sum_{\substack{p_1>\dots>p_d\ge 0,\ q_1>\dots>q_d\ge 0 \\ |p|+|q|+d=|l|+n}} \chi^{(p_1,\dots,p_d \mid q_1,\dots,q_d)}_{(l_1+1,\dots,l_n+1)} \\
&\quad\quad \times \frac{t^d}{(t)_{|p|+|q|+d}} \prod_{i=1}^{d} \frac{(z+1)_{p_i}(z'+1)_{p_i}(-z+1)_{q_i}(-z'+1)_{q_i}}{p_i!\, q_i!} \\
&\quad\quad \times \prod_{i,j=1}^{d} (p_i+q_j+1)^{-1} \prod_{1\le i,j\le d} (p_i-p_j)(q_i-q_j)
\end{aligned} \tag{3.2}$$

where $l_1, \dots, l_n = 0, 1, 2, \dots$ *and*

$$|p| = p_1 + \cdots + p_d, \quad |q| = q_1 + \cdots + q_d, \quad |l| = l_1 + \cdots + l_n.$$

Proof. This follows from Proposition 3.2 and Proposition 2.6. □

We close the section by indicating first applications of the controlling measures; we shall see that σ_1 and σ_2 control the location of the spectral measure P with respect to the faces (1.5)–(1.6) of the Thoma simplex.

Proposition 3.4. *A spectral measure P is concentrated on the face Ω_0 (see (1.6)) if and only if the first controlling measure σ_1 has no atom at* 0.

Proof. Let us regard the parameter γ as a function $\gamma(\omega)$ on Ω. This function is lower semicontinuous, hence a Borel function. By definition of σ_1 and of $\nu^{(\omega)}$,

$$\sigma_1(\{0\}) = \int_\Omega \gamma(\omega) P(d\omega).$$

Since $\gamma(\omega)$ is nonnegative, this expression is equal to 0 if and only if $\gamma(\omega) = 0$ almost surely with respect to (Ω, P), and the latter happens if and only if P is concentrated on the face Ω_0. □

Proposition 3.5. *Let P be a spectral measure and let σ_1, σ_2 be its first two controlling measures. Consider the subset*

$$\Delta = \{(x, x) \mid x \neq 0\} \subset [-1, 1]^2$$

and identify it with the punctured interval $I = [-1, 1] \setminus \{0\}$.
We always have

$$\sigma_2\big|_\Delta(dx) \ \geq \ |x| \cdot \sigma_1\big|_I(dx),$$

(the former measure majorates the latter), and both measures coincide if and only if, for P-almost all points $\omega \in \Omega$, there are no repetitions of the type $\alpha_i = \alpha_{i+1} \neq 0$ or $\beta_i = \beta_{i+1} \neq 0$.

Note that if such repetitions occur with a nonzero probability, then at least one of the faces $\Omega_{\pm i}$, $i \geq 1$, is not a P-negligible set.

Proof. For any $\omega \in \Omega$ we have

$$\begin{aligned}(\nu^{(\omega)} \times \nu^{(\omega)})\big|_\Delta &= \sum_{\substack{i,j \\ \alpha_i = \alpha_j > 0}} \alpha_i \alpha_j \, \delta(\alpha_i)\delta(\alpha_j) + \sum_{\substack{i,j \\ \beta_i = \beta_j > 0}} \beta_i \beta_j \, \delta(-\beta_i)\delta(-\beta_j) \\ &\geq \sum_{i:\, \alpha_i > 0} \alpha_i^2 \, \delta(\alpha_i) + \sum_{i:\, \beta_i > 0} \beta_i^2 \, \delta(-\beta_i) = |\cdot| \, \nu^{(\omega)}\big|_I,\end{aligned}$$

and equality holds if and only if there are no couples $i \neq j$ such that $\alpha_i = \alpha_j > 0$ or $\beta_i = \beta_j > 0$.

Integrating over ω with respect to P gives the desired claim. □

4 Point processes

We start with generalities about point processes and their correlation functions; our main reference here is the book [DVJ].

Let X be a standard Borel space equipped with a "bornology." The latter means that we know what subsets of X are "bounded." We assume that the family of the bounded subsets is closed under finite union and passage to a subset, and the whole space X can be represented as countable union of bounded subsets. Bounded Borel subsets will be called *test subsets*.

A *configuration* in X is a finite or countable system of points $\xi = (x_1, x_2, \dots)$ in I such that its intersection with any test set is finite. The word "system" is employed to emphasize that ξ is neither a subset of X (since repetitions are permitted) nor a sequence (since the points are not ordered); strictly speaking, ξ is a *multiset*. However, it is occasionally convenient to regard ξ as the image of a sequence of points.

The space of configurations in X will be denoted as Ξ. For a configuration $\xi \in \Xi$ and a test set A, let $|\xi \cap A|$ denote the number of points of ξ (counted with their multiplicities) that occur in A. By the assumption, this number is always finite. We equip Ξ with the Borel structure generated by the functions of the form $\xi \mapsto |\xi \cap A|$.

A *point process* in X is a (Borel) measurable map from a probability space (the *state space*) to Ξ. Equivalently, a point process defines a *random configuration* in X. A point process is called *simple* if the random configuration has no multiple points almost surely.

Given a test subset A, let N_A be the number of points in A for the random configuration (if the process is not simple, points are counted with their multiplicities); this is a random variable, which is defined on the state space and takes values in $\{0, 1, \dots\}$.

Assume that for any A, all the moments of N_A are finite. Then we assign to our point process a sequence $\rho_1, \rho_2, \dots$ of measures. The nth measure ρ_n lives on the n-fold direct product $X^n = X \times \cdots \times X$. In terms of the random configuration $\xi = (x_1, x_2, \dots)$, ρ_n is defined as the expectation

$$\rho_n = \mathbb{E}\Big\{ \sum_{i_1,\dots,i_n} \delta(x_{i_1}) \times \cdots \times \delta(x_{i_n}) \Big\},$$

where summation is taken over all n-tuples of *pairwise distinct* indices. Note that for any test subset A,

$$\rho_n(A^n) = \mathbb{E}\{N_A(N_A - 1)\dots(N_A - n + 1)\},$$

the nth *factorial* moment of N_A. The measure ρ_n is called the nth *correlation measure* or the nth *factorial moment measure*.

Clearly, the correlation measures are symmetric and take finite values on products of test subsets. The first correlation measure ρ_1 is also called the *density measure*. The value of ρ_1 on a test set A is the mean number of points (counted with multiplicities) occurring in A.

Assume X is equipped with a "reference" measure dx. When the process is simple and the measure ρ_n has a density $\rho_n(x_1, \dots, x_n)$ with respect to the measure $dx_1 \dots dx_n$ on X^n, this density is called the nth *correlation function*. Informally, $\rho_n(x_1, \dots, x_n)$ is equal to the probability that the random configuration intersects each of infinitesimal volumes $dx_1, \dots, dx_n$ around $x_1, \dots, x_n$, divided by $dx_1 \dots dx_n$. When X is a domain in an Euclidean space and dx is Lebesgue measure it is convenient to regard the measure ρ_n as a distribution and call the latter the correlation function, even if we do not know *a priori* that the measure is absolutely continuous with respect to $dx_1 \dots dx_n$.

From now on, we take as X, the punctured interval $I = [-1, 1] \setminus \{0\}$. This is a locally compact space in the natural topology, the point 0 playing the role of infinity. A subset of I will be called "bounded" if it is relatively compact in I, i.e., does not intersect a sufficiently small interval $(-\varepsilon, \varepsilon)$. Thus, for a configuration in I, the only possible accumulation point in $\mathbb{R}$ is 0.

Define a map $\Omega \to \Xi$ as

$$\omega \mapsto \xi = (\alpha_1, \alpha_2, \dots, -\beta_1, -\beta_2, \dots),$$

where all the α_i and β_i are assumed to be nonzero. In particular, the point $\omega = (\alpha, \beta, \gamma) = (0, 0, 1)$ is represented by the empty configuration. One can verify that $\omega \mapsto \xi$ is a Borel map. Thus, any probability measure P on Ω defines a point process in I with state space (Ω, Ξ); we shall denote this process by $\mathcal{P}$.

Propostion 4.1. *Any point process $\mathcal{P}$ determined by a probability measure P on Ω has the following special property: all the random variables of type N_A are bounded.*

Proof. Take $\varepsilon > 0$ such that A is contained in $[-1, -\varepsilon] \cup [\varepsilon, 1]$. Then, for any configuration ξ originating from a point $\omega \in \Omega$,

$$\sum_{x \in \xi \cap A} |x| \geq \varepsilon \, |\xi \cap A|.$$

Since the left-hand side does not exceed 1, we conclude that $|\xi \cap A| \leq \varepsilon^{-1}$. □

Consequently, $\mathcal{P}$ possesses correlation measures.

Proposition 4.2. *Consider a point process in I possessing correlation measures. If the diagonal in I^2 is a null set with respect to ρ_2 then the process is simple. Conversely, if the process is simple then, for any $n \geq 2$ and any couple $i \neq j$ of indices, the set*

$$\{(x_1, \dots, x_n) \in I^n \mid x_i = x_j\}$$

is a null set with respect to ρ_n.

Proof. This is verified by the same argument as in Proposition 3.5. Actually, the claim holds for general spaces X. □

Proposition 4.3. *Let P be a probability measure on Ω, σ_n its controlling measures, $\mathcal{P}$ the point process defined by P, and ρ_n the correlation measures. Set*

$$(I^n)' = \{(x_1, \dots, x_n) \in I^n \mid x_i \neq x_j \quad \text{for } i \neq j\}.$$

On $(I^n)'$, we have

$$\rho_n = |x_1 \dots x_n|^{-1} \sigma_n.$$

Proof. This is a direct consequence of the definitions of the measures σ_n and ρ_n. □

Thus, if $\mathcal{P}$ is simple (which can be tested with the help of Proposition 3.5), then the correlation measures are expressed in a simple way through the controlling measures. Indeed, on the subset $(I^n)' \subset I^n$ we can use the formula of Proposition 4.3; the complementary subset $I^n \setminus (I^n)'$ is negligible by virtue of Proposition 4.2.

Let $\Pi(n, r)$ be the set of partitions of the *set* $\{1, \dots, n\}$ consisting of r nonempty blocs, $1 \leq r \leq n$. An element $\pi \in \Pi(n, r)$ can also be regarded as an equivalence relation on $\{1, \dots, n\}$ with r equivalence classes. We assign to π a subset $I^n_\pi \subset I^n$ obtained by intersecting I^n with all hyperplanes of the form $x_i = x_j$ where $i \equiv j \mod \pi$. Thus, I^n_π is a "diagonal section" of I^n. Choosing a representative i in each equivalence class mod π we get natural coordinates $\{x_i\}$ in the section I^n_π by means of which we can identify I^n_π with I^r. Then we carry over the correlation measure ρ_r from I^r to $I^n_\pi \subset I^n$ and denote the resulting measure on I^n by $\rho_{r,\pi}$.

Proposition 4.4. *If $\mathcal{P}$ is simple, then*

$$\sigma_n \big|_{I^n} = |x_1 \dots x_n| \sum_{r=1}^{n} \sum_{\pi \in \Pi(n,r)} \rho_{r,\pi} .$$

Example. Let $n = 3$. There are 5 partitions π of the set $\{1, 2, 3\}$, and we have the following equality on I^3 (below $dx = dx_1 dx_2 dx_3$):

$$\begin{aligned}\sigma_3(x_1, x_2, x_3)dx = |x_1 x_2 x_3| \{ &\rho_3(x_1, x_2, x_3)dx + \rho_2(x_1, x_2)\delta(x_2 - x_3)dx \\ &+\rho_2(x_2, x_3)\delta(x_3 - x_1)dx + \rho_2(x_3, x_1)\delta(x_1 - x_2)dx \\ &+\rho_1(x_1)\delta(x_1 - x_2)\delta(x_1 - x_3)dx\}.\end{aligned}$$

Proof. Assume first that P is the delta measure concentrated at a point $\omega = (\alpha, \beta) \in \Omega$ such that both α and β have no repetitions. Let $\xi = (-\beta_1 < -\beta_2 < \dots < \alpha_2 < \alpha_1)$ be the corresponding subset of I. By the definition of the controlling measures and the correlation measures,

$$\sigma_n \big|_{I^n} = \sum_{x_1, \dots, x_n \in \xi} |x_1 \dots x_n| \delta(x_1) \otimes \dots \otimes \delta(x_n),$$

$$\rho_r = \sum_{\substack{y_1, \dots, y_r \in \xi \\ y_i \neq y_j}} \delta(y_1) \otimes \dots \otimes \delta(y_r).$$

To each n-tuple $(x_1, \dots, x_n)$ one can assign a partition π of the set $\{1, \dots, n\}$ as follows: two distinct indices i, j are in the same bloc of π if and only if $x_i = x_j$. Let r be the number of blocs of π. Then, in the sum over $(x_1, \dots, x_n)$, we group together summands corresponding to the same bloc π. To pass from $(x_1, \dots, x_n)$ to $(y_1, \dots, y_r)$ we order the blocs in an arbitrary way and assign to the ith bloc π_i the ith coordinate y_i (in other words, $y_i = x_j$ for any index $j \in \pi_i$). Then we get the desired relation.

It is worth noting that in the above reasoning, an enumeration of blocs was used for convenience only; actually, the coordinates on the section I_π^n correspond just to blocs of π.

Thus, we have verified the claim of the proposition in the particular case when P is a delta measure. In the general case, it remains to average over ω with respect to P. □

From Proposition 3.5 or Proposition 4.4 it is clear that for $n \geq 2$, the controlling measure σ_n can never be absolutely continuous relative to Lebesgue measure on the cube $[-1, 1]^n$ because the diagonal sections $x_i = x_j$ of the cube always have a nonzero mass. On the contrary, the correlation measures ρ_n can be absolutely continuous, at least, in the interior of I^n. In [P.II], it is shown that the correlation measures for $P = P_{zz'}$ have analytic densities.

Remark 4.5. Even if a process of the form $\mathcal{P}$ is not simple, its correlation measures can be expressed through the controlling measures. Conversely, each controlling measure σ_n can be expressed through the correlation measures ρ_i, $i = 1, \dots, n$. It follows, in particular, that *the initial measure P is uniquely determined by the correlation measures of $\mathcal{P}$*. Another proof of this fact can be obtained from a general result of the theory of point processes and Proposition 4.1.

Recall (see Section 1) that the Thoma simplex possesses a natural symmetry $\Omega \to \Omega$,

$$\omega = (\alpha, \beta) \mapsto \omega^t = (\beta, \alpha). \tag{4.1}$$

Let $\mathcal{P}_{zz'}$ be the point process defined by the spectral measure $P_{zz'}$ and let $\rho_n^{(zz')}$ be its correlation functions.

Proposition 4.6. *The symmetry map* (4.1) *takes $P_{zz'}$ to $P_{-z,-z'}$. Likewise, the symmetry $x \mapsto -x$ of I takes the process $\mathcal{P}_{zz'}$ to the process $\mathcal{P}_{-z,-z'}$. In particular, we have*

$$\sigma_n^{(-z,-z')}(x) = \sigma_n^{(zz')}(-x),$$
$$\rho_n^{(-z,-z')}(x) = \rho_n^{(zz')}(-x).$$

Proof. This follows from (1.8), (2.2) and the definition of the measures σ_n and ρ_n. □

5 The density function

The aim of this section is to calculate the first correlation measure (or the density measure) $\rho_1^{(zz')}$ of $\mathcal{P}_{zz'}$, the point process corresponding to $P_{zz'}$. It will be shown that $\rho_1^{(zz')}$ is absolutely continuous with respect to Lebesgue measure on I. Hence, one can speak about the first correlation *function* $\rho_1^{(zz')}(x)$ which is also called the density function.

By Proposition 4.3, we have

$$\rho_1^{(zz')} = |x|^{-1}\sigma_1^{(zz')}\big|_I ,$$

so that it suffices to calculate the measure $\sigma_1^{(zz')}$ on $[-1, 1]$.

Lemma 5.1. *The measure $\sigma_1^{(zz')}$ is the unique solution of the moment problem*

$$\int_{-1}^{1} x^l \sigma_1^{(zz')}(dx) = \sum_{\substack{p,q\ge 0\\ p+q=l}} \frac{(-1)^q\, t\,(z+1)_p(-z+1)_q(z'+1)_p(-z'+1)_q}{(t)_{p+q+1}\,(p+q+1)\,p!\,q!}, \tag{5.1}$$

where $l = 0, 1, 2, \dots$.

Proof. We apply Proposition 3.3. Since $n = 1$, the parameter d takes the only value of 1, and the summation is taken over the hook diagrams $\lambda = (p\,|\,q)$. It follows from the Murnaghan–Nakayama rule [M, §I.7, Ex. 5] that

$$\chi^{(p\,|\,q)}_{(p+q+1)} = (-1)^q.$$

This yields the desired formula. As was already mentioned, uniqueness holds because the support of the measure in question is bounded. □

We shall deal with the distributions

$$\phi_a(u) = \frac{u_+^a}{\Gamma(a+1)}, \qquad u \in \mathbb{R}, \quad a \in \mathbb{C},$$

concentrated on the right semiaxis $\mathbb{R}_+$. Here the numerator u_+^a coincides with the function u^a when $u > 0$ and vanishes when $u < 0$. When $\Re a > -1$, ϕ_a is a locally integrable function, and when $\Re a \le -1$, it is defined via analytic continuation.[4] For any $a \in \mathbb{C}$, the only singularity of the distribution ϕ_a may be at 0.

The product

$$\phi_{ab}(u) = \phi_a(u)\,\phi_b(1-u), \qquad a, b \in \mathbb{C}$$

[4]Here and in what follows, to simplify the notation, we shall write distributions as if they were ordinary functions.

is correctly defined because the possible singularities of the factors are at different points ($u=0$ and $u=1$, respectively). The result is a distribution concentrated on $[-1,1]$.

The following formula holds:

$$\int u^p(1-u)^q\phi_{ab}(u)=\frac{(a+1)_p(b+1)_q}{\Gamma(a+b+p+q+2)},\qquad p,q=0,1,2,\dots \quad (5.2)$$

Indeed, when $\Re a>-p-1$ and $\Re b>-q-1$, this formula is equivalent to the classical Euler beta integral formula, and for arbitrary $a,b\in\mathbb{C}$ the result holds by analytic continuation.

Set

$$\Phi(a+1;\zeta)=\int e^{\zeta u}\phi_{a,-a}(u)du,\qquad \zeta\in\mathbb{C},$$

i.e., $\Phi(a+1;\,\cdot)$ is the Laplace transform of $\phi_{a,-a}$. Expanding the exponential function and using the beta integral (5.2) with $q=0$, one sees that this is a special case of Kummer's hypergeometric function ${}_1F_1$ (also called the confluent hypergeometric function):

$$\Phi(a+1;\zeta)=\sum_{k=0}^{\infty}\frac{(a+1)_k}{(k+1)!\,k!}\zeta^k={}_1F_1(a+1;2;\zeta),$$

see [E, ch. 6]. Clearly, $\Phi(a+1;\zeta)$ is an entire function in ζ.

Theorem 5.2. *The density measure satisfies the following equation*

$$\begin{aligned}\int_{-1}^{1}\Phi(t+1;\zeta x)\,\sigma_1^{(zz')}(dx)&=\Phi(z+1;\zeta)\,\Phi(-z'+1;-\zeta)\\&=\Phi(z'+1;\zeta)\,\Phi(-z+1;-\zeta),\qquad \zeta\in\mathbb{C}.\quad(5.3)\end{aligned}$$

Proof. We rewrite formula (5.1) as

$$\begin{aligned}&\frac{\zeta^l(t+1)_l}{(l+1)l!}\int_{-1}^{1}x^l\sigma_1^{(zz')}(dx)\\&\quad=\sum_{\substack{p,q\ge0\\p+q=l}}\frac{(z+1)_p(-z+1)_q}{(p+q+1)!}\,\frac{(z'+1)_p(-z'+1)_q}{(p+q+1)!}\,\frac{\zeta^p(-\zeta)^q}{p!\,q!}.\end{aligned}$$

Next, we replace the first and the second ratios on the right by the corresponding beta integrals (5.2) and sum over l. The result will be as follows:

$$\begin{aligned}&\sum_{l\ge0}\int_0^1\frac{(\zeta x)^l(t+1)_l}{(l+1)!\,l!}\sigma_1^{(zz')}(dx)\\&\quad=\sum_{p,q\ge0}\iint\phi_{z,-z}(u)\phi_{z',-z'}(v)\frac{(\zeta uv)^p}{p!}\frac{(-\zeta(1-u)(1-v))^q}{q!}\,dudv.\end{aligned}$$

Interchanging summation and integration, we get

$$\int_0^1 \Phi(t+1;\zeta x)\sigma_1^{(zz')}(dx) = \iint e^{\zeta(uv-(1-u)(1-v))}\phi_{z,-z}(u)\phi_{z',-z'}(v)dudv.$$

Since

$$uv-(1-u)(1-v) = u+v-1 = u-(1-v) = v-(1-u),$$

the double integral on the right factorizes into a product of two one-dimensional integrals. Making a change of a variable, $(1-v)\mapsto v$ or $(1-u)\mapsto u$, and using the definiton of the function Φ, we obtain the first and the second variants of formula (5.3), respectively.

Note that the equivalence of both variants of formula (5.3) also follows from the identity

$$\Phi(a+1;\zeta) = e^{\zeta}\Phi(-a+1;\zeta),$$

which is a particular case of Kummer's transform for ${}_1F_1$, see [E, §6.3, (21)]. □

Remark 5.3. The idea of using the beta integral is due to Borodin. My initial proof of Theorem 5.2 was more complicated: I transformed the right-hand side of (5.1) to the form

$$\frac{l!\,(l+1)!\,t}{(t)_{l+1}} \sum_{p+q=l} \frac{(-z+1)_q(z'+1)_p}{p!\,(p+1)!\,q!\,(q+1)}$$

with the help of a Leibniz-type formula for the difference operator $f(z)\mapsto f(z)-f(z-1)$.

Set

$$\phi_{ab}^{(-)}(u) = \phi_{ab}(-u);$$

this is a distribution concentrated on $[-1,0]$.

Corollary 5.4. *If $t=1$, then the measure $\sigma_1^{(zz')}$ is the convolution product of two distributions, concentrated on* $[0,1]$ *and* $[-1,0]$:

$$\sigma_1^{(zz')} = \phi_{z,-z} * \phi_{-z',z'}^{(-)}.$$

Proof. When $t=1$, the function $\Phi(t+1;\zeta)$ degenerates to the exponential e^{ζ}. It follows that the left-hand side of (5.3) reduces to the Laplace transform of the measure $\sigma_1^{(zz')}$. On the other hand, the right-hand side of (5.3) is the product of the Laplace transforms of the distributions $\phi_{z,-z}$ and $\phi_{-z',z'}^{(-)}$. □

In the general case, to extract from the equation (5.3) an expresssion for $\sigma_1^{(zz')}$ we need a little formalism which will also be employed in §6.

Define an operation $\odot$ on distributions by the rule

$$(A\odot B)(x) = \iint \delta(x-y_1y_2)A(y_1)B(y_2)dy_1dy_2.$$

Or, in terms of a test function ψ,

$$\int (A \odot B)(x)\psi(x)dx = \int\int A(y_1)B(y_2)\psi(y_1 y_2)dy_1 dy_2.$$

The operation $\odot$ may be viewed as the convolution product on the semigroup $(\mathbb{R}, \cdot)$ of real numbers under multiplication; for this reason we shall call it *pseudoconvolution*.

Of course, in order that the pseudoconvolution $A \odot B$ be correctly defined, A and B must satisfy appropriate conditions. For example, it suffices that they be compactly supported; then $A \odot B$ will be compactly supported too; moreover, $(A \odot B) \subseteq \operatorname{supp} A \cdot \operatorname{supp} B$. We shall employ this operation for distributions concentrated on $[-1, 1]$; since $[-1, 1]$ is a subsemigroup of $(\mathbb{R}, \cdot)$, the result will always be a distribution of the same kind.

Remark 5.5. (i) Assume that A and B are integrable functions or (complex) measures of finite variance with no atom at 0. Then the point 0 may be neglected, and the pseudoconvolution reduces, in essence, to the conventional convolution on the multiplicative group $\mathbb{R}^* = \mathbb{R} \setminus \{0\}$.

(ii) The same is also true if both A and B are distributions whose supports do not contain 0. However, in the general case, the point 0 can cause complications.

(iii) Here is an illustrative example of what can happen in the extreme case when both A and B are supported at 0: Denoting by δ_0 the delta function and by $\delta_0^{(m)}$ its derivative of order m, we have

$$\delta_0^{(m)} \odot \delta_0^{(n)} = \begin{cases} m!\,\delta_0^{(m)}, & m = n \\ 0, & m \neq n. \end{cases}$$

Lemma 5.6. *Formula* (5.3) *of Theorem 5.2 is equivalent to*

$$\sigma_1^{(zz')} \odot \phi_{t,-t} = \phi_{z,-z} * \phi^{(-)}_{-z',z,}. \tag{5.4}$$

Proof. We claim that (5.3) coincides with the Laplace transform of the latter formula. Indeed, for the right-hand side this was already noted in the proof of Corollary 5.3, and for the left-hand side, this is readily verified by substituting the integral reprsentation of the function $\Phi(t+1; \cdot)$ to the left-hand side of (5.3). □

Lemma 5.7. *Denoting by δ_1 the Dirac mass at the point 1, we have*

$$\phi_{t,-t} \odot \phi_{1,t-2} = \frac{1}{\Gamma(t+1)}\delta_1.$$

Proof. Let us check that the nth moment of left-hand side (where $n = 0, 1, 2, \dots$) is equal to $(\Gamma(t+1))^{-1}$. By the definition of $A \odot B$, the nth moment of $A \odot B$ is equal to the product of the nth moments of A and B. Next, observe that

$$\int x^n \phi_{ab}(x)dx = \frac{(a+1)_n}{\Gamma(a+b+n+2)}.$$

Therefore, the nth moment in question is equal to

$$\frac{(t+1)_n}{\Gamma(n+2)}\,\frac{(2)_n}{\Gamma(t+n+1)}=\frac{1}{\Gamma(t+1)},$$

as was to be shown. □

Theorem 5.8. *We have*

$$\sigma_1^{(zz')}=\Gamma(t+1)\,(\phi_{z,-z}*\phi^{(-)}_{-z',z'})\odot\phi_{1,t-2}, \tag{5.5}$$

so that

$$\rho_1^{(zz')}(x)=\frac{\Gamma(t+1)}{|x|}\,\{(\phi_{z,-z}*\phi^{(-)}_{-z',z'})\odot\phi_{1,t-2}\},\qquad x\neq 0. \tag{5.6}$$

Proof. Take the pseudoconvolution of both sides of formula (5.4) of Lemma 5.6 with $\phi_{1,t-2}$ and then apply Lemma 5.7. □

The expression of Theorem 5.8 is a two-dimensional integral representation of the measure $\sigma_1^{(zz')}$. It can also be derived from the moment formula (5.1) by the same method as that employed in the proof of Theorem 5.2.

Remark 5.9. There is a somewhat different formula for $\sigma_1^{(zz')}$: set $D=x\frac{d}{dx}$; then

$$\sigma_1^{(zz')}(x)=\Gamma(t+1)\,(-D)\,[(\phi_{z,-z}*\phi^{(-)}_{-z',z,})\odot\phi_{0,t-1}].$$

Indeed, by virtue of Proposition 6.8 (see below), the right-hand side is equal to

$$\Gamma(t+1)\,(\phi_{z,-z}*\phi^{(-)}_{-z',z'})\odot(-D\phi_{0,t-1}).$$

It is readily verified that

$$-D\,\phi_{0,t-1}=\phi_{1,t-2}.$$

So the formula in question is equivalent to (5.5).

Theorem 5.10. *The restriction of the density measure $\rho_1^{(zz')}$ to $(-1,0)\cup(0,1)$ is absolutely continuous with respect to Lebesgue measure dx:*

$$\rho_1^{(zz')}(dx)=\rho_1^{(zz')}(x)dx,$$

and the density function $\rho_1^{(zz')}(x)$ is real analytic on $(0,1)$ and on $(-1,0)$.

When $x\in(0,1)$, the density function can be written in the following form

$$\rho_1^{(zz')}(x)=\frac{\Gamma(t+1)}{\Gamma(z+1)\Gamma(z'+1)}\,(1-x)^{-z-z'+t}\,\langle A,\Psi\rangle,\qquad x\in(0,1), \tag{5.7}$$

where A is a two-dimensional distribution concentrated on the triangle $u, v \geq 0$, $u + v \leq 1$, and Ψ is a test function, which is correctly defined and smooth in a neighborhood of that triangle:

$$A = A(u, v) = \phi_{-z}(u)\phi_{-z'}(v)\phi_{t-2}(1 - u - v),$$
$$\Psi = \Psi(u, v) = (1 - (1 - x)u)^{z}(1 - (1 - x)v)^{z'}(1 - (1 - x)(u + v))^{-t}.$$

When $x \in (-1, 0)$, the density function can be written in the same form, with the only modification: x is replaced by $|x|$ and the parameters z, z' are replaced by $-z, -z'$.

Proof. We start with a formal transformation of the expression (5.6) given by Theorem 5.8.

Assume that $x \in (0, 1)$. By (5.6) and the definition of $\odot$,

$$\sigma_1^{(zz')}(x) = \Gamma(t+1) \iiint \delta(x - w(u+v-1))\phi_{z,-z}(u)\phi_{z',-z'}(v)\phi_{1,t-2}(w)dudvdw,$$

where $\delta(\,\cdot\,)$ is the delta function. We can exclude the integration over w using the formula

$$\begin{aligned}
\int \frac{1}{x}\delta(x - w(u + v - 1))\phi_{1,t-2}(w)dw
&= \int \frac{1}{x(u + v - 1)}\delta\Big(w - \frac{x}{u + v - 1}\Big)\phi_{1,t-2}(w)dw \\
&= \frac{1}{x(u + v - 1)}\phi_{1,t-2}\Big(\frac{x}{u + v - 1}\Big) \\
&= \frac{1}{(u + v - 1)^{t}}\phi_{t-2}(u + v - 1 - x).
\end{aligned}$$

This gives

$$\begin{aligned}
\rho_1^{(zz')}(x) = \Gamma(t + 1) \iint & \phi_z(u)\phi_{-z}(1 - u)\phi_{z'}(v)\phi_{-z'}(1 - v) \\
& \times \phi_{t-2}(u + v - 1 - x)\frac{dudv}{(u + v - 1)^{t}}.
\end{aligned}$$

Make a change of variables,

$$1 - u = (1 - x)u_1, \quad 1 - v = (1 - x)v_1.$$

Then

$$\begin{aligned}
dudv &= (1 - x)^2 du_1 dv_1, \\
u &= 1 - (1 - x)u_1, \quad v = 1 - (1 - x)v_1, \\
u + v - 1 &= 1 - (1 - x)(u_1 + v_1), \quad u + v - 1 - x = (1 - x)(1 - u_1 - v_1).
\end{aligned}$$

Substituting these expressions into the integral and then renaming the variables u_1, v_1 to u, v, we come to formula (5.7). In the case $x \in (-1, 0)$ one can use exactly the same argument; an alternative possibility is to use the symmetry property indicated in Lemma 4.6.

To justify these formal transformations from (5.6) to (5.7), let us, for a moment, interpret $z, -z, z', -z'$, and $t-2$ as five independent complex parameters. If the real parts of these variables are strictly positive, then all the distributions become ordinary continuous functions, and our transformations are readily justified. On the other hand, the expressions (5.6) (with $x > 0$) and (5.7) are both correctly defined distributions in x, which depend holomorphically on the parameters. Hence, by the principle of analytic continuation, they are equivalent.

Finally, the expresion (5.7) is real analytic, because the test function Ψ is correctly defined for complex values of the parameters x in the strip $0 < \Re x < 1$ and depends on x analytically. □

Remark 5.11. By Lemma 6.10 (see below), the density measure $\rho_1^{(zz')}$ has no atom at 0, and one can show that $\rho_1^{(zz')}$ has no atoms at the endpoints ± 1. So it is completely determined by the density function on the open intervals $(0, 1)$ and $(-1, 0)$.

We shall express the density function through a multivariate hypergeometric series. Let

$$a = (a_1, \dots, a_n), \quad b = (b_1, \dots, b_n), \quad c$$

be complex parameters and

$$y = (y_1, \dots, y_n)$$

be n complex variables. The n-dimensional *Lauricella hypergeometric function* of type B is defined by the series

$$F_B^{[n]}(a, b; c \mid y) = \sum_{m_1, \dots, m_n \geq 0} \frac{(a_1)_{m_1} (b_1)_{m_1} \dots (a_n)_{m_n} (b_n)_{m_n}}{(c)_{m_1 + \dots + m_n}\, m_1! \dots m_n!} y_1^{m_1} \dots y_n^{m_n}$$

where the series is absolutely convergent for $|y_1| < 1, \dots, |y_n| < 1$, see [AK, Ex]. When $n = 1$, this is Gauss' hypergeometric function, and when $n = 2$, this is Appell's hypergeometric function F_3. Note also that the function F_B remains invariant when the couples $(a_1, b_1), \dots, (a_n, b_n)$ are permuted or the parameters in couples are interchanged.

Theorem 5.12. *Let*

$$a = (-z, -z', t-1), \quad b = (-z+1, -z'+1, t),$$
$$c = -z - z' + t + 1 = (-z+1)(-z'+1).$$

For $0 < x \leq 1$, *we have*

$$\rho_1^{(zz')}(x) = \frac{\Gamma(t+1) x^{-t}}{\Gamma(z+1)\Gamma(z'+1)} \phi_{c-1}(1-x)\, F_B^{[3]}\left(a, b; c \,\middle|\, 1-x, 1-x, \frac{x-1}{x}\right). \tag{5.8}$$

The same expression holds for $-1 \le x < 0$ *provided that* x *is replaced by* $|x|$ *and* z, z' *are multiplied by* -1.

Proof. First, note that the parameter c, as defined above, is a strictly positive real number (this follows from the fundamental assumptions on the parameters z, z'). It follows that $\phi_{c-1}(1-x)$ is integrable at $x = 1$. Consequently, the whole expression is integrable at $x = 1$, as it should be.

As before, the case of negative x can be reduced to that of positive x by symmetry. Next, by Remark 5.11, we may suppose $0 < x < 1$.

Set

$$\begin{aligned} u_1 &= u, \quad u_2 = v, \quad u_3 = 1-u-v, \\ d_1 &= -z, \quad d_2 = -z', \quad d_3 = t-2, \\ y_1 &= y_2 = 1-x, \; y_3 = \frac{x-1}{x}. \end{aligned}$$

By the binomial expansion,

$$\Psi = x^{-t} \sum_{m_1, m_2, m_3 \ge 0} \frac{(-z)_{m_1}(-z')_{m_2}(t)_{m_3}}{m_1!\, m_2!\, m_3!} (y_1u_1)^{m_1}(y_2u_2)^{m_2}(y_3u_3)^{m_3}.$$

Substituting this into (5.7) and employing the Euler beta integral

$$\begin{aligned} &\int_{\substack{u_1+u_2+u_3=1 \\ u_1,u_2,u_3 \ge 0}} \phi_{d_1}(u_1)\,\phi_{d_2}(u_2)\,\phi_{d_3}(u_3)\, u_1^{m_1}\, u_2^{m_2}\, u_3^{m_3}\, du \\ &\quad = \frac{(d_1+1)_{m_1}(d_2+1)_{m_2}(d_3+1)_{m_3}}{\Gamma(d_1+d_2+d_3+3)\,(d_1+d_2+d_3+3)_{m_1+m_2+m_3}} \end{aligned}$$

we obtain the desired formula. □

6 An application

In this section we shall prove the following result.[5]

Theorem 6.1. *All the spectral measures* $P_{zz'}$ *are concentrated on the face* Ω_0 *defined in* (1.5).

By Proposition 3.4, this is equivalent to the fact that $\sigma_1^{(zz')}$ has no atom at 0. The proof of the latter claim is divided into a series of lemmas.

Let $C_0(\mathbb{R})$ be the space of continuous compactly supported functions on $\mathbb{R}$ and let $C_0^\infty(\mathbb{R})$ be its subspace consisting of smooth compactly supported functions. Let a, b be complex parameters and let x be the coordinate on $\mathbb{R}$. Let

$$\mathcal{F}_a^+ = \phi_a \cdot C_0^\infty(\mathbb{R})$$

[5] I am grateful to Jean-Louis Clerc and Jacques Faraut for discussions related to the proof of Theorem 6.1.

denote the space of distributions formed by the products $\phi_a \cdot f$ with $f \in C_0^\infty(\mathbb{R})$, and let $\mathcal{F}_a^-$ denote the image of $\mathcal{F}_a^+$ under the reflection $x \mapsto -x$.

Lemma 6.2. *We have*

$$\frac{d}{dx}\mathcal{F}_a^\pm \subset \mathcal{F}_{a-1}^\pm + \mathcal{F}_a^\pm.$$

Proof. Indeed, this follows from the well-known formula

$$\frac{d}{dx}\phi_a(x) = \phi_{a-1}(x).$$

□

Lemma 6.3. *We have*

$$\mathcal{F}_a^\pm \subset \frac{d}{dx}\mathcal{F}_{a+1}^\pm + \mathcal{F}_{a+1}^\pm.$$

Proof. Indeed, for any $f \in C_0^\infty(\mathbb{R})$,

$$\phi_a f = (\phi_{a+1} f)' - \phi_{a+1} f' \in (\frac{d}{dx}\mathcal{F}_{a+1}^+ + \mathcal{F}_{a+1}^+).$$

The same argument works for the sign "−." □

Lemma 6.4. *We have*

$$\mathcal{F}_a^+ * \mathcal{F}_b^- \subset (\mathcal{F}_{a-1}^+ * \mathcal{F}_{b+1}^-) + (\mathcal{F}_a^+ * \mathcal{F}_{b+1}^-).$$

Proof. By Lemma 6.3,

$$\begin{aligned}\mathcal{F}_a^+ * \mathcal{F}_b^- \subset \mathcal{F}_a^+ * (\frac{d}{dx}\mathcal{F}_{b+1}^- + \mathcal{F}_{b+1}^-) &= (\mathcal{F}_a^+ * \frac{d}{dx}\mathcal{F}_{b+1}^-) + (\mathcal{F}_a^+ * \mathcal{F}_{b+1}^-)\\ &= (\frac{d}{dx}\mathcal{F}_a^+ * \mathcal{F}_{b+1}^-) + (\mathcal{F}_a^+ * \mathcal{F}_{b+1}^-).\end{aligned}$$

Next, we apply Lemma 6.2. □

Lemma 6.5. *Assume that one of the following conditions holds:*

(i) $\Re a > 0$, $\Re b < 1$;
(ii) $\Re a = \Re b = 0$.

Then $\mathcal{F}_a^+ * \mathcal{F}_{-b}^- \subset C_0(\mathbb{R})$.

Proof. (i) In this case, the elements of $\mathcal{F}_a^+$ are continuous functions with compact support while the elements of $\mathcal{F}_{-b}^-$ are integrable functions with compact support. Therefore, the result of convolution is continuous functions.

(ii) In this case, the elements of both spaces are bounded measurable functions with compact support. So they are square integrable functions. Therefore, the result of convolution are again continuous functions. □

Lemma 6.6. *Assume that the parameters* a, b *satisfy the following condition: there exist integers* $m \ge n \ge 0$ *such that either*

$$m < \Re a < m+1, \qquad n < \Re b < n+1$$

or

$$\Re a = m, \qquad \Re b = n.$$

Then

$$\mathcal{F}_a^+ * \mathcal{F}_{-b}^- \subset C_0(\mathbb{R}).$$

Actually, we need only the case $m = n$, but it will be convenient to check a slightly more general claim with $m \geq n$.

Proof. Assume first that $n \geq 1$. By Lemma 6.4,

$$\mathcal{F}_a^+ * \mathcal{F}_{-b}^- \subset (\mathcal{F}_{a-1}^+ * \mathcal{F}_{-(b-1)}^-) + (\mathcal{F}_a^+ * \mathcal{F}_{-(b-1)}^-),$$

which enables one to reduce the claim of the lemma for a given couple $m \geq n \geq 1$ to the same claims with (m, n) replaced by $(m-1, n-1)$ or by $(m, n-1)$. The reduction stops when $n = 0$, but then we can apply Lemma 6.5. □

Lemma 6.7. *Near* 0*, the distribution*

$$\phi_{z,-z} * \phi_{-z',z'}^{(-)}$$

is given by a continuous function.

Proof. By making use of an appropriate partition of unity one can represent both factors in the form

$$\phi_{z,-z} = A_0 + A + A_1, \qquad \phi_{-z',z'}^{(-)} = B_{-1} + B + B_0,$$

where A, B are smooth functions and A_0, A_1, B_{-1}, B_0 are certain distributions concentrated near the points $0, 1, -1, 0$, respectively (they are obtained by multiplying the initial distributions $\phi_{z,-z}$ and $\phi_{-z',z'}^{(-)}$ by appropriate functions from $C_0^\infty(\mathbb{R})$). It follows that

$$\phi_{z,-z} * \phi_{-z',z'}^{(-)} = (A_0 * B_0) + (A_1 * B_{-1}) + (A_0 * B_{-1}) + (A_1 * B_0) + (\ldots),$$

where $(\ldots)$ is a smooth function. Since the distributions $A_0 * B_{-1}$ and $A_1 * B_0$ are concentrated near the points -1 and 1, respectively, it remains to check that the distributions $A_0 * B_0$ and $A_1 * B_{-1}$ are actually continuous functions.

We have $A_0 \in \mathcal{F}_z^+$ and $B_0 \in \mathcal{F}_{-z'}^-$, so that

$$A_0 * B_0 \in (\mathcal{F}_z^+ * \mathcal{F}_{-z'}^-).$$

Next, the distrubution $A_1 * B_{-1}$ will not change if we shift both factors by 1, to the left and to the right, respectively. The resulting distributions will lie in $\mathcal{F}_{-z}^-$ and $\mathcal{F}_{z'}^+$, respectively, whence

$$A_1 * B_{-1} \in (\mathcal{F}_{z'}^+ * \mathcal{F}_{-z}^-).$$

Now, let us compare our fundamental assumptions on the parameters z, z' with the assumptions on the parameters a, b in Lemma 6.6. Without loss of generality, we may assume that $\Re z \geq 0$ and $\Re z' \geq 0$ (otherwise we may apply Proposition 4.6). Then the couples $(a, b) = (z, z')$ and $(a, b) = (z', z)$ will satisfy the assumptions of Lemma 6.6 with $m = n$. Application of this lemma concludes the proof. □

Lemma 6.8. *Set* $D = x\frac{d}{dx}$. *For any compactly supported distributions* A, B, *we have*

$$(DA) \odot B = D(A \odot B) = A \odot (DB).$$

Note that the claim is obvious when the supports of A and B do not contain 0, because then, replacing x by $s = \log|x|$, we can reduce the pseudoconvolution to the ordinary convolution and the operator D to $\frac{d}{ds}$.

Proof. Set $D' = \frac{d}{dx} \circ x = D + 1$. For an arbitrary smooth test function ψ, we have

$$\begin{aligned}
\langle (DA) \odot B, \psi \rangle &= \int\int DA(y_1)B(y_2)\psi(y_1 y_2)dy_1 dy_2 \\
&= \int B(y_2)dy_2 \int DA(y_1)\psi(y_1 y_2)dy_1 \\
&= \int B(y_2)dy_2 \int A(y_1)(-y_1\frac{d}{dy_1} - 1)\psi(y_1 y_2)dy_1 \\
&= \int B(y_2)dy_2 \int A(y_1)(-D'\psi)(y_1 y_2)dy_1 \\
&= \langle A \odot B, -D'\psi \rangle = \langle D(A \odot B), \psi \rangle.
\end{aligned}$$

This proves the first equality, and the second one is verified similarly. □

Lemma 6.9. *Assume that* A *is a distribution with a compact support not containing 0, and* B *belongs to a class* $\mathcal{F}_a^+$ *with* $\Re a > -1$. *Then the distribution* $A \odot B$ *is a function.*

Proof. Since the support of A is separated from 0, A can be represented in the form $A = f(D)\widetilde{A}$, where f is a polynomial, D is as in Lemma 6.8, and $\widetilde{A}$ is an ordinary (say, integrable) function. By Lemma 6.8,

$$A \odot B = (f(D)\widetilde{A}) \odot B = \widetilde{A} \odot (f(D)B).$$

On the other hand, we have

$$D\phi_a = a\phi_a, \qquad a \in \mathbb{C},$$

which implies

$$D\mathcal{F}_a^+ \subseteq \mathcal{F}_a^+, \qquad a \in \mathbb{C},$$

so that $f(D)B$ lies in $\mathcal{F}_a^+$ together with B. When $\Re a > -1$, the distributions of class $\mathcal{F}_a^+$ are ordinary (integrable) functions. Consequently, the pseudoconvolution of A and $f(D)B$ is, in essence, the ordinary convolution product on the multiplicative group $\mathbb{R}^*$ (see Remark 5.5), and its result is an ordinary function. □

The next lemma concludes the proof of Theorem 6.1.

Lemma 6.10. *The measure $\sigma_1^{(zz')}$ has no atom at* 0.

Proof. We shall use different arguments when $t < 1$, $t = 1$, and $t > 1$. When $t = 1$, the measure $\sigma_1^{(zz')}$ is given by Corollary 5.4, and the claim follows from Lemma 6.7.

Assume that $t < 1$ and examine the expression (5.4). Since $0 < t < 1$, the distribution $\phi_{t,-t}$ is actually a nonnegative integrable function. We can decompose the measure $\sigma_1^{(zz')}$ into the sum of two components: one is const $\cdot\, \delta_0$ (a multiple of the Dirac mass at 0) and another is a measure A with no atom at 0. The measure $A \odot \phi_{t,-t}$ cannot have an atom at 0 (see Remark 5.5). On the other hand,

$$\text{const}\delta_0 \odot \phi_{t,-t} = \text{const}\,\Big(\int \phi_{t,-t}(y)dy\Big)\,\delta_0,$$

where the integral is strictly positive. But the right-hand side of formula (5.4) cannot have an atom at 0, by virtue of Lemma 6.7. Hence, const $= 0$, so that $\sigma_1^{(zz')}$ has no atom at 0, too.

Finally, assume that $t > 1$ and look at formula (5.5). By Lemma 6.7, one can write

$$\Gamma(t+1)\,\phi_{z,-z} * \phi^{(-)}_{-z',z'} = A_0 + A_1,$$

where A_0 is a continuous function and A_1 is a distribution whose support does not contain 0. Next, since $t > 1$, the distribution $\phi_{1,t-2}$ can be written as

$$\phi_{1,t-2} = B_0 + B_1,$$

where $B_0 \in \mathcal{F}_1^+$ and B_1 is an integrable function concentrated near 1. Then we have

$$\begin{aligned}\sigma_1^{(zz')} &= (A_0 + A_1) \odot (B_0 + B_1)\\ &= (A_0 \odot B_0) + (A_1 \odot B_0) + (A_0 \odot B_1) + (A_1 \odot B_1).\end{aligned}$$

Let us examine the four summands of the latter expression. Since A_0, B_0, B_1 are ordinary (integrable) functions, the terms $A_0 \odot B_0$ and $A_0 \odot B_1$ are ordinary functions according to Remark 5.5. The term $A_1 \odot B_0$ is an ordinary function by Lemma 6.9. The term $A_1 \odot B_1$ is a distribution concentrated outside a neighborhood of 0. We conclude that the whole expression cannot have an atom at 0. □

7 Appendix: Correlation functions of the Poisson–Dirichlet processes

In this appendix, we briefly discuss a parallel (but simpler) theory. It is related to the Poisson–Dirichlet distributions, a remarkable one-parameter family of probability measures $\{PD(t) \mid t > 0\}$ which live on an infinite-dimensional subsimplex $\Delta \subset \Omega$. The measures $PD(t)$ determine point processes $\mathcal{PD}(t)$ on $(0, 1]$, called the Poisson–Dirichlet processes, and we shall calculate the correlation functions of $\mathcal{PD}(t)$ by using the general formalism of §§1–4.

The Poisson–Dirichlet distributions were studied in many papers from different points of view, see, e.g., [Ki2, W]. Our interest in them is due to the fact that they are one of the basic elements in the construction [KOV] of the generalized regular representations T_z. So both types of measures, the $PD(t)$'s and the $P_{zz'}$'s, are connected with the same construction — that of the representations T_z. But they appear at different levels of that construction: the former — at the "group" level (as they are responsible for certain quasiinvariant measures on $\mathfrak{S}$, see the introduction), and the later — at the "dual" level (the representation level). However, both $PD(t)$ and $P_{zz'}$ are spectral measures in the sense that they govern the decomposition of certain objects into indecomposable ones,[6] and both $PD(t)$ and $P_{zz'}$ can be interpreted as point processes.

For these reasons, it seems interesting to compare the point processes $\mathcal{PD}(t)$ and $\mathcal{P}_{zz'}$, and the main purpose of the present appendix is to prepare a foundation for such a comparison (see [P.III, §7]). Another purpose is to illustrate the formalism of Sections 1–4 on easier material.

We define the simplex Δ as the closed subset of Ω determined by $\beta_1 = \beta_2 = \cdots = 0$. In other words, Δ is the set of sequences $\alpha = (\alpha_1 \geq \alpha_2 \geq \cdots \geq 0)$ such that $\sum \alpha_i \leq 1$. The specialization (1.3) is replaced by the following:

$$p_1 \mapsto \widetilde{p}_1 = \widetilde{p}_1(\alpha; \gamma) := \sum_{i=1}^{\infty} \alpha_i + \gamma$$

$$p_n \mapsto \widetilde{p}_n = \widetilde{p}_n(\alpha; \gamma) := \sum_{i=1}^{\infty} \alpha_i^n, \qquad n \geq 0,$$

which is equivalent to

$$1 + \sum_{n=1}^{\infty} h_n u^n \mapsto 1 + \sum_{n=1}^{\infty} \widetilde{h}_n u^n := e^{\gamma u} \prod_{i=1}^{\infty} \frac{1}{1 - \alpha_i u}.$$

In what follows, we shall assume that α is a point of Δ and

$$\gamma = 1 - \sum_{i=1}^{\infty} \alpha_i .$$

---

[6]At the "group" level, the "objects" are K-invariant probability measures on $\mathfrak{S}$ and the "indecomposable objects" are ergodic ones, see [Ki1, KOV].

Then all $\widetilde{p}_n$ turn into continuous functions functions on Δ (note that $\widetilde{p}_1 \equiv 1$). Consequently, any element $f \in \Lambda$ is converted into a continuous function $\widetilde{f} = \widetilde{f}(\alpha)$ on Δ.

In place of the Schur functions s_μ we shall deal with the *monomial symmetric functions* $m_\lambda \in \Lambda$. Recall that in the standard realization of Λ as the algebra of symmetric functions in variables $x_1, x_2, \dots$, the function m_λ is the sum of all *distinct* monomials obtained from $x^\lambda = x_1^{\lambda_1} x_2^{\lambda_2} \dots$ by permuting variables, see [M].

The Pieri formula (1.3) for the Schur functions is replaced by its counterpart for the monomial symmetric functions,

$$m_\mu \cdot p_1 = \sum_{\lambda:\lambda \searrow \mu} \kappa_0(\mu, \lambda)\, m_\lambda,$$

where the coefficients $\kappa_0(\mu, \lambda)$ are positive integers defined as follows. Given $\lambda \searrow \mu$, there exists a unique $i \in \{1, 2, \dots\}$ such that $\lambda_i = \mu_i + 1$ (and $\lambda_j = \mu_j$ for all $j \neq i$). Then $\kappa_0(\mu, \lambda)$ is the multiplicity of the part λ_i in the partition λ.

In "exponential notation" for partitions,

$$\mu = (1^{r_1(\mu)} 2^{r_2(\mu)} \dots), \qquad \lambda = (1^{r_1(\lambda)} 2^{r_2(\lambda)} \dots),$$

$\lambda \searrow \mu$ means that there exists a unique k (equal to λ_i) such that

$$r_k(\lambda) = r_k(\mu)+1, \quad r_{k-1}(\lambda) = r_{k-1}(\mu)-1, \quad r_l(\lambda) = r_l(\mu) \quad \text{for } l \neq k,\ k-1,$$

and then

$$\kappa_0(\mu, \lambda) = r_k(\lambda).$$

The role of Young graph $\mathbb{Y}$ is played by the *Kingman graph* $\mathbb{K} = (\mathbb{Y}, \kappa_0)$: its vertices are the same as for $\mathbb{Y}$ (arbitrary partitions) but each edge $\mu \nearrow \lambda$ turns into $\kappa_o(\mu, \lambda)$ edges with the endpoints μ and ν. The grading $\mathbb{K} = \cup \mathbb{K}_n$ of the vertices remains unchanged. That is, the nth level $\mathbb{K}_n$ consists of partitions of n.

A new dimension function $\dim_0$ then arises: $\dim_0 \lambda$ is still defined as the number of oriented paths from $\varnothing$ to λ but we take into account edge multiplicities. The recurrence relation is modified as follows:

$$\dim_0 \lambda = \sum_{\mu:\mu \nearrow \lambda} \dim_0 \mu\, \kappa_0(\mu, \nu).$$

By definition, the numbers $\dim_0(\lambda)$ are nothing but the coefficients in the expansion

$$p_1^n = \sum_{\lambda:\in \mathbb{K}_n} \dim_0 \lambda \cdot m_\lambda.$$

The harmonicity condition is written as follows:

$$\varphi(\mu) = \sum_{\lambda:\lambda \searrow \mu} \kappa_0(\mu, \lambda)\, \varphi(\lambda).$$

A *coherent system of distributions on the graph* $\mathbb{K}$ is a sequence $M = (M_n)$ of probability distributions on the finite sets $\mathbb{K}_n$ (partitions of n), subject to the coherence condition

$$M_n(\mu) = \sum_{\lambda:\,\lambda\searrow\mu} \frac{\dim_0 \mu \cdot \kappa_0(\mu, \lambda)}{\dim_0 \lambda} M_{n+1}(\lambda), \qquad \mu \in \mathbb{K}_n\,,$$

equivalent to harmonicity of the function $M(\lambda)/\dim_0 \lambda$ (here it is convenient to write $M(\mu)$ in place of $M_n(\mu)$). This concept is equivalent to that of the *partition structure* in the sense of Kingman [Ki1].

The next relation establishes a bijective correspondence $M \leftrightarrow P$ between coherent systems M on Kingman's graph and probability measures P on the simplex Δ:

$$\frac{M(\lambda)}{\dim_0 \lambda} = \int_\Delta \widetilde{m}_\lambda(\alpha)\, P(d\alpha) \tag{7.1}$$

(recall that $\widetilde{m}_\lambda$ is the image of $m_\lambda \in \Lambda$ under the specialization $f \mapsto \widetilde{f}$ defined above). This result is *Kingman's theorem* [Ki1, Ke1, KOO].

The coherent *t-systems* $M^{(t)} = (M_n^{(t)})$ of distributions on Kingman's graph are defined as follows: $t > 0$ is a parameter and

$$M_n^{(t)}(\lambda) = \frac{t^{\ell(\lambda)} n!}{(t)_n\, z_\lambda}, \qquad \lambda \in \mathbb{K}_n, \tag{7.2}$$

where $\ell(\lambda)$ is the length of the partition λ (number of nonzero parts) and, in exponential notation,

$$z_\lambda = \prod_i \lambda_i \cdot \prod_k r_k(\lambda)! = \prod_k k^{r_k(\lambda)} r_k(\lambda)!\,. \tag{7.3}$$

The fact that $M^{(t)}$ is indeed a coherent system on $\mathbb{K}$ is verified directly using the above formulas for $\dim_0 \lambda$ and $\kappa_0(\mu, \lambda)$.

The systems $M^{(t)}$ are known as *Ewens partition structures* [Ki1, Ki2]. According to Kingman's theorem, they determine certain probability measures on Δ. The latter are called the *Poisson–Dirichlet distributions*, see [Ki2], and denoted as $PD(t)$. So, the link between $M^{(t)}$ and $PD(t)$ is as follows:

$$M^{(t)}(\lambda) = \int_\Delta \widetilde{m}_\lambda(\alpha)(PD(t))(d\alpha).$$

Given a probability measure P on Δ, we may regard it as a measure on Ω. Hence, the definitions of the *control measures* σ_n, of the *point process* $\mathcal{P}$ attached to P and of the *correlation measures* ρ_n of $\mathcal{P}$ make sense. As in the present situation there are no β's, the measure σ_n is concentrated on the cube $[0, 1]^n$ contained in $[-1, 1]^n$, the state space of the process $\mathcal{P}$ is the semiopen interval $(0, 1]$, and the nth correlation function ρ_n lives on $(0, 1]^n$.

Assume that P is related to M by (7.1). According to (3.1a), the moments of σ_n are given by

$$\int_{[0,1]^n} x_1^{l_1} \dots x_n^{l_n} \, \sigma_n(dx_1 \dots dx_n) = \int_\Delta \widetilde{p}_{l_1+1}(\alpha) \dots \widetilde{p}_{l_n+1}(\alpha) \, P(d\alpha),$$

which implies

$$\int_{[0,1]^n} x_1^{l_1} \dots x_n^{l_n} \, \sigma_n(dx_1 \dots dx_n) = \sum_\lambda [p_{l_1+1} \dots p_{l_n+1} : m_\lambda] \frac{M(\lambda)}{\dim_0 \lambda}. \qquad (7.4)$$

Here λ ranges over $\mathbb{K}_{l_1+\dots l_n+n}$ and, for $f \in \Lambda$, the symbol $[f : m_\lambda]$ denotes the coefficient of m_λ in the expansion of f into linear combination of monomial symmetric functions. It is worth noting that only partitions λ with $\ell(\lambda) \le n$ really contribute to this formula, cf. (3.1b).

Consequently, given M, we have again a collection of moment problems to determine the control measures σ_n, and from the measures σ_n one can get the correlation functions ρ_n. The procedure is exactly the same as for the Young graph. We shall apply it to $M = M^{(t)}$. The corresponding control measures will be denoted as $\sigma_n^{(t)}$ and the correlation measures as $\rho_n^{(t)}$. The point process attached to $PD(t)$ will be called the *Poisson–Dirichlet process* and denoted as $\mathcal{PD}(t)$.

We shall see that in the present situation, calculation of the correlation functions turns out to be much easier than in the case of the Young graph. A formal explanation is that the coefficients in the expansion of $p_{l_1+1} \dots p_{l_n+1}$ on monomial functions m_λ are given by much simpler expressions than when expanding on Schur functions s_λ.

To state the result we need some notation related to set partitions. Recall (see §4) that partitions of a set are the same as equivalence relations. We shall also need *ordered* partitions, i.e., partitions with a fixed enumeration of the blocs. As in Proposition 4.4, we denote the set of partitions of $\{1, \dots, n\}$ with r (nonempty) blocs by $\Pi(n, r)$; the set of ordered partitions with r blocs will be denoted as $\widetilde{\Pi}(n, r)$. There is a natural projection $\widetilde{\Pi}(n, r) \to \Pi(n, r)$ — forgetting enumeration. For each $\pi \in \Pi(n, r)$, there are $r!$ ordered partitions $\widetilde{\pi}$ over π.

Given $\pi \in \Pi(n, r)$, we define a "diagonal section" $[0, 1]^n_\pi$ of the cube $[0, 1]^n$ as the intersection with all hyperplanes of the form $x_i = x_j$ where $i \sim j \mod \pi$. To an ordered partition $\widetilde{\pi} = \widetilde{\pi}_1 \sqcup \dots \sqcup \widetilde{\pi}_r$ over π we assign a bijective map

$$[0, 1]^r \to [0, 1]^n_\pi \subset [0, 1]^n, \quad y = (y_1, \dots, y_r) \mapsto x = (x_1, \dots, x_n),$$
$$\text{where } x_j = y_i \text{ provided that } j \in \widetilde{\pi}_i. \qquad (7.5)$$

Define a measure $\widehat{\sigma}^{(t)}_{r,\widetilde{\pi}}$ on the cube $[0, 1]^r$ as follows:

$$\widehat{\sigma}^{(t)}_{r,\widetilde{\pi}}(dy) = t^r \cdot \prod_{i=1}^r y_i^{|\widetilde{\pi}_i|-1} \cdot (1 - \sum_{i=1}^r y_i)_+^{t-1} \, dy, \qquad (7.6)$$

and let $\sigma^{(t)}_{n,r,\tilde\pi}$ be its image under the above map $y \mapsto x$. The latter measure lives on $[0,1]^n_\pi$ and does not depend on the choice of $\tilde\pi$ over π, whereupon we may denote it as $\sigma^{(t)}_{n,r,\pi}$.

Example. For $n = 3$ the list of partitions and measures is as follows:

- $\Pi(3,1)$:

$$\pi = \{1,2,3\}, \quad \sigma^{(t)}_{3,1,\pi} = t^3(1 - x_1 - x_2 - x_3)_+^{t-1} dx$$

- $\Pi(3,2)$:

$$\pi = \{1\} \sqcup \{2,3\}, \quad \sigma^{(t)}_{3,2,\pi} = t^2 x_2 (1 - x_1 - x_2)_+^{t-1} \delta(x_2 - x_3) dx$$
$$\pi = \{2\} \sqcup \{3,1\}, \quad \sigma^{(t)}_{3,2,\pi} = t^2 x_3 (1 - x_2 - x_3)_+^{t-1} \delta(x_3 - x_1) dx$$
$$\pi = \{3\} \sqcup \{1,2\}, \quad \sigma^{(t)}_{3,2,\pi} = t^2 x_2 (1 - x_2 - x_3)_+^{t-1} \delta(x_1 - x_2) dx$$

- $\Pi(3,3)$:

$$\pi = \{1\} \sqcup \{2\} \sqcup \{3\}, \quad \sigma^{(t)}_{3,3,\pi} = t^3 x_1^2 (1 - x_1)_+^{t-1} \delta(x_1 - x_2)\delta(x_1 - x_3) dx.$$

Theorem 7.1. *In the above notation,*

$$\sigma^{(t)}_n = \sum_{r=1}^{n} \sum_{\pi \in \Pi(n,r)} \sigma^{(t)}_{n,r,\pi}.$$

Proof. We apply the general formula (7.4) to $M = M^{(t)}$. Let us abbreviate

$$|l| = l_1 + \cdots + l_n, \qquad r_k = r_k(\lambda).$$

Using the expressions for $M^{(t)}$ and z_λ given in (7.2) and (7.3), we get

$$\int_{[0,1]^n} x_1^{l_1} \dots x_n^{l_n} \sigma^{(t)}_n(x) dx$$
$$= \sum_{r=1}^{n} \sum_{\substack{\lambda:\, |\lambda| = |l| + n \\ \ell(\lambda) = r}} [p_{l_1+1} \cdots p_{l_n+1} : m_\lambda] \frac{t^r \prod_{i=1}^{r} (\lambda_i - 1)!}{(t)_{|l|+n} \prod_{k \ge 1} r_k!}. \tag{7.7a}$$

Remark that for any $f \in \Lambda$ and any λ with $\ell(\lambda) = r$,

$$[f : m_\lambda] = [f : x_1^{\lambda_1} \dots x_r^{\lambda_r}],$$

where the square brackets on the right denote the coefficient of $x_1^{\lambda_1} \dots x_r^{\lambda_r}$ in the expansion of f as linear combination of monomials. Therefore, we may rewrite (7.7a) as

$$\int_{[0,1]^n} x_1^{l_1} \dots x_n^{l_n} \sigma^{(t)}_n(x) dx$$
$$= \sum_{r=1}^{n} \sum_{\substack{\lambda:\, |\lambda| = |l| + n \\ \ell(\lambda) = r}} [p_{l_1+1} \cdots p_{l_n+1} : x_1^{\lambda_1} \dots x_r^{\lambda_r}] \frac{t^r \prod_{i=1}^{r} (\lambda_i - 1)!}{(t)_{|l|+n} \prod_{k \ge 1} r_k!}. \tag{7.7b}$$

Next, write λ as the n-tuple $\lambda_1 \geq \cdots \geq \lambda_r > 0$ and remark that each summand on the right-hand side of (7.7b) makes sense for any n-tuple of positive integers $\lambda_1, \ldots, \lambda_r$ and is symmetric with respect to their permutations. Moreover, remark that there are exactly $\frac{r!}{r_1!r_2!\ldots}$ distinct permutations of the numbers $\lambda_1, \ldots, \lambda_r$. It follows that we may drop the restriction $\lambda_1 \geq \cdots \geq \lambda_r$ and at the same time replace $\prod r_k!$ by $r!$. Thus, we can transform (7.7b) to

$$\int_{[0,1]^n} x_1^{l_1} \ldots x_n^{l_n} \sigma_n^{(t)}(x)dx = \sum_{r=1}^{n} \sum_{\substack{\lambda_1>0,\ldots,\lambda_r>0 \\ \sum \lambda_i = |l|+n}} [p_{l_1+1} \ldots p_{l_n+1} : x_1^{\lambda_1} \ldots x_r^{\lambda_r}] \frac{t^r \prod_{i=1}^r (\lambda_i - 1)!}{(t)_{|l|+n} r!}. \quad (7.8)$$

Write

$$p_{l_1+1} \ldots p_{l_n+1} = (x_1^{l_1+1} + x_2^{l_1+1} + \ldots) \ldots (x_1^{l_n+1} + x_2^{l_n+1} + \ldots)$$

and remove the parentheses. Then we will get a sum of monomials, each of which corresponds to a choice of a summand from the first, second, ..., nth parentheses. We are interested only in monomials of the form $x_1^{\lambda_1} \ldots x_r^{\lambda_r}$ with strictly positive $\lambda_1, \ldots, \lambda_r$. There is a bijective correspondence between such monomials and ordered partitions $\tilde{\pi} \in \tilde{\Pi}(n, r)$. Specifically, an index j belongs to the ith bloc of π if in the jth parentheses, the ith summand was chosen. This yields a correspondence

$$\tilde{\pi} \mapsto \lambda = \lambda(\tilde{\pi}), \qquad \lambda(\tilde{\pi})_i = \sum_{j \in \tilde{\pi}_i} (l_j + 1) = (\sum_{j \in \tilde{\pi}_i} l_j) + |\tilde{\pi}_i|.$$

Now we shall split the moment problem (7.8) into a collection of moment problems corresponding to various $\tilde{\pi}$. Specifically, assume that for any $r = 1, \ldots, n$ and any $\tilde{\pi} \in \tilde{\Pi}(n, r)$ we dispose of a measure $\sigma_{n,r,\tilde{\pi}}^{(t)}$ on $[0, 1]^n$ which solves the moment problem

$$\int_{[0,1]^n} x_1^{l_1} \ldots x_n^{l_n} \sigma_{n,r,\tilde{\pi}}^{(t)}(x)\, dx = \frac{t^r \prod_{i=1}^r (\lambda(\tilde{\pi})_i - 1)!}{(t)_{|l|+n}}. \quad (7.9)$$

Then the measure

$$\sigma_n^{(t)} := \sum_{r=1}^{n} \frac{1}{r!} \sum_{\tilde{\pi} \in \tilde{\Pi}(n,r)} \sigma_{n,r,\tilde{\pi}}^{(t)} \quad (7.10)$$

will solve the moment problem (7.8).

Looking at (7.9) we remark that the $(l_1, \ldots, l_n)$-moment depends only on the sums

$$m_i := \sum_{j \in \tilde{\pi}_i} l_j, \qquad i = 1, \ldots, r.$$

This indicates that the desired measure should live in the section $[0,1]^n_\pi$, where π stands for the (unordered) partition corresponding to $\widetilde{\pi}$. Then we identify $[0,1]^n_\pi$ with $[0,1]^r$ via the map $y \mapsto x$ defined in (7.5) and rewrite the moment problem (7.9) in terms of the coordinates $y_1, \dots, y_n$:

$$\int_{[0,1]^r} y_1^{m_1} \dots y_r^{m_r} \widehat{\sigma}_r(y)\, dy = \frac{t^r \prod_{i=1}^r (m_i + |\widetilde{\pi}_i| - 1)!}{(t)_{\sum (m_i + |\widetilde{\pi}_i|)}}, \tag{7.11}$$

where $\widehat{\sigma}_r$ stands for the unknown measure on $[0,1]^r$ and we have used the identity

$$|l| + n = \sum_{i=1}^r (m_i + |\widetilde{\pi}_i|).$$

On the other hand, consider the measure $\widehat{\sigma}_{r,\widetilde{\pi}} = \widehat{\sigma}^{(t)}_{r,\widetilde{\pi}}$ as defined in (7.6). This is a Dirichlet measure on a simplex (see [Ki2]) whose moments can be readily calculated (this is a multivariate version of the classical Euler beta integral). One verifies that this measure solves the moment problem (7.11).

Thus, we have shown that the measure (7.10) indeed solves the initial moment problem (7.7). Finally, we remark that the measure $\sigma^{(t)}_{n,r,\widetilde{\pi}}$ in (7.10) actually depends on the image $\pi \in \Pi(n,r)$ of $\widetilde{\pi}$, which concludes the proof. □

Theorem 7.1 yields new proofs of certain well-known properties of the Poisson–Dirichlet distributions $PD(t)$.

Consider the face $\Delta_0 = \Omega_0 \cap \Delta$ of the simplex Δ, i.e.,

$$\Delta_0 = \{(\alpha; \gamma) \mid \gamma = 1 - (\alpha_1 + \alpha_2 + \dots) = 0\}.$$

Corollary 7.2 (cf. [Ki2, 9.4–9.5]). *The Poisson–Dirichlet distribution $PD(t)$ is concentrated on the face Δ_0 of Δ.*

Proof. Applying Theorem 7.1 with $n = 1$, we get

$$\sigma_1^{(t)}(dx) = t(1-x)^{t-1} dx, \qquad 0 \le x \le 1.$$

As this measure has no atom at zero, we conclude, by Proposition 3.4, that $PD(t)$ lives on the face Ω_0 and hence on Δ_0. □

Corollary 7.3. *The Poisson–Dirichlet process $\mathcal{PD}(t)$ is simple (see the definition at the beginning of §4).*

Proof. Apply Theorem 7.1 with $n = 2$. Both $\Pi(2,2)$ and $\Pi(2,1)$ consist of a single element π: specifically, $\pi = \{1\} \sqcup \{2\}$ and $\pi = \{1,2\}$, respectively. So, we have

$$\sigma_2^{(t)} = \sigma^{(t)}_{2,2,\{1\}\sqcup\{2\}} + \sigma^{(t)}_{2,1,\{1,2\}}.$$

The first component on the right is an absolutely continuous measure on the square $[0,1]^2$, while the second one is a singular measure supported by the diagonal $x_1 = x_2 = y$ of $[0,1]^2$. According to Proposition 3.5, we have to examine only the second component. It is equal to

$$tx_1(1-x_1)_+^{t-1} \delta(x_1 - x_2) dx_1 dx_2$$

or, in terms of the coordinate y, to $ty(1-y)_+^{t-1}dy$. The latter expression is just $\sigma_1^{(t)}(dy)$ multiplied by y. From Proposition 3.5 we conclude that the process is simple. □

Note that this fact is evident from the construction of $PD(t)$ via a subordinator, see [Ki2].

Corollary 7.4. *The correlation functions of the Poisson–Dirichlet process* $\mathcal{PD}(t)$ *are given by the formula*

$$\rho_n^{(t)}(x_1,\dots,x_n) = \frac{t^n(1-x_1-\dots-x_n)_+^{t-1}}{x_1\dots x_n} \qquad n=1,2,\dots.$$

This result is due to Watterson [W] but our approach differs from that of [W]. One more proof can be obtained by making use of the fact that the so-called *size-biased sampling* from $PD(t)$ yields a sequence of *independent* random variables [Ki2, 9.6].

Proof. Since the process is simple (Corollary 7.3), all diagonal sections $[0,1]^n_\pi$ with $\pi \neq \{1\}\sqcup\dots\sqcup\{n\}$ are negligible sets with respect to the nth correlation measure $\rho_n^{(t)}$ (Proposition 4.3). Let

$$([0,1]^n)' = [0,1]^n \cap (I^n)' = \{x \in (0,1]^n \mid x_i \neq x_j, \quad i<j\}$$

be the complement to all proper diagonal sections. By Theorem 7.1, all the components $\sigma_{n,r,\pi}^{(t)}$ with $r<n$ are concentrated outside $([0,1]^n)'$, and the only component with $r=n$ is absolutely continuous with respect to dx with density $t^n(1-x_1-\dots x_n)_+^{t-1}$.

According to Proposition 4.3, the nth correlation measure is obtained from the latter measure by dividing it by $|x_1\dots x_n| = x_1\dots x_n$. □

Remark 7.5. Note that the structure of the nth controlling measure, as described by Theorem 7.1, is in perfect accord with the decomposition given in Proposition 4.4.

8 Appendix (A. Borodin): A proof of Theorem 2.1

In this appendix we present a simple direct proof of Theorem 2.1, which is due to A. Borodin.

We shall use the Frobenius notation $\lambda = (p_1,\dots,p_d \mid q_1,\dots,q_d)$ for a Young diagram λ, see (2.5). We start with the observation that the function $\varphi := M_{zz'}/\dim\lambda$ can be written in the determinantal form

$$\varphi(\lambda) = \frac{\det[m_{p_i q_j}]}{(t)_n}, \qquad n = |\lambda| = \sum(p_i+q_i+1), \qquad (8.1)$$

where

$$m_{p,q} = t\frac{(z+1)_p(z'+1)_p(-z+1)_q(-z'+1)_q}{p!q!(p+q+1)}; \quad t = zz'. \tag{8.2}$$

Indeed, this easily follows from the formulas (2.6), (2.7), and the Cauchy formula

$$\frac{\prod\limits_{1\le i,j\le d}(p_i - p_j)(q_i - q_j)}{\prod\limits_{1\le i\le d}\prod\limits_{1\le j\le d}(p_i + q_j + 1)} = \det\left[\frac{1}{p_i + q_j + 1}\right].$$

Let us introduce a class of functions on the Young graph. We fix a number $t \in \mathbb{C} \setminus \{0, -1, -2, \dots\}$, a sequence $\{m_{pq}\}_{p,q=0}^{\infty}$ and set

$$\varphi(\lambda) = \frac{\det[m_{p_i,q_j}]_{i,j=1}^{d}}{t(t+1)\cdots(t+n-1)}. \tag{8.3}$$

Theorem 8.1. *If a sequence* $\{m_{pq}\}_{p,q=0}^{\infty}$ *satisfies the relations*

$$m_{p+1,q} + m_{p,q+1} - (p+q+1)m_{p,q} = m_{p,0}m_{0,q}, \qquad p, q = 0, 1, \dots,$$

$$m_{0,0} = t \tag{8.4}$$

then the function φ *defined by* (8.1) *is harmonic. In other words,*

$$\varphi(\lambda) = \sum_{\nu\searrow\lambda}\varphi(\nu), \quad \lambda \in \mathbb{Y}. \tag{8.5}$$

Proof. For a $l \times l$ matrix $A = (a_{ij})$ we shall denote by $A\binom{i_1\dots i_k}{j_1\dots j_k}$ the determinant of the submatrix of A formed by the intersections of rows with numbers $i_1 \dots i_k$ and columns with numbers $j_1 \dots j_k$. We shall also denote by A_{ij} the cofactor of a_{ij}. That is,

$$A_{ij} = (-1)^{i+j} A\begin{pmatrix}1\dots\hat{i}\dots l\\ 1\dots\hat{j}\dots l\end{pmatrix}.$$

The transposed matrix to (A_{ij}) is equal to the inverse matrix of A, multiplied by $\det A$, so that

$$\sum_j a_{ij}A_{ij} = \sum_i a_{ij}A_{ij} = \det A.$$

It follows that for any two sequences of numbers $v_1, \dots, v_l$ and $w_1, \dots, w_l$

$$\sum_{i,j=1}^{l}(v_i + w_j)a_{ij}A_{ij} = \sum_{i=1}^{l}(v_i + w_i)\cdot\det A. \tag{8.6}$$

We proceed to verify the harmonicity relation (8.5). The Frobenius coordinates of a diagram $\nu \searrow \lambda$ in (8.5) are obtained from the Frobenius coordinates of the diagram λ by applying one of the following three operations:

1) $p_i \to p_i + 1$ for a certain $i = 1, \ldots, d$, which corresponds to creating a new box in the ith row above the diagonal;

2) $q_j \to q_j + 1$ for a certain $j = 1, \ldots, d$, which corresponds to creating a new box in the jth column below the diagonal;

3) adding a couple of coordinates $p_{d+1} = 0$, $q_{d+1} = 0$, which corresponds to creating a new box on the diagonal.

It may happen that creating a new box in a certain position is forbidden, because the resulting shape ν is not a Young diagram: this occurs exactly when the set of the coordinates for ν contains two equal p-coordinates or two equal q-coordinates. However, in such a case the formal application of formula (8.3) will give $\varphi(\nu) = 0$ as the determinant in the numerator of (8.3) will vanish. This makes it possible to sum up over all the operations of type 1), 2), 3), regardless of whether the corresponding shape ν is a Young diagram. Then the harmonicity relation (8.5) can be rewritten in the following form (below we set $M = (m_{pq})$)

$$\left(t + \sum_{i=1}^{d} (p_i + q_i + 1)\right) M \begin{pmatrix} p_1 \ldots p_d \\ q_1 \ldots q_d \end{pmatrix}$$
$$= \sum_{i=1}^{d} M \begin{pmatrix} p_1 \ldots p_i + 1 \ldots p_d \\ q_1 \ldots q_d \end{pmatrix} + \sum_{j=1}^{d} M \begin{pmatrix} p_1 \ldots p_d \\ q_1 \ldots q_j + 1 \ldots q_d \end{pmatrix}$$
$$+ M \begin{pmatrix} p_1 \ldots p_d & 0 \\ q_1 \ldots q_d & 0 \end{pmatrix}.$$

Expansion along the ith row gives

$$M \begin{pmatrix} p_1 \ldots p_i + 1 \ldots p_d \\ q_1 \ldots q_d \end{pmatrix} = \sum_{j=1}^{d} (-1)^{i+j} m_{p_i+1, q_j} M \begin{pmatrix} p_1 \ldots \hat{p}_i \ldots p_d \\ q_1 \ldots \hat{q}_j \ldots q_d \end{pmatrix}.$$

Similarly, expanding along the jth column, we get

$$M \begin{pmatrix} p_1 \ldots p_d \\ q_1 \ldots q_j + 1 \ldots q_d \end{pmatrix} = \sum_{i=1}^{d} (-1)^{i+j} m_{p_i, q_j+1} M \begin{pmatrix} p_1 \ldots \hat{p}_i \ldots p_d \\ q_1 \ldots \hat{q}_j \ldots q_d \end{pmatrix}.$$

Finally, expanding $M \begin{pmatrix} p_1 \ldots p_d & 0 \\ q_1 \ldots q_d & 0 \end{pmatrix}$ along the last row and column we get

$$M \begin{pmatrix} p_1 \ldots p_d & 0 \\ q_1 \ldots q_d & 0 \end{pmatrix}$$
$$= m_{0,0} M \begin{pmatrix} p_1 \ldots p_d \\ q_1 \ldots q_d \end{pmatrix} + \sum_{i,j=1}^{d} (-1)^{i+j+1} m_{p_i,0} m_{0,q_j} M \begin{pmatrix} p_1 \ldots \hat{p}_i \ldots p_d \\ q_1 \ldots \hat{q}_j \ldots q_d \end{pmatrix}.$$

Adding everything up, employing the assumption (8.4) and applying the relation (8.6), we get

$$\sum_{i=1}^{d} M\begin{pmatrix} p_1 \dots p_i+1 \dots p_d \\ q_1 \dots q_d \end{pmatrix} + \sum_{j=1}^{d} M\begin{pmatrix} p_1 \dots p_d \\ q_1 \dots q_j+1 \dots q_d \end{pmatrix}$$

$$+M\begin{pmatrix} p_1 \dots p_d & 0 \\ q_1 \dots q_d & 0 \end{pmatrix} = tM\begin{pmatrix} p_1 \dots p_d \\ q_1 \dots q_d \end{pmatrix}$$

$$+\sum_{i,j=1}^{d} (-1)^{i+j}(m_{p_i+1,q_j} + m_{p_i,q_j+1} - m_{p_i,0}m_{0,q_j}) M\begin{pmatrix} p_1 \dots \hat{p}_i \dots p_d \\ q_1 \dots \hat{q}_j \dots q_d \end{pmatrix}$$

$$= tM\begin{pmatrix} p_1 \dots p_d \\ q_1 \dots q_d \end{pmatrix} + \sum_{i,j=1}^{d} (-1)^{i+j}(p_i+q_j)m_{p_i,q_j} M\begin{pmatrix} p_1 \dots \hat{p}_i \dots p_d \\ q_1 \dots \hat{q}_j \dots q_d \end{pmatrix}$$

$$= \left(t + \sum_{i=1}^{d}(p_i+q_i+1)\right) M\begin{pmatrix} p_1 \dots p_d \\ q_1 \dots q_d \end{pmatrix},$$

which concludes the proof. □

Corollary 8.2. *The claim of Theorem 2.1 holds.*

Proof. It suffices to check that the sequence (8.2) satisfies the assumption (8.4) of Theorem 8.1. But this is easily verified. □

References

[A] D. J. Aldous, *Exchangeability and Related Topics*, Lecture Notes in Math. **1117**, Springer, 1985, pp. 2–199.

[AK] P. Appell, M. J. Kampé de Fériet, *Fonctions hypergéometriques et hypersphériques. Polynômes d'Hermite*, Gauthier-Villars, 1926.

[B1] A. Borodin, Characters of symmetric groups and correlation functions of point processes, *Funct. Anal. Appl.* **34**, No. 1 (2000), 12–28.

[B2] A. Borodin, Harmonic analysis on the infinite symmetric group and the Whittaker kernel, *Algebra i analiz* **12**, No. 5 (2000), 28–63 (Russian); English translation in *St. Petersburg Math. J.* **12**, No. 5 (2001), 733–759.

[B3] A. Borodin, Riemann–Hilbert problem and the discrete Bessel kernel, *Intern. Math. Research Notices*, No. 9 (2000), 467–494.

[BOO] A. Borodin, A. Okounkov, and G. Olshanski, Asymptotics of Plancherel measures for symmetric groups, *J. Amer. Math. Soc.* **13**, No. 3 (2000), 481–515.

[BO1] A. Borodin and G. Olshanski, Point processes and the infinite symmetric group, *Math. Research Letters* **5** (1998), 799–816.

[BO2] A. Borodin and G. Olshanski, Distributions on partitions, point processes, and the hypergeometric kernel, *Commun. Math. Phys.* **211** (2000), 335–358.

[BO3] A. Borodin and G. Olshanski, Harmonic functions on multiplicative graphs and interpolation polynomials, *Electronic J. Comb.* **7** (2000), paper #R28; `math/9912124`.

[BO4] A. Borodin and G. Olshanski, Z-Measures on partitions, Robinson–Schensted–Knuth correspondence, and $\beta = 2$ ensembles. In *Random Matrix Models and their Applications*, (P. M. Bleher, A. R. Its, eds.), MSRI Publications, Vol. 40, Cambridge Univ. Press, 2001, 71–94.

[DVJ] D. J. Daley and D. Vere-Jones, *An Introduction to the Theory of Point Processes*, Springer Series in Statistics, Springer, 1988.

[E] A. Erdelyi (ed.), *Higher Transcendental Functions*, Vol. 1, McGraw-Hill, 1953.

[Ex] H. Exton, *Multiple Hypergeometric Functions and Applications*, 1976.

[JK] G. James and A. Kerber, *The Representation Theory of the Symmetric Group*, Encyclopedia of Mathematics and Its Applications **16**, Addison-Wesley, 1981.

[Ke1] S. V. Kerov, Combinatorial examples in the theory of AF-algebras. In *Differential Geometry, Lie Groups and Mechanics X*, Zapiski Nauchnykh Seminarov LOMI, Vol. 172, 1989, pp. 55–67 (Russian); English translation in *J. Soviet Math.* **59**, No. 5 (1992), 1063–1071.

[Ke2] S. V. Kerov, Anisotropic Young diagrams and Jack symmetric functions, *Funct. Anal. Appl.* **34**, No. 1 (2000), 41–51.

[KOO] S. Kerov, A. Okounkov, and G. Olshanski, The boundary of Young graph with Jack edge multiplicities, *Intern. Math. Research Notices*, No. 4 (1998), 173–199.

[KOV] S. Kerov, G. Olshanski, and A. Vershik, Harmonic analysis on the infinite symmetric group. A deformation of the regular representation, *Comptes Rend. Acad. Sci. Paris, Sér. I* **316** (1993), 773–778.

[KV] S. Kerov and A. Vershik, The Grothendieck group of the infinite symmetric group and symmetric functions with the elements of the K_0-functor theory of AF-algebras. In *Representation of Lie Groups and Related Topics*, Adv. Stud. Contemp. Math. **7** (eds. A. M. Vershik and D. P. Zhelobenko), Gordon and Breach, 1990, pp. 36–114.

[Ki1] J. F. C. Kingman, Random partitions in population genetics, *Proc. Roy. Soc. London A.* **361** (1978), 1–20.

[Ki2] J. F. C. Kingman, *Poisson Processes*, Oxford University Press, 1993.

[M] I. G. Macdonald, *Symmetric Functions and Hall Polynomials*, 2nd edition, Oxford University Press, 1995.

[Ok1] A. Yu. Okounkov, Thoma's theorem and representations of infinite bisymmetric group, *Funct. Anal. Appl.* **28**, No. 2 (1994), 101–107.

[Ok2] A. Yu. Okounkov, $SL(2)$ and z-measures, In *Random Matrix Models and their Applications*, (P. M. Bleher, A. R. Its, eds.), MSRI Publications, Vol. 40, Cambridge Univ. Press, 2001, 407–420.

[Ok3] A. Yu. Okounkov, Infinite wedge and random partitions, *Selecta Math. (NS)* **7** (2001), 57–81.

[OV] A. Yu. Okounkov and A. Vershik, A new approach to representation theory of symmetric groups, *Selecta Mathematica, New Series* **2** (1996), 581–605.

[Ol1] G. I. Olshanskii, Unitary representations of (G, K)-pairs connected with the infinite symmetric group $S(\infty)$, *Algebra i Analiz* **1**, No. 4 (1989), 178–209 (Russian); English transl. in *Leningrad Math. J.* **1** (1990), 983–1014.

[Ol2] G. I. Olshanskii, Unitary representations of infinite-dimensional pairs (G, K) and the formalism of R. Howe. In *Representations of Lie groups and related topics*, Adv. Stud. Contemp. Math. **7** (eds. A. M. Vershik and D. P. Zhelobenko), Gordon and Breach, 1990, pp. 269–463.

[P.I] G. Olshanski, Point processes and the infinite symmetric group. Part I: The general formalism and the density function, preprint, 1998, `math/9804086`.

[P.II] A. Borodin, Point processes and the infinite symmetric group. Part II: Higher correlation functions, preprint, 1998, `math/9804087`.

[P.III] A. Borodin and G. Olshanski, Point processes and the infinite symmetric group. Part III: Fermion point processes, preprint, 1998, `math/9804088`.

[P.IV] A. Borodin, Point processes and the infinite symmetric group. Part IV: Matrix Whittaker kernel, preprint, 1998, `math/9810013`.

[P.V] G. Olshanski, Point processes and the infinite symmetric group. Part V: Analysis of the matrix Whittaker kernel, preprint, 1998, `math/9810014`.

[P] A. Postnikov, unpublished manuscript.

[R] N. A. Rozhkovskaya, Multiplicative distributions on Young graph. In *Representation Theory, Dynamical Systems, Combinatorial and Algorithmic Methods II* (A. M. Vershik, ed.), Zapiski Nauchnykh Seminarov POMI **240**, Nauka, St. Petersburg, 1997 pp. 246–257 (Russian); English transl. *J. Math. Sci.* (New York), **96** No. 5, (1999), 3600–3608.

[S] J. R. Stembridge, A characterization of supersymmetric polynomials, *J. Algebra* **95** (1985), 439-444.

[T] E. Thoma, Die unzerlegbaren, positive-definiten Klassenfunktionen der abzählbar unendlichen, symmetrischen Gruppe, *Math. Zeitschr.* **85** (1964), 40–61.

[V] A. M. Vershik, Asymptotic aspects of the representation theory of symmetric groups, *Selecta Math. Sov.* **11** (1992), 159–179.

[VK] A. M. Vershik and S. V. Kerov, Asymptotic theory of characters of the symmetric group, *Funct. Anal. Appl.* **15**, No. 4 (1981), 246–255; *J. Math. Sci.* **96** No. 5 (1999), 3600–3608.

[W] G. A. Watterson, The sampling theory of selectively neutral alleles, *Adv. Appl. Prob.* **6** (1974), 463–488.

Grigori Olshanski
Dobrushin Mathematics Laboratory
Institute for Information Transmission Problems
Bolshoy Karetny 19
101447 Moscow GSP–4, Russia
olsh@iitp.ru, olsh@online.ru

Some Toric Manifolds and a Path Integral

W. Rossmann

1 Introduction

Toric manifolds and path integrals look like an odd couple at first sight, but are in fact intimately related, a phenomenon having its origin in statistical mechanics and quantum mechanics. It is not my intention to discuss this topic in any depth — and I am ill-equipped to do so in any event— but only to suggest that this curiosity deserves study. It is in this spirit that I have mixed some physical terminology into what is really a mathematical exercise.

The toric manifolds in question are of a very special type. They were invented by Bott and studied by Grossberg and Karshon [1994]. Their intriguing results stimulated my interest in these matters, especially in the symplectic geometry of these manifolds. One thing new here is the construction of two-forms on these manifolds as curvature forms of holomorphic line bundles with unitary metrics. These two-forms are the objects of interest and their realization here leads naturally to an explicit formula from which everything else follows by simple calculations, things like canonical coordinates and positivity criteria.

The more important innovation, however, seems to me the method, using a construction of these toric manifolds as a special case of what I call "$\mathbb{P}^1$-chains," a construct for which I have some hope. The toric example was worked out as a test for the general case, to be discussed elsewhere. Here the methods are completely elementary, relying on direct calculations presented in calculus style.

A major reason for studying these manifolds comes from Kirillov's [1968] orbit theory for unitary representations of Lie groups and Kostant's [1970] theory of quantization. The point of view here is to look for the character of a unitary representation rather than for the index of a virtual representation. This requires some vanishing theorems for cohomology, which I deduce from a positivity result for the curvature by Kodaira's theorem. The unitary structure on the representation space depends on the curvature form and its positivity as well and the spectrum of the representation is determined with the help of the Kähler potential and its convexity.

Kirillov's universal formula for the character of representations constructed by quantization has inspired much of what I have to say on this topic here. In keeping with the physical origins of the subject, I have taken the opportunity to explain in some detail two formulas for the character in question — alias partition function

— one along the lines of statistical mechanics, the other in terms of a path integral. The first is very simple, but seems of some merit to me; the second, the path integral formula, seems of some interest as well, even though it may be commonplace in a formal way. The point is simply to write out rather carefully how the path integral can be defined in the situation at hand. The proof that it does give the right answer is then immediate in any case, although the reason why it does so, and not only here but also in many other situations, is far from clear.

The paper consists then really of two parts, one dealing with the very special toric varieties mentioned, the other with rather more general questions, for which these toric varieties serve as a vehicle of exposition. It seemed to me that each part benefits from the other.

2 $\mathbb{P}^1$ itself

Write G for SL(2, $\mathbb{C}$) acting by holomorphic automorphisms of $\mathbb{P}^1$,

$$z(g\xi) = \frac{az(\xi)+c}{bz(\xi)+d}, \qquad ad-bc=1.$$

The points ξ_0, ξ_∞ with $z=0,\infty$ are the fixed points of the subgroup T of diagonal matrices and $z=z(\xi)\in\mathbb{C}\cup\{\infty\}$ is the affine coordinate on $\mathbb{P}^1$ centered at ξ_0. The weights by which T acts on the cotangent spaces at these points ξ_0, ξ_∞, denoted $\alpha, -\alpha$, are twice the weights on the two coordinate lines $\mathbb{C}e_{\alpha/2}, \mathbb{C}e_{-\alpha/2}$ in $\mathbb{C}^2$ which correspond to the points ξ_0, ξ_∞ in $\mathbb{P}^1$. They are interchanged by the reflection $s=s_\alpha$, transforming the affine coordinate z centered at ξ_0 into the affine coordinate $z\circ s=-1/z$ centered at ξ_∞. Choose ξ_0 as base-point for $\mathbb{P}^1$ and let B denote its isotropy group in G, thus providing the realization of $\mathbb{P}^1$ as the coset space G/B. The triple (G, B, T) determines uniquely a basis $H_\alpha, E_\alpha, E_{-\alpha}$ of the Lie algebra of G so that T is generated by H_α, B by H_α together with E_α, and

$$[H_\alpha, E_{\pm\alpha}] = \pm 2E_{\pm\alpha}, \quad [E_\alpha, E_{-\alpha}] = H_\alpha.$$

Write N and N$^-$ for the subgroups generated by E_α and $E_{-\alpha}$, respectively. The affine coordinate $z \doteq z(g\mathrm{B})$ on G/B centered at ξ_0 is determined by $v=e^{z(g)E_{-\alpha}}$ if $g=vcn$ according to G$\doteq$ N$^-$TN. The symbol $\doteq$ indicates that the decomposition is valid only for $g\mathrm{B}\neq\xi_\infty$. The compact real form U:=SU(2) of G has $iH_\alpha, E_\alpha \pm iE_{-\alpha}$ as basis for its Lie algebra.

Homomorphisms T$\to$GL(1, $\mathbb{C}$) will be written exponentially as $h\mapsto h^\lambda$, λ being a linear functional on the Lie algebra, referred to as a weight of T. A homomorphism of T extends to B = TN so that $b^\lambda = h^\lambda$ if $b=hn$. $\mathbb{C}^2$ carries a symplectic form invariant under G and a unitary metric invariant under U. Write $g\mapsto g^*$ for the involution of G acting as $u^*=u^{-1}$ on U. The decomposition G= UB, $g=ub$, is valid everywhere and unique if $b\in\mathrm{B}=\mathrm{TN}$ is normalized so that its component in T is real and positive.

The complex-valued holomorphic functions $f(g)$ on G satisfying $f(gb) = f(g)b^\lambda$ are the holomorphic sections of a line bundle $\mathcal{L}_\lambda$ on $\mathbb{P}^1 = \mathrm{G/B}$. In this capacity $f(\xi)$ stands for the element in the line $\mathcal{L}_\lambda(\xi)$ at $\xi = g\mathrm{B}$. This line bundle has a U-invariant unitary metric given by

$$|f(\xi)|^2 := |f(u)|^2 \quad \text{if } \xi = u\mathrm{B},\ u \in \mathrm{U}.$$

One such function is $f(g) := \langle e_{-\alpha/2}, ge_{\alpha/2}\rangle^l$, the pointed brackets denoting the G-invariant symplectic pairing on $\mathbb{C}^2$. The curvature form σ_λ of $\mathcal{L}_\lambda$, normalized so that it represents the Chern class, is

$$\sigma_\lambda = \frac{-\lambda(H_\alpha)}{2\pi \mathrm{i}} \frac{d\bar{z} \wedge dz}{(1+\bar{z}z)^2}\,.$$

This $(1, 1)$-form σ_λ is *negative* when $\lambda(H_\alpha)$ is positive. These matters will be explained in a more general setting when needed.

In short, the triple (G, B, T) determines the basis $(H_\alpha, E_\alpha, E_{-\alpha})$, which allows the reconstruction of $\mathbb{P}^1$, complete with its affine coordinates z, the matrix representation of its automorphism group on $\mathbb{C}^2$, its line bundles $\mathcal{L}_\lambda$ with their unitary structures, etc. All of these $\mathbb{P}^1$-paraphernalia are thus available for any triple of groups known to be abstractly isomorphic with (G, B, T). Nevertheless, for the purpose of calculation in a more general setting it is preferable to work directly with the triple (G, B, T). Three of its properties are assembled here for reference.

Let $v(g)$, $c(g)$ be the components in N^-, T according to the decomposition $\mathrm{G} \doteq \mathrm{N}^-\mathrm{TN}$. Let $u(g)$, $b(g)$ be components in U, B according to G= UB, normalized so that $b(g) \in a(g)\mathrm{N}$ with $a(g) \in \mathrm{T}$ real and positive.

Lemma 2.1. *If* $l := \lambda(H_\alpha) = 0, 1, 2, \cdots$ *is a nonnegative integer, then* $c_s^\lambda(g) := c(sg)^\lambda$ *extends to a holomorphic function on all of* G and $c_s^\lambda(g) = 0$ iff $g \in s\mathrm{B}$.

Proof. In terms of the G-invariant pairing on $\mathbb{C}^2$,

$$c_s^{\alpha/2}(g) = \langle se_{\alpha/2}, ge_{\alpha/2}\rangle, \quad \langle se_{\alpha/2}, e_{\alpha/2}\rangle = 1.$$

If $l := \lambda(H_\alpha)$ then $\lambda = l(\alpha/2)$ and

$$c_s^\lambda(g) = \langle se_{\alpha/2}, ge_{\alpha/2}\rangle^l.$$

The assertion follows. □

Lemma 2.2. *If* $u = u(g)$, $v = v(g)$ *for some* g, *then* $c(u) = c^{-1}(b(v))$.

Proof. If $u = u(g)$, $v = v(g)$, then $u\mathrm{B} = v\mathrm{B}$ so $v = ub$ for some b necessarily equal to $b(v)$. Hence $u = vb^{-1}$ and $c(u) = c(b^{-1}) = c^{-1}(b)$. □

The third lemma is the essential one; its formulation may seem awkward at this point.

Lemma 2.3. *Let* $e^{zE_{-\alpha}} \in \mathrm{N}^-$, $z = re^{i\phi}$, *and* $a \in \mathrm{T}$ *real and positive. Then*

$$b(ae^{zE_{-\alpha}}) \in \tilde{a}\mathrm{N}, \quad \tilde{a} := a\mathrm{a}^{H_\alpha/2}(\tilde{r}), \quad \tilde{r} := a^{-\alpha}r$$

where

$$\mathrm{a}(r) := 1 + r^2.$$

and $\mathrm{a}^{H_\alpha/2} := e^{\frac{1}{2}(\log \mathrm{a})H_\alpha}$.

Proof. First take $a = 1$ and set $v = e^{zE_{-\alpha}}$. Then $ve_{\alpha/2} = e_{\alpha/2} + ze_{\alpha/2}$. Write $v = ub$ and $b \in \tilde{a}\mathrm{N}$ with $\tilde{a}$ real and positive. Then $ve_{\alpha/2} = \tilde{a}^{\alpha/2}ue_{\alpha/2}$. Hence

$$\tilde{a}^{\alpha/2}ue_{\alpha/2} = e_{\alpha/2} + ze_{\alpha/2}.$$

Taking square norms and using the normalization $\tilde{a}$ = real, positive, find

$$\tilde{a}^{\alpha} = 1 + r^2 = \mathrm{a}(r).$$

Hence $\tilde{a} = \mathrm{a}^{-H_\alpha/2}(z)$, which is the desired relation $b(e^{zE_{-\alpha}}) \in \mathrm{a}^{H_\alpha/2}(z)\mathrm{N}$ for $a = 1$. For general a,

$$b(ae^{zE_{-\alpha}}) = b(e^{a^{-\alpha}zE_{-\alpha}})a \in \mathrm{a}^{H_\alpha/2}(a^{-\alpha}z)a\mathrm{N}. \qquad \square$$

3 Definitions

Let $(G_\ell, B_\ell, T_\ell), \ldots, (G_1, B_1, T_1)$ be a sequence obtained from standard $\mathbb{P}^1$-triples $(\mathrm{G}_{\alpha_i}, \mathrm{B}_{\alpha_i}, \mathrm{T}_{\alpha_i})$ by simply enlarging the 1-tori T_i to bigger tori T_i through central extensions:

$$G_i = \mathrm{G}_{\alpha_i}T_i, \quad B_i = \mathrm{B}_{\alpha_i}T_i, \quad \mathrm{G}_{\alpha_i} \cap T_i = \mathrm{T}_{\alpha_i} = \mathrm{B}_{\alpha_i} \cap T_i\,.$$

All groups are required to be complex algebraic. The scheme is to denote by roman letters objects belonging to $\mathbb{P}^1$ itself, like the triples $(\mathrm{G}_\alpha, \mathrm{B}_\alpha, \mathrm{T}_\alpha)$. In addition, let $T_i \to T_{i-1}$, $h_i \mapsto h_i\gamma_{i\,i-1}$, be a sequence of homomorphisms of these tori, written on the *right*. The semidirect decomposition $\mathrm{B}_{\alpha_i} = \mathrm{T}_{\alpha_i}\mathrm{N}_{\alpha_i}$ extends to $B_i = T_i\mathrm{N}_{\alpha_i}$ and the homomorphisms extend to

$$B_i = \mathrm{N}_{\alpha_i}T_i \to T_i \xrightarrow{\gamma_{i\,i-1}} T_{i-1} \subset B_{i-1},$$

producing the sequence

$$G_\ell \overset{\subset}{\leftarrow} B_\ell \overset{\gamma_{\ell,\ell-1}}{\to} G_{\ell-1} \overset{\subset}{\leftarrow} \cdots \overset{\subset}{\leftarrow} B_2 \overset{\gamma_{21}}{\to} G_1\,.$$

The $\mathbb{P}^1$-*chain* associated to this sequence is the quotient

$$\mathcal{X} := G_\ell \times_{B_\ell} \times G_{\ell-1} \times \cdots \times_{B_2} \times G_1/B_1$$

of $G_\ell \times G_{\ell-1} \times \cdots \times G_1$ by the (free) right action of $B_\ell \times B_{\ell-1} \times \cdots \times B_1$ defined by

$$(p_\ell, p_{\ell-1}, \ldots, p_1) \mapsto (p_\ell b_\ell, b_\ell^{-1}\gamma_{\ell\ell-1}p_{\ell-1}, \ldots p_2 b_2, b_2^{-1}\gamma_{21}p_1).$$

The homomorphisms γ are the *connecting homomorphisms* of the $\mathbb{P}^1$-chain. The construction can proceed one step at a time, from right to left. Write s:= $(G_\ell, B_\ell, T_\ell, \gamma_{\ell\ell-1}, \cdots, \gamma_{21}, G_1, B_1, T_1)$ for the whole sequence of data which goes into the construction and $\mathcal{X} = \mathcal{X}_s$ for the result. Write ξ_0 for the element of $\mathcal{X}_s$ represented by $g_i = 1$ for all i, $\xi = g_\ell . g_{\ell-1} . \cdots . g_1 . \xi_0$ for the element represented by $(g_\ell, g_{\ell-1}, \cdots, g_1)$ and $G_\ell . G_{\ell-1} . \cdots . G_1 . \xi_0$ for $\mathcal{X}_s$ itself. For brevity, write also ξ as $g_{\ell-1}.g_\ell \cdots$ and $\mathcal{X}_s$ as $G_\ell . G_{\ell-1} \cdots$. Equality in the quotient $\mathcal{X}_s$ means

$$g_\ell b_\ell . g_{\ell-1} \cdots = g_\ell . b_\ell \gamma_{\ell\ell-1} g_{\ell-1} \cdots .$$

Omit $\gamma_{\ell\ell-1}$ and write $b_\ell g_{\ell-1}$ for $b_\ell \gamma_{\ell\ell-1} g_{\ell-1}$ when the meaning is clear from the indices. Let $G_s := \prod_s G_i$, $B_s := \prod_s B_i$ denote the direct products and $\mathcal{X}_s := G_s/_\gamma B_s$, the quotient by the right action of B_s on G_s defined above. The product torus $T_s := \prod_s T_i$ acts on $\mathcal{X}_s$ on the left, denoted $\xi \mapsto h\xi$ and given by

$$g_\ell . g_{\ell-1} . \cdots \mapsto h_\ell g_\ell . h_{\ell-1} g_{\ell-1} . \cdots .$$

T_s has an open dense orbit on $\mathcal{X}_s$, so that $(\mathcal{X}_s, T_s)$ is a smooth toric variety in the usual sense.

Associated to the sequence $(G_i, B_i, T_i, \gamma_{ii-1})$ is the sequence of roots α_i and a system of integers c_{ji}, $j > i$, defined as follows. Let $h_j \to h_j\gamma_{ji}$ be the composite of the connecting homomorphisms from T_j to T_i and let $\gamma_{ji}\alpha_i \leftarrow \alpha_i$ be the pullback of the root from T_i to T_j. The value of $\gamma_{ji}\alpha_i$ on the canonical generator H_{α_j} of the subtorus T_{α_j} of T_j is an integer denoted $c_{ji} := \langle H_{\alpha_j}, \gamma_{ji}\alpha_i \rangle$. The homomorphism γ_{ji} will be omitted if the meaning of $\langle H_{\alpha_j}, \alpha_i \rangle$ is clear from the indices.

Some comments

The example which motivated the definition is the case when the (G_i, B_i, T_i) are triples naturally associated to a sequence of roots α_i for a triple (G, B, T) consisting of a semisimple group G, a Borel subgroup B and a Cartan subgroup T. In this case $T_i = T$ for all i and one can take the identity map $T \to T$ to construct the connecting homomorphisms $B_i \to B_{i-1}$. If one replaces the triples (G_i, B_i, T) by triples (P_i, B, T) where the P_i are minimal (proper) parabolic subgroups of G containing B and takes the identity $B \to B$ for connecting homomorphisms one obtains the well-known Bott–Samelson manifolds. Other connecting homomorphisms, which are generally an essential ingredient, are of interest as well, e.g., those induced by conjugations in G for the triples (P_i, B, T). Manifolds isomorphic to the toric $\mathbb{P}^1$-chains defined above were first defined by Bott using a different procedure and studied by Grossberg and Karshon [1994], as mentioned in the introduction. Although it would not be worth while introducing $\mathbb{P}^1$-chains constructed by means of connecting homomorphisms just for the case discussed here, there is a generalization for which it is: each triple (G_i, B_i, T_i) can be taken to consist of an arbitrary algebraic groups G_i whose quotient G_i/B_i by its maximal solvable subgroup is isomorphic with $\mathbb{P}^1$. An extension of SL(2, $\mathbb{C}$) by a Heisenberg–Weyl group is an example, beyond those already mentioned.

4 The curvature formula

Let $\mathcal{X}_s$ be the $\mathbb{P}^1$-chain built from a sequence

$$\mathrm{s} := (G_\ell, B_\ell, T_\ell, \gamma_{\ell\,\ell-1}, \cdots \gamma_{21}, G_1, B_1, T_1).$$

Line bundles

Let $c(g_i)$ be the component in T_i according to the decomposition $G_i \doteq \mathrm{N}^-_{\alpha_i} T_i N_i$, defined for g_i in this open set. A weight λ of the product torus T_s is a sequence $\lambda = (\lambda_i)$. Define a function c^λ_s on the open set $\prod_\mathrm{s} B_i s_i B_i$ in G_s by the formula

$$c^\lambda_\mathrm{s}(g_\ell, \cdots, g_1) := \prod_\mathrm{s} c^{\lambda_i}_{s_i}(g_i), \quad c_{s_i}(g_i) := c(s_i g_i).$$

This function is defined on the domain where all the factors are defined, i.e., $g_i \in s_i N_i^- B_i = B_i s_i B_i$. We record a transformation property of the function c^λ_s under the right action of B_s on G_s. Let $\gamma_{i\,i-1}\lambda_{i-1}$ denote the pullback of a weight λ_{i-1} from B_{i-1} toB_i and set

$$\gamma\lambda := (\gamma_{\ell\,\ell-1}\lambda_{\ell-1}, \cdots \gamma_{21}\lambda_1, 0), \quad \mathrm{s}\lambda := (s_\ell\lambda_\ell, \cdots, s_1\lambda_1).$$

Also set $w_i := s_i \cdots s_1$ and $\mathrm{w}\lambda := (w_\ell\lambda_\ell, \cdots, w_1\lambda_1)$.

Lemma 4.1. *As a function of $g \in G_\mathrm{s}$, $c^\lambda_\mathrm{s}(g)$ transforms under the right action of b according to the rule $c^\lambda_\mathrm{s}(g \cdot b) = c^\lambda_\mathrm{s}(g) b^{\tilde\lambda}$ where*

$$\tilde\lambda := (1 - \mathrm{s}\gamma)\lambda, \ \lambda = (1 + \mathrm{w}\gamma)\tilde\lambda.$$

Proof. Omit the connecting homomorphisms from the notation to calculate

$$\begin{aligned} c^\lambda_\mathrm{s}(g \cdot b) &= c^\lambda_\mathrm{s}(g_\ell b_{\ell-1}, b_\ell^{-1} g_{\ell-1} b_{\ell-1}, \cdots b_2^{-1} g_1 b_1) \\ &= c^{\lambda_\ell}(s_\ell g_\ell b_\ell)\, c^{\lambda_{\ell-1}}(s_{\ell-1} b_\ell^{-1} g_{\ell-1} b_{\ell-1}) \cdots c^{\lambda_1}(s_1 b_2^{-1} g_1 b_1) \\ &= c^{\lambda_\ell}(s_\ell g_\ell) b_\ell^{\lambda_{\ell-1}} b_\ell^{-s_{\ell-1}\lambda_{\ell-1}} c^{\lambda_{\ell-1}}(g_{\ell-1}) b_{\ell-1}^{\lambda_{\ell-1}} \cdots b_2^{-s_1\lambda_1} c^{\lambda_1}(g_1) b_1^{\lambda_1} \\ &= c^\lambda_\mathrm{s}(g) b^{\tilde\lambda}. \end{aligned}$$

One checks that $\tilde\lambda := (1 - \mathrm{s}\gamma)\lambda$ is equivalent to $\lambda = (1 + \mathrm{w}\gamma)\tilde\lambda$. □

The locally defined holomorphic functions $f(g)$ on G_s satisfying $f(g \cdot b) = f(g) b^{\tilde\lambda}$ where defined are the local sections of a holomorphic line bundle $\mathcal{L}_\lambda$ on

$\mathcal{X}_s$. Write $f(\xi)$ for the value of the section f in the line $\mathcal{L}_\lambda(\xi) \approx \mathbb{C}$ attached $\xi \in \mathcal{X}_s$. This line $\mathcal{L}_\lambda(\xi)$ carries a unitary metric defined by $|f(\xi)|^2 := |f(u)|$ if $\xi = u.\xi_0$ with $u \in U_s := \prod_s U_i$, U_i being the compact from of G_i. To make use of fiber bundle technology is useful to note this reformulation of the lemma.

Corollary 4.2. *$\mathcal{L}_\lambda = G_s \times_{B_s} \mathbb{C}_{\tilde{\lambda}}$ is the holomorphic line bundle on $\mathcal{X}_s = G_s/_\gamma\, B_s$ associated to the 1-dimensional representation $\tilde{\lambda} = (1 - s\gamma)\lambda$ of the structure group B_s of the principal bundle $G_s \to \mathcal{X}_s$. On the underlying real manifold $\mathcal{X}_s = U_s/_\gamma T_{sU}$ ($T_{sU} := T_s \cap U_s$) the line bundle is $\mathcal{L}_\lambda = U_s \times_{T_{sU}} \mathbb{C}_{\tilde{\lambda}}$ and its unitary metric is associated to the T_{sU} invariant unitary metric on $\mathbb{C}_{\tilde{\lambda}}$.*

Any holomorphic line bundle $\mathcal{L}$ equipped with a unitary structure has a (1, 1)-form σ as curvature. Normalized so that it represents the Chern class, this form is given by

$$\sigma = -\frac{1}{2\pi \mathrm{i}}\bar{\partial}\partial \log |f|^2$$

for any local holomorphic section f, the formula being valid where $f \neq 0$ [Giffiths and Harris, 1978, p. 142]. It will be computed explicitly for the case at hand.

Affine coordinates

The equation

$$\xi := e^{z_\ell E_{\alpha_\ell}} s_\ell . e^{z_{\ell-1} E_{\alpha_{\ell-1}}} . s_{\ell-1} \cdots e^{z_1 E_{\alpha_1}} s_1 . \xi_0 .$$

defines holomorphic functions $z_\ell, z_{\ell-1}, \cdots, z_1$ on the image of the dense open set $\prod_s B_i s_i B_i$ under the quotient map $G_s \to \mathcal{X}_s$. Call $(z_\ell, \cdots, z_1)$ the *affine coordinates* on $\mathcal{X}_s$ centered at $\xi_\infty := s_\ell . s_{\ell-1} \ldots s_1 . \xi_0$. These functions $z_i = z_i(\xi)$ are well defined because $G_s \to \mathcal{X}_s$ maps $\prod_s N_i s_i$ isomorphically onto its image, but their definition requires that ξ be written in the form indicated.

Lemma 4.3. *Under the action of an element $h = (h_i)$ of T_s the affine coordinates (z_i) transform according to the rule*

$$z_i \circ h = h^{\varpi_i} z_i, \quad \varpi_i := (w_{\ell i}\alpha_i, \cdots, w_{i+1\, i}\alpha_i).$$

$w_{ji} := s_j \gamma_{j\, j-1} s_{j-1} \cdots \gamma_{i+1\, i} s_i$ is the homomorphism $T_j \to T_i$ composed of the reflections s and the connection homomorphisms γ.

Proof. Let $\xi := \exp(z_\ell E_{\alpha_\ell}) s_\ell . \exp(z_{\ell-1} E_{\alpha_{\ell-1}}) . s_{\ell-1} \cdots$. Omit the connection homomorphisms from the notation to calculate

$$\begin{aligned} h\xi &:= h_\ell \exp(z_\ell E_{\alpha_\ell}) s_\ell . h_{\ell-1} \exp(z_{\ell-1} E_{\alpha_{\ell-1}}) . s_{\ell-1} \cdots \\ &= \exp(h_\ell^{\alpha_\ell} z_\ell E_{\alpha_\ell}) h_\ell s_\ell . h_{\ell-1} \exp(z_{\ell-1} E_{\alpha_{\ell-1}}) . s_{\ell-1} \cdots \\ &= \exp(h_\ell^{\alpha_\ell} z_\ell E_{\alpha_\ell}) s_\ell . h_\ell^{s_\ell} h_{\ell-1} \exp(z_{\ell-1} E_{\alpha_{\ell-1}}) . s_{\ell-1} \cdots \\ &= \exp(h_\ell^{\alpha_\ell} z_\ell E_{\alpha_\ell}) s_\ell . \exp(h_\ell^{s_\ell \alpha_{\ell-1}} h_{\ell-1}^{\alpha_{\ell-1}} z_{\ell-1} E_{\alpha_{\ell-1}}) h_\ell^{s_\ell} h_{\ell-1} . s_{\ell-1} \cdots \end{aligned}$$

etc. This gives

$$z_i \circ h = (h_\ell^{w_{\ell i}\alpha_i} h_{\ell-1}^{w_{\ell-1 i}\alpha_i} \cdots h_{i+1}^{w_{i+1 i}\alpha_i} h_i^{\alpha_i}) z_i,$$

and hence the assertion. □

The points in the coordinate domain with all coordinates nonzero form a single orbit of T_s, namely its open dense orbit on the toric manifold $\mathcal{X}_s$.

Curvature

Let $\lambda = (\lambda_i)$ be a weight of T_s and $\mathcal{L}_\lambda$ the associated holomorphic line bundle on $\mathcal{X}_s$ with its unitary metric. The following formula gives its curvature.

Theorem 4.4. *The curvature form σ_λ of $\mathcal{L}_\lambda$ is given by*

$$\sigma_\lambda = +\frac{1}{2\pi \mathrm{i}} \sum_{ij} \frac{\partial^2 K_\lambda}{\partial \bar{z}_i \partial z_j} d\bar{z}_i \wedge dz_j, \quad K_\lambda := \log a_\lambda,$$

$$a_\lambda := \mathrm{a}(\tilde{r}_1)^{l_1} \cdots \mathrm{a}(\tilde{r}_\ell)^{l_\ell}, \quad l_i := \lambda_i(H_{\alpha_i}), \quad \mathrm{a}(\tilde{r}) := 1 + \tilde{r}^2.$$

$\tilde{r}_\ell, \cdots, \tilde{r}_1$ *are the functions of* $r_\ell, \ldots, r_1$ *and v.v. defined inductively by*

$$r_\ell = \tilde{r}_\ell, \quad r_i = \tilde{r}_i \prod_{j:j>i} \mathrm{a}^{c_{ji}/2}(\tilde{r}_j).$$

The exponents c_{ji} are the integers $c_{ji} = \langle H_{\alpha_j}, \alpha_i \rangle$.

The variables $\tilde{r}_i$ are defined where the $z_i = r_i e^{\sqrt{-1}\phi_i}$ are and $(\tilde{r}_i, \phi_i)$, may be used as real coordinates on $\mathcal{X}_s$ instead of (r_i, ϕ_i). Their recursive definition proceeds in the direction opposite to the construction of the $\mathbb{P}^1$- chain. Combine the $(\tilde{r}_i, \phi_i)$ into complex-valued variables $\tilde{z}_i := \tilde{r}_i e^{\sqrt{-1}\tilde{\phi}_i}$, considered as functions of the $z_i = r_i e^{\sqrt{-1}\phi_i}$ by the above equations supplemented by $\phi_i = \tilde{\phi}_i$, even though these functions $\tilde{z}_i$ are not holomorphic. The angle variables $\phi_i = \tilde{\phi}_i$ play no role in much of what is to follow and will then be suppressed. (The hoary notion of "variables," which leaves it to the reader and context to decide what is a function of what and in what way, is most convenient here and even more so later.)

Proof. To simplify the notation take a $\mathbb{P}^1$-chain of length three, and take (α, β, γ) as indices. Use the shorthand notation like $\xi = g_\alpha . g_\beta . g_\gamma . \xi_0$ etc, and omit the connecting homomorphisms. By definition

$$|c_s^\lambda(g_\alpha . g_\beta . g_\gamma . \xi_0)|^2 = |c^{\lambda_\alpha}(u_\alpha)|^2 |c^{\lambda_\beta}(u_\beta)|^2 |c^{\lambda_\gamma}(u_\gamma)|^2,$$

if $g_\alpha . g_\beta . g_\gamma . \xi_0 = u_\alpha . u_\beta . u_\gamma . \xi_0$. In terms of the components $u(g) \in \mathrm{U}, b(g) \in \mathrm{B}$ according to $\mathrm{G} = \mathrm{UB}$,

$$\begin{aligned} s_\alpha g_\alpha . s_\beta g_\beta . s_\gamma g_\gamma . \xi_0 &= u_\alpha b_\alpha . s_\beta g_\beta . s_\gamma g_\gamma . \xi_0 \quad [u_\alpha := u(s_\alpha g_\alpha), b_\alpha := b(s_\alpha g_\alpha)] \\ &= u_\alpha . u_\beta b_\beta . s_\gamma g_\gamma . \xi_0 \quad [u_\beta := u(b_\alpha s_\beta g_\beta), b_\beta := b(b_\alpha s_\beta g_\beta)] \\ &= u_\alpha . u_\beta . u_\gamma b_\gamma . \xi_0 \quad [u_\gamma := u(b_\beta s_\gamma g_\gamma), b_\gamma := b(b_\beta s_\gamma g_\gamma)] \\ &= u_\alpha . u_\beta . u_\gamma . \xi_0 . \end{aligned}$$

In terms of the components $v(g) \in \mathrm{N}^-$ according to $\mathrm{G} \doteq \mathrm{N}^-\mathrm{TN}$ (Lemma 2.2, $b^\lambda := c(b)^\lambda$),

$$|c^{\lambda_\alpha}(u_\alpha)|^2 = b^{-2\lambda_\alpha}(\tilde{v}_\alpha) \quad [\tilde{v}_\alpha := v(s_\alpha g_\alpha)]$$
$$|c^{\lambda_\beta}(u_\beta)|^2 = b^{-2\lambda_\beta}(\tilde{v}_\beta) \quad [\tilde{v}_\beta := v(b_\alpha s_\beta g_\beta)]$$
$$|c^{\lambda_\gamma}(u_\gamma)|^2 = b^{-2\lambda_\gamma}(\tilde{v}_\gamma) \quad [\tilde{v}_\gamma := v(b_\beta s_\gamma g_\gamma)].$$

Define $\tilde{z}_\alpha, \tilde{z}_\beta, \tilde{z}_\gamma$ by

$$e^{\tilde{z}_\alpha E_{-\alpha}} := \tilde{v}_\alpha,\ e^{\tilde{z}_\beta E_{-\beta}} := \tilde{v}_\beta,\ e^{\tilde{z}_\gamma E_{-\gamma}} := \tilde{v}_\gamma.$$

so that

$$|c_s^\lambda(g_\alpha . g_\beta . g_\gamma . \xi_0)|^2 = b^{-2\lambda_\alpha}(e^{\tilde{z}_\alpha E_{-\alpha}})\, b^{-2\lambda_\beta}(e^{\tilde{z}_\beta E_{-\beta}})\, b^{-2\lambda_\gamma}(e^{\tilde{z}_\gamma E_{-\gamma}}).$$

By Lemma 2.3 with $a = 1$ this becomes

$$|c_s^\lambda(g_\alpha . g_\beta . g_\gamma . \xi_0)|^2 = \mathrm{a}^{-l_\alpha}(\tilde{z}_\alpha)\mathrm{a}^{-l_\beta}(\tilde{z}_\alpha)\mathrm{a}^{-l_\gamma}(\tilde{z}_\gamma).$$

It remains to calculate the $\tilde{z}_\alpha, \tilde{z}_\beta, \tilde{z}_\gamma$ as functions of the coordinates $z_\alpha, z_\beta, z_\gamma$ of

$$\xi := e^{z_\alpha E_\alpha} s_\alpha . e^{z_\beta E_\beta} s_\alpha . e^{z_\gamma E_\gamma} s_\gamma \xi_0 .$$

Thus take

$$g_\alpha := e^{z_\alpha E_\alpha} s_\alpha,\ g_\beta := e^{z_\beta E_\beta} s_\alpha,\ g_\gamma := e^{z_\gamma E_\gamma}$$

or equivalently

$$s_\alpha g_\alpha = e^{-z_\alpha E_{-\alpha}},\ s_\alpha g_\beta = s_\alpha e^{-z_\beta E_{-\beta}},\ s_\gamma g_\gamma = e^{-z_\gamma E_{-\gamma}}.$$

The problem is this. Given $b_\alpha \in B_\alpha$ find $b_\beta \in B_\beta$ from $b_\beta = b(b_\alpha e^{-z_\beta E_{-\beta}})$ in order to determine $\tilde{v}_\gamma = e^{\tilde{z}_\gamma E_{-\gamma}} \in \mathrm{N}_\gamma^-$ from $\tilde{v}_\gamma = v(b_\beta e^{-z_\gamma E_{-\gamma}})$.

At the first step of the process determine b_α from $b_\alpha = b(g_\alpha)$. For $g_\alpha = e^{-z_\alpha E_{-\alpha}}$ the equation becomes

$$b_\alpha = b(e^{-z_\alpha E_{-\alpha}}).$$

By Lemma 2.3 (which is not affected by the minus sign in the exponent)

$$b_\alpha = b(e^{-z_\alpha E_{-\alpha}}) \in \tilde{a}_\alpha \mathrm{N}_\beta,\ \tilde{a}_\alpha = \mathrm{a}^{H_\alpha/2}(\tilde{z}_\alpha) \quad \tilde{z}_\alpha := z_\alpha$$

At the second step, determine b_β from the equation

$$b_\beta = b(b_\alpha e^{-z_\beta E_{-\beta}}), \quad b_\alpha \in \tilde{a}_\alpha \mathrm{N}_\alpha .$$

$b_\alpha e^{-z_\beta E_{-\beta}}$ is the action of b_α on B_β through the connecting homomorphism, which is trivial on N_α. Hence b_α may be replaced by $\tilde{a}_\alpha$ and same lemma applies:

$$\begin{aligned} b_\beta &= b(\tilde{a}_\alpha e^{-z_\beta E_{-\beta}}) \in \tilde{a}_\beta \mathrm{N}_\beta, \\ \tilde{a}_\beta &:= \tilde{a}_\alpha \mathrm{a}^{H_\beta}(\tilde{z}_\beta) = \mathrm{a}^{H_\alpha}(\tilde{z}_\alpha)\mathrm{a}^{H_\beta}(\tilde{z}_\beta), \\ \tilde{z}_\beta &:= (\tilde{a}_\alpha)^{-\beta} z_\beta = \mathrm{a}^{-\beta(H_\alpha)/2}(\tilde{z}_\alpha) z_\beta. \end{aligned}$$

At the third step,

$$b_\gamma = b(b_\beta e^{-z_\gamma E_{-\gamma}}), \quad b_\beta \in \tilde{a}_\beta \mathrm{N}_\beta.$$

This is

$$\begin{aligned} b_\gamma &= b(\tilde{a}_\beta e^{-z_\gamma E_{-\gamma}}) \in \tilde{a}_\gamma \mathrm{N}_\gamma, \\ \tilde{a}_\gamma &:= \tilde{a}_\beta \mathrm{a}^{H_\gamma}(\tilde{z}_\gamma) = \mathrm{a}^{H_\alpha}(\tilde{z}_\alpha)\mathrm{a}^{H_\beta}(\tilde{z}_\beta)\mathrm{a}^{H_\gamma}(\tilde{z}_\gamma), \\ \tilde{z}_\gamma &:= (\tilde{a}_\beta)^{-\gamma} z_\gamma = \mathrm{a}^{-\gamma(H_\alpha)/2}(\tilde{z}_\alpha)\mathrm{a}^{-\gamma(H_\beta)/2}(\tilde{z}_\beta)\mathrm{a}^{-\gamma(H_\beta)/2}(\tilde{z}_\gamma) z_\gamma. \end{aligned}$$

These are the desired formulas for these variables. The result is

$$|c_{\mathrm{s}}^\lambda(\xi)|^2 = \prod \mathrm{a}^{-\lambda_i(H_{\alpha_i})}(\tilde{z}_i) = a_\lambda(\xi)^{-1}$$

and

$$\sigma_\lambda = -\frac{1}{2\pi \mathrm{i}}\bar{\partial}\partial |c_{\mathrm{s}}^\lambda|^2 = +\frac{1}{2\pi \mathrm{i}}\bar{\partial}\partial a_\lambda. \qquad \square$$

5 Canonical coordinates

It will be convenient to change variables:

$$\begin{aligned} z_i &= e^{\zeta_i} \quad \zeta_i := \tau_i + \sqrt{-1}\phi_i \\ \tilde{z}_i &= e^{\tilde{\zeta}_i} \quad \tilde{\zeta}_i := \tilde{\tau}_i + \sqrt{-1}\tilde{\phi}_i\,. \end{aligned}$$

In these variables the formula for σ_λ reads

$$\sigma_\lambda = \frac{1}{2\pi \mathrm{i}}\sum_{ij}\frac{\partial^2 K_\lambda}{\partial \bar{\zeta}_i \partial \zeta_j} d\bar{\zeta}_i \wedge d\zeta_j = \frac{1}{2\pi}\sum_{ij}\frac{1}{2}\frac{\partial^2 K_\lambda}{\partial \tau_i \partial \tau_j} d\tau_i \wedge d\phi_j,$$

$$K_\lambda = \sum_i l_i \mathrm{K}(\tilde{\tau}_i), \quad \mathrm{K}(\tilde{\tau}) := \log \mathrm{a}(\tilde{\tau}), \quad \mathrm{a}(\tilde{\tau}) := 1 + e^{2\tilde{\tau}}.$$

Also introduce

$$J_i := -\frac{1}{2}\frac{\partial K_\lambda}{\partial \tau_i}, \quad \mathrm{J}(\tilde{\tau}) := \frac{1}{2}\frac{d\mathrm{K}(\tilde{\tau})}{d\tilde{\tau}} = \frac{e^{2\tilde{\tau}}}{1 + e^{2\tilde{\tau}}}.$$

The range of J is to be noted: $0 \le \mathrm{J} \le 1$. These formulas have the following consequence.

Theorem 5.1. *The curvature form of $\mathcal{L}_\lambda$ is given by*

$$\sigma_\lambda = \frac{1}{2\pi}\sum_i dJ_i \wedge d\phi_i ,$$

and the action of $e^H \in T_s$ is given by

$$(J_i, \phi_i) \circ e^H = (J_i, \phi_i + \varpi_i(H)).$$

Define linear combinations $L_1, \cdots, L_\ell$ of the J_i's by

$$L_1 := 0, \quad L_j := \sum_{i:j>i} c_{ji} J_i .$$

Then $J_1, \ldots, J_\ell$ are recursively determined by

$$J_1 = l_1 \mathrm{J}_1, \quad J_j = \mathrm{J}_j(l_j - L_j), \quad [\mathrm{J}_j := \mathrm{J}(\tilde{\tau}_j)]$$

and their range is determined by $0 \le \mathrm{J}_j \le 1$, i.e.,

$$0 \le \frac{J_1}{l_1} \le 1, \quad 0 \le \frac{J_j}{l_j - L_j} \le 1$$

with the understanding that a vanishing denominator means a vanishing numerator.

Proof. The formulas for σ_λ and the action of T_s in terms of the variables (J_i, ϕ_i) follows from the corresponding formulas in terms of the variables (z_i). The formula for K_λ reads

$$\frac{1}{2} dK_\lambda = \sum_i l_i \mathrm{J}_i d\tilde{\tau}_i, \quad \mathrm{J}(\tilde{\tau}) := \frac{1}{2}\frac{d\mathrm{K}(\tilde{\tau})}{d\tilde{\tau}} .$$

In terms of the τ_i's,

$$\frac{1}{2} dK_\lambda = \sum_i J_i d\tau_i, \quad J_i := \frac{1}{2}\frac{\partial K_\lambda}{\partial \tau_i} .$$

The relations between the $r_i = e^{\tau_i}$ and the $\tilde{r}_i = e^{\tilde{\tau}_i}$ are

$$r_\ell = \tilde{r}_\ell \quad \cdots \quad r_i = \tilde{r}_i \prod_{j:j>i} \mathrm{a}^{-\frac{1}{2}c_{ji}}(\tilde{r}_j).$$

In terms of the new variables these equations say

$$\tau_i = \tilde{\tau}_i - \sum_{j:j>i} \frac{1}{2} c_{ji} \mathrm{K}(\tilde{\tau}_j).$$

With some rearrangements this gives

$$d\tau_i = d\tilde{\tau}_i + \sum_{j:j>i} c_{ji} \mathrm{J}_j d\tilde{\tau}_j$$

with $\mathrm{J}_j := \mathrm{J}(\tilde{\tau}_j) := (1/2)d\mathrm{K}(\tilde{\tau})/d\tilde{\tau}$. Use these equations to compare the coefficients of the $d\tilde{\tau}_j$'s in the two expressions

$$\frac{1}{2}dK_\lambda = \sum_j l_j \mathrm{J}_j d\tilde{\tau}_j, \quad \frac{1}{2}dK_\lambda = \sum_i J_i d\tau_i .$$

Proceeding in the order $d\tilde{\tau}_1, d\tilde{\tau}_2, \cdots$ find

$$l_1\mathrm{J}_1 = J_1, \quad l_j\mathrm{J}_j = J_j + \sum_{i:j>i} c_{ji}\mathrm{J}_j J_i$$

as required. □

Some comments

(1) The theorem agrees with the results of Grossberg and Karshon [1994], but their two-form is constructed by a rather special procedure adapted to their setting. (2) If the form σ_λ is everywhere nondegenerate, the preceding theorem says that the canonical coordinates (J_i, ϕ_i) are action-angle variables for action of the compact real form of T_s on the symplectic manifold $\mathcal{X}_s, \sigma_\lambda$ [Arnold, 1989]. In this context σ_λ is the object of interest, and indeed as form, not only as a cohomology class. (3) In contrast to the construction of the variables $\tilde{r}_\ell, \ldots, \tilde{r}_1$, the inductive construction of $J_1, \cdots, J_\ell$ in terms of $\tilde{\tau}_1, \cdots, \tilde{\tau}_\ell$ proceeds in the same direction as the construction of the $\mathbb{P}^1$-chain. As a consequence, $J_j = \mathrm{J}_j(l_j - \sum_{i:j>i} c_{ji} J_i)$ depends only on the variables $\tilde{\tau}_i$ with $i \le j$ but on all of the variables τ_i.

For a curvature form of a holomorphic line bundle it is positivity which is of primary interest, rather than just nondegeneracy. This is the question to be addressed next.

6 Positivity Questions

Let $c_{ji}^{\pm} := c_{ji}$ if $\pm c_{ji} > 0$ and $= 0$ otherwise. Let $J_{j\,\min}, J_{j\,\max}$ be the min / max of $J_j := \mathrm{J}_j(l_j - \sum_{i:j>i} c_{ji} J_i)$ over $0 \le \mathrm{J}_j \le 1$ for fixed values of $J_{j-1}, \ldots, J_1$, inductively given by

$$J_{j\,\min} = \min\{0, l_j - \sum_{i:j>i} c_{ji}^{+} J_{i\,\max} + c_{ji}^{-} J_{i\,\min}\}$$

$$J_{j\,\max} = \max\{0, l_j - \sum_{i:j>i} c_{ji}^{+} J_{i\,\min} + c_{ji}^{-} J_{i\,\max}\}$$

$J_{j\,\min/\max} = J_{j\,\min/\max}(\lambda;\ J_{j-1}, \dots, J_1)$ are linear combinations of $J_{j-1}, \dots, J_1$ depending on λ.

Let $\min J_j$, $\max J_j$ be the min / max of J_j over its complete range, inductively given by

$$\min J_j = \min\{0,\ l_j - \sum_{i:j>i} c_{ji}^{+} \max J_i + c_{ji}^{-} \min J_i\}$$
$$\max J_j = \max\{0,\ l_j - \sum_{i:j>i} c_{ji}^{+} \min J_i + c_{ji}^{-} \max J_i\};$$

$\min / \max J_j = \min / \max J_j(\lambda)$ depends on λ only.

Theorem 6.1. *Assume $l_1 \neq 0, \cdots, l_\ell \neq 0$. The form $\sigma_{-\lambda}$ is positive if and only if λ satisfies*

$$\min J_1(\lambda) \geq 0 \quad \cdots \quad \min J_\ell(\lambda) \geq 0.$$

If so, the range of (J_i) is the cubic polytope given by the ℓ independent linear inequalities

$$\pi(\mathcal{X}): \quad 0 \leq J_j \leq l_j - L_j, \quad L_j := \sum_{i:j>i} c_{ji} J_i.$$

Remarks. (1) The conditions $\min J_j(\lambda) \geq 0$ are equivalent to $\min J_j(\lambda) = 0$ and amount to ℓ-piecewise linear inequalities in λ. They are certainly satisfied if $l_\ell \gg l_{\ell-1} \gg \cdots \gg l_1 > 0$; one only needs to insure that the first term in $J_j = \mathrm{J}_j(l_j - \sum_{i:j>i} c_{ji} J_i)$ dominates the rest.

(2) The action variables J_j are determined only up to an additive constant depending on λ by the equation $\sigma_\lambda = \frac{1}{2\pi} \sum_j dJ_j \wedge d\phi_j$. The J_j's used here are normalized so that $J_j|_{z_j=0} = 0$. Without this normalization the positivity condition reads $\min J_j(\lambda) = J_j|_{z_j=0}$. The polytope $\pi(\mathcal{X})$ is still given by $J_{j\,\min} \leq J_j \leq J_{j\,\max}$, provided 0 is replaced by $J_j|_{z_j=0}$ in the definition of $J_{j\,\min/\max}$.

The proof of the theorem needs some preparation. We have

$$J_j = \mathrm{J}_j(l_j - L_j), \quad L_j = \sum_{i:j>i} c_{ji} J_i.$$

L_j depends only on the variables $\tilde{\tau}_i$ with $j > i$ and

$$\mathrm{J}_i := \mathrm{J}(\tilde{\tau}_i), \quad \mathrm{K}(\tilde{\tau}) := \log \mathrm{a}(\tilde{\tau}), \quad \mathrm{J}(\tilde{\tau}) := \frac{1}{2} \frac{\mathrm{dK}(\tilde{\tau})}{\mathrm{d}\tilde{\tau}}.$$

Lemma 6.2. *The curvature form of $\mathcal{L}_\lambda$ on $\mathcal{X}_S$is given by*

$$\sigma_\lambda = \sum_{ji=1}^{\ell} \frac{1}{2} \frac{\partial^2 K_\lambda}{\partial \tau_i \partial \tau_j} \frac{1}{4\pi \mathrm{i}} \frac{d\bar{\zeta}_i \wedge d\zeta_j}{\bar{\zeta}_i \zeta_j}.$$

Its ℓth exterior power is

$$\sigma_\lambda^\ell = D_\lambda \prod_{j=1}^{\ell} \tilde{\sigma}_{-\alpha_j/2}, \quad D_\lambda := \prod_j (l_j - L_j), \quad \tilde{\sigma}_{-\alpha_j/2} := \prod_{j=1}^{\ell} \frac{\mathrm{K}''(\tilde{\tau}_j)}{2\pi \mathrm{i}} d\tilde{\tau}_j \wedge d\tilde{\phi}_j,$$

$\tilde{\sigma}_{-\alpha_j/2}$ being the curvature form on a copy of $\mathbb{P}^1$ itself with affine coordinate $\tilde{\zeta}_j = e^{\tilde{\tau}_j + \sqrt{-1}\tilde{\phi}_j}$.

Proof of the Lemma. Write $\sigma_\lambda = \frac{1}{2\pi \mathrm{i}} \bar{\partial}\partial K_\lambda$ in terms of the operators

$$\partial = \sum d\zeta_j \frac{\partial}{\partial \zeta_j}, \quad \bar{\partial} = \sum d\bar{\zeta}_j \frac{\partial}{\partial \bar{\zeta}_j} \quad [d = \bar{\partial} + \partial].$$

Use the relations

$$dK_\lambda = 2 \sum J_j d\tau_j, \quad d\tau_j = \frac{1}{2}(\partial \log \zeta_j + \bar{\partial} \log \bar{\zeta}_j) = \frac{1}{2}\left(\frac{d\zeta_j}{\zeta_j} + \frac{d\bar{\zeta}_j}{\bar{\zeta}_j}\right)$$

to compute

$$\sigma_\lambda = \frac{1}{2\pi \mathrm{i}} \bar{\partial}\partial K_\lambda = \frac{1}{\pi \mathrm{i}} \sum \bar{\partial} J_j \wedge \partial \tau_j = \frac{1}{4\pi \mathrm{i}} \sum \frac{\partial J_j}{\partial \tau_i} \frac{d\bar{\zeta}_i \wedge d\zeta_j}{\bar{\zeta}_i \zeta_j}.$$

This proves the first assertion. To prove the second one, note that

$$\det \frac{\partial J_j}{\partial \tau_i} = \det(\frac{\partial J_j}{\partial \tilde{\tau}_l}) \det(\frac{\partial \tilde{\tau}_l}{\partial \tau_i}) = \prod_j \frac{\partial J_j}{\partial \tilde{\tau}_j}$$

since both determinants are triangular, the second with diagonal entries $= 1$. Since L_j depends only on the variables $\tilde{\tau}_i$ with $j > i$,

$$\frac{\partial J_j}{\partial \tilde{\tau}_j} = \frac{\partial((l_j - L_j)\mathrm{J}_j)}{\partial \tilde{\tau}_j} = (l_j - L_j)\mathrm{J}'(\tilde{\tau}_j) = \frac{1}{2}(l_j - L_j)\,\mathrm{K}''(\tilde{\tau}_j).$$

Thus

$$\det \frac{\partial J_j}{\partial \tau_i} = \prod_j (l_j - L_j) \frac{1}{2} \mathrm{K}''(\tilde{\tau}_j), \quad \sigma_\lambda^\ell = D_\lambda \prod_{j=1}^{\ell} \frac{\mathrm{K}''(\tilde{\tau}_j)}{8\pi \mathrm{i}} \frac{d\bar{\zeta}_j \wedge d\zeta_j}{\bar{\zeta}_j \zeta_j}.$$

Since $\zeta_j = \tilde{\zeta}_j +$ terms depending only on $\tilde{\zeta}_{j+1}, \ldots, \tilde{\zeta}_\ell$, the ζ_j can be replaced by $\tilde{\zeta}_j$ in the last formula. In terms of $\tilde{\tau}_j, \tilde{\phi}_j$ this gives

$$\sigma_\lambda^\ell = D_\lambda \prod_{j=1}^{\ell} \tilde{\sigma}_{-\alpha_j/2}, \quad \tilde{\sigma}_{-\alpha_j/2} := \prod_{j=1}^{\ell} \frac{\mathrm{K}''(\tilde{\tau}_j)}{2\pi \mathrm{i}} d\tilde{\tau}_j \wedge d\tilde{\phi}_j.$$

This proves the second assertion. □

Proof of the Theorem. Generally, a (1,1) form $\frac{1}{2i}\sum K_{ij}d\bar{z}_i \wedge dz_j$ is positive iff the matrix (K_{ij}) is positive definite. So the form $\sigma_{-\lambda}$ is positive definite at a point ξ if and only if the matrix $\partial^2 K_{+\lambda}/\partial\tau_i\partial\tau_j$ is, and this happens if and only if all principal minors of this matrix are positive [Gantmacher, 1959, p.307]. Stated in terms of forms rather than matrices, this criterion reads as follows. *A* $(1,1)$ *form* σ *is positive at a point if and only if the sequence of exterior powers* $\sigma^\ell, \sigma^{\ell-1}, \dots, \sigma$ *is positive on a complete flag in the tangent space.* For a form of top degree, like σ^ℓ, positivity means $\sigma^\ell = D\prod \frac{1}{2i}d\bar{z}_j \wedge dz_j$ with $D > 0$, by definition.

To argue by induction on ℓ, write the $\mathbb{P}^1$-chain $\mathcal{X}_s$ as $\mathcal{X}_s = G_\ell \times_{B_\ell} \mathcal{X}_{s'}$ where $\mathcal{X}_{s'}$ is the $\mathbb{P}^1$-chain built from the sequence s′ obtained from w by omitting the last letter $i = \ell$. $\mathcal{X}_{s'}$ is embedded in $\mathcal{X}_s$ as the fiber of $\mathcal{X}_s \to G_\ell/B_\ell \approx \mathbb{P}^1$ over the base-point $1B_\ell$. The coordinate z_ℓ on $\mathcal{X}_s$ is the pullback of the affine coordinate on $\mathbb{P}^1$ centered at the point $s_\ell B$, also denoted z_ℓ. The submanifold $\mathcal{X}_{s'}$ of $\mathcal{X}_w$ is then given by the equation $z_\ell \circ s_\ell = 0$. Since $\tilde{\zeta}_\ell = \zeta_\ell$, the formula for σ_λ^ℓ shows that at such a point

$$\sigma_\lambda^\ell\Big|_{z_\ell \circ s_\ell = 0} = \left[(l_\ell - L_\ell)\frac{\mathrm{K}''(\tau_\ell)}{8\pi \mathrm{i}}\frac{d\bar{\zeta}_\ell \wedge d\zeta_\ell}{\bar{\zeta}_\ell\zeta_\ell}\right]_{z_\ell \circ s_\ell = 0} \wedge \sigma_{\lambda'}^{\ell-1}$$

where λ' is obtained by omitting the ℓth component and $\sigma_{\lambda'}$ is the corresponding form on $\mathcal{X}_{s'}$. The form in brackets is a pullback of this form G_ℓ/B_ℓ written in the coordinate ζ_ℓ. It is positive if and only if $l_\ell - L_\ell$ is positive. Since every point of $\mathcal{X}_s$ can be brought to $\mathcal{X}_{s'}$ by the action of the compact form U_ℓ of G_ℓ, which leaves σ_λ invariant, one finds that $\sigma_\lambda^\ell > 0$ on $\mathcal{X}_s$ if and only if

$$l_\ell - L_\ell > 0 \text{ and } \sigma_{\lambda'}^{\ell-1} > 0 \text{ on } \mathcal{X}_{s'}.$$

$l_\ell - L_\ell = l_\ell + \cdots$ is independent of $\tilde{\tau}_\ell$ and $J_\ell = \mathrm{J}_\ell(l_\ell - L_\ell)$, $0 \le \mathrm{J}_\ell \le 1$. Hence $l_\ell - L_\ell > 0$ amounts to $J_\ell \ge 0$ and not$\equiv 0$ on $\mathcal{X}_s$. Since $J_\ell = \mathrm{J}_\ell l_\ell + \cdots$ the latter condition is satisfied as soon as $l_\ell \ne 0$, which holds by hypothesis, so that we may replace $l_\ell - L_\ell > 0$ by $J_\ell \ge 0$. Thus $\sigma_{-\lambda} = -\sigma_\lambda$ is positive if and only if $J_1 \ge 0, \cdots, J_\ell \ge 0$ throughout, i.e., $\min J_j(\lambda) = 0$. If so, the range of the $J_j = \mathrm{J}_j(l_j - L_j), 0 \le \mathrm{J}_j \le 1$, is indeed given by $0 \le J_j \le l_j - L_j$. □

The condition for the positivity of σ_λ will be referred to as the *positivity condition* on λ.

7 Quantization

Projective embedding. Kodaira's theorems assert the following consequences of positivity [Griffiths and Harris, 1978, p. 156, 181].

Theorem 7.1. *Assume* λ *satisfies the positivity condition.*

(1) *Let $\Omega^q(\mathcal{L}_\lambda)$ be the sheaf of holomorphic q-forms with values in $\mathcal{L}_\lambda$.*

$$\mathrm{H}^p(\mathcal{X}_\mathrm{s}, \Omega^q(\mathcal{L}_\lambda)) = 0 \text{ for } p + q > \ell.$$

(2) *Let $\mathcal{O}(\mathcal{L}_\lambda)$be the sheaf of holomorphic sections of $\mathcal{L}_\lambda$. The natural map*

$$\mathcal{X}_\mathrm{s} \to \mathbb{P}(\mathcal{H}^*), \mathcal{H} := \mathrm{H}^0(\mathcal{X}_\mathrm{s}, \mathcal{O}(\mathcal{L}_{k\lambda}))$$

is an embedding for any sufficiently large multiple $k\lambda$ of λ.

Some comments

In general. the natural map $\mathcal{X} \to \mathbb{P}(\mathcal{H}^*)$ into the projective space dual to the space $\mathcal{H} := \mathrm{H}^0(\mathcal{X}, \mathcal{O}(\mathcal{L}))$ of global holomorphic sections is defined as follows. Given $\xi \in \mathcal{X}$, choose any nonzero element $e(\xi)$ in the line $\mathcal{L}(\xi)$. The map $f \mapsto f(\xi)/e(\xi)$ defines a linear functional $e_\xi \in \mathcal{H}^*$, which up to a scalar depends only on ξ and represents the image of ξ in $\mathbb{P}(\mathcal{H}^*)$. The line bundle $\mathcal{L}$ on $\mathcal{X}$ is the pullback of the hyperplane bundle on the ambient projective space $\mathbb{P}(\mathcal{H}^*)$, so its Chern class is the pullback of that of the hyperplane bundle. These Chern classes are represented by (normalized) curvature forms $\sigma_\mathcal{X}, \sigma_{\mathbb{P}^N}$ for any unitary metrics on these line bundles. As form, however, $\sigma_\mathcal{X}$ need not coincide with the pullback of the Fubini–Study form $\sigma_{\mathbb{P}^N}$; on the contrary, the relation between these forms is generally a delicate question. In particular, the unitary norm on the Hilbert space of sections, which is of interest for applications to representation theory, depends on the form $\sigma_\mathcal{X}$ itself.

Some notation

Throughout λ is assumed to satisfy the positivity condition and will remain fixed. It will be dropped as a subscript, as will be the subscript s, e.g., on σ and $\mathcal{X}$. Write T for the image of T_s as transformation group on $\mathcal{X}$. Like any complex torus, $T = T_\mathrm{U} T_\mathrm{P}$ has a unique decomposition into a 'unitary' real form $T_\mathrm{U} \approx \prod \mathrm{U}(1)$ and a 'positive' real form $T_\mathrm{P} \approx \prod \mathrm{GL}(1, \mathbb{R})_{\text{positive}}$. The weights ϖ_i by which T acts on the coordinates z_i via $z_i \circ h = h^{\varpi_i} z_i$ form a $\mathbb{Z}$-basis for a lattice $\mathfrak{t}_\mathrm{P}(\mathbb{Z})$ in the real dual of $\mathfrak{t}_\mathrm{P}^*$ of the Lie algebra of T_P. Let N_i be the dual basis for the dual lattice $\mathfrak{t}_\mathrm{P}(\mathbb{Z})$ in $\mathfrak{t}_\mathrm{P}(\mathcal{X}) \subset \mathfrak{t}_\mathrm{P}$, defined by $\langle \varpi_i, N_j \rangle = \delta_{ij}$. The N_i form an $\mathbb{R}$-basis for $\mathfrak{t}_\mathrm{P}$ and the $\mathrm{i}N_i$ for $\mathfrak{t}_\mathrm{U}$. The imaginary unit $\mathrm{i} := \sqrt{-1}$ will be written explicitly whenever $\mathfrak{t}_\mathrm{U}$ is concerned: $\mathfrak{t}_\mathrm{P}$ is taken as "the" real form of $\mathfrak{t}$ and $\mathfrak{t} = \mathfrak{t}_\mathrm{P} + \mathfrak{t}_\mathrm{U}$ as the decomposition of $\mathfrak{t}$ into "real" and "imaginary" parts. For example, write

$$\mathrm{i}H \in \mathfrak{t}_\mathrm{U} : H = \sum_i \epsilon_i N_i, \epsilon_i \in \mathbb{R}, \quad \mathrm{i}\eta \in \mathfrak{t}_\mathrm{U}^* : \eta = \sum_i n_i \varpi_i \, , \; n_i \in \mathbb{R}.$$

Thus i functions as the map $\mathfrak{t}_\mathrm{P} \to \mathfrak{t}_\mathrm{U}$. The pairing of $H \in \mathfrak{t}_\mathrm{P}$ and $\eta \in \mathfrak{t}_\mathrm{P}^*$ is written as $\langle \eta, H \rangle$ or $\langle H, \eta \rangle$.

The moment set

Any element $\mathrm{i}H = \mathrm{i}\sum \epsilon_i N_i$ of $\mathfrak{t}_{\mathrm{U}}$ has $J_{\mathrm{i}H}(\xi) := 2\pi \sum \epsilon_i J_i(\xi)$ as a Hamiltonian function, in the sense that $dJ_{\mathrm{i}H} = -i(\mathrm{i}H)\sigma$, the inner product of σ with the vector field on $\mathcal{X}$ induced by $-\mathrm{i}H \in \mathfrak{t}_{\mathrm{U}}$. This follows from the formulas $\sigma = \frac{1}{2\pi}\sum dJ_i \wedge d\phi_i$, $\iota(\mathrm{i}N_i)dJ_j = 0$, $\iota(\mathrm{i}N_i)d\phi_j = \delta_{ij}$. In particular the Hamiltonian function for $\mathrm{i}N_i$ is the action variable $J_i(\xi)$. Since $J_{\mathrm{i}H}(\xi)$ depends linearly on $\mathrm{i}H$, it defines a map $\mathcal{X} \to \mathfrak{t}_{\mathrm{U}}^*$, the *moment map* of the T_{U} action on $\mathcal{X}$. Transferred to $\mathfrak{t}_{\mathrm{P}}^*$ it becomes a map $\pi : \mathcal{X} \to \mathfrak{t}_{\mathrm{P}}^*$, defined by $\langle \pi(\xi), H\rangle := J_{\mathrm{i}H}(\xi)$, i.e., $\pi(\xi) = \sum J_i(\xi)\varpi_i$. The image $\pi(\mathcal{X})$ of π in $\mathfrak{t}_{\mathrm{P}}^*$ can evidently be identified with the range of the J_i and will be called the *moment set*.

The weight set

Let $\mathcal{H}$ be the space of holomorphic sections of $\mathcal{L}$, a finite-dimensional space with a positive definite norm defined by the formula

$$\|f\|^2 := \int_{\mathcal{X}} |f|^2 \sigma^{\ell}.$$

There is a natural representation of the complex torus on $\mathcal{H}$, defined by

$$(U_h f)(\xi) := f(h^{-1}\xi),$$

which is unitary on the real torus T_{U}.

Any $\mathrm{i}\,\eta \in \mathfrak{t}_{\mathrm{U}}^*(\mathbb{Z})$ defines a Laurent monomial $z^{\eta} := z_1^{n_1}\cdots z_{\ell}^{n_{\ell}}$ in the coordinates $z_i = e^{\tau_i + \mathrm{i}\phi_i}$ on $\mathcal{X}$ and hence a holomorphic section $f_\eta = z^\eta c^\lambda$ on the coordinate domain. (f_η transforms by $\tilde{\eta} := \eta + \tilde{\lambda}$ under T_{U} since c^λ transforms by $\tilde{\lambda}$.) The *weight set* Π is the set of weights $\eta \in \mathrm{i}^{-1}\mathfrak{t}_{\mathrm{U}}^*(\mathbb{Z})$ for which f_η extends to a holomorphic section on all of $\mathcal{X}$, a condition which may be reformulated in other ways.

Lemma 7.2. *Let $f = z^\eta c^\lambda$. The following conditions are equivalent.*

(1) *$f(\xi)$ extends to a global holomorphic section on $\mathcal{X}$.*
(2) *$|f(\xi)|$ is bounded on $\mathcal{X}$.*
(3) *$|f(\xi)|^2\sigma^\ell$ is integrable over $\mathcal{X}$.*

Proof. It suffices to prove the equivalence locally, i.e., with $\mathcal{X}$ replaced by a neighborhood of each of its points. Since the coordinates are meromorphic functions on $\mathcal{X}$, f is a meromorphic section on $\mathcal{X}$, represented locally by a meromorphic function, regular on the coordinate domain, and for such a function the local conditions are indeed equivalent. □

Theorem 7.3. *The weight set $\Pi \subset \mathfrak{t}_{\mathrm{P}}^*$ is $\Pi = \pi(\mathcal{X}) \cap \mathrm{i}^{-1}\mathfrak{t}_{\mathrm{U}}^*(\mathbb{Z})$, i.e., $f_\eta = z^\eta c^\lambda$ extends to a global holomorphic section on all of $\mathcal{X}$ if and only if η belongs to the image $\pi(\mathcal{X})$ of the moment map. These f_η, $\eta \in \Pi$, form a basis for $\mathcal{H}$.*

Proof. The coordinates $z_i = e^{\tau_i + \mathrm{i}\phi_i}$ on $\mathcal{X}$ provide a bijection

$$T_{\mathrm{U}}\backslash\mathcal{X} \leftrightarrow \mathfrak{t}_{\mathrm{P}}, \quad T_{\mathrm{U}}\xi \mapsto H := \sum \tau_i(\xi)N_i,$$

defined on the coordinate domain. Equivalently, $\xi = e^H\xi_1$ where ξ_1 has coordinates $z_1 = \cdots = z_\ell = 1$. The pointwise square norm $a(\xi) = |c^\lambda(\xi)|^2$ is really a function on $T_{\mathrm{U}}\backslash\mathcal{X}$. Use the bijection $T_{\mathrm{U}}\xi \mapsto H$ to write the equation $K(\xi) = \log a(\xi)$ as $K(H) = \log a(H)$ in terms of $H \in \mathfrak{t}_{\mathrm{P}}$.

By the lemma, f_η extends iff its pointwise square-norm is bounded. This square norm is

$$|f_\eta|(\xi)^2 = |z^\eta c^\lambda(\xi)|^2 = |z|^{2\eta}e^{-K(\xi)} = e^{2\langle\eta,H\rangle - K(H)}$$

where $H = \sum_i \epsilon_i N_i$, $\epsilon_i = \tau_i(\xi)$. Thus the condition is precisely that $2\langle\eta, H\rangle - K(H)$ be bounded from above as function of H. Since the curvature form

$$\sigma = \sum_{ji=1}^{\ell} \frac{1}{2}\frac{\partial^2 K(H)}{\partial\tau_i\partial\tau_j}\frac{1}{4\pi\mathrm{i}}\frac{d\bar\zeta_i \wedge d\zeta_j}{\bar\zeta_i\zeta_j}$$

is positive definite, so is the Hessian matrix of $K(H)$. Hence $K(H)$ *is strictly convex* as a function of $H \in \mathfrak{t}_{\mathrm{P}}$ [Rockafellar, 1970, p. 27]. For a smooth convex function $K(H)$, $2\langle\eta, H\rangle - K(H)$ achieves a finite maximum iff its partials vanish at some point X_o [loc. cit., p. 258], and this amounts to $2n_i = \partial K/\partial\tau_i = \pi_i$ at X_o, i.e., $\eta \in \pi(\mathcal{X})$. The elements f_η, $\eta \in \Pi$, are precisely the weight vectors for the representation of T_{U} on $\mathcal{H}$, hence form a basis. □

Corollary 7.4. *The function* $Z(\mathrm{i}H) :=$ trace$(e^{\mathrm{i}H} \mid \mathcal{H})$ *of* $e^{\mathrm{i}H} \in T_{\mathrm{U}}$ *is given by the formula*

$$Z(\mathrm{i}H) = \sum_{\eta\in\Pi} e^{\mathrm{i}\eta(H)}.$$

A comment

The determination of the weight set by convex calculus, as in the proof of the theorem, is as general as it is simple. Useful tools are available, especially the apparatus of conjugate functions. By definition, the *conjugate* of a convex function $K(H)$ is $K^c(\eta) := \max\{2\langle\eta, H\rangle - K(H) \mid H \in \mathfrak{t}_{\mathrm{P}}\}$ as function on the dual space $\mathfrak{t}_{\mathrm{P}}^*$. (In this context max $= \infty$ is allowed, by convention.) In the language of convex analysis, the weight set could be described as the set of integral points in the domain of the conjugate function $K^c(\eta)$, $\eta \in \mathfrak{t}_{\mathrm{P}}^*$ of the Kähler potential $K(H)$, $H \in \mathfrak{t}_{\mathrm{P}}$. To an analyst, convex conjugates will be familiar as asymptotic phase function according to the principle of stationary phase [Guillemin and Sternberg, 1977, p. 397].

Partition functions

Whether the function $Z(\mathrm{iH})$ is to be called *(quantum) partition function* or *character* is a matter of taste and tradition. The prejudice displayed here is adopted for this occasion only, because of a curious relation between toric geometry and statistical mechanics, as will now be explained.

Fix $\mathrm{i}H \in \mathfrak{t}_{\mathrm{U}}$ and let $J := J_{\mathrm{i}H}$ be its Hamiltonian function. The classical density associated to this Hamiltonian is the function in $\mathcal{X}$ defined by the formula

$$\rho_{\text{classical}}(\xi) := \frac{1}{Z_{\text{classical}}} \frac{e^{-\beta J(\xi)}}{\ell!}, \quad \int_{\mathcal{X}} \rho_{\text{classical}} \, \sigma^{\ell} = 1$$

[Honerkamp 1998, p. 28]. The denominator $Z_{\text{classical}}$ is a normalizing factor, depending only on H:

$$Z_{\text{classical}} := \int_{\mathcal{X}} e^{-\beta J} \frac{\sigma^{\ell}}{\ell!}.$$

It can be written in the form

$$Z_{\text{classical}} := \int_{\mathcal{X}} e^{-\beta J + \sigma}$$

the exponential being taken in the exterior algebra with the understanding that the integral picks out the component of the relevant degree $2\ell = \dim_{\mathbb{R}} \mathcal{X}$.

On the other hand, the quantum density is the density in the spectral resolution of $e^{-\beta \mathrm{i}H}$ as operator on $\mathcal{H}$:

$$\rho_{\text{quantum}}(\eta) := \frac{m(\eta)}{Z_{\text{quantum}}} e^{-\beta \mathrm{i}\eta}, \quad \sum_{\eta} \rho_{\text{quantum}}(\eta) = 1,$$

$m(\eta)$ being the multiplicity of the $\mathrm{i}\eta$ as eigenvalue of $\mathrm{i}H$ on $\mathcal{H}$. The normalizing factor Z_{quantum} is therefore

$$Z_{\text{quantum}} = \sum_{\eta} m(\eta) e^{-\beta \mathrm{i}\eta} = \mathrm{Tr}(e^{-\beta \mathrm{i}H} \mid \mathcal{H}).$$

[Honerkamp, 1998, p. 204]. To bring out the dependence on $\mathrm{i}H \in \mathfrak{t}_{\mathrm{U}}$ write $\rho(\mathrm{i}H; \xi)$ and $Z(\mathrm{i}H)$, if necessary.

The fundamental relation between the classical and quantum partition function is **Kirillov's Formula** [Berline, Getzler, and Vergne, p. 250]. In the present situation it states that for $\mathrm{i}H$ in a neighborhood of zero in $\mathfrak{t}_{\mathrm{U}}$, on which the exponential map is one-to-one, one has

$$\int_{\mathcal{X}} e^{J_{\mathrm{i}H} + \sigma} \widehat{A}(\mathrm{i}H) = \sum_{\eta} m(\eta) e^{\mathrm{i}\langle \eta, H \rangle}.$$

The sum on the right is the quantum partition function and the integral on the left is the classical partition function, except for the extra factor $\widehat{A}(\mathrm{i}H)$ under the integral, a characteristic class involving the Riemann curvature of $\mathcal{X}$, whose definition can be found loc. cit. (Its appearance here suggests that the classical definition of the partition function should be modified so as to incorporate this factor, a modification which is irrelevant in a flat phase space.) That $\mathrm{i}H$ has to remain in neighborhood of zero is a serious restriction.

Returning to the toric setting, assume that Kodaira's map $\psi : \mathcal{X} \to \mathbb{P}(\mathcal{H}^*)$, $\xi \mapsto \psi_\xi$, is an embedding. Let $\mathcal{X}'$ be the image of $\mathcal{X}$ and T' the image of T as a transformation group of $\mathcal{X}'$. Let $\mathcal{L}'$ be the line bundle on $\mathcal{X}'$ induced by the hyperplane bundle on $\mathcal{H}^*$. Let σ' be the curvature form for the unitary metric on $\mathcal{L}'$ induced by any T_{U}-invariant unitary metric on $\mathcal{H}^*$, for example the metric defined above.

Let $J'_{\mathrm{i}H'}(\xi)$ be a Hamiltonian function for $\mathrm{i}H' \in \mathfrak{t}'_{\mathrm{U}}$ acting on $(\mathcal{X}', \sigma')$, i.e., $dJ'_{\mathrm{i}H'} = -i(\mathrm{i}H')\sigma'$. This equation determines $J'_{\mathrm{i}H'}(\xi)$ up to a constant depending linearly on $\mathrm{i}H'$. Fix this constant as follows. Let ξ_1 be a base-point for $\mathcal{X}$ in the open T-orbit, say the point with all coordinates z_i equal to 1. Let ξ'_1 be it image in $\mathcal{X}'$, and require that $J'_{\mathrm{i}H'}(\xi'_1) = J_{\mathrm{i}H}(\xi_1)$ if $\mathrm{i}H' \in \mathfrak{t}'_{\mathrm{U}}$ is the image of $\mathrm{i}H \in \mathfrak{t}_{\mathrm{U}}$. Thus we have another set of data $\mathcal{X}', \sigma' \cdots$ like $\mathcal{X}, \sigma \cdots$.

Lemma 7.5. *Under the identification* $\mathfrak{t}' = \mathfrak{t}$,

$$Z'_{\mathrm{classical}}(\mathrm{i}H) = Z_{\mathrm{classical}}(\mathrm{i}H), \quad Z'_{\mathrm{quantum}}(\mathrm{iH}) = Z_{\mathrm{quantum}}(\mathrm{iH}).$$

Proof. The equality of the quantum partition functions is obvious, since the line bundles $\mathcal{L}$ and $\mathcal{L}'$ correspond under $\mathcal{X} = \mathcal{X}'$. The equality of the classical partition functions, however, is not, since the forms σ and σ' need not coincide. What is obvious is that their cohomology classes coincide, since both represent the Chern class of $\mathcal{L}' = \mathcal{L}$. Thus $\sigma' = \sigma + d\varphi$ for some smooth 1-form φ, which may be chosen T_{U}-invariant like σ and σ'. Then $d \circ \iota(\mathrm{i}H)\varphi = -\iota(\mathrm{i}H) \circ d\varphi$ for all $\mathrm{i}H \in \mathfrak{t}_{\mathrm{U}}$ and (suppressing H) the equations $dJ' + \iota\sigma' = 0$, $dJ + \iota\sigma = 0$, give $dJ' = dJ + d\iota\varphi$, i,e $J' = J + \iota\varphi + C$. The normalization $J'(\xi_1) = J(\xi_1)$ implies $C = 0$, so $J' + \sigma' = J + \sigma + (d + \iota)\varphi$. The rest of the proof uses three observations: (1) $(d + \iota)(J + \sigma) = 0$, (2) $(d + \iota)$ is a derivation, and (3) $(d + \iota)\Phi = d\Phi +$forms of lower degree. Armed with these one computes

$$\begin{aligned}\int_{\mathcal{X}} e^{J+\sigma} e^{(d+\iota)\varphi} &= \int_{\mathcal{X}} (e^{J+\sigma} + e^{J+\sigma}(d+\iota)\Phi)) \quad [e^{(d+\iota)\varphi} = 1 + (d+\iota)\Phi] \\ &= \int_{\mathcal{X}} (e^{J+\sigma} + (d+\iota)(e^{J+\sigma}\Phi)) \quad [\text{by (1) and (2)}] \\ &= \int_{\mathcal{X}} e^{J+\sigma} + 0 + 0 \quad [\text{by Stokes and (3)}].\end{aligned}$$

This equation says that $Z'_{\mathrm{classical}} = Z_{\mathrm{classical}}$ except for the factor β, which may be absorbed into J and J'. □

The image of the measure $|\sigma^\ell| = |\prod \frac{1}{2\pi} dJ_i \wedge d\phi_i|$ under $\eta = \pi(\xi)$ i.e., $n_i = J_i(\xi)$ is evidently $\rho(\eta)d\eta$, the Haar measure $d\eta = \prod dn_i$ multiplied by the indicator function $\rho(\eta)$ concentrated on $i\eta \in \pi(\mathcal{X})$. The partition function becomes

$$Z_{\mathrm{classical}}(\mathrm{i}H) = \int_{\eta \in \pi(\mathcal{X})} e^{-\beta\langle \eta, H\rangle} d\eta.$$

after an integration over the angle variables, the Laplace transform of $\rho(\eta)$. The lemma therefore implies that the moment sets for $\mathcal{X}'$ and $\mathcal{X}$ coincide, i.e., $\pi'(\mathcal{X}) = \pi(\mathcal{X})$ under the identification $\mathfrak{t}' = \mathfrak{t}$.

The main point is the topological nature of classical partition function: $Z_{\text{classical}}(H)$ *depends only on the Chern class of the line bundle.* (This phenomenon is evident in the setting of Kirillov's formula in the form presented loc. cit, where the lemma is understood in terms of equivariant cohomology. The calculation of Fourier transforms like one above was one of the first applications of that theory [Berline and Vergne, 1983].) Replacing σ by σ' allows one to apply the theory of toric varieties, which provides information to the moment map and the weight set. There is in particular an intriguing description of the moment set in terms of the maximum entropy principle of statistical mechanics, due to C. Lee [1990, p. 13, Ewald 1996, p. 298].

The path integral formula

The unitary representation of T on $\mathcal{H}$ is a quantization of the symplectic manifold $\mathcal{X}$, σ associated to the totally complex, positive polarization provided by its complex structure [Woodhouse, 1991]; the wave functions are the holomorphic sections $f(\xi)$ which make up $\mathcal{H}$. The quantum evolution operators are those representing T. On the other hand, the canonical variables $(J, \phi) := (J_i, \phi_i)$ provide a natural real polarization, at least on their domain of definition, whose wave functions are functions of ϕ. (It is natural to choose ϕ rather than J, because (J, ϕ) has an interpretation as coordinates on the cotangent bundle of T with ϕ as coordinate on the base.) We now leave the realm of geometric quantization for that of path integral quantization. The evolution operator generated by a Hamiltonian function $H(J, \phi)$ is then supposed to be an integral operator on wave functions $\psi(\phi)$,

$$U_H\psi(\phi'') = \int K_H(\phi'', \phi')\psi(\phi')d\phi',$$

whose kernel is given by a path integral, formally written as

$$K_H(\phi'', \phi') := \int e^{\frac{i}{\hbar}S_H[J(t),\phi(t)]}d\phi.$$

$S_H[J(t), \phi(t)]$ is the action along a path $(J(t), \phi(t) \mid t' \le t \le t'')$, defined by

$$S_H[J(t), \phi(t)] := \int_{t'}^{t''} [J d\phi - H(J, \phi)dt].$$

The path integral for $K_H(\phi'', \phi')$ is supposed to be taken over all such paths with $\phi(t') = \phi', \phi(t'') = \phi''$. The quantum partition function $Z(H)$, now written without subscript, is the trace of U_H, obtained from $K_H(\phi'', \phi')$ by setting $\phi'' = \phi'$ and integrating. The difficulties encountered when trying to define path integrals as true integrals are well known and need not be discussed here. Instead,

I shall give a recipe for evaluating $Z(H)$, which can be taken as a definition of this path integral in the special case we are concerned with, i.e., for a $\mathbb{P}^1$-chain $\mathcal{X}$, equipped with its curvature form σ and canonical coordinates (J, ϕ), and for Hamiltonians which are linear functions $H = H(J)$ of the J's only.

Consider first the case of $\mathbb{P}^1$ itself. The coordinates are (J, ϕ) with J ranging over an interval $\pi(\mathcal{X}) : J_{\min} \leq J \leq J_{\max}$ and ϕ over $\mathbb{R}/2\pi\mathbb{Z}$. The Hamiltonian is $H(J) := HJ$ for some constant also denoted H. *Define* the path integral

$$Z(H) = \int_{\phi(0)\equiv\phi(1)} e^{\mathrm{i}S_H[J(t),\phi(t)]} d\phi$$

by the following recipe.

(1) Fix N and replace $S_H[J(t), \phi(t)]$ by the sum

$$S^{(N)} := [J\phi]_0^1 - \sum_{i=0}^{N-1} \phi_i \Delta J_i + HJ_i\Delta t,$$

$$\Delta J_i := J_{i+1} - J_i, \qquad \Delta t := 1/N.$$

(2) Fix $n \in \mathbb{Z}$ and integrate over all sequences $(J_0\phi_0, \dots, J_N\phi_N)$

$$J_{\min} \leq J_i \leq J_{\max}, \qquad -\infty < \phi_i < \infty, \qquad \phi_N - \phi_0 = 2\pi n.$$

(3) Sum over $n \in \mathbb{Z}$.

(4) Take the limit $N \to \infty$.

Written as a single formula (not recommended), this amounts to the definition

$$Z(H) := \lim_{N\to\infty} \sum_{n=0}^{\infty} \int_{J_{\min}}^{J_{\max}} dJ_N \int_{-\infty}^{\infty} d\phi_N \cdots \int_{J_{\min}}^{J_{\max}} dJ_1 \int_{-\infty}^{\infty} d\phi_1$$
$$\times\, \delta(\phi_N - \phi_0 - 2\pi n) e^{i(J_N\phi_N - J_0\phi_0) - i\sum_{i=1}^{N-1} \phi_i (J_i - J_{i-1}) + HJ_i(\frac{1}{N})}.$$

The integrals in this formula must be interpreted in the sense of generalized functions, i.e., as limits of integrals against compactly supported test functions tending to the constant function $\equiv 1$. The integrals over $J_{\max} \leq J \leq J_{\min}$, in particular, are defined as limits against continuous compactly supported functions identically $\equiv 1$ on this closed interval. In (1), an integration by parts has been applied to the integral for S_H, replacing the form $Jd\phi$ by $-\phi dJ$ before replacing $J(t), \phi(t)$ by $(J_1\phi_1, \dots, J_N\phi_N)$. In (2), the variables ϕ_i replacing $\phi(t)$ are allowed to vary from $-\infty$ to $+\infty$, while the endpoint condition $\phi(0) \equiv \phi(1) \bmod 2\pi$ is interpreted as the *sum* over all such paths and *not* by requiring that the angle itself return to its initial value. This is an essential feature.

In the case of a general toric $\mathbb{P}^1$-chain the coordinate pair (J, ϕ) is replaced by n-coordinate pairs (J_i, ϕ_i) and the interval $\pi(\mathcal{X}) = \{J_{\max} \leq J \leq J_{\min}\}$ by the cubic polytope $\pi(\mathcal{X}) = \{J_{i\,\min} \leq J_i \leq J_{i\,\max}\}$. The path integral $Z(H)$ is then defined as an n-fold integral, each of which is given by the recipe above.

Theorem 7.6. *The path integral for $Z(H)$ exists and is given by*

$$Z(H) = \sum_{\eta \in \Pi} e^{\mathrm{i}\langle \eta, H \rangle}.$$

Proof. Follow the steps in the definition:

(1) Fix N. As indicated let

$$S_H = \int_0^1 (J d\phi - HJ dt) = [J\phi]_0^1 - \int_0^1 (\phi dJ + HJ dt),$$

$$S_H^{(N)} := [J\phi]_0^1 - \sum_{k=0}^{N-1} [\phi_k (J_{k+1} - J_k) + \frac{1}{N} H J_k].$$

(2) Fix n. For fixed k, the part of the integral involving $(J_k \phi_k)$ is of the form

$$\int_{J_{\min}}^{J_{\max}} dJ_k \int_{-\infty}^{\infty} d\phi_k \, e^{-\mathrm{i}\phi_k (J_{k+1} - J_k) - \frac{\mathrm{i}}{N} H J_k}.$$

Use the Fourier inversion formula on the real line in the form

$$\int_{-\infty}^{\infty} \frac{d\phi_k}{2\pi} e^{-\mathrm{i}\phi_k (J_{k+1} - J_k)} = \delta(J_{k+1} - J_k),$$

$$\int_{-\infty}^{\infty} dJ_k \, f(J_k) \delta(J_k - J_{k-1}) = f(J_{k-1}).$$

This gives

$$Z(H)^{(N,n)} := \int_{J_{\min}}^{J_{\max}} dJ_N \int_{J_{\min}}^{J_{\max}} dJ_0 \, e^{i(J_N \phi_N - J_0 \phi_0) - iHJ_0} \, \delta(J_N - J_0)$$

$$= \int_{J_{\min}}^{J_{\max}} d\eta \, e^{-\mathrm{i}\eta(\phi_N - \phi_0 + H)}.$$

(3)By Poisson summation,

$$\sum_{n \in \mathbb{Z}} Z(H)^{(N,n)} = \sum_{n \in \mathbb{Z}} \int_{J_{\min}}^{J_{\max}} d\eta \, e^{-\mathrm{i}\eta(2\pi n + H)} = \sum_{\eta \in \Pi} e^{\mathrm{i}\eta H}.$$

(4) The result of (3) is independent of N; the limit $N \to \infty$ is trivial.

The proof has been formulated so that it remains valid without change in the setting of $\mathbb{P}^1$-chains. □

Some comments

(1) The appearance of the delta function $\delta(J_N - J_0)$ in step (2) may be interpreted to mean that the path integral for $Z(H)$ is automatically concentrated on paths $(J(t), \phi(t))$ closing in J as well as in ϕ, which is not part of the input.

(2) The theorem gives $Z(\mathrm{i}H)$ again as a sum over the set of integral points in the range of the action variables (J_i). Since the proof is nothing but Poisson's summation formula applied to definition of the path integral, as specified in the recipe, the canonical coordinates themselves remain as the essential ingredient. The applicability of the path integral in many other situations seems to indicate that it does capture some basic feature, and not only in a formal way.

(3) In a more philosophical vein, it appears that the unitary-positive splitting $T = T_{\mathrm{U}} T_{\mathrm{P}}$ is reflected in the statistical mechanics vs. quantum mechanics setting of the two formulas for the partition functions. The odd nature of this juxtaposition is an old puzzle [Feynman and Hibbs, 1965, p. 296].

References

[1] V.I. Arnold, *Mathematical Methods of Classical Mechanics*, Second Edition, Springer Verlag, New York, 1989.

[2] M. Atiyah, Convexity and commuting Hamiltonians. *Bull. London Math. Soc*, **14**, (1982), 1–15.

[3] N. Berline, E. Getzler, and M. Vergne, *Heat Kernels and Dirac Operators*, Springer Verlag, New York, 1992.

[4] N. Berline, E. Getzler and M. Vergne, *Fourier transforms of orbits of the coadjoint represenation*; In Proceedings of the conference on representation theory of reductive groups (Park City, Utah, 1982), Progress in Mathematics, vol. 40, Birkhäuser, Boston, 1983.

[5] G. Ewald, *Combinatorial Convexity and Algebraic Geometry*, Springer Verlag, New York, 1996.

[6] R.P. Feynman and A.R. Hibbs, *Quantum Mechanics and Path Integrals*, McGraw-Hill Publishing Co., New York, 1965.

[7] F.R. Gantmacher, *The Theory of Matrices*, Volume I. Chelsea Publishing Co., New York, 1959.

[8] P. Griffiths and J. Harris, *Principles of Algebraic Geometry*, John Wiley & Sons, New York, 1978.

[9] M. Grossberg and Y. Karshon, Bott towers, complete integrability, and the extended character of representations, *Duke Math*.J., **76** (1994), 23–58.

[10] V. Guillemin and S. Sternberg. *Geometric Asymptotics*, American Math. Soc. Providence, 1977.

[11] J. Honerkamp, Statistical Physics, Springer Verlag, New York, 1998.

[12] A.A. Kirillov, The method of orbits in the theory of unitary representations of Lie groups. *Functional Anal. i Prilozhen*. **2** no 1, (1968), 96–98.

[13] B. Kostant, Quantization and unitary representations; In *Lectures in Modern Analysis and Applications III,* Lecture Notes in Mathematics, **170**, Springer Verlag, New York, 1970.

[14] C. Lee, Some recent results on convex polytoptes, *Contemporary Math*, **144** (1990), 3–19.

[15] R. T. Rockafellar, *Convex Analysis*, Princeton U. Press, Princeton, New Jersey, 1970.

[16] N. M. J. Woodhouse, *Geometric Quantization*, Clarendon Press, Oxford, 1991.

W. Rossmann
Department of Mathematics
University of Ottawa
Ottawa, Cananda
rossmann@uottawa.ca

Projective Schur Functions as Bispherical Functions on Certain Homogeneous Superspaces

Alexander Sergeev[1] [2]

ABSTRACT Here I show that the projective Schur functions can be interpreted as bispherical functions of either the triple $(\mathfrak{q}(n), \mathfrak{q}(n) \oplus \mathfrak{q}(n), \mathfrak{q}(n))$, where $\mathfrak{q}(n)$ is the "odd" (queer) analog of the general linear Lie algebra, or the triple $(\mathfrak{pe}(n), \mathfrak{gl}(n|n), \mathfrak{pe}(n))$, where $\mathfrak{pe}(n)$ is the periplectic Lie superalgebra which preserves the nondegenerate odd bilinear form (either symmetric or skew-symmetric). Making use of this interpretation I characterize projective Schur functions as common eigenfunctions of an algebra of differential operators, namely, the algebra of radial parts of either the center of $U(\mathfrak{gl}(n|n))$ or that of $U(\mathfrak{q}(n))$.

1 Introduction

1.1 History. Main result

Hereafter the ground field is $\mathbb{C}$.

I. Schur [Sch] introduced projective Schur functions to characterize projective representations of symmetric groups.

In [Se1] I showed that projective Schur functions can be also described as characters of tensor representations of the Lie superalgebra $\mathfrak{q}(n)$.

Stembridge [St] interpreted projective Schur functions as characteristics of spherical functions of a certain *twisted Gelfand pair* (for the definition see [St]). We recall the definition of bispherical functions and demonstrate that these Stembridge characteristics are precisely bispherical functions of the triple $(\mathfrak{pe}(n), \mathfrak{gl}(n|n), \mathfrak{pe}(n))$ for the two different ways of embedding $\mathfrak{pe}(n)$ into $\mathfrak{gl}(n|n)$: either as the diagonal or as a nondegenerate symmetric odd bilinear form or a skew-symmetric one.

We obtain one more realization of projective Schur functions by considering bispherical functions of the triple $(\mathfrak{q}(n), \mathfrak{q}(n) \oplus \mathfrak{q}(n), \mathfrak{q}(n))$ for both of the two different ways of embedding $\mathfrak{q}(n)$ into $\mathfrak{q}(n) \oplus \mathfrak{q}(n)$: either as the diagonal or as a "twisted diagonal."

Both ways of realizing projective Schur functions allow one to construct an al-

---

[1] **AMS Subject classification:** 05E05, 17B35.

[2] **Keywords and phrases:** Schur functions, Lie superalgebras, homogeneous superspaces.

gebra of differential operators for which the projective Schur functions are eigenfunctions. This algebra appears as the algebra of radial parts of Laplace operators for the Lie superalgebra $\mathfrak{gl}(n|n)$ or $\mathfrak{q}(n)$. (Although the center of $U(\mathfrak{gl}(n|n))$ is not isomorphic to that of $U(\mathfrak{q}(n))$, the algebras of *radial parts* of these centers are isomorphic.)

1.2 Differential operators and projective Schur functions

Let $I = \{1, \dots, n\}$, V an n-dimensional vector space, $\{e_i\}_{i\in I}$ a basis of V, and $\{\varepsilon_i\}$ the dual basis of V^*. Recall that $S(V)^*$ can be identified with the algebra of formal power series in n indeterminates; let e^l for $l \in V^*$ denote the homomorphism $S(V) \longrightarrow \mathbb{C}$ given by $e^l(v) = l(v)$ for any $v \in V$.

On $S(V)^*$, define a family of differential operators. Set

$$\partial_i^{(1)}(e^l) = l(e_i)e^l = \partial_i(e^l) \text{ for any } l \in V^*.$$

Set further $\varepsilon_{ij} = \varepsilon_i - \varepsilon_j$ for $i, j \in I,\ i \neq j$ and

$$\partial_i^{(k)} = \begin{cases} \partial_i\partial_i^{(k-1)} + \sum\limits_{j\neq i} \frac{2}{e^{\varepsilon_{ij}} - e^{\varepsilon_{ji}}}(\partial_i^{(k-1)} - \partial_j^{(k-1)}) & \text{for } k \text{ odd,} \\ (\partial_i - 1)\partial_i^{(k-1)} + \sum\limits_{j\neq i} \left(\frac{2}{e^{\varepsilon_{ij}} - e^{\varepsilon_{ji}}}\partial_i^{(k-1)} - \frac{2e^{\varepsilon_{ij}}}{e^{\varepsilon_{ij}} - e^{\varepsilon_{ji}}}\partial_i^{(k-1)}\right) & \text{for } k \text{ even.} \end{cases}$$

Finally, for k odd, set

$$\Omega_k = \sum_{i=1}^n \partial_i^{(k)}.$$

It is not difficult to verify directly that

$$\Omega_3 = \sum_{i=1}^n \partial_i^3 + \sum_{i<j} \frac{6}{e^{\varepsilon_{ij}} - e^{\varepsilon_{ji}}}(\partial_i^2 - \partial_j^2) - \sum_{i<j} \frac{6}{(e^{\frac{1}{2}\varepsilon_{ij}} + e^{\frac{1}{2}\varepsilon_{ji}})^2}(\partial_i + \partial_j)$$
$$+ 24 \sum_{i\notin\{j,k\}} \frac{1}{(e^{\varepsilon_{ij}} - e^{\varepsilon_{ji}})(e^{\varepsilon_{ik}} - e^{\varepsilon_{ki}})}\partial_i - \left(\sum_{i=1}^n \partial_i\right)^2,$$

where $\{j, k\} \subset I$ is any two-element subset.

Introduce new indeterminates: $x_i = e^{\varepsilon_i}$. Then $\partial_i = x_i \frac{\partial}{\partial x_i}$ and

$$\partial_i^{(k)} = \begin{cases} \partial_i\partial_i^{(k-1)} + \sum\limits_{j\neq i} \frac{2x_ix_j}{x_i^2 - x_j^2}\left(\partial_i^{(k-1)} - \partial_j^{(k-1)}\right) & \text{for } k \text{ odd,} \\ (\partial_i - 1)\partial_i^{(k-1)} + \sum\limits_{j\neq i} \left(\frac{2x_ix_j}{x_i^2 - x_j^2}\partial_i^{(k-1)} - \frac{2x_i^2}{x_i^2 - x_j^2}\partial_j^{(k-1)}\right) & \text{for } k \text{ even.} \end{cases}$$

We have

$$\Omega_3 = \sum_{i=1}^{n} \partial_i^3 + 6\sum_{i<j} \frac{x_i x_j}{x_i^2 - x_j^2}(\partial_i^2 - \partial_j^2) - 6\sum_{i<j} \frac{x_i x_j}{(x_i + x_j)^2}(\partial_i + \partial_j)$$
$$+24\sum_{i<j<k} \frac{x_i x_j x_k}{(x_i^2 - x_j^2)(x_i^2 - x_k^2)} \Big(x_i(x_j^2 - x_k^2)\partial_i - x_j(x_i^2 - x_k^2)\partial_j + x_k(x_i^2 - x_j^2)\partial_k\Big)$$
$$-\left(\sum_{i=1}^{n} \partial_i\right)^2.$$

For $i = 1, \dots, n$ and any k, define differential operators $\tilde{\partial}_i^{(k)}$ and $\tilde{\partial}_{\bar{i}}^{(k)}$ by setting

$$\tilde{\partial}_i^{(1)} = \tilde{\partial}_i = \tilde{\partial}_{\bar{i}}^{(1)} = \tilde{\partial}_{\bar{i}} = \partial_i$$

and

$$\tilde{\partial}_i^{(k)} = \tilde{\partial}_i \tilde{\partial}_i^{(k-1)} + \sum_{j \neq i} \frac{e^{\frac{1}{2}\varepsilon_{ij}}}{e^{\frac{1}{2}\varepsilon_{ij}} - e^{\frac{1}{2}\varepsilon_{ji}}}(\tilde{\partial}_i^{(k-1)} - \tilde{\partial}_j^{(k-1)})$$
$$- \sum_{j \neq i} \frac{e^{\frac{1}{2}\varepsilon_{ij}}}{e^{\frac{1}{2}\varepsilon_{ij}} + e^{\frac{1}{2}\varepsilon_{ji}}}(\tilde{\partial}_i^{(k-1)} + \tilde{\partial}_j^{(k-1)}),$$

$$\tilde{\partial}_{\bar{i}}^{(k)} = -\tilde{\partial}_i \tilde{\partial}_{\bar{i}}^{(k-1)} - \sum_{j \neq i} \frac{e^{\frac{1}{2}\varepsilon_{ij}}}{e^{\frac{1}{2}\varepsilon_{ij}} - e^{\frac{1}{2}\varepsilon_{ji}}}(\tilde{\partial}_{\bar{i}}^{(k-1)} - \tilde{\partial}_{\bar{j}}^{(k-1)})$$
$$+ \sum_{j \neq i} \frac{e^{\frac{1}{2}\varepsilon_{ij}}}{e^{\frac{1}{2}\varepsilon_{ij}} + e^{\frac{1}{2}\varepsilon_{ji}}}(\tilde{\partial}_{\bar{i}}^{(k-1)} + \tilde{\partial}_j^{(k-1)}).$$

Set further

$$\widetilde{\Omega}_k = \sum_{i=1}^{n} (\tilde{\partial}_i^{(k)} + \tilde{\partial}_{\bar{i}}^{(k)}).$$

Lemma 1.2.1. *The algebra Ω generated by the operators Ω_k for $k = 1, 3, 5, \dots$ coincides with the algebra generated by the operators $\widetilde{\Omega}_k$ for $k = 1, 2, 3, 4, \dots$.*

Set

$$\Delta_i^{(k)+} = \tilde{\partial}_i^{(k)} + \tilde{\partial}_{\bar{i}}^{(k)}, \qquad \Delta_i^{(k)-} = \tilde{\partial}_i^{(k)} - \tilde{\partial}_{\bar{i}}^{(k)}.$$

Now, it is not difficult to verify the following identities:

$$\Delta_i^{(k)+} - \Delta_i^{(k-1)+} = \partial_i \Delta_i^{(k-1)-} + \sum_{j \neq i} \frac{2}{e^{\varepsilon_{ij}} - e^{\varepsilon_{ji}}}(\Delta_i^{(k-1)-} - \Delta_j^{(k-1)-}),$$

$$\Delta_i^{(k)-} = (\partial_i - 1)\Delta_i^{(k-1)+} + \sum_{j \neq i} \left(\frac{2}{e^{\varepsilon_{ij}} - e^{\varepsilon_{ji}}} \Delta_i^{(k-1)+} - \frac{2e^{\varepsilon_{ij}}}{e^{\varepsilon_{ij}} - e^{\varepsilon_{ji}}} \Delta_j^{(k-1)+} \right).$$

Now observe that

$$\Delta_i^{(2k)+} = f_{2k}(\partial_i^{(1)}, \dots, \partial_i^{(2k-1)})$$

and

$$\Delta_i^{(2k+1)+} = f_{2k+1}(\partial_i^{(1)}, \dots, \partial_i^{(2k+1)}),$$

where f_{2k} and f_{2k+1} are certain linear functions; as it is easy to verify by induction, these functions are interrelated:

$$\begin{aligned} f_{2k+1} &= f_{2k} + f_{2k-1}(\partial_i^{(3)}, \dots, \partial_i^{(2k+1)}), \\ f_{2k} &= f_{2k-1} + f_{2k-2}(\partial_i^{(3)}, \dots, \partial_i^{(2k-1)}). \end{aligned} \tag{1.2.1}$$

Now, it is easy to verify that

$$\tilde{\Omega}_1 = 2\Omega_1 = \tilde{\Omega}_2, \quad \tilde{\Omega}_3 = 2\Omega_3 + 2\Omega_1, \quad \tilde{\Omega}_4 = 4\Omega_3 + 2\Omega_1$$

and, by induction, (1.2.1) implies that $\tilde{\Omega}_k$ is a linear combination of the Ω_{2l+1} for $2l + 1 \leq k$. One can show that, the other way round, Ω_{2l+1} can be expressed via $\tilde{\Omega}_k$. This proves Lemma 1.2.1. □

1.2.2. Projective Schur functions as common eigenfunctions of an algebra of differential operators

Let $x_1, \dots, x_n, t$ be indeterminates. Introduce polynomials $q_k(x_1, \dots, x_n)$ from the equation

$$\sum_{k=0}^{\infty} q_k(x_1, \dots, x_n)t^k = \prod_{i=1}^{n} \frac{(1 + x_i t)}{(1 - x_i t)}.$$

Further, set

$$Q_{k,l} = q_k q_l + 2\sum_{p=1}^{l} q_{k+p} q_{l-p}.$$

We see that $Q_{k,0} = q_k$ and $Q_{k,l} = -Q_{l,k}$ for $k + l > 0$. If $\lambda = (\lambda_1, \dots, \lambda_m)$ is a strict partition, i.e., $\lambda_1 > \lambda_2 > \dots > \lambda_m > 0$, then (Pf is the Pfaffian of a skew-symmetric matrix)

$$Q_\lambda(x_1, \dots, x_n) = \begin{cases} \mathrm{Pf}(Q_{\lambda_i,\lambda_j}) & \text{if } m \text{ is even,} \\ Q_{(\lambda,0)} & \text{if } m \text{ is odd.} \end{cases}$$

Lemma. *Set* $\delta = \prod_{1\leq i<j\leq n} \left(\dfrac{e^{\frac{1}{2}\varepsilon_{ij}} + e^{\frac{1}{2}\varepsilon_{ji}}}{e^{\frac{1}{2}\varepsilon_{ij}} - e^{\frac{1}{2}\varepsilon_{ji}}} \right)$. *Then*

i) $\delta^{-1}\Omega_3\delta = \sum_{i=1}^{n} \partial_i^3 - \left(\sum_{i=1}^{n} \partial_i\right)^2.$

ii) *The polynomials Q_λ are common eigenfunctions of the operators Ω_k for $k = 1, 3, 5, \dots$.*

iii) *Let $P(x_1, \dots, x_n)$ be a polynomial symmetric with respect to $x_1, \dots, x_n$ which, after substituting $x_i = t$, $x_j = -t$, becomes independent of t. If P is a common eigenfunction of the operators Ω_k, $k = 1, 3, 5, \dots$, then, up to a scalar multiple, P coincides with one of the Q_λ.*

A proof of Lemma 1.2.2 will be given in Subsection 3.3.

1.3 Bispherical functions

Let $\mathfrak{g}$ be a finite-dimensional Lie superalgebra. Its enveloping algebra $U(\mathfrak{g})$ possesses a canonical antiautomorphism $t : u \mapsto {}^t u$ (the *principal antiautomorphism*) given by the formula $t(x) = -x$ for any $x \in \mathfrak{g}$ and extended to the whole algebra via the following formula:

$$^t(uv) = (-1)^{p(u)p(v)}\, {}^t(v)^t(u).$$

Convention. For brevity, when it cannot cause misunderstanding, we sometimes write $(-1)^u$ instead of $(-1)^{p(u)}$.

The left and right coregular representations of $U(\mathfrak{g})$ are defined for any $l \in U(\mathfrak{g})^*$ and $v, u \in U(\mathfrak{g})$ by the formulas

$$(L^*(u)l)(v) = (-1)^{ul} l({}^t u \cdot v) \text{ and } (R^*(u)l)(v) = (-1)^{u(l+v)} l(vu)\,.$$

Let $\mathfrak{b}_1$ and $\mathfrak{b}_2$ be subalgebras of $\mathfrak{g}$. A functional $l \in U(\mathfrak{g})^*$ is said to be *two-sided invariant* if

$$l(x_1 u) = l(u x_2) = 0 \text{ for any } x_1 \in \mathfrak{b}_1, x_2 \in \mathfrak{b}_2 \text{ and } u \in U(\mathfrak{g}).$$

Let V be a $\mathfrak{g}$-module containing a nonzero $\mathfrak{b}_2$-invariant vector $v \in V$; suppose also that there exists a $\mathfrak{b}_1$-invariant vector $v^* \in V^*$. The matrix coefficient $\Theta(v^*, v) \in U(\mathfrak{g})^*$ defined by the formula

$$\Theta(v^*, v)(u) = (-1)^{uv} v^*(uv) \text{ for } u \in U(\mathfrak{g})$$

is a *bispherical function associated with the triple* (V, v^*, v).

Observe that if $z \in Z(\mathfrak{g})$, then $L^*(z)l$ is two-sided invariant if l is as well. Therefore, on the space of invariant functionals, every $z \in Z(\mathfrak{g})$ determines a linear operator $\Omega_{(z)} : l \mapsto L^*(z)l$.

1.4 Another $\mathfrak{g}$-action on $U(\mathfrak{g})$

In addition to the usual $\mathfrak{g}$-module structure on $U(\mathfrak{g})$, the extension of the adjoint action, consider the $\mathfrak{g}$-module structure on $U(\mathfrak{g})$ with respect to the following action ([Se4], [G]):

$$x * u = xu - (-1)^{x(u+\bar{1})} ux. \tag{1.4.1}$$

It is easy to verify that for a finite-dimensional $\mathfrak{g}$-module V the functional $u \mapsto \mathrm{tr}_V(u)$ is invariant with respect to this action.

In $\mathfrak{g} \oplus \mathfrak{g}$, consider two subalgebras:

$$\mathfrak{g}_1 = \{(x, (-1)^x x) \mid x \in \mathfrak{g}\} \text{ and } \mathfrak{g}_2 = \{(x, x) \mid x \in \mathfrak{g}\}.$$

Lemma. *The algebra of functionals on* $U(\mathfrak{g} \oplus \mathfrak{g})$ *biinvariant with respect to* $\mathfrak{g}_1$ *and* $\mathfrak{g}_2$ *is isomorphic to the algebra of functionals on* $U(\mathfrak{g})$ *invariant with respect to the action* (1.4.1).

It is sufficient to prove the following assertion:

Let $\mathfrak{g}$, $\mathfrak{g}_1$, $\mathfrak{g}_2$ *be selected as in Section 1.4, let* I *be the left ideal in* $U(\mathfrak{g}\oplus\mathfrak{g})$ *generated by* $\mathfrak{g}_2$, *and let* $M = U(\mathfrak{g}\oplus\mathfrak{g})/I$. *Let* $\sigma : \mathfrak{g} \longrightarrow \mathfrak{g}\oplus\mathfrak{g}$ *be the embedding in the first summand, i.e.,* $\sigma(x) = (x, 0)$. *Let* $\tilde{\sigma} : U(\mathfrak{g}) \longrightarrow M$ *be the map induced by the homomorphism* $U(\mathfrak{g}) \longrightarrow U(\mathfrak{g}\oplus\mathfrak{g})$ *that extends* σ *and* $\rho(x) = (x, (-1)^x x)$ *an isomorphism of* $\mathfrak{g}$ *with* $\mathfrak{g}_1$. *Then* $\tilde{\sigma}(x * u) = \rho(x)\tilde{\sigma}(u)$.

Proof.

$$\begin{aligned}
\tilde{\sigma}(x * u) &= \tilde{\sigma}(xu - (-1)^{x(u+\bar{1})}ux) = xu\otimes 1 - (-1)^{x(u+\bar{1})}ux\otimes 1\\
&= xu\otimes 1 - (-1)^{x(u+\bar{1})}ux\otimes 1 - \rho(x)\tilde{\sigma}(u) + \rho(x)\tilde{\sigma}(u)\\
&= \rho(x)\tilde{\sigma}(u) + xu\otimes 1 - (-1)^{x(u+\bar{1})}ux\otimes 1\\
&\qquad -(x\otimes 1 + (-1)^x 1\otimes x)(u\otimes 1)\\
&= \rho(x)\tilde{\sigma}(u) - (-1)^{x(u+\bar{1})}(ux\otimes 1 + u\otimes x)\\
&= \rho(x)\tilde{\sigma}(u) - (-1)^{x(u+\bar{1})}(u\otimes 1)(x\otimes 1 + 1\otimes x)\\
&\equiv \rho(x)\tilde{\sigma}(u) \qquad (\text{mod } I).
\end{aligned}$$

This proves the lemma. □

1.4.1. Convention

Let $I = I_{\bar{0}} \cup I_{\bar{1}} = \{1, \dots, n\} \cup \{\bar{1}, \dots, \bar{n}\}$ be the union of the "even" and "odd" indices. Parity of *all* elements from $I_{\bar{0}}$ is equal to $\bar{0}$ and that of *all* elements from $I_{\bar{1}}$ is equal to $\bar{1}$.

Let $\dim V = (n|n)$ and $\{e_i\}_{i\in I}$ be a basis of V such that the parity of each vector of the basis is the same as that of its index. Set $\bar{\bar{i}} = i$ and define the odd operator $\Pi \in \mathfrak{gl}(V)$ by setting

$$\Pi(e_i) = (-1)^{p(i)} e_{\bar{i}} \text{ for any } i \in I.$$

Define the parity operator $P \in \mathfrak{gl}(V)$ by setting

$$P(e_i) = (-1)^{p(i)} e_i \text{ for any } i \in I.$$

We denote the superspace of operators in V by End(V); the superspace of supermatrices (in the standard format with respect to the basis in which all the even vectors come first) by Mat(V) and the Lie superalgebra structures in these isomorphic superspaces is denoted by $\mathfrak{gl}(V)$ and $\mathfrak{gl}(\dim V)$, respectively.

1.4.2. Queer superalgebras $\mathfrak{q}(V)$

Let

$$\mathfrak{q}(n) = \{X \in \mathfrak{gl}(V) \mid [X, \Pi] = 0\}. \qquad (1.4.2)$$

It is easy to verify that $\mathfrak{q}(n) = \text{Span}(e_{ij}, f_{ij} \mid i, j \in I_{\bar{0}})$, where

$$e_{ij} = e_i \otimes e_j^* + e_{\bar{i}} \otimes e_{\bar{j}}^*, \quad f_{ij} = e_i \otimes e_{\bar{j}}^* + e_{\bar{i}} \otimes e_j^* \text{ for any } i, j \in I_{\bar{0}}$$

and where $\{e_i^*\}_{i\in I}$ is the left dual of the basis $\{e_i\}_{i\in I}$; observe that we have identified $\mathrm{End}(V)$ with $V \otimes V^*$.

Set $\mathfrak{g} = \mathfrak{q}(n) \oplus \mathfrak{q}(n)$ and select $\mathfrak{g}_1$ and $\mathfrak{g}_2$ as at the beginning of Section 1.4. Let $\sigma : \mathfrak{q}(n) \longrightarrow \mathfrak{g}$ be the embedding into the first summand, $\mathfrak{h}_{\bar{0}} = \mathrm{Span}(e_{ii} \mid i \in I_{\bar{0}})$ be the even part of the Cartan subalgebra of $\mathfrak{q}(n)$. Let us define inductively the following elements of $U(\mathfrak{q}(n))$:

$$\begin{aligned} e_{ij}^{(1)} &= e_{ij},\ f_{ij}^{(1)} = f_{ij}, \\ e_{ij}^{(p)} &= \sum_{l=1}^{n} e_{il} e_{lj}^{(p-1)} + (-1)^{p-1} \sum_{l=1}^{n} f_{il} f_{lj}^{(p-1)}, \\ f_{ij}^{(p)} &= \sum_{l=1}^{n} e_{il} f_{lj}^{(p-1)} + (-1)^{p-1} \sum_{l=1}^{n} f_{il} e_{lj}^{(p-1)}. \end{aligned} \tag{1.4.3}$$

The following relations are subject to straightforward verification:

$$\begin{aligned} [e_{ij}, e_{kl}^{(p)}] &= \delta_{jk} e_{il}^{(p)} - \delta_{il} e_{kj}^{(p)}, \\ [e_{ij}, f_{kl}^{(p)}] &= \delta_{jk} f_{il}^{(p)} - \delta_{il} f_{kj}^{(p)}, \\ [f_{ij}, e_{kl}^{(p)}] &= (-1)^{p+1} \delta_{jk} f_{il}^{(p)} - \delta_{il} f_{kj}^{(p)}, \\ [f_{ij}, f_{kl}^{(p)}] &= (-1)^{p+1} \delta_{jk} e_{il}^{(p)} + \delta_{il} e_{kj}^{(p)}. \end{aligned} \tag{1.4.4}$$

As easily verified, see [Se3], the elements $z_k = \sum_{i=1}^{n} e_{ii}^{(k)}$, $k = 1, 3, 5, \ldots$, are central if we embed $\mathfrak{q}(n)$ into $\mathfrak{g} = \mathfrak{q}(n) \oplus \mathfrak{q}(n)$ as the first summand, i.e., strictly speaking, we identify z_k with $\sigma(z_k) \in U(\mathfrak{g})$.

Let $\{\varepsilon_i\}_{i\in I_{\bar{0}}}$ be the basis of $\mathfrak{h}_{\bar{0}}$ dual to $\{e_i\}_{i\in I_{\bar{0}}}$. Let $\lambda = (\lambda_1, \ldots, \lambda_m)$ be a strict partition of k, i.e., $\lambda_1 > \lambda_2 > \ldots > \lambda_m > 0$. Let V^λ be an irreducible submodule of $V^{\otimes k}$ corresponding to λ, see [Se1]. Then the $\mathfrak{g}$-module $W^\lambda = V^\lambda \otimes (V^\lambda)^*$ is irreducible and contains a unique even $\mathfrak{g}_2$-invariant vector ω_λ corresponding to the identity operator under identification $V^\lambda \otimes (V^\lambda)^* = \mathrm{End}(V^\lambda)$. The dual module $(V^\lambda)^* \otimes V^\lambda$ is also irreducible and contains an even $\mathfrak{g}_1$-invariant vector ω_λ^* corresponding to the parity operator P under identification $(V^\lambda)^* \otimes V^\lambda = \mathrm{End}(V^\lambda)$.

Theorem 1.5. *Let $\mathfrak{q}(n)$ be embedded into $\mathfrak{g} = \mathfrak{q}(n) \oplus \mathfrak{q}(n)$ as the first summand. Then*

i) *Every bi-invariant with respect to $\mathfrak{g}_1$ and $\mathfrak{g}_2$ functional on $\mathfrak{g}$ is uniquely determined by its restriction to $S(\mathfrak{h}_{\bar{0}}) \subset \mathfrak{q}(n)$.*

ii) *Every $z \in Z(\mathfrak{q}(n))$ uniquely determines a differential operator $\Omega_{(z)}$ on the space of restrictions of left-invariant functionals $S(\mathfrak{h}_{\bar{0}})^{*inv}$.*

iii) *$\Omega_{(z_k)} = \Omega_k$ for $z = z_k$.*

iv) *Let $\varphi_\lambda = \Theta(\omega_\lambda^*, \omega_\lambda)$. Then the restriction of φ_λ to $S(\mathfrak{h}_{\bar{0}})$ coincides up to a scalar multiple with $Q_\lambda(e^{\varepsilon_1}, \ldots, e^{\varepsilon_n})$.*

A proof of Theorem 1.5 will be given in Section 3.4.

1.6 Periplectic superalgebras $\mathfrak{pe}(V)$

Let $\mathfrak{g} = \mathfrak{gl}(\dim V)$, where $\dim V = (n|n)$. Let $I = I_{\bar{0}} \cup I_{\bar{1}}$, where $I_{\bar{0}} = \{1, \dots, n\}$ and $I_{\bar{1}} = \{\bar{1}, \dots, \bar{n}\}$. Let $\{e_i\}$ be a basis of V such that the parity of each vector coincides with that of its index. Denote by $\mathfrak{pe}_1(n)$ and $\mathfrak{pe}_2(n)$ the Lie subsuperalgebras in $\mathfrak{g}$ preserving the respective tensors:

$$\omega_1 = \sum_{i \in I} e_i^* \otimes e_{\bar{i}}^*, \quad \omega_2 = \sum_{i \in I} (-1)^{p(i)} e_i^* \otimes e_{\bar{i}}^*.$$

Let ψ_1 and ψ_2 be involutive antiautomorphisms of $\mathfrak{g}$ that single out $\mathfrak{pe}_1(n)$ and $\mathfrak{pe}_2(n)$, respectively, i.e.,

$$\mathfrak{pe}_i = \{X \in \mathfrak{gl}(V) \mid \psi_i(X) = -X\} \text{ for } i = 1, 2.$$

It is easy to see that the restrictions of ψ_1 and ψ_2 to $\mathfrak{g}_{\bar{0}}$ coincide. Set

$$\mathfrak{h}^+ = \{X \in \mathfrak{h} \mid \psi_1(X) = \psi_2(X) = X\},$$

where $\mathfrak{h}$ is a Cartan subalgebra in $\mathfrak{g}$.

Let λ be a partition and let V^λ be the corresponding $\mathfrak{g}$-submodule in the tensor algebra of the identity representation, see [Se1]. It follows from [Se4] that, (in Frobenius' notation [Ma])

$$V^\lambda \text{ contains a } \mathfrak{pe}_1\text{-invariant vector if } \lambda = (\alpha_1, \dots, \alpha_p | \alpha_1 + 1, \dots, \alpha_p + 1)$$

and

$$V^\lambda \text{ contains a } \mathfrak{pe}_2\text{-invariant vector if } \lambda = (\alpha_1 + 1, \dots, \alpha_p + 1 | \alpha_1, \dots, \alpha_p).$$

Similarly,

$$(V^\lambda)^* \text{ contains a } \mathfrak{pe}_1\text{-invariant vector if } \lambda = (\alpha_1 + 1, \dots, \alpha_p + 1 | \alpha_1, \dots, \alpha_p)$$

and

$$(V^\lambda)^* \text{ contains a } \mathfrak{pe}_2\text{-invariant vector if } \lambda = (\alpha_1, \dots, \alpha_p | \alpha_1 + 1, \dots, \alpha_p + 1).$$

Now, let $\lambda = (\alpha_1 + 1, \dots, \alpha_p + 1 | \alpha_1, \dots, \alpha_p)$, let $v_\lambda^* \in (V^\lambda)^*$ be a $\mathfrak{pe}_1$-invariant vector, $v_\lambda \in V^\lambda$ a $\mathfrak{pe}_2$-invariant vector, and let $\varphi_\lambda = \Theta(v_\lambda^*, v_\lambda)$ be the corresponding bispherical function. In $\mathfrak{h}^+$, select the basis $\{e_{ii}^+ = \frac{1}{2}(e_{ii} + \psi_1(e_{ii})) \mid i \in I_{\bar{0}}\}$ and let ε_i be the left dual vectors.

Let e_{ij} be the basis of matrix units in $\mathfrak{gl}(V)$. Set

$$e_{ij}^{(1)} = e_{ij}, \quad e_{ij}^{(k)} = \sum_{i \in I} (-1)^p e_{ip} e_{pj}^{(k-1)}.$$

It is easy to verify that

$$[e_{ij}, e_{pq}^{(k)}] = \delta_{jp} e_{ip}^{(k)} - (-1)^{(i+j)(p+q)} \delta_{iq} e_{pj}^{(k)}.$$

This easily implies that $z_k = \sum_{i \in I} e_{ii}^{(k)} \in Z(\mathfrak{g})$.

Theorem 1.7. i) *Each two-sided invariant with respect to* $\mathfrak{pe}_1$ *and* $\mathfrak{pe}_2$ *functional on* $U(\mathfrak{g})$ *is uniquely determined by its restriction to* $S(\mathfrak{h}^+)$.

ii) *Every* $z \in Z(\mathfrak{g})$ *uniquely determines a differential operator* $\Omega_{(z)}$ *on the space* $S(\mathfrak{h}^+)^*$ *of restrictions of invariant functionals.*

iii) *If* $z = z_k$, *then* $\Omega_{(z_k)} = \tilde{\Omega}_k$.

iv) *The functional* φ_λ *coincides, up to a scalar multiple, with* $Q_\lambda(e^{\varepsilon_1}, \dots, e^{\varepsilon_k})$.

A proof of Theorem 1.7 will be given in Section 4.3.

2 The algebra dual to the enveloping algebra

In this section we follow Dixmier's book [Dix] applied *mutatis mutandis* to Lie superalgebras.

2.1 $U(\mathfrak{g})^*$ as a coalgebra

Let $\mathfrak{g}$ be a Lie superalgebra. We endow $U(\mathfrak{g})^*$ with a coalgebra structure by setting

$$c : \mathfrak{g} \longrightarrow U(\mathfrak{g}) \otimes U(\mathfrak{g}), \quad c(x) = x \otimes 1 + 1 \otimes x \ \text{ for any } x \in \mathfrak{g},$$

so that ${}^t c(x) = c({}^t x)$ where the first t is the principal antiautomorphism of the superalgebra $U(\mathfrak{g}) \otimes U(\mathfrak{g}) = U(\mathfrak{g} \oplus \mathfrak{g})$, see Section 1.3.

Lemma. *Let* $\dim \mathfrak{g} = (n|m)$. *Then* $U(\mathfrak{g})^*$ *is isomorphic to the supercommutative superalgebra of formal power series in n even and m odd indeterminates.*

Proof. Let $\mathfrak{g}_{\bar 0} = \mathrm{Span}(e_1, \dots, e_n)$, $\mathfrak{g}_{\bar 1} = \mathrm{Span}(e_{\bar 1}, \dots, e_{\bar m})$, $I_{\bar 0} = \{1, \dots, n\}$, $I_{\bar 1} = \{\bar 1, \dots, \bar m\}$.

Denote:

$$M = \{\nu = (\nu_1, \dots, \nu_n, \nu_{\bar 1}, \dots, \nu_{\bar m}) \mid \nu_i \in \mathbb{N} \text{ if } i \in I_{\bar 0} \text{ and } \nu_i \in \{0, 1\} \text{ if } i \in I_{\bar 1}\}.$$

For $\nu \in M$, set

$$e_\nu = \frac{e_1^{\nu_1}}{\nu_1!} \cdots \frac{e_n^{\nu_n}}{\nu_n!} \frac{e_{\bar 1}^{\nu_{\bar 1}}}{(\nu_{\bar 1})!} \cdots \frac{e_{\bar m}^{\nu_{\bar m}}}{(\nu_{\bar m})!}$$

and let $t_1, \dots, t_n, t_{\bar 1}, \dots, t_{\bar m}$ be the set of even and odd supercommuting indeterminates. The correspondence

$$U(\mathfrak{g})^* \ni L \mapsto \sum_{\nu \in M} L(e_\nu) t_{\bar m}^{\nu_{\bar m}} \dots t_{\bar 1}^{\nu_{\bar 1}} t_n^{\nu_n} \dots t_1^{\nu_1}$$

determines the desired homomorphism. □

2.2 Left and right coregular representations

For any $u, v \in U(\mathfrak{g})$ and $L \in U(\mathfrak{g})^*$ set

$$(L^*(u)L)(v) = (-1)^{uL} L({}^t uv) \text{ and } (R^*(u)L)(v) = (-1)^{u(L+v)} L(vu).$$

The following statements are easy to verify:

i) $u \longrightarrow L^*(u)$ is a representation of $U(\mathfrak{g})$ in $U(\mathfrak{g})^*$ (we call it *left regular*);

ii) $u \longrightarrow R^*(u)$ is a representation of $U(\mathfrak{g})$ in $U(\mathfrak{g})^*$ (we call it *right regular*);

iii) If $x \in \mathfrak{g}$, then $L^*(x)$ and $R^*(x)$ are superderivations of the superalgebra $U(\mathfrak{g})^*$.

Observe also that algebra $U(\mathfrak{g})^*$ possesses an automorphism

$$L \mapsto L^T : L^T(u) = L({}^t u) \text{ for any } u \in U(\mathfrak{g}),\ L \in U(\mathfrak{g})^*.$$

2.3 Matrix coefficients

Let V be a $\mathfrak{g}$-module, let $\pi : \mathfrak{g} \longrightarrow \mathfrak{gl}(V)$ be the corresponding representation, and V^* the dual module. Let $v \in V$ and $v^* \in V^*$. Denote by $\Theta^\pi(v^*, v)$ the linear form on $U(\mathfrak{g})^*$:

$$\Theta^\pi(v^*, v)(u) = (-1)^{uv} v^*(\pi(u)v).$$

Finally, the subalgebra in $U(\mathfrak{g})^*$, generated by $\Theta^\pi(v^*, v)$, is denoted by $C(\pi)$ or $C(V)$ for all $v^* \in V^*$ and $v \in V$.

Lemma 2.3.1.

$$\Theta^{\pi_1 \otimes \pi_2}(v_1^* \otimes v_2^*, v_1 \otimes v_2) = (-1)^{v_1 v_2^*} \Theta^{\pi_1}(v_1^*, v_1) \otimes \Theta^{\pi_2}(v_2^*, v_2).$$

Proof. We have

$$\begin{aligned}
&\Theta^{\pi_1 \otimes \pi_2}(v_1^* \otimes v_2^*, v_1 \otimes v_2)(u_1 \otimes u_2)\\
&= (-1)^{(u_1+u_2)(v_1+v_2)} v_1^* \otimes v_2^*(\pi_1(u_1) \otimes \pi_2(u_2)(v_1 \otimes v_2))\\
&= (-1)^{(u_1+u_2)(v_1+v_2)+v_1u_2+v_2^*(u_1+v_1)} v_1^*(\pi_1(u_1)v_1) v_2^*(\pi_2(u_2)v_2)\\
&= \Theta^{\pi_1}(v_1^*, v_1) \otimes \Theta^{\pi_2}(v_2^*, v_2)(u_1 \otimes u_2)\\
&= (-1)^{u_1(v_2+v_2^*)} \Theta^{\pi_1}(v_1^*, v_1)(u_1) \Theta^{\pi_2}(v_2^*, v_2)(u_2)\\
&= (-1)^{u_1(v_2+v_2^*)+v_1u_1+v_2u_2} v_1^*(\pi_1(u_1)v_1) v_2^*(\pi_2(u_2)v_2). \qquad \square
\end{aligned}$$

Lemma 2.3.2. 1.) *The map $V^* \otimes V \longrightarrow U(\mathfrak{g})$ given by $v^* \otimes v \mapsto \Theta^\pi(v^*, v)$ is a $\mathfrak{g} \oplus \mathfrak{g}$-module homomorphism; here we consider $U(\mathfrak{g})^*$ as a $\mathfrak{g} \oplus \mathfrak{g}$-module with respect to the left coregular representation of the first summand and the right coregular representation of the second summand.*

2) *If V is irreducible, the above map is an isomorphism.*

Proof. Clearly, there exists a linear map $\varphi : V^* \otimes V \longrightarrow U(\mathfrak{g})^*$ such that $\varphi(v^* \otimes v) = \Theta^\pi(v^*, v)$. Let $x \in \mathfrak{g}$. Then

$$\varphi((x \otimes 1)(v^* \otimes v))(u) = \varphi(xv^* \otimes v)(u)$$
$$= \Theta^\pi(xv^*, v)(u) = (-1)^{vu}(xv^*)(u) = (-1)^{vu+\bar{1}+xv^*}v^*(xu).$$

On the other hand,

$$L^*(x)\varphi(v^*, v)(u) = (-1)^{x(v+v^*)+\bar{1}}\Theta^\pi(v^*, v)(xu) =$$
$$(-1)^{x(v+v^*)+\bar{1}+(x+u)v}v^*(xuv) = (-1)^{uv+\bar{1}+xv^*}v^*(xu).$$

Further,

$$\varphi((1 \otimes x)(v^* \otimes v))(u) = (-1)^{xv^*}\varphi(v^* \otimes xv)(u)$$
$$= (-1)^{xv^*}\Theta^\pi(v^*, xv)(u) = (-1)^{xv^*+u(x+v)}v^*(uxv).$$

On the other hand,

$$R^*(x)\varphi(v^* \otimes v)(u) = (-1)^{x(v^*+v+u)}\varphi(v^* \otimes v)(ux) =$$
$$(-1)^{x(v^*+v+u)}\Theta^\pi(v^* \otimes v)(ux) =$$
$$(-1)^{x(v^*+v+u)+(u+x)v}v^*(uxv) = (-1)^{xv^*+u(x+v)}v^*(uxv).$$

This proves the first statement. The second one is obvious. □

Lemma 2.3.3. *Let ρ be the tensor product of representations π_1 and π_2. Then*
i) $\Theta^\rho(v_1^* \otimes v_2^*, v_1 \otimes v_2) = (-1)^{v_1 v_2^*}\Theta^{\pi_1}(v_1^*, v_1)\Theta^{\pi_2}(v_2^*, v_2)$.
ii) $C(\rho) = C(\pi_1)C(\pi_2)$.
iii) *If π is finite dimensional, then* $(\Theta^\pi(v^*, v))^T = (-1)^{vv^*}\Theta^{\pi^*}(v, v^*)$.

Proof. i)

$$(-1)^{v_1 v_2^*}(\Theta^{\pi_1}(v_1^*, v_1)\Theta^{\pi_2}(v_2^*, v_2))(u)$$
$$= (-1)^{v_1 v_2^*}(\Theta^{\pi_1}(v_1^*, v_1) \otimes \Theta^{\pi_2}(v_2^*, v_2))(C(u))$$
$$= \Theta^{\pi_1 \otimes \pi_2}(v_1^* \otimes v_2^*, v_1 \otimes v_2)(C(u))$$
$$= (-1)^{(v_1+v_2)u}(v_1^* \otimes v_2^*)(\pi_1 \otimes \pi_2) \cdot C(u)(v_1 \otimes v_2)$$
$$= (-1)^{(v_1+v_2)u}(v_1^* \otimes v_2^*)(\rho(u)(v_1 \otimes v_2))$$
$$= \Theta^\rho(v_1^* \otimes v_2^*, v_1 \otimes v_2)(u).$$

ii) follows from i).
iii) One has $(\Theta^\pi(v^*, v))^T(u) = \Theta^\pi(v^*, v)({}^t u)(-1)^{uv}v^*(\pi({}^t u)(v))$. On the other hand,

$$\Theta^{\pi^*}(v, v^*)(u) = (-1)^{v^* u}v(\pi^*(u)v^*)$$
$$(-1)^{v^* u+v(u+v^*)}(\pi^*(u)v^*)(v) = (-1)^{v^* u+v(u+v^*)+uv^*}v^*(\pi({}^t u)v). \qquad \square$$

Lemma 2.3.4. *Let V be a $\mathfrak{g}$-module. Consider the map*

$$\varphi : V^* \otimes V \otimes V^* \otimes V \longrightarrow U(\mathfrak{g})^*$$
$$\varphi(v_1^* \otimes v_1 \otimes v_2^* \otimes v_2) = (-1)^{v_1^* v_2^* + v_2^* v_1 + v_1^* v_1} (\Theta^\pi(v_2^*, v_1))^T \cdot \Theta^\pi(v_1^*, v_2).$$

If we consider $V^* \otimes V \otimes V^* \otimes V = (V^* \otimes V) \otimes (V^* \otimes V)$ as a $\mathfrak{g} \oplus \mathfrak{g}$-module in such a way that the first two factors are one $\mathfrak{g}$-module and the last two are the other module, then φ is a $\mathfrak{g} \oplus \mathfrak{g}$-module homomorphism. (Recall that we consider $U(\mathfrak{g})^*$ as a $\mathfrak{g} \oplus \mathfrak{g}$-module under simultaneous coregular actions of the first summand from the left and the second summand from the right.)

Proof. Let $x \in \mathfrak{g}$. Then

$$\begin{aligned}
&\varphi((x \otimes 1)(v_1^* \otimes v_1 \otimes v_2^* \otimes v_2)) = \\
&\quad = \varphi((x v_1^* \otimes v_1 + (-1)^{x v_1^*} v_1^* \otimes x v_1) \otimes v_2^* \otimes v_2) \\
&\quad = (-1)^{v_2^*(x + v_1^*) + v_2^* v_1 + (x + v_1^*) v_1} (\Theta^\pi(v_2^*, v_1))^T \Theta^\pi(x v_1^*, v_2) \\
&\qquad + (-1)^{v_1^* v_2^* + v_2^*(x + v_1) + v_1^*(x + v_1) + x v_1^*} (\Theta^\pi(v_2^*, x v_1))^T \Theta^\pi(v_1^*, v_2) \\
&\quad = (-1)^{v_1^* v_2^* + v_2^* v_1 + v_1^* v_1} L^*(x) (\Theta^\pi(v_2^*, v_1))^T \Theta^\pi(v_1^*, v_2) \\
&\qquad + (-1)^{v_1^* v_2^* + v_2^* v_1 + v_1^* v_1 + x(v_2^* + v_1)} (\Theta^\pi(v_2^*, v_1))^T L^*(x) \Theta^\pi(x v_1^*, v_2) \\
&\quad (-1)^{v_1^* v_2^* + v_2^* v_1 + v_1^* v_1} L^*(x) [(\Theta^\pi(v_2^*, v_1))^T \Theta^\pi(v_1^*, v_2)] \\
&\quad = L^*(x) \varphi(v_1^* \otimes v_1 \otimes v_2^* \otimes v_2).
\end{aligned}$$

The identity

$$\varphi((1 \otimes x)(v_1^* \otimes v_1 \otimes v_2^* \otimes v_2)) = R^*(x) \varphi(v_1^* \otimes v_1 \otimes v_2^* \otimes v_2)$$

is similarly verified. □

Lemma 2.3.5. *Let V be a finite-dimensional $\mathfrak{g}$-module, let $\{v_i\}_{i \in I}$ be a basis, and let $\{v_i^*\}_{i \in I}$ be the left dual basis of V^*. Then*

$$\sum_i (\Theta^\pi(v^*, v_i))^T \Theta^\pi(v_i^*, v) = v^*(v) \cdot \varepsilon,$$

where $v \in V$, $v^* \in V^*$ and $\varepsilon \in U(\mathfrak{g})^*$ is the counit.

Proof. The functional ε is uniquely, up to a scalar multiple, characterized by its invariance with respect to the right coregular representation. Further,

$$\omega = \sum_{i \in I} (-1)^{p(i)} v_i^* \otimes v_i$$

is an invariant of the $\mathfrak{g}$-module $V^* \otimes V$. Hence, by Lemma 2.3.4,

$$\varphi\left(\sum_{i \in I} (-1)^{p(i)} v_i^* \otimes v_i \otimes v^* \otimes v\right)$$

is an invariant with respect to the right coregular representation on $U(\mathfrak{g})$, i.e., φ is an invariant functional on $U(\mathfrak{g})$. Hence, $\varphi(\omega \otimes v^* \otimes v) = \alpha\varepsilon$.

On the other hand, $\varphi(\omega \otimes v^* \otimes v) = \sum\limits_{i \in I}(-1)^{p(i)}\varphi(v_i^* \otimes v_i \otimes v^* \otimes v) = \sum\limits_{i \in I}(\Theta^\pi(v^*, v_i))^T \cdot \Theta^\pi(v_i^*, v)$. Hence,

$$\sum_{i \in I}(\Theta^\pi(v^*, v_i))^T \cdot \Theta^\pi(v_i^*, v) = \alpha\varepsilon.$$

To find α, let us substitute $u = 1$ into both sides of the identity. We obtain $\alpha = \alpha\varepsilon(1) = \sum\limits_{i \in I}((\Theta^\pi(v^*, v_i))^T \cdot \Theta^\pi(v_i^*, v)(1) = \sum\limits_{i \in I} v^*(v_i)v_i^*(v) = v^*(v)$. □

3 Bispherical functions and radial parts of Laplace operators for the triple $(\mathfrak{q}(n), \mathfrak{q}(n) \oplus \mathfrak{q}(n), \mathfrak{q}(n))$

3.1 Modules over $\mathfrak{q}(n)$

Let $\mathfrak{g} = \mathfrak{q}(n)$, let $Z(\mathfrak{g})$ be the center of $U(\mathfrak{g})$ and let $\mathfrak{h} = \operatorname{Span}(e_{ii}, f_{jj} \mid i, j \in I_{\bar{0}})$ be the Cartan subalgebra (for the definition of e_{ij} and f_{ij} see Section 1.4.2). Let $\mu \in \mathfrak{h}_{\bar{0}}^*$ and let $M(\mu)$ be the corresponding Verma module with the highest weight with respect to the decomposition $\mathfrak{g} = \mathfrak{n}_- \oplus \mathfrak{h} \oplus \mathfrak{n}_+$, where $\mathfrak{n}_- = \operatorname{Span}(e_{ij}, f_{ij} \mid i > j)$ and $\mathfrak{n}_+ = \operatorname{Span}(e_{ij}, f_{ij}) \mid i < j)$, see [Se4], [Pe]. The module $M(\mu)$ has a central character, i.e., there exists a homomorphism

$$\chi_\mu : Z(\mathfrak{g}) \longrightarrow \mathbb{C}, \quad zv = \chi_\mu(z)v \text{ for any } v \in M(\mu) \text{ and } z \in Z(\mathfrak{g}).$$

Let ε_i be the left dual vector to e_{ii}. Set $R^+ = \{\varepsilon_i - \varepsilon_j \mid i < j\}$ and

$$B(\mu) = \{\nu \mid \nu = \lambda - \sum_{\alpha \in R^+} n_\alpha \alpha \text{ for any } n_\alpha \in \mathbb{Z}_{\geq 0}, \text{ where } \alpha \in R^+, \text{ and } \chi_\nu = \chi_\mu\}.$$

Lemma. *Let $L(\mu)$ be an irreducible $\mathfrak{q}(n)$-module with highest weight μ. Then*

$$\operatorname{ch} L(\mu) = \sum_{\nu \in B(\mu)} C_{\mu\nu} \operatorname{ch} M(\nu), \tag{3.1.1}$$

where for any fixed $\tau \in \mathfrak{h}_{\bar{0}}^*$ only a finite number of summands on the right-hand side contains τ as a weight (for the definition function ch, see [Dix]).

Proof. Let $\alpha_1, \dots, \alpha_{n-1}$ be a base (system of simple roots) in the root system R. Set, by definition,

$$ht(\nu) = \sum_1^{n-1} k_i \quad \text{for any } \nu = \mu - \sum_{i=1}^{n-1} k_i\alpha_i.$$

By induction on t we prove that

$$\mathrm{ch}L(\mu) = \sum_{\nu\in B(\mu),\ ht(\nu)\le t} C_{\mu\nu}\,\mathrm{ch}M(\nu) + \sum_{i=1}^{m(t)} \mathrm{ch}V_i \tag{3.1.2}$$

for any $\tau \in \mathrm{Supp\,ch}V_i$, $ht(\tau) > t$, $\chi_{V_i} = \chi_\mu$ and $\mathrm{Supp\,ch}V_i \subset B(\mu)$.

Let $t = 0$. Then we obtain an exact sequence

$$0 \longrightarrow N \longrightarrow M(\mu) \longrightarrow L(\mu) \longrightarrow 0$$

with $N \subset \bigoplus_{\nu\neq\mu} M(\mu)_\nu$. Therefore, $ht(\nu) > 0$ and $\mathrm{ch}L(\mu) = \mathrm{ch}M(\mu) - \mathrm{ch}(N)$; moreover, $\mathrm{Supp\,ch}N \subset B(\mu)$.

Let (3.1.2) hold. Consider the set $ht(\nu)$, where $\nu \in \mathrm{Supp\,ch}V_i$, and let k be the least element of this set. Clearly, $k > t$. There exists then a finite number of vectors $v_1^i, \dots, v_l^i \in V_i$ with weights $\mu_{1i}, \dots, \mu_{li}$ such that $ht(\mu_{1i}) = ht(\mu_{2i}) \dots = k$. For all other weights ν of V_i, we have $ht(\nu) > k \ge t+1$. The vectors $v_1^i, \dots, v_l^i$ are, obviously, the highest weights, so we have an exact sequence

$$0 \longrightarrow N \longrightarrow \bigoplus_{j=1}^{l} M(\mu_{ji}) \longrightarrow V_i \longrightarrow K \longrightarrow 0.$$

This implies that

$$\mathrm{ch}V_i = \sum_{j=1}^{l} \mathrm{ch}M(\mu_{ji}) + \mathrm{ch}K - \mathrm{ch}N$$

and if $\nu \in \mathrm{Supp\,ch}K \cup \mathrm{Supp\,ch}N$, then $ht(\nu) > t+1$. This proves (3.1.2).

Let $\nu \in \mathfrak{h}_{\bar 0}^*$ be any weight. Set $t = ht(\nu)$; apply (3.1.2) to see that $\nu \notin \cup\mathrm{Supp\,ch}V_i$. This proves (3.1.1). □

Lemma 3.2. *Let* $z_3 = \sum_{i=1}^{n} e_{ii}^{(3)}$ *(see Section* 1.4.2*) and let* v_μ *be the highest weight vector of the* $\mathfrak{g}$*-module* $M(\mu)$*. Then*

$$z_3 v_\mu = \left(\sum_{i=1}^{n} \mu_i^3 - \left(\sum_{i=1}^{n} \mu_i\right)^2\right) v_\mu.$$

Proof. First, observe that

$$e_{ii}^{(3)} = \sum_{j=1}^{n} e_{ij}e_{ji}^{(2)} + \sum_{j=1}^{n} f_{ij}f_{ji}^{(2)} = \sum_{i<j} e_{ij}e_{ji}^{(2)} + \sum_{i<j} f_{ij}f_{ji}^{(2)}$$
$$+ \sum_{i>j} e_{ij}e_{ji}^{(2)} + \sum_{i>j} f_{ij}f_{ji}^{(2)} + e_{ii}e_{ii}^{(2)} + f_{ii}f_{ii}^{(2)}.$$

But, as it is easy to verify that for $i < j$ $e_{ij}^{(2)} v_\mu = f_{ij}^{(2)} v_\mu = 0$. Therefore,

$$\begin{aligned} z_3 v_\mu &= \left(\sum_{i=1}^n e_{ii} e_{ii}^{(2)} + \sum_{i=1}^n f_{ii} f_{ii}^{(2)}\right) v_\mu + \left(\sum_{i<j} e_{ij} e_{ji}^{(2)} + \sum_{i<j} f_{ij} f_{ji}^{(2)}\right) v_\mu \\ &= \left(\sum_{i=1}^n e_{ii} e_{ii}^{(2)} + \sum_{i=1}^n f_{ii} f_{ii}^{(2)} + \sum_{i<j}(e_{ii}^{(2)} - e_{jj}^{(2)}) + \sum_{i<j}(e_{jj}^{(2)} - e_{ii}^{(2)})\right) v_\mu \\ &= \left(\sum_{i=1}^n e_{ii} e_{ii}^{(2)} + \sum_{i=1}^n f_{ii} f_{ii}^{(2)}\right) v_\mu. \end{aligned}$$

Further, it is easy to verify that

$$e_{ii}^{(2)} v_\mu = \left(\mu_i^2 - \mu_i - 2\sum_{k>i} \mu_k\right) v_\mu \text{ and } f_{ii}^{(2)} v_\mu = 0.$$

Hence, $z_3 v_\mu = (\sum_{i=1}^n \mu_i^3 - (\sum_{i=1}^n \mu_i)^2) v_\mu$. □

Corollary. *Let* $\delta = \prod_{\alpha \in R^+} \frac{e^{\alpha/2} + e^{-\alpha/2}}{e^{\alpha/2} - e^{-\alpha/2}}$ *(see Section 1.4.2) and let* $\Omega_3^* = \sum_{i=1}^n \partial_i^3 - (\sum_{i=1}^n \partial_i)^2$, *where* $\partial_i e^l = l(e_i) e^l$. *Then* $\Omega_3^* \delta^{-1} = 0$ *and* $(\sum_{i=1}^n \partial_i^3)\delta^{-1} = 0$.

Proof. Let us prove a more general statement; namely, let φ_μ be the character of an irreducible $\mathfrak{g}$-module with highest weight μ. Then

$$\delta^{-1}\varphi_\mu \text{ is an eigenfunction of } \Omega_3^* \text{ with eigenvalue } \sum_{i=1}^n \mu_i^3 - (\sum_{i=1}^n \mu_i)^2.$$

Indeed, by Lemma 3.1

$$\varphi_\mu = \sum_{\nu \in B(\mu)} C_{\mu\nu} \mathrm{ch} M(\nu).$$

By multiplying both sides of the inequality by δ^{-1} we obtain

$$\delta^{-1}\varphi_\mu = \sum_{\nu \in B(\mu)} \tilde{C}_{\mu\nu} e^\nu.$$

If $z \in Z(\mathfrak{g})$, then $z v_\mu = \chi_\mu(z) v_\mu$. Since $\nu \in B(\mu)$, it follows that $\chi_\mu(z) = \chi_\nu(z)$. But by Lemma 3.2

$$\chi_\mu(z_3) = \sum_{i=1}^n \mu_i^3 - \left(\sum_{i=1}^n \mu_i\right)^2 = \chi_\nu(z_3) = \sum_{i=1}^n \nu_i^3 - \left(\sum_{i=1}^n \nu_i\right)^2.$$

So for $\Omega_3^* = \sum_{i=1}^{n} \partial_i^3 - (\sum_{i=1}^{n} \partial_i)^2$, we have

$$\Omega_3^*\delta^{-1}\varphi_\mu = \sum_{\nu\in B(\mu)} \Omega_3^*(\tilde{C}_{\mu\nu}e^\nu) = \sum_{\nu\in B(\mu)} \tilde{C}_{\mu\nu}\left(\sum_{i=1}^{n} v_i^3 - \left(\sum_{i=1}^{n} v_i\right)^2\right)e^\nu$$
$$= \left(\sum_{i=1}^{n} \mu_i^3 - \left(\sum_{i=1}^{n} \mu_i\right)^2\right)\delta^{-1}\varphi_\mu.$$

In particular, applying this statement to the trivial module we obtain $\Omega_3^*\delta^{-1} = 0$, implying $(\sum_{i=1}^{n} \partial_i^3)\delta^{-1} = 0$. □

3.3 Proof of Lemma 1.2.2

Let us prove i). Let $R = \{\varepsilon_i - \varepsilon_j \mid i \neq j\}$ and $R^+ = \{\varepsilon_i - \varepsilon_j \mid i < j\}$. Set $\Delta_\alpha^+ = e^{\frac{1}{2}\alpha} + e^{-\frac{1}{2}\alpha}$ and $\Delta_\alpha^- = e^{\frac{1}{2}\alpha} - e^{-\frac{1}{2}\alpha}$ for any $\alpha \in R$. Then

$$\partial_i(\Delta_\alpha^+) = \frac{1}{2}\alpha(e_i)\Delta_\alpha^-, \quad \partial_i(\Delta_\alpha^-) = \frac{1}{2}\alpha(e_i)\Delta_\alpha^+.$$

We also set

$$\varphi_i = \sum_{\alpha\in R^+} \frac{\alpha(e_i)}{\Delta_\alpha^+\Delta_\alpha^-}, \qquad \psi_i = \sum_{\alpha\in R^+} \frac{\alpha^2(e_i)}{(\Delta_\alpha^+)^2}, \qquad \Theta_i^\pi = \sum_{\{\alpha,\beta\}\subset R^+} \frac{\alpha(e_i)\beta(e_i)}{\Delta_\alpha^+\Delta_\alpha^-\Delta_\beta^+\Delta_\beta^-},$$

where the last sum runs over the two-element subsets of R^+. It is not difficult to verify that one can express the operator $\hat{\Omega}_3 = \Omega_3 + (\sum \partial_i)^2$ in the form

$$\widehat{\Omega}_3 = \sum_{i=1}^{n} \partial_i^3 + 6\sum_{i=1}^{n} \varphi_i\partial_i^2 - 6\sum_{i=1}^{n} \psi_i\partial_i + 24\sum_{i=1}^{n} \Theta_i^\pi\,\partial_i.$$

It is easy to verify that $\delta^{-1}\partial_i\delta = \partial_i - 2\varphi_i$; hence,

$$\delta^{-1}\partial_i^2\delta = \partial_i^2 - 4\varphi_i\partial_i + 4\varphi_i^2 - 2\partial_i(\varphi_i);$$
$$\delta^{-1}\partial_i^3\delta = \partial_i^3 - 6\varphi_i\partial_i^2 + 3(4\varphi_i^2 - 2\partial_i(\varphi_i))\partial_i - 8\varphi_i^3 - 2\partial_i^2(\varphi_i) + 12\varphi_i\partial_i(\varphi_i).$$

Therefore,

$$\delta^{-1}\hat{\Omega}\delta = \sum_i[\partial_i^3 + (24\Theta_i^\pi - 6\psi_i - 12\varphi_i^2 - 6\partial_i(\varphi_i))\partial_i$$
$$+ (16\varphi_i^3 - 2\partial_i^2(\varphi_i) - 48\Theta_i^\pi\varphi_i + 12\psi_i\varphi_i)].$$

Direct calculations show that

$$24\Theta_i^\pi - 6\psi_i - 12\varphi_i^2 - 6\partial_i(\varphi_i) = 0.$$

Hence,

$$\delta^{-1}\hat{\Omega}\delta = \sum_{i=1}^{n} \partial_i^3 + f.$$

But, $\hat{\Omega}(1) = 0$, so

$$(\delta^{-1}\hat{\Omega}\delta) \cdot (\delta^{-1}) = \left(\sum_{i=1}^{n} \partial_i^3\right)(\delta^{-1}) + f\delta^{-1} = 0.$$

But due to Corollary 3.2, $(\sum_{n-1}^{n} \partial_i^3)(\delta^{-1}) = 0$; hence, $f\delta^{-1} = 0$ and $f = 0$.

ii) immediately follows from our proof of Corollary 3.2.

Let us prove iii). It is not difficult to see that the algebra Ω generated by all the Ω_k, $k = 1, 3, 5, \ldots$, is the image of $Z(\mathfrak{q}(n))$ under the homomorphism

$$r : Z(\mathfrak{q}(n)) \longrightarrow \Omega, \quad r(z)Q_\lambda = \varphi(z)(\lambda)Q_\lambda,$$

where φ is the Harish-Chandra homomorphism, see [Se3]. Therefore, the statement of iii) can be reformulated as follows:

Let R_n be the algebra of symmetric polynomials $r(t_1, \ldots, t_n)$ that do not depend on s after substituting $t_i = s$, $t_j = -s$. Define an R_n-action on the algebra generated by the Q_λ by setting

$$r * Q_\lambda = r(\lambda)Q_\lambda.$$

If P is an eigenvector for R_n and belongs to the subalgebra generated by the Q_λ, then P coincides, up to a scalar multiple, with one of the Q_λ.

Indeed, let $P = \sum_{\lambda\in\Lambda} C_\lambda Q_\lambda$, where Λ is the set of partitions of length $\leq n$. Since $r * P = C(r)P$, it follows that $r(\lambda) = const$ for any $\lambda \in \Lambda$. Let $\lambda, \mu \in \Lambda$ and $\lambda \neq \mu$; then $r(\lambda) = r(\mu)$ for any $r \in R_n$. Hence,

$$\prod_{i=1}^{n} \frac{t - \lambda_i}{t + \lambda_i} = \prod_{i=1}^{n} \frac{t - \mu_j}{t + \mu_j}.$$

We may assume that $\lambda_i > 0$ and $\mu_i > 0$ for all i and j. Then the identity

$$\prod_{i=1}^{n}(t - \lambda_i)\prod_{i=1}^{n}(t + \mu_j) = \prod_{i=1}^{n}(t + \lambda_i)\prod_{i=1}^{n}(t + \mu_j)$$

implies $\prod_{i=1}^{n}(t - \lambda_i) = \prod_{i=1}^{n}(t - \mu_j)$, hence, $\lambda = \mu$. Contradiction. □

3.4 Proof of Theorem 1.5

Let us prove i). Due to [Se4] and [G] we have an isomorphism of $\mathfrak{g}$-modules with respect to the action (1.4.1):

$$U(\mathfrak{g}) \simeq \mathrm{Ind}_{\mathfrak{g}_{\bar{0}}}^{\mathfrak{g}}(U(\mathfrak{g}_{\bar{0}})).$$

Therefore, there exists a bijection between the space of the $\mathfrak{g}$-invariant with respect to the action (1.4.1) functionals on $U(\mathfrak{g})$ and the space of $\mathfrak{g}_{\bar{0}}$-invariant functionals on $U(\mathfrak{g}_{\bar{0}})$. Moreover, any $\mathfrak{g}$-invariant functional is uniquely determined by its restriction to $U(\mathfrak{g}_{\bar{0}})$. On the other hand, every $\mathfrak{g}_{\bar{0}}$-invariant functional is uniquely determined by its restriction to $U(\mathfrak{h}_{\bar{0}}) = S(\mathfrak{h}_{\bar{0}})$.

It is easy to verify ii) of the theorem.

Let us prove iii). Let $l \in (U(\mathfrak{g})^*)^{\mathfrak{g}}$ and let φ_l be the generating function of its restriction to $S(\mathfrak{h}_{\bar{0}})$, i.e.,

$$\varphi_l(t_1, \dots, t_n) = \sum \frac{l(e_{11}^{\nu_1} \dots e_{nn}^{\nu_n})}{(\nu_1)! \dots (\nu_n)!} t_1^{\nu_1} \dots t_n^{\nu_n}$$

where the sum is taken by the integers $\nu_1 \geq 0, \dots, \nu_n \geq 0$. On $S(\mathfrak{h}_{\bar{0}})^*$, define the following operators by setting for any $f \in S(\mathfrak{h}_{\bar{0}})$:

$$(\partial_i^{(k)} l)(f) = l(e_{ii}^{(k)} f), \quad (\delta_i^{(k)} l)(f) = L(f_{ii}^{(k)} f),$$
$$(D_{ij}^{(k)} l)(f) = l(e_{ij} e_{ji}^{(k)} f), \quad (\Delta_{ij}^{(k)} l)(f) = L(f_{ij} f_{ji}^{(k)} f).$$

Lemma. *Let* $\alpha = \varepsilon_i - \varepsilon_j$. *Then* i) $D_{ij}^{(k)} = \dfrac{e^\alpha}{e^\alpha - 1}(\partial_i^{(k)} - \partial_j^{(k)})$.

ii) $\Delta_{ij}^{(k)} = \dfrac{e^\alpha}{e^\alpha + 1}(\partial_j^{(k)} + (-1)^{k+1}\partial_i^{(k)})$.

iii) $l(f \cdot f_{ii} f_{ii}^{(k)}) = l(f e_{ii}^{(k)})$ *for* k *odd.*

iv) $l(f \cdot f_{ii} f_{ii}^{(k)}) = 0$ *for* k *even.*

Proof. i) and ii) are similarly proved. Consider i):

$$\begin{aligned}(D_{ij}^{(k)} l)(f) &= l(f e_{ij} e_{ji}^{(k)}) = L(e_{ij} f(h + \alpha(h)) e_{ij}^{(k)} = l(f(h + \alpha(h)) e_{ij}^{(k)} e_{ij}) \\ &= l(f(h + \alpha(h)) e_{ij} e_{ji}^{(k)}) - l(f(h + \alpha(h))[e_{ij}, e_{ji}^{(k)}]) \\ &= (e^\alpha D_{ij}^{(k)} l)(f) - l(f(h + \alpha(h))(e_{ii}^{(k)} - e_{jj}^{(k)})) \\ &= (e^\alpha D_{ij} - e^\alpha(\partial_i^{(k)} - \partial_j^{(k)}))(l)(f).\end{aligned}$$

This proves i).

iii) We have

$$f_{ii} * (f \cdot f_{ii}^{(k)}) = f_{ii} f \cdot f_{ii}^{(k)} - f \cdot f_{ii}^{(k)} f_{ii}$$

$$= f(f_{ii} f_{ii}^{(k)} - f_{ii}^{(k)} f_{ii}) = f(2 f_{ii} f_{ii}^{(k)} - [f_{ii}, f_{ii}^{(k)}]) = 2 f f_{ii} \cdot f_{ii}^{(k)} - 2 e_{ii}^{(k)} f$$

or $f \cdot f_{ii} \cdot f_{ii}^{(k)} = \frac{1}{2} f_{ii} * (f \cdot f_{ii}^{(k)}) + f \cdot e_{ii}^{(k)}$. This proves iii). iv) is similar. □

This proves the lemma and item ii) of Theorem 1.5. □

Let us prove iv) of the theorem. Let $V^\lambda = \mathrm{Span}(v_p)$, where the $\{v_p\}_{p\in P}$ form a basis of V^λ, let the v_p^* form the left dual basis; set $\omega_\lambda^* = \sum\limits_{p\in P} v_p^* \otimes v_p$, and $\omega_\lambda = \sum_{p\in P} v_p \otimes v_p^*$.

If $\mathfrak{g}$ is embedded in $\mathfrak{g}\oplus\mathfrak{g}$ as the first summand, and $u \in U(\mathfrak{g})$, then

$$\varphi_\lambda(u\otimes 1) = \Theta^\pi(\omega_\lambda^*, \omega_\lambda)(u\otimes 1) = \omega_\lambda^*(u\otimes 1(\omega_\lambda))$$

$$= \omega_\lambda^*(\sum_{p\in P} uv_p \otimes v_p^*) = \sum_{p\in P} v_p^*(uv_p) = \mathrm{tr}(u).$$

In other words, the matrix coefficient $\Theta^\pi(\omega_\lambda^*, \omega_\lambda)$ coincides with the functional $\mathrm{tr}_{V^\lambda}(u)$ after restriction to $U(\mathfrak{g})$. But due to [Se1] we have, up to a scalar multiple,

$$\mathrm{tr}_{V^\lambda}|_{\mathfrak{h}_{\bar 0}} = Q_\lambda(e^{\varepsilon_1}, \dots, e^{\varepsilon_n}). \qquad \square$$

4 Bispherical functions and the radial parts of Laplace operators for the triple $(\mathfrak{pe}(n), \mathfrak{gl}(n|n), \mathfrak{pe}(n))$

4.1 Periplectic superalgebras $\mathfrak{pe}(V)$ (continued from Section 1.6)

In terms of Sections 1.4 and 1.6, the supertransposition antiautomorphism in $\mathfrak{gl}(V)$ is of the form

$$e_{ij}^t = (-1)^{p(i)(p(j)+\bar 1)} e_{ji}.$$

Define two antiautomorphisms ψ_1 and ψ_2 by setting

$$\psi_1(x) = (-1)^x \Pi(x^t)\Pi \quad\text{and}\quad \psi_2 = P\psi_1 P.$$

Direct calculations show that

$$\psi_1(e_{ij}) = (-1)^{p(j)(p(i)+\bar 1)} e_{\bar j\bar i}, \quad \psi_2(e_{ij}) = (-1)^{p(i)(p(j)+\bar 1)} e_{\bar j\bar i}.$$

Observe that $\psi_1(x) = \psi_2(x)$ if $p(x) = \bar 0$ and $\psi_1(x) = -\psi_2(x)$ if $p(x) = \bar 1$. Therefore,

$$\mathfrak{pe}_1(V) = \mathfrak{gl}(V)_{\bar 0}^- \oplus \mathfrak{gl}(V)_{\bar 1}^-, \quad \mathfrak{pe}_2(V) = \mathfrak{gl}(V)_{\bar 0}^- \oplus \mathfrak{gl}(V)_{\bar 1}^+,$$

where

$$\begin{aligned}
\mathfrak{gl}(V)_{\bar 0}^- &= \{x \in \mathfrak{gl}(V)_{\bar 0} \mid \psi_2(x) = -x\},\\
\mathfrak{gl}(V)_{\bar 1}^- &= \{x \in \mathfrak{gl}(V)_{\bar 1} \mid \psi_1(x) = -x\},\\
\mathfrak{gl}(V)_{\bar 1}^+ &= \{x \in \mathfrak{gl}(V)_{\bar 1} \mid \psi_1(x) = x\}.
\end{aligned}$$

For every $x \in \mathfrak{gl}(V)$, set

$$x^+ = \frac{1}{2}(x + \psi_1(x)), \quad x^- = \frac{1}{2}(x - \psi_1(x)).$$

Also set

$$\mathfrak{h}^+ = \mathrm{Span}(e_{ii}^+ \mid i \in I_{\bar{0}}).$$

Lemma 4.2. *For $f \in S(\mathfrak{h}^+)$ and $\alpha = \varepsilon_i - \varepsilon_j$ set*

$$R_{ij}^- f = \tfrac{1}{2}[f(h - \alpha(h)) - f(h + \alpha(h))],$$
$$R_{ij}^+ f = \tfrac{1}{2}[f(h - \alpha(h)) + f(h + \alpha(h))].$$

Then the following identities hold:

i) $e_{ij}^- f = R_{ij}^+ f e_{ij}^- + R_{ij}^- f \cdot e_{ij}^+$.

ii) $R_{ij}^- f e_{ij} e_{ji}^{(k)} - (R_{ij}^- - R_{ij}^+) f \cdot [e_{ij}^-, e_{ji}^{(k)}] \in \mathfrak{pe}_1 U(\mathfrak{g}) + U(\mathfrak{g})\mathfrak{pe}_2$ *if* $p(i)+p(j) = \bar{0}$.

iii) $R_{ij}^+ f e_{ij} e_{ji}^{(k)} - (R_{ij}^+ - R_{ij}^-) f \cdot [e_{ij}^+, e_{ji}^{(k)}] \in \mathfrak{pe}_1 U(\mathfrak{g}) + U(\mathfrak{g})\mathfrak{pe}_2$ *if* $p(i)+p(j) = \bar{1}$.

iv) $[e_{ij}^-, e_{ji}^{(k)}] = \dfrac{1}{2}[e_{ii}^{(k)} - e_{jj}^{(k)}]$ *if* $p(i) + p(j) = \bar{0}$.

v) $[e_{ij}^+, e_{ji}^{(k)}] = \dfrac{1}{2}[e_{ii}^{(k)} + e_{jj}^{(k)}]$ *if* $\bar{i} \neq j$ *and* $p(i) + p(j) = \bar{1}$.

vi) $[e_{i\bar{i}}^+, e_{\bar{i}i}^{(k)}] = e_{ii}^{(k)} + e_{\bar{i}\bar{i}}^{(k)}$ *if* $p(i) = \bar{1}$.

vii) $[e_{i\bar{i}}^+, e_{\bar{i}i}^{(k)}] = 0$ *if* $p(i) = \bar{0}$.

Proof. It is reduced to direct verification. □

4.3 Proof of Theorem 1.7

Let us prove i). It suffices to prove that $U(\mathfrak{g}) = S(\mathfrak{h}^+) + \mathfrak{pe}_1 U(\mathfrak{g}) + U(\mathfrak{g})\mathfrak{pe}_2$. Indeed, any element of $U(\mathfrak{g})$ can be represented in the form

$$u = u_0 + u_1 u_0' u_2, \text{ where } u_1 \in U(\mathfrak{g}_{\bar{1}}^-),\ u_2 \in U(\mathfrak{g}_{\bar{1}}^+),\ u_0, u_0' \in U(\mathfrak{g}_{\bar{0}}).$$

Therefore, we may assume that

$$u = f e_{\alpha_1}^+ \dots e_{\alpha_k}^+,$$

where $f \in S(\mathfrak{h}^+)$, and $e_{\alpha_1}^+, \dots, e_{\alpha_k}^+ \in \mathfrak{g}_{\bar{0}}$ are the weight vectors.

Now, let us do induction on k. If $k = 1$, then $R_{ij}^- f e_{ij}^+ = e_{ij}^- f - R_{ij}^+ f e_{ij}^-$ by Lemma 4.2.i). Hence, $f e_{ij}^+ \in \mathfrak{g}_{\bar{0}}^- U(\mathfrak{g}_{\bar{0}})\mathfrak{g}_{\bar{0}}^-$. Let $k > 1$. Then

$$\begin{aligned} R_{\alpha_1}^- f e_{\alpha_1}^+ \dots e_{\alpha_k}^+ &= e_{\alpha_1}^- f e_{\alpha_2}^+ \dots e_{\alpha_n}^+ - R_{\alpha_1}^+ f e_{\alpha_1}^- e_{\alpha_2}^+ \dots \\ &\equiv \ {\scriptstyle (\mathrm{mod}\ \mathfrak{g}_{\bar{0}}^- U(\mathfrak{g})\mathfrak{g}_{\bar{0}}^-)} - R_{\alpha_1}^+ f e_{\alpha_1}^- e_{\alpha_2}^+ \dots e_{\alpha_n}^+ \\ &\equiv R_{\alpha_1}^+ f \cdot [e_{\alpha_1}^-, e_{\alpha_2}^+ \dots e_{\alpha_k}^+] \\ &= -R_{\alpha_1}^+ f \cdot [e_{\alpha_1}^-, e_{\alpha_2}^+] e_{\alpha_3}^+ \dots e_{\alpha_k}^+ \\ &\quad - R_{\alpha_1}^+ f e_{\alpha_2}^+ [e_{\alpha_1}^-, e_{\alpha_3}^+] \dots e_{\alpha_n}^+ + \dots \in \mathfrak{g}_{\bar{0}}^- U(\mathfrak{g}_{\bar{0}})\mathfrak{g}_{\bar{0}}^-. \end{aligned}$$

The statement of ii) is obvious.

iii) follows from Lemma 4.2 and the fact that $z_k = \sum_{i\in I} e_{ii}^{(k)}$.

Let us prove iv) of Theorem 1.7. It is easy to verify that $\Theta^* = \sum_{i\in I} e_i^* \otimes e_{\bar{i}}^*$ is $\mathfrak{pe}_1$-invariant, whereas $\Theta = \sum_{i\in I} e_i \otimes e_{\bar{i}}$ is $\mathfrak{pe}_2$-invariant. According to [Se2] the linear hull of all the $\mathfrak{pe}_1$-invariants in $V^{\otimes 2k}$ is isomorphic to

$$\mathrm{Ind}_{H_k}^{\mathfrak{S}_{2k}}(\varepsilon) = \bigoplus_{\lambda=(\alpha_1+1,\dots,\alpha_p+1|\alpha_1,\dots,\alpha_p)} S^\lambda, \tag{4.3.1}$$

where $H_k = \mathfrak{S}_k \ltimes \mathbb{Z}_2^k$ is the semidirect product, and $h((\Theta^*)^{\otimes k}) = \varepsilon(h)(\Theta^*)^{\otimes k}$ for any $h \in H_k$.

Similarly, the module of all $\mathfrak{pe}_2$-invariants in $V^{\otimes 2k}$ is of the form (4.3.1). This implies that

$$\varphi_\lambda(u) = \Theta(v_\lambda^*, v_\lambda)(u) = \Theta(e_\lambda\Theta^{*2k}, e_\lambda\Theta^{2k})(u),$$

where e_λ is the minimal idempotent in the Hecke algebra $H(\mathfrak{S}_{2k}, H_{2k}, \varepsilon)$ corresponding to the partition λ, see [St].

If $\sigma \in \mathfrak{S}_{2k}$, then the map

$$\sigma \mapsto \varphi_\sigma, \quad \text{where } \varphi_\sigma(u) = (-1)^{\frac{1}{2}k(k-1)}\Theta^{*\otimes k}(\sigma u \Theta^{\otimes k}),$$

satisfies

$$\varphi_{h_1\sigma h_2} = \varepsilon(h_1h_2)\varphi_\sigma \quad \text{for any } h_1, h_2 \in H_k.$$

Therefore, we obtain a map $H(\mathfrak{S}_{2k}, H_k, \varepsilon) \longrightarrow U(\mathfrak{g})^{*inv}$. By restricting this map to $S(\mathfrak{h}_{\bar{0}})$ we obtain a map

$$\mathrm{ch} : H(\mathfrak{S}_{2k}, H_k, \varepsilon) \longrightarrow S(\mathfrak{h}_{\bar{0}})^*. \tag{4.3.2}$$

Let $K = K_1 \cup K_2$, where $K_1 = \{1, \dots, 2k_1\}$ and $K_2 = \{2k_1+1, \dots, 2k_1+2k_2\}$. Let σ_1 permute the elements of K_1 and let σ_2 permute the elements of K_2; let $\sigma = \sigma_1\sigma_2$. Now, one can verify that

$$\mathrm{ch}(\sigma_1\sigma_2) = \mathrm{ch}(\sigma_1)\mathrm{ch}(\sigma_2).$$

Let now k be odd and $\sigma = (1, \dots, 2k)$ a cycle. Let us calculate $\mathrm{ch}(\sigma)$. We have

$$\Theta^{\otimes k} = \sum e_\psi, \quad \Theta^{*\otimes k} = \sum e_\psi^*,$$

where the sums run over all the maps $\psi : \{1, \dots, 2k\} \mapsto \{1, \dots, n, \bar{1}, \dots, \bar{n}\}$ such that $\psi(2i) = \overline{\psi(2i-1)}$ for $i = 1, \dots, k$ and where $e_\psi = e_{\psi(1)} \otimes \dots \otimes e_{\psi(2k)}$ and e_ψ^* is similarly defined. Therefore,

$$\mathrm{ch}(\sigma)(u) = (-1)^{\frac{1}{2}k(k-1)}\Theta^{*\otimes k}(u \sum \sigma e_\psi),$$

where in the sum one has to take into account only the summands e_ψ for which $\sigma\psi$ possesses the same property as does ψ, i.e., $\psi(2i) = \overline{\psi(2i-1)}$. Therefore, we may assume that

$$\psi(1) = \overline{\psi(2)} = \psi(3) = \overline{\psi(4)} = \ldots = \psi(2k-1) = \overline{\psi(2k)},$$

i.e., $e_\psi = (e_i \otimes e_{\bar{i}})^{\otimes k}$ for $i \in \{1, \ldots, n, \bar{1}, \ldots, \bar{n}\}$. But for such ψ and $u \in S(\mathfrak{h}_{\bar{0}})$ we have

$$\begin{aligned} ch(\sigma)(u) &= (-1)^{\frac{1}{2}k(k-1)}\Theta^{*\otimes k}(u \sum_{i\in I} \sigma(e_i \otimes e_{\bar{i}})^{\otimes k}) \\ &= (-1)^{\frac{1}{2}k(k-1)}\Theta^{*\otimes k}(u \sum_{i\in I} (e_i \otimes e_{\bar{i}})^{\otimes k}) \\ &= (-1)^{\frac{1}{2}k(k-1)}\Theta^{*\otimes k}(\sum_{i\in I_{\bar{0}}} 2e^{\varepsilon_i}(u)(e_i \otimes e_{\bar{i}})^{\otimes k}) \\ &= 2\sum_{i\in I_{\bar{0}}} 2e^{k\varepsilon_i}(u). \end{aligned}$$

Hence, $\mathrm{ch}(\sigma) = 2\sum_{i=1}^{n} 2e^{k\varepsilon_i}$. Therefore, if $\sigma = \sigma_{\nu_1} \ldots \sigma_{\nu_p}$ is the product of independent cycles of odd lengths, then

$$\mathrm{ch}(\sigma) = 2^{l(\nu)} P_\nu, \text{ where } P_\nu = P_{\nu_1} \ldots P_{\nu_p} \text{ and } P_l = \sum_{i=1}^{n} e^{l\varepsilon_i}.$$

We see that the map (4.3.2) coincides with the characteristic map Stembridge constructed in [St], p. 85. Therefore, by Theorem 5.2 from [St] we have

$$\mathrm{ch}(e_\lambda) = \Theta_\lambda(e^{\varepsilon_1}, \ldots, e^{\varepsilon_n}) \cdot 2^{n-e(\lambda)} \cdot g_\lambda,$$

where g_λ is the number of shifted standard tableaux of shape λ. □

Acknowledgment. I wish to thank G. Olshanskii who raised the problem and D. Leites for help.

References

[Dix] Dixmier J., *Enveloping algebras.* Revised reprint of the 1977 translation. Graduate Studies in Mathematics, 11. American Mathematical Society, Providence, RI, 1996. xx+379 pp.

[G] Gorelik M., Annihilation theorem and separation theorem for basic classical Lie superalgebras, math/RT0008143.

[Ma] Macdonald I., *Symmetric functions and Hall polynomials*, with contributions by A. Zelevinsky, Oxford Mathematical Monographs, Oxford Science Publications, Clarendon Press, Oxford University Press, New York, 1995. x+475 pp.

[Pe] Penkov I., Characters of typical irreducible finite-dimensional $\mathfrak{q}(n)$-modules, *Funktsional. Anal. i Prilozhen* **20** no. 1 (1986), 37–45; (Russian; English translation: *Functional Anal. Appl.* **20** no. 1 (1986), 30–37).

[Sch] Schur I., Über die darstellung der symmetrischen und der alternierden Grouppe durch gebrochene lineare Substitutionen, *J. Reine. Angew. Math.* **139** (1911), 155–250.

[Se1] Sergeev A., The tensor algebra of the identity representation as a module over Lie superalgebras $\mathfrak{gl}(m|n)$ and $\mathfrak{q}(n)$, *Math. USSR Sbornik* **51** (1985), 419–427.

[Se2] Sergeev A., An analog of the classical invariant theory for Lie superalgebras. I, II. mathRT/9810111, 9904079; *Michigan J. Math.* **49** no. 1 (2001), 113–146, 147–168.

[Se3] Sergeev A., The centre of enveloping algebra for the Lie superalgebra $\mathfrak{q}(n; \mathbb{C})$, *Lett. Math. Phys.* **7** no. 3 (1983), 177–179.

[Se4] Sergeev A., The invariant polynomials on simple Lie superalgebras, *Representation Theory*, **3** (1999), 250–280.

[St] Stembridge J., On Schur's Q-functions and the primitive idempotents of a commutative Hecke algebra, *J. Alg. Combinatorics* **1** no. 1 (1992), 71–95.

Alexander Sergeev
Department of Mathematics
University of Stockholm
Roslagsv. 101, Kräftriket hus 6
SE-106 91, Stockholm, Sweden
mleites@matematik.su.se
(on leave of absence from
Balakovo Institute of Technique, Technology and Control
Russia)

Maximal Subalgebras of the Classical Linear Lie Superalgebras

Irina Shchepochkina

ABSTRACT Dynkin's classification of maximal subalgebras of simple finite-dimensional complex Lie algebras is generalized to *linear* Lie superalgebras. The maximal nonsimple irreducible subalgebras of $\mathfrak{gl}(p|q)$, $\mathfrak{q}(n)$, $\mathfrak{sl}(p|q)$, $\mathfrak{osp}(m|2n)$, $\mathfrak{pe}(n)$, and $\mathfrak{spe}(n)$ are classified.

Introduction

0.1 Dynkin's result

In 1951 Dynkin published two remarkable papers somewhat interlaced in their theme: on the classification of semisimple [D1] and maximal [D2] subalgebras of simple (finite-dimensional) Lie algebras. These classifications are of interest *per se*; they also proved to be useful in the studies of integrable systems and in representation theory.

A. L. Onishchik and D. Leites asked me to generalize Dynkin's results to "classical" Lie superalgebras. In this paper (partly preprinted in [Sh1], [Sh3]) I try to give the answer in a form similar to that of Dynkin's result. Let me first recall Dynkin's result. In what follows the adjective "linear" describes a Lie (super)algebra $\mathfrak{g}$ whose elements are realized as linear operators in a linear space V.

Let $\mathfrak{g}$ be a simple complex linear Lie algebra, e.g., $\mathfrak{g} = \mathfrak{sl}(V)$, $\mathfrak{o}(V)$, or $\mathfrak{sp}(V)$, $\mathfrak{h} \subset \mathfrak{g}$ its maximal subalgebra. In this setting only the following 3 cases can occur:

1) the representation of $\mathfrak{h}$ in V is irreducible; $\mathfrak{h}$ is not simple. Dynkin's list is as follows (on the next page); here the cases $\dim V_2 = 4$ in the second and fourth lines, as well as $\dim V_2 = \dim V_1 = 2$ in the third one, are exceptional because $\mathfrak{o}(4) \cong \mathfrak{sp}(2) \oplus \mathfrak{sp}(2)$).

2) $\mathfrak{h}$ is simple and irreducible; i.e., $\mathfrak{h}$ irreducibly acts on V. *For practically every irreducible representation of a simple $\mathfrak{h}$ in a linear space V the image is a maximal subalgebra in one of the three classical simple linear algebras $\mathfrak{sl}(V)$, $\mathfrak{o}(V)$ or $\mathfrak{sp}(V)$.* Dynkin proved this and listed the exceptional cases.

AMS Subject classification: 17A70
Keywords and phrases: maximal subalgebras, Lie algebras, Lie superalgebras

	$\mathfrak{h}$	$\mathfrak{g}$	condition
1	$\mathfrak{sl}(V_1) \oplus \mathfrak{sl}(V_2)$	$\mathfrak{sl}(V_1 \otimes V_2)$	$\dim V_2 \geq \dim V_1 \geq 2$
2	$\mathfrak{sp}(V_1) \oplus \mathfrak{o}(V_2)$	$\mathfrak{sp}(V_1 \otimes V_2)$	$\dim V_1 \geq 2$, $\dim V_2 \geq 3$, $\dim V_2 \neq 4$ or $\dim V_1 = 2$ and $\dim V_2 = 4$;
3	$\mathfrak{sp}(V_1) \oplus \mathfrak{sp}(V_2)$	$\mathfrak{o}(V_1 \otimes V_2)$	$\dim V_2 \geq \dim V_1 \geq 2$ except $\dim V_1 = \dim V_2 = 2$,
4	$\mathfrak{o}(V_1) \oplus \mathfrak{o}(V_2)$	$\mathfrak{o}(V_1 \otimes V_2)$	$\dim V_2 \geq \dim V_1 \geq 3$ and $\dim V_1$, $\dim V_2 \neq 4$.

(0.1)

3) $\mathfrak{h}$ is reducible. Then $\mathfrak{h}$ can be described as the collection of all operators from $\mathfrak{g}$ that preserve a subspace $W \subset V$. Here W can be arbitrary for $\mathfrak{g} = \mathfrak{sl}$, whereas for $\mathfrak{g} = \mathfrak{o}(n)$ and $\mathfrak{sp}(2n)$ the bilinear form ω on V, preserved by $\mathfrak{g}$, must either be nondegenerate or identically vanish on W. Such algebras $\mathfrak{h}$ are certain *parabolic* subalgebras of $\mathfrak{g}$.

Passing to subsuperalgebras we encounter the same cases but difficulties in their superization range widely. Additionally, there is case 4): maximal subalgebras that are not subsuperalgebras.

Superization of case 3) to Lie superalgebras with Cartan matrix; linear ones is more or less straightforward, see [ZZO].

Superization of case 2) requires a more or less explicit description of finite-dimensional irreducible modules over simple Lie superalgebras. There are two types of such modules: *typical* and *atypical* ones, cf. [K1], [K2], [PS]. Observe that Lemma 4.4 demonstrates that, unlike Lie algebra case, the images of Lie superalgebras $\mathfrak{sl}(m|n)$, $\mathfrak{osp}(2|2n)$, $\mathfrak{pe}(n)$, $\mathfrak{spe}(n)$, $\mathfrak{vect}(0|n)$, $\mathfrak{svect}(0|n)$, $\mathfrak{h}(0|n)$, $\mathfrak{sh}(0|n)$ in the typical modules are not maximal, except for the case considered in Lemma 3.4.1.

The case of other algebras and atypical modules constitutes an open problem. For a partial result see [J1]. With a general character formula, even conjectural, [PS], [Se] one can now hope to be able to derive the complete result.

Superization of case 1) is what is done in this paper: the description of the irreducible nonsimple maximal subsuperalgebras of linear complex Lie superalgebras either simple or "classical," i.e., certain algebras closely related to simple types.

Regrettably, a precise description of maximal Lie superalgebras of type 1) is more involved than Dynkin's description (0.1) above. Indeed, there are too many exceptional cases having small dimensions. Nevertheless, in the Main Theorem, I distinguish four main types of subalgebras such that any type 1 linear Lie superalgebra is contained in one of these four types of Lie superalgebras. Thus, the subalgebras distinguished in the Main Theorem are the main candidates for the roles of maximal subalgebras.

Observe that two of these four types are similar to Dynkin's types, whereas the other two are of a totally different nature; the picture is similar to the one over fields of prime characteristic.

In Section 1 I describe the main constructions and give a precise formulation of the Main Theorem. Statements describing when the subalgebras from the Main Theorem are indeed maximal are collected in Tables 1–3. Section 2 is devoted to a proof of maximality of Dynkin-type subalgebras, and in Section 3 the other two types are considered. In Section 4 I prove the Main Theorem.

0.2 Comparison with the case of prime characteristic

Our results resemble Ten's result in prime characteristic [T]. To formulate this, recall that a subalgebra of a (finite-dimensional) Lie algebra is called *regular* if it is invariant with respect to a maximal torus. Here, as well as in our nonstandard cases, we need semidirect sums of Lie (super)algebras denoted by $\mathfrak{i} \subplus \mathfrak{b}$, the ideal is on the left.

Theorem. *Let $\mathbb{K}$ be an algebraically closed field of characteristic $p > 3$.*

1) *Any non-semisimple maximal subalgebra of $\mathfrak{sl}(n)$ for $n \not\equiv 0 \mod p$, $\mathfrak{o}(n)$ or $\mathfrak{sp}(2n)$ is regular.*

2a) *Let $\dim V = np^m$, $(n, p) = 1$, $n > 1$; let $x = (x_1, \dots, x_m)$. If $\mathfrak{m}$ is an irreducible maximal subalgebra in $\mathfrak{sl}(V)$ such that $\mathfrak{pm} = \mathfrak{m}/\mathrm{Span}(\mathrm{id})$ is not semisimple, then $V = U \otimes O_m$, where $O_m = \mathbb{K}[x]/(x_1^p, \dots, x_m^p)$ and $\mathfrak{m} = \mathfrak{gl}(U) \otimes O_m \subplus \mathfrak{vect}(m)$.*

2b) *If $n = 1$, then in addition to the above examples 2a) the algebra $\mathfrak{m} = \mathfrak{sp}(2m) \supplus \mathfrak{hei}(2m|0)$ (where $\mathfrak{hei}(2m|0)$ is an even version of the Heisenberg Lie algebra, see sec. 2.7) is also maximal in $\mathfrak{sl}(V)$.*

3) *Any maximal subalgebra in $\mathfrak{g}_2$ is regular except $\mathfrak{vect}(1)$ for $p = 7$ and $\mathfrak{sl}(2)$ for $p > 7$.*

0.3 Other related results

Maximal solvable Lie subsuperalgebras

Such subalgebras for $\mathfrak{gl}(m|n)$ and $\mathfrak{sl}(m|n)$ are classified in [Sh2]. A bizarre series of subalgebras was discovered. Maximal solvable subalgebras — Borel subalgebras — of simple Lie superalgebras are important in representation theory (e.g., for constructing Verma modules). In the supersetting, the maximal solvable subalgebras can be larger than what is used for constructing Verma modules and what Penkov suggested calling Borel subalgebras. Conjecturally, these larger algebras are related with atypical representations.

Superization: case 4) A nonhomogeneous with respect to parity subalgebra $\mathfrak{h}$ of the Lie superalgebra $\mathfrak{g}$ is called a *Volichenko algebra*. A list of simple finite dimensional Volichenko subalgebras in simple Lie superalgebras is obtained under a technical condition by Serganova [S]. (For motivations and the infinite-dimensional case see [LS], [KL].) Simple Volichenko algebras are one more, new, type of maximal subalgebras of simple Lie superalgebras: Volichenko subalgebras are not Lie subsuperalgebras.

1 Notation, background and main statements

Before we formulate our result, we need to fix several notation and constructions. Linear algebra type equations are generalized to the linear superalgebra setting by means of linearity and the Sign Rule, as for example, are the definitions for a supercommutator and a Lie superalgebra. There are, however, notions, e.g., that of supertrace, which although they follow from the Sign Rule, are not obvious direct corollaries, cf. [D]. Some facts, like existence of (at least) two analogs of the general linear Lie algebra, or two types of bilinear forms (even and odd) are even less familiar.

1.1 Basics

Throughout the paper the ground field is $\mathbb{C}$. A *superspace* is a $\mathbb{Z}/2$-graded linear space $V = V_{\bar 0} \oplus V_{\bar 1}$, where $\mathbb{Z}/2 = \{\bar 0, \bar 1\}$ and $p(v) = \bar i$ if $v \in V_{\bar i}$. The *superdimension* of V is a pair $N = m|n$, where $m = \dim V_{\bar 0}$, $n = \dim V_{\bar 1}$. The usual equation $\dim V \otimes W = \dim V \cdot \dim W$ becomes manifest if we introduce a formal symbol ε such that $\varepsilon^2 = 1$ and set $\dim V = \dim V_{\bar 0} + \dim V_{\bar 1}\varepsilon$.

For a superspace $V = V_{\bar 0} \oplus V_{\bar 1}$ denote by $\Pi(V)$ another copy of the same superspace — with the shifted parity, i.e., $(\Pi(V))_{\bar i} = V_{\bar i + \bar 1}$. The *subsuperspace* $U \subset V$ is a subspace such that $U_{\bar i} = U \cap V_{\bar i}$. A superspace structure in V induces the superspace structure in the space $\mathrm{End}(V)$.

The *Lie superalgebra* of all linear operators in V is called the *general Lie superalgebra*. It is denoted by $\mathfrak{gl}(V)$ or $\mathfrak{gl}(\dim V)$. Having selected a homogeneous basis of V, we can represent operators by supermatrices; in this paper I only need supermatrices in the standard format, i.e., when the even basis vectors of V are collected together and come first.

The space of operators with zero supertrace constitutes the *special linear* Lie subsuperalgebra $\mathfrak{sl}(V)$, also denoted as $\mathfrak{sl}(\dim V)$.

There are, however, at least two, not one, super versions of $\mathfrak{gl}(V)$. Another version is called the *queer* Lie superalgebra and is defined as the one that preserves the complex structure given by an *odd* operator J, i.e., is the centralizer $C(J)$ of J:

$$\mathfrak{q}(V) = C(J) = \{X \in \mathfrak{gl}(V) \mid [X, J] = 0\}, \text{ where } J^2 = -\mathrm{id}.$$

It is clear that by a change of basis we can reduce J to the form $J_{2n} = \begin{pmatrix} 0 & 1_n \\ -1_n & 0 \end{pmatrix}$.

Then in matrix form we have

$$\mathfrak{q}(n) = \left\{ X \in \mathfrak{gl}(n|n) \mid X = \begin{pmatrix} A & B \\ B & A \end{pmatrix} \right\}.$$

On $\mathfrak{q}(n)$, the *queer trace* is defined as qtr : $\begin{pmatrix} A & B \\ B & A \end{pmatrix} \mapsto \operatorname{tr} B$. Denote by $\mathfrak{sq}(n)$ the Lie superalgebra of *queer-traceless* operators.

Observe that the identity representations of $\mathfrak{g} = \mathfrak{q}(V)$ and $\mathfrak{sq}(V)$ in V, although irreducible in the supersetting, are not irreducible in the nongraded sense: take linearly independent vectors $v_1, \dots, v_n$ from $V_{\bar{0}}$; then $\operatorname{Span}(v_1 + J(v_1), \dots, v_n + J(v_n))$ is a $\mathfrak{g}$-invariant subspace of V which is not a subsuperspace.

A representation is called *irreducible of G-type* if it has no invariant subspace; it is called *irreducible of Q-type* if it has no invariant sub*super*space, but has an invariant subspace.

1.2 *The action of* $\mathfrak{g}_1(V_1) \oplus \mathfrak{g}_2(V_2)$ *in* $V_1 \otimes V_2$

Given two irreducible (of G- or Q-type) linear Lie superalgebras $\mathfrak{g}_i \subset \mathfrak{gl}(V_i)$, we obtain a representation of the Lie superalgebra $\mathfrak{g}_1 \oplus \mathfrak{g}_2$ in the superspace $V = V_1 \otimes V_2$, the *tensor product* of the given representations:

$$X_1 + X_2 \mapsto X_1 \otimes 1 + 1 \otimes X_2 \quad \text{for } X_1 \in \mathfrak{g}_1,\ X_2 \in \mathfrak{g}_2. \tag{1.1}$$

G-construction: the $\mathfrak{g}_i$ *are both not of Q-type*

If both $\mathfrak{g}_1$ and $\mathfrak{g}_2$ contain the identity operators, the representation (1.1) has a 1-dimensional kernel. By $\mathfrak{g}_1 \odot \mathfrak{g}_2$ we will mean the image of the direct sum $\mathfrak{g}_1 \oplus \mathfrak{g}_2$ under the representation (1.1). Observe that $\mathfrak{g}_1 \odot \mathfrak{g}_2$ is irreducible in this case.

It is convenient to retain the notation $\mathfrak{g}_1 \odot \mathfrak{g}_2$ even in the absence of the kernel, i.e., when $\mathfrak{g}_1 \odot \mathfrak{g}_2 \cong \mathfrak{g}_1 \oplus \mathfrak{g}_2$.

Q-construction: both $\mathfrak{g}_i$ *are irreducible of Q-type*

In this case the standard action of $\mathfrak{g}_1 \oplus \mathfrak{g}_2$ in $V_1 \otimes V_2$ is reducible. That is why the construction of $\mathfrak{g}_1 \odot \mathfrak{g}_2$ is different, as follows.

Let U be a fixed $(1|1)$-dimensional superspace. The operators $J_U = \begin{pmatrix} 0 & 1 \\ -1 & 0 \end{pmatrix}$ and $I_U = \begin{pmatrix} 0 & i \\ i & 0 \end{pmatrix}$ from $\operatorname{End}(U)$ supercommute and $J_U^2 = I_U^2 = -1$. Let us represent the $n_1|n_1$-dimensional superspace V_1 as a tensor product $V_1 = (V_1)_{\bar{0}} \otimes U$ and $n_2|n_2$-dimensional superspace V_2 as a tensor product $V_2 = U \otimes (V_2)_{\bar{0}}$. Set $J = 1 \otimes J_U \in \operatorname{End}(V_1)$ and $I = I_U \otimes 1 \in \operatorname{End}(V_2)$. It is clear that

$$C(J) = \operatorname{End}(V_1)_{\bar{0}} \otimes \operatorname{Span}(1, I_U) \cong \mathfrak{q}(V_1);$$
$$C(I) = \operatorname{Span}(1, J_U) \otimes \operatorname{End}(V_2)_{\bar{0}} \cong \mathfrak{q}(V_2).$$

Define the *Q-tensor product* by setting

$$V_1 \otimes^Q V_2 := (V_1)_{\bar{0}} \otimes U \otimes (V_2)_{\bar{0}}.$$

The map

$$X_1 + X_2 \mapsto X_1 \otimes 1_{(V_2)_{\bar{0}}} + 1_{(V_1)_{\bar{0}}} \otimes X_2 \quad \text{for } X_1 \in \mathfrak{g}_1,\ X_2 \in \mathfrak{g}_2$$

determines an irreducible representation of the Lie superalgebra $\mathfrak{g}_1 \oplus \mathfrak{g}_2$ in the space $V_1 \otimes^Q V_2$. We will denote the image of $\mathfrak{g}_1 \oplus \mathfrak{g}_2$ in this superspace by $\mathfrak{g}_1 \odot \mathfrak{g}_2$. The usual tensor product $V_1 \otimes V_2$ considered as a $\mathfrak{g}_1 \oplus \mathfrak{g}_2$-module is a direct sum of two submodules isomorphic to $V_1 \otimes^Q V_2$. (When $\mathfrak{g}_1 = \mathfrak{g}_2 = \mathfrak{g}$ and, consequently, $V_1 = V_2$, Sergeev and his followers denote the diagonal action of $\mathfrak{g}$ in $V \otimes^Q V$ by $2^{-1}(V \otimes V)$, cf. [Ser2].)

1.3 Projectivization

If $\mathfrak{g} \subset \mathfrak{gl}(n|n)$ is a Lie subsuperalgebra containing scalar operators, then the *projective* Lie superalgebra of type $\mathfrak{g}$ is $\mathfrak{pg} = \mathfrak{g}/\mathbb{C}$. Lie superalgebras $\mathfrak{g}_1 \odot \mathfrak{g}_2$, described in Section 1.2, are projective.

Projectivization sometimes leads to new Lie superalgebras, for example to $\mathfrak{pgl}(n|n)$, $\mathfrak{psl}(n|n)$, $\mathfrak{pq}(n)$, $\mathfrak{psq}(n)$, whereas $\mathfrak{pgl}(p|q) \cong \mathfrak{sl}(p|q)$ if $p \neq q$.

1.4 Lie superalgebras that preserve bilinear forms

We will often use a general notation $\mathfrak{aut}(\omega)$ for the Lie superalgebra that preserves the nondegenerate bilinear form ω in the superspace V, i.e., $\mathfrak{aut}(\omega) =$

$$\{X \in \mathfrak{gl}(V) \mid \omega(Xv_1, v_2) + (-1)^{p(X)p(v_1)}\omega(v_1, Xv_2) = 0 \text{ for any } v_1, v_2 \in V\}.$$

If the form ω is even and supersymmetric, then the Lie superalgebra $\mathfrak{aut}(\omega)$ is called *orthosymplectic* and denoted $\mathfrak{osp}(V) = \mathfrak{osp}(\dim V)$. Observe that the passage from V to $\Pi(V)$ sends the supersymmetric forms to superskew-symmetric ones. This is why we use the notation $\mathfrak{osp}^{sk}$ for the Lie superalgebra that preserves the *superskew*-symmetric form. We have an isomorphism $\mathfrak{osp}(V) \cong \mathfrak{osp}^{sk}(\Pi(V))$, but matrix representations of elements from $\mathfrak{osp}(V)$ and $\mathfrak{osp}^{sk}(\Pi(V))$ are different.

If the form ω is odd, then the Lie superalgebra $\mathfrak{aut}(\omega)$ is called, as A. Weil suggested, *periplectic* and denoted $\mathfrak{pe}(n) := \mathfrak{pe}^{sk}(n)$ or $\mathfrak{pe}^{sy}(n)$, in accordance with the symmetry of ω. The passage from V to $\Pi(V)$ sends the supersymmetric forms to superskew-symmetric ones and establishes an isomorphism $\mathfrak{pe}^{sy}(n) \cong \mathfrak{pe}^{sk}(n)$.

The *special periplectic* superalgebra is $\mathfrak{spe}(n) = \{X \in \mathfrak{pe}(n) \mid \mathrm{str} X = 0\}$.

Observe that the map $\chi_\lambda : X \mapsto \lambda \cdot \mathrm{str} X$, where $\lambda \in \mathbb{C}$, $\lambda \neq 0$, determines a nontrivial character of $\mathfrak{pe}(n)$. Denote by $\mathfrak{pe}_\lambda(n)$ the image of $\mathfrak{pe}(n)$ in the representation $\mathrm{id} \otimes \chi_\lambda$.

1.5 Sergeev Lie superalgebra

A. Sergeev proved that there is just one nontrivial central extension of $\mathfrak{spe}(n)$. It exists only for $n = 4$ and is denoted by $\mathfrak{as}$. Let us represent an arbitrary element $X \in \mathfrak{as}$ as a pair $X = x + d \cdot z$, where $x = \begin{pmatrix} A & B \\ C & -A^t \end{pmatrix} \in \mathfrak{spe}(4)$ ($\operatorname{tr} A = 0$, $B = B^t$, $C = -C^t$), $d \in \mathbb{C}$ and z is the central element. The bracket in $\mathfrak{as}$ is

$$[x + d \cdot z, x' + d' \cdot z] = [x, x'] + \operatorname{tr} CC' \cdot z.$$

1.6 Heisenberg Lie superalgebra

Denote by $\mathfrak{hei}(0|m)$ the Heisenberg Lie superalgebra with m odd generators of creation and annihilation, i.e., the Lie superalgebra with odd generators $\xi_1, \dots, \xi_n; \eta_1, \dots, \eta_n$ if $m = 2n$ or $\xi_1, \dots, \xi_{n-1}; \eta_1, \dots, \eta_{n-1}$ and θ if $m = 2n - 1$ and an even generator z satisfying the relations

$$[\xi_i, \eta_j] = \delta_{i,j} \cdot z, \; [\xi_i, \xi_j] = [\eta_i, \eta_j] = [z, \mathfrak{hei}(0|2n)] = 0;$$
$$[\theta, \theta] = z, \; [\theta, \xi_i] = [\theta, \eta_j] = 0.$$

The irreducible finite-dimensional representation of $\mathfrak{hei}(0|m)$ was first described in [K]; see also [Ser1].

Every irreducible finite-dimensional representation of $\mathfrak{hei}(0|m)$ is scalar on z and the only one that sends z to the identity operator is realized in the superspace $\Lambda(n) = \Lambda(\xi)$ for $m = 2n$ or $\Lambda(n) = \Lambda(\xi, \theta)$ for $m = 2n - 1$ by the equations (where $i \leq n$ for $m = 2n$ and $i \leq n - 1$ for $m = 2n - 1$):

$$z \mapsto 1, \quad \xi_i \mapsto \xi_i, \quad \eta_i \mapsto \partial_{\xi_i}; \quad \theta \mapsto \theta + \partial_\theta.$$

This representation is irreducible of G-type if $m = 2n$ and irreducible of Q-type if $m = 2n - 1$.

The normalizer of $\mathfrak{hei}(0|m)$ in $\mathfrak{gl}(\Lambda(n))$ is $\mathfrak{g} = \mathfrak{hei}(0|m) \oplus \mathfrak{o}(m)$; it acts in the spinor representation of $\mathfrak{o}(m)$, or, in terms of differential operators: $\mathfrak{g} = \operatorname{Span}(1, \xi, \partial, \xi\partial, \partial\partial, \xi\xi)$ for $m = 2n$ and $\mathfrak{g} = \operatorname{Span}(1, \xi, \partial, \theta + \partial_\theta, \xi(\theta + \partial_\theta), \partial(\theta + \partial_\theta), \xi\partial, \partial\partial, \xi\xi)$ for $m = 2n - 1$.

Let $n = 3$. Then $\mathfrak{g}$ is contained in $\tilde{\mathfrak{g}} = (\mathfrak{hei}(0|6) \oplus V) \oplus \mathfrak{o}(6)$ (sum as $\mathfrak{o}(6)$-modules), where the highest weight of the $\mathfrak{o}(6) \simeq \mathfrak{sl}(4)$-module V is $(2, 0, 0)$, i.e., $\tilde{\mathfrak{g}}$ is isomorphic to the nontrivial central extension $\mathfrak{as}$ of $\mathfrak{spe}(4)$. (Observe that the $\mathfrak{sl}(4)$-module $\mathfrak{hei}(0|6)$ is the direct sum of the trivial module and the exterior square of the dual to the standard 4-dimensional $\mathfrak{sl}(4)$-module.)

1.7 Densities

Let $\mathfrak{vect}(0|n) = \mathfrak{der}\Lambda(n)$ be the Lie superalgebra of vector fields on an $(0|n)$-dimensional superspace. Irreducible representations of $\mathfrak{vect}(0|n)$ are described in [BL]. The most important for us will be a one-parameter family of representations

T^λ of $\mathfrak{vect}(0|n)$ in the superspace $Vol^\lambda = \Lambda(\xi)vol^\lambda(\xi)$ of λ-densities. We define it by the equation

$$T^\lambda(D)(f(\xi)vol^\lambda) = (D(f) + \lambda \mathrm{div} D \cdot f)vol^\lambda.$$

The representations T^λ are irreducible if $\lambda \neq 0, 1$. The representation T^0 determines the action of the Lie superalgebra of vector fields $\mathfrak{vect}(0|n)$ in the space of functions; the constants form an invariant 1-dimensional subspace. The representation T^1 is the dual representation in the space of volume forms, Vol; it contains an irreducible subspace of codimension ε^n spanned by the volume element vol. Therefore, $T^0 \cong (\Pi^n(T^1))^*$.

Define the form $\omega_{1/2}$ on $\sqrt{Vol}$ by the equation

$$\omega_{1/2}(f\sqrt{vol}, g\sqrt{vol}) = \int fg \cdot vol.$$

It is clear that $T^{1/2}(\mathfrak{vect}(0|n))$ preserves $\omega_{1/2}$.

1.8 The Poisson superalgebra

On $\Lambda(m) = \Lambda(\Theta_1, \dots, \Theta_m)$, define a Lie superalgebra structure by setting (the extra minus is convenient for calculations with weights)

$$\{f, g\}_{P.b.} = -(-1)^{p(f)} \sum_{j \leq m} \frac{\partial f}{\partial \Theta_j} \frac{\partial g}{\partial \Theta_j}.$$

Sometimes it is more convenient to denote the Θ's differently and set (here $i^2 = -1$ and $[m + 1/2]$ is the integer part):

$$\xi_j = \frac{1}{\sqrt{2}}(\Theta_j - i\Theta_{n+j}); \ \eta_j = \frac{1}{\sqrt{2}}(\Theta_j + i\Theta_{n+j})$$

for $j \leq n = [m + 1/2]$ $\theta = \Theta_{2n+1}$. In new indeterminates the Poisson bracket is defined by equation (the summand with θ only exists for m odd):

$$\{f, g\} = -(-1)^{p(f)} \left(\sum_{i \leq n} \left(\frac{\partial f}{\partial \xi_i} \frac{\partial g}{\partial \eta_i} + \frac{\partial f}{\partial \eta_i} \frac{\partial g}{\partial \xi_i}\right) + \frac{\partial f}{\partial \theta} \frac{\partial g}{\partial \theta} \right).$$

This Lie superalgebra is denoted $\mathfrak{po}(0|m)$; it is a finite-dimensional analog of the Poisson algebra, and it turns into the general matrix superalgebra under quantization, see [LSh].

On $\mathfrak{po}(0|m)$, define the grading to be $\mathrm{gr}(f) = \deg f - 2$, where $\deg f$ is the degree of f as an element of $\Lambda(m)$: $\deg \Theta_i = 1$ for any i. Thus, $\mathfrak{po}(0|m) = \bigoplus_{i=-2}^{m-2} \mathfrak{po}_i$, where $\mathfrak{po}_0 \cong \mathfrak{o}(m)$ and where $\bigoplus_{i<0} \mathfrak{po}_i$ is isomorphic to $\mathfrak{hei}(0|m)$.

Since the supertrace and the queer trace are quantum versions of the integral, $\mathfrak{po}(0|m)$ possesses an ideal $\mathfrak{spo}(0|m)$, the special Poisson superalgebra, of codimension ε^m, and a 1-dimensional center, the space of constant functions.

The classical isomorphisms of Lie algebras $\mathfrak{sl}(4) \cong \mathfrak{o}(6)$ and their modules $\Lambda^2(\mathrm{id}_{\mathfrak{sl}(4)}) \cong \mathrm{id}_{\mathfrak{o}(6)}$ show that $\mathfrak{as}$ described in Section 1.5 can be embedded in $\mathfrak{po}(0|6)$. The embedding sends the central element $z \in \mathfrak{as}$ into $1 \in \mathfrak{po}(0|6)$.

1.9 Simplicity

The Lie superalgebras $\mathfrak{sl}(m|n)$ for $m > n \geq 1$, $\mathfrak{psl}(n|n)$ for $n > 1$, $\mathfrak{psq}(n)$ for $n > 2$, $\mathfrak{osp}(m|2n)$ for $mn \neq 0$ and $\mathfrak{spe}(n)$ for $n > 2$ are simple, see [K] (as well as [Kap], [FK], [SNR]).

1.10 Almost simplicity

We say that a Lie superalgebra $\mathfrak{g}$ is *almost simple* if it can be sandwiched (not-strictly) between a simple Lie superalgebra $\mathfrak{s}$ and the Lie superalgebra $\mathfrak{ders}$ of the derivations of the latter: $\mathfrak{s} \subset \mathfrak{g} \subset \mathfrak{ders}$.

Theorem 1.11 (Main Theorem). 1° *Let $\mathfrak{g}$ be an irreducible linear Lie superalgebra which is neither almost simple nor a central extension of an almost simple Lie superalgebra.*

Then $\mathfrak{g}$ is contained in one of the following four major types of Lie superalgebras:

1) $\mathfrak{gl}(V_1) \odot \mathfrak{gl}(V_2)$;
2) $\mathfrak{q}(V_1) \odot \mathfrak{q}(V_2)$;
3) $\mathfrak{gl}(V) \otimes \Lambda(n) \subset\!\!\!+ \mathfrak{vect}(0|n)$;
4) $\mathfrak{hei}(0|2n) \subset\!\!\!+ \mathfrak{o}(2n)$.

2° *Let, in addition to conditions of* 1°, $\mathfrak{g}$ *be a subalgebra of* $\mathfrak{q}(V) = C(J)$ *for some J. Then $\mathfrak{g}$ is contained in one of the following Lie superalgebras (numbered as in* 1°*):*

1q) $\mathfrak{q}(V_1) \odot \mathfrak{gl}(V_2)$ *and* $J = J_1 \otimes 1$, *where* $\mathfrak{q}(V_1) = C(J_1)$;
3q) $\mathfrak{q}(V_1) \otimes \Lambda(n) \subset\!\!\!+ \mathfrak{vect}(0|n)$ *and* $J = J_1 \otimes 1$, *where* $\mathfrak{q}(V_1) = C(J_1)$;
4q) $\mathfrak{hei}(0|2n-1) \subset\!\!\!+ \mathfrak{o}(2n-1)$.

3° *Let, in addition to conditions of* 1°, $\mathfrak{g}$ *preserve a nondegenerate homogeneous form ω, either symmetric or skew-symmetric. Then $\mathfrak{g}$ is contained in one of the following Lie superalgebras (numbered as in* 1°*):*

1ω) $\mathfrak{aut}(\omega_1) \odot \mathfrak{aut}(\omega_2)$ *and* $\omega = \omega_1 \otimes \omega_2$;
3ω) $\mathfrak{aut}(\omega_1) \otimes \Lambda(n) \subset\!\!\!+ T^{1/2}(\mathfrak{vect}(0|n))$ *and* $\omega = \omega_1 \otimes \omega_{1/2}$.

1.11 Maximal subalgebras from the Main Theorem

Tables 1–2 describe cases when subalgebras of the form $\mathfrak{g}_1 \odot \mathfrak{g}_2$ are maximal in a linear Lie superalgebra $\mathfrak{g}$. They are similar to those that Dynkin described.

Table 1

In this table we assume that $\mathfrak{g}_i \subset \mathfrak{gl}(V_i)$, and $\dim V_i = N_i = m_i + n_i\varepsilon \neq 1$ or ε.

$\mathfrak{g}_1$	$\mathfrak{g}_2$	$\mathfrak{g}$	conditions
$\mathfrak{gl}(N_1)$	$\mathfrak{gl}(N_2)$	$\mathfrak{gl}(N_1N_2)$	$N_i \neq 1+\varepsilon$; $m_1 \neq n_1$ or $m_2 \neq n_2$
$\mathfrak{gl}(N_1)$	$\mathfrak{gl}(N_2)$	$\mathfrak{sl}(N_1N_2)$	$N_i \neq 1+\varepsilon$; $m_1 = n_1$ and $m_2 = n_2$
$\mathfrak{sl}(N_1)$	$\mathfrak{sl}(N_2)$	$\mathfrak{sl}(N_1N_2)$	$N_i \neq 1+\varepsilon$; $m_1 \neq n_1$ or $m_2 \neq n_2$
$\mathfrak{q}(n_1)$	$\mathfrak{q}(n_2)$	$\mathfrak{sl}(n_1n_2\vert n_1n_2)$	$n_1n_2 > 1$
$\mathfrak{q}(n_1)$	$\mathfrak{gl}(m_2+\varepsilon n_2)$	$\mathfrak{q}(n_1(m_2+n_2))$	$n_1 \geq 1$; $m_2 \neq n_2$
$\mathfrak{q}(n_1)$	$\mathfrak{gl}(m_2+n_2\varepsilon)$	$\mathfrak{sq}(n_1(m_2+n_2))$	$n_1 \geq 1$; $m_2 = n_2 > 1$

The case $N_i = 1+\varepsilon$ is exceptional because if we identify the $(1+\varepsilon)$-dimensional superspace V with $\Lambda(1)$, then $\mathfrak{gl}(V) \cong \Lambda(1) \subset\!\!\!\!+ \mathfrak{vect}(0|1)$. Therefore,

$$\mathfrak{g}_1 \odot \mathfrak{gl}(V) \subset \mathfrak{g}_1 \otimes \Lambda(1) \subset\!\!\!\!+ \mathfrak{vect}(0|1)$$

for any Lie superalgebra $\mathfrak{g}_1$. The case $n_1n_2 = 1$ in the fourth line is exceptional due to the fact that $\mathfrak{q}(1) \odot \mathfrak{q}(1) \cong \mathfrak{sl}(1|1)$.

Table 2

Table 2 describes maximal subalgebras of the form $\mathfrak{aut}(\omega_1) \oplus \mathfrak{aut}(\omega_2)$ of the Lie superalgebra $\mathfrak{g}$ that preserves a nondegenerate bilinear form $\omega = \omega_1 \otimes \omega_2$. It is clear that if both forms ω_1 and ω_2 are supersymmetric or skew, then ω is symmetric, while if one of the forms ω_1 or ω_2 is symmetric and the other is skew, then ω is skew. We take into account isomorphisms $\mathfrak{osp}(V) \cong \mathfrak{osp}^{sk}(\Pi(V))$ and $\mathfrak{pe}^{sy}(V) \cong \mathfrak{pe}^{sk}(\Pi(V))$, see Section 1.4, and skip the types of symmetry just to save space.

In this table $N_i = n_i + 2m_i\varepsilon$, and $n_im_i \neq 0$. The conditions are established by the fact that the identity representations of $\mathfrak{o}(2)$ and $\mathfrak{pe}(1)$ are reducible; $\mathfrak{pe}(2)$ is an exception because

$$\mathfrak{pe}(2) \cong \mathfrak{sp}(2) \otimes \Lambda(1) \subset\!\!\!\!+ T^{1/2}(\mathfrak{vect}(0|1)). \tag{1.2}$$

If we identify the $2|2$-dimensional superspace of the identity representation of $\mathfrak{pe}(2)$ with $\mathbb{C}^2 \otimes \Lambda(1)$, we obtain an embedding

$$\mathfrak{aut}(\omega_1) \odot \mathfrak{pe}(2) \subset \mathfrak{aut}(\omega_1 \otimes \omega_2) \otimes \Lambda(1) \subset\!\!\!\!+ T^{1/2}(\mathfrak{vect}(0|1)),$$

where ω_2 is the standard form in $\mathbb{C}^2$ preserved by $\mathfrak{sp}(2)$.

$\mathfrak{g}_1$	$\mathfrak{g}_2$	$\mathfrak{g}$	conditions
$\mathfrak{osp}(N_1)$	$\mathfrak{osp}(N_2)$	$\mathfrak{osp}(N_1 N_2)$	$---$
$\mathfrak{o}(n)$	$\mathfrak{osp}(N_2)$	$\mathfrak{osp}(nN_2)$	$n > 2, n \neq 4$
$\mathfrak{sp}(2n)$	$\mathfrak{osp}(N_2)$	$\mathfrak{osp}(2nN_2)$	$n \geq 1$
$\mathfrak{pe}(n_1)$	$\mathfrak{pe}(n_2)$	$\mathfrak{osp}(2n_1n_2\|2n_1n_2)$	$n_1,\ n_2 > 2$
$\mathfrak{osp}(n_1\|2m_1)$	$\mathfrak{pe}(n_2)$	$\mathfrak{pe}(n_1n_2 + 2m_1n_2)$	$n_2 > 2,\ n_1 \neq 2m_1$
$\mathfrak{osp}(2m\|2m)$	$\mathfrak{pe}(n)$	$\mathfrak{spe}(4mn)$	$n > 2$
$\mathfrak{osp}(n_1\|2m_1)$	$\mathfrak{pe}_\lambda(n_2)$	$\mathfrak{pe}_\mu(n_1n_2 + 2m_1n_2)$, for $\mu = \frac{\lambda}{n_1 - 2m_1}$	$n_2 > 2,\ n_1 \neq 2m_1$
$\mathfrak{o}(n)$	$\mathfrak{pe}(m)$	$\mathfrak{pe}(nm)$	$n, m > 2, \quad n \neq 4$
$\mathfrak{sp}(2n)$	$\mathfrak{pe}(m)$	$\mathfrak{pe}(2nm)$	$m > 2, \quad n \geq 1$

1.12 Table 3: Non-Dynkin type of subalgebras

The maximal subalgebras considered in Tables 1–2 are similar to those considered by Dynkin. There are, however, maximal subalgebras of linear superalgebras of a totally different nature. We represent them in Table 3: $\mathfrak{g}_1$ is a maxiaml subalgebra in $\mathfrak{g}$; we set $\dim V_1 = N_1 = m_1 + n_1\varepsilon$. In this case we allow $m_1 n_1 = 0$ but, of course, exclude $m_1 = n_1 = 0$. In lines 5–7 we assume that n_1 is even and in lines 8–9 we assume that $m_1 = n_1$. As in Table 2 we do not mention types of symmetry of bilinear forms.

	$\mathfrak{g}_1$	$\mathfrak{g}$	conditions
1	$\mathfrak{gl}(V_1) \otimes \Lambda(n) ⊄ \mathfrak{vect}(0\|n)$	$\mathfrak{sl}(V_1 \otimes \Lambda(n))$	$N_1 \neq 1 + \varepsilon$; either $n \neq 1$ or ($n = 1$ and $m_1 = n_1 > 1$)
2	$\mathfrak{gl}(V_1) \otimes \Lambda(1) ⊄ \mathfrak{vect}(0\|1)$	$\mathfrak{gl}(V_1 \otimes \Lambda(1))$	$m_1 \neq n_1;\quad N_1 \neq 1$ or ε
3	$\mathfrak{gl}(V_1) \otimes \Lambda(1) ⊄ \mathbb{C} \cdot \partial$	$\mathfrak{sl}(V_1 \otimes \Lambda(1))$	$m_1 \neq n_1;\quad N_1 \neq 1$ or ε
4	$\mathfrak{q}(V_1) \otimes \Lambda(n) ⊄ \mathfrak{vect}(0\|n)$	$\mathfrak{sq}(V_1 \otimes \Lambda(n))$	$m_1 = n_1 \geq 1;\quad n \geq 1$
5	$\mathfrak{osp}(V_1) \otimes \Lambda(2k) ⊄ T^{1/2}(\mathfrak{vect}(0\|2k))$	$\mathfrak{osp}(V_1 \otimes \Lambda(2k))$	$N_1 \neq 1, 2;\quad k > 0$
6	$\mathfrak{osp}(V_1) \otimes \Lambda(2k+1) ⊄ T^{1/2}(\mathfrak{vect}(0\|2k+1))$	$\mathfrak{spe}(V_1 \otimes \Lambda(2k+1))$	$N_1 \neq 1, 2$; either $k > 0$ or ($k = 0$ and $m_1 = n_1 > 1$)
7	$\mathfrak{osp}(V_1) \otimes \Lambda(1) ⊄ T^{1/2}(\mathfrak{vect}(0\|1))$	$\mathfrak{pe}(V_1 \otimes \Lambda(1))$	$N_1 \neq 1, 2; m_1 \neq n_1$
8	$\mathfrak{pe}(V_1) \otimes \Lambda(2k) ⊄ T^{1/2}(\mathfrak{vect}(0\|2k))$	$\mathfrak{spe}(V_1 \otimes \Lambda(2k))$	$m_1 = n_1 > 2;\quad k > 0$
9	$\mathfrak{pe}(V_1) \otimes \Lambda(2k+1) ⊄ T^{1/2}(\mathfrak{vect}(0\|2k+1))$	$\mathfrak{osp}(V_1 \otimes \Lambda(2k+1))$	$m_1 = n_1 > 2;\quad k \geq 0$
10	$\mathfrak{hei}(2n) ⊄ \mathfrak{o}(2n)$	$\mathfrak{sl}(\Lambda(n))$	$n \geq 2, \quad n \neq 3.$
11	$\mathfrak{hei}(2n-1) ⊄ \mathfrak{o}(2n-1)$	$\mathfrak{sq}(\Lambda(n))$	$n > 2.$

The exceptional cases:

1) If $\dim V_1 = 1$ or ε, then $\mathfrak{gl}(V_1) \otimes \Lambda(1) \subsetplus \mathfrak{vect}(0|1) \cong \mathfrak{gl}(1|1)$.

2) If $\dim V_1 = 1 + \varepsilon$, then having identified V_1 with $\Lambda(1)$, we obtain

$$\mathfrak{gl}(V_1) \otimes \Lambda(n) \subsetplus \mathfrak{vect}(0|n) \cong (\Lambda(1) \subsetplus \mathfrak{vect}(1)) \otimes \Lambda(n) \subsetplus \mathfrak{vect}(0|n) =$$
$$\Lambda(1) \otimes \Lambda(n) \subsetplus (\mathfrak{vect}(1) \otimes \Lambda(n) \subsetplus \mathfrak{vect}(0|n)) \subset \Lambda(n+1) \subsetplus \mathfrak{vect}(0|n+1).$$

3) The isomorphism (1.2) induces the inclusion

$$\mathfrak{pe}(2) \otimes \Lambda(n) \subsetplus T^{1/2}(\mathfrak{vect}(0|n)) \subset \mathfrak{sp}(2) \otimes \Lambda(n+1) \subsetplus T^{1/2}(\mathfrak{vect}(0|n+1)).$$

4) $n = 3$ in line 10; for motivation see Section 1.6.

2 Irreducible maximal subalgebras of the form $\mathfrak{g}_1 \odot \mathfrak{g}_2$

In this section we will prove theorems summarized in Tables 1 and 2.

Theorem 2.1. *Let* $\dim V_i = N_i = m_i + n_i\varepsilon$ *and* $N_i \neq 1$ *or* ε *or* $1 + \varepsilon$. *Let* $V = V_1 \otimes V_2$. *Then*

1) $\mathfrak{gl}(V_1) \odot \mathfrak{gl}(V_2)$ *is a maximal subalgebra in* $\mathfrak{gl}(V)$ *if* $n_1 \neq m_1$ *or* $n_2 \neq m_2$; *otherwise it is maximal in* $\mathfrak{sl}(V)$.

2) *The following subalgebras are maximal in* $\mathfrak{sl}(V)$:

a) $\mathfrak{sl}(V_1) \oplus \mathfrak{sl}(V_2)$ *if* $n_1 \neq m_1$ *or* $n_2 \neq m_2$;

b) $\mathfrak{gl}(V_1) \odot \mathfrak{gl}(V_2)$ *if* $n_1 = m_1$ *and* $n_2 = m_2$.

Proof. 1) Let $\mathfrak{g} = \mathfrak{gl}(V_1) \odot \mathfrak{gl}(V_2)$. Consider $\mathfrak{gl}(V)$ as a $\mathfrak{g}$-module. Then

$$\begin{aligned} &\mathfrak{gl}(V) \cong \mathfrak{gl}(V_1) \otimes \mathfrak{gl}(V_2) \quad \text{with the bracket} \\ &[A \otimes B, C \otimes D] = \\ &(-1)^{p(B)p(C)}[A, C] \otimes BD + (-1)^{p(A\otimes B)p(C)} CA \otimes [B, D]. \end{aligned} \tag{2.1}$$

Observe that as a $\mathfrak{gl}(V_i)$-module, $\mathfrak{gl}(V_i)$ contains only two nontrivial submodules: $\mathbb{C} \cdot \mathrm{id}$ and $\mathfrak{sl}(V_i)$. Thus, the minimal $\mathfrak{g}$-submodule W of $\mathfrak{gl}(V)$ larger than $\mathfrak{g}$ is of the form

$$W = \mathfrak{g} + \mathfrak{sl}(V_1) \otimes \mathfrak{sl}(V_2). \tag{2.2}$$

If $m_1 \neq n_1$ and $m_2 \neq n_2$, then the sum in (2.2) is direct and $W = \mathfrak{gl}(V)$, i.e., $\mathfrak{g}$ is maximal in $\mathfrak{gl}(V)$.

Let $\mathbb{C} \cdot \mathrm{id}_i \cong \mathfrak{gl}(V_i)/\mathfrak{sl}(V_i)$ be a trivial $\mathfrak{gl}(V_i)$-module. If $m_1 \neq n_1$ but $m_2 = n_2$, then $\mathfrak{gl}(V)/W \cong \mathfrak{sl}(V_1) \otimes \mathbb{C} \cdot \mathrm{id}_2$. If $m_1 = n_1$ and $m_2 = n_2$, then

$$\mathfrak{gl}(V)/W \cong \mathfrak{sl}(V_1) \otimes \mathbb{C} \cdot \mathrm{id}_2 + \mathbb{C} \cdot \mathrm{id}_1 \otimes \mathfrak{sl}(V_2) + \mathbb{C} \cdot \mathrm{id}_1 \otimes \mathbb{C} \cdot \mathrm{id}_2.$$

Since the spaces $\mathfrak{sl}(V_i)$ are not closed with respect to the operator product, equation (2.1) demonstrates that in both cases any subalgebra strictly containing $\mathfrak{g}$ must contain $\mathfrak{sl}(V_1)\otimes\mathfrak{gl}(V_2)+\mathfrak{gl}(V_1)\otimes\mathfrak{sl}(V_2)=\mathfrak{sl}(V)$. To complete the proof it suffices to observe that in the first case the subalgebra $\mathfrak{g}$ is not contained in $\mathfrak{sl}(V)$.

□

Theorem 2.2. *Let* $\dim V_i = N_i = m_i + n_i\varepsilon$, *where* $m_1 = n_1 \geq 1$ *and* $N_2 \neq 1$, *or* ε, *or* $1+\varepsilon$. *Let* $V = V_1 \otimes V_2$. *Then the Lie subsuperalgebra* $\mathfrak{g} = \mathfrak{q}(V_1) \odot \mathfrak{gl}(V_2)$ *is maximal in* $\mathfrak{q}(V)$ *if* $m_2 \neq n_2$; *it is maximal in* $\mathfrak{sq}(V)$ *if* $m_2 = n_2$.

Proof. The proof of this theorem largely repeats that of the previous one. Consider $\mathfrak{q}(V)$ as a $\mathfrak{g}$-module. Then $\mathfrak{q}(V) \cong \mathfrak{q}(V_1) \otimes \mathfrak{gl}(V_2)$ and the bracket in $\mathfrak{gl}(V)$ is defined via (2.1).

Note that for $m_1 > 2$ the space $\mathfrak{q}(V_1)$ considered as a $\mathfrak{q}(V_1)$-module, contains two nontrivial submodules, $\mathbb{C}\cdot \mathrm{id}$ and $\mathfrak{sq}(V_1)$. Therefore, the minimal $\mathfrak{g}$-submodule $W \subset \mathfrak{q}(V)$ is of the form $W = \mathfrak{g} + \mathfrak{sq}(V_1) \otimes \mathfrak{sl}(V_2)$. Since $\mathfrak{sq}(V_1)$ is not closed with respect to the operator product and taking into account equation (2.1), we see that any subalgebra $\mathfrak{h} \subset \mathfrak{q}(V)$ strictly containing $\mathfrak{g}$ should satisfy

$$\mathfrak{h} \supset \mathfrak{sq}(V_1) \otimes \mathfrak{gl}(V_2) + \mathfrak{q}(V_1) \otimes \mathfrak{sl}(V_2) = \mathfrak{sq}(V). \tag{2.3}$$

For $m_1 = 2$ the space $\mathfrak{q}(V_1) = \mathfrak{q}(2)$, considered as a $\mathfrak{q}(2)$-module contains one more nontrivial submodule, $\mathbb{C}\cdot \mathrm{id} \oplus \mathfrak{sq}(2)_{\bar{1}}$. Therefore, the minimal $\mathfrak{g}$-module $W \subset \mathfrak{q}(V)$ is in this case of the form $W = \mathfrak{g} + \mathfrak{sq}(V_1)_{\bar{1}} \otimes \mathfrak{sl}(V_2)$. Nevertheless, even in this case equation (2.1) leads to inclusion (2.3). To complete the proof for $m_1 \geq 2$ it only remains to observe that, for $m_2 \neq n_2$, the Lie superalgebra $\mathfrak{g}$ is not contained in $\mathfrak{sq}(V)$.

The case $m_1 = 1$ is even simpler: $\mathfrak{q}(V)/\mathfrak{g} \cong \mathfrak{pgl}(V_2)$. □

2.3 Embedding $\mathfrak{q}_1 \odot \mathfrak{q}_2 \subset \mathfrak{sl}(V_1 \otimes_Q V_2)$

First, describe $\mathfrak{gl}(V_i)$ as a $\mathfrak{q}(V_i)$-module.

As in Section 1.2, we introduce a $(1|1)$-dimensional superspace U and consider the following basis of $\mathrm{End}(U)$:

$$1_U, \quad J_U = \begin{pmatrix} 0 & 1 \\ -1 & 0 \end{pmatrix}, \quad I_U = \begin{pmatrix} 0 & i \\ i & 0 \end{pmatrix}, \quad D_U = \begin{pmatrix} 1 & 0 \\ 0 & -1 \end{pmatrix}.$$

Let us realize the superspaces V_1 and V_2 of dimension $n_1|n_1$ and $n_2|n_2$, respectively, as tensor products

$$V_1 = (V_1)_{\bar{0}} \otimes U, \quad V_2 = U \otimes (V_2)_{\bar{0}}.$$

Set $J = 1 \otimes J_U$ and $I = I_U \otimes 1$. It is clear that

$$C(J) = \mathrm{End}(V_1)_{\bar{0}} \otimes \mathrm{Span}(1, I_U) \cong \mathfrak{q}(V_1);$$
$$C(I) = \mathrm{Span}(1, J_U) \otimes \mathrm{End}(V_2)_{\bar{0}} \cong \mathfrak{q}(V_2).$$

Then $\mathfrak{gl}(V_i)$, as a $\mathfrak{q}(V_i)$-module, is a direct sum of two reducible but indecomposable modules:

$$\mathfrak{gl}(V_1) = \mathfrak{gl}(V_1)_{\bar{0}} \otimes \mathrm{Span}(1, I_U) \oplus \mathfrak{gl}(V_1)_{\bar{0}} \otimes \mathrm{Span}(D_U, J_U); \tag{2.4}$$

$$\mathfrak{gl}(V_2) = \mathrm{Span}(1, J_U) \otimes \mathfrak{gl}(V_2)_{\bar{0}} \oplus \mathrm{Span}(D_U, I_U) \otimes \mathfrak{gl}(V_2)_{\bar{0}}. \tag{2.5}$$

Observe that the first summands in (2.4) and (2.5) are isomorphic to $\mathfrak{q}(V_i)$, whereas the second ones, as $\mathfrak{q}(V_i)$-modules, are isomorphic to $\Pi(\mathfrak{q}(V_i))$.

Theorem 2.3.1. *If* $n_1 n_2 > 1$, *then* $\mathfrak{g} = \mathfrak{q}(V_1) \bigodot \mathfrak{q}(V_2)$ *is maximal in* $\mathfrak{sl}(V_1 \otimes^Q V_2)$.

Proof. Thus, the minimal $\mathfrak{g}$-submodule $W \subset \mathfrak{sl}(V_1 \otimes^Q V_2)$ containing $\mathfrak{g}$ is, thanks to equations (2.4)–(2.5), of the form

$$W = \mathfrak{g} + \mathfrak{sl}(V_1)_{\bar{0}} \otimes \mathfrak{gl}(U) \otimes \mathfrak{sl}(V_2)_{\bar{0}}.$$

When we close W with respect to the bracketing we obtain $\mathfrak{sl}(V_1 \otimes^Q V_2)$. □

2.4 The case with the billinear form

Let a nondegenerate homogeneous (with respect to parity) supersymmetric or superskew-symmetric bilinear form ω be given in a superspace V. Consider two objects associatted with ω:

1) The Lie superalgebra $\mathfrak{aut}(\omega)$ of operators preserving ω:

$$\mathfrak{aut}(\omega) = \{A \in \mathfrak{gl}(V) \mid \omega(Ax, y) + (-1)^{p(A)p(x)}\omega(x, Ay) = 0\}.$$

2) The space $\mathrm{sym}(\omega)$ of operators supersymmetric with respect to ω:

$$\mathrm{sym}(\omega) = \{A \in \mathfrak{gl}(V) \mid \omega(Ax, y) = (-1)^{p(A)p(x)}\omega(x, Ay)\}.$$

It is clear that, as a linear space, $\mathfrak{gl}(V) = \mathfrak{aut}(\omega) \oplus \mathrm{sym}(\omega)$.

Set

$$\mathfrak{saut}(\omega) = \mathfrak{aut}(\omega) \cap \mathfrak{sl}(V), \qquad \mathrm{ssym}(\omega) = \mathrm{sym}(\omega) \cap \mathfrak{sl}(V).$$

Note that if $p(\omega) = \bar{0}$, then $\mathfrak{aut}(\omega) = \mathfrak{saut}(\omega)$, and if $p(\omega) = \bar{1}$, then $\mathrm{sym}(\omega) = \mathrm{ssym}(\omega)$.

If ω is even, it determines a canonical isomorphism $V \cong V^*$. In this case,

$$\mathfrak{g} = \mathfrak{aut}(\omega) = \begin{cases} \mathfrak{osp}(V) \cong \Lambda^2(V) \text{ and } \mathrm{sym}(\omega) = S^2(V) & \text{if } \omega \text{ is symmetric} \\ \mathfrak{osp}^{sk}(V) \cong S^2(V) \text{ and } \mathrm{sym}(\omega) = \Lambda^2(V) & \text{if } \omega \text{ is skew.} \end{cases}$$

In both cases the space $\mathrm{sym}(\omega)$ contains a 1-dimensional subspace $\mathbb{C}\omega$ corresponding to scalar operators.

If, moreover, $\dim V_{\bar{0}} \neq \dim V_{\bar{1}}$, then $\mathrm{sym}(\omega) = \mathbb{C}\omega \oplus \mathrm{ssym}(\omega)$, and the $\mathfrak{g}$-module $\mathrm{ssym}(\omega)$ is irreducible.

If $\dim V_{\bar 0} = \dim V_{\bar 1}$, then $\mathbb{C}\omega \subset \mathrm{ssym}(\omega)$ and $\dim(\mathrm{sym}(\omega)/\mathrm{ssym}(\omega)) = 1$ and $\mathrm{ssym}(\omega)/\mathbb{C}\omega$ is irreducible.

If ω is odd, a canonical isomorphism $V^* \cong \Pi(V)$ implies that $V \otimes V^* \cong V \otimes \Pi(V) \cong \Pi(V \otimes V)$. In this case,

$$\mathfrak{g} = \mathfrak{aut}(\omega) = \begin{cases} \mathfrak{pe}^{sy}(V) \cong \Pi(\Lambda^2(V)) \supset \mathfrak{saut}(\omega) = \mathfrak{spe}^{sy}(V) & \text{if } \omega \text{ is symm.} \\ \mathfrak{pe}^{sk}(V) \cong \Pi(S^2(V)) \supset \mathfrak{saut}(\omega) = \mathfrak{spe}^{sk}(V) & \text{if } \omega \text{ is skew} \end{cases}$$

and

$$\mathrm{sym}(\omega) = \mathrm{ssym}(\omega) = \begin{cases} \Pi(S^2(V)) \supset \mathbb{C}\omega & \text{if } \omega \text{ is symmetric} \\ \Pi(\Lambda^2(V)) \supset \mathbb{C}\omega & \text{if } \omega \text{ is skew.} \end{cases}$$

In both cases the space $\mathrm{sym}(\omega)/\mathbb{C}\omega$ is an irreducible $\mathfrak{g}$-module if $\mathfrak{saut}(\omega)$ is simple, i.e., if $\dim V = n|n$ and $n > 2$.

Lemma 2.4.1. 1) $[\mathrm{sym}(\omega), \mathrm{sym}(\omega)] \subset \mathfrak{aut}(\omega)$, *but* $[\mathrm{sym}(\omega), \mathrm{sym}(\omega)]$ *is not contained in* $\mathfrak{saut}(\omega)$ *if* $p(\omega) = \bar 1$.

2) *Set* $\{A, B\} = AB + (-1)^{p(A)p(B)}BA$. *Then*

$$\begin{aligned} &\{\mathfrak{aut}(\omega), \mathfrak{aut}(\omega)\},\ \{\mathrm{sym}(\omega), \mathrm{sym}(\omega)\} \subset \mathrm{sym}(\omega); \\ &\{\mathfrak{aut}(\omega), \mathrm{sym}(\omega)\} \subset \mathfrak{aut}(\omega); \\ &\mathfrak{aut}(\omega_1 \otimes \omega_2) = \mathfrak{aut}(\omega_1) \otimes \mathrm{sym}(\omega_2) + \mathrm{sym}(\omega_1) \otimes \mathfrak{aut}(\omega_2); \\ &\mathrm{sym}(\omega_1 \otimes \omega_2) = \mathfrak{aut}(\omega_1) \otimes \mathfrak{aut}(\omega_2) + \mathrm{sym}(\omega_1) \otimes \mathrm{sym}(\omega_2). \end{aligned} \tag{2.6}$$

If $p(\omega) = \bar 0$, *then the subspace* $\mathrm{ssym}(\omega)$ *is not closed with respect to* $\{\cdot, \cdot\}$.

$$\begin{aligned} 3)\ [A_1 \otimes B_1, A_2 \otimes B_2] = \tfrac{1}{2}((-1)^{p(A_2)p(B_1)}[A_1, A_2] \otimes \{B_1, B_2\} + \\ (-1)^{p(A_2)(p(A_1+p(B_1))}\{A_2, A_1\} \otimes [B_1, B_2]). \end{aligned} \tag{2.7}$$

Proof. Direct calculations. □

Theorem 2.4.2 (Cf. Table 2). *For simple Lie superalgebras* $\mathfrak{saut}(\omega_1)$ *and* $\mathfrak{saut}(\omega_2)$ *the Lie subalgebra* $\mathfrak{g} = \mathfrak{aut}(\omega_1) \oplus \mathfrak{aut}(\omega_2)$ *is maximal in* $\mathfrak{aut}(\omega_1 \otimes \omega_2)$ *if either* $p(\omega_1) + p(\omega_2) = \bar 0$ *or if* $p(\omega_i) = \bar 0$, $p(\omega_j) = \bar 1$ *and* $\dim(V_i)_{\bar 0} \neq \dim(V_i)_{\bar 1}$ *for* $i, j \in \{1, 2\}$. *If* $p(\omega_i) = \bar 0$, $p(\omega_j) = \bar 1$ *and* $\dim(V_i)_{\bar 0} = \dim(V_i)_{\bar 1}$, *then* $\mathfrak{g}$ *is maximal in* $\mathfrak{saut}(\omega_1 \otimes \omega_2)$.

Proof. Equation (2.6) and the description of $\mathfrak{aut}(\omega_i)$ and $\mathrm{sym}(\omega_i)$ as $\mathfrak{aut}(\omega_i)$-modules immediately imply that any subalgebra of $\mathfrak{h} \subset \mathfrak{aut}(\omega_1 \otimes \omega_2)$ containing $\mathfrak{g}$ must also contain at least one of the submodules $\mathfrak{saut}(\omega_1) \otimes \mathrm{ssym}(\omega_2)$ or $\mathrm{ssym}(\omega_1) \otimes \mathfrak{saut}(\omega_2)$. But then, by (2.7) we see that $\mathfrak{h}$ must contain both of these modules, hence,

$$\mathfrak{h} \supset \mathfrak{saut}(\omega_1) \otimes \mathrm{sym}(\omega_2) + \mathrm{sym}(\omega_1) \otimes \mathfrak{saut}(\omega_2). \tag{2.8}$$

If $p(\omega_i) = \bar{0}$ for $i = 1, 2$ then the rhs of (2.8) coincides with $\mathfrak{aut}(\omega_1 \otimes \omega_2)$. If $p(\omega_i) = \bar{1}$ for $i = 1, 2$, then by bracketing the elements from distinct summands and taking into account that $\{\mathfrak{saut}(\omega_i), \mathrm{sym}(\omega_i)\} = \mathfrak{aut}(\omega_i)$ we again obtain that $\mathfrak{h} = \mathfrak{aut}(\omega_1 \otimes \omega_2)$.

Finally, if $p(\omega_2) = \bar{0}$ and $p(\omega_2) = \bar{1}$, then the first summand in (2.8) coincides with $\mathfrak{aut}(\omega_1) \otimes \mathrm{sym}(\omega_2)$. By bracketing the elements from distinct summands, we see that apart from inclusion (2.8), there is an inclusion $\mathfrak{h} \supset \mathrm{ssym}(\omega_1) \otimes \mathfrak{aut}(\omega_2)$, i.e., $\mathfrak{h} \supset \mathfrak{saut}(\omega_1 \otimes \omega_2)$.

This completes the proof when $\dim(V_1)_{\bar{0}} = \dim(V_1)_{\bar{1}}$. But, for $\dim(V_1)_{\bar{0}} \neq \dim(V_1)_{\bar{1}}$ it suffices to observe that $\mathfrak{g}$ is not contained in $\mathfrak{sl}(V)$. □

3 Irreducible linear maximal subalgebras of non-Dynkin form

Theorem 3.1. *If $m = 2n$, where $n \geq 2$, $n \neq 3$, then $\mathfrak{g} = \mathfrak{hei}(0|m) \in\!\!\!\!+ \mathfrak{o}(m)$ is a maximal subalgebra in $\mathfrak{sl}(\Lambda(n))$.*

If $n = 3$, then $\mathfrak{g} \subset \mathfrak{as} \subset \mathfrak{sl}(\Lambda(3))$, and $\mathfrak{as}$ is maximal in $\mathfrak{sl}(\Lambda(3))$.

If $m = 2n - 1$, $n > 1$, then $\mathfrak{g}$ is a maximal subalgebra in $\mathfrak{sq}(\Lambda(n))$.

Proof. Let $\mathfrak{hei} = \mathfrak{hei}(0|m)$. Consider the image of the universal enveloping algebra $U(\mathfrak{hei})$ in the irreducible representation from Section 1.6. Due to irreducibility this image, considered as a Lie superalgebra, coincides with $\mathfrak{gl}(\Lambda(n))$ for $m = 2n$ and with $\mathfrak{q}(\Lambda(n))$ for $m = 2n - 1$.

Consider $U(\mathfrak{hei})_L$ as a filtered Lie superalgebra with respect to the filtration induced by the natural filtration of the enveloping algebra, and consider the associated graded one, $\mathrm{gr}(U(\mathfrak{hei})_L)$. As is known ([LSh]), $\mathrm{gr}(U(\mathfrak{hei})_L)$ is isomorphic to the Poisson superalgebra with its standard grading described in Section 1.8, $\mathfrak{po}(0|m) = \oplus_{i=-2}^{m-2}\mathfrak{po}_i$. The graded image $\mathrm{gr}(\mathfrak{g})$ of the Lie superalgebra $\mathfrak{g}$ coincides with the nonpositive part of $\mathfrak{po}$:

$$\mathrm{gr}(\mathfrak{g}) = \oplus_{i=-2}^{0}\mathfrak{po}_i.$$

Since $\mathfrak{po}(0|m)$ is transitive, $\mathfrak{po}_1$ is (for $m \neq 6$) an irreducible $\mathfrak{o}(m) \cong \mathfrak{po}_0$-module and generates $\bigoplus\limits_{i=1}^{m-3} \mathfrak{po}_i$, the positive part of the special Poisson subalgebra, $\mathfrak{spo}(0|m)$, we see that $\mathrm{gr}(\mathfrak{g})$ is maximal in $\mathfrak{spo}(0|m)$ for $m \neq 6$.

To complete the proof, observe that $\mathfrak{spo}(0|2n) = \mathrm{gr}(\mathfrak{sl}(\Lambda(n)))$ and $\mathfrak{g} \subset \mathfrak{sl}(\Lambda(n))$, whereas $\mathfrak{spo}(0|2n-1) = \mathrm{gr}(\mathfrak{sq}(\Lambda(n)))$ and $\mathfrak{g} \subset \mathfrak{sq}(\Lambda(n))$. □

3.2 $V = V_1 \otimes \Lambda(n)$

Let $\dim V_1 = (m_1, n_1)$, $\mathfrak{n} = \mathfrak{gl}(V_1) \otimes \Lambda(n)$ and let $\mathfrak{g}$ be the semidirect sum of the ideal $\mathfrak{n}$ and the subalgebra $\mathfrak{vect}(0|n)$ with the natural action on the ideal. The Lie superalgebra $\mathfrak{g}$ has a natural faithful representation ρ in $V = V_1 \otimes \Lambda(n)$, defined by the equations for any $A \otimes \varphi \in \mathfrak{n}$, $D \in \mathfrak{vect}(0|n)$, and $v \otimes \psi \in V$, we have

$$\begin{aligned} \rho(A \otimes \varphi)(v \otimes \psi) &= (-1)^{p(\varphi)p(v)} Av \otimes \varphi\psi, \\ \rho(D)(v \otimes \psi) &= (-1)^{p(D)p(v)} v \otimes D(\psi). \end{aligned}$$

In the following, we will always identify elements of $\mathfrak{g}$ with their images under ρ. Therefore, we will consider $\mathfrak{g}$ embedded in $\mathfrak{gl}(V)$ which coincides, as a linear space, with $\mathrm{End}(V) \cong \mathrm{End}(V_1) \otimes \mathrm{End}(\Lambda(n))$.

Theorem 3.2.1. 1) *The Lie superalgebra* $\mathfrak{g} = \mathfrak{gl}(V_1) \otimes \Lambda(n) \subset\!\!\!\!+ \mathfrak{vect}(0|n)$ *is a maximal Lie subsuperalgebra of* $\mathfrak{sl}(V)$, *where* $V = V_1 \otimes \Lambda(n)$, *in all cases except*

a) $\dim V_1 = 1 + \varepsilon$ *or*
b) $n = 1$ *and* $\dim V_1 = m_1 + n_1\varepsilon$ *with* $m_1 \neq n_1$.

2) *If* $n = 1$ *and* $\dim V_1 = m_1 + n_1\varepsilon$, *where* $m_1 \neq n_1$ *and* $m_1 + n_1 > 1$, *then* $\mathfrak{g}$ *is maximal Lie subsuperalgebra of* $\mathfrak{gl}(V)$. *In this case the Lie superalgebra* $\mathfrak{sg} := \mathfrak{gl}(V_1) \otimes \Lambda(1) \subset\!\!\!\!+ Span(\partial)$ *is maximal in* $\mathfrak{sl}(V)$.

Let us first prove a particular case of Theorem 3.2.1.

Lemma 3.2.2. *If* $n > 1$, *then* $\mathfrak{g} = \Lambda(n) \subset\!\!\!\!+ \mathfrak{vect}(0|n)$ *is maximal in* $\mathfrak{sl}(\Lambda(n))$.

Proof. We will make use of the arguments used in the proof of Theorem 3.1. Observe that $\mathfrak{g}$ contains a G-type irreducible subalgebra $\mathfrak{hei} = \mathfrak{hei}(0|2n) = Span(1, \xi, \partial)$. This means that the image of $U(\mathfrak{hei})$ in $\mathrm{End}(\Lambda(n))$ coincides with $\mathrm{End}(\Lambda(n))$, and the graded Lie superalgebra associated with $U(\mathfrak{hei})_L$, is isomorphic to $\mathfrak{po}(0|2n)$. Clearly, subalgebra $\mathfrak{sl}(\Lambda(n))$ corresponds to $\mathfrak{spo}(0|2n)$.

Let us realize the elements of $\mathfrak{po}(0|2n)$ by means of generating functions in ξ, η:

$$\xi_i \longleftrightarrow \xi_i, \quad \partial_i \longleftrightarrow \eta_i.$$

The image $\mathrm{gr}(\mathfrak{g}) = \bigoplus_{i \geq -2} \mathfrak{g}_i$ of $\mathfrak{g}$ in $\mathfrak{po}(0|2n)$ is the linear space of functions of degree ≤ 1 in η's. In particular, $\mathfrak{g}_0$ consists of the elements of $\mathfrak{po}(0|2n)_0 \cong \mathfrak{o}(2n)$ that preserve the space $Span(\xi_1, \dots, \xi_n)$. By Dynkin's theorem $\mathfrak{g}_0$ is a maximal subalgebra in $\mathfrak{o}(2n)$.

On the other hand, it is clear that

$$\mathfrak{g}_i = \{X \in \mathfrak{g}_{i-1} \mid [X, \mathfrak{g}_{-1}] \subset \mathfrak{g}_{i-1}\} \text{ for all } i \geq 1,$$

i.e., $\mathfrak{g}$ is a maxiaml subalgebra in $\mathfrak{po}(0|2n)$ with a given nonpositive part. Therefore, any subalgebra $\mathfrak{h} \subset \mathfrak{po}(0|2n)$ containing $\mathfrak{g}$ must contain the whole of $\mathfrak{po}_0$. Since for $n \neq 3$ the $\mathfrak{o}(2n)$-action in $\mathfrak{po}_1$ is irreducible, $\mathfrak{h}$ must contain the whole component $\mathfrak{po}_1$, hence, the whole of $\mathfrak{spo}$.

If $n = 3$, then $\mathfrak{g}_1$ does not lie in any of the ireducible $\mathfrak{o}(6)$-submodules of $\mathfrak{po}_1$. Therefore, again, $\mathfrak{h} \supset \mathfrak{spo}$.

In both cases we see that $\mathfrak{g}$ is maximal in $\mathfrak{sl}(\Lambda(n))$. □

3.2.3. Proof of Theorem 3.2.1

Let $\dim V_1 \neq 1$ or ε or $1+\varepsilon$ and let $\mathfrak{h} \subset \mathfrak{gl}(V)$ be a subalgebra containing $\mathfrak{g}$. Clearly, to prove both items of the theorem, it suffices to show that $\mathfrak{h} \supset \mathfrak{sl}(V)$.

Observe that $\mathfrak{g}$ contains two subalgebras, $\mathfrak{g}_1 = \mathfrak{gl}(V_1) \otimes 1$ and $\mathfrak{g}_2 = 1 \otimes \Lambda(n) \subsetplus \mathfrak{vect}(0|n)$. Since $\mathrm{End}(V_1)$ only contains two nontrivial $\mathfrak{gl}(V_1)$-submodules, $\mathbb{C}\cdot\mathrm{id}$ and $\mathfrak{sl}(V_1)$, we deduce that $\mathfrak{h}$ must contain a subspace of one of the two types: either $1 \otimes W$ or $\mathfrak{sl}(V_1) \otimes W$ for a $\mathfrak{g}_2$-invariant subspace $W \subset \mathfrak{gl}(\Lambda(n))$.

In the first case, by Lemma 3.2.2, $\mathfrak{h} \supset 1 \otimes \mathfrak{sl}(\Lambda(n))$ and, by Theorem 2.1, $\mathfrak{h} \supset \mathfrak{sl}(V_1 \otimes \Lambda(n))$.

In the second case, let us realize the elements of $\mathfrak{gl}(\Lambda(n))$ by differential operators acting on $\Lambda(n)$. It is clear that by bracketing with ξ_i and ∂_i we can reduce any differential operator to the form ∂_j, i.e., $W \supset \partial_1, \dots, \partial_n$. Applying equation (2.1) to elements of the form $A \otimes \partial_j$ and $B \otimes D$, where $D \in \Lambda(\partial)$ is a differential operator with constant coefficients, we see that $W \supset \Lambda(\partial)$. Therefore, by $\mathfrak{g}_2$-invariance, $W \supset \mathfrak{sl}(\Lambda(n))$ and, therefore, $\mathfrak{h} \supset \mathfrak{sl}(V_1 \otimes \Lambda(n))$. □

Theorem 3.3. *The Lie superalgebra $\mathfrak{g} = \mathfrak{q}(V_1) \otimes \Lambda(n) \subsetplus \mathfrak{vect}(0|n)$ is a maximal Lie subsuperalgebra of $\mathfrak{sq}(V)$, where $V = V_1 \otimes \Lambda(n)$.*

The proof is similar to that of Theorem 3.2.1. □

3.4 Lie superalgebras preserving a nondegenerate bilinear form

We use notation from Sections 1.4 and 1.7. Observe that for $n \leq 2$ we have the isomorphisms

$$T^{1/2}(\mathfrak{vect}(0|1)) \cong \mathfrak{pe}(1); \quad T^{1/2}(\mathfrak{vect}(0|2)) \cong \mathfrak{osp}(2|2).$$

Lemma 3.4.1. *For $n > 2$ the Lie superalgebra $\mathfrak{g} = T^{1/2}(\mathfrak{vect}(0|n))$ is maximal in $\mathfrak{saut}(\omega_{1/2})$.*

Proof. Let us realize the space $\mathfrak{gl}(\Lambda(\xi))$ by differential operators. Observe that any function $\varphi \in \Lambda(\xi)$ is an element from the space $\mathrm{sym}(\omega_{1/2})$, and any differential operator with constant coefficients $D \in \Lambda^k(\partial)$ belongs to $\mathfrak{aut}(\omega_{1/2})$ for k odd and to $\mathrm{sym}(\omega_{1/2})$ for k even. Since $\{A, B\} = 2AB - (-1)^{p(A)p(B)}[A, B]$, it follows that

$$\mathfrak{aut}(\omega_{1/2}) = \mathrm{Span}(\varphi, D \mid \varphi \in \Lambda(\xi),\ D \in \bigoplus_k \Lambda^{2k+1}(\partial));$$

$$\mathrm{sym}(\omega_{1/2}) = \mathrm{Span}(\varphi, D \mid \varphi \in \Lambda(\xi),\ D \in \bigoplus_k \Lambda^{2k}(\partial)).$$

Set

$$V_l = \bigoplus_{k \leq l,\ k \equiv l \pmod 2} \{\Lambda(\xi),\ \Lambda^k(\partial)\}.$$

Clearly, each V_l is $\mathfrak{g}$-invariant, and as follows from [BL], the $\mathfrak{g}$-action in V_l / V_{l-2} is irreducible for $l \neq 0, n$ (we assume that $V_{-2} = V_{-1} = 0$). The explicit equation for the bracket of $A = \{\xi_1\xi_2, \partial_1\} \in \mathfrak{g}$ and $B = \{\xi_1\xi_2, \partial_1\partial_2\partial_{i_3}\dots\partial_{i_k}\} \in V_k$ is

$$[A, B] = -\xi_2\partial_{i_3}\dots\partial_{i_k} \in V_{k-2};$$

this shows that the representation of $\mathfrak{g}$ in V_k is not completely reducible and the minimal $\mathfrak{g}$-invariant subspace W such that $\mathfrak{g} \subset W \subset \mathfrak{aut}(\omega_{1/2})$ is V_3 for $n > 3$ and $\mathfrak{saut}(\omega_{1/2}) \cong \mathfrak{spe}(4)$ for $n = 3$.

Finally, the equation

$$[\partial_1\partial_2\partial_3, \ \xi_1\partial_{i_1}\partial_{i_2}\dots\partial_{i_k}] = \partial_2\partial_3\partial_{i_1}\dots\partial_{i_k} \in V_{k+2}$$

shows that V_3 generates the whole Lie superalgebra $\mathfrak{saut}(\omega_{1/2})$. □

A corollary from the proof of Lemma 3.4.1 is the following statement: *the minimal $\mathfrak{g}$-invariant subspace W such that $\Lambda(n) \subset W \subset \mathrm{sym}(\omega_{1/2})$ is $W = V_2$.*

Theorem 3.5 (Cf. Table 3). *Let ω_1 be a nondegenerate supersymmetric or skew bilinear form in V_1 of dimension $N_1 = m_1 + n_1\varepsilon$, where $N_1 \neq 1, \ 2$ and n_1 is even if $p(\omega_1) = \bar{0}$, and where $m_1 = n_1 > 2$ if $p(\omega_1) = \bar{1}$.*

Then $\mathfrak{g} = \mathfrak{aut}(\omega_1) \otimes \Lambda(n) \oplus\!\!\!\!\!\subset T^{1/2}(\mathfrak{vect}(0|n))$ is a maximal subalgebra in $\mathfrak{saut}(\omega_1 \otimes \omega_{1/2})$ except for the case when $n = 1$, $p(\omega_1) = \bar{0}$ and $m_1 \neq n_1$; in the exceptional case $\mathfrak{g}$ is maximal in $\mathfrak{aut}(\omega_1 \otimes \omega_{1/2})$.

Lemma 3.4.1 and its corollary make it possible, essentially, to combine proofs of Theorem 3.2.1 and 2.4.2.

4 Proof of the Main Theorem

4.1 Superization of a proposition by Dixmier

In this section $\mathfrak{g}$ is an irreducible linear Lie superalgebra, ρ its standard representation in a finite-dimensional superspace V (in particular, ρ is faithful). Let $\mathfrak{i}$ be an ideal of $\mathfrak{g}$. We will assume that $\dim V > 1$ (because the case $\dim V = 1$ is trivial). Our proof of the Main Theorem is largely based on the following constructions and statements (Theorems 4.1.1 and 4.1.2), a superization of statements well known to the reader from Dixmier's book [Di] (Proposition 5.5.1).

Let τ be an irreducible subrepresentation of the restriction $\rho|_{\mathfrak{i}}$ in the subspace $U \subset V$ and let $V_1 \subset V$ be the sum of all $\mathfrak{i}$-submodules of $\rho|_{\mathfrak{i}}$, isomorphic to either τ or $\Pi(\tau)$.

The *stabilizer* $\mathfrak{st}(\tau)$ of τ is the set of $Y \in \mathfrak{g}$ such that there exists an $A \in \mathrm{End}(U)$ for which $\tau([Y, X]) = [A, \tau(X)]$ for any $X \in \mathfrak{i}$.

Theorem 4.1.1. *1)* $\mathfrak{h} = \mathfrak{st}(\tau) \supset \mathfrak{g}_{\bar{0}}$;

2) *V_1 is $\mathfrak{h}$-invariant;*

3) *If $\mathfrak{h} \neq \mathfrak{g}$, then $\rho \equiv \mathrm{Ind}_{\mathfrak{h}}^{\mathfrak{g}}\sigma$, where σ is a representation of $\mathfrak{h}$ in V_1.*

Theorem 4.1.2. *Let $\mathfrak{g}$ be an irreducible linear Lie superalgebra, ρ its standard representation (in particular, ρ is faithful). Let $\mathfrak{i}$ be a nontrivial ideal of $\mathfrak{g}$. Then* 3 *cases are possible:*

A) *$\rho|_{\mathfrak{i}}$ is irreducible;*

B) *$\rho|_{\mathfrak{i}}$ is a multiple of an irreducible $\mathfrak{i}$-module τ and the multiplicity of τ is* > 1*;*

C) *there exists a proper subalgebra $\mathfrak{h} \subset \mathfrak{g}$ such that $\mathfrak{i} \subset \mathfrak{h}$ and $\rho \equiv \mathrm{ind}_{\mathfrak{h}}^{\mathfrak{g}} \sigma$ for an irreducible $\mathfrak{h}$-module σ.*

The proof largely follows the lines of [Di] with one novel case: irreducible modules of Q-type might occur. Fortunately, their treatment is rather straightforward and not necessary to prove.

We will say that the *representation ρ is of type A, B, or C with respect to the ideal* $\mathfrak{i}$ if the corresponding case holds. Lemmas 4.6, 4.2 and 4.4 deal with types A, B and C, respectively.

Lemma 4.2. B) 1) *If ρ is of type B with respect to the ideal $\mathfrak{i}$, then either $\mathfrak{g} \subset \mathfrak{gl}(V_1) \odot \mathfrak{gl}(V_2)$ or $\mathfrak{g} \subset \mathfrak{q}(V_1) \odot \mathfrak{q}(V_2)$ for some V_1 and V_2.*

2) *If, moreover, $\mathfrak{g} \subset \mathfrak{q}(V)$, then $\mathfrak{g} \subset \mathfrak{q}(V_1) \odot \mathfrak{gl}(V_2)$ for some V_1, V_2.*

3) *If ω is a nondegenerate 2-form on V and $\mathfrak{g} \subset \mathfrak{aut}(\omega)$, then $\mathfrak{g} \subset \mathfrak{aut}(\omega_1) \odot \mathfrak{aut}(\omega_2)$, where $\omega = \omega_1 \otimes \omega_2$.*

The proof of Lemma 4.2 follows, except statements involving $\mathfrak{q}$, the standard scheme of the proof of a similar statement for Lie algebras, and the exception is also easy to consider, so I am skipping it.

Remark 4.3. If ρ is of type B with respect to the (nontrivial) ideal $\mathfrak{i}$ and $\dim \tau = 1$ or ε, then due to the faithfulness of ρ, the ideal $\mathfrak{i}$ should be a 1-dimensional center of $\mathfrak{g}$.

Lemma 4.4. C) 1) *If ρ is of type C with respect to the ideal $\mathfrak{i}$, then $\mathfrak{g} \subset \mathfrak{gl}(V_1) \otimes \Lambda(n) \subset\!\!\!\!+ \mathfrak{vect}(0|n)$ for some V_1 and n and $V = V_1 \otimes \Lambda(n)$.*

2) *If, moreover, $\mathfrak{g} \subset \mathfrak{q}(n)$, then $\mathfrak{g} \subset \mathfrak{q}(V_1) \otimes \Lambda(n) \subset\!\!\!\!+ \mathfrak{vect}(0|n)$.*

3) *If ρ is of type C with respect to the ideal $\mathfrak{i}$ and $\mathfrak{g} \subset \mathfrak{aut}(\omega)$ for a nondegenerate form ω, then $\mathfrak{g} \subset \mathfrak{aut}(\omega_1) \otimes \Lambda(n) \subset\!\!\!\!+ T^{1/2}(\mathfrak{vect}(0|n))$, where $\omega = \omega_1 \otimes \omega_{1/2}$ and the representation $T^{1/2}$ of $\mathfrak{vect}(0|n)$ in the superspace $\sqrt{Vol}$ of half-densities is defined in Section* 1.7.

Proof. 1) By the definition of type C, $\rho = \mathrm{ind}_{\mathfrak{h}}^{\mathfrak{g}} \sigma$. Let

$$\mathfrak{h}_1 = \{X \in \mathrm{Ker}\sigma \mid [X, \mathfrak{g}] \subset \mathfrak{h}\}.$$

Clearly, $\mathfrak{h}_1$ is a subalgebra in $\mathfrak{g}$ and ideal in $\mathfrak{h}$. For each integer $i > 1$ define inductively a subalgebra

$$\mathfrak{h}_i = \{X \in \mathfrak{h}_{i-1} \mid [X, \mathfrak{g}] \subset \mathfrak{h}_{i-1}\}.$$

We obtain a decreasing filtration in $\mathfrak{g}$:

$$\mathfrak{h}_{-1} = \mathfrak{g} \supset \mathfrak{h}_0 = \mathfrak{h} \supset \mathfrak{h}_1 \supset \dots$$

Since $\dim \mathfrak{g} < \infty$, the filtration stabilizes, i.e., $\mathfrak{h}_k = \mathfrak{h}_{k+1} = \dots$ for some k. This means that $\mathfrak{h}_k$ is an ideal in $\mathfrak{g}$ lying in the kernel of σ. But then $\mathfrak{h}_k \subset \operatorname{Ker}\rho$ and, therefore, $\mathfrak{h}_k = 0$ because ρ is faithful.

Consider the associated graded Lie superalgebra

$$\operatorname{gr}(\mathfrak{g}) = \bigoplus_{i=-1}^{k-1} \mathfrak{g}_i, \text{ where } \mathfrak{g}_i = \mathfrak{h}_i/\mathfrak{h}_{i+1}.$$

Observe that we have two homomorphisms of $\mathfrak{h}$:

$$\sigma : \mathfrak{h} \longrightarrow \mathfrak{gl}(V_1) \quad \text{and} \quad \operatorname{ad}_{\mathfrak{g}/\mathfrak{h}} : \mathfrak{h} \longrightarrow \mathfrak{gl}(\mathfrak{g}/\mathfrak{h}),$$

and $\mathfrak{h}_1 = \operatorname{Ker}\sigma \cap \operatorname{Ker}(\operatorname{ad}_{\mathfrak{g}/\mathfrak{h}})$. Set $W = \mathfrak{g}/\mathfrak{h}$, and let $\dim W = 0|n$. We have obtained an embedding

$$\mathfrak{g}_0 \subset \mathfrak{gl}(V_1) \oplus \mathfrak{gl}(W) \cong \mathfrak{gl}(V_1) \oplus W^* \otimes W,$$

which for every $i > 0$ induces an embedding

$$\mathfrak{g}_i \subset \mathfrak{gl}(V_1) \otimes \Lambda^i(W^*) \oplus \Lambda^{i+1}(W^*) \otimes W,$$

which add up to an embedding of the whole $\mathfrak{g}$:

$$\operatorname{gr}(\mathfrak{g}) \subset \mathfrak{gl}(V_1) \otimes \Lambda(W^*) \niplus \mathfrak{vect}(W^*).$$

It remains to observe that since $\mathfrak{vect}(W^*)$ contains a grading operator, the embedding exists not only for $\operatorname{gr}(\mathfrak{g})$, but for $\mathfrak{g}$ itself. Under the embedding the space V is identified with $V_1 \otimes \Lambda(W) \cong V_1 \otimes \Lambda(W^*)^*$. As we observed in Section 1.7, the representation in $\Lambda(W^*)^*$, as a $\mathfrak{vect}(W^*)$-module, is isomorphic to $\Pi^n(T^1)$. Since we are not interested in the representation itself but only in the image of $\mathfrak{g}$ in $\mathfrak{gl}(V)$, we can replace for convenience $V_1 \otimes \Lambda(W)$ with $\tilde{V}_1 \otimes \Lambda(W^*)$ for some $\tilde{V}_1$. This completes the proof of the first item of the lemma.

2) Observe that if $\mathfrak{g} \subset \mathfrak{q}(V) = C(J)$, then the subspace V_1 is J-invariant and, therefore, $\sigma(\mathfrak{h}) \subset \mathfrak{q}(V_1) = C(J|_{V_1})$.

3) If $\mathfrak{g}$ preserves a nondegenerate bilinear form ω in V, consider its restriction $\tilde{\omega}$ on V_1. Since σ is irreducible, $\tilde{\omega}$ is either nondegenerate or vanishes identically. Denote by $V_1^{\perp}$ the subspace orthogonal to V_1 with respect to ω. Clearly, $V_1^{\perp}$ is $\mathfrak{h}$-invariant.

If $\tilde{\omega}$ is nondegenerate, then $V = V_1 \oplus V_1^{\perp}$. This means that $\mathfrak{g} \subset \mathfrak{aut}(\tilde{\omega}) \otimes 1 \oplus 1 \otimes \mathfrak{vect}(0|n)$. Therefore, the $\mathfrak{g}$-action in V is reducible; a contradiction.

If $\tilde{\omega} \equiv 0$, then by $\mathfrak{g}$-invariance of ω, the space $V_1^{\perp}$ must contain the image of V under $\rho(\operatorname{Ker}\sigma)$. Hence, the $\mathfrak{h}$-module $V/V_1^{\perp} \cong V_1^*$ should be of the form $V_1 \otimes \Lambda^n(W)$. This implies that $\omega = \omega_1 \otimes \omega_{1/2}$ for some ω_1 and $\mathfrak{g} \subset \mathfrak{aut}(\omega_1) \otimes \Lambda(n) \niplus T^{1/2}(\mathfrak{vect}(0|n))$. □

4.5 What remains to be proved

To complete the proof of the Main Theorem, it suffices to consider the case when ρ possesses the following property: for any ideal $\mathfrak{i} \subset \mathfrak{g}$, either ρ is of type A with respect to $\mathfrak{i}$ or $\rho|_{\mathfrak{i}}$ is the multiple of a character. Due to Remark 4.3 the second possibility means that $\mathfrak{i}$ is a 1-dimensional center of $\mathfrak{g}$. In particular, any nontrivial commutative ideal of $\mathfrak{g}$ coincides with its 1-dimensional center.

Let $\mathfrak{r}$ be the radical of the linear Lie superalgebras $\mathfrak{g}$. We see that either $\dim \mathfrak{r} \leq 1$ or $\mathfrak{r}$ is not commutative. In the case when $\mathfrak{r}$ is not commutative, consider the derived series of $\mathfrak{r}$:

$$\mathfrak{r} \supset \mathfrak{r}_1 \supset \cdots \supset \mathfrak{r}_k \supset \mathfrak{r}_{k+1} = 0, \text{ where } \mathfrak{r}_{i+1} = [\mathfrak{r}_i, \mathfrak{r}_i].$$

Clearly, each $\mathfrak{r}_i$ is an ideal in $\mathfrak{g}$ and the last ideal, $\mathfrak{r}_k$ is commutative. Hence, $\dim \mathfrak{r}_k = 1$ and $\mathfrak{r}_k$ is the center of $\mathfrak{g}$.

Lemma 4.6. **A)** 1) *If ρ is of type A with respect to $\mathfrak{r}_{k-1}$ and $\rho|_{\mathfrak{r}_k}$ is scalar, then either $\mathfrak{r}_{k-1} \cong \mathfrak{hei}(0|2n)$ or $\mathfrak{r}_{k-1} \cong \mathfrak{hei}(0|2n-1)$; $V \cong \Lambda(n)$ or $\Pi(\Lambda(n))$ and in either case $\mathfrak{g} \subset \mathfrak{hei}(0|2n) \ltimes \mathfrak{o}(2n)$.*

2) *If additionally $\mathfrak{g} \subset \mathfrak{q}(V)$, then $\mathfrak{r}_{k-1} \cong \mathfrak{hei}(0|2n-1)$, and $\mathfrak{g} \subset \mathfrak{hei}(0|2n-1) \ltimes \mathfrak{o}(2n-1)$.*

3) *Under the assumptions of item* 1) $\mathfrak{g}$ *does not preserve any nondegenerate bilinear form on V.*

Proof. We will prove items 1 and 2 simultaneously.

1) Since $\mathfrak{r}_k$ is the center of $\mathfrak{g}$ and $\dim \mathfrak{r}_k = 1$ we have $\mathfrak{r}_{k-1} = \mathfrak{hei}(0|m)$ for some m.

2) $\rho|_{\mathfrak{r}_{k-1}}$ is irreducible and faithful, so it can be realized in the superspace of functions, $\Lambda(n)$, or in $\Pi(\Lambda(n))$, where $n = [\frac{m+1}{2}]$; observe that $\rho|_{\mathfrak{r}_{k-1}}$ is irreducible of G-type for $m = 2n$ and it is irreducible of Q-type for $m = 2n-1$, see Section 1.6.

3) $\mathfrak{g}$ is contained in the normalizer of $\mathfrak{hei}(0|m)$ in $\mathfrak{gl}(\Lambda(n))$. This means that $\mathfrak{g} \subset \mathfrak{hei}(0|m) \ltimes \mathfrak{o}(m)$. □

Lemma 4.7. *Let* $\dim \mathfrak{r} \leq 1$, *i.e.,* $\mathfrak{g}$ *is either semisimple or a nontrivial central extension of a semisimple Lie superalgebra but NOT an almost simple or a central extension of an almost simple Lie superalgebra. Then we can always choose an ideal* $\mathfrak{i} \neq \mathfrak{r}$ *such that* ρ *is of type B or C with respect to* $\mathfrak{i}$.

4.8 Proof of Lemma 4.7 for semisimple Lie superalgebras

By definition $\mathfrak{g}$ is *semisimple* if its radical is zero. By analogy with a description of semisimple Lie algebras over fields of prime characteristic, V. Kac [K] described semisimple finite-dimensional Lie superalgebras as follows. Let $\mathfrak{s}_1, \dots, \mathfrak{s}_k$ be simple Lie superalgebras, let $n_1, \dots, n_k$ be nonnegative integers, let $\Lambda(n_j)$ be the supercommutative Grassmann superalgebra, and $\mathfrak{s} = \oplus \mathfrak{s}_j \otimes \Lambda(n_j)$. Then $\mathfrak{ders} = \oplus\left((\mathfrak{ders}_j) \otimes \Lambda(n_j) \ltimes 1 \otimes \mathfrak{vect}(n_j)\right)$. Let $\mathfrak{g}$ be a subalgebra of $\mathfrak{ders}$ containing $\mathfrak{s}$.

1) If the projection of $\mathfrak{g}$ on $1 \otimes \mathfrak{vect}(n_j)_{-1}$ coincides with $\mathfrak{vect}(n_j)_{-1}$ for each $j = 1, \dots, k$, then $\mathfrak{g}$ is semisimple.

2) All semisimple Lie superalgebras arise in the manner indicated.

Let $\dim \mathfrak{r} = 0$, i.e., $\mathfrak{g}$ is semisimple. Since $\mathfrak{g}$ is not almost simple, then due to Kac's description, the alternative arises: either $\mathfrak{g}$ contains an ideal $\mathfrak{i}$ of the form

$$\mathfrak{i} = \mathfrak{s} \otimes \Lambda(n) \quad \text{with simple } \mathfrak{s} \text{ and } n > 0 \tag{4.1}$$

or $\mathfrak{g} = \bigoplus_{j \leq k} \mathfrak{s}_j$, where each $\mathfrak{s}_j$ is almost simple and $k > 1$.

Lemma 4.8.1. *If* $\mathfrak{g} = \bigoplus_{j \leq k} \mathfrak{s}_j$ *and* $k \geq 2$, *then any irreducible faithful representation of* $\mathfrak{g}$ *is of type B with respect to any of its ideal* $\mathfrak{s}_j$.

Proof. Since the stabilizer of any irreducible representation of $\mathfrak{s}_j$ is the whole $\mathfrak{g}$, the type of any irreducible representation of $\mathfrak{g}$ with respect to $\mathfrak{s}_j$ can be either A or B. Due to the faithfulness case A is ruled out. □

Lemma 4.8.2. *Let* $\mathfrak{s}$ *be a simple Lie superalgerba and* $\mathfrak{i} = \mathfrak{s} \otimes \Lambda(n)$, $n > 0$. *Then* $\mathfrak{i}$ *has no faithful irreducible finite-dimensional representations.*

For the proof see Section 4.8.4.

Corollary 4.8.2.1. *If* $\mathfrak{g}$ *contains an ideal* $\mathfrak{i}$ *of the form* (4.1), *then* $\mathfrak{g}$ *cannot have any faithful irreducible finite-dimensional representation of type A with respect to the ideal* $\mathfrak{i}$.

Corollary 4.8.2.2. *Lemmas* 4.8.1, 4.8.2 *and Corollary* 4.8.2.1 *prove Lemma* 4.7 *for semisimple Lie superalgebras.*

Recall the following well-known and simple statement.

Lemma 4.8.3. *If* $\mathfrak{s}$ *is a simple Lie superalgebra, then*

$$[\mathfrak{s}_{\bar{1}}, \mathfrak{s}_{\bar{1}}] = \mathfrak{s}_{\bar{0}} \quad \textit{and} \quad [\mathfrak{s}_{\bar{0}}, \mathfrak{s}_{\bar{1}}] = \mathfrak{s}_{\bar{1}}\ .$$

4.8.4. Proof of Lemma 4.8.2

For $n > 0$ the Lie superalgerba $\mathfrak{i}$ contains the supercommutative ideal

$$\mathfrak{n} = \mathfrak{s} \otimes \xi_1 \dots \xi_n.$$

Let us show that $\mathfrak{n}$ is contained in the kernel of any irreducible representation ρ of $\mathfrak{i}$. Consider a nilpotent ideal

$$\mathfrak{m} = \mathfrak{s} \otimes \bigoplus_{i \geq 1} \Lambda^i(\xi) \subset \mathfrak{i}.$$

As follows from [K], [Ser1], any irreducible finite-dimensional representation of $\mathfrak{m}$ is given by a character $\lambda \in \mathfrak{m}^*$ that vanishes on $[\mathfrak{m}_{\bar 0}, \mathfrak{m}_{\bar 0}] \oplus \mathfrak{m}_{\bar 1}$.

For $n = 2k$ we have

$$[\mathfrak{m}_{\bar 0}, \mathfrak{m}_{\bar 0}] \supset [\mathfrak{s}_{\bar 1} \otimes \Lambda^1(\xi), \mathfrak{s}_{\bar 1} \otimes \Lambda^{2k-1}(\xi)] = [\mathfrak{s}_{\bar 1}, \mathfrak{s}_{\bar 1}] \otimes \Lambda^{2k}(\xi) = \mathfrak{n}_{\bar 0}$$

and $\mathfrak{m}_{\bar 1} \supset \mathfrak{s}_{\bar 1} \otimes \Lambda^{2k}(\xi) = \mathfrak{n}_{\bar 1}$. For $n = 2k + 1 > 1$ we have

$$[\mathfrak{m}_{\bar 0}, \mathfrak{m}_{\bar 0}] \supset [\mathfrak{s}_{\bar 1} \otimes \Lambda^1(\xi), \mathfrak{s}_{\bar 0} \otimes \Lambda^{2k}(\xi)] = [\mathfrak{s}_{\bar 1}, \mathfrak{s}_{\bar 0}] \otimes \Lambda^{2k+1}(\xi) = \mathfrak{n}_{\bar 0}$$

and $\mathfrak{m}_{\bar 1} \supset \mathfrak{s}_{\bar 0} \otimes \Lambda^{2k+1}(\xi) = \mathfrak{n}_{\bar 1}$. Thus, $\mathfrak{n}$ is contained in the kernel of any irreducible finite-dimensional representation of the ideal $\mathfrak{m}$, hence, in the kernel of any irreducible finite-dimensional representation of $\mathfrak{i}$.

If $n = 1$ and τ is an arbitrary irreducible subrepresentaiton of $\rho|_{\mathfrak{n}}$, then $\dim \tau = 1$ or ε because $\mathfrak{n}$ is supercommutative. Hence, $\tau|_{\mathfrak{n}_{\bar 1}} = 0$. Item 1) of Theorem 4.1.1 implies that the restriction of τ onto $[\mathfrak{g}_{\bar 0} \otimes 1, \mathfrak{n}] \supset [\mathfrak{s}_{\bar 0}, \mathfrak{s}_{\bar 1}] \otimes \xi = \mathfrak{s}_{\bar 1} \otimes \xi = \mathfrak{n}_{\bar 0}$ must vanish. Thus, $\tau|_{\mathfrak{n}} = 0$ and by item 3) of Theorem 4.1.1, $\mathfrak{n} \subset \ker \rho$.

Lemma 4.8.2 is proved. □

4.9 Proof of Lemma 4.7 for central extensions of semisimple Lie superalgebras

In this section we assume that $\dim \mathfrak{r} = 1$, hence (see Remark 4.3), $\mathfrak{r}$ is the center of $\mathfrak{g}$, and $\tilde{\mathfrak{g}} = \mathfrak{g}/\mathfrak{r}$ is semisimple.

First, consider the case when $\tilde{\mathfrak{g}} = \bigoplus\limits_{i \leq k} \mathfrak{s}_i$, where the $\mathfrak{s}_i$ are almost simple and $k > 1$. Let $\pi : \mathfrak{g} \longrightarrow \tilde{\mathfrak{g}}$ be the natural projection. The Lie superalgebra $\pi^{-1}(\mathfrak{s}_1) = \mathfrak{i}$ is an ideal in $\mathfrak{g}$ and $\dim \mathfrak{i} > 1$.

Lemma 4.9.1. *If $k > 1$, then ρ cannot be irreducible of type A with respect to $\mathfrak{i}$.*

Proof. Assume the contrary. Let $\rho|_{\mathfrak{i}}$ be irreducible. Set $\mathfrak{g}_+ = \bigoplus\limits_{i>1} (\mathfrak{s}_i)_{\bar 0}$ and let $\mathfrak{g}_+$ be a Lie algebra. Then $[\mathfrak{s}_1, \mathfrak{g}_+] = 0$. Since $\rho(\mathfrak{r})$ acts by scarlar operators and ρ is a finite-dimensional representation, it follows that $[\pi^{-1}(\mathfrak{g}_+), \mathfrak{i}] = 0$. Since $\rho|_{\mathfrak{i}}$ is irreducible by the hypothesis, this means that $\rho(\pi^{-1}(\mathfrak{g}_+)) = \mathbb{C} \cdot 1$ and, since ρ is faithful, this implies $\mathfrak{g}_+ = 0$. □

4.9.2. $\tilde{\mathfrak{g}}$ contains an ideal $\mathfrak{i}$ of the form (4.1)

The central extension is defined by a cocycle $c : \tilde{\mathfrak{g}} \times \tilde{\mathfrak{g}} \longrightarrow \mathbb{C}$. The cocycle condition is

$$c(f, [g, h]) = c([f, g], h) + (-1)^{p(f)p(g)} c(g, [f, h]) \quad \text{for any} f, g, h \in \tilde{\mathfrak{g}}. \tag{4.2}$$

As earlier, we assume that $\mathfrak{g}$ has a faithful finite-dimensional representation; so the restriction of c to $\tilde{\mathfrak{g}}_{\bar 0} \times \tilde{\mathfrak{g}}_{\bar 0}$ is trivial. Besides, $c|_{\tilde{\mathfrak{g}}_{\bar 0} \times \tilde{\mathfrak{g}}_{\bar 1}} = 0$ by parity considerations. Therefore, nonzero values of the cocycle c are only possible on $\tilde{\mathfrak{g}}_{\bar 1} \times \tilde{\mathfrak{g}}_{\bar 1}$.

Lemma. *Let, as above,* $\mathfrak{n} = \mathfrak{s} \otimes \Lambda^n(\xi)$. *Then* $c|_{\mathfrak{n}\times\tilde{\mathfrak{g}}} = 0$.

Proof. First, let us prove that $c|_{\mathfrak{n}_{\bar{1}}\times\mathfrak{n}_{\bar{1}}} = 0$. If $n = 2k + 1$, it suffices to check condition (4.2) for the triple $f = f_{\bar{0}} \otimes \xi_1 \cdot \dots \cdot \xi_n$, $g = g_{\bar{1}} \otimes \xi_1 \cdot \dots \cdot \xi_n$ and $h = h_{\bar{1}} \otimes 1$, where $f_{\bar{0}} \in \mathfrak{s}_{\bar{0}}$ and $g_{\bar{1}}, h_{\bar{1}} \in \mathfrak{s}_{\bar{1}}$. With Lemma 4.8.3, the left-hand side of (4.2) gives us the values of c on an arbitrary pair of elements from $\mathfrak{n}_{\bar{1}}$, whereas the right-hand side vanishes because $[f, g] = 0$ and $c(g, [f, h]) = 0$ since $p(g) = \bar{0}$.

If $n = 2k$, similar arguments are applicable to the triple $f = f_{\bar{1}} \otimes \xi_1 \cdot \dots \cdot \xi_n$, $g = g_{\bar{0}} \otimes \xi_1 \cdot \dots \cdot \xi_n$ and $h = h_{\bar{1}} \otimes 1$, where $f_{\bar{1}}, h_{\bar{1}} \in \mathfrak{s}_{\bar{1}}$ and $g_{\bar{0}}, \in \mathfrak{s}_{\bar{0}}$.

Now set $\mathfrak{L}^k = \mathfrak{s} \otimes \Lambda^k(\xi_1, \dots, \xi_n)$ and let us prove that $c|_{\mathfrak{n}\times\mathfrak{L}^k} = 0$. We will perform the inverse induction on k. For $k = n$ we have already proved this.

Let the statement be true for all $k > k_0$. Let us show that it is true for $k = k_0$ as well. Observe that due to the description of semisimple Lie superalgebras, $\tilde{\mathfrak{g}}$ contains n elements η_i such that $\operatorname{ad}\eta_i|_{\mathfrak{s}\otimes\Lambda(n)} = \partial_{\xi_i} + D_i + X_i \otimes \alpha_i$, where $D_i \in \mathfrak{vect}(0|n)$ and $D_i(0) = 0$, $X_i \in \mathfrak{ders}$, $\alpha_i \in \Lambda(\xi)$. Let $\varphi \in \Lambda^{k_0+1}(\xi_1, \dots, \xi_n)$, $\psi \in \Lambda^n(\xi_1, \dots, \xi_n)$; let $g, h \in \mathfrak{s}$ and $p(g \otimes \varphi) = \bar{0}$ while $p(h \otimes \psi) = \bar{1}$. Then we have

$$c([\eta_i, g \otimes \varphi], h \otimes \psi) = (-1)^{p(g)}\left(c(g \otimes \frac{\partial\varphi}{\partial\xi_i}, h \otimes \psi) + c(g \otimes D_i\varphi, h \otimes \psi)\right) + (-1)^{p(g)p(\alpha_i)}c([[X_i, g] \otimes \alpha_i\varphi, h \otimes \psi).$$

As $D_i\varphi$ and $\alpha_i\varphi \in \bigoplus_{l\geq k_0+1} \Lambda^l(\xi)$, the last two summands vanish by the inductive hypothesis.

On the other hand, due to (4.2) we have

$$c([\eta_i, g \otimes \varphi], h \otimes \psi) = c(\eta_i, [g \otimes \varphi, h \otimes \psi]) + c([\eta_i, h \otimes \psi], g \otimes \varphi).$$

As $\deg\varphi + \deg\psi = k_0 + 1 + n > n$, the bracket in the first summand above vanishes. The second summand vanishes by parity considerations. So $c(g \otimes \frac{\partial\varphi}{\partial\xi_i}, h \otimes \psi) = 0$ for arbitrary $g \otimes \frac{\partial\varphi}{\partial\xi_i} \in \mathfrak{L}^{k_0}_{\bar{1}}$ and $h \otimes \psi \in \mathfrak{L}^n_{\bar{1}} = \mathfrak{n}_{\bar{1}}$. □

Lemma 4.9.3. $\mathfrak{i}$ *has no faithful irreducible finite-dimensional representations.*

Proof. Word-for-word, the proof of Lemma 4.8.2 with the help of Lemma 4.9.2. □

Corollary. ρ *cannot be of type A with respect to the ideal* $\mathfrak{i}$.

5 Summing up

Theorem 4.1.2 put together with Lemmas 4.2 (Lemma B), 4.4 (Lemma C), 4.6 (Lemma A), sec. 4.5 and Lemma 4.8 prove Main Theorem. □

Acknowledgment. I am greateful to D. Leites for help; RFBR grants 99-01-00245 and 01-01-00490a, and NFR for partial financial support; Stockholm University for hospitality.

References

[BL] Bernstein J., Leites D., Invariant differential operators and irreducible representations of Lie superalgebras of vector fields, *Sel. Math. Sov.* **1** no. 2 (1981), 143–160.

[D] Deligne P. et al., eds., *Quantum fields and Strings: A Course for Mathematicians*. Vols. 1, 2. Material from the Special Year on Quantum Field Theory held at the Institute for Advanced Study, Princeton, NJ, 1996–1997. AMS, Providence, RI; Institute for Advanced Study (IAS), Princeton, NJ, 1999. Vol. 1: xxii+723 pp.; Vol. 2: pp. i–xxiv and 727–1501.

[Di] Dixmier J., *Algèbres enveloppantes*, Gauthier-Villars, Paris, 1974; *Enveloping algebras*. Revised reprint of the 1977 translation. Graduate Studies in Mathematics, 11. American Mathematical Society, Providence, RI, 1996. xx+379 pp.

[D1] Dynkin E.B., Semisimple subalgebras of semisimple Lie algebras, *Matem. Sbornik* **30** (1952) (72), 349–462 (Russian) = English transl. in Moscow Math. Soc. Translations Ser. 2, v.6, 111–244.

[D2] Dynkin E.B., Maximal subgroups of classical groups, *Trudy Moskovskogo Matem. obshchestva* **1** (1952), (Russian) = English transl. in: Moscow Math. Soc. Translations Ser. 2, v.6, 245–378.

[FK] Freund P., Kaplansky I., Simple supersymmetries, *J. Math. Phys.* **17** no. 2 (1976), 228–231.

[J1] Jeugt J. van der., Principal five-dimensional subalgebras of Lie superalgebras, *J. Math. Phys.* **27** (12) (1986), 2842–2847.

[K] Kac V. G., Lie superalgebras, *Adv. Math.* **26** (1976), 8–96.

[K1] Kac V.G., Characters of typical representations of classical Lie superalgebras, *Comm. Algebra* **5** (1977), no. 8, 889–897.

[K2] Kac V.G., Representations of classical Lie superalgebras, *Differential geometrical methods in mathematical physics*, II (Proc. Conf., Univ. Bonn, Bonn, 1977), pp. 597–626, Lecture Notes in Math., 676, Springer, Berlin, 1978.

[Kap] Kaplansky I., Superalgebras, *Pacific J. Math.* **86** (1980), no. 1, 93–98.

[KL] Kotchetkov Yu., Leites D., Simple Lie algebras in characteristic 2 recovered from superalgebras and on the notion of a simple finite group. In: Kegel O. et. al. (eds.) *Proc. Intern. algebraic conference*, Novosibirsk, August 1989, Contemporary Math. AMS, 1992, (Part 2), v. 131, 59–67.

[LS] Leites D., Serganova V., Metasymmetry and Volichenko algebras, *Phys. Lett. B* **252** no.1 (1990), 91–96; id., Symmetries wider than supersymmetries. In: S. Duplij and J. Wess (eds.), *Noncommutative Structures in Mathematics and Physics*, Proc. NATO Advanced Research Workshop, Kiev, 2000. Kluwer, 2001, 13–30.

[LSh] Leites D., Shchepochkina I., Howe's duality and Lie superalgebras, In: S. Duplij and J. Wess (eds.), *Noncommutative structures in Mathematics and Physics*, Proc. NATO Advanced Research Workshop, Kiev, 2000. Kluwer, 2001, 93–112.

[PS] Penkov I., Serganova V., Generic irreducible representations of finite dimensional Lie superalgebras. Internat. J. Math. 5, 1994, 389–419.

[SNR] Scheunert M., Nahm W., Rittenberg V., Classification of all simple graded Lie algebras whose Lie algebra is reductive. I, II. Construction of the exceptional algebras. J. Mathematical Phys. 17 (1976), no. 9, 1626–1639, 1640–1644.

[Se] Serganova V., Characters of irreducible representations of simple Lie superalgebras. *Proceedings of the International Congress of Mathematicians*, Vol. II (Berlin, 1998). Doc. Math. 1998, Extra Vol. II, 583–593 (electronic).

[S] Serganova V., Simple Volichenko algebras. In: Kegel O. et.al. (eds.) *Proc. Intnat. algebraic conference, Novosibirsk, August* 1989, Contemporary Math. AMS, 1992, (Part 2), v. 131, 50–58.

[Ser1] Sergeev A., Irreducible representations of solvable Lie superalgebras. In: Leites D. (ed.) *Seminar on supermanifolds*, n. 22. Reports of Dept. of Math. of Stockholm Univ. 1988, n. 4, 1–12; math.RT/9810109 and Represent. Theory 3 (1999), 435–443.

[Ser2] Sergeev A., An analog of the classical invariant theory for Lie superalgebras. I. math.RT/9810113; Michigan J. Math, v. 48, 2001; II, math.RT/9904079; Michigan J. Math, v. 48, 2001.

[Sh1] Shchepochkina I., On maximal and maximal solvable subalgebra of simple Lie superalgebras. In: Onishchik A. (ed.), *Questions of group theory and homological algebra*, Yaroslavl Univ., Yaroslavl, 1987, 156–161 (in Russian).

[Sh2] Shchepochkina I., Maximal solvable subalgebra of the Lie superalgebras $\mathfrak{gl}(m|n)$ and $\mathfrak{sl}(m|n)$. Funktsional. Anal. i Prilozhen. 28 (1994), no. 2, 92–94; English translation in Functional Anal. Appl. 28 (1994), no. 2, 147–149.

[Sh3] Shchepochkina I., Maximal subalgebras of matrix Lie superalgebras In: Leites D.(ed.) *Seminar on Supermanifolds* v. 32, Reports of Stockholm University, 1992, 1–43; hep-th/9702122, 1997, 18 pp.

[T] Ten O.K. Semisimple maximal subalgebras of Lie p-algebras of classical type. (Russian) Vestnik Moskov. Univ. Ser. I Mat. Mekh. 1987, no. 4, 69–71, 103. id., Letter to the editor, Vestnik Moskov. Univ. Ser. I Mat. Mekh. 1988, no. 2, 103. 17B50; id., On nonsemisimple maximal subalgebras of Lie p-algebras of classical type. (Russian) Vestnik Moskov. Univ. Ser. I Mat. Mekh. 1987, no. 2, 65–67, 103.

[ZZO] Zharov O., Zolotaryev B., Onishchik A., Reducible maximal subalgebra of classical Lie superalgebras. In: Onishchik A. (ed.), *Questions of group theory and homological algebra*, Yaroslavl Univ., Yaroslavl, 1987, 161–163 (in Russian).

Irina Shchepochkina
Independent University of Moscow
Bolshoj Vlasievsky per.
RU-121002 Moscow, Russia
Ira@Paramonova.mccme.ru

Progress in Mathematics

Edited by:

Progress in Mathematics is a series of books intended for professional mathematicians and scientists, encompassing all areas of pure mathematics. This distinguished series, which began in 1979, includes authored monographs and edited collections of papers on important research developments as well as expositions of particular subject areas.

We encourage preparation of manuscripts in some form of TEX for delivery in camera-ready copy which leads to rapid publication, or in electronic form for interfacing with laser printers or typesetters.

Proposals should be sent directly to the editors or to: Birkhäuser Boston, 675 Massachusetts Avenue, Cambridge, MA 02139, USA or to Birkhauser Verlag, 40-44 Viadukstrasse, Ch-4051 Basel, Switzerland.

143 GONZALEZ-VEGA/RECIO (eds). Algorithms in Algebraic Geometry and Applications
144 BELLAÏCHE/RISLER (eds). Sub-Riemannian Geometry
146 JARDINE. Generalized Etale Cohomology
147 DIBIASE. Fatou Theorems
148 HUANG. Two-Dimensional Conformal Geometry and Vertex Operator Algebras
149 SOURIAU. Structure of Dynamical Systems: A Symplectic View of Physics
150 SHIOTA. Geometry of Subanalytic and Semialgebraic Sets
151 HUMMEL. Gromov's Compactness Theorem for Pseudo-holomorphic Curves
152 GROMOV. Metric Structures for Riemannian and Non-Riemannian Spaces. Sean Bates (translator)
153 BUESCU. Exotic Attractors: From Liapunov Stability to Riddled Basins
154 BÖTTCHER/KARLOVICH. Carleson Curves, Muckenhoupt Weights, and Toeplitz Operators
155 DRAGOMIR/ORNEA. Locally Conformal Kähler Geometry
156 GUIVARC'H/JI/TAYLOR. Compactifications of Symmetric Spaces
157 MURTY/MURTY. Non-vanishing of L-functions and Applications
158 TIRAO/VOGAN/WOLF (eds). Geometry and Representation Theory of Real and *p*-adic Groups
159 THANGAVELU. Harmonic Analysis on the Heisenberg Group
160 KASHIWARA/MATSUO/SAITO/SATAKE (eds). Topological Field Theory, Primitive Forms and Related Topics
161 SAGAN/STANLEY (eds). Mathematical Essays in Honor of Gian-Carlo Rota
162 ARNOLD/GREUEL/STEENBRINK. Singularities: The Brieskorn Anniversary Volume
163 BERNDT/SCHMIDT. Elements of the Representaiton Theory of the Jacobi Group
164 ROUSSARIE. Bifurcations of Planar Vector Fields and Hilbert's Sixteenth Problem
165 MIGLIORE. Introduction to Liaison Theory and Deficiency Modules
166 ELIAS/GIRAL/MIRO-ROIG/ZARZUELA (eds). Six Lectures on Commutative Algebra
167 FACCHINI. Module Theory
168 BALOG/KATONA/SZA'SZ/RECSKI (eds). European Congress of Mathematics, Budapest, July 22-26, 1996. Vol. I
169 BALOG/KATONA/SZA'SZ/RECSKI (eds). European Congress of Mathematics, Budapest, July 22-26, 1996. Vol. II
168/169 Sets Vols I, II

170 PATERSON. Groupoids, Inverse Semigroups, and their Operator Algebras
171 REZNIKOV/SCHAPPACHER (eds). Regulators in Analysis, Geometry and Number Theory
172 BRYLINSKI/BRYLINSKI/NISTOR/TSYGAN/ XU (eds). Advances in Geometry
173 DRÄXLER/MICHLER/RINGEL (eds). Computational Methods for Representation of Groups and Algebras: Euroconference in Essen
174 GOERSS/JARDINE. Simplicial Homotopy Theory
175 BAÑUELOS/MOORE. Probabilistic Behavior of Harmonic Functions
176 BASS/LUBOTZKY. Tree Lattices
177 BIRKENHAKE/LANGE. Complex Tori
178 PUIG. On the Local Structure of Morita and Rickard Equivalence Between Brauer Blocks
179 MORALES RUIZ. Differential Galois Theory and Non-Integrability of Hamiltonian Systems
180 PATERNAIN. Geodesic Flows
181 HAUSER/LIPMAN/OORT/QUIRÓS (eds). Resolution of Singularities: In Tribute to Oscar Zariski
182 BILLEY/LAKSHMIBAI. Singular Loci of Schubert Varieties
183 KAPOVICH. Hyperbolic Manifolds and Discrete Groups
184 du SAUTOY/SEGAL/SHALEV (eds). New Horizons in pro-p Groups
185 FARAUT/KANEYUKI/KORÁNYI/LU/ROOS. Analysis and Geometry on Complex Homogeneous Domains
186 KLEINERT. Units in Skew Fields
187 BOST/LOESER/RAYNAUD. Courbes Semi-stables et Groupe Fondamental en Géométrie Algébrique
188 DOLBEAULT/IORDAN/HENKIN/SKODA/ TRÉPREAU (eds). Complex Analysis and Geometry: International Conference in Honor of Pierre Lelong
189 ANDRÉ/BALDASSARI. DeRham Cohomology of Differential Modules on Algebraic Varieties
190 van den ESSEN. Polynomial Automorphisms and the Jacobian Conjecture.
191 KASHIWARA/MIWA (eds). Physical Combinatorics
192 DEHORNOY. Braids and Self-Distributivity
193 KNUDSON. Homology of Linear Groups
194 JUHL. Cohomological Theory of Dynamical Zeta Functions
195 FABER/van der GEER/OORT (eds). Moduli of Abelian Varieties
196 AGUADÉ/BROTO/CASACUBERTA (eds). Cohomological Methods in Homotopy Theory
197 CHERIX/COWLING/JOLISSAINT/JULG/ VALETTE (eds). Groups with the Haagerup property
198 LANDSMAN/PFLAUM/SCHLICHENMAIER (eds). Quantization of Singular Symplectic Quotients
199 PEYRE/TSCHINKEL. Rational Points on Algebraic Varieties
200 GOLUBITSKY/STEWART. The Symmetry Perspective
201 CASACUBERTA/MIRÓ-ROIG/VERDERA/ XAMBÓ-DESCAMPS (eds). European Congress of Mathematics, Volume 1
202 CASACUBERTA/MIRÓ-ROIG/VERDERA/ XAMBÓ-DESCAMPS (eds). European Congress of Mathematics, Volume 2
203 BLAIR. Riemannian Geometry of Contact and Symplectic Manifolds
204 KUMAR. Kac–Moody Groups, their Flag Varieties, and Representation Theory
205 BOUWKNEGT/WU. Geometric Analysis and Applications to Quantum Field Theory
206 SNAITH. Algebraic K-groups as Galois Modules
207 LONG. Index Theory for Symplectic Paths with Applications
208 TURAEV. Torsions of 3-dimensional Manifolds
209 UNTERBERGER. Automorphic Pseudodifferential Analysis and Higher Level Weyl Calculi
210 KNAPP. Lie Groups Beyond an Introduction, Second Edition
211 HERNÁNDEZ-LERMA/LASSERE. Markov Chains and Invariant Probabilities
212 LUBOTZKY/SEGAL. Subgroup Growth
213 DUVAL/GUIEU/OVSIENKO (eds). The Orbit Method in Geometry and Physics. In Honor of A.A. Kirillov
214 DUNGEY/TER ELST/ROBINSON. Analysis on Lie Groups with Polynomial Growth

MIX
Papier aus verantwortungsvollen Quellen
Paper from responsible sources
FSC® C105338

If you have any concerns about our products,
you can contact us on
ProductSafety@springernature.com

In case Publisher is established outside the EU,
the EU authorized representative is:
Springer Nature Customer Service Center GmbH
Europaplatz 3, 69115 Heidelberg, Germany

Printed by Libri Plureos GmbH
in Hamburg, Germany